PATHOLOGY OF TREES AND SHRUBS

WITH SPECIAL REFERENCE TO BRITAIN

BY

T. R. PEACE

CHIEF RESEARCH OFFICER
FORESTRY COMMISSION

OXFORD
AT THE CLARENDON PRESS
1962

Reprinted by Trollius Publications
2001

Prefatory Note

Of all the published texts on the diagnosis of disease in trees and shrubs none has stood the test of time better than T R Peace's classic descriptive work. Unavailable for many years, generations of students and arboricultural workers have had little or no access to the mine of information which it contains.

We make no apology, therefore, despite present-day diffrerences of approach, for publishing this unabridged reprint of "Pathology of Trees and Shrubs", in the hope and belief that it will be of continued service to all those interested in the care and conservation of trees.

Jon Atkins
Trollius Publications
August 2001

ISBN 0-9539718-1-3

Published by Trollius Publications
2001

Printed and bound by Antony Rowe Limited, Eastbourne

PREFACE

IT is certainly unnecessary to justify the production of a British textbook on tree diseases. Its lack has been apparent for many years to all interested in forestry. The practical forester wishes for guidance on the cause and treatment of the troubles which he sees afflicting his trees, the university student requires a background to his studies of forest pathology, and the research worker needs a book to which he can turn for basic information and more particularly for an introduction to the work which has been done on any particular disease. I may have been too ambitious in attempting to cater for so wide a range of needs. This preface is, therefore, also my apologia for certain inadequacies and inconstancies in my treatment of the subject. In the hope of making this book usable by the practical forester, the use of botanical terms has been cut to a minimum, and those that have been employed are explained in a glossary. For the sake of the research worker the number of literature references given is larger than any degree student would require.

As first envisaged, the book was to be restricted to British diseases, with only a very brief mention of the more important foreign ones. I soon realized that this was a difficult position to maintain. Some foreign diseases may eventually reach this country, and in any case, work done on foreign diseases often throws light on related diseases present in Britain. For these reasons the book now includes fungal and bacterial diseases of trees over the whole of the temperate regions, though more emphasis has been placed on their occurrence and behaviour in Britain. In descriptions of damage due to non-living agencies, foreign experience has been drawn upon only in so far as it affects what may happen in Britain. Even this has involved the inclusion of a great deal of work done in other countries.

In one other aspect, by the inclusion of diseases of ornamental trees and shrubs, I have allowed the book to exceed the scope originally envisaged for it. At the outset 'Forest Pathology' was the proposed title, obviously involving a limitation to forest trees. This limitation raised two queries. Firstly, should I include diseases of forest shrubs such as hazel or dogwood? Secondly, were trees such as *Liriodendron* or *Aesculus*, normally grown for ornament, but potentially timber producers, to be brought in? In any case no existing British book covers diseases of ornamental trees and shrubs, even to the very limited extent that the lack of available information allows, though diseases of fruit trees and of roses have been well covered. Leaving these out, there is still enough known on diseases of ornamentals to make the longest chapter, despite a serious lack of basic information on many of them.

Several people have expressed regret that the book contains no keys as an aid to identification. The overwhelming reasons against any attempt to provide such keys are outlined in the introduction to Chapter 2. The identification of some tree diseases is difficult, while other diseases are due to the complex and interdependent action of two or more agencies. I should be doing no good service if I provided deceptive aids which might appear to make identification easier, but would certainly make it very inaccurate.

The most vexed point, however, is that of nomenclature. For many years now practising botanists and foresters have been plagued with constant name changes. Many of these have removed names with which we have all become familiar. Others have substituted names which are less reasonable than the old ones. *Picea abies* instead of *P. excelsa* is an obvious case in point. Some of these changes result from the application of the rules of botanical nomenclature, but others, and this applies particularly to the fungi, arise from further study leading to the splitting of genera or to the transfer of a fungus from one genus to another. It appears to me that these changes bring to light two serious faults in the attitude of taxonomic botanists towards the science which they practise. Firstly, do they realize that taxonomy, which is certainly one of the bases of biological science, is also its handmaiden? When a name is altered, do they envisage the confusion created in card catalogues and indices, let alone in the minds of foresters and botanists, to whom names are merely convenient descriptive tools? Secondly, do they still believe that the concepts of genus and species have some exact significance beyond the fact that they are useful and, within rough limits, definable divisions? In many instances the creation of a new genus could be avoided merely by indicating in specialist publications that the species within the existing genus fell into two groups sufficiently different to be made separate genera, but that such a course was not being taken.

It seems clear to me that, if taxonomic botanists were really determined to avoid name changes, they could do it in the majority of cases by creating lists of conserved names, by rejecting some of the more ridiculous recent changes, and by restricting generic transfers, even if this meant a severe stretching of their genus concept. The fact that the rules of botanical nomenclature remain unchanged and almost every week brings forth a new paper playing havoc with fungal generic names, suggests to me that they are even now more concerned with the ramifications of their own science than with the well-being of biology as a whole.

That being my opinion, it is perhaps logical that in nearly every case, both for trees and for fungi, I have used the names that are most familiar to me and, I believe, to other foresters. In some cases I am probably wrong, for I do not take the view that all name changes can be avoided. To have decided every case on its merits, however, would have consumed far too much time. In a few instances I have given specific reasons for rejecting a change, but in most cases the use of old names is based on the general arguments set out above.

In so doing I may merely be building one of the last hummocks to remain above water in the rising tide of taxonomic autocracy. I have taken this action, however, in the belief that those waters will one day ebb, allowing, by the application of more adaptable rules and by a saner taxonomic outlook, the use of many of the names I have retained.

The use of capital letters for English names is a matter of much less importance, but about which, nevertheless, there is a good deal of disagreement. Taking the view that a name, being a name, deserves at least one capital, I have used not merely Scots pine, but also Red pine, White pine blister rust, or Oak wilt, but when talking of a genus or group, merely pine, oak, or rust fungus. Metric measures have been used throughout, to make the book more easily used abroad, and in the hope that they may one day be adopted generally.

For the use of research workers and others who may wish to follow up a disease in some detail a fairly large bibliography has been provided.

Even had I possessed the linguistic ability, reading all the references quoted would have been an impossible task. In many cases, therefore, I have depended on abstracts. While I should like to pay a most sincere tribute to the usefulness and accuracy of *Forestry Abstracts* and the *Review of Applied Mycology*, I am bound to admit that the use of abstracts introduces considerable possibilities of error and misinterpretation. I can only hope that the inaccuracies arising from this cause are not sufficiently serious to mar the book.

Pathology, as I have treated it in this book, is an immensely wide subject, and I cannot claim to have a proper knowledge of all the aspects I have included. In the hope of avoiding some of the errors of ignorance, I have asked for various chapters to be read and criticized by specialists in particular fields. Even now I fear some erroneous interpretations may remain. My deepest thanks go to my colleagues, Dr. S. A. B. Batko and Mr. J. S. Murray, whose untiring work on all the chapters has brought in much additional information, and has done a great deal to compensate for my mycological and other deficiencies. Without their help the book would hardly have been possible. The entire text has also been critically corrected by Mr. C. W. T. Young, to whom I am deeply grateful. The writing of the book was only possible because Mr. Murray and Mr. J. Jobling kindly spared me the time by taking on much of my normal work.

My thanks are particularly due to the following, among many, who criticized individual chapters or parts of chapters: Mr. J. R. Aldhous, Dr. A. Biraghi, Mr. G. Buszewicz, Mr. M. V. Edwards, Mr. R. J. Gladman, Mr. D. E. Green, Dr. J. L. Harley, Dr. W. H. Hinson, Mr. G. D. Holmes, Dr. E. W. Jones, Dr. J. E. Kuntz, Mr. A. D. C. Le Sueur, Mr. R. Lines, Dr. I. Levisohn, Dr. L. Leyton, Mr. J. D. Low, Dr. J. D. Ovington, Dr. R. F. Patton, Dr. R. G. Pawsey, Dr. A. J. Riker, Dr. W. S. Rogers, Mr. K. G. Stott, Dr. E. P. Van Arsdel, and Dr. G. van den Ende. I have purposely not mentioned the particular chapters involved, feeling that it would be unfair if these persons should, as a result, be assumed to have approved all I have said. Others too numerous to mention assisted with criticism and advice, and I should particularly like to thank Dr. R. W. G. Dennis, who has helped Dr. Batko on numerous occasions, and Mr. W. C. Moore who allowed me to use the manuscript of his book *British Parasitic Fungi* before its actual publication.

The book owes much to the skill and application of the three illustrators, Miss Janet Chandler (now Mrs. A. F. Dyer), Miss Janet Nimmo (now Mrs. R. G. F. Worsley), and Mr. A. Coram. I am particularly grateful to Professor Sir Harry Champion and Mr. A. C. Hoyle of the Oxford School of Forestry for allowing me to make use of Miss Chandler's services. The photographs are mostly from the Forestry Commission collection, the others are individually acknowledged. The onerous task of typing has fallen on the willing shoulders of Mrs. Joyce Cam, and much of the checking and other secretarial work on those of Mr. J. G. Jackman; most of the bibliography was checked and arranged by Mrs. Janet Morgan and Mrs. Edna Jeffers; Mr. P. G. Beak gave invaluable assistance with Slavonic names and titles. I am full of gratitude to them and to the many others who have helped in various ways.

The whole project was made possible only by the kind and patient forbearance of the Forestry Commission, who allowed me to use official time over a much longer period than I had originally requested.

T. R. P.

TO THE LATE

F. T. BROOKS

WHO LED ME TO THE

PATHOLOGICAL FOREST

AND TO

W. R. DAY

WHO FIRST GUIDED ME

THEREIN

CONTENTS

LIST OF PLATES

(*at end*)

1

INTRODUCTION

A FEW words of general explanation are required for the full understanding of this book. Pathology has been taken to include all diseases and damage to trees arising from non-animal causes. This, of course, excludes damage by insects and animals, except where their actions affect diseases due to other causes, a matter dealt with briefly in a special section of the book.

Examination of the table of contents will show that the book is divided into two main parts. Following two introductory chapters, Section I covers damage caused by non-living agencies. This is dealt with first, because such damage often precedes fungal or bacterial diseases. Following a short section on the influence of man, animals, and insects, comes Section III, in which diseases caused by living agencies are considered, first in their general aspects and then in their specific ones, on the basis of the host tree genera attacked.

In these chapters in Section III a number of standard headings and sub-headings have been used. These are:

Diseases due to non-living agencies. This is only a brief recapitulation of any important points already dealt with in Section I, but particularly affecting the genus under consideration.
Nursery diseases.
Root diseases.
Stem diseases—wilts. Pathogens infecting the internal tissues of the stem and causing wilting.
Stem diseases—bark and cambium. Pathogens attacking the outer parts of the stem causing dieback and canker.
Diseases of leaves and shoots.

The omission of any of these headings from a chapter merely indicates that there are no important diseases to be included under it. Information on the more important fungal and bacterial diseases is divided under four sub-headings:

Pathogen,
Symptoms and Development,
Distribution and Damage,
Control.

These are probably self-explanatory, but it is worth pointing out that 'Distribution' includes occurrence on different species and provenances within species, as well as purely regional considerations.

Some further limitations on the scope of Section III may require explanation. Descriptions are limited to macroscopic characters. Nothing is described which is too small to be seen with a good hand lens. The reasons for this are more fully explained in the introduction to Chapter 2. Decay is considered only as it affects the standing tree. Decay in felled timber and in forest products has already been adequately covered in other books, references to which are given in Chapter 19.

Naturally consideration of diseases outside, as well as in Britain, results in possibilities of confusion in the text. However, it can generally be assumed that any disease dealt with is present in Britain unless the contrary is stated, and that all remarks on the development and treatment of a disease apply to Britain, unless some qualification is made. In this context it may be useful to explain that the word 'Continent' has been used throughout the book to denote Europe, excluding the British Isles. In the bibliography the papers fall into four, often overlapping, classes:

(*a*) Papers giving a good general account of a disease or of some important aspect of it.
(*b*) Recent papers, with good bibliographies of their own, which can be used as a start in tracking down all the literature available.
(*c*) Papers supporting some statement in the text, which might otherwise be considered to be purely the author's own opinion or experience.
(*d*) Papers referring to the occurrence or progress of the disease in Britain.

With reference to (*c*) above, most statements unaccompanied by any reference are either culled from general textbooks or from the author's own experience. In some cases, where it seemed desirable to do so, the author's own contributions have been made more definite by inserting (Peace unpubl.).

No attempt has been made to give a complete bibliography. This would obviously have been an impossible task. Apart from many other omissions, authorities for records of fungal or bacterial occurrence in a particular country have been given only when they were of outstanding importance, or contained other matter of more general interest. Within the general groups outlined above, first preference has been given to papers referring to Britain, and second preference to papers in the English language. It is hoped that the abstract references given may be an aid to those wishing to consult papers in some of the more obscure English language, and all the foreign language, journals.

A number of textbooks and other works have been in such frequent use that the information culled from them could not be made the subject of continual references in the text. Books and papers used in this general way are listed below:

General information on fungal or bacterial diseases

Hartig 1894, 1900; Massee 1910; Neger 1924; Ferdinandsen and Jørgensen 1938–9; Dodge and Rickett 1948; Boyce, J. S. 1948; Pirone 1948; Butler and Jones 1949; Viennot-Bourgin 1949; Baxter 1952; Gram and Weber 1952; Brooks 1953; Buckland, Redmond and Pomerleau 1957; Dowson 1957.

General information on non-living agencies

Boyce, J. S. 1948; Hawley and Stickel 1948; Pirone 1948; Baxter 1952; Edlin and Nimmo 1957.

Information on special groups of fungi

Grove (Uredinales) 1913; Rea (Basidiomycetes) 1922; Darker (Hypodermataceae) 1932; Grove (Sphaeropsidales and Melanconiales) 1935; Petch (Hypocreales) 1938; Bisby and Mason (Pyrenomycetes) 1940; Wakefield and Bisby (Hyphomycetes) 1941; Boyce, J. S. (Uredinales) 1943; Wakefield and Dennis (Larger Basidiomycetes) 1950; Ramsbottom and Balfour Brown (Discomycetes) 1951; Wilson and Bisby (Uredinales) 1954.

Fungal nomenclature

Brooks 1953; Ainsworth and Bisby 1954.

British and other fungal records

Ainsworth 1937; Moore, W. C. 1948; Spaulding 1956, 1958; Foister 1959; Moore, W. C. 1959.

Nomenclature and behaviour of host trees and other plants

Rehder 1949; Bean 1950–1; Clapham, Tutin and Warburg 1952; Macdonald *et al.* 1957.

The Index is so arranged that it is possible to find immediately the principal references in the text to any particular pathogen, disease, or non-living agency. Living pathogens are separated according to species. In the case of host trees the breakdown is only to the generic level, but at that level diseases due to different pathogens and agencies are indexed separately. As a partial compensation for the lack of host species in the Index, a list of these, with English names and important synonyms, is given as an appendix. This list is the only place where authorities are given for host names.

2

THE SYMPTOMS AND DIAGNOSIS OF TREE DISEASES

THE symptoms of tree disease follow general patterns according to the part or parts of the tree which are injured, and it is the purpose of this chapter to describe these rather than to discuss the precise symptoms produced by different pathogenic agencies. The detailed symptoms associated with particular diseases are described more fully elsewhere in this book, where, however, the descriptions of symptoms have generally been limited to those which can be seen with the naked eye or with a hand lens, so that, for instance, only the gross characters of fungal fruit bodies have been described. Most fungi can only be identified firstly by reference to related genera and secondly in comparison with other members of the same genus. Details of spore-sizes or of the microscopic anatomy of fruit-bodies, if given in a general textbook such as this, may therefore do more harm than good by leading to hasty and incorrect identifications. For this reason in particular, but also on the grounds of space, they have been omitted. In some instances, unfortunately, it has proved impossible to provide proper descriptions, even of the macroscopic symptoms of diseases of minor importance. Authors, primarily mycological in interest, have often described the causal fungus without paying much attention to the symptoms it produced. This is one reason why no keys have been provided for the recognition of diseases. Another strong argument against the use of such keys is that two quite different agencies can produce exactly the same set of symptoms. If the causal agent is a living one, the confusion thus arising can sometimes be resolved by the appearance or discovery of its fruit-bodies, or by culturing the pathogen from the diseased tissues. If non-living agencies are involved the true cause can sometimes be ascertained by a careful study of the past and present environment. The position may be still further complicated because the symptoms are produced by two or even more pathogenic agencies, usually acting together. All these complications tend to make diagnosis on the basis of symptoms alone often very difficult and sometimes quite impossible.

Leaf symptoms

In general leaf symptoms tend to progress from discoloration and death of small areas of tissue towards the involvement of the whole leaf, followed by withering and sometimes, though not always, by shedding. Many fungi cause small discoloured spots on leaves, followed by local necrosis (Cunningham 1928). Eventually a large proportion of the leaf area may be involved, either by increase in size of a limited number of big spots, or by coalescence of a larger number of small ones. Leaf spots are not invariably fungal or bacterial; some non-living agencies, such as toxic fumes or hail, can cause spotting. On the narrow needles of conifers development of damage is naturally rather different; a small lesion can soon isolate the distal portion of the needle; so that spotting is usually a very short-lived phase resulting almost immediately in tip death. Both in leaves and in needles, gum or resin barriers are sometimes laid down as a reaction to fungal invasion. These are particularly noticeable in pine needles attacked by *Lophodermium pinastri.*

The distribution of fungal and bacterial lesions sometimes bears a relation to the structure of the leaf (Fig. 75), but often they are scattered haphazardly (Fig. 98). Leaf damage by non-living agencies on the other hand is often distributed in rough conformity with the anatomy of the leaf on which it occurs; interveinal and marginal discoloration and marginal withering are common symptoms for such pathogenic agencies as mineral deficiency, drought, or industrial fumes (Figs. 19, 26). Yellowing of the whole leaf blade or needle, usually described as chlorosis, is primarily associated with deficiency disease. Variegation, in which part of the leaf blade is yellow or white and part green, the division often being abrupt and sometimes very regular, is often due to genetic mutation, not to disease at all, but it is also quite a common symptom of virus attack on broadleaved trees. Superficial mycelium, such as is formed by the mildews on succulent leaves and shoots, obviously assists in diagnosis.

All these forms of local attack may and often do lead to the death of the whole leaf or needle. If the leaf has a rigid structure it may retain its general shape, though possibly curling slightly, but soft-structured leaves, especially if they die when young, collapse to a varying degree. Large-scale withering of leaves is one of the forms of disease sometimes rather loosely referred to as 'Blight' (Fig. 83).

Collapse and death of leaves can be caused indirectly by injury to the lower parts of the tree. Anything that cuts off the water-supply to the leaves, such as drought, death of roots, girdling of the stem, or blockage of the vessels will lead eventually to the death of foliage. In some wilt diseases this process may be hastened by the production of toxins by the causal agent.

Reduction in size of leaves or shortening of needles on conifers is almost always a sign of damage to the stem or more particularly to the roots, or else of some serious disturbance in the nutritional balance of the tree. Reduction in needle-length of conifers often follows defoliation the previous year. If hardwoods produce a second crop of leaves following premature defoliation, they are always abnormally small. Reduction in size does not indicate any direct damage to the leaves themselves. Fusion of needles in conifers (Fig. 21), which is often accompanied by shortening, is also an indirect effect, having been variously associated with drought, with mineral deficiency, and with lack of mycorrhiza.

The whole subject of the abscission of leaves has been reviewed by Addicott and Lynch (1955). Leaf or needle diseases may result either in premature defoliation or in retention of the dead foliage beyond its normal span. For instance, frost injury on larch may kill the needles, which then still adhere to the plant, whereas attack by the fungus *Meria laricis* results in premature needle fall. In general rapid death of the leaves results in their retention, whereas slow death, which is the more normal outcome of fungal attack, ends in premature shedding. Campbell and Vines (1938) have described in detail the effect of *Lophodermium macrosporum,* which does sometimes cause needle retention, on the abscission of spruce needles.

The effect of defoliation on a tree depends not only on its degree, but also on the time of its occurrence. Its effect on the general health of the tree is greater in conifers than in hardwoods and greater in evergreen than in deciduous trees. The most obvious result is a reduction in growth, which is roughly proportional to the degree of defoliation (Bruce 1956; Mott, Nairn, and Cook 1957; Stark and Cook 1957; Blais 1958). In evergreens, loss of the young foliage is generally more serious than loss of the old (Craighead 1940). Apart from any other considerations loss of new leaves or needles will affect the tree for a longer period than loss of old. During a single year, however, Linzon (1958) found that loss of year-old needles of White pine had a greater effect than loss of those of the current year. Most trees, even conifers, can survive a single complete defoliation with no worse effects than a reduction in growth, but two or three such occurrences in conifers normally result in death (Stark and Cook 1957). Broadleaved trees can stand repeated yearly defoliation in the late summer or early autumn with little effect. Defoliation earlier in the season normally results in the production of a second crop of leaves. There is evidence that repeated early defoliation in broadleaved trees eventually weakens the tree and leads to dieback. There is also evidence, particularly in the case of poplar, that loss of foliage towards the end of the summer retards ripening of the defoliated shoots, so that they remain abnormally susceptible to frost. In this way defoliation may lead indirectly to dieback. Defoliation, as well as reducing the width of the annual ring, can also affect the nature of the wood by reducing the thickness of the cell walls, and therefore making the summer wood less clearly defined (Harper 1913).

Shoot symptoms

It is not possible exactly to separate the symptoms on young shoots and twigs from those on the older wood. Possibly the clearest division as far as symptoms are concerned is at the point where the shoot hardens and becomes more or less rigid. This hardening is a gradual process proceeding up the stem, so that the whole of the shoot produced in any one year has lost its succulence shortly after growth ceases. If succulent tips remain for long after growth has ceased, they are often damaged by frost or attacked by fungi such as *Botrytis cinerea.* As long as the shoot or shoot-tip is succulent it is capable of wilting and withering, and is often subject to attack by some of the same fungi that infect leaves. On the other hand, many fungi which attack shoot-tips can also extend into older, harder stems. Diseases involving the leaves and succulent shoots are often known as 'Shoot blights' or 'Anthracnoses', those involving the shoots and twigs as 'Twig blights' or 'Twig diebacks'. In wilt diseases, involving either interruption of the water-supply or the presence of toxins in the sap stream, the succulent shoot-tips and the leaves are both damaged.

Shoots may be deformed by damage which does not actually kill them. Such deformations are particularly striking if the injury occurs while the shoot is still in the process of elongation. This often occurs in attack by *Melampsora pinitorqua* on Two-needled pines (Fig. 61). Many fungi distort and often enlarge the growing parts of their host plants. Such distortion is particularly apt to occur in developing shoots and leaves. Fungi belonging to the *Taphrinales* are particularly apt to behave in this way (Figs. 94, 97).

Buds are, of course, inevitably involved, if the twig on which they are borne is diseased or dies, but they can also die without any immediate injury to the twig. A few fungi attack only or mainly buds, for instance *Cucurbitaria piceae* on spruce and *Pycnostysanus azaleae* on rhododendron. Frost will often kill buds which have started to become active before the cambium of the twigs on which they are borne. The fate of a shoot deprived of buds depends on its ability to produce new ones. A shoot left permanently budless will eventually die (Onaka 1950).

Bud infection may sometimes lead to the production of abnormal growths such as witches' brooms; but many of these abnormalities also arise from bud alterations which are not attributable to invasion by any organism (Chapter 20).

The flushing in the same year of buds formed during the summer leads to the production of lammas shoots, which may be particularly prone to fungal infection, as are those of oak to *Microsphaera alphitoides*, or to autumn frost injury. Lammas shoots may also occasionally result in deformed growth, as exemplified by 'Extra-seasonal growth' in pine (Chapter 22).

Stem symptoms

The woody aerial parts of the tree provide a very wide variety of externa and internal symptoms. Damage can be to the bark, to the cambium, or to the wood. Injury to the bark alone is usually very difficult to distinguish from that to the bark and cambium together. In its simplest form such damage involves the death of a patch of bark, the dead tissue of which may extend to the cambium. If this lesion surrounds the stem it will result in girdling and death of all the upper portion of the stem. If the stem is not girdled development will usually involve successive annual regrowths of healing tissue, possibly interrupted by extensions of the lesion, either by the original, or possibly by other, pathogenic agencies. Such wounds surrounded by ridges of healing tissue are generally known as 'cankers'. They vary greatly in form, sometimes being highly irregular and erumpent, and sometimes forming regular ridges of tissue around the focus of the wound, the so-called target cankers (Fig. 91). Ring counts in the callus tissue can be used to date the initial injury, or the death of a branch (Andrews and Gill 1939). On slow-growing trees, for instance on old beech, or on lesions which are extending rapidly, such as those which *Endothia parasitica* sometimes causes on *Castanea*, little healing takes place, and there is no real canker formation. Such lesions are best described as bark dieback (Plate XV, fig. 81). Zycha (1955*b*) has tried to establish a clear distinction between canker and bark dieback, but in fact there are many marginal examples which cannot clearly be placed in either category. Bark dieback, canker, shoot dieback, and stem dieback are all closely related symptoms, since the last two will occur as soon as a lesion or canker has extended right round the stem.

Lesions on the stem are often accompanied by the production of resin or gum from the host tissues. Resin bleeding occurs chiefly in conifers, while gum production is found particularly in the genus *Prunus*. Exudations of this sort are often, though not always, signs of disease, so that trees displaying them are usually worth investigation. As a means of diagnosis they are only of limited value, since on some trees they are an almost universal reaction to wounding of any kind.

Wilt diseases show externally by the collapse of the leaves and succulent shoots, but many display very distinctive internal symptoms, visible if infected wood is cut across. This is the case with Elm disease (*Ceratocystis ulmi*), *Verticillium* wilt of maple, and Watermark disease of Cricket-bat willow. The dark markings in the wood in these cases are mainly due to the production of tyloses and wound gum in the vessels. The formation of these bodies, with especial reference to Elm disease, has been studied by Broekhuizen (1929).

According to the nature of the invading fungus infection of the wood often, though not always, leads to staining and possibly to decay. The symptoms associated with these are dealt with in Chapter 19. The wood is also subject to a number of mechanical defects mostly arising from climatic stresses. These include frost crack, drought crack, and partial failures of the wood due to wind, or to the weight of ice or snow. The heartwood of older trees often contains radial or circumferential cracks of rather obscure origin, known as shakes (Chapter 20).

Other abnormalities in the wood result from injury to the cambium and the subsequent formation of abnormal cells. These cells usually lie in the form of a ring, being those formed while the cambium was still in an injured condition. Abnormal rings of this kind can be caused by frost, lightning, drought, and locally by wounding or by fungal attack (Nylinder 1951; Manners 1957) (Figs. 12, 17, Plate II. 3), and are described in more detail in appropriate chapters. They are particularly valuable in diagnosis, not only because they may disclose in conjunction with a knowledge of the environment of the tree, the cause of the abnormal ring and of other injuries associated with it, but because, by their position in the annual ring, they give some measure of the time of their occurrence. Fig. 1 illustrates diagrammatically how the position of the injury makes it possible to date within broad limits the occurrence of the causal catastrophe. Natural frost rings have been used to date other occurrences in the tree (Glock 1951), and those caused artificially by the local application of solid carbon dioxide to give permanent dating 'tags' on experimental trees (Studhalter and Glock 1948).

Disease often reduces the growth rate of a tree, which is reflected not only in the narrow width of the annual rings, but also by a reduction in the length of the new shoots. On the other hand, annual rings in the neighbourhood of wounds are often abnormally wide, and decayed trees sometimes exhibit increased diameter growth near the base, giving the lower stem a slightly bottle-shaped appearance.

Root symptoms

Root symptoms by virtue of their position are naturally harder to observe, and therefore less often recorded than those on the above-ground parts of the tree. The death of fine roots appears to be a normal part of a tree's function, and it is very difficult to decide at what level this phenomenon becomes pathological. In a seedling with a very limited number of root-tips it certainly has a pathological significance. Death of the ends of larger roots from circumstances such as a rise in the water-level is also a very common phenomenon, but is probably sufficiently abnormal and damaging to be regarded as a disease symptom. Death of whole roots, which is often the result of fungal attack, is very definitely a matter of disease. When death has reached the collar it may extend up the lower part of the stem, sometimes internally in the form of decay, as happens with *Fomes annosus*, sometimes

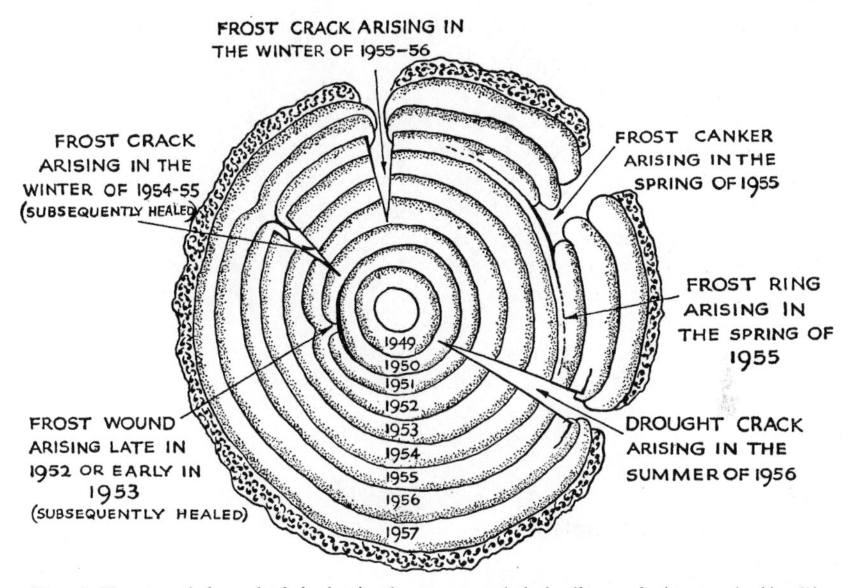

FIG. 1. Frost and drought injuries in the stem, and their diagnosis (see text). (A. C.)

externally as death of the cambium, for example the tongue-shaped watery dead patches extending up the stem above roots killed by *Phytophthora* on *Castanea* and other trees. Resin flow often occurs from diseased conifer roots.

The initial infection of roots is, of course, seldom observed. Sometimes it is through dead tips killed by other agencies, sometimes by root contact or root-grafts between diseased and healthy trees, as with *Fomes annosus*. Root grafts, which display their presence in *Pseudotsuga* and some other species by the healing of cut stump surfaces because they are attached by grafts to still living trees, are certainly much commoner than is generally realized (Yli-Vakkuri 1954; Kobendza 1955; Šafar 1955; Molotkov 1956), and may have considerable pathological significance. With a root graft the fungus is able to stay in the internal tissues while passing from tree to tree and there is no need for it to grow from surface to surface, where it would be subject to the sometimes controlling influence of other organisms in the rhizosphere. Infection may also take place directly, through the soil, as is the case with *Armillaria mellea* and *Pythium* spp. In such cases excavation of roots at an early stage will disclose necrotic spots around the points of infection. As with shoots, once a lesion has extended right round a root, the distal portion dies. Eventually dead roots become decayed. Sometimes, where the initial damage was fungal in origin, the fungus causing decay may be different from the original invader.

Recovery symptoms

If damage to a tree is the result of a single catastrophe, or if a progressive attack lessens in intensity or ceases, even if only for a period, the tree, provided it has retained sufficient vigour, will start to produce recovery tissue. Those

FIG. 2. Juvenile recovery shoots following drought or frost injury to *Pinus halepensis*, Madrid, Spain, April. (J. C.)

parts of the tree with an active cambial layer are obviously best able to do this. Damaged buds and leaves are hardly able to make any efforts towards recovery, beyond the production of gum barriers.

Stems, on the other hand, produce healing tissue more easily. This tissue may succeed in covering the lesion, or if further damaged may form the basis of canker development. If buds have been killed or if dieback has taken place, dormant buds already present develop or new adventitious buds may be produced, sometimes in very large quantities. Production of very large numbers of recovery buds and sometimes of resultant shoots is a very common reaction to the pruning effect of frost injury in some species. This is particularly striking in larch where all the needles in an initially single-budded short-shoot may produce axillary recovery buds, if the centre bud is killed. The ability to produce recovery buds varies enormously between different trees. Some genera such as *Corylus*, *Castanea*, or *Quercus* have the ability to initiate buds on quite old wood, which in practice renders them suitable species for growth as coppice. Pathologically it enables them to recover from very severe dieback. Recovery shoots on conifers sometimes bear juvenile leaves, such as are normally found on seedlings (Fig. 2).

Roots also display considerable powers of recovery, adventitious roots being produced from the live portions of diseased ones. Under favourable circumstances a whole new root system, sufficient to support and maintain the tree, may be formed above the original roots lost by disease. Minor regenerations of dead roots are an almost normal part of the life of any tree.

The final outcome of disease on a tree, whether it be caused by living or non-living agencies, may often depend more on the recuperative powers of the host than on the initial severity of attack. Some trees have much greater powers of recovery than others. Japanese larch, for instance, displays much greater ability to recover from branch dieback than does European; loss of buds is much more serious on *Abies* than on *Picea*, and so on.

Evaluation of symptoms

For the purposes of comparing the severity of disease between different trees or different stands, or as a measure of the progress or decline of a disease in a given area over a period of time, methods of evaluating symptoms are an essential. The mere size of a tree makes some agricultural methods, such as the counting of lesions, normally of little value for trees, except in the nursery. In comparison with agriculture very little work has been done on this aspect of forest pathology, though Buchanan, Harvey and Welch (1951) elaborated a system for numerical evaluation of the symptoms of Pole blight in Western white pine. In general an approximate method applied to a large number of trees is of more value than an accurate method applied to a few. Sometimes it may be possible to count the number of dead branches or dead upper roots. With diseases such as *Lophodermium pinastri*, which on pole-stage pine tends to result in premature killing of the lower branches, an estimate of the number of whorls of live branches may be of value. The survey of Elm disease in Britain (Chapter 31) has been done mainly by a visual evaluation of symptoms based roughly on the proportion of active disease and of death in the crown. Inoculations with *Pseudomonas syringae fm. populea*, the cause of Bacterial canker of poplar, have been assessed on visual estimate of the size of the lesion produced. The severity of *Fomes annosus* attack causing death of pines has been assessed on shortening of the annual shoot growth, on diminutions in the size of the needles and on reduction in their number, and the same symptoms are used in such stands for the selection of trees for thinning. The presence of stain or decay in the cut stumps of thinnings can be used as the basis of an approximate estimate both of the disease condition of the roots and of the decay condition of the stems in the remaining standing trees. Other diseases have been dealt with on even vaguer personal estimates of severity classes. The great advantage of such methods is that they are extremely rapid, and can therefore be applied to very large numbers of individuals. Unfortunately they lose much of their value with any change of observer. If periodical observations are going to be made by different people, a much more objective method of estimation, depending on one or more easily countable or estimatable characters, must be devised. In this matter it is quite impossible to make suggestions of general application. It is far better to build up the most workable and appropriate method for each case where the evaluation of symptoms is necessary.

The diagnosis of tree diseases

It has already been pointed out that the diagnosis of tree diseases is too difficult a matter to be expressed in any kind of key, however complex. Nevertheless, it can best be approached in a methodical manner and it can be considered, both in theory and in practice, under a number of separate heads.

(i) **The reaction of the tree.** The symptoms on the affected tree or trees should be studied in an endeavour to decide exactly what parts are affected, and in particular what parts were affected first. In some cases symptoms may be specific to a certain disease or a certain pathogen, in other cases they may indicate the form of attack rather than the cause. Where a number of trees are concerned it may be possible to get an idea of the progress of the disease in the tree by studying different individuals in various stages of injury. Where a group of trees is concerned the distribution on the ground of trees in different damage classes may also throw light on the direction and rate of progress of the disease, and indeed on whether it is still spreading.

(ii) **The presence of pathogens.** A search should be made for signs of fungal mycelium or fruit bodies, for bacterial slime, and so on. The discovery of a pathogen should never lead to the assumption that it is the cause of the disease. Its presence must be considered in relation to the symptoms on the tree and to the conditions under which the tree is growing.

(iii) **The environment of the tree.** The surroundings of the tree should be studied as far as possible in all their aspects. It is of particular value if they can be compared with the surroundings of otherwise similar, undiseased trees. If this is not done there is a risk that an adverse factor, which is operative over a large area, may be blamed for a local outbreak of disease occurring in only part of its range. Small changes in the environmental conditions studied in relation to the distribution or severity of the disease may often provide valuable clues.

(iv) **The environmental history of the tree.** The surroundings of the tree are important, not only in their present condition, but also in their past history. Directly damaging factors such as frost or wind, or indirect influences such as periods of damp weather favouring fungal infection, are naturally not permanent parts of the environment and information on them must be obtained from past records.

Much of the data under the four headings given above can be acquired on the site. Laboratory investigation may be required for a more exact interpretation of the symptoms, and microscope examination, possibly accompanied by culturing, for the discovery of all the potential pathogens present. Examination of the environment may involve such investigations as soil analysis, and the consultation of climatic records. In practice, of course, diagnostic examinations must always be in some measure incomplete. Indeed it would not be reasonable to investigate all aspects, if for instance the presence of a known pathogen and its typical symptoms clearly indicated the cause of the trouble. Investigation of the present and past environment should always be limited to those aspects which a preliminary examination suggests may be relevant to the symptoms on the tree.

The fact that older trees are very large organisms, parts of which may be out of easy reach, and which have a long history of exposure to varying influences, often makes the diagnosis of their diseases particularly difficult. In all but the most obvious cases it is necessary to regard the problem primarily as one of detection. A gradual sifting of the useless and building up of the significant evidence will usually give a workable theory, if not always a definite answer, on the cause of any particular disease.

It is difficult to be more exact on the matter of tree-disease diagnosis without referring to actual examples. In practice such examples would be so varied that they would not form the basis of any general scheme. The few suggestions made above should not therefore be taken as a list of observations to be made, some of the most obvious having been omitted, but as isolated reminders of a few of the lines of investigation which are worth following up under certain circumstances. There should be no need, for instance, to call attention to the value of abnormal rings in the wood as an aid in diagnosis of frost and other damaging agencies, but annual rings are also valuable in dating the occurrence of wounds, or the death of branches (Andrews and Gill 1939). The dating of annual rings themselves is often aided by the occurrence, in the same ring of all the trees concerned, of abnormalities, not in themselves of any pathological significance, but which can be related to some known climatic occurrence (Dobbs 1951, 1952). One cannot overstress the importance of observing the distribution both of the symptoms on the diseased tree, and of the diseased trees in relation to their surroundings. For instance, lesions consistently on the south side of a tree would almost certainly have some connexion with the sun, while the clear demarcation of a diseased area by a drain obviously indicates the presence of a pathogen which cannot cross it, or else an abrupt change in soil conditions, sufficient to influence tree health, between its two sides. It is a general rule, which is often overlooked, that diseased trees which are still alive are much more valuable as a means of diagnosis than trees which are dead. The latter are rapidly invaded by fungi which may mask or even destroy the symptoms of the original disease. In culture work the most likely place to isolate the pathogen is at the joint of live and dead tissue in affected parts.

The evaluation of losses from tree diseases

This subject has been thoroughly discussed and reviewed by Baxter (1952) and by Chester (1950), both of whom have stressed its extreme complexity. Losses in tree crops, because of the long growth periods involved, present one of the most difficult cases in a generally difficult subject.

Estimation of loss must take much more into account than the mere reduction in present market value of the tree or crop involved. The time already occupied in growing the crop and the money spent on it, as well as potential future losses, must all be considered. In some cases loss of amenity value may also have to be assessed and this may occasionally be greater than the loss in timber value. Disease may affect not only the value of the crop, but also the value of the site. For instance, soil may deteriorate under the very open cover of a plantation seriously damaged by disease. This loss of value may even extend into the next rotation, taking, for instance, the form of a carry-over of infection. In the case of *Fomes annosus* (Chapter 16), light infection in one crop is mainly important because it provides a scattering of infected stumps to start an early and much more serious outbreak in the next rotation. An estimate of loss may also need to take into account the increased risk of disease to adjoining crops, where, for example, a high concentration of spores of the pathogen in the air will encourage the establishment of new infections nearby. Gaps in the canopy caused by disease may form foci for windblow during subsequent gales.

Even the estimation of immediate loss is not an easy matter. A dead tree is not necessarily a total loss, it may still be possible to market it. Some diseased trees may be perfectly saleable if they are felled in time. The economic effects of Elm disease in Britain have been greatly mitigated by the existence of a ready market for elm timber even from diseased trees. The visible damage in a tree is not by any means proportional to the loss in sale value. Severe crown dieback, for instance, may have no effect at all on the value of the main stem, provided the tree is marketed before stain and decay fungi have penetrated from the dead limbs into the trunk and before there has been a marked reduction in growth. Even the percentage of decay in a stem is not necessarily the same as the percentage loss, because its distribution, as well as its amount, will influence conversion and consequent value.

Many diseases result in a reduction in growth rate and therefore in volume increment. In the case of *Fomes* root-rot this may sometimes represent a greater monetary loss than that caused by the actual decay (Petrini 1946; Arvidson 1954; Rattsjö and Rennerfelt 1955). Growth-rate reduction is an aspect of loss which must be considered in the evaluation of any kind of damaging disease, not merely those on roots.

In some cases the degree of loss depends on the market for which the timber is destined. Cricket-bat willows visibly infected with Watermark disease are quite unsaleable for bats, buyers fearing that the stain may extend into the butt. In this instance the mere presence of the disease gives the tree a much lower value. Seedling pine defoliated by *Lophodermium pinastri* can often be safely transplanted and will make a complete recovery, but their immediate sale value is greatly diminished by the defoliation.

Forecasts of future losses, based not only on the damage already done by the disease, but also on its anticipated future development, are notably hard to make. In the author's experience few tree diseases have advanced at the rate or in the way that was expected of them from the evidence provided by their initial attacks.

In view of all these complications it is hardly surprising that, relative to agriculture, very little work has been done on losses due to disease on trees, with the exception of those due to wood-rotting fungi, an aspect of decay which is briefly considered in Chapter 19. The losses caused by White pine blister rust (*Cronartium ribicola*)(Chapter 22) have probably been investigated more carefully than those of any other tree disease not involving decay (Snell 1931; Filler 1933).

It has been pointed out elsewhere that expenditure on control measures in forestry is strictly limited by the relatively low annual increment in value of the crop. That being so, it is obviously important to know the extent of the actual and potential loss, before measures are taken. It is surprising in view of this how little work has been done.

3

TREES IN RELATION TO THEIR ENVIRONMENT

IT is obvious that the proper study of any living organism involves consideration of its surroundings. The limitations which the environment imposes on it may be so severe as to prevent its very existence. On the other hand, the environment may be sufficiently favourable to allow it the maximum development which is genetically possible. The concept of disease is discussed more fully below, but it is usual to speak of disease when environmental limitations reach a level at which parts of the tree are seriously abnormal in growth, are dying or are dead.

The remainder of this book is almost entirely taken up with a consideration of the pathological effects on the tree of its environment, for pathogenic fungi, bacteria, and viruses are part of that environment.

The climate of Britain has been described by Bilham (1938) and Manley (1952); while a very useful summary with particular reference to forestry is provided by Wood (Macdonald *et al.* 1957). The limitations which the climate places on tree growth in Britain, and in particular on the species we can use, have been summarized by J. Macdonald (1951). In various ways climate exercises a far greater limitation than any other factor in the trees' environment. Forests have been planted in most of the different climates found in Britain, except of course in those which are too harsh for tree growth. The direct effect of climate as a cause of disease is covered in this book, by Chapter 4, 'The Effect of Frost on Trees', Chapter 5, 'Damage to Trees by Other Climatic Agencies and by Fire', and, as far as salt spray is concerned, by part of Chapter 8, 'Damage to Trees by Toxic Substances'. Rainfall, influencing the water-supply of the tree, and the damaging effects of dry winds are both dealt with in Chapter 6, 'Damage to trees by unsuitable Soil and Soil Water Conditions'.

British forests are in the main relegated to soils unsuitable for agriculture. Thus as regards tree growth a general survey of British soils is of much less value than a general survey of the British climate. A certain amount has been published on individual forest soil types, and a number of local investigations of soils in relation to forestry have been carried out. Little has been published on British forest soils as a whole, except that Macdonald has given a useful summary (Macdonald *et al.* 1957).

The relegation of forestry in Britain to soils marginal or unsuitable for agriculture inevitably means that much of our planting is on sites edaphically very far from ideal for tree growth. The limitations imposed by such sites may sometimes be sufficiently serious in their effects to be considered pathological. W. R. Day (1949) suggested that the planting of some soils may prove uneconomic. Since we have attempted to extend our planting limits, using sites which were considered unplantable in the past, and species whose site limitations are not fully known, it is inevitable that in some instances the trees should prove to be out of balance with the site conditions, so that unhealthy crops result. By and large, however, these new pioneer plantings have been remarkably successful.

In this book the air and water conditions of the soil, in so far as they influence or cause disease, are dealt with in Chapter 6, diseases due to lack of mineral elements in the soil in Chapter 7, 'Deficiency Diseases of Trees', and soil toxicities in Chapter 8.

The effect of most of these non-living agencies is to some extent relative. One tree may grow quite healthily under conditions which would cause disease or even death in another tree, even of the same species, if the latter were suddenly subjected to them. In many cases the degree and rate of change in the environment are more important than the actual level which the factors of the environment reach. For instance, a tree growing under cold conditions may be unharmed by a frost which would kill a similar tree that had been growing in warm conditions. A tree planted in a hole in a pavement may grow quite well, but an established tree may die for lack of soil oxygen when a pavement is laid over its roots. This question of degree and rate of environmental change must be kept in mind in all considerations of injury due to non-living agencies.

It is perhaps not so immediately obvious that living agencies such as fungi, bacteria, and other plants, as well as viruses, growing in, on or around trees and sometimes parasitizing them, are also part of the environment. They are dealt with in Section III of this book. Equally environmental to the tree are insects and animals, whose influence on tree diseases, though not the direct damage they do, is discussed in Chapter 10, 'The Influence of Man, Animals and Insects on Tree Diseases'. Man directly and indirectly greatly influences the conditions under which trees grow. Indeed as a forester or arboriculturist it is his duty so to do. Man's activities are particularly considered in Chapter 10, but one special aspect of them is dealt with in Chapter 8, 'Damage to Trees by Toxic Substances'.

It is clear, therefore, that non-living and living agencies, acting singly or in concert and often in succession, form the environment of the tree. Under certain circumstances, and usually in a very complex manner, this environment can become damaging. Baxter and Wadsworth (1939) have described a complex pathological succession in the natural forests of Alaska. The importance of the environment has been considerably and rightly stressed in recent years, in its narrower definition of climatic and edaphic surroundings by W. R. Day (1955) and many Continental silviculturists (Dengler 1930; Rubner, K. 1953) and in its broader sense by Baxter (1952). There are certainly many cases, dealt with later in this book, where the non-living factors of the environment greatly influence the initiation and subsequent development of fungal and bacterial diseases. It is entirely wrong to ignore such factors when considering any disease. It is equally wrong to assume that their substantial intervention is necessary for the successful invasion and development of all living pathogens. Many fungi and bacteria are alarmingly successful on trees which would have satisfied any reasonable definition of good health before they were attacked. It is important that the wide range of factors influencing the health of trees should not lead us, with our present limited knowledge, into drawing broad conclusions. Each case of disease should be, and is here, studied on its own merits. Some climatic and edaphic factors are certainly concerned, though not always in the same way, with a wide range of tree diseases. A study of one disease may well throw light on another, but any effort to impose broad generalizations on a subject so imperfectly understood as forest pathology is to court disaster. We can only learn to recognize the influences involved, and the pathological symptoms they cause, and with this knowledge study each disease without trying to fit it into any general pattern of our own or another's making.

The concept of disease

One obvious difficulty in the study of forest pathology is to decide what constitutes disease. The tree makes its maximum growth only on an optimum site, but it is hardly reasonable to consider any falling away from this maximum as disease. At the opposite end of the scale, disease obviously includes death or even deterioration of large parts of the tree. Death of small parts is not necessarily disease, for leaves, shoots, and small roots all die and are shed on occasion as part of the normal behaviour of the tree. The presence of a fungus or bacterium in the tissues can hardly be regarded as disease, unless it produces some pathological effect. Slow growth is not usually considered a matter of disease, unless it is accompanied by more definite pathological symptoms. These may be defined as abnormal development, deterioration or death of tissue. W. R. Day (1953) has referred to the death of fine roots on spruce as 'root-disease'. Probably this term should be used only if the amount of root death is appreciably greater than that found on similar trees growing at a satisfactory rate on the same or on other sites. In this matter one possible yardstick may even be the managerial one, by which measure we should regard a tree as diseased only if it were suffering sufficient loss of increment or timber, to lessen its sale value. In this connexion it is necessary to remember that slow growth may sometimes enhance the value of timber.

It is fairly certain that trees do not just die of 'old age' (Salisbury, E. J. 1947). Parts of even the oldest tree, for instance the leaves, the shoots, the fine roots, and the outer wood, are constantly renewed. As a tree becomes older it becomes less and less adaptable by its mere increase in size to changes in environment, and has a progressively larger area subject to adverse influences and to injuries. Fungal infections, particularly by wood-rotting fungi, become more and more extensive. There may well be gradual accumulation of the by-products of assimilation until they reach a toxic level (Salisbury, E. J. 1947). Fortunately in forestry, though not of course in arboriculture, we are hardly concerned with the senescent tree, since we aim to fell and replace a tree before it has reached this condition. Nevertheless, the deterioration of old trees is as much a part of pathology as the diseases of young and middle-aged ones.

The natural forest

In Britain we are not primarily concerned with the natural forest. It exists here largely as a concept, which is considered by some to be the ideal basis for forest practice, both from a pathological and silvicultural point of view (Anon. 1939*b*; Anderson, M. L. 1956; Köstler 1956*a*). The dangers in this idea of the inherent desirability of the natural forest have been discussed by Peace (1957*b*). Some natural forests have probably reached a state of pathological equilibrium, in which damage by disease is no longer particularly apparent; but the behaviour of introduced diseases such as Chestnut blight, *Endothia parasitica*, in the mixed hardwood forests of the eastern United States, or even of Elm disease, *Ceratocystis ulmi*, in semi-natural elm woods in Britain, shows how easily this balance can be disturbed.

As practical foresters we are concerned with growing utilizable trees in the climates and the soils at our disposal. For this the trees must maintain both a reasonable level of health, and an economic growth rate, and also reach a utilizable size. Whether we look at the matter pathologically or silviculturally, there is no reason for us to assume that this can only be done, or even best be done, by reference to the natural forest. Even in initially undisturbed forest the intrusion of the forester, felling and regenerating, immediately produces abnormal conditions, so that no managed forest can be truly natural. Some would argue that forestry should try to keep these interferences to a minimum, and to bring them as close as possible to normal forest behaviour. It is certainly desirable that we should try to discover in each instance what balance, if any, we are disturbing, and whether the disturbance is damaging or beneficial to the future of the forest. We should then have a better basis for our forest practice than any unreasoning appeal to 'natural laws'.

The artificial forest

In Britain we are almost entirely concerned with artificial forests, and it is often suggested that we ought to have made a sustained effort to bring these into a semi-natural condition. Plantation forestry, particularly as it has been practised in Britain during the last forty years, is subject to two particular criticisms, one directed against the use of exotics and the other against the use of pure stands. It is curiously often assumed that a tree growing as an exotic is inevitably subject to conditions less favourable to it than are found in its native range. Since many plant distributions are a matter of geography, topography, and chance, it would be rash indeed to assume that a plant's natural range comprises the only conditions entirely suitable for its growth. In some cases trees grow better outside their native range than they do within it. In many cases, as has clearly been illustrated in Britain, exotics grow better on some sites than native trees. The information on the encouraging growth and in many cases comparative freedom from disease of exotic trees in Britain, given by Macdonald *et al.* (1957), should go far to refute the gloomy views of those who consider that exotics are more often than not a failure (Boyce, J. S. 1941, 1954*a*; Grimal 1956). There are, of course, numerous examples of the failure of introduced species, though in many such cases the exotic nature of the tree has no bearing on the disease. In any case it can hardly be supposed that introduced trees would always be moved to environments where they would thrive and remain free from disease. Many such failures are much less complete than they appear at first sight, and there are plenty of examples to show that exotics can be just as successful as, and no more prone to disease than, native trees (Moulds 1957; Peace 1957*b*).

Pure crops are also a matter of constant criticism, especially on the Continent (Anon. 1939*b*; Leibundgut 1947; Frohlich 1949; Boyce, J. S. 1954*a*). It must be admitted that pure stands provide ideal conditions for the build-up of a pathogen, since infection can pass directly from tree to tree. A mixed stand is usually less dangerous, provided the pathogen is not one affecting all the species in the mixture. It is very important to bear in mind, however, that the ideal stand would be a pure one of a resistant species. Many failures have been attributed to stands being pure that were in reality due to entirely different causes (Champion 1933). We should certainly experiment further with mixed plantings, but we should not assume at the start that they are pathologically more desirable under all circumstances. Agriculture is largely based on monoculture, and to secure full production, forestry will almost certainly have to be run on some of the same lines (Peace 1957*b*). To attain more productive and healthier crops we can use choice of provenance, selection, and breeding to improve disease resistance, albeit more slowly than in agriculture, and we can elaborate control methods, though they may have to be cheaper and less direct than agricultural ones.

Selection and breeding lead to consideration of the use of clones. The risks sometimes involved in pure stands are certainly enhanced in pure clonal ones, where the trees should display no variation in resistance to a pathogen (Hartley 1939). It would be rash to assume that any particular clone was resistant to all the pathogens that could attack it.

It is also sometimes suggested that clones propagated vegetatively over long periods eventually become senescent (Rohmeder 1957; Schröck 1957). Senescence has sometimes been put forward as a basis for deterioration in species, even though in that case propagation is by seed. The idea of species senescence can almost certainly be discounted (Cain 1944), and there is little foundation for the supposed senescence of clones (Salisbury, E. J. 1947). The idea of clonal degeneration has been based largely on the progressive increase of virus diseases in vegetatively propagated stocks, which is, of course, a very different matter. There has also been confusion because of the reduced rooting ability of material from the older parts of trees as compared with that of juvenile shoots.

With most of our forest trees propagation from seed ensures the use of mixed stocks. In the future when most of our seed sources will be the result of selection and breeding, we can use seed orchards in which a sufficient number of selected clones have been planted. Where vegetative propagation is practised, as with poplars, a sufficient number of clones must be used, certainly over the country as a whole, and preferably in each forest or estate, to do away with the serious disease risk involved in the widespread use of one or two clones.

4

THE EFFECT OF FROST ON TREES

THERE are many small areas in Britain which experience temperatures below freezing in every month of the year. There are many larger ones where July and August are the only two months which are regularly and completely frost free. Most of Britain, however, enjoys a long period free from damaging frosts from May to September or even from April to October inclusive. But in some years this is interrupted by spring frosts which may occur as late as the beginning of June, and by autumn frosts any time after mid-September. The start and finish of the summer growing period in Britain is often subject to alternations of warm and cold weather, which can be extremely damaging to vegetation. The winter climate in Britain is mitigated by the nearness of the sea, so that we never suffer temperatures as low as those which occur, for instance, in the northern part of the continent of Europe, or in the interior of the United States. Nevertheless, there are occasional winters when cold air from the Arctic regions or from the big land masses of northern Europe brings hard, prolonged frost, accompanied by dry winds, and by substantial freezing of the soil, in fact conditions very conducive to winter damage to vegetation. Indeed frost plays a much greater part in forestry and arboriculture than the generally favourable nature of our climate would lead one to expect. In fact it is one of the most serious causes of damage to trees in Britain. Frost, winter cold, and other related aspects of the British climate have been fully discussed by Bilham (1938) and by Manley (1952).

In describing the effect of frost and cold weather, winter cold is considered separately from autumn and spring frost. In the autumn, of course, the typical frosts of that season tend to merge into winter conditions, and likewise early in the year it is impossible to make a definite separation between winter and spring frosts. Despite this it is preferable to deal with them separately, because they differ so much in the way they develop and in the way they affect plants. It is necessary first to mention the ways in which plants are damaged by frost, and the reasons for their varying resistance to it.

Hardiness and the mechanism of frost injury

This extremely complex subject has been reviewed by Chandler (1954), and particularly fully by J. Levitt (1956). The latter considers that the damage caused by excessively low temperatures depends mainly on the formation of ice crystals in the plant. This may happen inside the cells, in which case the protoplasm is invariably killed by mechanical disruption, or in the intercellular spaces, in which case the protoplasm may be damaged by the withdrawal of water through the cell wall. Hardiness in plants, therefore, depends particularly on an increase in the amount of 'bound' water in the protoplasm, i.e. water held by the proteins of the protoplasm so that it cannot be withdrawn on freezing. Its presence increases the ability of the protoplasm to resist dehydration and therefore raises the frost resistance of the cell. An increase in bound water seems to follow a rise in soluble sugars, which in itself lowers the freezing-point of the cell sap. This normally takes place in the autumn. In the spring the sugars are reconverted into starch again, and the water bound by the cell proteins decreases. The freezing-point of the cell sap lies below that of water, so that most plants, even when they have reached their maximum susceptibility, can survive temperatures just below freezing-point. Damage to trees very seldom occurs at temperatures higher than $-2°$ C.

Thus the cold resistance of a plant rises during the autumn to a winter maximum and falls again in the spring to a summer minimum. This variation can be illustrated graphically by plotting the temperatures at which plants are damaged against the time of year (Day and Peace 1934; Ulmer 1937). The rate, the timing, and the extent of this annual variation in susceptibility vary for different species, for varieties and provenances of the same species, and even for individuals within a provenance. This accounts for the strikingly wide differences in injury between trees, which can be observed whenever frost damage occurs. There is also variation in hardiness between different parts of the same tree. These variations in frost resistance at different times of year, and between different trees, can be demonstrated by subjecting trees to artificial as well as to natural frost (Winkler 1913; Day and Peace 1934; Parker, J. 1955).

The kinds of differences that exist are illustrated diagrammatically in Fig. 3. Norway spruce, which is a typical winter-hardy but spring-susceptible conifer, shows a very rapid rise of susceptibility in the spring, and a fairly early fall in the autumn to its winter level of hardiness. Norway spruce can be sorted into early and late flushing strains, the latter being definitely the hardier. In the diagram this shows by its later development of susceptibility in the spring. Japanese larch, which behaves very similarly to early flushing spruce in the spring, continues growth later in the autumn, so that its hardening is delayed, and it is more susceptible than spruce to autumn frost. Scots pine is definitely hardier than early spruce or larch in the spring, being comparable with late flushing spruce; but the severe damage it suffered in the exceptional frost of 30 June 1954 (Murray, J. S. 1955*b*) suggests that its susceptibility continues to rise in the late spring until it reaches at least the same level as in most other conifers. The other example chosen, Monterey cypress, *Cupressus macrocarpa*, is native of a cool coastal belt, where the variations between summer and winter temperatures are slight. Probably for this reason there is a moderate difference between its winter-frost susceptibility, which is high, and its spring susceptibility, which is low. The winter susceptibilities of the other species are not shown on the diagram, since they have never been properly ascertained. They are certainly much lower than that of Monterey cypress. Parker (1955) found that needles of *Abies grandis* and *Pinus ponderosa* were able to withstand temperatures as low as $-55°$ C. in January. We shall see below that much of the winter damage which does occur, at any rate on evergreens, is a matter of excessive water loss, rather than of direct damage by low temperature.

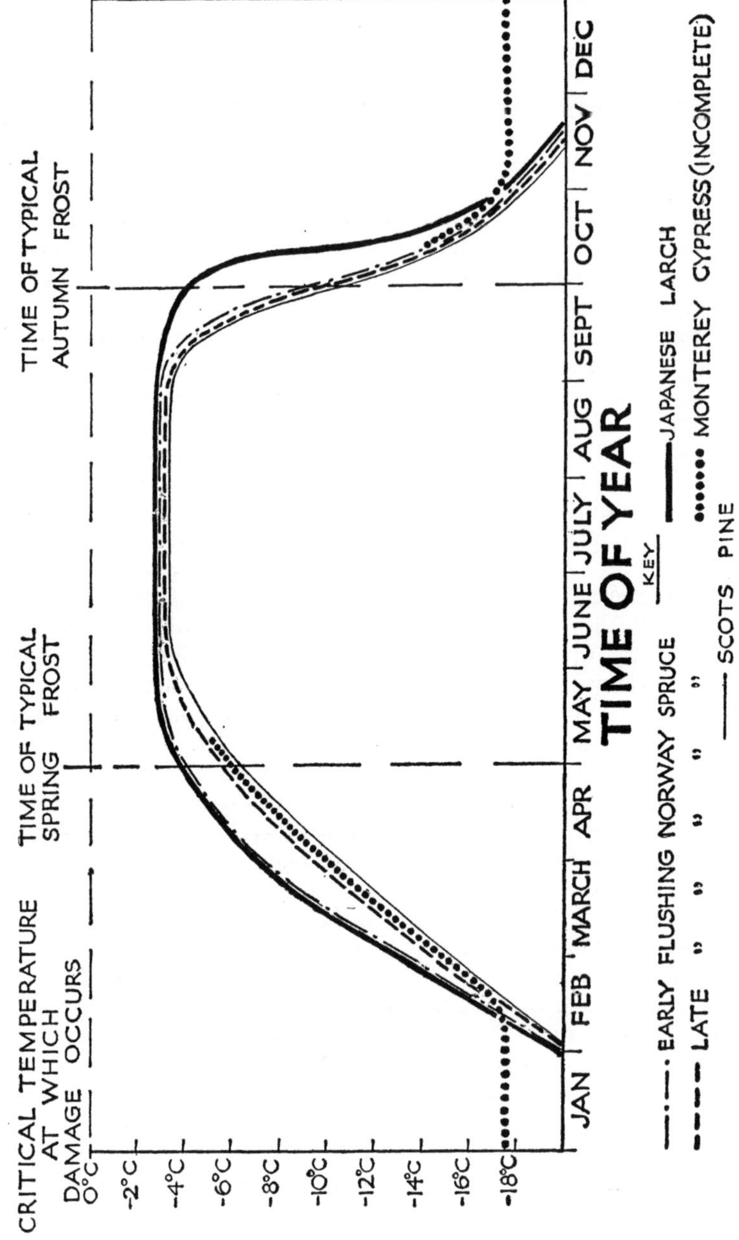

FIG. 3. Variation in frost susceptibility of four species of conifer throughout the year (see text). (A. C.)

The spring and autumn changes in susceptibility seem to be brought about by changes mainly in temperature, but probably also in day length. While these changes are operative, they are very easily influenced by changes in temperature. Exposure to low temperature increases hardiness and exposure to high temperature lessens it. But this can only happen within certain limits. Exposure of a growing plant to prolonged low temperatures will certainly make it hardier, but it can never make it able to withstand really low temperatures. Nevertheless, the temperature at which plants were growing before a period of damaging frost has a big influence on their hardiness and therefore on the amount of damage they suffer (Harvey, R. B. 1930; Day and Peace 1937*a*). Frost following a warm spell is much more damaging than frost following a cold one. One of the reasons why frost damage is so serious in Britain is that in our fickle climate damaging spells of frost are often preceded by warm weather.

Various other factors influence the amount of damage caused by any given freezing temperature (Day and Peace 1937*a*, 1937*b*; Levitt 1956). The most important of these are probably the rate of freezing and the rate of thawing. The more quickly plant tissue freezes or thaws, the worse it is damaged (Davis, Macarthur and Williams 1955; Levitt 1956). In nature the onset of frost is usually gradual. In a typical radiation frost, for instance, the temperature falls steadily throughout the night, but thawing in the first rays of the morning sun can be very rapid. For this reason rate of thawing is likely to be of more practical importance than rate of freezing. Garcia and Rigney (1914) have pointed out that the nearer the minimum temperature is to sunrise, the worse the damage. Chandler (1954) and others, however, did not find any increase of frost damage, which might be due to more rapid thawing, on the east side of trees. Nevertheless, anything which slows up thawing, such as artificial shelter, or an aspect protected from the morning sun, may lessen injury.

The duration of the damaging temperature can also affect the degree of injury. If the temperature is very much below the critical point, damage can be caused by an exposure as short as five minutes, but under more normal conditions, where the temperature is just low enough to cause injury, there is an increase of damage with exposure up to one hour, but little increase beyond that (Day and Peace 1937*a*). This means that quite a short period of frost during the night is enough to cause the maximum damage possible at that degree of cold.

Naturally, since freezing and thawing have such a marked effect on injury, one would expect repeated freezing and thawing to increase damage. Day and Peace (1937*a*) found that they did in fact do more damage than prolonged exposure. This suggests that freezing nights and warm days are worse than a sustained period of temperatures below freezing-point. Field (1942), however, found that continuous freezing of apple blossoms caused more damage than discontinuous; but his periods of discontinuous freezing may have been too short to cause the maximum damage.

Other influences affecting frost injury are not so definite in their effects. There is evidence that trees which are wet, when they freeze, are more damaged than dry ones (Korstian 1924; Day and Peace 1937*b*; Field 1942).

In nature this is probably only of slight importance. It must not be confused with the protective effect, discussed later, of sprinkling trees with water during the frost.

Manuring may increase or decrease hardiness according to substance, time of application, and so on (Anon. 1945*b*; Levitt 1956). There is, however, a general consensus of opinion that nitrogen manuring increases susceptibility, if it is applied at such a time that it increases growth during a season of frost danger (Smith and Tingley 1940; Sudds and Marsh 1943; Edgerton 1957). This is most likely, of course, to happen in the autumn. The effect of soil moisture is rather obscure. Under some conditions plants growing in dry soil seem to become hardier than those growing under moist conditions, but this may be merely a question of retarded growth. Day and Peace (1937*b*) were unable to detect any soil-moisture influence, but Fraser and Farrer (1957) found pine and spruce worse damaged in dry soil than in wet. At any rate there is no reason to suppose that sprinkling or irrigation, both used as protective measures, will have any marked effect on damage, owing to the increase of soil moisture which they cause.

WINTER COLD

The coldest periods in the winter are normally associated with anticyclones over the Continent and their associated dry easterly or north-easterly air streams. During the calmer nights the lowest temperatures recorded will develop locally owing to additional loss of heat by radiation; but the over-ruling factor is the presence of moving, cold, dry air. Damage due primarily to low temperature alone certainly occurs in any really cold winter, especially on species which are not adapted to our winter climate, for instance *Cupressus macrocarpa*, *Eucalyptus* species, and many ornamental trees and shrubs. But most of the winter damage to the commoner evergreen species, and especially to our forest conifers, is due primarily to loss of water from the foliage, under conditions where its replacement from the soil is difficult or even impossible, owing to the freezing of the stems and of the ground. This condition is known as 'physiological or winter drought'. Mikhin (1928) found that even dormant buds of deciduous broadleaved trees could lose more water during the winter than they could replace, so that injury to buds during the winter was roughly proportional to their rate of transpiration. In any case absorption of water from the soil shows a marked reduction with lowering of temperature, even though the soil is not frozen (Kramer 1949). Bethlahmy (1952) maintained that a frozen soil is physiologically a dry soil, while Fries (1943) found that water transport was impossible in frozen stems. It follows that winter injury is particularly likely to occur in places exposed to east or north-east winds, and more particularly on hill-tops and on the easterly margins of plantations. This is in direct contrast to the occurrence of spring and autumn frost particularly in sheltered hollows and valleys. Increased damage occurs if the days are warm enough to promote transpiration without any compensatory thawing of the ground (Felt 1943; Curry and Church 1952).

Evergreen trees and shrubs planted in the autumn may sometimes be damaged by the cold winds of the subsequent winter. In this case the inability to extract water from the soil is not so much due to frost, but to the lack of absorbing roots, which do not develop in the cold soil. This type of damage happens frequently in the forest with Douglas fir and *Tsuga* and occasionally with spruce and other evergreen conifers, and may result in quite widespread death of newly planted trees. For exposed and windy places, or for delicate evergreens in the garden, early autumn planting, which gives a chance for some root development to take place while the soil is still warm, or spring planting are advisable.

In some countries snow helps to protect trees from drying winds as well as from extreme cold, so that the worst damage occurs when the snow cover is inadequate (Ernstson and Hadders 1948). In Britain snowfall is normally too slight and short-lived to have much effect. Snow covering the ground alone may increase damage to trees projecting through it, not only because exceptionally low temperatures often occur over snow, owing to its blanketing effect on heat radiation from the ground, but because surprisingly low humidities may be found above it (Baldwin 1928).

Symptoms of winter injury

It is clear, therefore, that winter injury, on all but the more susceptible species, is really a form of drought damage. The symptoms are likely therefore to resemble those directly due to drought. This in fact they do, showing first as a browning of the margins of leaves and of the ends of needles, and developing later, in more serious cases, into complete withering and eventual defoliation. Typical needle injury may be observed after any cold winter on *Sequoia sempervirens*, which is obviously on the margin of its winter tolerance in Britain. In North America winter injury on conifers tends to occur in belts along the contours, and is often known as 'Red belt' (Boyce, J. S. 1948; Henson 1952). The damage is sometimes serious, and, apart from actual dieback and death, may also cause appreciable loss of increment in surviving trees (Blyth, A. W. 1953).

Two special forms of winter injury, frost crack and sunscorch, are discussed separately below. Frost lift, which occurs mainly in the winter, is dealt with at the end of the chapter. There are, however, types of winter damage other than these and the drought-type symptoms described above. Injury directly due to cold produces death of foliage on evergreen species, and in severe cases death of buds and twig dieback on deciduous and evergreen trees and shrubs. Frost rings (see below) may be produced in the wood by winter as well as by spring and autumn frost, but the association of these rings with damage is not so invariable in the winter as it is in the spring. V. R. Gardner (1944) made the interesting observation that on juniper the juvenile foliage was less damaged by winter cold than adult foliage. Minderman (1949) records the breaking and shedding of spruce twigs rendered excessively brittle by freezing. Roots are also sometimes damaged by winter cold; Hord, Van Groenewoud and Riley (1957) found winter injury of roots a serious cause of damage to elms planted outside their proper climatic range. On the Continent, but not in America, extreme winter cold has been associated with a non-fungal staining of the heartwood of beech. This was particularly noticeable after the winter of 1928–9 (Custer 1934; Larsen, P. 1944). There are no definite records of this damage in beech in Britain, though unexplained staining of the heartwood is not uncommon.

Winter injury, whether it be due to cold or physiological drought, is notably affected by local conditions. It is commonplace that trees which have survived in the nursery without damage may be subject to winter injury as soon as they have been planted in the forest.

Some conifers during the winter show needle discoloration, which is certainly the result of cold, but does not rank as injury, since it is reversible, the normal green colour reappearing in the spring. Young plants of *Thuja plicata* often turn bronze in the winter, indeed sometimes they are so nearly brown as to cause considerable apprehension. Winter purpling of the needles is sometimes very noticeable on young pine. In the United States the winter coloration of *Pinus banksiana* varies according to provenance (Stoeckeler and Rudolf 1956).

The degree of recovery from real injury depends naturally on its extent. Defoliation alone is seldom fatal, even if nearly complete, though Rudolf (1949) found that pines with more than 75 per cent. of the foliage injured usually died. Stoeckeler (1948) described the recovery of a large number of species, and stressed their extreme variability in reaction. Since the damage happens during the dormant season a considerable time will elapse before there are any visible signs of recovery. It may be necessary to wait till May or even June before the damage can be properly assessed. It is a common mistake to cut back winter-injured shrubs before the exact extent of the injury can be appreciated, or to uproot trees which in fact are still alive.

Distribution of winter injury

The distribution and severity of winter injury depends not only on the distribution of low temperatures and cold dry winds, but also on the species grown. In areas where cold winters are normal it is impossible to grow the less hardy trees and shrubs. The damage after exceptional winters is therefore less striking there than in countries where the comparative infrequency of winter cold renders the cultivation of less hardy species marginally possible. Britain is such a country, so that every time it has a really cold winter there are reports of damage to trees and shrubs introduced from warmer climates. In the winter of 1955–6 severe cold extended unusually far south in Europe, and damage to trees and shrubs on the Mediterranean coast of France and Italy was much more striking than farther north, where only hardy trees are grown (Dugelay 1957). In Britain the warming influence of the Gulf Stream, particularly on the west coast, provides a coastal belt wherein half-hardy exotics can be grown under nearly frost-free conditions. Their existence in this belt, and the fact that in Britain there may sometimes be periods of ten years or more between cold winters, provides a constant temptation to try half-hardy trees and shrubs in the colder parts of our country, although they are likely then to be damaged every time there is an exceptional cold winter.

It is important to realize that winter cold, despite its very limited effect on the hardier trees that we grow on a large scale, is nevertheless one of the biggest limitations on the species that can be grown successfully in Britain. Whether milder winters, unaccompanied by other climatic alterations such as hotter summers, would enable us to grow many more trees well enough for forest use is an open question. Such a change would certainly widen enormously the range of trees and shrubs that could be planted in Britain, even if their rate of growth in our cool summers was too slow for commercial purposes.

Reaction of different species to winter cold

There is great variability in the reactions of different trees and shrubs to winter cold. Damage, as we have seen, can be caused by cold dry winds or by actual low temperatures, and regardless of whether one or both of these are operative, injury will vary in extent according to the condition of the tree at the time. Thus a species conspicuously damaged in one cold winter will not necessarily be noticeably so in the next. There is little evidence in Britain of winter damage, other than frost crack and bark injury, on the commoner deciduous trees, though twig dieback of Japanese larch, which may have resulted from winter injury, has been recorded. Among the commoner evergreen conifers, Scots pine is probably the most frequently browned, but the damage, though striking, is seldom really serious. Most of the other conifers, with the possible exception of Sitka spruce, occasionally show some needle-browning after exceptional winters. In Douglas fir this is frequently associated with scattered twig dieback, which may, however, be due partially to the fungus *Phomopsis pseudotsugae* (Chapter 25). *Sequoia sempervirens* in Britain regularly shows needle-browning after severe winters. But the damage seldom has any appreciable effect on its general health. However, MacGinitie (1933) has suggested that winter cold may be responsible for the northern limit of its native range on the coast of Oregon. In Britain some of the marginal forest species, for instance *Pinus radiata, Cupressus macrocarpa,* and *Nothofagus* spp., in particular *N. procera,* are frequently damaged in the winter, and are probably only really suitable for the coastal parts of Britain. Nevertheless, it would generally be true to say that winter injury is of little importance in British forestry.

On the Continent, presumably owing to the lower temperatures experienced, winter damage to forest species is commoner and more important, but even there, as in Britain, the most serious injury is to rarer trees and shrubs in parks and gardens. Extensive lists of the degrees of winter injury have resulted from most of the very cold winters of the present century (Höfker 1919; Fabricius 1930; Melzer 1931; Wróblewski, Korczyńska and Wilusz 1952; Dugelay 1957).

In Britain useful lists have been provided by Balfour (1941) and by Harrow (1948). Certain important decorative genera are prominent in these lists. They include *Buddleia, Camellia, Ceanothus, Cistus, Escallonia, Eucryphia, Magnolia, Nothofagus,* and *Olearia.* Many of the rarer conifers also appear. Nevertheless, despite such periodic damage or even death, most of the woody plants listed remain in cultivation in Britain.

In a number of instances, pronounced differences in winter injury have been noted between different provenances. Ernstson and Hadders (1948) found marked variation in Norway spruce and Chiba (1955*a*) in *Cryptomeria japonica.* Krahl-Urban (1955), who found similar provenance differences in oak, considered Sessile oak on the whole hardier than Pedunculate. In the United States provenance variations in winter resistance to cold have been recorded for *Pinus resinosa* (Bates, C. G. 1930) and for various Southern pines (Minckler 1951).

Frost crack

Cracking of stems of trees during cold winters is sufficiently distinct from other types of winter injury to justify separate consideration. The exact mechanism of frost cracking is still a matter of some doubt, though it is probably due to unequal expansion and contraction of the wood during rapid freezing and thawing. The cracks are normally radial and occur particularly on the south or south-west sides of the trunk (Fergus 1956*c*), though Lamprecht (1950) found them on all aspects on oak in Switzerland, except on slopes when they tended to be on the downhill side. The larger cracks are open to the surface, but smaller ones may be entirely internal (Orr 1925). The actual splitting is said to occur suddenly and to make a noise like that of a pistol shot. As the temperature rises again the cracks tend to close, the rate at which this happens varying according to the species concerned (Schulz 1957). Those cracks which reach the surface tend to reopen in subsequent winters, though eventually most of them heal over, leaving a pronounced ridge of callus tissue up the stem.

Pirone (1948) and Ishida (1950) both attribute frost crack to the presence of excessive water in the heartwood. Indeed Ishida associates it with 'wet-wood' (see under 'Slime Flux', Chapter 20). High water content may well be important when actual freezing is the cause of the injury, but H. J. Lutz (1952) found cracks in spruce in Alaska, which he attributed to winter drought, water being withdrawn from the wood under conditions where it could not be replaced from the frozen soil. Such cracks, which may well occur elsewhere, would be generally of the same nature as those caused more typically by summer drought (Chapter 6).

Pirone (1948) associated frost crack with badly drained soils, Pecrot (1956) with shallow soils and those with a variable water table, and Lamprecht (1950) with certain ecological types. Igmándy (1956) considered that local climate as affected by topography was a decisive factor, cracks occurring much more frequently in valleys and plains, particularly on moist sites. More research is needed before any general relationship with edaphic conditions can be established. Removal of shelter will sometimes bring about frost crack in the same way that it encourages 'sunscorch'. Isolated trees are thus more affected than those in plantations, so that standards in coppice with standards are in more danger than trees in high forest (Lamprecht 1950; Malaisse 1957). More than 50 per cent. frost cracking has been observed in a poplar plantation following the removal of high hazel coppice from around the stems.

Frost crack is generally commoner on hardwoods than on conifers. Both oak and poplar are frequently affected. However, the fact that isolated trees are more liable to crack than those in plantations may partially account for the relative frequency of cracks in these genera, which are both fairly common hedgerow trees. But the fact that species does matter as well as exposure is clearly shown by the relative freedom from frost crack of elm, which is our commonest hedgerow tree. Sessile oak, according to Lamprecht (1950), develops slightly longer cracks than Pedunculate, but the number occurring on the two species is roughly the same. *Quercus rubra* equals Pedunculate oak in susceptibility (Malaisse 1957), and *Q. cerris* is more susceptible (Igmándy 1956).

Frost crack in Britain is not a serious source of loss, though the initial fall in value, owing to the crack itself, usually becomes greater as a result of fungal invasion. Cracked trees should always be removed as thinnings, even if the crack has healed over, and sudden exposure of stems should be avoided wherever possible.

Sun scorch

Death of patches of bark on the south or south-west side of the tree, which is usually known as sun scorch, is fully described in Chapter 5. It has usually been attributed to the lethal effect of the heat of the sun on thin-barked trees in the summer. In some instances this certainly is the cause of injury, but it has also been suggested that it may result from frost injury to the bark in the winter (Harvey, R. B. 1923; Hubermann 1943; Frerich 1957). This might come about because of the wide variation in day and night temperatures on the south side of the tree, involving very rapid freezing and thawing. Ferkl (1951) has found that the south side of a trunk may be 20° C. warmer than the north on a sunny winter day. Alternatively the heat on the south side might stimulate the cambium, initiating changes which would lower its hardiness.

For fruit trees, whitewashing the trunk to lower the temperature of the south side has been advocated to mitigate winter as well as summer sunscald (Aichele 1952; Karnatz 1957). In the forest nothing can be done beyond efforts to avoid the sudden exposure of thin-barked trees to the sun.

Prevention of winter injury

In the forest, very little indeed can be done to prevent or even mitigate winter frost injury. Winter-susceptible species should not be planted in exposed sites or unduly exposed later by the removal of sheltering crops. In the nursery, shelter over the beds, either of laths or of evergreen branches, lessens damage to susceptible species, partly by sheltering them from dry winds, partly by maintaining the temperature of the beds on cold nights and partly by lowering the rate of thawing after the frost. Shelter is also valuable against frost lift (see below) in nurseries. In the garden, half-hardy shrubs can be protected by evergreen branches stuck in the ground around and over them, or even by complete wrapping with straw or bracken. It is advisable to have materials ready against a sudden onset of cold weather. Such protection should always be applied as late as possible, and removed as soon as danger is over.

SPRING AND AUTUMN FROST

Spring and autumn frosts, sometimes known as late and early frosts, are rather different from winter cold, though in both cases the conditions concerned result from the arrival over Britain of cold continental or arctic air.

In the spring and autumn, however, the initial temperature of that air is above freezing, and it is only subsequent loss of heat by radiation that produces damagingly low temperatures. Thus the most severe spring and autumn frosts occur on clear, still nights during generally cold periods. Under these conditions the climate immediately over the ground is very different from that a few feet above it, and the climate in a sheltered valley from that over the locality as a whole. In the study of radiation frost, therefore, micro-climatology becomes of outstanding importance.

The conditions leading to the formation of radiation frost and the way in which it subsequently develops have been dealt with briefly by Day and Peace (1946), in rather more detail by Bush (1946), and very fully by Geiger (1950). Briefly, on clear nights the surface ground and other solid objects lose heat by radiation, and in consequence cool the air in contact with them. This cooling often continues until the ground and the air above it are well below freezing-point. Naturally this condition is reached more quickly if the air is initially cold. This explains why severe radiation frost only occurs when the temperature of the general air mass is already low. Presence of clouds or fog will prevent loss of heat by radiation, and wind also prevents or lessens radiation frost by continually moving in slightly warmer air. Only on very clear, still nights do really serious frosts develop. The cold air near the ground, being heavier than the warmer air above, has no tendency to rise, and a so-called temperature inversion develops. Under normal conditions, of course, the air near the ground is warmer and is constantly rising by convection. Under inversion conditions, on the other hand, temperatures rise with distance above the ground up to a varying level, above which the normal fall of temperature with height again operates. Smoke or fog, if it develops, usually remains trapped within the inversion layer. Temperature inversion is easily demonstrated by thermometer readings at increasing heights above the ground (Day and Peace 1946; Geiger 1950).

On level ground the layer of cold air simply increases in depth as cooling proceeds, but on slopes, owing to its greater density, it starts to move down-hill, producing quite appreciable 'katabatic' winds (Cornford 1938). The cold air tends to collect in hollows, in valleys, and even on shelves on hill-sides, in which places the lowest temperatures occur. Very often the tops of trees project above this layer of cold air, so that only the lower branches are damaged by frost (Plate I). A frost-line on trees, below which the shoots are brown and above which they are green, is a common phenomenon following spring frost.

In its flow, cold air is more like thick syrup than water, so that it can pool to a considerable depth on a flat surface without immediately running over the edges. For this reason flat hill-tops and hill-side shelves are often frosty sites. It does not very rapidly penetrate perforated barriers, such as hedges or belts of trees, though if it is deep enough it may flow over them. In passing over a young, dense plantation it tends, as it were, to get entangled in the branches and to descend to the ground. Hedges and shelterbelts can thus be used to divert cold air, for instance from orchards below; but appreciable pooling of cold air, and consequent low temperatures, can take place on the upper side of a belt or plantation (Fig. 7).

The lowest temperatures of all occur in fully enclosed hollows with no natural above-ground drainage. In such places freezing temperatures are likely to occur throughout the year. There are many such hollows in Britain. Sometimes, even in temperate countries, the local climate and therefore the

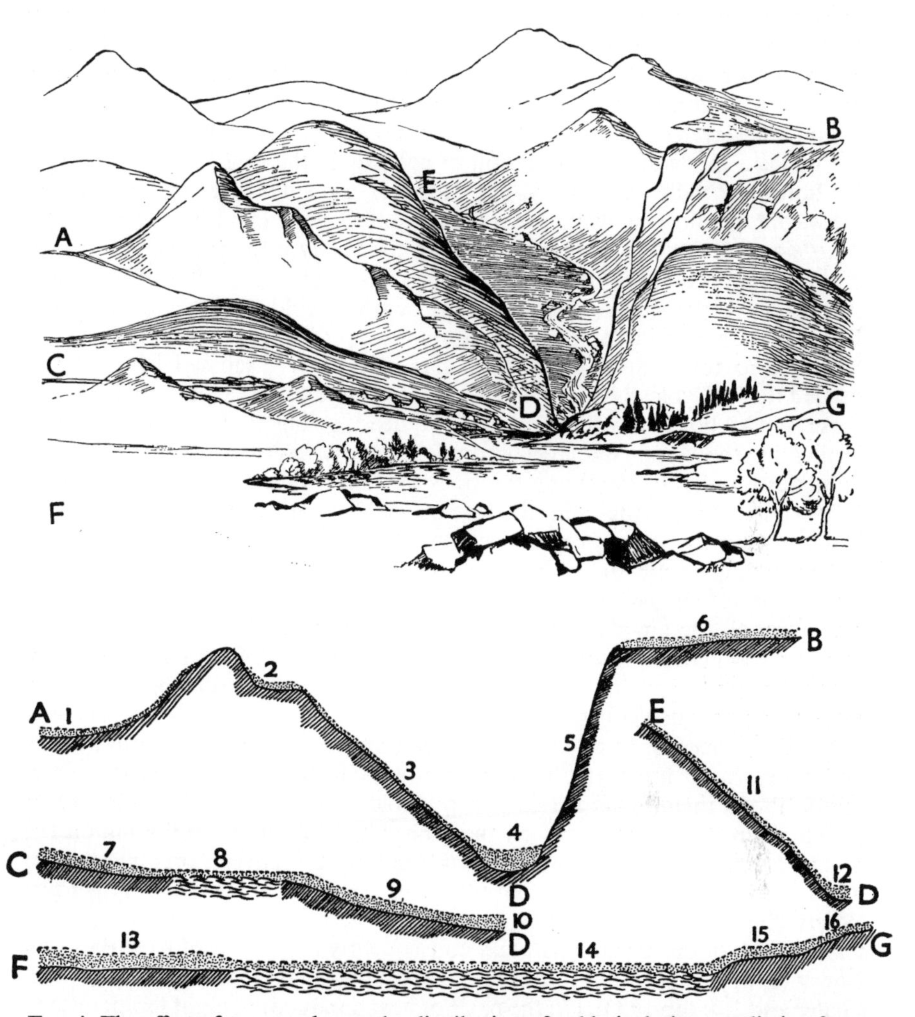

FIG. 4. The effect of topography on the distribution of cold air during a radiation frost; shown semi-diagrammatically, and in cross-section between various points on the diagram. (A. C.)

vegetation of a real frost hollow may be truly arctic (Hawke 1944). But many valleys with natural water drainage are nevertheless very frosty places, and are sometimes loosely called frost hollows. They are especially frosty if they are narrowed near the mouth, so that air flow is restricted, if they are long and flat-bottomed, so that the cold air does not flow out of them, or if they are fed by a very large catchment area, which will produce a lot of cold air to flow into them. Frost valleys may be quite limited in extent or may cover many square miles. The location of a meteorological station in such a valley may give very abnormal night temperature readings (Hawke 1944). Fig. 4 illustrates diagrammatically the depth of cold air and therefore the probability of damaging frost in different topographical positions. As we move from A to B there is a slight accumulation of cold air on a col (1) and on a shelf (2), very little cold air on the slope (3), but substantial pooling in the deep valley (4). The very steep slope (5) has no accumulation of cold air, but there is a moderate layer on the flat top of the adjacent hill (6). Moving from C to D there is a moderate accumulation of cold air in the shallow, but relatively flat-bottomed valley (7 and 9), but less over the small lake (8), because of the warming influence of the water. On the steep slope from E to D (11) there is little accumulation, but a good deal in the valley bottom (12). From F to G there is a deep accumulation of cold air in the plain (13), but less over the lake (14). Up the slope (15 and 16) the depth of cold air tends to decrease. The influence of topography on the development of spring frost, with especial reference to forest areas in Britain, has been fully discussed by Day and Peace (1946). Typically when a frost valley or hollow is planted with trees, those at the bottom will be stunted or dead owing to frost injury, while farther up the slopes damage becomes progressively less (Aichinger 1932; Pomerleau and Ray 1957). As the trees on the slope reach the thicket stage they tend to entrap the cold air, less of which then reaches the bottom, so that the valley becomes progressively less frosty (Fig. 8).

The cooling of the air is considerably affected by the condition of the soil and the nature of the soil cover (Franklin 1919–20). At night bare soil loses more heat to the air above than a dense cover of vegetation. A thick grass mat is particularly effective in preventing conduction of heat from the soil below, so that very low temperatures and very severe damage are often associated with such cover. Temperatures are not so low over heather as over grass, and a deep open cover such as bramble or gorse is even better. Snow and deep, loose litter are both dangerous soil coverings as regards frost injury above them. This suggests a possible danger inherent in the use of mulches. Freshly fallen snow, on the other hand, while it is still lying on the plants, definitely acts as a protection to the foliage it covers. It is later, when it has melted off them but still covers the ground, that it becomes dangerous. Where the soil is bare both its condition and moisture content affect the temperature of the air above it, and consequently the degree of frost damage. Lower temperatures are found more over loose than over compact soil (Slater and Ruxton 1955). On the basis of heat conduction, night temperatures should be higher over wet soil than over dry and this is sometimes the case, so that irrigation is one of the methods suggested for frost control (Kikuchi 1955). On the other hand, wet ground tends to be colder than dry ground at the end of the day and this initial difference is sometimes great enough to make the final minimum temperature during the night lower over wet soil. The frost effect of the wet peat, which now provides large areas of forest ground in Britain, is at present unknown. Large areas of open water certainly have an ameliorating influence on spring and autumn, as well as on winter, frost (Fig. 4). There is always less damage near the coast or on the shores of large lakes. Even a network of irrigation ditches can have an appreciable effect (Eimern and Loewel 1953).

Overhead cover, even if very open, lessens radiation from plants and from the ground beneath (Umann 1930). Where such cover is available it can sometimes be used to enable frosty sites to be planted with a frost-susceptible species. In the nursery, of course, shelter can be artificially provided. It serves a dual purpose, not only raising the night minimum, but beneficially slowing up the rate of thawing in the morning.

It is easy to see that aspect may often affect frost damage, but the various ways in which it acts have never been evaluated properly and may in some cases be contradictory. A slope exposed to the morning sun may suffer more, because of rapid thawing, than one which the sun does not reach till later in the day. A slope facing south-west or west will warm up in the afternoon, if the day is sunny, which is quite likely since the existence of the frost postulates a clear night. Such a warmed slope starts the night with a definite advantage over a cold aspect, and the air temperature above it, other things being equal, will not fall so low. On the other hand, the warmth on a southern slope may stimulate activity in the tree and thus render it more frost-susceptible. Under some conditions damage on southern slopes is certainly greater (Zieger, Pelz, and Hornig 1958). A northern slope, where average temperatures will be lower, may hasten hardening in the autumn or delay its disappearance in the spring, though there is some evidence that the lack of sunshine in the summer and autumn in such positions interferes with the hardening of some species.

It is thus clear that the evaluation of the frost status of any particular site is a matter of considerable difficulty, involving as it does so many different and sometimes conflicting considerations. Nevertheless, really bad frost sites are usually fairly easy to detect, even if the condition of the vegetation on them gives no aid. In this connexion bracken, which is very frost-susceptible, though it will continue to grow in frosty sites, is a valuable indicator.

The more exposed parts of the plant itself are usually slightly cooler during a radiation frost than the air surrounding them (Ehlers 1915; Grainger and Allen 1936). The difference, which is due to cooling by evaporation as well as radiation, may be as little as 0·5° C., but even that could have a considerable influence on the amount of frost injury.

Leaving abnormal frost hollows out of account, damaging frosts can occur as late as the end of May in most parts of Britain and well into June in the north or in frost valleys elsewhere. In 1954 a damaging frost was experienced on 30 June over a large area in central East Anglia (Murray, J. S. 1955*b*). In 1958 Scots pine were damaged by frost at Culbin Forest on the Moray Firth on 25 July. Damaging frosts are likely again any time from mid-September onwards. Some years are relatively very frost free, but few pass without frost injury in some part of Britain. In 1958 *Thuja* plants were examined from two frost valleys, one in Sussex and the other in Yorkshire; both lots showed frost rings in every annual ring for the five to seven years since they were planted.

Damage by autumn frosts is nearly always less widespread, affects fewer species, does less damage to them, and is therefore less striking. Nevertheless, few years are entirely free from autumn damage, and it is a pathological factor of considerable importance.

Symptoms of spring-frost injury

The symptoms of the damage caused by spring frost have been fully described by Day and Peace (1934, 1946). The spring change from hardiness to susceptibility has already been discussed, and it is clear that the nature and amount of damage to the tree will depend on the stage which has been reached in this change. The process does not necessarily start at the same time or proceed at an equal rate all over the tree, so that one part may be damaged more than another simply because it had already become more susceptible. Secondly, one part of the tree may be subjected to lower temperatures than another. For instance, the top of a tree may project above the cold air layer, so that it is less damaged than the bottom (Plate I), or the base of a tree may be sheltered by tall grass or bracken so that it is less damaged than the top. Variations in exposure to day warmth between the two sides of a tree may affect the initiation and rate of development of susceptibility and thus indirectly influence frost damage. The actual visible state of the tree, whether the buds are swelling, bursting, or in leaf, naturally moves in harmony with the internal changes and the consequent increase in susceptibility. Nevertheless, this increase starts well before the tree has flushed, and differences in susceptibility, which are not detectable by any corresponding differences in visible development, often exist between species, between individuals of the same species, or even between different parts of the same tree.

If frost occurs before flushing, damage is often confined to the buds. These tend to be in a more advanced state than the shoots on which they are borne, and can therefore be killed by frost without there necessarily being any shoot injury. It is doubtful if, even in large-budded species, the bud scales afford any appreciable protection against low temperatures (Genèves 1957). Destruction of buds, and their subsequent replacement by a larger number of dormant or adventitious buds, is one of the reasons why frosted trees tend to become so bushy. The destruction of short-shoot buds in larch often results in the production of rings of new buds, each in the axil of one of the short-shoot needles (Plate II. 1). Such clusters of buds are a common sign of frost injury in this genus.

After flushing, damage, which is quickly detectable, occurs particularly on the leaves and soft succulent shoots. In cases of slight injury the edges of the leaves (Fig. 5) or the tips of the needles are withered. Leaves killed by frost do not usually fall, because the sudden death prevents the formation of an abscission layer (Mühldorf 1928). It is much more usual for the whole new shoot to be involved. Withered, and sometimes live, permanently bent, shoots hanging on the tree are very typical. Often a large number of recovery buds develop, with consequent bushy growth. The leaves of broadleaved trees, produced after frost injury, are sometimes abnormal in shape and slightly distorted (Mišič 1956). Some conifers, such as spruce, which can stand repeated loss of shoots for many years, finally become almost cypress-shaped as a result of the repetitive frost pruning, when growing in very frosty sites.

Damage to the cambium is less immediately obvious. It may range from mere injury, in which case a frost ring is subsequently formed, to death, in which case canker (Plate II. 2) or dieback develops according to the amount killed (Day and Peace 1934). Frost rings arise when the cambium is injured

FIG. 5. Spring-frost injury to the foliage of *Populus trichocarpa*, June (× ½). (J. C.)

but not killed; the cells that are formed for a period after injury are abnormal, being usually larger, thinner-walled, and of much more irregular shape than normal wood cells (Plate II. 3). If the frost occurs early in the spring the abnormal ring will be at the beginning of the annual ring. If the frost is in late spring, there will be some production of normal wood before the frost ring is formed. The position of the ring in the wood gives therefore a very rough idea of the time of the damaging frost. A frost ring forms a definite line of weakness in the wood (Fig. 6), and may be one of the factors leading eventually to ring shake (Chapter 20). Cases have been found where this weakness has developed very rapidly. Young Japanese larch, severely frosted in the spring, have broken partially through in autumn gales of the same year, because of poor adhesion between the wood of the current year, and the inner core of earlier wood. Death of small areas of bark, resulting in an irregular, cracked appearance (Fig. 6), is also quite a common result of spring frost damage to smooth-barked stems. It is particularly common on *Thuja plicata*.

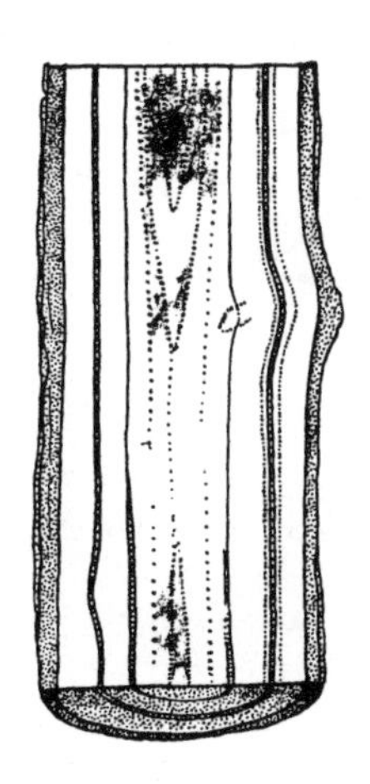

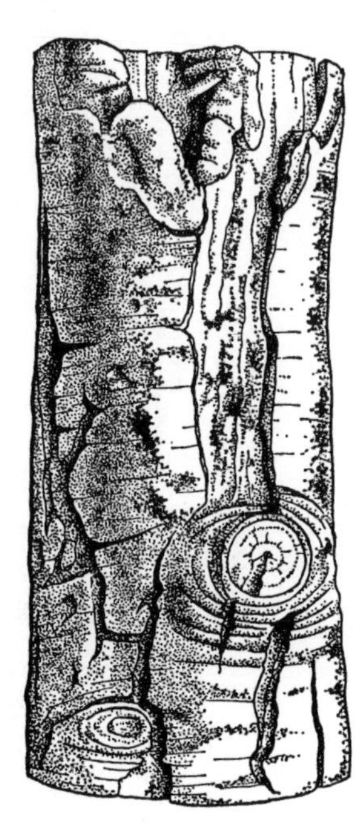

FIG. 6. Bark-cracking and frost rings in a young Douglas fir stem as a result of spring-frost injury ($\times \frac{2}{3}$). (J. C.)

If the cambium is killed right round the stem, dieback will take place, but if the injury is only local, the tree will start to heal over the dead patch. Such lesions are particularly liable to further frost injury in subsequent years, and of course also to fungal infection. The healing tissue around their edges tends to become active early in the spring, and also by its position lacks bark-protection; both are reasons why it is likely to be damaged, even by frosts slighter than that which caused the initial injury. Once a lesion is present repeated frost damage and healing soon produce a real canker.

In general, the greater the distance from the top of the tree the more backward is the cell activity in the spring (Priestley 1932), and the thicker is the protective bark. Thus the ends of twigs are more liable to injury than their lower parts, the twigs than the branches, and the branches than the trunk. The cambium on a main branch, however, first becomes active where a minor branch joins it, and consequently the cambium at this point becomes susceptible before that surrounding it. For this reason frost cankers often develop around side branches. R. B. Harvey (1923) has clearly demonstrated the protective effect of bark against low temperature.

Recovery from spring injury rests largely on the ability of the tree to form buds, and to produce vigorous shoots from them. *Picea* has an advantage over *Abies*, for instance, because it is better able to produce new buds, Japanese larch over European, because its recovery shoots are so much more vigorous. In many cases the final damage may depend as much on recovery ability as on the initial injury.

Moving a species from one climate to another may have a considerable effect on its reaction to spring frost in its new home. Trees from places with sustained cold winters, such as *Larix siberica*, *L. gmelini*, or *Populus koreana*, are tempted into precocious activity by warm spells in the late winter, and are almost invariably severely damaged later by frost.

Symptoms of autumn-frost injury

Autumn injury is confined almost entirely to succulent shoots, or more often to the succulent tips of shoots. It is thus found on species which are still growing or at any rate still internally active in the autumn. Its effect is, of course, to cause withering of the shoot-tips. In many cases the shoots affected are 'Lammas shoots', being the result of renewed growth in the late summer. Burger (1944) has pointed out that the more vigorous larches, for instance Japanese, Silesian, and Polish, are liable to autumn damage, because of their habit of renewed growth. Unusually severe autumn frost may produce dieback of older, hardened shoots, but this is uncommon.

Anything that encourages renewed or sustained autumn growth, or delays hardening, may lead to frost injury. Irrigation can sometimes increase frost damage in this way. Late defoliation of poplars by *Melampsora* rusts apparently prevents the proper ripening of the defoliated, leafless shoot-tips and leads to frost dieback. This phenomenon may well occur in other cases of autumn defoliation. Changes in length of day can also have a profound effect on autumn damage. Kramer (1937) increased the autumn frost susceptibility of *Abelia* by subjecting it to increased day length. Trees moved north into regions of longer day length tend to suffer from delayed cessation of growth in the autumn and are therefore subject to fost injury to their still-active shoots (Pauley and Perry 1954; De Boer 1957). This type of damage consistently occurs on poplar selections made in Italy, when they are planted in Britain.

Autumn-frost injury frequently leads to subsequent invasion by *Botrytis cinerea*, Grey mould (Chapter 15), which often spreads down the shoot and increases the initial injury. Autumn is the time when *Botrytis* is particularly active, since it is favoured by high humidities and undeterred by low temperatures.

Recovery reactions to autumn frost do not take place until the following spring. Likewise frost rings resulting from autumn injury do not develop till the following year. Thus by position they cannot be distinguished from rings developing as a result of frosts in the late winter or early spring.

Distribution of spring and autumn frost

It will be already clear that the occurrence of radiation frost is much more a matter of local conditions than of general climate. Spring and autumn frosts of damaging intensity are likely to occur, therefore, wherever local topographical conditions lead to the pooling of cold air. Such localities may be found in almost all parts of Britain, except on the immediate seaboard. The worst areas tend to be in hilly districts, but nevertheless some of the wide valleys and plains in the lowlands are in themselves frost hollows of great extent. Possibly the worst general area in Britain for early and late frosts is the Breckland region of East Anglia.

Reaction of different species to spring frost

Severe spring frosts have usually led to the publication of lists of damage to different species (Ramsey 1871–2; Ohlweiler 1912; Roth, J. 1920; Hunter Blair 1945; Day and Peace 1946). The list given by J. S. Murray (1955*b*) is of particular interest because it refers to an exceptionally late frost on 30 June 1954.

A full list, including many ornamental trees and shrubs, was given by Day and Peace, and, although this was based primarily on the exceptionally severe frosts of May 1935, there is every reason to think that it is generally applicable. In the summary list below only some of the commoner trees are mentioned:

	HARDWOODS	CONIFERS
VERY SUSCEPTIBLE	Walnut Ash Sweet chestnut Oak Beech	*Abies grandis* Sitka spruce Norway spruce European larch Japanese larch Douglas fir *Tsuga heterophylla* *Thuja plicata*
MODERATELY SUSCEPTIBLE	Sycamore Horse chestnut Some poplars	*Pinus contorta* Corsican pine *Picea omorika*
HARDY	Birch Hazel Hornbeam Lime Elm Most of the commonly planted poplars	Scots pine *Cupressus macrocarpa*

The number of ornamental trees and shrubs which are liable to spring-frost injury is enormous.

In many species there are large variations in frost susceptibility between different provenances, or even between individuals of the same provenance.

Differences in European larch are dealt with by W. R. Day (1951), and are discussed more fully under Larch canker (Chapter 24). Edwards (1953) has described provenance differences in Sitka spruce, plants from United States seed being much more susceptible than those from the Queen Charlotte Islands. Sylvén (1944) found provenance differences in oak. In Norway spruce there is wide variation in date of flushing between individuals, and it is possible to divide any batch of plants into early and late-flushing groups (Münch 1928; Reuss 1928). In some very frosty sites pure crops of the late-flushing strain have resulted from the elimination by frost of the early-flushing trees.

Despite the fact that many of our most important forest species, and some of our native trees, are highly susceptible to spring frost, it has surprisingly little influence on what we can plant and on what grows naturally. There are probably a few small areas where highly susceptible species, such as ash, have been suppressed by frost. There are certainly other places where the initial use of susceptible species for forest planting on bare ground is almost impossible; though in many instances it is possible to raise them later under the shelter of hardier species, or to plant them when the afforestation of the slopes has lessened the downflow of cold air.

Reaction of different species to autumn frost

Since autumn frost is much less important than spring frost, there is less available evidence on the species affected. Japanese larch frequently suffers damage; Veen (1954) has suggested that the active autumn growth that makes this species so susceptible may be the result of the day length in northern Europe being greater than that over most of Japan. Nienstaedt (1958) found that southern origins of *Tsuga canadensis* continued growth longer in the autumn and therefore made more growth, but were more susceptible to frost injury than northern origins. Height growth could therefore be used as a measure of frost susceptibility. Application of maleic hydrazide, a growth-reducing substance, has been found to lessen autumn-frost damage (Satô, Ôta, and Shoji 1955*a*). In nurseries, though not in the forest, Sitka spruce and Douglas fir are fairly often injured. Corsican pine is sometimes affected, despite the fact that it has already formed its terminal buds. Damage to hardwoods, with the exception of poplars from southern regions of longer day length, is not common in Britain.

Prevention of spring- and autumn-frost injury

A good deal can be done to prevent, or at any rate lessen, spring radiation frost injury by proper forest management. Given a knowledge of the frost potentialities of a site, a suitable choice of species or of provenances can be made. Exceptionally bad frost sites may demand the use of an uneconomic species in order to establish forest conditions. Pillichody (1936) and Lachaussée (1941) advocated Mountain pine to plant frost hollows. In view of the inter- and intraspecific variations already known to exist, there are obvious opportunities for selection and breeding for frost resistance.

Existing forest cover on the ground can be thinned and used as shelter (Umann 1930), although it is difficult in practice to get overhead shelter dense enough to protect, without cutting off too much light. This is especially so because frosts often occur when broadleaved deciduous shelter is leafless, while light values for the undercrop are critical after it has come into leaf. Shelter in a variety of forms can be provided in the nursery. All types of shelter, whether in forest or nursery, are valuable not only because they raise the minimum temperature but because they tend to retard the rate of thawing.

The forest can be manipulated in a number of ways to lessen injury. The slopes of frost hollows can be afforested first, to hold up the flow of cold air, before the bottoms are dealt with (Lachaussée 1941) (Fig. 7). Felling prior to replanting can be done in such a way that there is never any pooling of cold air against the parts not yet felled (Fig. 8 top). Belts of trees can be left to interrupt or divert the flow of cold air. Each area should be studied with regard to the local conditions before planting and later on before felling is started.

For nurseries, selection of site and its subsequent treatment, are of paramount importance. From the working point of view nurseries can hardly be put on steep slopes, although that would provide good cold-air drainage. Even if level ground is a necessity, it should still be possible to avoid the worst frost sites. Plantations round a nursery will sometimes act as barriers causing pooling of cold air within it. This is particularly apt to happen if trees are planted on the edge of a hill-side shelf, the rest of which is occupied by a nursery. As the trees grow up they increase the pooling of cold air over the nursery, making it seriously frosty (Fig. 8 centre and bottom). If there is free air drainage in any direction, it should be left open.

Frost control in horticulture, particularly for fruit crops, has become quite an elaborate matter. The methods used have been very fully reviewed (Anon. 1945*b*). The comparatively high annual value of a fruit crop, and the fact that a single severe frost may cause its complete loss for that particular year, justify fairly high expenditures on apparatus and on the use of it. Smoke production to cut down radiation is now more or less obsolete (Anon. 1945*b*), though it can be combined with heating by burning a smoke-producing fuel, such as peat (Jacobsen 1956). Direct air heating has probably been used more than any other method (Anon. 1945*b*; Bush 1946; Mursell 1949). Crude-oil burning in very simple heaters is the usual method, these being lighted when the temperature falls to a certain minimum, but even simpler fuels, such as straw bales, can be of value (Baker, C. 1953). Infra-red heating has also been tried with success (Farrall, Sheldon, and Hansen 1946). Indirect heating can be provided by drawing warmer air from above to replace the cold air near the ground by means of large-bladed, slow-moving fans (Bush 1946; Frith 1951; Hasen 1951); but this method has never been widely adopted. Sprinkling with water throughout the danger period is a valuable method for areas with a readily available water supply, and is particularly applicable to low-growing crops (Rogers *et al.* 1954). Because of the release of latent heat by the formation of ice on the sprayed plants, their temperature remains roughly at freezing-point, whereas a fall of several degrees below freezing is required to damage even the most susceptible material. This should not be confused with wetting the plants before the frost and then letting them freeze, which is damaging, or spraying the plants in the morning, the so-called 'washing off'

FIVE YEARS AFTER PLANTING

UNRESTRICTED FLOW OF COLD AIR.

POOLING OF COLD AIR

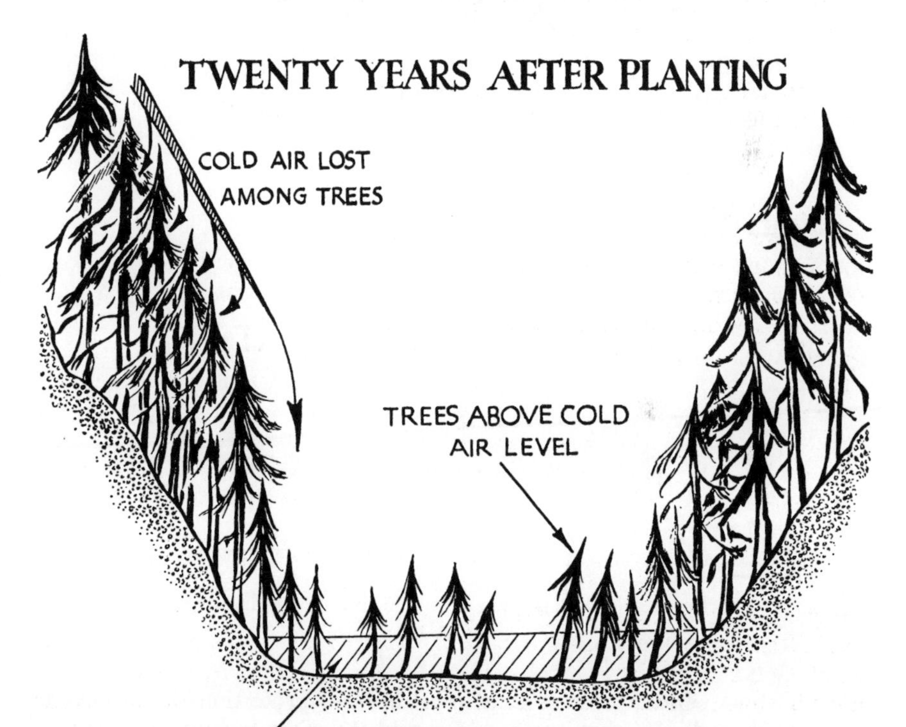

FIG. 7. The effect of tree growth on the pooling of cold air in a valley. (A. C.)

of the frosts, which may sometimes be slightly beneficial, but is generally valueless and occasionally harmful (Modlibowska 1953). All these methods could be applied to forest nurseries, if the value of the crop and the anticipated degree of damage justified them; but fruit normally requires protection over

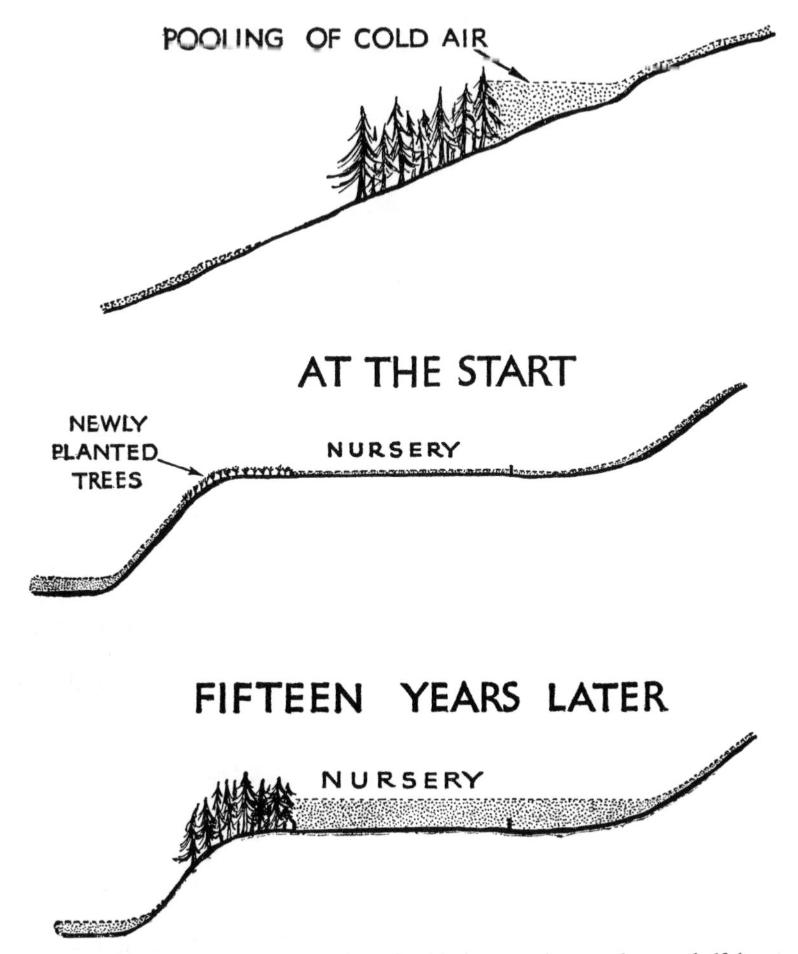

FIG. 8. The effect of trees on the pooling of cold air on a slope and on a shelf (see text). (A. C.)

quite a short period, whereas in a forest nursery protection would have to start much earlier in the season and would therefore involve a greater expenditure on labour and fuel. If a nursery has been properly sited the degree of damage that is likely to result, even from a severe radiation frost, would hardly justify the expenditure and trouble required for heating or spraying, though there is possibly a case for a trial of sprinkling in nurseries equipped for irrigation.

FROST LIFT

The ejection of small plants from the soil by winter frost, a process usually known as frost lift or frost heaving, can be a serious source of loss in nurseries. The same phenomenon is also very destructive to road surfaces, and from both these aspects it has been studied by a number of workers (Bouyoucos 1923; Taber 1929, 1930; Kokkonen 1933), who explained in some detail the mechanism of its occurrence. In brief, when the soil freezes layers of ice crystals are formed in it. The type and extent of these, which have been described and illustrated by Kokkonen (1926), depend on the structure and water content of the soil. During the formation of these ice layers, which naturally force up the soil above them, water is drawn from the soil beneath be capillary action. Thus the extent of the lift is determined not only by the water content of the soil near the surface, but also by the ability of that soil to draw up water from the subsoil and by the availability of water there. Taber states that whereas water increases by 9 per cent. of its volume in freezing, frost lift can be as much as 60 per cent. of the depth of frozen soil. Owing to this rise of water from below, the frozen layer contains an excess of water after the frost, whereas the lower layers are drier than before. It is this high surface-water content that makes the use of forest roads and rides during or immediately after a thaw so destructive to them.

In the first stages of freezing, seedlings and other small objects are firmly gripped by the frozen surface soil. When the surface is lifted by the formation of ice crystals beneath it the seedlings move up with it. Some of the deeper-going fine roots may be broken as the lift takes place. Larger plants with deeper and stronger root anchorage remain unmoved. In their case the frozen surface soil slides up their stems, and then down again when it thaws. The seedlings, on the other hand, remain at the new level when thawing takes place, because their lower roots are still resting on frozen soil. The soil, however, falls back, and may eventually reach a lower level than it formerly occupied, owing to the breakdown of lumps in it by the action of the frost. The final effect is to leave the upper part of the seedlings' root systems exposed (Fig. 9). Repeated freezings and thawings result in the complete ejection of the seedlings from the ground. Stones, garden labels, and other small unanchored objects are ejected in the same way. In addition seedlings are sometimes damaged or even girdled by the abrasive action of the moving frozen soil (Tyron 1943).

Observation under natural conditions and experiments have shown that frost lift is much greater in some soils than in others (Haasis 1923; Taber 1930; Jones and Peace 1939). In general, clay soils give the greatest lift and sandy soils the least. The frost-lift status of peat soils does not seem to have been investigated. The amount of lift also depends on the water available in the soil, both in the surface and in the deeper layers, at the time of the frost. A deep-rooting vegetation mat greatly reduces lift, since it holds the soil down (Ijjäsz 1933). Where roots in the soil are dense they also interfere with the formation of the ice crystals which play a major part in frost lift (Nikki, Ushiyama, and Tomii 1957). Apart from this, any kind of soil cover has a

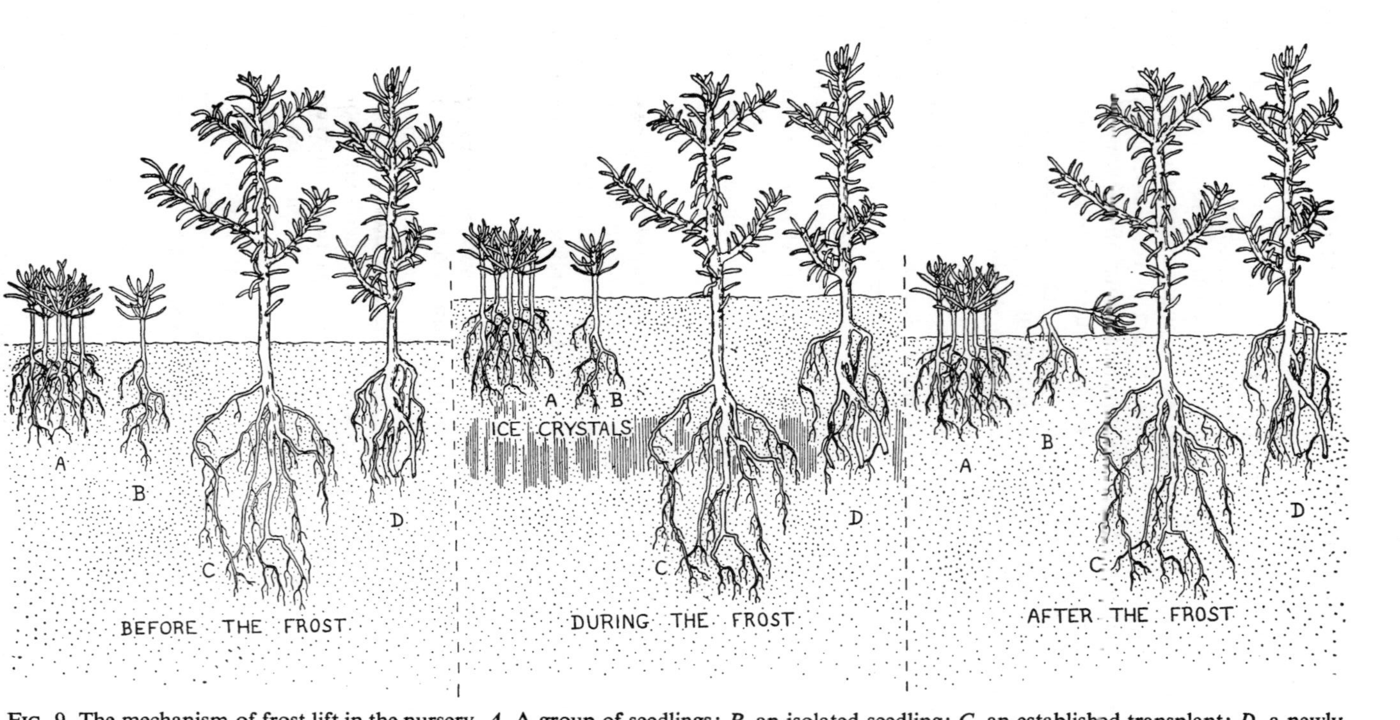

FIG. 9. The mechanism of frost lift in the nursery. *A*, A group of seedlings; *B*, an isolated seedling; *C*, an established transplant; *D*, a newly moved transplant. (J. N.)

reducing effect, because it protects the soil from loss of heat, and therefore prevents it from reaching as low a temperature as it would if it had been bare. McQuilkin (1946) found that mulching was essential for the establishment of young trees on bare clay sites.

Jones and Peace (1939) found that in the British climate frost lift can take place on as many as fifty separate occasions during an average winter, the earliest being in November and the latest in April. In a fairly heavy clay-loam with a rather high water table lifts as high as 27·0 mm. were registered, though the average for the winter was around 6 mm. The total lift for the whole winter was in the neighbourhood of 28 cm. These figures for maximum, average, and total lift would almost certainly be exceeded in a really cold winter with prolonged periods of frost. The figures refer to the actual lifting of the soil surface on each occasion. The ejection rate of objects embedded in the soil would naturally be less. Nevertheless, even if a seedling is only lifted a few millimetres on each occasion, the cumulative effect of up to fifty such lifts would be very substantial.

Generally when the seedling size and rooting depth is the same, dense beds are less damaged than sparse ones, because dense clumps of seedlings tend to lift and settle again together with the soil among their roots (Fig. 9). For this reason drill sowing is often regarded as preferable to broadcast in soils subject to frost lift. But Wahlenburg (1929) found that seedlings grown in dense beds suffered more from frost lift in the transplant lines than seedlings given greater seed-bed space and therefore more chance of developing good root systems. Naturally small plants are more damaged than older, larger ones, so that the risk of injury from frost heaving decreases with increase in size and age. In Britain it is hardly ever of importance outside the nursery.

Control of frost lift

There can be few causes of damage to trees which have so many possible lines of control. For many reasons other than that of frost lift it is a mistake to site a nursery on a heavy soil. But if a nursery has been established on a soil that lifts, and some otherwise satisfactory soils do possess this undesirable property, there are various methods of control which can be followed. Drill sowing or a sufficient increase in density of broadcast sowing to give a complete ground cover will help. Late autumn weeding should be avoided, probably because a certain amount of weed growth in the beds will enable the soil, weeds, and seedlings to lift and fall as a unit. Seedbeds raised well above the alleyways between them appear to give less lift than level beds, presumably because they tend to be drier (Lantelmé 1951).

McQuilkin (1946) found that the application of fertilizers materially reduced frost lift. This he attributed to the effect of the chemicals in lowering the freezing-point of the soil water. On the basis of this one instance, the use of fertilizers can hardly be recommended as a direct method of control; but fertilization, irrigation, early sowing, soil sterilization, or any other method likely to produce larger, better-rooted seedlings will naturally reduce lift. Lantelmé (1951) found that spruce were less lifted when sown in mixture with deeper rooted pine.

A large number of substances have been tried as seed-bed mulches. In parts of America where seedlings are under snow cover for much of the winter, frost heaving risk is limited to short periods in the autumn and spring. For these short periods quite heavy coverings of straw, chopped straw, or 'shingle tow', a kind of coarse wood wool, can be put over the seedlings. In Britain mulches must leave the whole top of the seedling exposed. There are considerable discrepancies between the effects of various materials used in different nurseries (Jones and Peace 1939). This strongly suggests that differences in climate and soil may influence success or failure. A mulch of granulated peat, and stone chippings used as a seed-bed cover, have both been successful in lessening lift on a number of occasions. Protection of the beds by lath shelter or by conifer branches stuck in the ground to form a canopy lessens lift, though it may fail to prevent it.

If seedlings have been lifted it is better to reline them on fresh ground than to try to push them back. The extent to which transplants are lifted naturally depends on their depth of rooting. Established two-year transplants are seldom lifted, but seedlings lined out in the autumn may suffer considerable lift, especially if their roots were small or the lining-out shallow (Fig. 9). Transplants which have been heaved by frost can usually be pushed down again by careful treading along the rows. Slight ridging, akin to earthing up potatoes, is another way of covering the exposed roots. Wherever possible, lining out should be left to the late winter and early spring in nurseries subject to frost lift. By then most of the frosts causing heaving will have occurred.

Heaving in plantations is seldom of importance in Britain, but does occasionally do damage on bare soils in some other countries. Deters (1939) found that mulching and firm planting were both of value in preventing the lifting of newly planted trees.

5

DAMAGE TO TREES BY OTHER CLIMATIC AGENCIES AND BY FIRE

DAMAGE BY WIND

WIND is certainly the most important climatic factor affecting the growing of trees in Britain. Not only do occasional gales cause considerable, and sometimes spectacular, damage, but much more important are the frequent high winds which limit the sites, rate of growth, form and final height of trees. Gales produce mainly mechanical damage. The blast effect of lesser winds is caused mainly by excessive transpiration, and consequent desiccation of the foliage and succulent twigs. This is more appropriately dealt with in Chapter 6; here only the mechanical damage caused by gales is considered.

Wind damage takes a variety of forms; at its worst the trees may be uprooted; they may break off; they may bend over, sometimes permanently, but more often eventually assuming a fairly upright position; they may rock about, chafing the base of the stem against stones or even hard-packed earth, a form of damage which normally affects only young trees. In addition persistent high winds mainly from one direction have a big effect on the form of trees and on their timber, making the growth one-sided, and often causing them to lean away from the wind, or to form a bend near the base. These forms of damage are dealt with separately, but, first, there are a number of relevant general factors to be considered.

It is outside the scope of this book to deal with the meteorological causes of high winds; but inevitably from their position the British Isles are particularly subject to such winds, which here normally blow from the south-west (Bilham 1938). Exceptionally gales may come from other quarters, and these are usually the more damaging because they assault trees from a direction in which they have developed no especial strength or anchorage. Trees exposed on one side or subject to winds mainly from one direction develop both their roots and their stem in a way that offers greater wind resistance.

The mechanical action of the wind on trees has been dealt with very fully by a number of workers (Fritzsche 1933; Jacobs, M. R. 1936; Woelfle 1936–7; Curtis, J. D. 1943; Mergen 1954), all of whom have realized the complexity of the problems involved in deciding the stresses exerted by an erratic force on an asymmetrical structure to some extent pliable in all its parts, and irregularly anchored in a substratum of extremely variable properties. Certain points emerge quite clearly; the turning moment exerted at the base of the tree is proportional both to the extent of the crown and to its height above ground, though as soon as a tree is moved out of the vertical its own weight and therefore the height of its centre of gravity above the ground also begin to influence the turning moment. If the tree breaks, the initial failure is normally on the leeward side, where the wood is compressed, rather than on the windward, where it is stretched. But the possibility of a tree breaking or uprooting is certainly not directly proportional to the turning moment. In fact, of course, the wind does not exert a steady pressure, the crown of the tree will often swing in a large ellipse with its long axis in the same direction as the wind, so that it becomes very difficult to produce exact mathematical data on the behaviour of what is in fact a large inverted pendulum, subject to innumerable irregularities in its form, suspension, and anchorage. In many trees the crowns are asymmetrical, so that the main stem is subjected to torque as well as to more direct pressure. Individual branches are even more subject to torque and often show obvious signs of having been twisted before being broken in the wind.

A certain amount of gale damage is an inevitable accompaniment of forestry in Britain, but the storm that occurred at the end of January 1953, and which resulted in extensive damage to trees, particularly in north-east Scotland, as well as serious sea-flooding along the southern parts of the North Sea coast, naturally attracted special attention (Lines 1953; Andersen 1954). There is also a vast literature on the effect of other gales, particularly on the Continent and in the New England States of America. Some of these papers are quoted below, where they have a bearing on particular types of injury.

Topography obviously has a very big influence on damage by wind (Woelfle 1936–7; Andersen 1954), but in view of the possibility of gales from unexpected directions, it is seldom possible to estimate with any certainty the wind susceptibility of a particular site, though south-east gales are so exceptional that trees exposed only in that direction can be considered safe. After a gale it is often possible to explain the blowing of one area or the escape of another on the grounds of topography and its influence on the direction and force of the wind; but it is much harder to prophesy what will occur in the future, even if a certain wind direction is assumed. In old forest areas past experience of gale damage is obviously a valuable guide in the treatment of sites, but in afforestation of course no such aid is available. Trees normally accustomed to exposure are less prone to damage than those usually sheltered, so that windblow on exposed hill-tops is seldom as great as one might expect, while damage in apparently sheltered situations may be much greater. Topography, especially in relation to the direction of prevailing gales, should certainly be considered in elaborating precautionary measures against windblow, but it should not be given too great a weight.

Inevitably in a sea-girt island, gales are likely to carry in salt, and especially in spring and early summer, when the foliage is developing, this salt can cause severe scorching. This source of injury is dealt with in more detail in Chapter 8.

Windthrow

In most gales, uprooting of trees is far more frequent than stem breakage or any of the other forms of mechanical injury due to wind. The mechanical complexities involved in the overthrowing of a tree by the wind have already been briefly considered. For instance, it cannot be assumed that trees with large high crowns will suffer most, although the turning moment is proportional both to the size and height of the crown. J. D. Curtis (1943), who considered this matter in some detail, was driven to make a number of exceptions, where this rule did not hold. In fact there is no general agreement as to whether, under similar wind conditions, large- or small-crowned trees of the same height are the most susceptible.

It has been observed with young trees in a few instances that the liability to wind damage was increased by heavy pruning. This may well be due in part to the smaller diameter growth and root development of the trees with reduced crowns, but it also suggests that the pendulum effect of a top-heavy tree swaying in the wind may influence its final uprooting or breaking. Uprooting is seldom the result of steady pressure by the wind, but is rather the outcome of lashing to and fro, whereby various stresses are exerted in several directions.

If the general height of a plantation is considered, however, it is clear that taller crops are likely to be worse damaged. Lines (1953) found that there was a marked increase in damage in the 9–12 metres height class, plantations below that height seldom being severely blown, while nearly all the disastrous windfalls were in plantations above it. Werner and Ärmann (1955) gave very similar results. There is disagreement, however, on whether a stand with a uniform canopy is less likely to be damaged than an uneven one (Woelfle 1936–7; Curtis, J. D. 1943; Macdonald, J. M. 1952; Werner and Ärmann 1955).

In considering windfall the development of the roots must be taken into account. M. R. Jacobs (1936) found small trees blown more frequently than large, and considered this to be mainly a reflection of their inferior root systems. Croker (1958) found that liability to windthrow decreased steadily with increasing soil depth. The closely interrelated effects of the soil and the root system have quite as large an effect on windthrow as have crown size and height. The depth of rootable soil, which may be limited either by the underlying strata or by drainage impedence, certainly has a profound effect on windblow (Day, W. R. 1950*b*, 1953; Lines 1953; Andersen 1954). Many of the most disastrous windblows occur on sites with naturally shallow soils or in places where the periodic or permanent presence of water in the soil limits depth of rooting. Thus the maintenance of the best possible drainage conditions is a big step towards the control of wind damage. The drains themselves unfortunately form lines of weakness, since they cut roots and disturb the continuity of the root mat. The number of drains therefore should be kept to the minimum necessary for the satisfactory removal of water.

Apart from rooting depth the general structure of the root system greatly affects windthrow (Pryor 1937). A number of authors state that windthrow is considerably influenced by the strength of the leeward roots, and that the failure of these roots under compression finally leads to the uprooting of the tree (Fritzsche 1933; Mergen 1954). Though breakage of this kind has been observed on a number of occasions in Britain, it was not commonly found after the 1953 gale, where the majority of the blown trees remained alive, because the leeward roots, now at the bottom of the upturned root plate, were only bent, not broken, and continued to supply the tree with enough water to support life, even if not active growth. It seems probable that, under shallow rooting conditions, the stability of the stand may greatly depend on the formation of a continuous interlaced root-mat (Wendelken 1955), and that much of the force needed to uproot the individual tree may be expended on tearing apart this structure. There is evidence (Yeatman 1955; Lines unpubl.) that interruption of this mat by actions such as drainage or deep ploughing at the time of planting can lead to windfall when the gale is at right angles to the direction of disturbance. In the case of ploughing, the roots tend to develop along the ridges and not to cross the furrows to any extent. It seems possible, however, that as plantations formed in this way get older, the furrows will fill with humus and the root distribution become more uniform. The soil may also affect windblow in another way; in highly fertile, well-watered soils, trees may form very large crowns and comparatively small root systems, and thus become liable to damage. This has happened with Douglas fir on some sites, and also with poplar.

Naturally anything that leads to a defective root system may increase liability to windthrow. It has been suggested that planting trees with the roots curled up like the handle of a walking stick, by forcing them into a shallow slit, a defect which certainly leads to wind-rocking in young trees (see below), may also result in a permanent maldistribution of the roots and subsequent instability. It is quite certain that fungi play a big part in windthrow (Hubert 1918; Wallis 1954; Gratkowski 1956). *Fomes annosus*, *Armillaria mellea*, and in older stands various other root fungi, greatly assist windthrow by killing roots; while the windblow of isolated trees, which one would normally expect to remain windfirm, is almost invariably preceded by substantial fungal attack on the roots (Campbell and Miller 1952). Gratkowski (1956) suggested that felling faces should never be made through or up to stands known to be affected by root-rotting fungi.

The relative resistance of different species of windblow is very difficult to assess, because so many of the factors contributory to wind damage affect different species in different ways. For this reason, although lists of relative susceptibility have been published, they are probably mainly of local significance. The spruces are often damaged, probably in part because they are naturally shallow-rooted; but they are very commonly planted on areas where deep rooting would be impossible and where any other species would be liable to windfall. Douglas fir suffered severely in the gales in south-west England and Wales in 1954 (Lines unpubl.), but this was partly due to its being commonly planted on rich, loose, loamy soils, which were sodden at the time of the gale and offered a very poor root-hold. The heavy damage to Scots pine in the 1953 gale was mainly due to its preponderance both in the affected area and in the height classes most liable to injury. Among the conifers, larch generally has a good record, and among the hardwoods, oak. Beech was the worst affected hardwood in the 1953 gale in Scotland, but it was also the commonest one in the area affected, occurring frequently in under-thinned shelterbelts.

Obviously thinning greatly affects windblow. Trees grown in open conditions from youth are generally much more stable than plantation trees. It may be that a similar stability could be achieved by strip or group planting (Macdonald, J. M. 1952), certainly heavy thinning in youth or extremely wide-spaced planting are likely to increase wind resistance. Thinning may precipitate windthrow primarily due to other causes such as root rot (Wallis 1954). The period between the operation of thinning and the gale is probably more important than the actual degree of thinning. The trees left standing after the operation require time to become accustomed to the additional exposure, and if a gale occurs before they are stabilized damage is likely to be severe (Werner and Ärmann 1955). For this reason it is safest to thin stands where windblow is likely to be troublesome, or plantations of especial value, in late spring so as to give them a period of stabilization before renewed gales may be expected. There is some evidence that under-thinned plantations are more liable to be injured than normally or heavily thinned ones. However, completely unthinned plantations have sometimes proved very stable (Werner and Ärmann 1955), but the longer they remain in this condition the harder it becomes to thin them without rendering them liable to subsequent wind damage.

Windblow is greatly influenced by the original layout of the forest and by its subsequent treatment. It enters into any discussion of silvicultural systems or of felling methods (Hawley 1946; Hawley and Stickel 1948; Troup 1952). Wind stability has been advanced as one of the reasons in favour of uneven-aged stands (Werner and Ärmann 1954; Meyfarth 1955; Köstler 1956*b*). Admixture of species, especially of broadleaved with conifers, is also frequently advocated as a means of increasing wind stability (Werner and Ärmann 1955; Hirata and Maezewa 1956). Very large areas of a single species and of uniform age are particularly dangerous, while relatively restricted blocks of different ages and species provide boundaries against which damage may stop in the event of extensive windfall. It is important to avoid the unnecessary creation of exposed faces in hitherto sheltered plantations. In clear felling, of course, such exposure is inevitable, but a great deal can be done by felling up to a more resistant species, or up to a crop which has not yet reached a dangerous height. Felling faces should never be left across an ill-drained site, or in a place known to be subject to wind-funnelling (Gratkowski 1956). Quite small indentations in a felling face are often foci for further damage, as are quite small gaps in an otherwise undisturbed stand. Heavily thinned plots in a normally thinned plantation tend to blow down, not directly as a result of the heavy thinning, but because they form a gap in the canopy into which the wind can eddy. In the same way gaps caused by diseases such as *Fomes annosus*, *Rhizina inflata*, or even lightning often start extensive windfalls. Indeed their greatest danger lies in the wind damage to which they may lead. The danger of windthrow should certainly influence every decision on forest layout and treatment, but it is doubtful whether wind is sufficiently important alone to determine the choice of silvicultural system.

Fellings made in connexion with road construction often lead to windfall, not only because they expose trees hitherto sheltered, but also because road work is often accompanied by considerable severance of the roots, and consequent instability, of the trees left standing. The road system in a forest should be planned before trees are put in, and ample widths left unplanted for subsequent construction. In established forests opportunities should be taken after felling and before replanting to make such new roads as are necessary on the area felled and to improve old ones. Gratkowski (1956) also stressed the desirability of avoiding damage to roots during the extraction of thinnings (Chapter 10), because such injury increases the risk of windfall.

Inevitably, large-scale windthrow raises the question of deterioration of the trees before they can be salvaged. Insect infestations are important more because of the damage the increased population may do to crops still standing, or to the young trees used for replanting, than because of the direct effect they have on the blown trees, but fungal deterioration is a serious possibility. If the trees are only partly uprooted and remain alive, as in the 1953 gale, surprisingly little stain or decay takes place. However, survival was high following the 1953 gale, because the summers of that year and of 1954 were generally cool and moist. A hot dry summer following a gale might kill many of the trees, despite their root attachment, and deterioration would then become a much more urgent problem. All that can be done to prevent loss is the removal first of those species most subject to decay (Chapter 19).

Windbreak

Windbreak is much less common in Britain than windthrow, though it is of course the result of the same forces. Breaking may occur at any point up the tree, even just below the crown. Normally it develops from an initial compression failure on the leeward side of the tree, rather than from a tension failure on the windward side (Hocking 1949; Mergen and Winer 1952). Trees which are damaged, but fail to break, may develop swellings over the partial failures in the stem. These swellings also occur as a result of bending by ice or snow (see below). Windbreak seems to occur especially on sites where there are strong eddies, which result in the tree being subjected to heavy wind buffeting from more than one direction, with the consequent production of compression failures on more than one side of the tree. Butt-rotted trees frequently suffer windbreak, and in this case the break is usually in the basal two metres.

Breaking of branches by wind, especially on large-crowned hardwoods, occurs frequently. Branches are often subject to twisting forces, which of course increase the possibilities of fracture. Gale-force winds during the growing season can damage trees by breaking off leaves and young shoots, though this is usually much less serious than 'blast' (Chapter 6). A. E. Moss (1940) reported almost complete defoliation of broadleaved trees in a September hurricane in New England, some of the leaves being torn away from the mid-rib and others broken off at the petiole. This was followed by dieback and by death of some of the resting buds. This can probably be attributed to failure of the defoliated shoots to ripen and become winter-hardy, a phenomenon which has been also noted in the case of trees defoliated by fungal disease. There is no record of such complete defoliation due to wind in Britain; though some green leaves are torn off in any summer gale, the degree of defoliation is probably seldom damaging.

Windbend and windlean

Trees may be bent over by the wind, though this type of damage is much more often caused by snow or ice deposition. Often the initial displacement is due to wind, but further bending is caused by the weight of the displaced crown. This type of damage is more likely to occur on trees with slender pliable stems. It is thus a frequent form of wind injury in seriously under-thinned plantations, and has also been observed by the author on very fast-growing young conifers on a very fertile, loose soil. In this instance the wind caused partial uprooting so that the trees developed a moderate lean. The subsequent bending appeared to be the result of the trees' own weight. Pronounced bending without displacement of the roots has been observed in fast-grown poplar, which showed strong clonal differences in degree of bending, presumably dependent on differences in wood strength. Spegazzini (1925) has reported spectacular bending in poplars with decayed heartwood, the tops actually resting on the ground. In many cases windbend is permanent, while in others the tree may eventually resume an upright position.

Trees constantly exposed to wind may, however, lean away from it without any apparent effort to resume a vertical position. This is a distinguishing characteristic of the so-called Black Italian poplar, *Populus 'serotina'*. It also occurs in Britain with larch, *Pinus contorta*, and Scots pine, but not with the spruces or Silver firs. M. R. Jacobs (1936) and Fielding (1940) describe large areas of Monterey pine in Australia leaning with such regularity as to pre-determine the direction of felling. Several other species of pine were similarly affected, but Corsican pine remained upright.

If trees are pushed over by the wind when young, it usually involves movement of the roots as well as the stem. Trees in this position often become stabilized and later resume upright growth. To do this, however, they develop near the base a permanent curvature, the concave side of which is towards the prevailing wind. This happens commonly with larch, Maritime pine, and Douglas fir (Goetz 1951; Stecki and Rada 1953), and may seriously reduce the value of the butt length.

Leaning trees, and indeed upright trees if frequently subject to wind pressure from a certain direction, tend to develop special reaction wood. In conifers this takes the form of compression wood on the leeward side (Jacobs, M. R. 1954; Scott and Preston 1955). In broadleaved trees tension wood is formed on the windward side. The properties of reaction woods differ materially from those of normal wood, so that they cause differential shrinkage during seasoning, irregularities in veneer peeling, and so on. On the other hand, tension wood in poplar, some varieties of which are markedly affected by wind, has higher cellulose content than normal wood (Jayme, Harders-Steinhäuser, and Mohrberg 1951). This is obviously advantageous for cellulose extraction. Jacobs found that swaying in the wind had a profound effect on diameter growth, and that trees guyed to prevent free movement put on less wood.

Windshake

Wind certainly plays a big part in the production of shakes, which are small splits in the heartwood. These are discussed in more detail in Chapter 20.

Windrock

Wind can damage young trees, particularly those with heavy tops and restricted root development, by rocking them about. This chafes the bark at the collar against stones or even against compressed soil (Moore, B. 1933; Day, W. R. 1934). Such injury is particularly frequent with Maritime and Corsican pine (Fig. 10). It may sometimes lead to infection by Honey fungus, *Armillaria mellea*. It is often suggested that wind-rocking results from bad planting, the roots having been forced in at an angle or even bent upwards, so that they fail to develop an even anchorage. Clarke (1956) found that young natural regeneration of *Pinus radiata* was much more windfirm than planted stock of the same age and size. The superiority of natural regeneration in this matter may, however, be due mainly to its initially slower growth, and the resultant better balance between crown and root development. Sometimes staking may be necessary to save a plantation that is being rocked by the wind; the best method is to attach the live branches of the trees themselves to the stakes (Chapter 10).

FIG. 10. Damage to roots of Corsican pine by wind-rocking. (J. C.)

It has been suggested that decline in vigour in older trees may result from the breakage of the finer roots by windsway. This is most likely to happen when the roots grow down from a loose movable medium, such as a peat layer, into a firm immovable one, such as clay. Redmond (1954), however, did not find root breakage to Balsam fir and Yellow birch growing in sandy loam in Nova Scotia. There is no evidence that it is a serious factor in tree disease. Severe damage to the roots of older trees, 5–10 metres high, has been recorded on a shallow soil over chalk with flints. The species concerned was *Cupressus macrocarpa*. The trees were wide-spaced and had developed large crowns. Many of the abraded roots had died, which had increased the liability of the trees to windthrow.

DAMAGE BY FLOODS AND BY EROSION

Injury due to the effect of flood water on the oxygen supply to tree roots, a type of damage that occurs only when flooding is prolonged, is considered in Chapter 6. The toxic effect of salt water on tree roots is dealt with in Chapter 8. Boulders and driftwood carried by fast-moving flood water may damage the stems of trees on the river bank, causing wounds near the base on the upstream side. On the Continent, similar damage has been caused by ice floating in flood water. Young poplars have even been broken off in this manner (Günzl 1953). The occurrence of floating ice of sufficient weight to cause injury is unlikely in Britain.

Trees growing on banks and on the edges of cuttings often have part of their root system exposed by the erosion of the soil. This process tends to be gradual, so that the tree can adapt itself to the altered conditions by developing deeper feeding roots and by increasing its anchorage. Such erosion may lead eventually to windthrow, but seldom to fungal infection or to any marked deterioration in the general health of the tree. Where a cutting is being made alongside a tree, erosion can be avoided by building a retaining wall (Chapter 37).

Erosion can also be a serious problem in some nurseries, especially those on light soils with a moderate humus or silt content. These encourage surface compaction and consequent run-off rather than absorption during heavy rain, and the light soil easily washes away. The most serious damage is to seed-beds. Seed may be washed away, or seedlings uprooted. Granite chippings certainly assist in preventing erosion from seed-beds; Bedford sand washes off rather easily. With proper orientation most of the water can be directed down the alleys between the beds. Those parts of the nursery most subject to erosion should be reserved for transplants.

DAMAGE BY HAIL

Hail is a well-known cause of damage to fruit, and is regarded with particular dread by vine-growers on the Continent, but the injury it causes to trees, in Britain at any rate, is much less familiar. Hailstones may perforate the leaves and cause small lesions on the exposed surfaces of young soft-barked twigs. Young shoots of conifers may be severed; fortunately leaders, because of their vertical position, are less liable to this type of injury than side shoots.

Hailstorms, and consequently hail damage, are normally very local in occurrence. In Britain hail is seldom violent enough to cause permanent damage to established trees, though Sampson (1901) reported very serious injury on a number of tree species due to a local hailstorm in Cheshire. The hailstones in this instance averaged 2·5 cm. in diameter. Certainly hail damage to forest crops in Britain would not justify any of the efforts that are being made experimentally on the Continent to precipitate the clouds in other less damaging forms. Severe hail injury has been reported both from America and the Continent, conifers and broadleaved trees being defoliated, many trees dying back, and wounds being formed which allowed the entry of fungi (Engel 1939; Anon. 1948*b*; Riley 1953; Tanner 1953; Grayburn 1957).

Hail damage is comparatively easy to diagnose, especially *in situ* when the orientation of the injuries on the tree and on the twigs can be observed. The lesions are mainly on the upper side of the twigs, and on one side of the tree, depending on the direction of the wind at the time of the storm. Sectioning of the twigs will show that all the injuries occurred at the same stage in the annual ring. Damage to the shoots is likely to be more severe on trees with thick rigid twigs, such as poplar, than on those with thin pliant twigs, such as birch; but the available information on relative susceptibility is somewhat contradictory. Pechmann (1949) has reported alterations in the annual rings following severe hail injury, namely reduction in the summer wood of the year of injury and reduction in width of the ring of the next year. Similar ring reductions have been described in trees defoliated by fire (see below).

DAMAGE BY THE DEPOSITION OF ICE

The meteorological conditions required for a glazed frost are too complex for discussion here (Brooks and Douglas 1956); but briefly it can only occur when there is a substantial layer of air above freezing lying over a lower layer of air below freezing-point. Under these conditions precipitation becomes supercooled during its passage through the lower cold layer. On touching any solid object the supercooled droplets turn instantly to ice. Such meteorological conditions are unstable, and therefore they rarely last for any length of time.

In January 1940, however, these conditions persisted for three or four days mainly along a belt stretching from North Wales, through Gloucestershire to Kent. In general to the north east of this belt the precipitation fell as snow and to the south-west of it as ordinary rain. In the affected area, especially on high ground and on aspects exposed to the prevailing wind, ice deposition up to 5 cm. thick occurred on twigs, branches and other exposed surfaces. Twenty-three telephone wires between two posts were estimated to be carrying over 11 tons of ice, and twigs might bear as much as sixty times their own weight. Under such a weight of ice, breakage of branches and crowns was inevitable. Damage naturally took a wide variety of forms, from the breaking of the main stem to severe bending of the branches. Some trees were pushed right over; this happened particularly on slopes where the crowns tended to be one-sided and the weight of ice therefore unevenly distributed. Breaking of branches was particularly serious on hardwoods (Plate III. 1), while conifers suffered more from loss of leaders or the upper parts of the crown and from being pushed over. McKellar (1942) found that the stems of three species of pine tended to break where they were approximately 5 cm. in diameter. On older hardwoods broken branches left many wounds eminently suitable for the entry of decay fungi (Campbell and Davidson 1940). The bases of pressed-down branches on conifers showed cracking and resin exudation, though there is no proof that this led to fungal infection. In many cases the slow increase in weight allowed young stems to bend gradually without breaking. On the other hand, very pliable stems sometimes bent too far, too easily, and reached positions from which they could not rise again. Many conifers had several metres of the crown in an inverted position, while others with normally upward-pointing branches, such as the pines, had all their branches pointing towards the ground. Recovery from these deflexions was slow, but in many cases eventually complete. Many trees did not reach a completely normal position until late in the summer (Gardner, R. C. B. 1940; Sanzen-Baker and Nimmo 1941). In some instances the timber of trees which were severely bent suffered a loss of elasticity, and trees permanently deflected produced compression wood (Rendle, Armstrong, and Nevard 1941). Thus the damage caused may extend beyond immediate breakage and the subsequent entry of decay. It is still possible in some areas to detect the breakages that occurred during this storm.

Similar storms have very occasionally been reported on the Continent (Noël 1941; Leroy, P. 1956) and rather more frequently in the eastern United States (Downs 1938; Spaulding and Bratton 1946). In America glazed frost is of sufficient importance for some work to have been done on its avoidance. Downs (1943) suggested that thinning, so as to produce well-spaced stands of fast-growing trees, was the best answer; however, in a series of plots in Pennsylvania thinned to different degrees, it was found that only the most heavily thinned suffered any glaze damage (Anon. 1950*a*). This result was later confirmed by Carvell, Tryon, and True (1957), who considered that the greater exposure of the crown in heavy thinning opened a larger area of the tree to ice deposition. McCulloch (1943) found worse damage among the residual trees in stands from which a low percentage of the dominants had been removed than in stands from which a higher percentage of the sub-dominants had been cut.

In 1940 in Britain damage was particularly severe in existing underthinned plantations or in underthinned plantations which had recently been thinned. In plantations of uneven height-growth the tallest trees often suffered very severe injury. Fortunately in Britain glazed frosts are far too infrequent for precautions to be necessary. Roberts and Clapp (1956) have described the effect of exceptionally heavy live-crown pruning on trees bent over by ice. This measure caused the trees to resume an upright position much more rapidly than unpruned trees.

Information on the relative damage to various species is so confused by differences due to age and location that it is often extremely contradictory. Among conifers the spruces seem to escape lightly, possibly because of the downward sweep of their branches and their normally well-balanced crowns.

In the British storm of 1940 isolated hardwoods, particularly poplar, but also birch, ash, sycamore, and alder, were generally worse injured than isolated conifers. Among the hardwoods elm in particular escaped with relatively little damage. This is curious because *Ulmus americana* was very near the top of a long list of relative damage given by Croxton (1939) following a glazed frost in the United States. In plantations, conifers were usually worse damaged than hardwoods, but the former probably occupied most of the sites most favourable for ice deposition.

Two unusual types of injury associated with glazed frost have been described. Ladefoged (1943) described failures on the compression side of the stem caused by the weight of the ice and the force of the wind. In some cases the bark was forced away from the wood. Subsequently the space between the bark and the wood was filled with weak woody tissue and appeared externally as a lump on the stem. In subsequent gales trees often broke at these points of weakness. H. J. Lutz (1936) described a much milder form of injury. The coating of ice on young, pliable, thin-barked stems cracked as the trees swayed, and the broken edges of the ice damaged the outer bark, leaving a series of parallel scars across the stem. The injury was too superficial to do any permanent damage.

DAMAGE BY SNOW

The amount of snow that will adhere to branches depends mainly on its temperature and consistency, but also on the slope and position of the branch, the force and direction of the wind, and various other factors (Anon. 1952*b*). Very cold snow usually falls in small particles which do not readily cohere, whereas snow nearer its melting-point falls in larger flakes which can build up on twigs and branches to a surprising degree. Unseasonable late spring snowfalls nearly always consist of this type of snow.

The damage caused by snow is usually very similar to, but less severe than, that caused by ice (see above). If the weight of snow causes any marked downward movement of the branches, this often brings about a shedding of the load, so that acute bend, such as occurs with ice, is rare. When very deep snow falls among young conifers, it may result in breakage of the branches below the snow level (Day, M. W. 1940). The most serious result of snow is certainly the pushing over of trees, especially on slopes where the crowns are ill balanced (Curtis, J. D. 1936). One tree falling over may give the final push to the one below, so that snow damage of this kind often extends in strips down a slope.

In countries subject to prolonged and heavy winter snowfall, types of conifers have developed which by virtue of the short length and downward snow shedding sweep of their branches are particularly suited to withstand heavy snow; *Picea engelmannii* and *P. omorika* are good examples of this. In relatively snow-free regions types occur that are singularly ill-fitted to withstand snow, in particular fastigiate conifers, such as *Cupressus macrocarpa* or *C. sempervirens stricta* which tend to fall apart when snow-weighted. Provenance differences and even local variations in reaction to snow certainly exist in some species (Hess 1933; Rubner, N. 1942), but do not appear to have been systematically studied. Hardwoods are normally much less damaged by snow than conifers. The worst injury, severe breakage of limbs, results only from very late snowfalls occurring after the trees are in leaf (Galoux 1951).

Breaking of branches on hardwoods is likely to lead to decay, and it has been suggested that cracks at the base of conifer branches resulting from their depression by snow may lead to fungal infection. Snow sometimes causes breakage at points of weakness resulting from other causes. Decayed stems, for instance, often break under snow. The fracture of weak joints between side-branches and the main stem in pine, resulting originally from premature development of some of the side buds, is described in Chapter 20.

The amount of damage caused by snow to trees in Britain does not justify special measures being taken against it. On the Continent, however, the adoption of particular silvicultural systems is often advocated as a means of avoiding snow damage. Even in Britain particular care should be exercised on slopes to remove trees with asymmetrical crowns, or thin-stemmed trees with very high crowns, during thinning. Both of these types are particularly liable to snow injury (Rosenfeld 1944). Leaning trees should never be left on slopes lest they prove the start of a chain of snow-felled trees.

In mountainous countries with very heavy snowfalls avalanches sometimes cause serious damage to high-altitude forests, sweeping away quite large areas of trees. On the other hand, trees themselves are often used for stabilizing snow on steep slopes. In Britain we are virtually free of avalanche danger, but properly sited tree planting could sometimes be used to prevent snow drifting on to roads and railways.

An unusual form of snow damage sometimes occurs in very exposed places owing to ice crystals blowing over the snow surface. Small trees less than 1 metre high may be bent over by the wind, and the bark scoured from the exposed parts by the moving ice (Fig. 11). The worst damage is on the windward side of the stem, and on the under sides of the windward branches. Less severe damage occurs on the upper side of leeward branches not buried in the snow. The base of the main stem and the lower branches, protected in the snow layer, are uninjured. After the snow has melted and the trees have resumed an upright position, damage of this nature can prove very puzzling.

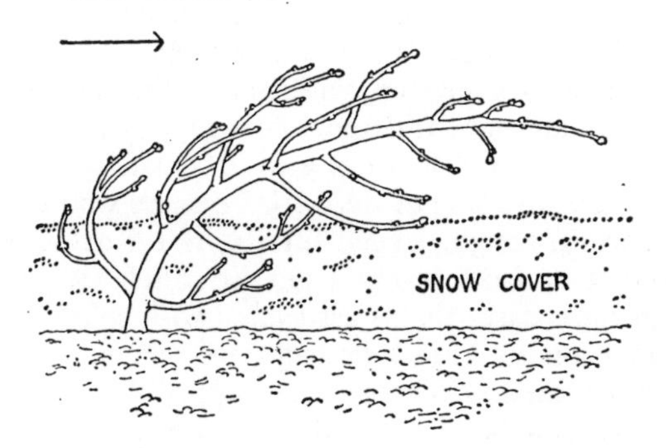

FIG. 11. Damage to European larch by blown snow crystals (see text). (J. N.)

DAMAGE BY LIGHTNING

Lightning is a familiar cause of damage to individual trees, on which its effect is often spectacular, though the total damage over the country as a whole is negligible. Lightning is much less familiar as a cause of death of groups of trees, an occurrence which happens often enough to justify attention, though again the total damage done is slight.

The mechanism of lightning discharge is too complex for discussion here. In brief, lightning is a vast electric spark between a cloud and the earth which tends to discharge through tall objects because this lessens the distance it has to travel. It is simply because they are tall objects that trees tend to act as channels of discharge (Dodge, A. W. 1936; Müller, K. M. 1938; Bruce and Golde 1949). The effect on the tree varies according to the intensity of the discharge and the condition of the tree. Damage may vary from a single scar down the trunk to complete shattering of the main stem. The long scars often involve only the stripping of the bark from the wood. If they penetrate the wood, they form a rough, jagged furrow, not a definite fissure such as occurs with drought or frost crack. The scars often follow a spiral course down the tree; in some cases this is associated with spiral grain in the wood (Chapter 20). Occasionally visible damage may extend along the roots, the soil being thrown off and one or more of the upper roots exposed. It has been suggested that trees can be struck without receiving any injury at all. In group strikes trees can be killed without displaying any visible external injury. A. W. Dodge (1936) describes damage to the twigs alone due to comparatively slight discharges. Lightning killing the top few feet of a Sequoia in Britain has been confirmed by the presence of lightning rings (Murray unpubl.). The diagnosis of slight injury by electrical discharges (such minor phenomena can hardly be described as lightning) is obviously difficult, since the actual causation is unlikely to be observed. It is impossible, therefore, to say how frequent or how important such minor damage is, but it may well account for more cases of crown dieback in trees than we at present realize.

A good deal of information is available on the relative frequency of lightning damage to different species (Stahl 1912; Dark 1935; James 1939; Thompson, A. R. 1943), but the order of frequency is not necessarily the same as the order of susceptibility. It is almost certain that oak, elm, and poplar, which generally head the lists of broadleaved trees, and pine which heads the lists of conifers, do so because they are the commonest isolated trees, and are therefore more likely to be struck than other species. Despite this it does seem possible that there are specific differences in susceptibility to lightning strike. Beech is certainly not damaged as often as its frequency of occurrence would suggest. Indeed in Germany 'Eichen sollst du weichen, Buchen sollst du suchen', 'Oaks you should avoid, beeches you should seek' is a traditional saying. The view is often taken (Walter, B. 1932; Dodge, A. W. 1936) that the escape of beech is due to its smooth bark, which, when wet, provides a continuous film of water, which may act as a harmless discharge path for the electric current. Stahl (1912) considered that it was the water content that determined the liability of a tree to lightning, by affecting its electrical conductivity. These ideas are contested by Szpor (1945), who states that the electrical resistance of a tree has no direct bearing on its liability to lightning damage. It is obvious that the mechanism and action of electrical discharges through trees are only imperfectly understood. This is inevitable since information can only be deduced from the damage and not from observation of the actual phenomenon.

Some sites are more liable to lightning strike than others. This depends on the general distribution of thunderstorms, local air conditions with particular reference to upward movements of hot air, and the nature of the soil and underlying rock. In the United States lightning conductors are occasionally affixed to ornamental trees of particularly high value (Dodge, A. W. 1936, 1937). It is doubtful whether the risk anywhere in Britain is great enough to justify this.

The death of groups of trees as a result of lightning has frequently been reported in Europe and elsewhere (Tubeuf 1905; Hoepffner 1910; Peace 1940; Entrican 1954; Wijbrans 1957; Murray, J. S. 1958). Such groups are usually less than an acre in extent, but J. S. Boyce (1948) reported damage, admittedly only browning of needles, over an area of nearly 500 acres, and more severe damage, though only some of the trees were killed, over an area of 25 acres. These groups have been reported mainly in conifers, but records of groups in oak (Grossh 1906) and the death of a number of small hardwoods in the Douglas fir group described by Peace (1940) prove that they can occur also among broadleaved trees. Group strikes usually occur in trees in the pole-stage or older, but they can happen in young crops, not only of trees (Jackson, L. W. R. 1940*a*; Rhoads 1943), but also of agricultural plants such as cabbages or potatoes (Spence 1956). J. S. Murray (1958) found lightning damage to small groups of *Tsuga* transplants in a nursery. The roots appeared to have been killed first and the stems of those plants that were still alive showed typical abnormal rings.

The groups in high forest are normally roughly circular and may include one or more trees showing typical furrows in the bark or sapwood; but most and sometimes all of the trees in the group die without showing external injuries. Marginal trees often have dieback on the branches which point into the group, and occasionally of the top of the crown. Wijbrans (1957) found that injured *Hevea* trees were rapidly invaded by secondary fungi. Sometimes the main damage to those trees that are not killed outright is to the deeper roots.

It is often possible to find in the live parts of injured trees a ring of abnormal tissue (Tubeuf 1906). These rings sometimes consist largely of resin ducts, but they are often indistinguishable from 'frost rings' (Chapter 4). They also occur in small trees struck by lightning in the nursery. Abnormal tissue following lightning does not, however, occur with the same consistency as do frost rings. Thus, while the presence of abnormal tissue, agreeing by its position in the annual ring with the known date of a thunderstorm, is valuable evidence in favour of lightning as the cause of injury, its absence must not be taken to prove that lightning was not involved.

Abnormal rings have been produced in young trees by the application of electrical currents (Peace unpubl.) (Fig. 12). Damage could be caused either by low amperage currents of high voltage from a coil, or by higher amperage currents from electric mains; but the balance between complete freedom from injury on the one hand and death of the tree on the other was very delicate, so that abnormal rings occurred in only a small proportion of the trees tested. In the same experiments it was found that small trees were easily killed by electric currents without any visible scarring or burning. In the field the nature of damage to trees probably depends on the intensity, the distribution, and possibly the duration of the discharge.

Damage can also occur as a result of trees touching overhead electric wires, particularly those carrying a high voltage (Stone, G. E. 1914), but this should seldom happen under modern conditions, where a proper clearance is usually provided.

The total damage by lightning in Britain is very small, though valuable ornamental trees may occasionally be destroyed. It is more significant in some areas in Europe, and in parts of America (Reynolds, R. R. 1940; Wadsworth 1943) it can be a serious source of loss, sometimes directly, sometimes by leading to attacks by bark beetles and fungi on lightning-damaged

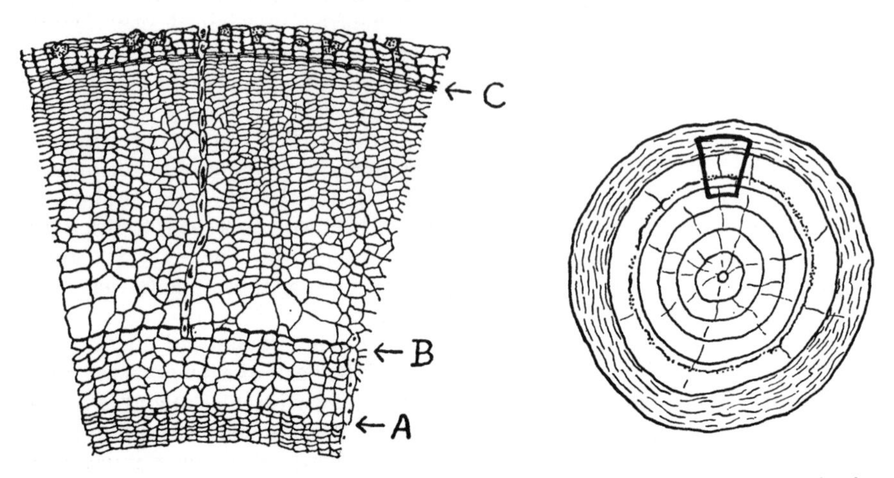

FIG. 12. Abnormal rings caused by giving an electric shock to a small European larch (highly magnified). *A*, Beginning of annual ring; *B*, abnormal ring; *C*, end of annual ring. The right-hand diagram shows the position of the section in the annual ring. (J. N.)

trees. In America lightning is also a very common cause of forest fires. Fires started by lightning have been recorded in Britain, but only rarely.

DAMAGE BY FIRE

Forest fires are a regrettably frequent cause of injury to our forests in Britain; but fire hazard never reaches the same level as it does in places such as North America or Australia, so that here fires tend to spread more slowly and burn less furiously. In addition, the scattered layout of forests in Britain tends to limit the areas burnt. Nevertheless, the usual outcome of forest fires here, as in other countries, is complete destruction of the aerial portions of most of the trees within the burnt area. It is not appropriate to deal with the causes or control of fires in this book. Only the damage done by fire to trees that survive it will be considered here.

Normally, unless the fire is on dry peat which readily catches fire and burns away, the underground portions of the trees are unharmed. Only the top 5 cm. or so of the soil reaches a lethal temperature (Beadle 1940). Thus any trees which can coppice will survive the fire, though the timber is lost. Most hardwoods, but few conifers, sprout again following the death of the main stem. Notable exceptions are *Sequoia sempervirens* and *Taxodium distichum*, which coppice readily; a few pine species, for instance *Pinus muricata*, can also produce coppice shoots.

In older woods in Britain, ground fires, burning vegetation and brash but never extending to the crowns of the trees above, are quite common. The extent of the damage these cause to trees depends on the heat of the fire and the protection offered by the bark. Thick bark offers a very substantial protection against fire, so that trees like *Sequoia* or *Pinus pinaster* often survive fires that kill thinner-barked species. In many cases, however, trees escape death, not because of the thickness of their bark, but because the heat of the fire was intense only on the windward side, so that the bark and cambium on that side alone was killed. Such damage regrettably often occurs around injudiciously large bonfires, lit among standing timber. Basal fire wounds are an easy means of entry for decay-causing fungi. In countries such as the United States, where very hot ground fires occur frequently, consequent decay losses can be very serious (Hepting 1935, 1941; Garren 1941; Gustafson 1946; Toole and Furnival 1957), though the wounding does not appear directly to result in any loss of increment (Jemison 1944). It is important, therefore, that ground fires should not be allowed to run unchecked.

Variation between different trees in their ability to withstand fire appears to depend on their bark thickness, on the presence of dead inflammable matter on the tree itself, such as dead needles in the lower crown, and on the inflammability of the living foliage, rather than on differences in the lethal temperatures for the living cells (Byram and Nelson 1952). Inoue and Masuda (1956) classified a large number of trees and shrubs according to the inflammability of their leaves or needles. Conifers were generally more inflammable than broadleaved trees, but one of the most resistant was Japanese larch, which was therefore recommended for firebreaks. Pryor (1940), who considered that resistance could be based mainly on bark thickness and on ability to produce new foliage, arranged a number of conifers, mostly pines, in order of resistance. *Pinus pinaster* was the only resistant species commonly planted in Britain. *P. radiata*, *P. contorta*, Scots and Corsican pine were all considered susceptible; while Weymouth pine and Douglas fir were among the very susceptible conifers.

Craighead (1927) found that defoliation by fire caused reduction of the summer wood of the annual ring of the year when the fire occurred, and reduction of the total width of several subsequent annual rings. In cases of severe injury a ring or two might be totally omitted. There was no production of abnormal tissue, such as is associated with frost injury.

In the United States controlled burning, done when the condition of the vegetation is such as to produce a fire of moderate intensity, is used to control needle blight of Longleaf pine caused by *Scirrhia acicola* (Chapter 22). Such a fire will burn off the infected needles without permanently damaging the pine (Siggers 1944). Bonfires and other small fires in the forest may form centres of spread for the fungus *Rhizina inflata*, which attacks the roots of trees (Chapter 16).

DAMAGE DUE TO THE HEAT OF THE SUN

Heat injury to trees takes two main forms: the first is direct scorching of bark by the rays of the sun; in the second the damage is indirect, for the sun heats the soil which in its turn damages the stems of seedlings or occasionally the superficial fine roots of trees growing in it. The mechanism of heat injury is still a matter for argument, but it seems clear that resistance to frost, drought, and heat are closely related (Levitt, J. 1956). Cracking of stems in hot dry summers is probably more a question of water-supply than of direct high temperature, while leaf-scorch in hot weather is also normally due to excessive water loss, though Mielke and Kimmey (1942) attributed damage to the foliage of Black oak in California to high temperatures rather than to excessive transpiration; drought crack and leaf scorch are therefore discussed in Chapter 6.

Sun scorch

Death of the bark and sometimes of the cambium on the south and south-west sides of thin-barked trees occurs quite commonly when they are suddenly exposed to full insolation. Normally a tree grown in full sunlight does not suffer from sunscorch, though Havelik (1935) attributed failure in Silver fir (*Abies alba*) partially to pure planting without a broadleaved admixture, and consequent exposure of the stems to the sun. The surface of the stem of a fully exposed tree can reach temperatures far above that of the surrounding air (Harvey, R. B. 1923). Hartig (1894) found that the south-west side of spruces fully exposed to the sun were at 55° C., when the shade temperature was 36° C. Nevertheless, it is usually only when some action is taken which exposes the stem of a hitherto shaded tree to the sun that injury occurs. This may be brought about by felling neighbouring trees, by thinning, by the removal of undergrowth (Havelik 1935), or even by pruning (Fryer 1947). It can also occur on the edges of gaps caused by gales.

Injury of this nature has been most often recorded on beech (Schwartz 1932), but also occurs frequently on sycamore and other maple species. Hartig (1894) considers Weymouth pine, Norway spruce, and hornbeam particularly susceptible. It has been found on poplar, cherry, and a number of other thin-barked trees. The rise of temperature is undoubtedly slower below thick bark (Harvey, R. B. 1923), so that it protects the cambium. In Britain sunscorch is rare on conifers. Since the damage occurs on those parts of the stem most exposed to the sun, it is usually found near the centre of the internodes on conifers (Fryer 1947). It was extremely prevalent in Britain on young beech left in areas felled during the Second World War. Trees too small for utilization were often left in the hope that they would become pioneers of the next crop, but almost invariably they developed long lesions on the sunward side. Beech should always be thinned with great circumspection, particularly on the southern edge of a plantation.

A number of authors have suggested that the damage is not due directly to heat, but to the repeated rapid freezing and thawing in the winter of the side of the tree exposed to the sun (Schwartz 1932; Huberman 1943). This is discussed further in Chapter 4.

Lesions due to heat can lead to fungal disease. Riley (1948) found that *Hypoxylon pruinatum* colonized heat injuries after thinning of aspen in Canada. Sommer (1955) associated attacks of *Hendersonula toruloidea* with heat injury to the bark of walnut. In Britain heat injuries on beech almost invariably lead to rapid invasion by a large number of fungi, including *Nectria*, and decay fungi such as *Ganoderma applanatum*.

It is very noticeable that many trees tend to develop a paler bark when growing in a hotter climate. This phenomenon sometimes makes identification based on bark colour rather difficult. It may help to protect the trees against high temperatures. Indeed whitening of the stems has been suggested as a protective measure (Brichet 1950). A coating of whitewash for two or three years might well be tried on the sunward side of the stem when a valuable thin-barked tree, hitherto shaded, has to be exposed to full sunlight; though the practice would not be feasible on a large scale.

It has often been suggested that drops of water on the surface of leaves exposed to the sun lead to injury. This is used as an argument against overhead watering in hot weather. Kramer (1939), however, found that water drops on leaves markedly lowered their temperature, and that there was no evidence that the water droplets were acting as lenses concentrating the rays of the sun. Indeed if they did, the focal point would be well behind the leaf. J. Levitt (1956), on the other hand, states that local water-saturation of the tissues, which might well occur beneath water droplets, increases the liability to heat injury. In practice, however, trees can safely be watered in full sunlight. Spraying with fungicides, insecticides, or weed-killers under such conditions is dangerous, because the chemical in the spray may be phytotoxic, if concentrated on the leaf surface by evaporation. Spraying therefore should be carried out in the evening or in dull weather. In the same way mist contaminated with industrial fumes may prove damaging if it is rapidly evaporated on the leaf surface (Chapter 8).

Damage due to hot soil

Injury to seedlings owing to excessive heating of the surface soil occurs quite commonly, even in the comparatively cool climate of Britain. It is most frequent on sandy soils or where rock chips have been used as a seed-bed covering, presumably because, owing to their poor conductivity, less heat is lost to the soil below, so that the surface layers reach a higher temperature. Tourney and Neethling (1924) found that a temperature of about 50° C. at the soil surface was lethal to small coniferous seedlings, even if it persisted for only 15–30 minutes. The fact that seedlings remained undamaged in the shadow of poles put up to support overhead lath shelter, when those in between were severely injured, clearly illustrates that injury can take place in a very short time. In this instance the damage occurred because a hot day followed soon after the mid-season removal of the shelter. Heat injury is more likely in dark soils than in light ones, because of the greater heat-absorbing capacity of the former (Isaac 1930). This may occasionally account for the loss of young trees planted on the black charcoal areas of old fire sites, though attack by *Rhizina inflata* (Chapter 16) is also possible. Vaartaja (1949, 1954*a*) reported high surface temperatures in humus, as well as in sand, so that heat injury may be one of the difficulties in the regeneration of dry peat sites, or in the use of litter seed-beds. J. Levitt (1956), in his review of the subject of heat-resistance, states that plants grown under moist conditions are more liable to heat injury than those grown under dry ones. This may account for its rather frequent occurrence in nurseries in western Britain, where, except during actual hot periods, the seedlings are likely to have ample water.

Heat injury occurs most commonly on seedlings during their first or second year; transplants are more rarely affected, and only very exceptionally does it happen in plantations. However, Scots pine transplants have been damaged in this way in East Anglia, when planted in bare furrows running north and south, so that the midday sun was able to heat the soil without any shading by vegetation. In hotter climates hot-soil injury has attracted attention mainly because of its effect on natural regeneration (Tourney and Neethling 1924; Baker, F. S. 1929). Damage has been recorded on many species including spruce, larch, Scots pine, and Douglas fir (Bates and Rosser 1924; Petrie and Mackay 1948). It is uncommon on broadleaved species. Injury is likely to be less severe in dense beds, where the soil is more shaded than in sparse ones, and seedlings with many or large cotyledons, short stems, or rapid development of true leaves are obviously less liable to injury than those with sparse foliage and slow development (Baker, F. S. 1929).

Damage takes a variety of forms according to its time of occurrence and the size of the affected plants. It can happen at any time from May until August, but is probably commonest in June and July. On very small seedlings complete and immediate collapse at soil level usually occurs, but larger seedlings or transplants, though girdled, often remain upright and continue to grow (Hartley 1918). In this girdled condition water can still pass from the root to the shoot, which continues to develop (Fig. 13), but the roots gradually deteriorate because, owing to the destruction of the phloem, they do not receive carbohydrates elaborated by the leaves. Eventually the death of the roots leads to yellowing, wilting, and dieback of the shoot. This may not take place till several weeks after the actual heat injury, so that diagnosis may be rather difficult. On plants which are not girdled, small lesions are sometimes formed at soil-level, usually on the south side of the stem (Fig. 14). These are valuable in diagnosis, especially if the plants are examined on the spot, so that the orientation of the lesions can be observed. Often these small wounds heal over so that the plant recovers, but more extensive lesions form points of weakness for a considerable period after their actual formation, and some affected seedlings break off at soil level during heavy rain. This also may happen weeks after the heat injury occurred.

It is easy to confuse heat injury, especially in the later stages when the bark on girdled seedlings has decayed and fallen off, with damage caused by cutworms, a group of Lepidopterous insects, which often chew the bark of young plants at soil-level. In the early stages heat injury can also be mistaken for damping off (Chapter 15), though it naturally occurs under very different conditions. Damping off tends to affect dense seed-beds more than sparse ones and to occur in patches, whereas heat injury occurs more on isolated seedlings, because the soil at their base is not shaded by their neighbours.

On older seedlings damage by the fungus *Pestalozzia hartigii* (Chapter 15) is often very similar to heat injury. Indeed there is a suspicion that this and other fungi may occasionally colonize plants initially damaged by heat.

Heat injury in seed-beds is probably most easily avoided by the use of lath or other shelters. Such shelters markedly reduce extreme temperatures (Papajoannou 1934). In all but the hottest weather, however, seedlings are

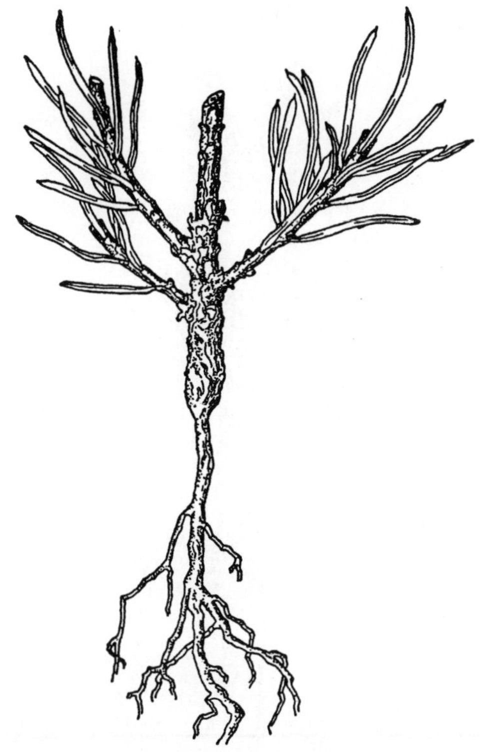

FIG. 13. European larch seedling girdled by hot soil (nat. size). (J. N.)

likely to grow best unshaded, and it may well be considered that the expense of shading against damage that may occur only occasionally in very hot summers is not justified. If shelter has been applied, it should not later be removed during hot weather. Seedlings raised under shelter are more susceptible to heat injury than those raised in the open. Early sowing, so that the seedlings are larger and less succulent by the time the hot weather starts, is certainly beneficial. Watering during hot weather markedly reduces the temperature of the soil surface (Maguire 1955), so that nurseries equipped for irrigation could adopt periodic watering as a means of prevention. One instance has been recorded, however, where spraying with water in hot weather seemed to lead to increased heat injury. It was suggested that drops of water running down the stems and resting on the soil surface were absorbed by the young stem tissue, so increasing its susceptibility to heat damage. Increase of water content in plant tissue is known to have this effect (Levitt, J.

1956). The amount of water applied in this case was insufficient to wet the surface soil. A heavier watering would almost certainly have been beneficial.

FIG. 14. Lesion on young *Thuja* caused by hot soil ($\times 2\frac{1}{2}$). (J. N.)

Damage has also been observed after seed-beds had been sprayed with an oil weedkiller. Oil droplets against the stems may have raised the susceptibility of the seedlings. At any rate the risk of such injury, as well as the possibility of foliage scorching, renders the application of weedkillers to germinated seed-beds dangerous in hot weather.

Despite the high temperatures which have been recorded in dry peat, a vegetable mulch over bare soil has been suggested (Anon. 1952*c*) as a means of preventing heat injury. Inclination of transplants to the south at the time of planting, so that the tops shade the roots more efficiently, has been suggested (Korstian and Fetherolf 1921). In Mexico pieces of waste veneer encircling the lower stems have been used to protect newly planted Spanish cedar (*Cedrela*) (Miller, Perry, and Borlaug 1957). Damage to transplants in Britain, however, is too rare to justify any special measures. Regeneration fellings in hot climates should be so arranged that the seed trees give partial shade to the young seedlings.

THE EFFECT OF LIGHT ON TREE HEALTH

Direct injury to plants due to light can hardly be separated from the effect of heat, so that though the former may exist it cannot be especially considered. Nevertheless, Collaer (1940) has suggested that high light intensity plays a part in determining the upper limits of forests. But light affects the health of the tree in a number of indirect ways.

Lack of light is the main reason for the death of the lower branches on trees in forests and plantations. The suppression and death of such branches is a necessary part of the production of high-quality timber, but nevertheless they are important, both as a reservoir for fungi and as a means of entrance to the main stem for facultative parasites. For plantations of equal density, the live crown is much deeper in countries of high light intensity and more vertical sunshine than in regions farther north. Excessive shading, together with root competition, causes suppression of growth and possibly death. Suppressed trees often fall an easy prey to fungal attack, but are not very significant, since they would have died eventually anyway. A large proportion of dead crown in a tree, or of dead and dying suppressed trees in a plantation, is undesirable from a pathological point of view, and should be avoided by appropriate silvicultural measures. Planting under cover, if the amount of overhead shade is excessive, often results in weak, etiolated growth or even in death. It is often very difficult to strike a balance between the benefits of overhead shade, such as protection from frost and exposure, and the damaging effect of lack of light. Some trees, such as European larch or Hybrid black poplar, are intolerant of any shading; others such as beech, *Abies grandis*, or yew grow quite well in low light intensities. The remainder are intermediate as regards their shade tolerance.

Photoperiodicity also has a profound effect, not only on the rate of growth of trees, but also on their health (Wareing 1948; Pauley and Perry 1954; Vaartaja 1954*b*; Wassink and Wiersma 1955). In general trees moved south to shorter summer day-lengths grow poorly, sometimes so poorly as to appear diseased, while trees moved northward into longer summer day-lengths suffer from delayed cessation of growth in the autumn, and consequent frost injury. Both these considerations are important in the selection of provenances of exotic trees for cultivation in Britain.

Photoperiodic effects occasionally appear on street trees as a result of artificial lighting (Matzke 1936; Taylor 1958). The artificial lengthening of the 'day' period may lead to delayed leaf fall, with consequent autumn frost injury. It is worth considering the position of street lamps in relation to any localized injuries to roadside trees.

6

DAMAGE TO TREES BY UNSUITABLE SOIL AND SOIL WATER CONDITIONS

SOIL TEMPERATURE

THE damage done by hot soil, which is an indirect effect of the heat of the sun, is dealt with in Chapter 5. The effects of frozen soil, as a cause both of direct damage to roots and interference with water absorption, have been covered in Chapter 4. Apart from these extremes, soil temperature is chiefly important because a reasonably warm soil promotes root growth and thus aids recovery from disease, from injury, or merely from transplanting. Newly transplanted trees may be unable to make fresh root growth because the soil is too cold. Some species, such as *Cupressus macrocarpa*, are quite unable to establish themselves in cold soil. If the difficulties of higher transpiration can be overcome, summer transplanting may be more suitable for this and other species requiring warm soil for root development. Cold soil conditions are often associated with a high water content. It is sometimes difficult, therefore, to separate the damage done by low soil temperature from that due to high water content.

SOIL AERATION

A certain concentration of oxygen in the soil is required for the maintenance of life in tree roots. The amount needed varies for different tree species and at different times of year. Willow, poplar, and alder roots, for instance, can all survive under conditions of poor aeration. The quantity available depends on the nature of the soil and in particular on the extent to which it allows diffusion of oxygen from the surface. In general soils of loose texture and dry soils have a better oxygen supply than compact or wet ones. The concentration of oxygen naturally tends to be less in the deeper layers, so that in badly aerated soils rooting is usually shallow, because only near the surface is there a sufficient supply for root growth. If the oxygen supply falls below a certain level, root death occurs, resulting, if the damage is sufficiently extensive, in crown symptoms and possibly death of the tree. In badly aerated soils it is not only difficult for oxygen to reach the roots, it is also difficult for carbon dioxide to diffuse away. The toxic action of the resultant carbon dioxide accumulation, as well as the lack of oxygen, is damaging to the roots. Cannon (1932) and Crescini (1940–1) found that many plants, including some trees, were able to supply part of the roots' oxygen requirement by above-ground absorption and translocation down through the stem and roots. Varying ability to do this is certainly one of the influences affecting tolerance to poor soil aeration. Adaptation to a low oxygen content in the soil may depend partly on the ability of the tree to increase the amount of oxygen supplied to the roots in this way. General outlines of this subject have been given by Hutchins (1947) and Kramer (1950).

The causes of poor soil aeration

Soil compaction

Compaction of the surface layers of the soil, preventing ingress of air, can come about in a number of ways. It is much more likely to occur in heavy soils than in light ones, because by their nature the former are more easily compacted. The problem of soil compaction, which is particularly important for ornamental and roadside trees, has been briefly discussed by Haddock and Gessel (1951).

Probably the commonest cause of soil compaction is trampling either by man or animals (Burger 1940; Read 1957), but in recent years the use of heavy machinery has affected some agricultural soils. Machinery in forest nurseries can lead to compaction, and it is known that the very heavy tractors used to extract timber from forests can do appreciable injury to soil structure (Steinbrenner and Gessel 1955). Damage following soil compaction by the feet of man and domestic animals is probably commoner than is generally realized. It has attracted particular attention in California because of the effect of tourist traffic on the spectacular groves of *Sequoia sempervirens* (Meinecke 1928).

For ornamental trees, loosening the surface or even injection of compressed air into the soil are possible remedial measures. In the forest it is best to avoid the initial damage, for instance by restricting the concentration of grazing animals in an area or limiting heavy haulage to well-defined tracks. In some forest nurseries lining out of transplants in wet weather may lead to serious compaction. It is best to postpone work till the soil is drier, but if necessary a layer of peat spread on the surface before use will greatly lessen the damage to the soil structure, and prevent subsequent poor growth on the compacted areas.

Surface sealing of the soil

It has sometimes been claimed that the accumulation of unrotted conifer needles or raw humus on the surface of the soil prevents the passage of oxygen to the lower layers and leads to ill health in trees (Graebner 1906; Růžička 1938*a*). On some moorland sites the superficial growth of lower plants, such as algae, lichens, etc., which forms a 'skin' on the surface of the soil, may have the same effect. However, these methods of surface sealing have never been critically examined, and the most definite cases of damage to trees due to this cause are all the result of man's activities.

The commonest cause is the laying down of impervious coverings over the soil in the course of road or building construction. A deep covering of soil placed over the roots of a tree can have the same effect, especially if the soil is heavy and relatively impervious. Knowlson (1939) found that large oak trees covered with 6 metres of rock and sand survived unharmed; but under these conditions aeration through the loose soil would still be relatively good. Crown dieback as a result of root damage can be expected if more than half the root system of a tree is suddenly covered with an oxygen-excluding surface. The gradual covering of the roots, on the other hand, would lead to less catastrophic damage; while a tree that is planted in a hole in a roadway will often grow quite well, despite the fact that nearly all its roots are under the impervious cover.

Where earth fills are involved, the damage can be lessened or even prevented by the use of broken rock instead of soil. The spaces between the rocks allow the ingress of oxygen, at any rate until the tree roots have become adapted to their new position. Adaptation probably depends more on the extension of the root system, nearer to the surface if necessary, into better aerated soil, than on any change in oxygen requirement, though increased transference of oxygen to the roots from above-ground organs probably plays a part. Descriptions of the various methods of using rock fills around trees are given in many American publications (Thompson, A. R. 1939; Fowler, Gravatt, and Thompson 1945; Pirone 1948).

Excess water in the soil

The commonest cause of bad soil aeration is excess water. Flowing water may carry enough dissolved oxygen for the roots of some trees. P. W. Zimmerman (1930) found that cuttings of willow rooted freely, and particularly near the base, in aerated water, in contrast to those in stagnant water where all root formation was near the surface. Willows, poplars, and alders often root into streams. Stagnant water, however, soon loses most of its dissolved oxygen and thus reduces to a very low level indeed the oxygen supply of the tree roots it covers. If soil is permanently waterlogged, the growth of most woody species is prevented. If it becomes waterlogged when trees are already present, they will deteriorate or die.

Waterlogging of the soil can come about in a variety of ways, by flooding, by the gradual or sudden blockage of drains, by land subsidence, or even by the removal of water-absorbing vegetation. For instance, felled areas often become very swampy for a period following the removal of the water-demanding tree crop (Koščeev 1952). In general the more sudden the change to wet conditions the greater the damage to the trees.

Short periods of flooding by fresh water, especially during the dormant season, do not usually cause damage, provided that the water is able to flow off and is not left to evaporate. In longer periods of flooding the amount of damage depends much more on the duration and on the rate of flow than on the depth of the water. If the roots are all covered the damage is roughly the same whatever the depth of water (Green, W. E. 1947). Flooding by salt water, where the toxic action of the salt greatly increases the damage, is dealt with in Chapter 8.

The symptoms and development of injury due to poor aeration

Lack of oxygen and accumulation of carbon dioxide in the soil, whatever the cause, lead directly to death of roots. In some cases only the fine roots are involved; Güssow (1929) attributed Needle blight of *Pinus strobus* to destruction of the root hairs during an excessively wet summer, and the consequent inability of the trees to absorb sufficient water when the weather became dry. But root death of trees suffering from poor aeration is normally more extensive than this. Since the trees are usually unable to take up sufficient water to meet the demands of transpiration, the symptoms are essentially similar to those caused directly by drought (Heinicke 1932; Hulshof and Zegers 1951). Typically there is withering of the margins of the leaves or of the ends of the needles, followed by defoliation and by twig or even branch dieback. Where the damage develops rapidly, the upper parts of the crown are most affected, giving in hardwoods a typical stag-headed effect (Plate IV). In chronic cases, however, where the development is more gradual, most of the dieback is on the lower branches, thus reducing the live crown depth below normal.

Where the injury is slow to develop, stunted growth is very typical. Damage of this kind has been described on *Pinus resinosa* growing in poorly drained sites (Stone, Morrow, and Welch 1954). Death of the root collar of fruit trees has been attributed to waterlogged soil (Furneaux and Kent 1937; Cooley 1948). In these cases there was some association with the water collecting in the funnel made around the stem when a tree rocks in the wind. There are no definite records of this type of injury on forest trees.

In nurseries a wet summer always causes unusually high losses among newly moved transplants. These are unable to initiate new root growth in the cold, wet soil. Plants in this condition may linger on alive until there is a period of hot weather, when the conditions of increased transpiration lead to withering and death. Under these circumstances the true cause of the damage may be overlooked. Root fungi may play a part in the later stages of this disease. Heeling-in of plants into very wet soil is undesirable because the badly aerated conditions may lead to root-dieback. Waterlogging in nurseries may also lead to purpling of the foliage of conifers, a common symptom which may result from several other causes. The discoloration is reversible and the trees usually resume their normal green colour when the ground has been drained.

The effect of lack of aeration on different species

Most of the information on this subject has been collected after floods, sometimes natural, sometimes artificial. Numerous lists of trees showing their susceptibility to damage by flooding have been published (Vlad 1944; Yeager 1949; Traunmüller 1954; Hall and Smith 1955; Ahlgren and Hansen 1957). Willows, poplars, and alders are notably resistant. *Fraxinus americana* is also very resistant, but the European ash, *F. excelsior*, is much more affected. Swamp cypress, *Taxodium distichum*, is a notable exception among the conifers. Indeed Bjallovič (1957), who made a very carefully classified list, placed it alone, as showing the maximum tolerance to flooding. Most conifers are very susceptible to injury by flooding. Ahlgren and Hansen (1957) found that a wide range of conifers were worse affected than the hardwoods occurring naturally with them. Norway spruce and Scots pine can both survive under wet soil conditions, but react by notably stunted growth, and superficial rooting. Elm, beech, elder, and oak have all been mentioned as susceptible to flooding. Tree seedlings have been grown under anaerobic conditions to investigate the effects of aeration. Using this method Leyton and Rousseau (1958) found that the roots of willow (*Salix atrocinerea*) required much less oxygen than those of a number of conifers, all of which had approximately the same oxygen requirements. Huikari (1955) found birch much more resistant to bad aeration than pine and spruce.

In Britain some species of *Salix* and *Rhamnus*, alder, birch, and sometimes poplar can persist in nearly waterlogged soil, but even those species that are normally capable of reaching timber size will not do so under such poor conditions. It is impossible to raise a forest crop in Britain on swampy ground without drainage.

Conifers, such as the spruces, which are capable of making very shallow root systems, can grow quite well in the very limited oxygenated layer of litter and humus which often exists above waterlogged peat or clay. Under such circumstances, however, growth is restricted and the risk of disease is enhanced (Day, W. R. 1953).

VARIABLE SOIL WATER CONTENT

In many soils the water content is very variable, usually being high in the winter and low in the summer. In some cases this may result from seasonal variations in the depth of the true water table below the surface of the soil. In other cases a temporary or perched water table develops during the winter, for instance above an impermeable clay layer, and this disappears altogether during the summer (Plaisance 1956). Such conditions are quite unsatisfactory for root growth. During the winter, when the soil is saturated, the roots tend to die back; whilst during the summer, when the soil is dry, the restricted root system is inadequate to absorb the water required for transpiration, and crown-dieback often ensues. Thus the final result of soil waterlogging may be drought damage. It has been pointed out that damage in a dry summer may best indicate which parts of an orchard or plantation are badly drained (Anon. 1955*b*). The sudden drainage of a wet site may leave the trees therein with root systems inadequate for the drier conditions to which they are then subjected. Alterations in water conditions should never be made too abruptly, if carried out while a crop is on the ground. Deficiencies in drainage should be corrected before planting.

Trees growing on soils with a variable water table tend to develop flat-bottomed root systems, in which all the sinker roots end in truncated tufts, resulting from the annual production and subsequent destruction of roots penetrating into the winter waterlogged zone (Fig. 15). It is a debatable point whether such a root system is itself pathological. It is certainly not ideal, but a tree with a root system of this kind will make merchantable growth under many conditions. Such trees are certainly less wind-stable than fully rooted ones, and may be more prone to fungal root disease, though in conifers there is no proven connexion between this type of root injury and attack by *Fomes annosus*. In the case of *Castanea*, however, there is a definite correlation between soils with variable water table and attack by *Phytophthora* species, producing the so-called Ink disease (Chapter 29). This association of variable soil water content and disease almost certainly operates in some other instances of *Phytophthora* root attack, and probably extends to some root diseases associated with other fungal genera. Damage to roots resulting from a variable water table, leading to attack by *Armillaria mellea*, by other fungi and by insects, is one of the explanations put forward to account for widespread deterioration of oak (Georgescu, Teodoru, and Badea 1945). It is possible that the stagheadedness of oak, and more particularly of ash, which occurs commonly in hedgerows on the clay soils in Britain, is associated with root damage arising basically from wide variations in soil water content,

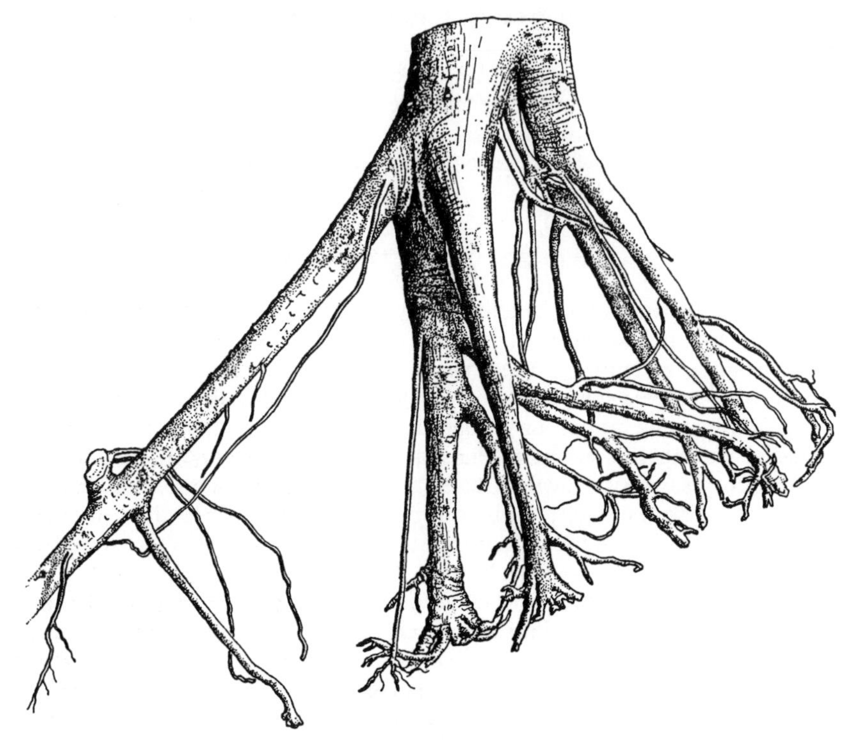

FIG. 15. Restricted root system of Sitka spruce from a site with a variable water table. The roots tend to die back during the winter. (J. C. after W. R. Day.)

possibly, though not certainly, followed by fungal invasion. It would be a grave mistake, however, to believe that root dieback due to variable soil water invariably leads to crown dieback; often it produces no visible effect on the tree above ground. It is also clear that damaging soil conditions are not an essential precursor to fungal attack on roots, nor does such attack necessarily follow their development.

DROUGHT

Drought can be considered either as an inadequate supply of water to the roots, or as an excessive loss of water from the leaves. These two aspects are not entirely separable in practice, since in many cases drought injury may result from their combined action. Yet drought due mainly to dry soil behaves very differently from drought due mainly to dry air, and they are dealt with separately here under the headings of 'Soil Drought' and 'Atmospheric Drought'.

The physiological basis of drought hardiness has been very fully covered by J. Levitt (1956) and J. Parker (1956*a*). Since dehydration of the cells is primarily concerned both in drought and in frost injury, one would expect some correlation between drought and frost hardiness, and this does in fact exist. Nevertheless, the conditions under which the two types of injury occur are very different, and it would be quite wrong to assume that species or individuals resistant or susceptible to frost at any particular time of year would show the same reaction to drought.

Soil drought

Soil drought occurs whenever for any reason the ground becomes abnormally dry. The criterion is not so much absolute dryness as dryness in relation to the normal condition of the soil. A tree usually growing under moist conditions may suffer more in a drought than a tree normally growing in dry soil, and certainly would do so if the soil water content dropped to the same level for the two trees. It is very important to keep this relative effect in mind when considering drought. Liese (1953), for instance, attributed the dieback of elms on low-lying ground to the cessation of the floods which periodically used to saturate the soil. Naturally, soil drought conditions are most likely to develop during the summer, though in Britain the most serious losses are usually to newly planted trees in the spring; April in particular often being a very dry month, and the plants in a more susceptible condition. Drought injury can also occur in the winter if the temperature rises sufficiently to cause a high rate of transpiration (Voigt 1951).

Drought damage depends among other things on the nature of the soil. Some soils hold more water than others, some yield it more readily to plant roots. The nature of the soil influences the extent of the tree's root system, and therefore affects its water-absorbing capacity. The influence of soil is obviously very complex, but drought is usually most serious on the extreme types, on gravels and sands which hold very little water, and on clays which hold more but are very water-retentive. Shallow soils over rock, over porous gravel, or over chalk also provide potential drought sites, particularly in the higher rainfall areas where under normal conditions they are moist enough to support more or less unrestricted tree growth. Shallow patches of soil over rock show as brown areas amid the general green, because of the death of the young trees on them, every time there is a damaging drought in the Highlands of Scotland or in the wetter parts of Wales (Fig. 16).

Competition with vegetation and with other trees greatly increases drought injury. In this connexion dense grass is particularly damaging; there is a possibility that it may have a toxic action on the tree roots (Chapter 8), but its main effect is almost certainly that of successful competition for water and nutrients (Howard, A. 1925; Albertson and Weaver 1945; Lane and McComb 1948). Drought damage to young Japanese larch frequently occurs when they have been planted in a grass mat. In the forest, experiments involving trenching round small plots have proved that competition for water, as well as for light, restricts the growth and survival of tree seedlings (Fabricius 1929; Kramer, Costing, and Korstian 1952; Ackerman 1957).

Dew or rain too light to penetrate the soil, but absorbed by the leaves, can lessen or prevent drought injury (Stone, E. C. 1957). Heavy rain naturally relieves drought and stops the development of injury, though cracking, following rain and apparently due to a sudden resumption of growth, has been observed on the stems of young conifer seedlings previously checked by prolonged drought.

It has already been pointed out that drought injury is in some measure

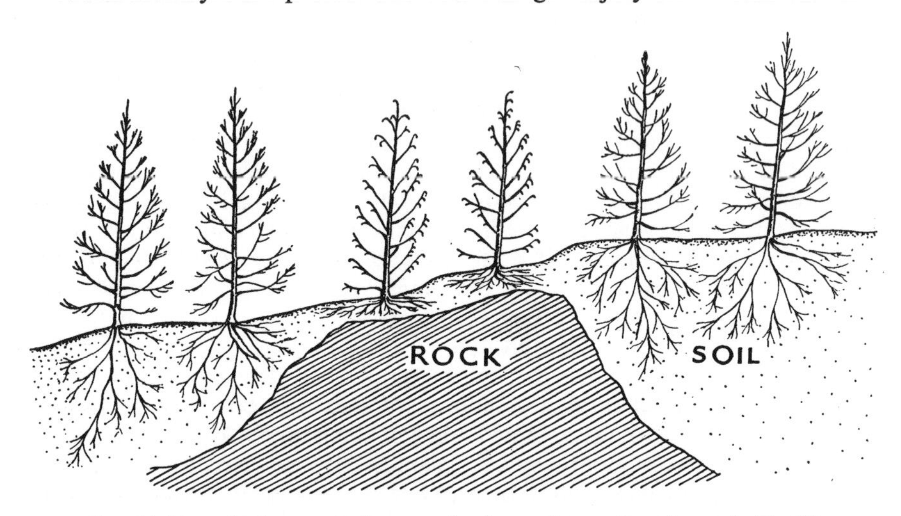

FIG. 16. Drought damage to Japanese larch growing on top of a rock. (A. C.)

relative to the soil water conditions experienced by the tree before the drought. One would expect, therefore, that watering would have an effect on subsequent drought resistance. This certainly seems to be the case, but the information available is somewhat contradictory. Stoeckeler (1951) found that *Pinus ponderosa* became more resistant following moderate watering, while Shirley and Meuli (1939*a*) increased the drought resistance of conifers by subjecting them first to a non-damaging degree of drought. One would also expect that nutrients which increase shoot growth, in particular nitrogen, would affect drought resistance, but here also there is contradiction. Bensend (1943) found that seedlings of *P. ponderosa* given sufficient nitrogen for optimum growth had a maximum drought resistance, whereas Shirley and Meuli (1939*b*) found that increased nitrogen application consistently decreased drought resistance in *P. resinosa*. In the presence of high nitrogen applications, however, heavy watering increased drought resistance, because it leached out the chemical from the soil before the tree could absorb it. No doubt the differences rest mainly on differential development of roots and shoots, the one favouring drought resistance, the other increasing susceptibility. In any case, drought damage in Britain is not serious enough to affect nursery practice as regards manuring.

Drought damage is one of the commonest predisposing factors to fungal attack. There is evidence that cells which are not fully turgid may be more easily invaded by some fungi. Drought has attracted particular attention in connexion with *Diplodia pinea* on pine (Chapter 22) and with *Dothichiza populea* on poplar (Chapter 32). Permanent lowering by pumping of the water table in the dunes in Holland has led to birch dieback associated with *Polyporus betulinus* and other fungi (Vliet, Verbrugge, and Boer 1955) (Chapter 34).

Symptoms caused by soil drought

The most typical symptoms of drought injury are wilting, and if the water content falls too low, withering of leaves usually starting at the margins. Conifer needles turn brown, beginning at the tips. Succulent young shoots, especially on conifers, wilt and wither as a whole. Sometimes rain may bring about a restoration of cell turgor before the wilting becomes fatal to the whole shoot, but after damage to some of the cells has allowed the shoot to drop into a partially pendulous position. Shoots bent near the base, but thereafter growing normally, are the result of this type of injury. Damage of the same general nature, involving partial collapse of the needles, has been described on *Pinus resinosa* under the name of Needle droop by Patton and Riker (1954). Severe drought injury involves shoot and branch dieback, or even the complete death of the tree. Repeated yearly droughts or prolonged non-lethal drought greatly reduce the size and density of the foliage and restrict shoot growth (Albertson and Weaver 1945; Copeland 1955). Any drought tends to reduce diameter growth; even a single severe drought can produce a very narrow annual ring and it may be several years before the normal growth rate is resumed (Wagener 1940; Albertson and Weaver 1945; Sipkens 1952; Copeland 1955).

On young conifers drought is one of the influences which sometimes causes reversible discoloration, usually reddening or purpling of the foliage. If the drought continues, this may be succeeded by typical wilting and withering, but if the drought ceases the needles gradually turn green again.

Curling of pine needles, which fail to expand or to separate completely, has been ascribed to drought by Neilson Jones (1952) on *Pinus palustris* and by L. W. R. Jackson (1948) on *P. echinata*. The symptoms in these cases are appreciably different from the 'Needle fusion' attributed both to lack of mycorrhiza and to mineral deficiency (Chapter 22), and in any case drought can act indirectly by affecting the absorption of minerals.

True and Tryon (1956) have reported drought as the most probable cause of stem cankers on oak, from which they were unable to isolate any causal organisms. The possibility that drought may be able to kill isolated patches of bark or cambium is of interest, in connexion with similar, so far unexplained, damage on other trees. Drought has already been put forward as one of the factors concerned with bark-dieback of beech (Chapter 28). The suggestion that the cambium can be killed by drought gains support from the comparatively frequent occurrence of drought rings in the wood, resulting from damage to the cambium in dry periods. These rings are less spectacular than those caused by frost or lightning (Fig. 17). They consist mainly of flattened cells, sometimes with the longer, circumferential walls distinctly bowed. These rings are quite different from the reduced ring width also associated with drought, though the two may occur together. J. M. Harris (1953) found that drought retarded heartwood formation of *P. radiata*.

Cracking of the stems of conifers, as a result of drought, has been fully described by W. R. Day (1954) (Fig. 18). It occurred particularly widely during the dry summer of 1947, but to a lesser extent in 1955 and other drought years. D. K. Barrett (1958) has described a tree of *Abies nobilis* with cracks, now healed over, in 1899, 1904, 1911, 1919, 1925, 1933, 1946, and 1947. The cracking in 1946, which had a wet summer, seems to have been due to

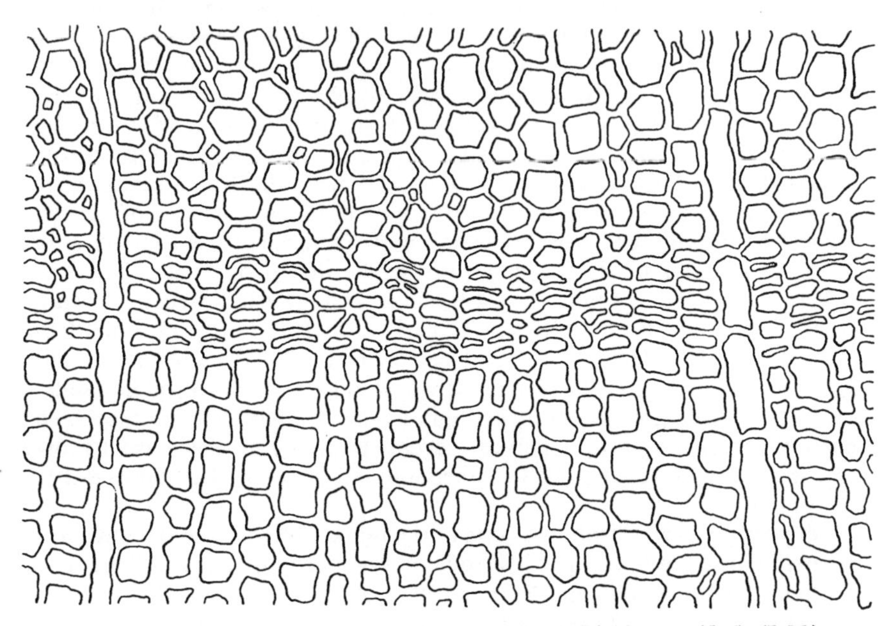

FIG. 17. Drought ring in the wood of Japanese larch (highly magnified). (J. N.)

an October drought. Drought crack is found particularly on sites of limited rooting capacity. In Britain it occurs most commonly on Sitka spruce and *Abies* spp. Rapidly grown and spiral-grained trees are particularly prone to cracking. Day suggested a connexion between cracking and poor wood structure, resulting from unsuitability of climate. It appears that water can be withdrawn from the wood into the transpiration stream during drought (Chalk and Bigg 1956). This may be one of the underlying causes of cracking.

There are other records of drought crack, both in Britain and abroad (Flander 1913; Moreillon, M. 1929; Lutz, H. J. 1952), but most Continental records refer to Norway spruce. Some cracks resulting from drought may have been attributed to frost, which is better known as a cause of cracking. Nevertheless, it is clear that this type of damage does not take place on a wide scale every time there is a hot dry summer, and that when it does occur it is limited to certain sites.

In nurseries there is sometimes confusion between drought and heat damage, since the two causative factors so often occur together. However, the symptoms of heat injury on young plants are usually solely at soil-level (Chapter 5), whereas drought affects the plant as a whole.

Species affected by soil drought

Less information is available than would be expected on the relative susceptibility of trees to drought, possibly because its effects are less immediate and

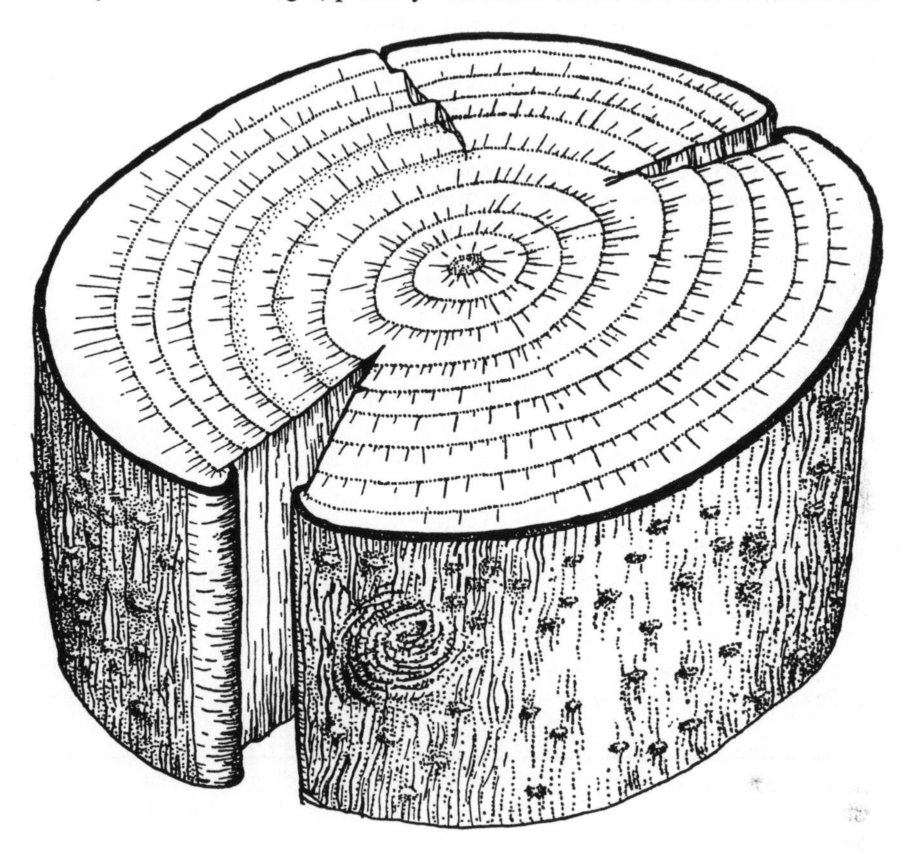

FIG. 18. Drought crack in Sitka spruce from Pitfichie Forest, Aberdeenshire (greatly reduced). (J. C.)

clear-cut than those of frost or even flooding. Experience with shelterbelt species in the dry interior of the United States has been summarized by George (1936). Here *Ulmus parvifolia* and *Fraxinus pennsylvanica* were particularly resistant. Various species of poplar and willow, while less resistant, were still good enough to be used regularly for shelterbelt planting, indicating that the natural habitat of a tree is not necessarily a guide to its drought resistance. Ability to form a deep and extensive root system is probably a very important factor in drought resistance. In Czechoslovakia, Pfeffer, Škoda, and Zlatuška (1948) found Douglas fir resistant and Norway spruce susceptible, particularly when in competition with other trees. Drought injury to

Norway spruce in competition with ash has been noted in Britain. Some of the damage to Norway spruce described by J. S. Murray (1957) under the name of Top dying was caused by soil drought, though in most cases it was probably due to excessive transpiration from trees suddenly exposed. Ow (1948) in Germany noted particularly the resistance of *Robinia*, oak, elm, and *Prunus padus*, while maple, alder, and ash were particularly susceptible. In contrast to Pfeffer *et al.*, he considered Norway spruce drought resistant. Since both were reporting on the same drought, that of 1947, it seems likely that local conditions must affect the relative susceptibility of different species. In Holland Japanese larch was particularly affected in 1947 (Veen 1954), while in Britain it suffered both in 1947 and in 1955. The injury was confined to particularly dry sites and to places with locally shallow soils, and there is no suggestion that Japanese larch is generally unsuited to the Dutch or British climates.

Variations in drought resistance have been observed between different provenances (Meuli and Shirley 1937; Zobel and Goddard 1955), as well as between individuals of the same species (Ow 1948). Oksbjerg (1956) found drought injury more serious on late-flushing than on early-flushing Norway spruce. The period of shoot development of the former coincided with drought periods, whereas the shoots of the latter were developed before drought conditions intervened. There is no doubt that there are ample opportunities for selection and breeding for drought resistance; but drought damage is not sufficiently serious in Britain for this aspect to be given primary consideration in tree breeding programmes.

The mitigation of soil drought

The most obvious counter-measure to drought is, of course, watering; but apart from limited areas in drier regions abroad, where regular irrigation is necessary to maintain tree growth, such treatment is normally limited to nurseries equipped for watering, or where water is locally available for pumping. In many nurseries, and certainly in the forest, other less direct means have to be employed. In the nursery these include shelter, mainly to lessen evaporation both from the soil and from the trees, autumn or early spring planting, so that the trees have a chance to develop some absorbing roots before spring droughts set in, and mulching to conserve soil moisture. Transplanting, which involves loss of water-absorbing roots and sometimes damaging exposure of the root system, naturally carries a high risk of drought injury, even though it is normally carried out when the plant is dormant. Newly planted trees benefit by the removal of competing vegetation, either by ploughing, screefing, or mulching (Ferguson 1957). In the case of ploughing, additional benefit both in shelter and soil moisture may sometimes accrue from planting in the furrow.

In established forest little can be done, though the initial choice of species has a marked effect on subsequent drought injury. There are in fact few sites in Britain where drought is a controlling factor, though Japanese larch in particular should be avoided on shallow soils over rock or in heavy grass, unless the soil it covers is known to be deep and moist. Stickel (1933) has pointed out that stands kept open are less liable to drought injury than dense stands, since the amount of water available for each tree is greater. On the other hand, sudden thinning can bring about atmospheric drought injury (see below), probably by increasing exposure and consequently transpiration beyond the supply capabilities of the roots, despite the additional soil water available. It is necessary therefore to proceed with caution in this matter.

Atmospheric drought

This type of drought, which is sometimes rather erroneously termed 'physiological drought', results from the tree losing water from its leaves faster than it can replace it, despite a sufficient supply being available in the soil. There is, of course, no clear dividing-line between this type of drought and soil drought. A marginal case is winter damage, when the soil is frozen, and therefore physiologically in a dry condition, while irreplaceable water loss causes damage to the leaves or needles of evergreens (Chapter 4).

Essentially the kind of drought damage with which we are concerned here arises under two different sets of conditions; it is found firstly on trees growing in exposed positions, where, during periods of high, dry wind, transpiration becomes too rapid, and secondly on trees growing in relatively sheltered conditions, which as a result of felling, thinning, or for some other reason are exposed to water losses greater than their roots or conduction systems can replace.

The scorching or blast effect of wind on exposed trees is probably the most serious limitation on tree growth in Britain (Plate III. 2). It is exposure rather than temperature that governs the upper limits to our forests (Macdonald, J. 1951). Near the coast, salt drift plays a big part in the scorching of tree foliage (Chapter 8), but over most of Britain it is excessive water loss, aided by some mechanical damage from the threshing of the twigs in the wind, that shears and stunts exposed trees, and in some places completely prevents their growth.

In plantations, however, the trees tend to give mutual shelter as they grow older. If they are able to pass from the initial stage, where they suffer as individuals, to the next stage, where one tree to some extent shelters the next, the formation of a proper crop with only the windward trees badly blasted is only a matter of time. Such a plantation will eventually have a long sloping face towards the prevailing wind, each tree being slightly higher than the one immediately to the windward of it.

Damaging water loss from the leaves or needles of trees growing under moist soil conditions is clearly illustrated by the sorry state of many exposed trees, which are obviously not short of water at the roots. Severe summer gales periodically cause marginal withering or death of leaves of trees, particularly those of beech. Such damage occurs on exposed trees more or less regardless of the soil moisture conditions and too far from the sea for salt to be involved. Experimental evidence was provided by Beilmann (1938), who caused a lime tree to wilt in the air current from a large fan, despite the fact that air temperatures were low and soil moisture abundant. Kramer (1949) considered that the resistance of the root cells to the passage of water might limit the rate of absorption and thus cause a water deficiency in the foliage, regardless of the moisture content of the soil.

In hotter, drier climates than that of Britain, damage due primarily to excessive transpiration may sometimes occur on relatively sheltered trees. Leaf scorch arising from this cause is quite common in the Mid-Western States (Anon. 1956*c*). Spaulding and Hansbrough (1943) attribute Needle blight of *Pinus strobus* to a deficiency of active absorbing roots, and a consequent insufficient supply of water to the needles during hot, dry weather.

In Britain damage to broadleaved trees and conifers frequently follows heavy thinning or selective felling. This has been noted on a wide variety of species in different parts of the country, and is certainly one of the main reasons for the failure of trees left for seed production, as shade for underplanting, or to get further increment after the partial felling of a crop. J. S. Murray (1957) has described very serious and widespread injury to Norway spruce under the name of Top dying, and attributes this particularly to atmospheric, but also to soil drought. He mentions heavy thinning and exposure by removal of shelter as two of the principal causes of this disease. Graham (1943) has described death of *Tsuga* in the United States following selective felling. Over-exposure may well be one factor concerned in the widespread death of birch in partially logged stands in the north-eastern United States (Hansbrough *et al.* 1950).

Symptoms caused by atmospheric drought

The damage resulting from atmospheric drought is essentially similar to that arising from soil drought (see above). On broadleaved trees withering of the margins of the leaves is a very typical symptom (Fig. 19). Damage to trees growing permanently in exposed positions tends to be chronic, the tree making growth during favourable periods, and dying back or suffering loss of foliage during periods of high, dry wind.

In the case of Top dying of Norway spruce the initial damage is to the current year's needles, many of which wither and fall. This process continues till the effects of defoliation bring about death of shoots, and, by lack of nourishment, of the smaller roots (Plate V). Once this stage is reached the death of the tree as a whole is comparatively rapid, the base of the trunk and the main roots being the last portions left alive (Murray, J. S. 1957). The dying needles are often invaded by *Rhizosphaera kalkhoffii*, which is the commonest, and probably the most important, of the secondary invaders (Chapter 23).

Species affected by atmospheric drought

Most of the British lists of wind-tolerant trees and shrubs refer in the main to coastal areas and are therefore influenced by salt injury. However, W. A. Cadman (1953), writing with particular reference to shelterbelts, gives a useful summary of the commoner species, while Arnold-Forster (1951) deals with a wider range of ornamentals. Cadman suggests the following species as particularly suitable for conditions of severe exposure.

Sitka spruce	*Pinus contorta*	Beech
Picea alba	Corsican pine	Sycamore
Abies nobilis	Austrian pine	
	Mountain pine	

In the warmer parts of the country *Pinus radiata* and *Cupressus macrocarpa* could be added to this list. Although sycamore is a tree capable of growing under severely exposed conditions, it is nevertheless one of the species quite commonly damaged by sudden exposure when it has been growing in shelter.

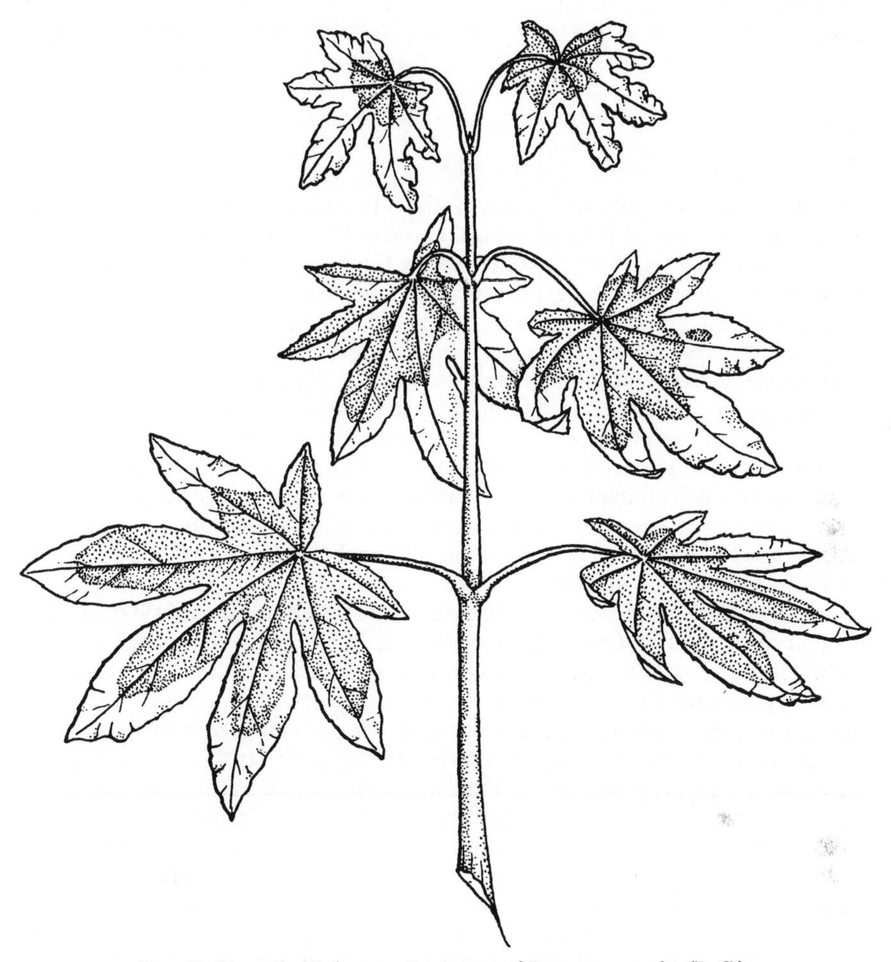

FIG. 19. Dry wind injury to the leaves of Japanese maple. (J. C.)

Damage to Norway spruce as a result of suddenly increased crown exposure is alarmingly common in Britain (Murray 1957), but, unlike sycamore, this species is unsuitable for planting under initially exposed conditions.

The inclusion of *Pinus contorta* in the list above and in fact its rather general use in exposed situations in Britain (Macdonald *et al.* 1957) is interesting in view of the work of Gail and Long (1935), who found that wind increased the transpiration rate of *P. contorta* much more rapidly than that of *P. ponderosa*. This combined with its shallower and less extensive root-system led to its restriction to sheltered sites. Possibly this may be a matter of provenance. Some of the inland races of *P. contorta* are known to do badly in Britain, and intolerance to exposure might well be one of the operative factors.

The fact that there are provenance differences in relation to atmospheric drought is clearly illustrated by the varied reactions of Scots pine of different origins in exposed plantings. There is no doubt that careful selection could greatly increase the ability of this species to withstand atmospheric drought. With Norway spruce, however, there is no obvious evidence of variation in atmospheric drought resistance with provenance, and even differences between individual trees seem more likely to be environmental than genetical.

The mitigation of atmospheric drought

For permanently exposed areas, choice of resistant species, at any rate for the windward edge of the plantation, is the only possible measure. In a sense it is necessary to establish a shelterbelt on the exposed side of the plantation. In exposed places collections of rarer trees and shrubs can often be established only in the shelter of commoner, but more resistant, species. The climatological effects of shelterbelts have been fully covered by Caborn (1957), and their construction and maintenance by W. A. Cadman (1953).

Temporary shelter can sometimes be given to newly planted trees by retention of part of the previous crop, where such exists. Reafforestation of many exposed areas, the initial planting of which was made very difficult by wind blast, should prove much easier if careful use is made of the existing shelter.

For the afforestation of open areas, planting in the bottom or on the side of plough furrows may provide desirable early shelter from wind (Rennie 1956). Atmospheric drought damage is naturally very apt to affect newly planted trees which have few absorbing roots. Some species are slow to make new absorbing roots after transplanting, especially if the soil is wet or cold. Douglas fir is particularly apt to suffer in this way. During the late spring and early summer newly planted trees may die off at intervals, because they have not made any appreciable quantity of absorbing roots. In a wet summer, and under conditions of heavy shade, newly planted Douglas fir have died as late as August, primarily because of their failure to develop absorbing roots.

A properly sited nursery should not require shelter from wind, except for exceptionally delicate species. There are, however, some nurseries where hedges are desirable, despite their interference with large-scale mechanical cultivation and the rather limited protection they provide. Drying out of nursery plants, while they are out of the soil, can be a serious cause of loss in forest nurseries, and during transport from the nursery to the planting site. The roots are the parts most likely to suffer, and any loss of fine roots lessens the chances of survival after planting. The length of exposure required to produce damage varies widely according to the species concerned, and of course according to the weather conditions. Any method of keeping the roots damp will obviously lessen damage, and puddling the roots in wet clay has been suggested as one method of doing this (Slocum and Maki 1956). The recently introduced use of polythene bags for plant transport should greatly mitigate such losses (Aldhous 1958). Waxy sprays have also been suggested as a protective measure to reduce water loss (Miller, Neilson, and Bandemer 1937; Marshall and Maki 1946; Allen 1955), but they have not been universally successful.

Reduction of the live crown by judicious pruning has been suggested in North America as a preventative measure against atmospheric drought injury (Anon. 1956*c*). Such a measure is not, of course, feasible in the forest. There care must be taken to guard against over-sudden exposure of the crowns. With species such as Norway spruce, which are susceptible to atmospheric drought, thinnings must be frequent and light, and great care must be taken to avoid the removal of adjacent sheltering plantations. In selective felling, species susceptible to atmospheric drought should be removed at the first cutting, for if they are left standing they will almost certainly deteriorate. If some individuals of a sensitive species have to be left, dominants, the crowns of which were previously exposed above those of their neighbours, should be selected to stand.

7

DEFICIENCY DISEASES OF TREES

THE growth and eventually the health of all plants are affected if they are not supplied with adequate quantities of major and minor (trace) elements. Substantial quantities of the major elements are required, but only minute amounts of the minor elements. Deficiencies are usually accompanied by fairly definite symptoms, the severity of which depends on the shortage of the element involved. It is certainly not possible to diagnose all deficiencies by the symptoms they cause. Nevertheless, proper comparative descriptions of the symptoms can be a very useful aid in diagnosis. Such descriptions, accompanied by coloured illustrations, are available for a large number of agricultural and horticultural crops, including fruit trees (Hambridge *et al.* 1941; Wallace 1951). The symptoms on fruit trees can be used as a guide as to what may be expected on deciduous broadleaved trees, but only in a few cases have the symptoms caused by the lack of different elements been properly compared for either broadleaved trees or conifers. Such descriptions as do exist are mentioned below.

Deficiencies in the forest

There is always a tendency for forestry to be driven by agriculture on to the less fertile soils. Only where the terrain is too steep or rocky for cultivation or in very remote places can forestry hope to remain on the better ground. In a highly developed country such as Britain forestry must inevitably occupy large areas of poor soil, parts of which are likely to be deficient in some elements to the point of causing disease. It has even been suggested that deficiencies, rendered more acute by periodic removal of the standing crop, may eventually prove a serious limitation in British forestry (Rennie 1957), though it is generally believed that this can be avoided by appropriate aerial manuring. Our upland peat soils, for instance, are locally deficient in some minor elements and notably lacking in available phosphorus (Mitchell 1954; Zehetmayr 1954). It is easy to quote examples of deficiency, causing at least defective growth, if not actual disease. Lime-induced chlorosis on chalk and limestone soils is often serious. Some nurseries have serious nutritional problems, and in many of them mild deficiency symptoms appear widely in wet seasons. Check in growth following planting is often associated with deficiencies of nitrogen or phosphorus. Lesser deficiencies, leading to reductions in growth rather than pathological symptoms, or masked by other diseases to which they have been a primary or contributing cause, may well be commoner than we suspect. Many deficiencies are far from simple. Excess of an element in the soil may prevent the absorption of others which are present in lesser quantities. Lime-induced deficiencies on limestone and chalk soils are of this nature. In the case of minor elements the margin between deficiency and toxicity is often comparatively narrow, so that efforts to correct deficiencies may result in toxicity symptoms (Chapter 8).

The symptoms of deficiency diseases vary according to the element and species involved, but they include purpling, yellowing, or withering of the leaves, from the tip downward on conifers and from the margin inwards on broadleaved trees, often accompanied by dieback of the tips of the shoots and probably, though this is less often observed, of the roots. The yellowing of the leaves and needles is typically chlorotic, and this rather unfortunately has led to the use of the term 'chlorosis' as if it was synonymous with deficiency.

The effect of environment on deficiencies

The extent to which deficiencies produce visible symptoms depends among other things on the conditions under which the tree is growing. Lack of a minor element, for instance, may become obvious only if other conditions for growth are favourable, so that the increase in size of the tree is rapid with consequent high nutritional demands. Drought may sometimes produce deficiency symptoms, because in the absence of available water the nutrient intake is restricted. Conversely, in very wet weather symptoms may develop because the constantly moist trees transpire little and thus absorb little water from the soil. Excess water in the soil can affect deficiencies by interfering with proper absorption by the roots. It is possible that heavy continuous rain may also contribute to deficiencies by washing nutrients out of plant foliage (Tamm 1951; Will 1955; Anon. 1956*b*). Cuttings in the more heavily misted parts of a mist propagation frame have developed deficiency symptoms almost certainly because the nutrients were leached from the foliage. Rain can also act by leaching nutrients from the upper soil layers, so that they are beyond the reach of the roots of young trees. Other vegetation may compete with trees for the available nutrients. This is certainly one of the reasons why a grass mat makes the establishment of trees so difficult (Howard 1925; Richardson, S. D. 1953). All these phenomena are more likely to affect small trees than large ones, and are therefore more noticeable in the nursery than in the forest.

The diagnosis of deficiencies

There are a number of ways of diagnosing deficiency diseases. They are very largely complementary, and it is usually unsafe to base a decision on the results of one method alone. The usual methods are:

1. Chemical analysis of parts of the tree, usually the foliage.
2. Field and pot trials to determine the effect of withholding or adding nutrients to the soil. Properly carried out this method is critical in diagnosis, and certainly indicates whether a cure is possible.
3. Soil analysis.
4. Addition of mineral nutrients directly into affected plants by injection or spraying. Properly carried out this method gives a definite diagnosis.
5. Diagnosis based on visual symptoms.

All these methods can be used in forestry. In dealing with large organisms such as trees, however, care must be taken in sampling for foliar analysis.

Only with small plants is it usual to analyse the whole tree. With older trees irregular distribution of the nutrient may give erroneous results, unless the samples are well distributed. Stimpfling and Shubert (1950) found very irregular distribution of iron in *Acer saccharinum*, and J. Parker (1956*b*) found very wide variations in content of copper, boron, and manganese between different healthy individuals of *Pinus ponderosa*. It would therefore be rash to arrive at any conclusions unless a number of healthy and unhealthy trees had been properly sampled. The methods outlined above have been described in detail by Wallace (1951). Goodall and Gregory (1946) have dealt particularly with the chemical composition of plants as an index of their nutritional status. Roach and Roberts (1945) have dealt with injection methods of diagnosis, which are of particular value on broadleaved trees, since they can be used on individual twigs or even leaves, but are not satisfactory for conifers. Diagnosis in forest crops has been particularly considered by Leyton (1957).

With the second and fourth methods great care must be taken to see that the chemicals used are pure. It is quite possible to supply the element actually responsible for the deficiency as an impurity in some other substance, lack of which might then be regarded erroneously as the cause of the trouble.

Curative methods

Deficiencies can be corrected either by the addition of the element that is lacking, or by promoting its natural release by the adjustment of other soil conditions, such as the correction of excessive acidity or alkalinity. The former can be achieved by addition to the soil, by injection, or by spraying followed by absorption through the leaves. At the moment the practical use of injection and spraying is generally limited to ornamental trees of individual value, but has occasionally been extended on a forest scale. Curative injections into fruit trees have been dealt with in some detail by Roach and Roberts (1945) and by Levy (1946). The methods and dosages suggested for fruit trees should be generally applicable for other broadleaved trees suffering from similar disorders.

Making good shortages of the major elements is usually only a matter of the application to the soil of the correct amount of the appropriate fertilizer, though interactions with other elements may sometimes produce difficulties. In forestry manuring with major elements has so far been largely confined to nurseries, and to the initial boosting of young trees at or soon after planting, but aerial application may make forest manuring much more feasible.

Injection and spraying with minor elements have been used commercially on ornamental trees, particularly in the United States, and on a forest scale against zinc deficiency in pine. Soil application of minor elements was difficult in the past because the minute quantities involved were either leached out or altered, by combination with other chemicals in the soil, into forms which could not be used by the plant. The recent introduction of chelates, where iron and zinc are applied as stable organic compounds (Bear *et al.* 1957), and fritted trace elements in which the elements are mixed with glass and applied as very small particles, have rendered soil application rather less difficult.

No doubt both these methods, which have been briefly described by Holdsworth (1956), will play a part in the treatment of deficiencies in trees, at any rate in nurseries. The appropriate remedial measures must depend on the local circumstances in each case. It is quite impossible, therefore, to make here specific suggestions for the treatment of soil deficiencies. In any case the consideration of curative treatment leads directly to the subject of forest manuring, which is obviously beyond the scope of this book.

Detailed consideration of deficiencies

In the sections below the information available on deficiencies of forest and ornamental trees is considered under the headings of the elements involved. It is important to realize that the descriptions of the symptoms, which in most cases are very incomplete, and mostly refer to young plants, have been taken from a large number of different papers and are not strictly comparable. Valuable descriptions of deficiencies, covering in each case only a limited number of nutrients, have been given by Ingestad (1957) for birch, by Worley, Lesselbaum, and Matthews (1941) for *Ulmus*, *Ailanthus*, and *Catalpa*, by Walker, Gessel, and Haddock (1955) for *Thuja plicata*, by M. E. Smith (1943) for *Pinus radiata*, and by Hobbs (1944) for four pine species. Walker (1956) dealt with the effect of potassium deficiency on various trees and shrubs native of New York State. Becker-Dillingen (1939) has also described and illustrated a number of deficiencies, particularly on pine and spruce. The symptoms developed by fruit trees such as apple and pear, and described by Wallace (1951), give some guidance to the probable behaviour of other broadleaved trees and are therefore quoted under the different elements below.

Trees differ not only in the symptoms caused by specific deficiencies, but also in their susceptibility. In fact the nutrient requirements of each species, and probably of individuals within that species, are different. On our present knowledge, however, we cannot single out any particular tree as suitable for potentially deficient soils. Certainly some trees, such as poplar or ash, appear particularly demanding, and others, such as Scots pine and *Pinus contorta*, much less so, but poor soil conditions are usually associated with so many other climatic and edaphic factors that it is seldom possible to draw definite conclusions. Nevertheless, it is possible on observational evidence to recommend, for instance, beech for chalk soils or *P. contorta* for some forms of poor peat.

Considerable knowledge is available as to the actual behaviour of the various nutrient elements in the plant, and therefore on the reasons why their presence is desirable. These considerations are too fundamental to be dealt with here. They have been summarized by Hambridge *et al.* (1941) and by Wallace (1951).

Boron

Deficiency of the minor element boron in trees has attracted little attention except in Australia. It does occur on agricultural plants in parts of Britain, but has not so far been reported as a forest disorder. In Australia it was considered as one possible explanation of some of the widespread and obscure disorders of *Pinus radiata* and other pines (Ludbrook 1942). Boron deficiency in pines shows mainly as stunted growth, accompanied by death of the growing tips of the shoots and roots. The needles near the apex of the shoots die, and there is resin exudation from the buds. Needles tend to be shortened, and may become fused, though it was not contended that boron deficiency was the main cause of this rather common pathological symptom in pines in Australia. The symptoms of boron deficiency are rather slow to develop (Ludbrook 1940; Smith, M. E. 1943). In *Thuja plicata* boron deficiency results in restricted growth of both roots and shoots, the needles becoming closely bunched; the young shoots also become weak and floppy, and the newly formed needles are bronze-purple (Walker, Gessel, and Haddock 1955). On fruit trees boron deficiency causes defoliation and dieback.

Calcium

The major element calcium is particularly important because it exerts such a large influence on other nutrients both in the soil and in the plant. Disturbances of the balance of calcium with other elements are the basis of many of the most serious deficiency diseases found in the forest.

Calcium deficiency has attracted far less attention than the effect of excess calcium in the soil on the absorption of other elements. On *Pinus taeda* lack of calcium produced yellowing, followed by browning of the needles, which became twisted and stiffened; there was also death of root tips (Davis, D. E. 1949). On *Thuja plicata* death of the tips of the shoots and roots was the most obvious symptoms of calcium deficiency (Walker, Gessel, and Haddock 1955). Chlorosis followed by dieback of shoots and roots occurred on three broadleaved species (Worley, Lesselbaum, and Matthews 1941). In fruit trees calcium deficiency causes inward rolling of the leaf margins and shoot dieback. On the highly acid soils, where naturally calcium deficiency occurs, there is nearly always confusion with other deficiencies, possibly with manganese or aluminium toxicities and with the effect of the very low pH on the soil flora.

But the real importance of calcium is that when present in excess it reduces the availability of other elements. In this role it is responsible for some of the most striking instances of forest deficiency in Britain. Iron and manganese are the elements most commonly involved, but excess calcium can affect the absorption or availability of many other nutrients. It can act in this way even when the other nutrients are present in adequate quantity. Björkman (1953) has reported lime-induced potassium deficiency in spruce. Alben and Boggs (1936), studying the 'Rosette' disease of *Carya pecan*, found that soils of high lime content interfered with the uptake of zinc. One of the most serious cases of deficiency in Britain, affecting Scots pine on oolite limestone, was thought to be associated with lack of manganese. Lime-induced deficiencies usually occur on naturally alkaline soils, but they may be induced or their severity increased by the use of irrigation water with a high lime content (Korstian *et al.* 1921). Typically symptoms of lime-induced deficiencies do not appear until the trees are beginning to close canopy or even later. Probably during the early period of normal growth the trees use the elements in the surface layers which, owing to leaching, generally have an appreciably lower lime content than the deeper layers of the soil.

Despite the common occurrence of lime-induced chlorosis in forests in

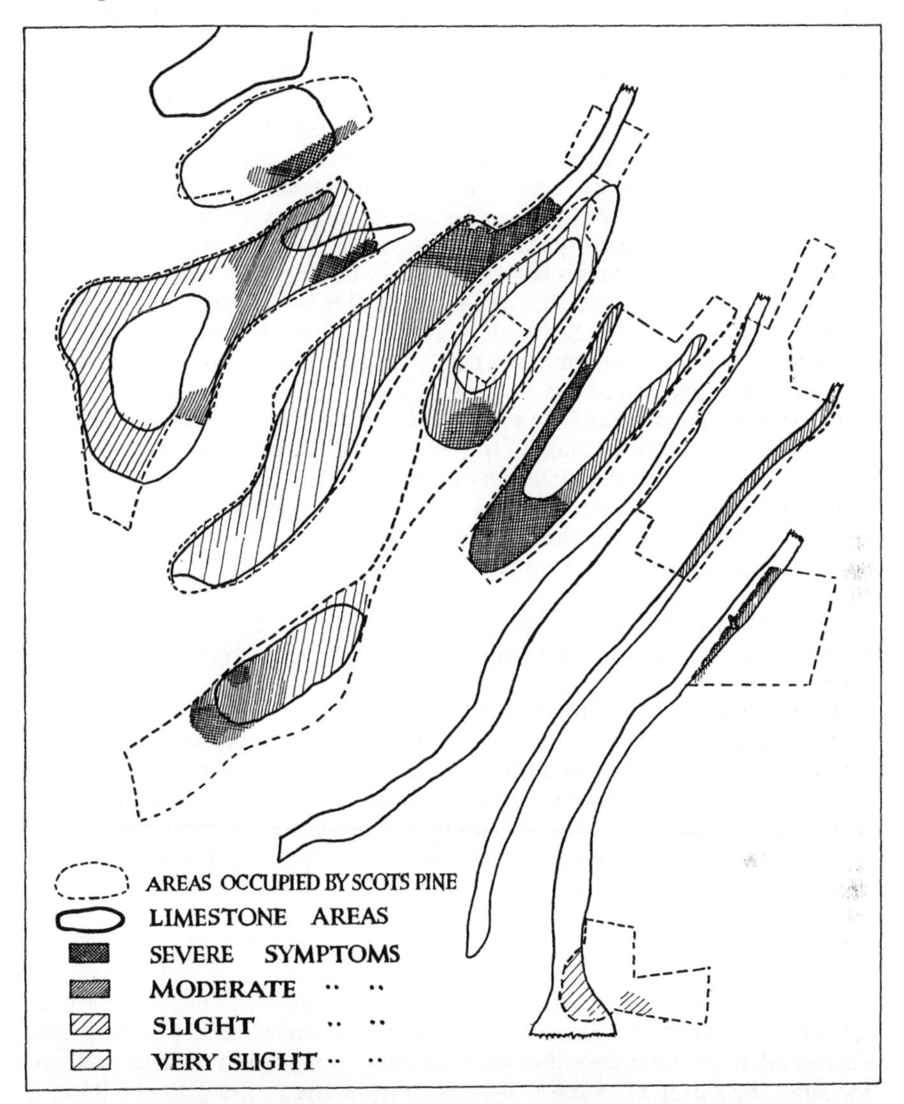

FIG. 20. Map showing the distribution of lime-induced chlorosis on Scots pine at Allerston Forest, Yorkshire (scale *c.* 2 in. to 1 mile). (A. C.)

Britain, references to it are few, though W. R. Day (1946) associated chlorosis of beech on shallow chalk soils with lime-induced iron deficiency. Severe lime-induced chlorosis of Scots pine, possibly acting through manganese deficiency, has occurred on the oolite limestone in Yorkshire. Fig. 20 illustrates clearly how well the boundaries of the disease coincide with those of the limestone outcrop. The symptoms were first observed when the pine were twelve years old and thereafter they developed rapidly. The needles turned yellow then brown and fell, while subsequent growth was both stunted and short-needled; many of the trees died. The disease reached its peak when the trees were about twenty years old, over half the trees dying in some parts of the area. Since then the death rate has declined and many of the affected trees have recovered. This apparent recovery, which has been observed in other instances of lime-induced deficiency, is possibly associated with the formation of a humus layer. The worst patches all occurred where the limestone was near the surface, but there was no simple relationship between soil depth and symptom severity. Corsican pine, sometimes mixed with the Scots, was unaffected. The same specific resistance has been observed elsewhere on chalk, and there are numerous instances of Corsican pine growing healthily on shallow limestone and chalk soils. Nevertheless, apparent lime-induced deficiency of this species has been recorded on both soil types, and in one instance was worse than on Scots pine. Austrian pine is even better able to withstand alkaline conditions.

Other serious cases of lime-induced deficiencies have been recorded on Scots pine, on Japanese larch, where the symptoms were yellowing and browning of the needles followed by tip dieback, on European larch, and on Douglas fir where it caused striking chlorosis. The spruces are generally less affected, possibly because they can form shallow root systems in the surface soil. The very selective nature of these deficiencies is clearly illustrated by the behaviour of various poplars in a chalk soil nursery. A large range of varieties grew healthily and as fast as the rather dry nature of the site allowed, with the exception of the hybrid *Populus koreana X trichocarpa*, which suffered severe chlorosis and stunting.

No doubt careful investigation of a wide range of alkaline soils in Britain would disclose that nearly all coniferous and many broadleaved genera are affected under extreme conditions. Nevertheless, there are large areas of relatively successful coniferous planting on calcareous soils where deficiency, if it exists, is not visibly damaging. On our present knowledge, apart from the obvious danger of very shallow soils, we cannot prophesy where on the limestone or chalk such deficiencies are likely to occur.

Copper

Deficiency of copper, a minor element, is comparatively rare even in agriculture in Britain. Where it does occur it is often associated with zinc deficiency. It has been described on fruit trees on dry sandy soils in southern England (Bould *et al.* 1953*b*). Benzian and Warren (1956) found it on a similar soil in Dorset causing tip-burn of the needles of seedling Sitka spruce. The symptoms normally appear during hot dry spells in the summer, mainly on the larger, faster-growing plants. The tips of the upper needles turn pale yellow, there being a very clear demarcation between the affected tip and the green remainder of the needle. The symptoms develop mostly in plots subject solely to inorganic fertilization; hop waste, which is an integral part of the manuring régime recommended for infertile soils, apparently contains enough copper to prevent the occurrence of deficiency. The disease could be virtually cured by foliar applications of copper. In the same nursery copper deficiency produced black discoloration of the leaves of poplars (Benzian 1958). Penningsfeld (1957) has recorded dieback of azaleas grown in copper-deficient peat. M. E. Smith (1943) found that copper deficiency caused withering of the tips of the needles of *Pinus radiata* grown in water culture; the symptoms were rather slow in developing. Rademacher (1940) thought copper deficiency of considerable importance in the moorland soils of north-west Germany. The status of copper in British forest soils might well repay investigation.

Iron

Iron, a minor element, plays a major part in the formation of chlorophyll in the plant. Chlorosis is therefore a characteristic symptom of iron deficiency. In iron-deficient *Thuja plicata* the young foliage is conspicuously yellow in contrast to the green of the older foliage (Walker, Gessel, and Haddock 1955). In fruit trees chlorosis is sometimes followed by the dieback of shoot tips, whole shoots, or even quite large branches. No doubt some of the very striking chlorosis, observed on various species on chalk and limestone soils, is due to lack of iron, which is generally considered the commonest of the lime-induced deficiencies. The presence of excess manganese in the soil also produces iron deficiency (Twyman 1946*a*).

Magnesium

Magnesium, usually regarded as a major element, is a constituent of chlorophyll, so that chlorosis is one of the signs of its deficiency. It is said to be very mobile in the plant, and when in short supply may be transferred from the older tissues to the growing parts. For this reason magnesium-deficiency symptoms sometimes but not always appear first in the older tissues. Magnesium deficient *Thuja plicata* show yellowing of the older needles, followed by browning and death. Badly affected plants are finally left with a long bare stalk carrying a tuft of green shoots at the top (Walker, Gessel, and Haddock 1955). Magnesium deficiency is very apt to occur on lime-deficient soils and can be brought about by heavy applications of potassium fertilizers.

Typical chlorotic symptoms have been recorded for a number of species of pine. Becker-Dillingen (1939) and Hobbs (1944) have described yellowing of the needles, while E. L. Stone Jr. (1953*a*) noted the same symptom, followed by death of the tips and sometimes by needle-fall. Yellow needles were also found by Jessen (1939) to be typical of Scots pine and Norway spruce suffering from magnesium deficiency. On young spruce the yellow needles usually lie at a less acute angle with the stem than in normal plants. On broadleaved trees interveinal chlorosis and stunting are typical (Worley, Lesselbaum, and Matthews 1941). Poplar shows a very marked interveinal chlorosis followed by necrosis (Hinson unpubl.). Interveinal necrosis, sometimes accompanied by chlorosis, also occurs on apples, a crop frequently affected by magnesium deficiency in Britain.

Manganese

The minor element manganese, like iron, is associated with chlorophyll formation, so that symptoms of its deficiency almost always include chlorosis. Manganese deficiencies occur quite widely in Britain on agricultural soils of high pH. The presence of excess iron in the soil sometimes leads to manganese deficiency (Twyman 1946*a*). There is no information on its importance on forest soils in Britain. The development of suspected lime-induced manganese deficiency in Scots pine in Yorkshire has already been described under 'Calcium'; the symptoms included very marked chlorosis. Chlorosis due to this cause has been described on Scots pine by Becker-Dillingen (1940), and accompanied by dieback on *Pinus radiata* by M. E. Smith (1943). Walnuts suffering from manganese deficiency exhibit chlorotic mottling, followed by browning and withering of the leaves (Braucher and Southwick 1941). Fruit trees usually show varying degrees of interveinal chlorosis.

Molybdenum

Deficiency of this minor element appears to be of little importance, even in agriculture. M. E. Smith (1943), using water culture, found that lack of molybdenum produced slight blueing of the foliage and thickening of the root tips in *Pinus radiata*, but the symptoms were not at all striking.

Nitrogen

Nitrogen, a major element, is required in large quantities for plant growth, and its availability in the soil is very dependent on the presence of organic matter and the activities of micro-organisms therein. Nitrogen, like magnesium, moves freely in the plant and can be transferred from the older to the younger tissues. Nitrogen-deficiency symptoms are apt therefore to appear first on the older organs. The older foliage dies first in nitrogen deficient *Thuja plicata* (Walker, Gessel, and Haddock 1955). In the nursery dense beds of pine and spruce often show nitrogen deficiency by the yellow-green coloration of the older needles and reduced growth of the plants away from the seed-bed edges. In general, lack of nitrogen shows more as reduced growth than as definite deficiency symptoms. In broadleaved species markedly reduced growth is often accompanied by premature autumn coloration. In beech the colour of the foliage, from light to dark green, can be used as a measure of nitrogen deficiency (Holstener-Jørgensen and Klubien 1957). This relationship would only hold in the absence of other deficiencies causing chlorosis. Many trees, which initially do not appear 'unhealthy', will give marked responses to nitrogenous fertilization, both in growth and in deepening of the green colour of the leaves.

Competition with other vegetation, particularly grass, is a frequent cause of nitrogen deficiency in trees (Howard 1925; Richardson, S. D. 1953). Heather acts in the same way (Duchaufour 1957).

Phosphorus

Phosphorus, a major element, is required by plants in comparatively large amounts. It has been suggested that the movement of phosphates in the soil is very limited, and they are therefore of little use to the plant unless they are near its roots. Thus an older tree with a widespread root system would be much better able to get the phosphorus it needs than a young tree, which can only tap a very restricted volume of soil. For this reason 'boost' manuring with phosphatic fertilizers has proved of particular value in overcoming initial growth check on soils of low phosphorus content, such as the peats (Zehetmayr 1954).

In young conifers phosphorus deficiency is characterized by purple discoloration and later by withering of the older foliage. This has been described for various American pines by Hobbs (1944) and for Scots pine and Norway spruce by Jessen (1938), but European larch and slightly affected spruce showed only a bluish-green discoloration of the needles. Němec (1936) described the discoloration on Scots pine as dull violet. In *Thuja plicata* the stems and older foliage become reddish or purplish, later turning brown and dying, but usually remaining attached to the plant. The youngest foliage remains a normal green (Walker, Gessel, and Haddock 1955).

Purple discoloration, particularly of pines, and reddish-purple discoloration, particularly of spruces, are both common phenomena in nurseries. They tend to appear during drought, in badly drained beds, and with the onset of cold weather in the autumn. All these influences could accentuate phosphorus deficiency. It would, however, be very rash to suggest that lack of phosphorus is the sole cause of these symptoms, and in most cases, lacking further evidence, it is best to associate them merely with the obvious cause of their appearance, and in the case of drought or bad drainage to remedy it.

Phosphorus deficiency is certainly associated with the widespread and complex disorders of pines which have occurred in Australia, though other elements, including zinc and boron, are also involved (Stoate 1951; Stoate and Bednall 1957). Ludbrook (1942) found that phosphatic manuring greatly reduced needle-fusion symptoms on *Pinus taeda* and *P. caribaea*, but not on *P. radiata*. The same has been found in Britain for *P. contorta*. It seems quite clear that the stunting and partial adhesion of the needles in the fascicle, which are the symptoms usually described as 'fused-needle' (Fig. 21), can be caused by various deficiencies and probably by other pathogenic influences.

Potassium

Potassium, a major element, which is required in large quantities by plants, is highly mobile within them. It is therefore one of the elements that moves from the older parts to the growing points in times of shortage, so that deficiency symptoms are usually first seen and develop most seriously in the older organs.

Walker (1956) has described the symptoms of potassium deficiency on a number of trees and shrubs native to New York State, using them as a guide to the location of deficient areas. *Pinus strobus* showed chlorosis and stunting of the needles, which often went brown at the tips. Hardwoods developed various forms of chlorosis, usually primarily marginal. The leaves of *Prunus serotina* showed bright-red margins. Tamm (1953) described yellowing of the needle tips in Norway spruce. Goor (1956) found similar damage on Scots and Corsican pine. More general chlorosis followed by browning and withering of the needles has been reported on species of *Pinus* and *Picea* in the United States (Hobbs 1944; Heiberg and White 1951) and on *Thuja plicata* (Meier 1937; Walker, Gessel, and Haddock 1955). Jessen (1939) and Benzian (1955) have both described reddish-purple or mauve discolorations of the needles of seedling spruce as a symptom preceding chlorosis. Benzian has also drawn attention to the difficulty of distinguishing the symptoms of potassium deficiency from those of magnesium deficiency in spruce. On

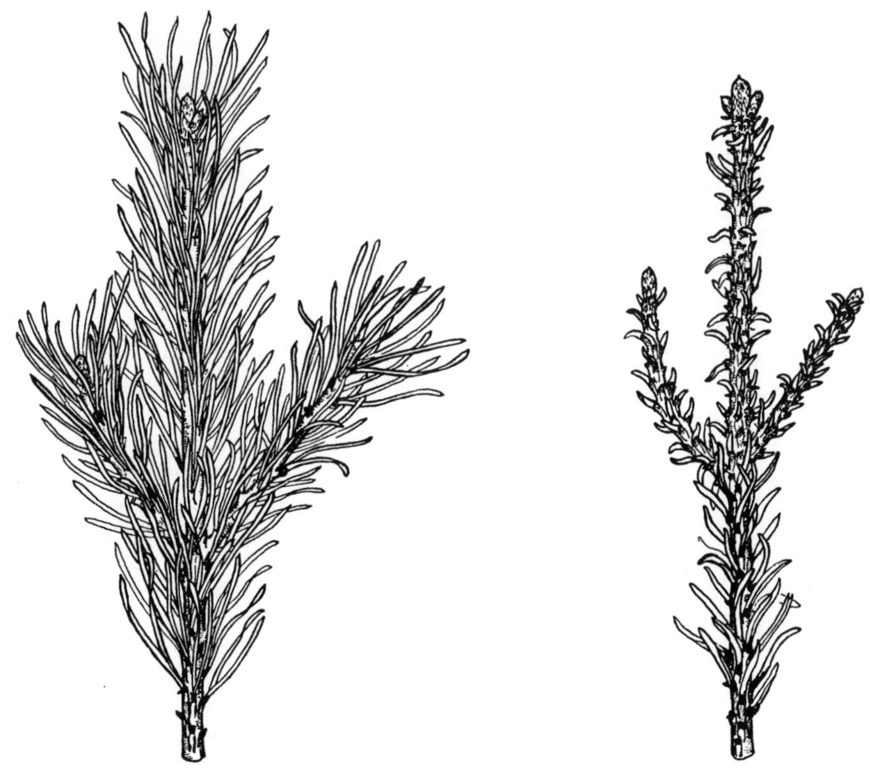

FIG. 21*a*

broadleaved trees discoloration of the foliage is often followed by marginal or apical withering (Worley, Lesselbaum, and Matthews 1941).

Potassium deficiency occurs fairly widely on agricultural soils in Britain, especially in the south and east. It certainly occurs in forest nurseries, though its extent there has never been investigated. It is now attracting some attention in British forests.

Sulphur

Deficiency of sulphur causes a reduction in growth and pronounced yellowing of adult needles of Corsican pine in pot cultures; it may be one of the deficiencies occurring on sand dunes at Culbin in Morayshire (Keay unpubl.). Otherwise it has not been observed in Britain. Walker, Gessel, and

Haddock (1955) found that lack of sulphur produced chlorosis on *Thuja plicata*, the symptoms being more pronounced on the younger than on the older growth. Worley, Lesselbaum, and Matthews (1941) produced chlorosis and reduced growth on three broadleaved species by withholding sulphur.

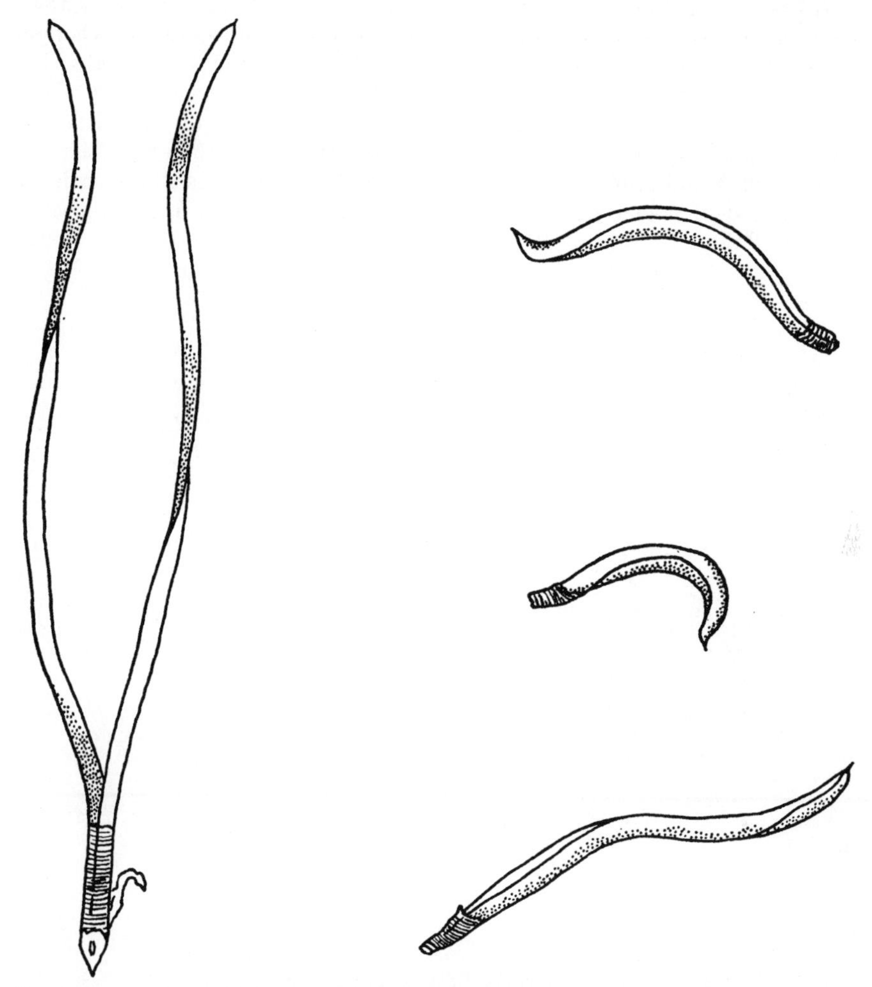

FIG. 21*b*. Fused-needle disease of *Pinus contorta*, Wareham Forest, Dorset. Normal and affected shoots, and detail of fused needle (×½, ×2½). (J. N.)

Zinc

Zinc deficiency is often associated with that of copper; both appear to be comparatively rare in Britain. Bould *et al.* (1953*a*) have recorded zinc deficiency in fruit trees in southern England, but so far there are no forest records.

C. C. Wilson (1953) found that lack of zinc caused chlorosis and stunting of *Pinus taeda* and *P. palustris*, M. E. Smith (1943) found chlorosis accompanied by red or purple spotting on the needles of *P. radiata*.

Zinc deficiency has attracted most attention in Australia, where it is one of the elements associated with 'Needle fusion' (Fig. 21), and other disorders of *P. radiata* and other pines (Stoate 1951). Spraying with zinc salts has proved beneficial.

It has also been associated with stunted and bunched shoot growth, known as 'Rosette', on cultivated *Carya pecan* in California. This was at first thought to be an iron deficiency because it responded to injection of iron salts; but it was found that these contained appreciable quantities of zinc (Alben, Cole, and Lewis 1932). In some places where the soils were naturally zinc deficient, the disease did not develop because there was sufficient zinc in the irrigation water (Finch and Kinnison 1933). This is one of the diseases where chelates, in this case of course chelated zinc, have rendered soil treatment possible (Alben 1955).

8

DAMAGE TO TREES BY TOXIC SUBSTANCES

INJURIES to trees by poisonous substances can be classified in a number of ways. Here they are divided into those due to man's activities, e.g. industrial fumes, and those due to natural agencies, e.g. salt spray from the sea.

Many diseases, particularly those known as 'wilts', are due to the production of toxins by pathogenic organisms in the tree, for instance Elm disease caused by *Ceratocystis ulmi* (Feldman, Caroselli, and Howard 1950) (Chapter 31), or Oak wilt caused by *C. fagacearum* (White, I. G. 1955) (Chapter 27). The subject of such toxins has been dealt with fully by Dimond and Waggoner (1953), Gäumann (1954), and Dimond (1955) and will not be discussed here.

DAMAGE DUE TO MAN'S ACTIVITIES

Damage by industrial pollution

A great deal has been written, particularly in Germany, on smoke and fumes from factories. We are here concerned mainly with publications that pay attention to the effect of air pollution on plants and in particular on trees. Among the books dealing fully with the subject are Stoklasa (1923), Haselhoff, Bredemann, and Haselhoff (1932), and Haselhoff (1932). Meetham (1952) deals fully with the causes, measurement, and prevention of pollution in Britain, but only very briefly with injury to plants. In America the work done on the damage caused by the Trail smelter on the borders of the United States and Canada (Anon. 1939*a*) remains the largest single investigation of fume injury ever undertaken. More recently the whole subject of pollution has been covered in a symposium sponsored by a United States Interdepartmental Committee (Anon. 1952*e*), and a brief review with particular reference to grasses has been given by Bleasdale (1957). There are also a number of shorter reviews of the subject referring particularly to tree damage (Boyce, J. S. 1948, pp. 51–55; Peace 1952; Zieger 1953–4, 1955; Grayson 1956). Unfortunately, there are still many gaps in our knowledge and, worse, there are still many difficulties in applying existing knowledge to the problem of growing trees in industrially polluted areas.

The nature and causes of pollution

An enormous number of industrial and chemical installations emit toxic fumes or dust, either in smoke from fuel or as part of the manufacturing process. Although screening and extraction methods are constantly being improved, the steady expansion of industry has meant, in Britain at any rate, that the potentially polluted area is tending to increase, and there has as yet been no marked improvement in the degree of damage within that area.

The main causes of injury in Britain have been reviewed by Meetham (1952) and by Thring (1957). In the neighbourhood of large industrial towns it is often impossible to separate the effects of various toxic agents, though most estimates of pollution are based on measurements of sulphur dioxide, sometimes with additional data on the deposition of soot. Fluorine, probably in the form of hydrogen fluoride or silicon fluoride, is also a very important toxicant (Sertz 1921). It is responsible for most of the damage around some isolated industrial plants such as brickworks (Peace 1958*a*) and aluminium smelters (Anon. 1948*d*). Even where it occurs mixed with sulphur dioxide and other toxicants it is still sometimes considered the primary damaging agent (Shaw *et al.* 1951; Adams *et al.* 1952). In Los Angeles both hydrogen fluoride and sulphur dioxide play a part in fume injury to plants, but oxidized hydrocarbons arising from motor fuel and excess ozone are also important (Middleton *et al.* 1956). Numerous special cases of damage can be quoted: Hiksch (1934) described injury to trees by arsenic fumes; Gerlach (1928) attributed tree damage to smoke from locomotives; waterglass dust was found to be very damaging owing to its extreme alkalinity by Czaja (1951); magnesium oxide dust not only formed a hard impenetrable crust over the soil, but was also highly toxic (Sievers 1924); zinc poisoning arising from the manufacture of brass has also been reported (Anon. 1921*a*); cement dust appears to do little damage to trees, although it has proved troublesome as a deposit on soft fruit. Many other substances, besides those mentioned above, are liable to damage trees. Unfortunately the continuing elaboration of chemical processes may well produce further hazards in the future.

A great deal of information has been collected on the concentrations of different substances, either in the air or in the leaf, required to produce visible symptoms of injury (Anon. 1939*a*; Romell 1941; Adams *et al.* 1952; Scheffer and Hedgcock 1955; Adams, Shaw, and Yerkes 1956). The variety of conditions under which these figures were measured makes summarization almost impossible. It is also difficult to correlate concentration in the air with injury, since the amount of damage depends on the state of the tree and therefore on the time of year, as well as on meteorological conditions. Damaging concentrations in the leaf vary for leaves of different ages and at different times of year. Since toxic substances may accumulate in the leaf, at least for short periods (Romell 1941), a relatively small amount of atmospheric pollution, if persistent or recurrent, may raise the concentration in the leaf beyond the danger point. These difficulties have been analysed and discussed by Mühlsteph (1942). The methods used in the estimation of pollution outside and inside the plant will not be discussed here.

It is generally agreed that pollution is worst in the winter months, particularly near towns where domestic fires play an appreciable part (Bleasdale 1953; Scurfield 1955). Also at that time of year high humidities favour deposition. But the worst injury generally occurs in the spring, when the young foliage is being produced (Anon. 1939*a*; Katz 1949; Vulterin 1952). During the winter little absorption takes place, even by evergreens, whereas the young growth in the spring is both absorptive and highly susceptible.

Damage can be acute or chronic (Anon. 1939*a*; Ceccarelli 1950), according to whether it arises from a single damaging deposition or from several lesser depositions leading to a gradual concentration of the toxic substance in the leaves. There has been considerable argument on the question of so-called 'invisible injury', i.e. that concentrations too low to produce visible injury may interfere with the metabolism of the tree sufficiently to reduce its growth rate or even affect its health. Several workers have denied its existence (Anon. 1939*a*; Katz 1949), but Thomas (1955) considered that interference with the metabolism can take place without visible symptoms; but pointed out that if no death of tissue occurred the process should be reversible and complete recovery should take place when the source of fumes was removed. Bleasdale (1952) found a reduction in the dry weight of grass grown under conditions of air pollution, although there were no visible injury symptoms. Linzon (1958*b*) found a definite correlation between the diminution of sulphur dioxide concentration, frequency, and duration with increasing distance from the source of fumes and improvement in the growth of *Pinus monticola.* Reckendorfer (1952) has described a very gradual chronic effect of fluorine, which may almost be regarded as invisble injury.

Evergreen trees are generally more damaged than deciduous ones, partly because their foliage is exposed to fumes for longer periods, particularly during the winter when pollution is worst. However, absorption is then at a minimum and it may be more important that polluted air hastens the senescence of leaves, a factor that would certainly have more effect on evergreens (Bleasdale 1953).

Smoke particles can have quite a serious influence on plant growth. It was at one time considered that the soot blocked the stomata and interfered with gaseous exchange, but this view has now been disproved (Rhine 1925). The reduction in the light reaching the leaves, due both to the deposit of soot on the leaves and to the smoke in the air, is the serious factor (Jennings 1934). The effect of dirt in lowering the rate of photosynthesis has been investigated by Eršov (1957). This restriction of light is likely, however, to have more effect on the rate of growth of the tree than on its visible 'health'.

The effect of pollution on the soil

Most of the damage by fumes is due to their deposition on and absorption by the leaves, but they can also produce undesirable changes in the soil. Acidification of the soil by sulphur dioxide has been put forward as a major reason for deterioration in the health of trees (Wieler 1922; Ewert 1924; Miyazaki, Okinaga, and Harata 1954), and liming has been suggested as a curative measure. Daines *et al.* (Anon. 1952*e*) have described the absorption of fluorine from the soil by plant roots. Zinc released in the manufacture of brass was said to exert its toxic effect as zinc sulphate in the soil water (Anon. 1921). Dust deposits are often more serious by their action on the soil than by their direct action on foliage (Sievers 1924).

McCool and Mehlich (1938), however, found that the soil around industrial areas producing sulphur dioxide had not been materially altered, while MacIntire (1957) found no measurable soil effect from the deposition on the soil of air-borne fluorides. In the Trail investigation (Anon. 1939*a*), although acidification of the soil was found some miles from the smelter, this effect was not considered to be inimical to tree health; indeed it was pointed out that under some circumstances the additional sulphur in the soil might be beneficial. It has even been suggested that other nutrients may reach the soil as a result of pollution (Bleasdale 1957).

The effect of pollution on fungi, lichens, and mosses

It has frequently been noted that some fungi, especially those growing superficially on leaves, do not occur in regions of heavy pollution (Scheffer and Hedgcock 1955). Notable examples are Oak mildew, *Microsphaera alphitoides* (Köck 1935), and Tar spot of sycamore, *Rhytisma acerinum*. Lichens and arboreal mosses are also noticeably absent from industrial districts (Wheldon and Travis 1915; Richards 1928; Jones, E. W. 1952), so their presence can be used as an indication of freedom from serious pollution.

Meteorological conditions for pollution damage

One set of conditions ideal for fume damage has been described by Peace (1958*a*). It involved mist, fog, or very fine rain to deposit the substance on the leaves, followed by drier conditions to concentrate it. The formation of an 'inversion layer' (Chapter 4) to hold the fumes near the ground increases injury (Okanoue 1958). Heavy rain, by washing the material out of the air and off the leaves, markedly lessens the possibility of injury. Wind should disperse the fumes and lessen injury, but Lines (unpubl.) has recorded higher sulphur dioxide readings on exposed gauges than on sheltered ones. Topography certainly plays a part in the distribution of fumes as well as meteorology. Setterstrom and Zimmermann (1939) found that injury from sulphur dioxide became greater with increases both in the humidity of the air and in the water supply.

On the average pollution is worst in the winter, the maximum concentrations of sulphur dioxide in the air occurring when the air is humid (Vulterin 1952), and the greatest deposition taking place when the leaf surfaces are damp. On the other hand, as has already been pointed out, absorption is low during the winter, even in evergreen trees, and the worst damage occurs to trees in active growth. Setterstrom and Zimmermann (1939) found that plants growing at temperatures below 40° F. were markedly more resistant to sulphur dioxide injury.

The distribution of damage

Naturally damage is worst near the source of the fumes, and becomes less serious and less apparent with distance. It is usually impossible to define with exactitude the limits at which pollution ceases to affect tree health. Normally damage extends much farther down the prevailing wind, but this does not prevent the occurrence of injury in another direction should favourable conditions occur when the wind is in that quarter (Peace 1958*a*). The theory of distribution has been dealt with by Sutton (1947). In practice there are many anomalies. Certain areas may escape damage owing to their topographical position, whereas winds concentrated in valleys may carry the fumes farther than would otherwise be expected. Acute damage is often limited to distances of 1–3 kilometres from the source of the fumes (Hiksch 1934; Adams *et al.* 1952); but in the case of the Trail smelter (Anon. 1939*a*), where a very large source was involved, damage extended as far as 50 kilometres, and all the conifers within 20 kilometres were killed.

Chronic injury is usually found over a large and comparatively constant area round the source, but acute injury, depending on local high concentrations of the toxicant and suitable meteorological conditions for its deposition, may be much more local and restricted. It is uncertain whether acute injury is more serious than chronic, certainly the importance of the latter must not be underrated. The former may only happen on a very small number of days during the year, so that any attempt to evaluate the practicability of planting trees in a given area should involve daily observations on pollution over a long period and in as many places as possible. In this way the short period occurrences of high concentration can be noted as well as the general level of deposition of the toxic agent.

Much of the ground subject to pollution in Britain is at comparatively high elevations and is therefore subject also to exposure and windblast. However, in areas of general pollution, damage is often worse at high elevations, possibly because mist and very fine rain, which lead to the greatest deposition, occur more commonly there than lower down. In addition, many polluted areas have naturally poor soils. In some places, also, the water table has been lowered by pumping or raised by the subsidence of undermined ground, either of which is likely to damage the health of existing tree crops. With so many confusing factors, it is often very difficult to separate the adverse effects of fume injury from those due to exposure, and unfavourable soil and water conditions. In some cases plantation failures due to totally unrelated causes have been attributed to pollution, merely because the forest happened to lie in an industrial district. In other places, of course, smoke and fumes may only act as the final limiting factors preventing tree growth, where other conditions are also acting adversely. Trees on otherwise good sites may have the necessary vitality to recover from fume injury, while those on poor sites show persistent damage, or succumb.

The symptoms and effect of pollution injury

The typical symptoms of fume injury, as caused for instance by sulphur dioxide or fluorine, have been particularly well described and illustrated in the case of the Trail smelter (Anon. 1939*a*) and by Stoklasa (1923). On broadleaved trees, marginal and interveinal yellowing is usually followed by browning of the same parts (Solberg and Adams 1956). Damage may extend to the whole leaf surface, but the leaves often remain in a partially withered condition. On conifers the tips of the needles turn reddish-brown and wither. It has been found in the case of fluorine damage to *Pinus ponderosa* that young needles are more damaged than old needles, irrespective of the amount of toxicant absorbed (Adams, Shaw, and Yerkes 1956). Sometimes brown spots appear on the needles below the dead tips or in the case of broadleaved trees on otherwise uninjured areas (Fig. 22). Frequently the needles or leaves of affected trees are smaller than normal and are shed earlier, so that the foliage always appears sparse (Ferda 1954). According to Antipov (1957*a*) this may result in a reduction of the total vegetative period by as much as forty days. The distribution of injury on leaves and needles has never been completely explained, though marginal and interveinal damage suggest that concentrations may be greatest farthest from the rising sap stream. The occasional occurrence of necrotic spots irrespective of the anatomy of the leaf does suggest the evaporation of droplets on the leaf surface. Repeated foliar injury either in broad-leaved trees or conifers leads to restricted growth and often to dieback. Biraghi (1938) reported dead patches and bark peeling on the branches of walnut, which he attributed to fumes, and Linzon (1958*b*) associated roughen-

FIG. 22. Damage to Scots pine and Douglas fir from aluminium smelter fumes, Glen Nevis, Scotland ($\times \frac{3}{5}$). (J. C.)

ing and discoloration of the bark of *Pinus monticola* with smelter fumes; but foliar symptoms are far commoner than injury to other parts of the tree. Even when the visible damage is very slight or negligible, exposure to fumes can result in reduction in growth (Güde 1954), which appears to support the conception of invisible injury discussed above. Observations around the Trail smelter (1939*a*) showed that recovery from such a depression in growth rate is rapid once the source of pollution has been removed. In addition to their slower growth, trees subject to slight fume injury over long periods often have yellowish foliage of abnormally small size and a generally unhealthy appearance (Reckendorfer 1952). The increased rate of senescence brought about by exposure to fumes (Bleasdale 1953) certainly leads some evergreen species, for instance *Ligustrum lucidum*, a normally fully evergreen privet, or holly, to become wholly or partially deciduous (Scurfield 1955).

Less is known about the injuries caused by substances other than sulphur dioxide and fluorine. Damage by oxidized hydrocarbons takes the form of glazing and silvering of the lower leaf surfaces, sometimes followed by a bronze or reddish discoloration; while ozone, which is present in toxic quantities in Los Angeles 'smog' and presumably also in that of other towns, causes bleaching and chlorosis, sometimes followed by withering (Middleton *et al.* 1956). The symptoms caused by a wide range of other toxicants have been reviewed by P. W. Zimmermann (1955).

Young trees are generally more susceptible than older trees. The high susceptibility of seedlings and the reduction in cone crops on older trees suffering from fume injury both adversely affect regeneration in polluted areas (Scheffer and Hedgcock 1955), though Antipov (1957*b*) found that seed from several broadleaved species growing under polluted conditions showed higher germination and less fungal infection than that from unpolluted trees. Established trees will often linger on in places where the establishment of young trees of the same species proves impossible. For instance, because of pollution it is now difficult to raise young conifers in Kew Gardens, though many of the older trees still appear to be in quite good health. Nevertheless, damage to old trees occurs very frequently and very large trees can sometimes be killed.

The relative resistance of trees and shrubs

Numerous lists of susceptibility and resistance have been made either from widespread observations or in relation to one particular source of injury. The variations which occur between different lists probably depend partly on differences in the nature of the toxicants concerned, and partly on local differences in climate and soil. Comprehensive lists have been given by Janson (1925), Schimmler (1935), Krüger (1951), Scurfield (1955), Scheffer and Hedgcock (1955), Morling (1956), and Glocker and Krüssman (1957). Conder (1957) deals only with forest trees, and Peace (1958*a*) with trees damaged by fluorine in a single episode.

Any effort to summarize such lists immediately reveals discrepancies. There is not even general agreement that evergreens are more damaged than deciduous trees and shrubs. Schimmler (1935) considered evergreens more resistant to industrial fumes, as indeed they often are to sea spray, on account of their thicker cuticle. Possibly where damage is chronic, or where soot deposits are heavy, evergreen plants may suffer more because of the longer period for which they carry their leaves, whereas under other conditions intermittent damage may do worse injury to the more delicate foliage of deciduous species. There is, however, fairly general agreement on the trees and shrubs given in the list below:

RESISTANT

BROADLEAVED	CONIFERS
Ailanthus	*Pinus nigra*
Betula	*Pinus nigra calabrica*
Laburnum	*Pinus montana*
Liriodendron	*Taxus*
Platanus	*Ginkgo biloba*
Populus	*Taxodium distichum*
Prunus	

RESISTANT

BROADLEAVED	CONIFERS
Pyrus	
Robinia	
Tilia	
Ulmus	

SUSCEPTIBLE

Fagus silvatica	*Tsuga*
Quercus (moderately susceptible)	*Abies*
	Larix (Japanese is certainly much better than European larch)
	Picea (*P. pungens* and *P. omorika* are sometimes quoted as resistant)
	Pinus sylvestris
	Pinus strobus
	Pseudotsuga taxifolia

In most lists *Fraxinus* and *Acer* (in particular the sycamore and Norway maple) are classed as resistant, but there are reports of severe injury. Hiksch (1934), for instance, found ash the most susceptible of a number of broad-leaved species to arsenic fumes. In the author's experience, ash is generally resistant to fumes, but often affected by the soil difficulties involved in town cultivation. Sycamore, on the other hand, frequently suffers leaf injury, but quickly recovers. It is certainly a successful town tree; indeed a map of the areas where sycamore is the dominant species in mixed woodlands in Britain might almost be a map of the industrial areas, because under pollution sycamore can compete successfully with species which would suppress it under more normal conditions. In most lists *Sorbus* species are regarded as good town trees; but here again there is disagreement, possibly owing to exceptional local conditions. Most of the poplars grow well in polluted air, and it is curious that a variety of *Populus nigra*, the so-called Manchester poplar, should have achieved a particular reputation for its suitability to town conditions. However, size of crown (except in Lombardy poplar and a few other varieties) and root-spread in all varieties limit the usefulness of poplars for planting in built-up areas.

Much less attention has been given to variation in susceptibility within a species. There is a tradition in the north of England that *Fraxinus excelsior monophylla*, a variety in which a single blade replaces the normal pinnate leaf, is more resistant to fumes than the normal form of ash. Pelz (1956) reported individual differences in the reaction of spruce to industrial fumes in Germany. Hendrix (1956) and H. J. Walter (1956) found marked variations in susceptibility to atmospheric fluorides among *Pinus ponderosa* seedlings; while Peace (1958*a*) noted individual variations among Scots pine in their reaction to fluorine from brickworks. There would appear to be opportunities for selection and breeding in this field.

Prevention of pollution injury

The real answer to injury by fumes lies, of course, in their extinction at the source. In two instances in Britain injunctions to restrain the emission of fumes damaging to trees have been obtained. The first of these affected an iron smelter at Glencorse (Lauder 1909; Newbigin 1909), the second an aluminium smelter near Fort William (Anon. 1948*d*). Once toxicants are present in the atmosphere little can be done to mitigate their effects. Choice of species is of great importance, and any action that can be taken to improve general growing conditions will increase the chances of recovery from injury.

Spraying plants with calcium oxide will prevent fluorine injury (Shaw *et al.* 1951), while fungicides of the dithiocarbamate, benzothiazole, and thiuram-sulphide groups will protect plants exposed to oxidized hydrocarbons from fuel oil (Kendrick, Middleton, and Darley 1954; Kendrick *et al.* 1956). These protectants become less effective as the concentration of the toxicant increases, and their effect naturally fades away with time. It can hardly be suggested that protective methods of this kind have any practical application.

Escapes of illuminating gas

Leaks of illuminating gas occur occasionally as a result of ageing of pipes, or their fracture by soil disturbance or subsidence. When this happens trees and other vegetation in the immediate vicinity are damaged. Trees are usually affected first because of their deeper roots. The damage is mainly due to phytotoxic substances in the gas, but may be partially the result of the replacement by the gas of the oxygen required by the roots. The chemicals concerned almost certainly vary with the kind of coal used for distillation and the exact nature of the process. Ethylene, carbon monoxide, and hydrogen cyanide are the three constituents of coal gas usually considered to be responsible for damage to plants; this part of the subject has been fully reviewed by M. D. Thomas (1951), and will not be discussed here, since the symptoms, as far as is known, are generally similar whatever the chemical concerned.

The visible symptoms, which result of course from damage to or the death of the roots, can vary from a slow decline of the tree to rapid wilting, withering, and dieback, according to the amount of gas escaping (Crocker 1931; Deuber 1936). If a large escape occurs during the winter affected trees may fail to come into leaf in the spring, or come into leaf very slowly and then wilt. The foliage on damaged trees is often abnormally small and pale green in colour. Injury in the crown is sometimes one-sided, since the leak is bound to be on one side of the tree. This is particularly noticeable if trees are planted along the edge of a road; the roots under the impervious road surface, where the gas accumulates, are killed, while those in open ground away from the road remain alive. This distribution of root injury is often reflected in the crown symptoms. Sometimes when root damage is very restricted, death of bark at the base of the tree on the affected side takes place before the crown symptoms become obvious. In lower concentrations, coal gas stimulates cork formation (Woffenden and Priestley 1924). In some cases this may produce visible proliferations of discoloured spongy tissue above ground. Symptoms of this nature have been described on plane by May, Walter, and Mook (1941).

Gas injury can be suspected particularly if trees of different kinds, known to be in the neighbourhood of a gas main, simultaneously develop symptoms of the type described above. Major leaks are comparatively easily confirmed by methods based on differential diffusion rates through a porous membrane of gas as compared with normal air. Smaller concentrations can be detected by testing air samples for carbon monoxide, very small quantities by testing for benzene residues in the soil. The last method gives results for a considerable time after the actual leak has ceased (Anon. 1945*a*). The gas does not usually travel far horizontally, unless it is underneath an impervious road surface. It may, however, occasionally enter electric conduits, disused drains, etc., and cause damage at a considerable distance from the actual leak.

Nothing is known about varietal susceptibility, but where mixed plantings are affected, differences in injury seem to depend on depth of rooting, larger and deeper rooting trees being more affected than smaller, shallow-rooted ones. If the gas is trapped under a road surface, trees with widespread root systems, parts of which are under the road, may be more affected than those with compact roots not so situated.

The first point in control is obviously the detection and repair of the leak. Severely damaged trees are unlikely to recover and should be removed. Forced aeration of the soil, followed by repeated drenchings with water, may be used to save less severely damaged trees of particular value, or to render the soil fit for replanting. Heavy watering is a useful method in any case. With street trees it is desirable to examine the condition of the major roots. If all the roots on one side were dead, it would be dangerous to leave a tree, even if its crown was still alive.

Escapes of natural gas

If natural gas is collected and piped, leaks can cause damage to trees, in the same way as escapes of manufactured coal gas, though it is generally considered that natural gas is appreciably less poisonous. Typical fume injury to foliage can often be observed in the neighbourhood of hot springs. For instance, *Pinus contorta* growing in the vicinity of the geyser basins in Yellowstone National Park in the United States shows browning of the tips of the needles, the damage fading out rapidly with distance from the sources of contaminated steam. The substance mainly concerned is presumably sulphur dioxide, though Vergnano (1953) reported boron injury to poplar and elm in the neighbourhood of boriferous steam vents in Italy. There is a possibility in this case, however, that the soil water was contaminated rather than the air.

Damage by irrigation water

There is very little evidence that irrigation water is likely to be toxic to trees in nurseries, parks, gardens, or other irrigated sites. Zimmermann and Berg (1934) found that the concentration of chlorine required to cause injury to plants was well above that normally found in chlorinated water supplies. It is possible, however, to raise the alkalinity of soil sufficiently adversely to affect the growth of lime-intolerant plants such as rhododendron or even conifers by continual use of irrigation water of high lime content. Unfortunately many water supplies in the south of England, where irrigation is most required, do originate from limestone or chalk rocks. The same effect may result from the use of alkaline chippings as a cover for coniferous seed-beds. That damage due to other elements in the water is possible, however, is illustrated by injuries due to excess boron in irrigation water used in Californian orchards (Eaton, McCallum, and Mayhugh 1941), and by the absorption of sodium chloride by the leaves of citrus bushes subjected to sprinkler irrigation (Harding, Miller, and Fireman 1956). Iron deficiency in *Araucaria*, resulting from the high salt concentration of irrigation water, has also been reported (Simmonds 1940).

Damage by weedkillers

During the last decade there has been an enormous increase in the use of weedkillers, particularly in agriculture, but also in forestry. This has brought in its train problems of damage to trees in and around the treated areas. The materials, the methods of application, and the mechanism and type of damage all vary with the purpose for which the weedkillers are being used. The nature of the toxic action of the various substances employed has been dealt with by Woodford (1957) and will not be discussed here. The legal aspect of accidental damage involved in the use of weedkillers is covered in the Weed Control Handbook (Anon. 1957*a*). The mechanism of damage, the type of injury caused, and the species affected are dealt with below according to the purpose of application.

Weedkillers used for the total eradication of vegetation

Simple weedkillers such as sodium chlorate or sodium arsenite are usually used either in powder form, or more commonly dissolved in considerable quantities of water, to clear all weeds from paths, fallow ground, etc. Drift is seldom serious and the substances do not vaporize, but damage may occur to trees or shrubs with roots near the surface in the treated area. Marginal or total withering of the foliage takes place, mainly as a result of the destruction of the roots. This may be followed by dieback or, as with salt-water injury, by a second flush of much smaller leaves. The damage is often restricted to that side of the tree on which the weedkiller application was made. If weedkillers are being put on above tree roots, the minimum quantity should be used. In some cases hormone weedkillers might prove to be a safer alternative.

Weedkillers used to kill individual trees

It is sometimes desirable, in connexion with tree-disease eradication or control, to poison individual diseased trees. This is often done by frill girdling and pouring a strong poison solution into the frill. Similar methods have been used to kill thinnings in hardwood and conifer crops while they were still standing. 'Seasoned on the stump' in this way they are much lighter to extract and the bark may be more easily removed. Simple substances such as copper sulphate and sodium arsenite have been used in the past, but more recently hormone weedkillers, such as 2,4,5-T, have been increasingly employed.

Rushmore (1956) found that suckers could be damaged by poisons applied to the parent tree, while Kuntz and Riker (1950*b*) observed the same effect through root grafts of oak. Cook and Welch (1957) found that sodium arsenite applied to individual trees of *Pinus strobus* may kill others round them. There is therefore a risk that damage may be done to surrounding trees of the same species as those poisoned, especially if they are suckers arising from a common root system. The concentrated method of application to the individual tree makes damage by any other means than root transmission unlikely. It has been noted, however, that *Rhizina inflata* (Chapter 16) does not develop on the sites of fires made from the brush of trees poisoned with sodium arsenite (Murray and Young 1959). This suggests that other biological effects may arise from the use of such potent long-lasting poisons.

Hormone weedkillers

Hormone weedkillers, most of which are selective in their action, are now widely used in agriculture, mainly because they kill many dicotyledonous weeds without seriously damaging cereals or grasses. They are also being used, particularly in the United States, to release conifers from competition with unwanted hardwoods (Hawkes 1953; Arend 1956; Atkins 1956). Agricultural usage carries dangers of drift both of droplets and vapour, and less frequently percolation through the soil to damage roots (Jacobs, H. L. 1950). Use in the forest usually involves some damage to the species which it is desired to favour, as well as more serious injury to those it is desired to destroy.

Injury normally takes the form of extreme distortion of those parts of the tree which are in active growth at the time of poisoning or which are formed shortly afterwards (Holly 1954). In cases of severe damage dieback also occurs. In some broadleaved trees the symptoms are suggestive of virus infection (Fig. 23). On pine the damage could be confused with that caused by the fungus *Melampsora pinitorqua* or the insect *Evetria buoliana* (Fig. 24), but apart from the absence of any living pathogen, the large number of shoots affected and the distribution of the damage in the crop would normally suggest the real cause of the injuries.

Examples of injury due to drift of droplets and vapour are becoming regrettably frequent (Hubert, McCubbin, and Wheeler 1952; Bartholomew 1955), and the symptoms, which are particularly striking on pine (Fig. 24), are becoming familiar. In slight cases recovery is usually complete the year following spraying, though the distorted shoots remain as a perpetual disfigurement. Andersson (1953–4), however, reported residual effects in the second year after poisoning, minor distortion being produced when the trees came again into active growth.

Information on the relative susceptibility of different tree species is still somewhat contradictory (Wurgler 1955; Bartholomew 1955; Bylterud 1956). The work on the selective eradication of broadleaved species from among conifers indicates that the latter are generally more resistant than the former. On the other hand, it is well known that under some conditions conifers can be damaged. Arend (1955) found that conifers could only safely be sprayed after the end of July. Wurgler (1955) found less damage to juniper growing in the shade than in the open. Obviously, since a considerable number of substances applied in different concentrations and under varying conditions at different times of year are involved, great care is needed to preserve from

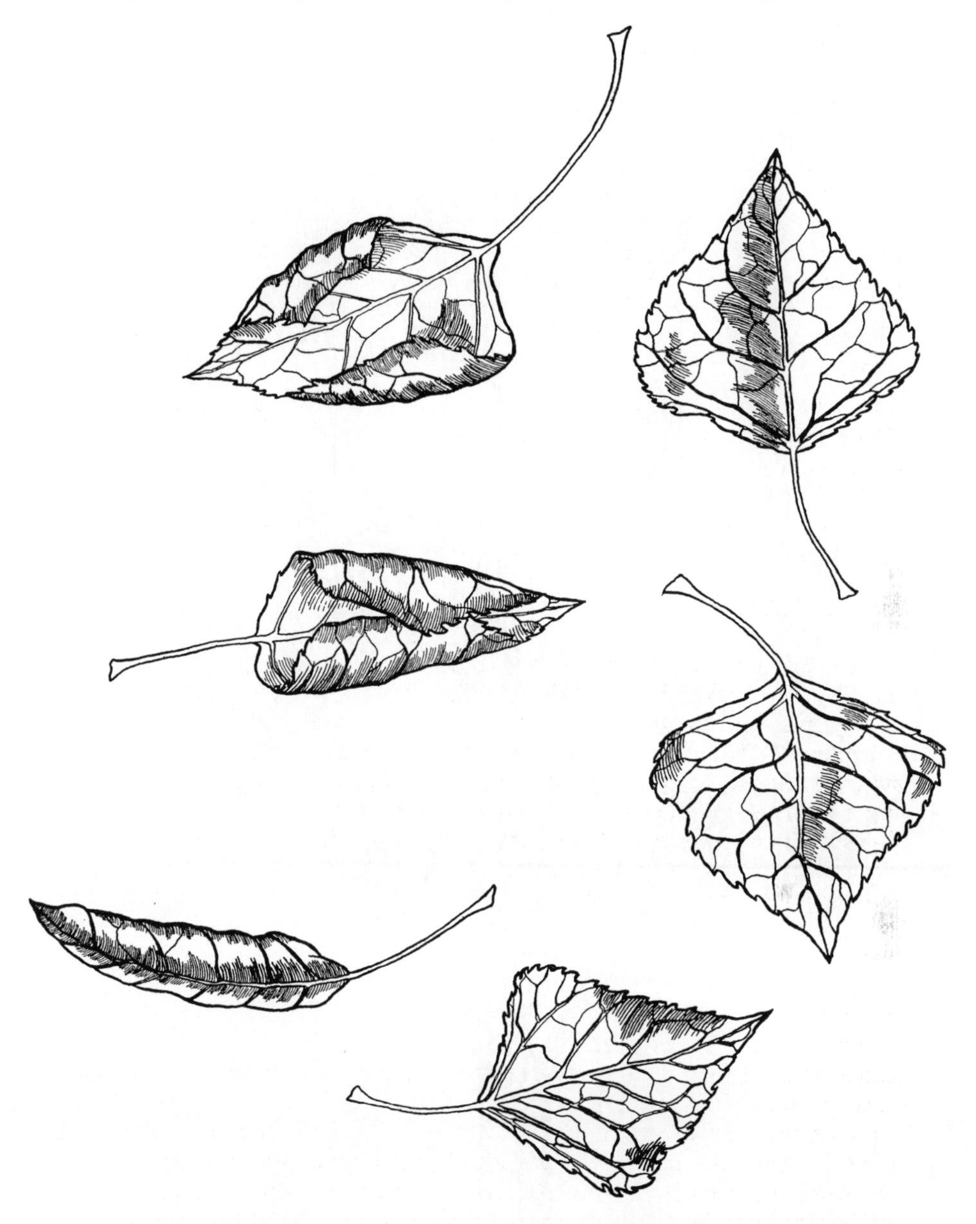

FIG. 23. Hormone weedkiller injury to *Populus 'robusta'*, Bentley, Hants, June ($\times \frac{1}{2}$). (J. N.)

injury those trees which it is desired to retain, whether the weedkiller is being applied in neighbouring fields or to kill unwanted species in the forest.

FIG. 24. Hormone weedkiller injury to Scots pine ($\times \frac{2}{3}$). The damage shows in the two-year-old shoots. (J. N.)

Little information is yet available on the distances over which drift is likely to take place. According to Hawkes (1953) aeroplanes flying 60 metres above the ground can give a drift of 0·8 kilometres down wind. From ground applications distances are likely to be much less than this, and much of the mist will be intercepted by the outer trees. Nevertheless, damage can extend 40–50 metres into a young plantation.

Weedkillers in the nursery

The use of oil weedkillers, both before and after seedling emergence, is now accepted practice in forest nurseries (Holmes, G. D. 1953). Both times of application, but particularly the latter, carry possibilities of damage to the seedling crop. Pre-emergence sprays applied too late may scorch the roots and cotyledons of the developing seedlings or even prevent germination. Post-emergence spraying carries a greater risk, since slight errors in strength or dosage, application in hot weather, or treatment of beds from which shelter has recently been removed, can all result in damage. This usually takes the form of scorching of the foliage, but one case is known where post-emergence spraying appeared in some unexplained way to have aggravated injury due primarily to the heating of the surface soil (Chapter 5).

Some information is available (Anon. 1957*a*, p. 56) on the relative susceptibility of tree seedlings to the light mineral oils normally used for post-emergence spraying. Many pines, including Scots and Corsican, the spruces, *Thuja*, and Lawson cypress are resistant; *Pinus contorta*, the larches, Douglas fir, *Tsuga*, and *Abies* are less resistant; while most hardwoods are so susceptible as to render selective spraying impossible.

Phytotoxicity of fungicides and insecticides

Many fungicides and some insecticides damage plants if they are applied in too high a concentration or under the wrong conditions. Fungicides containing copper, or polysulphides, and insecticidal oil emulsions are particularly liable to be phytotoxic. Fungicidal and insecticidal dusts can also cause damage under certain circumstances. Surprisingly little attention has been devoted to this subject in textbooks on fungicides and insecticides, though Voigt (1954) has recorded the effects of a number of fungicides on *Pinus radiata* seedlings. Naturally care must be taken that strong solutions of particularly phytotoxic substances, intended for use when deciduous trees are dormant, are applied only when the trees are really in that condition. Care must be taken, when such materials are used, that the spray does not drift on to evergreens or on to trees which are in a more advanced condition than those being sprayed. There may, for instance, be difficulties in growing spruce as a Christmas tree undercrop in orchard if winter washes are used on the fruit trees. For instance, dinitro-ortho-cresol is known to damage Norway spruce. Damage can generally be avoided by careful observation of recommended strengths or, with new substances, by experiments prior to general use. Instances of particular liability to spray injury are mentioned in other parts of this book where specific diseases are discussed, but generally the subject, like that of fungicidal sprays themselves, is too large and too complex for discussion here. It has been dealt with fully by Horsfall (1945), though he does refer mainly to agricultural and horticultural crops.

The condition of the plants and of the weather at the time of spraying greatly affects spray injury. Small and delicate plants, or those softened by undue shading, are more liable to injury than older plants, or those grown under more normal conditions. If shaded beds are sprayed, the shading material should always be replaced for a period of several days after the operation; likewise, several days should elapse between weeding and spraying of nursery beds which carried a heavy weed cover. Hot sun and low humidity, both of which tend to concentrate the spray on the leaves, are the conditions under which spray injury is most likely to occur. Dull weather should always be preferred for spraying, especially when potentially phytotoxic substances are being used. Cram and Vaartaja (1956) called attention to the possibility of injury to seeds as a result of surface sterilization, especially if this is done after stratification.

Damage by fungicides and insecticides is essentially similar to that caused by industrial fumes, though the distribution of the injury may be influenced to some extent by the method of application and the nature of the spray deposit.

Damage by excess of fertilizers

Most fertilizers damage plants if they are applied in excessive quantities. Damage happens much more readily if they come into contact with the foliage, and this is obviously a limitation on the foliar application of fertilizers, which has recently become an accepted method. In the deliberate application of fertilizers avoidance of damage is largely a question of care over dosage and placement.

Injury often occurs from unexpected concentrations, particularly of nitrogen, over or in the rooting zones of trees. This type of injury can occur if manure heaps are made near trees, or if a large number of fertilizer bags are opened, with inevitable spilling, over their roots. Escapes of sewerage can sometimes have the same effect; a case has been reported where a beech was damaged because some of its roots had penetrated into a cess-pit (Le Sueur 1931). There are probably differences in susceptibility, since poplars and willows normally grow extremely well immediately beside sewage beds. A case has been noted where the deliberate feeding of cattle under a group of maples resulted in death of the trees from excess of nitrogen in the manure deposited over their roots. Use by dogs is said to cause the death of many young trees in New York City, and may well do the same in other towns. The symptoms of this type of damage are normally scorching of the foliage and dieback of the twigs. Control lies in removal of the offending source and repeated heavy waterings to leach the excess out of the soil.

Metallic toxicants

Some metals can prove toxic to trees. T. M. Harris (1946) noted damage to wild herbaceous plants and particularly to lichens and mosses growing under galvanized wire-netting cages, and similar zinc poisoning has been found on tree seedlings caged for protection from birds. The zinc is presumably dissolved by small quantities of acids present in the rain. In Australia damage to the roots of *Callitris* and *Eucalyptus* seedlings grown in galvanized containers was noted; it could be nearly prevented by the addition of lime to the soil

(Worsnop 1955). It has been suggested by Pirone (1948) that rain-water, which has passed down copper pipes, may eventually damage trees if it is allowed continually to wet their roots. Gourlay (1951) observed damage due to the drip from copper telegraph wires, causing stunted growth and premature defoliation in *Prunus cerasifera pissardi*, but no appreciable injury to other species of *Prunus* growing under the same wires. The damage was restricted to those parts of the trees immediately beneath the wires.

Damage by other chemicals on the soil

There are various ways in which toxic chemicals, other than weedkillers and fertilizers in excess, can reach the soil and the roots of trees growing therein. In factories damage to trees due to leaky drums of chemical stacked on the root-area has been observed. Leakages of fuel oil sometimes fail to penetrate far into the soil, and may form an airtight seal preventing proper aeration of the roots; on the other hand, where sufficient quantities are spilt to cause deep penetration, they can cause rapid and apparently direct death of tree roots (Fuller and Leadbeater 1935).

Salt-treated grit for use on icy roads in the winter can damage trees if it is piled above their roots. Death of moderate-sized elms from this cause has been observed in Britain. Grit of this nature should always be placed well away from trees. Direct salting of icy roads has caused damage to trees growing along them (Wysong 1952). Even the careless emptying of brine from ice-cream freezers over the roots of trees has been known to damage them. In America calcium chloride, because of its hygroscopic properties, is sometimes used for laying dust on 'dirt' roads. This can result in damage to trees, especially if they are growing where the run-off rain-water soaks away (Strong 1944).

DAMAGE DUE TO NATURAL CAUSES

Damage by salt spray

In Britain oceanic influences have a large effect on tree growth. One influence, the action of salt spray on plants, is often very difficult to separate from the direct effect of the wind which bears the salt. However, deposition of salt sometimes produces more rapid and more striking damage to foliage than can be caused by wind alone, and with summer off-sea gales in particular it is possible to get fairly definite evidence on the extent, severity, and nature of salt injury. It is often possible to see salt spray being whipped by the wind from the tops of breaking waves, but these relatively large visible droplets do not travel far. Woodcock (1955) found that bursting air bubbles on the tops of waves produce aerosols (near stable suspensions of minute droplets in the air), and it is these which travel farthest and do most damage. Injury, attributed to salt rather than to direct effect of the wind, has been recorded as far as 80 kilometres from the sea by A. E. Moss (1940), who checked chemically the presence of salt on the damaged twigs, and by Edlin (1957), who was able to taste the salt on affected foliage. Edlin's observations, which were made partly at Bedgebury in Kent, are supported by the observations of Madgwick and Ovington (1959), who recorded abnormally high salt content in the rainfall at the time the damage occurred. At such distances it is usually only found at fairly high elevations on trees exposed to the off-sea winds.

Most of the injury occurs within 15 to 20 kilometres of the coast, and acute damage within a belt 5 to 6 kilometres wide. Topography has a large effect on injury. In the gale of July 1956, for instance, little damage was found on the north slope of the South Downs (Evison 1957) (Fig. 25).

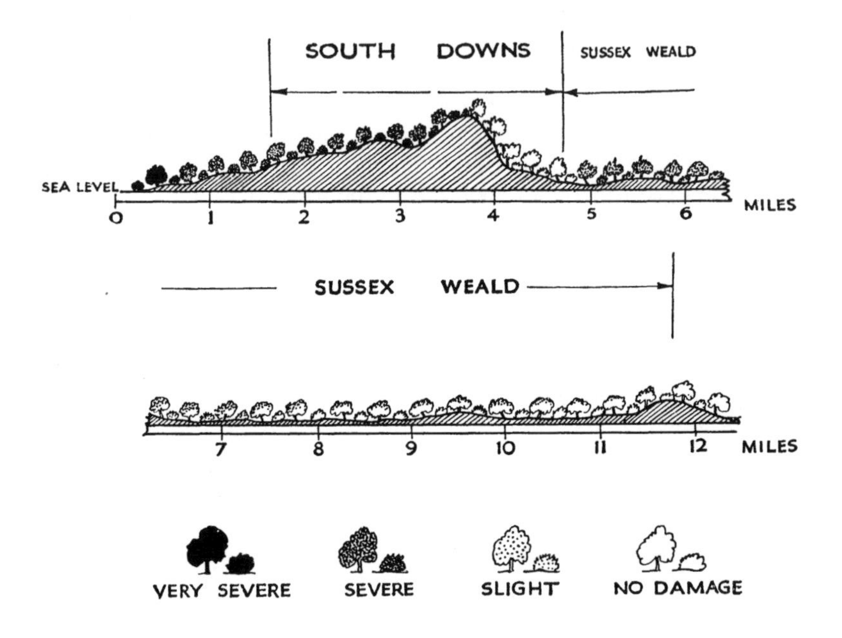

FIG. 25. Transect northwards from Brighton, showing the distribution of damage by salt spray. (A. C. after J. R. B. Evison.)

Symptoms of salt spray damage closely resemble those caused by industrial fumes. On broadleaved trees there is marginal and interveinal scorching of leaves not completely withered (Fig. 26). Frequently defoliation is followed by a fresh crop of leaves and even of flowers later in the summer (Edlin 1957; Evison 1957). With conifers the tips of the needles are browned.

There are wide differences in susceptibility between different trees. In general, evergreen are more resistant than deciduous trees. Kurauchi (1956) divided leaves into two classes: (*a*) those, mainly evergreens, where the salt is deposited on the surface but does not penetrate, and (*b*) those where the salt penetrates and damages the leaf. The first group also contains trees such as *Populus alba*, the hairy leaf surface of which probably lessens salt penetration. Relative amounts of injury are also affected by the times of occurrence of the salt-bearing winds. Describing a storm which occurred at the end of May, Edlin (1943) found that some trees had escaped damage because they were still in bud. It has frequently been suggested that the amount of damage

FIG. 26. Injury by salt spray to leaves of sycamore, oak, beech, hawthorn, and wild plum (×½); Tor Point, Cornwall, September. (J. C.)

is affected by the weather immediately following the salt-bearing wind. Though there is no definite evidence on this matter, it seems very probable that rain washes off the salt, thus lessening damage, and that hot sun concentrates it and increases the injury. Little, Mohr, and Spicer (1958) found that salt-spray damage to *Pinus taeda* followed storms with high wind and little rain, though this combination of circumstances inevitably suggests that excessive transpiration may have been responsible for part of the damage.

In the absence of any action by man, salt-bearing winds may exercise a limiting effect on tree growth and produce a local salt-controlled vegetation climax (Wells 1939). It has occasionally been suggested (Edlin 1957) that sodium chloride, as well as other elements included in 'sea-salt', may act beneficially, when they are not deposited in toxic quantities, either through direct foliar penetration, or through the soil after being washed off the foliage by rain. On the other hand, Woods (1955) reported death of *P. radiata* in low-lying areas of shallow soil in Southern Australia owing to gradual accumulation of wind-borne oceanic salt in the soil.

A number of authors have given lists of relative susceptibility based on observations made after salt-bearing storms (Wallace and Moss 1939; Wyman 1939; Edlin 1943, 1957; Evison 1957). These lists are not exactly the same, but there is general agreement that the highest resistance is shown by a number of shrubs, notably *Tamarix*, *Escallonia*, and *Olearia*, that among broadleaved trees *Quercus ilex* is outstandingly resistant, and that beech is particularly susceptible. Among the conifers *Cupressus macrocarpa*, Austrian pine, and Monterey pine have perhaps the best records. Wyman (1939) was particularly impressed by the resistance of the Japanese *Pinus thunbergii*, and Wallace and Moss (1939) by that of *Picea pungens*. Larch is often severely damaged. There are in existence also numerous lists of trees and shrubs recommended for seaside planting (Gaut 1907; Webster 1918; Bean 1950–1; Evison 1954); but the basis for these is, of course, somewhat different. Ability to withstand wind as well as salt spray is a vital factor in seaside survival, and many species which are quite severely damaged by salt may still find a place in lists of this kind because of their ability to make rapid recovery. Sycamore is often noticeably successful near the sea, but its leaves are easily salt-damaged. Sitka spruce, which is only moderately resistant to salt, is very successful near the sea in its native range in western North America, and on the more humid western side of the British Isles.

Some protection against salt deposition can be given by planting shelterbelts of resistant species (Iizuka, Tamate, Takakuwa, and Satô 1950). Such belts not only break the force of the wind, but act as filters, reducing the salt content of the air passing through them. In solid blocks of forest near the sea the effect of salt diminishes rapidly with distance from the sea. Great restraint must be exercised in the removal of salt damaged trees; even when dead or nearly so they will to some extent shelter and protect those behind them. Constant tidying up of the damaged seaward edge of a shelterbelt or plantation will often increase the exposure of, and consequently the probability of damage to, the interior trees.

Damage by flooding with salt water

Fortunately salt-water flooding does not often occur, and when it does can affect, in Britain, only a relatively small area largely devoted to agriculture. However, sea-banks do occasionally break, leading to salt-water flooding of land and trees normally subject to fresh-water conditions. In the North Sea gale of January 1953 this occurred on a large scale in Britain and much more catastrophically in Holland.

The basic causes of damage by salt-water flooding have been discussed by S. D. Richardson (1955), but the physiological and chemical basis of the damage does not appear to have been investigated. Undoubtedly some of the damage is due to the presence of water above the roots and the consequent restriction of air-supply, such as occurs in the case of fresh-water flooding (Chapter 6). That much of the damage is due to the concentration of salt is clearly indicated by the greater severity of injury in salt-water than in fresh-water floods, and by the deleterious effects of the salt remaining in the soil after the flood-waters have receded. Salt flooding, by replacing some of the calcium in the soil by sodium, damages the soil structure, particularly of alkaline clays (Bower 1954) and raises the pH to a dangerously high level. It is not known how important this is to the health of trees.

The symptoms of salt injury have mostly been described in relation to the floods of January 1953 (Linde and Meiden 1953; Brett 1954; Richardson, S. D. 1955). They refer, therefore, to trees which were dormant at the time of the flooding. Generally trees came into leaf, but the leaves rapidly withered from the margins inwards, and were succeeded by a second crop of leaves; in some cases a succession of crops of leaves, each smaller and more feeble than the one preceding it, was produced throughout the summer after the floods. In more susceptible species, or in areas where the salt concentration was high, the defoliation was accompanied by progressive dieback, which often resulted in the death of the whole tree. Some trees remained apparently healthy until the onset of hot weather in August, when their decline presumably followed increased transpiration. Flooding is most likely to occur during the winter, but if a summer inundation did happen the symptoms would probably be similar, except that they would start with the withering of the existing foliage and, since the highest salt concentration would occur when the trees were in active growth, the final damage might well be worse.

A number of lists of relative susceptibility have been drawn up (Wyman 1939; Buxton 1942; Linde and Meiden 1953; Brett 1954). These lists have been compared by S. D. Richardson (1955). There was close agreement between those of Buxton and Brett, both based on floods in Norfolk, but in different years. There were some variations between these English and the Dutch list, which might be explained on differences in soil or on the longer periods of actual flooding experienced in Holland. Oaks, White poplars (*Leuce* group), and some species of willow (especially those of the Sallow group) appear to be relatively resistant, while sycamore, beech, ash, and the Black poplars (*Aigeiros* group) are particularly susceptible. All the conifers observed in the Dutch and English lists were badly damaged by salt water, despite the fact that some of them are so well able to tolerate salt-laden winds. Wyman (1939), however, found that a few species of conifers were able to withstand twenty-four hours' flooding with salt-water under conditions where the soil had been previously saturated by heavy rain. The Dutch list is headed as might be expected by *Tamarix* and the less well-known shrub *Lycium halimifolium*, both typical coastal plants. In China *Tamarix* and *Ailanthus glandulosa* were both found by experiment to be especially resistant to salt (Tsing, Fang, and Wang 1956).

Rehabilitation of salt-flooded areas depends mainly on the leaching of the salt from the soil. On some heavier soils gypsum can usefully be applied to lower the high pH and to preserve the soil structure (Bower 1954). Any improvements in sub-soil drainage will hasten the removal of the salt and Werff (1955) has suggested irrigation with fresh water for the same purpose. The time necessary for recovery depends on the initial amount of salt, on the drainage, and on the rainfall, but several years are likely to elapse before salt-flooded soil is again fit for tree planting, except with the more salt-resistant species.

Toxic soils

Some soils are naturally toxic to plant growth. To some extent this is so with all highly calcareous soils; but since the mechanism in this case is the prevention of absorption of some minor elements by the excess of calcium in the soil, and the effect on the tree is typical deficiency chlorosis, such soils have been discussed under 'Soil Deficiencies' in Chapter 7. Serpentine soils, derived from the igneous rocks of that name, are sometimes toxic to plant growth. Němec (1951) associated stunting of pine on Serpentine soils in Czechoslovakia with excess of chromium, nickel, and cobalt. It would be rash, however, to assume that all unhealthy tree growth on these soils was due to soil toxicity, for such soils are also frequently deficient in some of the essential elements.

Boron, lead, manganese, magnesium, nickel, cobalt, copper, iron, and other elements are associated with soil toxicity damage to agricultural crops. Such toxicities are often very complex. They certainly occur in forests, but little information is available on their frequency or on the specific symptoms they produce on trees. They are certainly not widespread or important in Britain. Stone and Baird (1956) have investigated boron toxicity in pines. The symptoms are browning of the tips of the current year's needles. Where damage is chronic, growth becomes stunted.

True alkali or saline soils do not occur in Britain. In these soils salts accumulate in toxic quantities in the surface layers, owing to excess of evaporation over rainfall. Under these circumstances, of course, damage may be due to the effect of high salt concentration on the uptake of other substances, as well as to a direct toxic effect. In the hotter regions, where saline soils occur, conditions are often too dry for the growth of most trees and shrubs, even if the soils were not toxic. *Tamarix* is one of the most successful shrubs under these conditions (Krupenikov 1947, 1951). Stoeckeler (1946) has investigated the relative tolerance of various species mainly to sodium sulphate and sodium carbonate. He found *Elaeagnus angustifolia*, *Fraxinus pennsylvanica*, *Ulmus pumila*, *Caragana arborescens*, *Gleditsia triacanthos*, and *Morus alba* the most resistant species of a considerable number tested. Oganesjan (1953) also found *Elaeagnus* particularly tolerant to high salt concentrations.

Tree planting has often been adopted, and could be used more frequently, to utilize and beautify colliery spoil heaps. In a very few cases, however, these are toxic. This seems to be connected with high acidity, perhaps brought about by the oxidation of sulphur compounds.

Toxic effect of tree roots

It has often been suggested that the roots of walnut, and possibly of other trees such as *Robinia*, are toxic to those of other plants. Loehwing (1937) pointed out that most of these supposedly toxic effects are due to competition for water or nutrients, or to derangement by one species of the biological or nutritional balance of the soil, so that it becomes unsuitable for the growth of another. There seems little doubt, however, that walnut can be toxic to other plants growing in the same soil (Pirone 1938). This damage is probably due to a substance juglone, which also has fungicidal properties (Gries 1943). Massey (1925) was able to reproduce the effects found on tomato and potato plants around walnuts in the field by planting them in pots containing portions of walnut root. Schreiner (1950) reported damage to young pines within a radius of 13 metres of *Juglans nigra* trees some 10 metres high, and Schneiderhan (1927) to apples sometimes as much as 25 metres from the trunks of the walnuts. It is curious, however, that this toxic effect is not invariable. Under some conditions walnut trees can be grown without any more effect on the plants around them than would be expected from normal root competition.

Laing (1932) and Braathe (1950) have both produced evidence for the toxic effect of *Calluna* on young spruce. There is also evidence that grass roots, or organisms such as bacteria associated with them, have an inhibiting effect on the growth of young trees. This effect is not wholly one of competition for nutrients and moisture, though under natural conditions there are obvious difficulties in distinguishing it from such competition (personal communication from E. J. Schreiner, based on unpublished data; Rubin *et al.* 1952). Nevertheless, grass probably acts chiefly by competing for water and nitrogen (Howard, A. 1925; Joachim 1957*a*).

The possibility of minor reactions between the roots of different species of trees has also been suggested (Lyubich 1955). These may result in stimulation, or restriction by the production of toxins, according to the species involved, but probably have only a minor pathological significance.

9

THE AVOIDANCE AND MITIGATION OF DAMAGE BY NON-LIVING AGENCIES

THE cost of forest-tree protection must always be considered in relation to the rather low annual increase in value of the crop. Careful thought must be taken before any specific measures are introduced, and methods based on the adaptation of normal silvicultural techniques are particularly attractive. In considering control measures the following factors must be taken into account:

(*a*) The value of the crop. It is desirable here to consider future as well as present value. Trees in streets, parks, and gardens may, of course, have a high amenity value, which may justify the adoption of control practices which would be quite out of place in the forest.

(*b*) The degree of damage anticipated. Some causes of damage, for instance hail, may be quite frequent, without the aggregate causing enough damage to justify any expense on control or protection measures. The possibility and extent of recovery must be brought into the picture. The less obvious effects of injury, such as reduction in the growth rate, or enhanced liability to the entry of decay, must not be overlooked.

(*c*) The probability of damaging conditions occurring. For instance, measures designed to mitigate glazed-frost injury would not be justified in view of its rare occurrence in Britain.

(*d*) The total area likely to be affected. If this is relatively small, protection may not be justified.

(*e*) The effectiveness of the proposed measures both in limiting the occurrence of damaging conditions and in mitigating their effect.

(*f*) The possibility of the control measure having additional undesirable effects on the crop. For instance, mustard sown to protect Douglas fir regeneration from hot-soil injury damaged the conifer seedlings by competing for water and had to be destroyed by the use of a selective weedkiller (McKell and Finnis 1957).

(*g*) The cost of the proposed measures. If protection takes the form of adaptation of a normal silvicultural practice, or if the same measure serves more than one purpose, it may be permissible to charge only a proportion of the cost against protection. In some cases one measure may give protection against more than one damaging agency.

Within certain limits we can anticipate from the start the hazards of soil and climate to any particular plantation. Exposure, liability to frost, possibilities of flooding, and drainage conditions can all be evaluated by a proper examination of the site and a study of climatic records. There is thus generally no need to await the actual damage before taking protective measures. A preliminary decision as to what measures are justified can be made before the area is even planted, and indeed may well affect the choice of species. Protection against non-living agencies should therefore become a part of the forest working plan, and should be embodied in the basic design and layout of parks and gardens.

In considering protection, it is vital to realize that the degree of damage in many cases of injury by non-living agencies depends on the rate of change of the environment. Frost following warm weather does more damage than if it comes at the end of a cold spell; a sudden change in soil water conditions is more damaging than a slow one, and so on. Indeed, the severity of damage done by some of the agencies considered in Chapter 5 depends on the suddenness of their occurrence. Chronic damage, of course, also occurs, so that we cannot render environmental changes entirely harmless merely by making them slow in action. Nevertheless, measures designed primarily to slow up the rate of change arc very valuable. Apparently desirable remedial measures may have damaging results if they are so applied as to bring about large and rapid changes in the tree's environment.

It would be wrong to suggest that the vigorous tree is normally less affected by damaging non-living agencies than the weak one. In some cases this is certainly so, for instance vigorous, deep-rooted seedlings will better resist frost lift. In other cases, however, vigour may be a disadvantage; drought crack occurs frequently on large dominant spruce; excessive transpiration is particularly liable to damage large, heavy-canopied trees. There is, however, no doubt that the vigorous tree has an enormous advantage after the damage has occurred and when recovery starts. Measures designed primarily to increase vigour and growth rate, though they may have no protective intention at all, will usually be of value in lessening the ultimate effects of injury.

Details of measures suggested for the avoidance and mitigation of damage have already been described in the individual chapters. Here it is only necessary to allude to the most generally used methods and to point out some of their limitations. A general review of the subject has been given by Hawley and Stickel (1948).

Choice of site

In forestry there are strict limitations on site avoidance. It has already been pointed out that forests in Britain are inevitably relegated in the main to inferior sites. Areas too exposed, unplantably rocky, or undrainable must be rejected, but many plantations have to be established under conditions far from ideal. However, certain species, such as poplar, demand specially favourable conditions and should be limited to suitable sites. Trees in parks and gardens, and along roadsides, like forest trees, usually have to put up with the local conditions, whether they be good or bad.

The area required for nurseries is relatively small, and it should be possible to site most of them under more nearly ideal conditions. This is possible as far as we can define such conditions—there are still, of course, gaps in our knowledge—and so far as the various conditions are compatible. For instance, a frost-free site on a slope may be too steep for economic cultivation, or soil entirely free from any tendency to frost lift may be too dry in the summer.

Choice of species

Since information on the major factors of the environment should be available before planting, there is every opportunity to select the species to be planted on the basis of anticipated conditions. It may sometimes be necessary to modify pathological requirements in face of economic ones, and there is no objection to this, provided the disease loss on the apparently more valuable species does not outweigh the direct loss of using the less valuable one. On the other hand, use of a less economic or even uneconomic species may be justified in order to establish forest conditions, which may eventually allow more profitable planting. The use of Mountain pine in the afforestation of frost hollows is a case in point. It is sometimes suggested that marketing considerations make it desirable to plant as few species as possible in any one area. To do so would certainly mean serious neglect of pathological considerations, and involve greater risks of damage by non-living, as well as living, agencies.

Possibly the main difficulty in the selection of species lies in the complexity of the environment. There will obviously have to be compromise, since a tree resistant to one adverse influence may be susceptible to another. Viewing the wide range of British forest trees, it is easy to feel that we have at least one species suitable for every condition. This may be true, but we have not got species suitable for every combination of conditions that occurs.

This makes the discovery of provenances, and the selection and breeding of strains of increased resistance to specific adverse factors or combinations of factors, a very important aspect of protection. In many cases we already know of provenance or individual differences in resistance. In view of the very large number of factors involved, however, selection and breeding must obviously be limited to those factors of outstanding importance. It would certainly be valuable to select or breed trees of increased frost or wind resistance; it would be a waste of time to search for resistance to lightning or hail.

Silvicultural techniques

There are a great many silvicultural techniques which can be adapted to give protection against non-living agencies. Indeed, in some cases protection is afforded merely by doing a normal and necessary operation in the best possible manner. As an instance of this one can quote care in plant transport and in particular the use of polythene bags, which should overcome the risk of damage by water loss during that critical period in a tree's life. In some cases the damage itself may be directly caused by avoidable carelessness. Injury by weedkiller spilling or drift and excessive application of fertilizers both fall in this category. The possibilities of adaptation are far too numerous to list, but as instances there are early or late transplanting, and the avoidance of autumn seed-bed weeding to prevent frost lift, delayed removal of an overstory, to give frost protection, and so on.

It has already been pointed out that disease in some of its aspects is merely an extreme case of poor growth. This being so, silvicultural treatments designed to improve growth should normally lessen the chances of disease. Irrigation, which may be applied primarily to increase germination and growth, may also prevent drought losses. Mulches, which are most likely to be put on as a growth-improving measure, will also lessen damage by frost lift, hot soil, and drought, and indeed may sometimes be applied specifically for protection against one or more of these agencies. Correct fertilization, applied primarily to increase growth, should also correct deficiency symptoms, though its effect on the damage by some other agencies, for instance drought, is more problematical.

Some of the silvicultural techniques which are most important from the protection point of view are considered separately below.

Drainage

Drainage can have a profound effect not only on the ability of ground to carry a commercial tree crop at all, but on the growth, health, and development of that crop. Drainage provides, however, an outstanding example of the need for care in altering the existing conditions around established trees. As far as possible major alterations in drainage should be made between rotations, rather than during them. If they have to be carried out while the trees are present, they will be least damaging when the trees are comparatively young and adaptable.

Thinning

Thinning influences the effects of various adverse agencies, though its influence is not always consistent. Wind damage is greatly influenced by thinning, as is injury by snow and by ice deposition. Thinning should lessen drought injury, since it lowers the root demand on the soil moisture, but in fact it may increase atmospheric drought injury, a condition where the transpiration from the foliage exceeds the absorbing capacity of the roots, or the transmitting capacity of the stem. This happens because the crown becomes more exposed to drying winds as a result of the operation. Pathologically it is quite certain that sudden drastic thinnings are undesirable. If wide spacing is required for silvicultural, pathological, or economic reasons it should be achieved gradually, or established at the time of planting.

Shelter in the forest

Shelter in the forest is normally provided by the retention of part of a previous crop. It can provide a measure of protection against wind, frost, sunscorch, and even salt spray. In practice, however, it is not always easy to achieve a proper balance between the amount of shelter required for protection and the degree of light or exposure required for the proper growth of the new crop. It is particularly difficult to maintain overhead shelter heavy enough to protect against radiation frost, without cutting off too much light. However, manipulation of part of the existing crop as shelter from exposure can greatly aid reafforestation and increase the range of possible species, as compared with the original afforestation of the same ground.

In some cases it may be desirable to establish part of the initial planting in advance to provide shelter for the remainder. This influences not only exposure to wind, but also the flow of cold air during radiation frost. Advance afforestation of the slopes of a valley will impede the downward flow of cold air and make the initially frosty bottom warmer and therefore easier to plant. On the other hand, cold air can pool up behind standing trees, so that on some sites removal of shelter may be required to render an area sufficiently frost-free for planting.

Shelter does not affect only wind and cold-air movements, it may protect trees from the sun, or occasionally shade-bearing species from excessive light. Some of the Japanese maples, for instance, appear to need partial shade in the sunnier parts of Britain.

Shelter in the nursery

The desirability of shelter in the nursery is a debatable point. The shelter itself, and particularly the semi-permanent supports usually used to carry it, gravely hamper other nursery operations such as weeding, or weedkiller spraying, especially if they are being done mechanically. The influences, to mitigate which shelter is applied, are of erratic occurrence, so that for several seasons the application of shelter may be completely unnecessary. Local experience may show clearly that a certain cause of damage, say frost, frost lift, or heat, occurs with sufficient frequency to justify shelter at any rate on certain susceptible species. But in most cases a long series of careful records of occurrences and damages would be required as the basis for a decision, and these seldom exist. Experiments would not give reliable results unless they were carried out over a long period, and even then losses in the damaging years would have to be set against the cost of the shelter and the inconvenience it causes. It seems likely that the use of nursery shelter will remain a matter of local practice and prejudice for a long time to come.

Protection against constructional damage

Construction work on roads, buildings, etc., is one of the commonest causes of damage to trees, not only directly, but even more frequently indirectly, by altering the soil conditions or the exposure. In such cases the urgency of the work seldom allows alterations in the environment of trees to be made gradually. There are some instances, such as the use of rock fills to allow the access of air to the covered roots of trees, where changes can be made less suddenly. In all such work, where existing trees are involved, every effort should be made to keep environmental alterations to a minimum. When new plantings are envisaged, the planning should include provision for any roads, building sites, drains, etc., which are likely to be required during the rotation. These can then be constructed, when they are required, without disturbance of the tree crop. Strip cutting for the construction of roads, or even for the widening of existing roads in an established crop, is pathologically a very dangerous practice.

10

THE INFLUENCE OF MAN, ANIMALS, AND INSECTS ON TREE DISEASES

THE INFLUENCE OF MAN

SINCE there is practically no natural forest in Britain, a great deal of this book is inevitably concerned with the diseases of trees deliberately planted and maintained for profit or pleasure. That being so, the influence of man is a factor affecting, by his choice of site and species, by his silvicultural treatment, and by his decision to fell, all the diseases being considered. This chapter, however, will deal only with certain ways, not covered elsewhere in this volume, whereby man, usually unintentionally, though often carelessly, injures trees.

Direct damage by man

Damage by deliberate wounding

Some damage, either in the form of small branches broken by children or the carving of initials on smooth-barked trees, is perhaps inevitable. The broken branches are usually too small, and the wounds made by carving too shallow to permit the entry of heart-rotting fungi. Except in cases of repeated injury, the damage affects the appearance of the tree rather than its general health.

Breakage of tree shoots sometimes follows injury from stray shot. Such damage is often difficult to diagnose, because the pellets usually pass through the shoots.

Decay fungi can enter through blaze markings on trees, especially if they extend into the cambium (Weir 1920; Ekbom 1928), and even timber-scribe marks may result in fungal infection (Oppliger 1932; Silvén 1944). It is important therefore that blazes on trees which are to be left standing should be confined to the outer bark, and that thinnings should not be blaze-marked, if the marking is to be subject to later correction. The use of spiked climbing irons is also potentially damaging to trees.

Increment borings are also occasionally a source of decay and frequently of stain in trees (Lorenz 1944; Hepting, Roth, and Sleeth 1949). Borings should always be sealed with wooden plugs or small corks, and, if possible, the instrument sterilized before use in a fresh tree. In Horse chestnut (*Aesculus*) bacterial infection and slime-flux so frequently follow increment borings that the use of the instrument is best avoided.

Wade (1953) described damage to thin-barked trees, such as beech, by blows from an axe head made in the course of sounding for possible internal decay. Weddell (1942) found that cankers caused by *Hypoxylon rubiginosum* (Pers.) Fr. (Ascomycetes, Sphaeriales) developed on Catalpas as a result of fishermen hitting the trunks with clubs to dislodge insects for bait. These cases are supported by the frequency with which *Phomopsis pseudotsugae* invaded experimental hammer wounds on Japanese larch (Chapter 24).

Damage during felling and extraction

The wounds caused by felling and by moving logs and vehicles in the course of timber extraction are much more serious than those discussed in the previous section. The stripping of bark and cambium caused by a felled tree sliding down a standing one will probably heal over fairly rapidly, but may remain exposed long enough to become infected by wood-destroying fungi, and in any case will give rise to a defect in the wood. Broken branches may also create a definite decay hazard. The injuries caused by the careless towing or 'tushing' of logs can be much more serious. Along extraction routes the wounding tends to be repetitive, producing much deeper injuries than would result from a single blow. A wound at the base of the stem is very well placed for fungal infection, and such infection can be very serious (Wright and Isaac 1956; Příhoda 1957*a*). An open wound can be infected by the many fungi which normally enter by wounds or broken branches. Being in contact with the litter layer, it may also be infected by fungi which normally enter through roots. In some cases, if the injury where the root joins the stem is severe, the whole root may die, making infection even easier. Worse still, the fungus is in a position immediately to decay the most valuable part of the tree, the butt length. It is essential, therefore, that extraction should be carried out so as to keep such wounding to a minimum. If a route is to be used regularly for extraction it should be marked out carefully and a trial log brought down at a slow pace. This should disclose which trees are likely to be hit. These should either be protected by driving heavy stakes on the danger side, or, if they can be spared from the crop, removed. It may be even better practice in some instances to leave such trees standing to protect others behind them, and to remove them when extraction is completed. Deitschman and Herrick (1957), in the course of a thorough study of the influence of different machinery and extraction practices on damage, found that tree-length logging using a wheeled arch caused appreciably less damage than dragging out log-lengths along the ground.

Treating extraction wounds with preservatives is not generally worth while. The wood structure is sometimes shattered and there is then no smooth wood surface on which to establish a protective covering. Preservatives with a penetrating power, such as creosote, may kill the cambium, and would certainly delay healing. Where, however, a valuable tree has received only superficial damage, use of a suitable wound protectant (Chapter 37) may be fully justified.

Olson (1952) pointed out that power extraction is leading to increasing damage to roots, as well as to the butts of standing trees. With more powerful extraction machinery it is possible to drag out roughly trimmed logs, or bundles of poles so large that the butts of the lower ones cut deep furrows in the soil. In areas of superficial rooting, which of course are very common, considerable injury may occur due to the tearing of roots by snags or the butts of logs moving like ploughs through the ground. The effect may be both direct by reducing the spread of feeding roots and, possibly more important, indirect by allowing fungal infection of the weakened trees. In the case of *Fomes annosus*, damage to roots appears to be a more important means of entry than stem wounding. The use of heavy machinery can also alter the physical properties of the soil, in particular reducing its permeability (Steinbrenner and Gessel 1955), which is very likely to produce indirect damage to tree growth (Chapter 6).

Very occasionally mechanical wounds may assume the appearance of target cankers (Plate VI). This happens where the cause of damage is periodic, as when a road is used for extraction for a short period at one particular season of the year. During the remainder of the growing season the wound will start to heal, only to be reopened when traffic is once more resumed.

In Britain we are not often faced with the problem of damage to natural regeneration by the felling and removal of timber, but where regeneration exists it is likely to be injured during the extraction of the seed trees. Such damage can be minimized by proper planning of felling and extraction (Wales 1931), in just the same way as can injury to the stems and roots of older trees.

Damage by pruning, pollarding, and coppicing

The question of pruning and dressings for pruning wounds is dealt with in Chapter 37. Coppicing and pollarding, where the tree is cut back either at ground-level or higher with the object of producing periodic crops of poles for special purposes, inevitably leave a large number of wounds, many of which by their position are slow to heal. Almost invariably pollard or coppice stools eventually become infected with decay fungi. Methods by which this form of attack can be delayed and whereby infection of the new crop of poles can be avoided are also discussed in Chapter 37.

Damage due to road-construction work

Construction work, particularly road-making, is a frequent source of injury to trees. Apart from mechanical damage by road machinery and by rocks hurled in blasting, injuries to the roots, caused by their severance or by their being covered with soil, occur very commonly. The question of over-burden on roots is largely one of aeration and water-supply, and is dealt with in Chapter 6. The severance of roots may act by depriving the tree of part of its water-supply, by destroying its stability or by providing a means of entry for fungi. The first two factors are probably of most importance. In the case of valuable trees these effects can be to some extent mitigated if the construction work can be anticipated and the root severance done in two or three operations, spaced over a year or even longer. This should enable the tree to produce recovery roots on the mutilated part of the root system before another section is cut back. Stone banking to hold earth around the severed roots and to prevent erosion will improve the chances of recovery and re-anchorage (Fig. 101). Severance should naturally be made as far as possible from the bole, and certainly never nearer than 3 to 4 feet, except on very small trees. If more than one-third of the root system has to be removed, crown reduction is desirable, both to restore the balance between the roots and the foliage and to improve stability. These methods are discussed more fully by Fowler, Gravatt, and Thompson (1945) and Le Sueur (1949).

Obviously such careful work can be undertaken only for trees of special value. Thus in the forest it can be done only in a very small percentage of cases. Apart from these, trees should be felled if more than one-third of the root would have to be removed. The maximum distance from the trunk to the point of severance should be allowed in those left standing. The damage to the roots of trees entailed in forest road construction makes an early decision on road plans essential, so that routes can be marked out and left unplanted.

Damage caused by bad planting

It has frequently been suggested that bad planting, in particular pushing the roots into a shallow hole so that they are bent up again into a 'walking stick' shape, leads to root disease. There is some evidence that occasionally *Armillaria mellea* attacks particularly trees so handicapped, but whether this is the direct effect of the bad planting or whether it follows injury due to wind rocking (Chapter 5) consequent on the unstable root-hold is not clear.

A considerable amount of work has been done on the effect of different methods of planting on survival and early growth, but naturally few of these experiments have included deliberate bad planting as one of the treatments. Some rather inadequate work carried out by Cheyney (1927) indicated that under nursery conditions, and over a period of only four years, plants with the roots bundled into a ball grew quite as well as those with the roots spread out. In an unpublished Forestry Commission experiment, conifers planted with the roots bent at right angles or in a 'U' shape grew quite as well for six years as properly planted ones. There was evidence, however, that the 'U' planting produced a shallower and less evenly distributed root system. Later Schantz-Hansen (1945) used bad planting, by forcing the roots into an inadequate slit, as one treatment in an experiment comparing methods of planting of a number of coniferous species. Over a period of two years it gave slightly lower survival than more orthodox methods, but had little effect on growth. Neither of these experiments continued long enough to disclose the pathological possibilities of bad planting. There is in fact little evidence that bad planting does have any major effect on the subsequent health of the tree, apart from immediate post-planting losses or wind-rock at a slightly later stage.

Most silvicultural textbooks have advocated careful planting without producing any concrete evidence in its favour. The observed association of bad planting with subsequent disease is sometimes produced as evidence, but is of little value since it cannot be compared with the proportion of cases of similar disease among well planted trees, or the proportion of badly planted ones which do not become diseased. Nevertheless, it would obviously be wrong to advocate bad planting, and it is probably right to suggest that plants should be put in as carefully as can be done without unduly raising the cost of the operation.

It is probable that failure to spread the roots at the time of planting may result occasionally in the partial girdling of the tree by one of its own misplaced roots (Fig. 27). This is a phenomenon occasionally met with in ornamental trees; it is unlikely to be noticed in the forest, unless the afflicted tree blows over as a result of the strangulation (Watson, H. 1937). It can usually be remedied by the removal of the offending root.

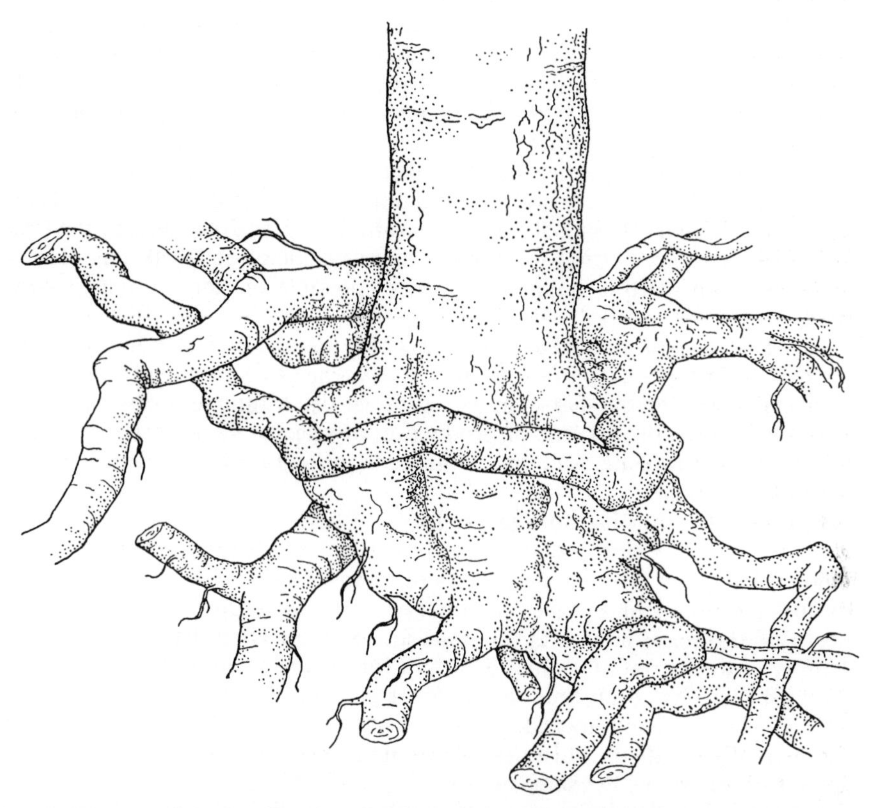

FIG. 27. A young beech with the soil removed to show a root which may later girdle it ($\times \frac{3}{8}$). (J. N.)

The effect of girdling on tree diseases

Girdling has occasionally been practised on a large scale as a cheap means of killing the stems of unwanted scrub to allow the establishment of a productive crop. The work of R. Leach (1937, 1939) on *Armillaria mellea*, which is discussed more fully in Chapter 16, suggests that girdled trees may be less dangerous, as sources of root infection to the new crop, than their stumps would have been if they had been felled. Further experiments are required before it can be assumed that this is always the case.

Partial girdling or strangulation is one of the methods used by the forest geneticist to increase seed production. Inevitably it weakens the growth of the tree, but so far there is no evidence that it leads to disease. Trees treated in this way are normally grown on selected sites and at wide spacing. On poorer sites and under conditions of heavy competition the effects might be very different.

Grafting as a cause of disease in trees

Grafting is one of the most convenient methods of propagating certain trees and shrubs. It is particularly valuable for clonal stocks, nearly always providing large plants more quickly than they can be raised from cuttings. Where cuttings do not strike, grafting may be the only method of raising clonal plants. Certain difficulties, however, are inherent in the method. Incompatabilities, which may take various forms, and which are sometimes very slow in developing, may arise between scion and stock. Except where cultivation is very intensive, the mere fact that all shoots arising below the graft have to be removed is a very big drawback. Where grafting has been done on a freely-suckering stock, this can make the maintenance of the tree almost impossible. During the present century, therefore, there has been a recession from grafting in favour wherever possible of plants on their own roots.

However, as a result of the interest now taken in the breeding and selection of trees, grafting has again come into prominence as a means of raising the plants required for tree-banks and seed-orchards. Although it can hardly be expected to provide a means of production of selected clones for forest purposes, grafting is being used on a large scale in the raising of clonal trees for roadside planting.

Incompatibility in its more severe forms usually declares itself at a fairly early stage, but it does not necessarily prevent the formation of an initial union. In poplars, for instance, grafts of almost any species on another may be initially successful, but incompatibilities often show up by the wilting and death of the scion late in the first season or even after the resumption of growth in the second. Gradual deterioration of the scion, quite a long time after grafting, has been reported in other genera. This happens particularly with *Syringa* (Lilac) grafted on *Ligustrum* (Privet) (Chester 1930, 1931; Cadman, C. H. 1940), but does not occur if *Syringa* is grafted on another species or variety of the same genus.

Occasionally, despite apparently normal growth for many years, wind breakage of grafted trees occurs at the graft union, disclosing a curiously discontinuous arrangement of the wood of the stock and scion. Such breaks are not infrequent in fruit trees, but are sufficiently rare in other trees to be somewhat puzzling when they are encountered. The tissue of the separated faces at the break shows a radial fan-like arrangement, and gives the appearance of complete lack of vertical continuity between the two sides of the break (Plate VII. 1).

Indirect damage by man

Tree ties and guards as a cause of injury

In the forest where trees of a very small size are normally planted, staking and tying are emergency measures adopted only following windblow, if it occurs a few years after planting when the trees are 4 to 15 feet in height. But in positions such as roadsides, where larger trees are planted, support may be necessary if the trees are to remain upright until they are established. Unfortunately both the stake and the tie are liable to cause injury when the tree moves in the wind (Fig. 28). If the tree is bound sufficiently tightly to the stake to prevent movement, it is likely to be strangled. The ideal material and method of tying have yet to be evolved. A soft tie will generally do less

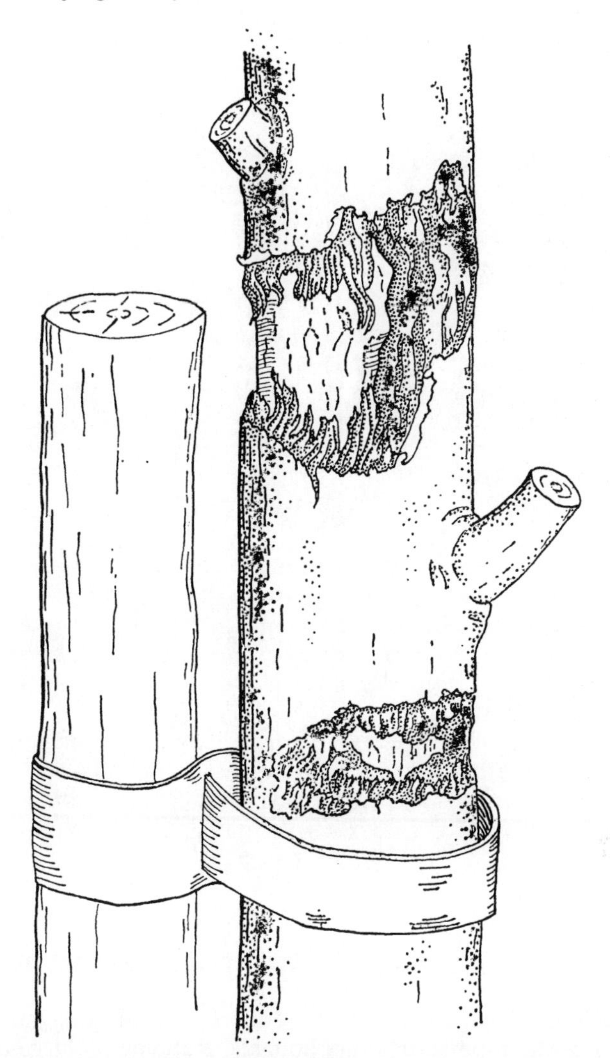

FIG. 28. Damage to a young poplar caused by chafing against the stake to which it was tied ($\times \frac{1}{3}$). (J. N.)

damage than a hard one, a broad tie less damage than a narrow. Chamfering the top of the stake, so as to remove projecting corners, is some help and the stake should always be put on the windward side of the tree, so that in the prevailing wind the tree will pull away from the stake instead of rubbing against it. For conifers, or for broadleaved trees if suitable twigs are available, the best method is really the tethering of the tree by its own pliable branches.

Two branches arising at the right level, and initially pointing in the general direction of the stake, are pulled towards it, crossed, passed round the stake, crossed again and tied at the second point of crossing (Fig. 29). This gives a springy tie, which naturally cannot rub the tree, and which keeps the tree

FIG. 29. Method of tying a tree by its own branches ($\times \frac{1}{12}$) (see text). (J. N.)

from contact with the stake. Though it appears inelegant, it is extremely effective.

Considerable damage can also be done by guards designed to prevent damage by cattle, rabbits, or other animals. Rubbing on large cattle guards is easily prevented by tying the tree loosely to the guard in three directions, at an angle of roughly 120°. Injury from netting sleeves used for protection against rabbits is more difficult to prevent. Hay or straw packed in the top will sometimes stop the netting from injuring the stem. Where individual tree protection against rabbits is necessary the use of cardboard or waterproof paper sleeves should be considered. Chemical repellents, which also act against other animals, are useful, but require periodic renewal.

If any form of metallic guard is used, and this applies to wire netting sleeves as well as larger constructions, it must be removed before it starts to interfere with the growth of the roots. Most guards will tend to be slightly buried in the soil, and it should be borne in mind that the diameter of the tree just below ground is considerably greater than that above. It is necessary therefore to remove guards long before they begin to restrict the tree above ground-level. Wire-netting guards tend to get interwoven below the soil with the roots and runners of herbaceous plants and grass, so that there is a temptation to cut the wire at soil-level, leaving the buried section in place. If this is done the detached portion will become embedded in the base of the tree, causing lines of weakness along which the tree may break in high winds.

Fences as a cause of injury

Fences should be kept well away from trees. Young trees may rub against any form of fence, and with older trees there is always a temptation to fasten the fence to the tree. Nails or staples in a tree quickly become embedded, and may later, if the tree is felled and utilized, cause damage to saws. Even substantial iron railings can be partially overgrown by a tree if they remain in close contact with it. Wire fences stapled to a tree greatly increase the danger to livestock if the tree is struck by lightning, for the current will travel from the tree along the wire, and may injure animals in contact with it.

It has recently been suggested (Jørgensen, E. 1955) that fence posts made from wood infected with the decay-causing fungus *Fomes annosus* (Chapter 16) may act as centres of spread to neighbouring trees.

Damage by removal of litter

In some countries it is common practice to remove the surface litter from forests in order to use it for mulching and fertilizing agricultural ground, for bedding for domestic animals, or for other purposes (Krauss 1956). This may result in reductions of as much as 30 per cent., but often less, in the rate of growth of the trees on the robbed sites, though not in actual disease (Wiedeman 1935; Jemison 1943). Němec (1956) found appreciable lowering of the nutrient content of the soil of Scots pine stands from which litter had been removed. There is thus a possibility that nutrient deficiencies might arise from this practice. Though the removal of litter would not have a significant effect on all sites, the possibility of damage should be borne in mind when collecting litter for Dunnemann beds, or for mulching.

The effect of thinning on tree disease

Occasional mechanical injury by a falling tree hitting a standing one has been mentioned above. But thinning is known to have a far more important indirect effect, in that it leaves stumps which can be used as centres of infection by root fungi such as *Fomes annosus*. This occurrence, its significance, and the measures that can be taken are discussed in Chapter 16.

Thinning can, of course, affect disease in other ways. The remaining trees have greater space and should therefore grow more vigorously. In the case of fungal diseases badly diseased trees are likely to be removed so that the amount of infective material present is decreased. On the other hand, the remaining trees are more exposed and this may affect their water relationships (Chapter 6).

The transmission of diseases on tools

Diseases may be transmitted by the use of contaminated tools. This is particularly likely in the pruning of trees suffering from wilts, where infected sap may exude on to the tools from the cut branch ends. It was found to be the most important method of transmission for Canker stain of plane (*Endoconidiophora* sp.) (Walter, J. M. 1946) (Chapter 35). It has also been suggested as a possible means of transmission for Oak wilt (*Ceratocystis fagacearum*). It has not, however, been proved to occur in practice, though infection is possible with artificially contaminated tools (Jones and Bretz 1955).

Certainly pruning tools used in connexion with disease control, or on trees known to be diseased, should be sterilized (Chapter 37) before they are used on healthy ones.

THE INFLUENCE OF ANIMALS

It is no part of the purpose of this book to discuss in detail the damage done by animals and birds to trees, or the methods of preventing it. However, many animals feed on trees, leaving wounds which may act as means of entry for damaging fungi, and in some cases they have been thought to be responsible for disease transmission.

Browsing, particularly by sheep or rabbits, can be an additional cause of trouble on frosty sites. The effect of the animals and of the frost are curiously similar. Both remove the tips of the shoots, and produce bushy restricted growth and slow height increase. Without browsing, two or three years free of spring frost may enable the trees to make enough height growth to get their leaders above the normal frost-level, and thereafter to grow away even in frosty years (Chapter 4). Without frost, a few years of freedom from sheep or rabbit trespass may enable the trees to achieve a height where the animals cannot reach the leaders, for many trees have remarkable powers of recovery (Pearson 1931). But frost and browsing together, one affecting the tree when it is free of the other, can end in complete crop failure.

Various animals, ranging from voles to domestic stock, feed on the bark of trees, leaving injuries which are readily colonized by wood-rotting fungi, especially if they are at the base of the tree. The wounds are usually much less irregular and the wood surface much less broken than in wounds caused during extraction, so that on valuable trees at any rate it may be feasible to apply a preservative covering (Chapter 37). Damage higher up the trunk, especially on timber for special purposes, such as Cricket bat willow, or on trees destined for veneering, is particularly serious because despite subsequent healing it remains as a defect in the wood.

Voles occasionally feed on the bark of branches of young beech, and it is possible that they provide means of entry for *Nectria* and other pathogens. *Cryptostroma corticale* has been found growing saprophytically on sycamore girdled by grey squirrels, but there was no evidence that this initial infection allowed the fungus to extend to the uninjured parts of the tree (Chapter 30). Galloy (1925) found that decay entered pines through wounds made by deer.

Voles have also been thought to assist in the spread of *Fomes annosus*. Fruit bodies of the fungus, which are frequently produced on roots in cavities in the litter or upper soil layers, sometimes occur thus in vole runs. It is possible that the spores adhering to the fur of the voles may be carried to roots of other trees, especially those gnawed by the voles themselves. But there is no proof that this does occur, and it is certainly not an important means of transmission (Chapter 16).

A point of interest is that voles find the soft spongy bark of the cankerous swellings caused by White pine blister rust (*Cronartium ribicola*) on Weymouth and other Five-needled pines particularly edible, so that the presence of the disease is sometimes first detected by the extensive vole gnawing of the diseased limbs (Chapter 22). It has been suggested by Mielke (1935) that this may appreciably reduce the production of aeciospores.

Squirrels and voles are a possible means of transmission of Oak wilt (*Ceratocystis fagacearum*) (Himelick, Schein, and Curl 1953; Himelick and Curl 1955). But generally the role of animals in the transmission of tree diseases appears to be a minor one.

Woodpeckers are probably the only birds that actually wound the bark of trees. However, their activities are mostly restricted to trees already decayed and insect infested, so that woodpecker wounds are of little importance as a means of entry for fungi.

Birds may be carriers of a number of tree diseases. For instance, they may take a major part in the long-distance spread of Chestnut blight (*Endothia parasitica*) (Heald and Studhalter 1914) (Chapter 29). Tits have been suggested as a means of spread of Watermark disease of Cricket bat willow (Burtt Davy and Day 1957) (Chapter 33). Birds may also be vectors of Oak wilt (*Ceratocystis fagacearum*) (Tiffany, Gilman, and Murphy 1955). In spite of such cases, however, birds are much less important than insects in the transmission of tree diseases.

THE INFLUENCE OF INSECTS

The transmission of tree diseases by insects

Insects are known to play an important part in the transmission of plant diseases, particularly those due to viruses and bacteria (Leach, J. G. 1940; May and Baker 1952). Owing to difficulties of observation, and to the fact that many tree diseases have not yet been intensively studied, exact information on the transmission of tree-disease pathogens is still very scarce, and a great deal of surmise still exists.

Perhaps the best-known example is the transmission of *Ceratocystis ulmi*, the cause of Elm disease, by *Scolytus* beetles (Chapter 31). More recently Nitidulid beetles have been associated with the spread of *C. fagacearum*, the cause of Oak wilt (Hepting 1955*a*) (Chapter 27). Insects, in America at any rate, are important vectors of Chestnut blight (*Endothia parasitica*)

(Studhalter and Ruggles 1915) (Chapter 29). Gipsy moth larvae have been suggested as agents in the dissemination of White pine blister rust (Gravatt and Posey 1918) (Chapter 22). A leaf-hopper, *Graphocephala coccinea* Forst., appears to be the chief means of dissemination of the Rhododendron bud blast (*Pycnostysanus azaleae*) (Baillie and Jepson 1951).

Insects are almost certainly the chief vectors of bacterial diseases of trees such as Watermark disease of willow; indeed the beetle *Cryptorrhynchidus lapathi* L. has been associated in Holland with the transmission of a similar disease of willows caused by *Pseudomonas saliciperda* (Lindeijer 1932). Bacterial canker of poplar and ash are also probably insect carried, but no definite associations have been discovered.

Insects are known to be the most important vectors of virus diseases (Chapter 12), so that we may expect tree diseases of this nature to be insect carried. This is certainly the case with Phloem necrosis of elm (Baker, W. L. 1948, 1949, 1950). Sucking insects, such as the aphids, are frequently involved in virus transmission. In view of the multiplicity of such pests on trees, there would appear to be no lack of possible vectors.

The association of wood-decaying and wood-staining fungi with bark and wood-boring beetles, and their introduction into trees by these insects, has been the subject of a good deal of research (Cartwright and Findlay 1959). In the case of Ambrosia beetles the fungus growing in the borings in the wood is actually eaten by the insects (Bakshi 1950; Fisher, Thompson, and Webb 1953). Wood wasps of the genus *Sirex* have been found to introduce the serious wood-rotting fungus *Stereum sanguinolentum* into conifers, both in Britain (Cartwright 1938) and in New Zealand (Clark 1933). Francke-Grosmann (1939) has found other wood-rotting fungi living in close association with *Sirex* spp., which was certainly introduced by them into the attacked trees. In Britain what appears to be the conidial stage of *Fomes annosus* has been found fruiting in the tunnels of Ambrosia beetles (Bakshi 1950), though it is not known how commonly it occurs there, or whether its presence has any pathological significance. But insects of this nature occur most commonly in dead or recently felled trees, and very seldom, if ever, in really healthy ones. They are therefore, as far as living trees are concerned, significant mainly in introducing fungi into trees already weakened by other causes.

Insects as predisposing factors

Insects can serve as predisposing factors in a number of tree diseases without actually transmitting the pathogen concerned. This they do by providing wounds which serve as a means of entry, by weakening the tree so that it is more easily infected by a feeble parasite, or by altering its growth rhythm so that it is in proper condition for infection at an unusual time, when the infective capacity of the pathogen is at a high level.

Suggestions that insect wounds lead to fungal infection have been made for a number of tree diseases. For instance, infection by *Dasyscypha willkommii* through insect wounds has been put forward as one of the important factors in Larch canker (Chapter 24). Possibly the best documented case is the infection of beech in the United States by *Nectria coccinea* through punctures made by the Beech coccus (*Cryptococcus fagi* Baer.) (Ehrlich 1934). This association of aphid and fungus has been found also on beech in Denmark (Thomsen, Buchwald, and Hauberg 1949), but in this case *Nectria galligena* was the fungus concerned. This complex does not seem to occur in Britain, despite the frequent presence of the insect and both species of *Nectria* on beech (Chapter 28).

In Italy bark wounds made by the larvae of the moth *Gracillaria simploniella* F.R. serve as a means of entry for Chestnut blight (*Endothia parasitica*) (Orsenigo 1949) (Chapter 29). In America 95 per cent. of the *Hypoxylon pruinatum* cankers on poplar examined by Graham and Harrison (1954) originated in injuries caused by various insects (Chapter 32). In Canada root wounding by larvae of the weevil *Hypomolyx* is considered a major factor in infection of live trees of White spruce (*Picea alba*) by various decay fungi (Warren and Whitney 1951; Whitney 1952). Injuries by *Hylobius* weevils serve as infection entries for butt-rotting fungi in *Abies balsamea* in Canada (Smerlis 1957).

Insect wounds, of course, are usually small and inconspicuous. Therefore, unless the early stages of a disease are carefully observed, infection through insect injuries may well be overlooked. It is probable that this avenue of infection is used more commonly, and by a larger variety of fungi, than have yet been recorded.

Initial weakening of the tree by insects, and subsequent attack by fungi, is not a common phenomenon. It probably happens occasionally with fungi such as *Armillaria mellea*, but more frequently death of a tree from Honey fungus following insect attack merely indicates that the insect damage tipped the balance in favour of the fungus in a tree already infected.

The classic example of interference with growth rhythm is provided by the spring defoliation of oak by Lepidopterous larvae. This results in a fresh flush of foliage in summer, at a time when spores of the Oak mildew (*Microsphaera alphitoides*) are abundantly present. Since newly developed leaves are much more susceptible than old ones, the new flush is severely attacked and the tree may be partially defoliated for the second time. Insect and fungus defoliation together can have a very depressing effect on the general health of the tree (Chapter 27).

Insects as secondary parasites to fungi

It is much more usual to find insects attacking trees weakened by fungi than the reverse. Many bark beetles have a liking for trees which are unhealthy, though they most commonly attack the parts which are still alive; for instance, the base of the trunk. In this way insects often hasten the death of trees initially attacked by fungal or bacterial pathogens, or damaged by non-living agencies. Indeed in many instances they may bring about the death of trees which otherwise would have recovered. The bark beetle, *Dendroctonus micans* Kug., acts in this way in Schleswig and Denmark on Sitka spruce weakened by *Fomes annosus* (Francke-Grosmann 1948). In the same way Wichmann (1953) has associated attacks on spruce by *Ips typographus* L. and *Polygraphus polygraphus* L. with prior infection by *Armillaria mellea*.

Scolytus beetles are often responsible for the final death of elms attacked by *Ceratocystis ulmi*; indeed the appearance of the emergence 'shot holes' of the beetle in the main stem is one of the surest indications that a tree is doomed (Chapter 31). There are numerous other instances where insects, especially bark beetles, give the *coup de grâce* to trees initially diseased by other agencies.

11

THE ORGANISMS ASSOCIATED WITH DISEASES OF TREES

SOME of the organisms associated with tree diseases have been considered in other chapters and need not be dealt with here. Parasitic flowering plants, in particular the mistletoes, and climbing flowering plants which do mechanical damage by twining or smothering are covered in Chapter 13. The superficial flora of trees, which includes *Algae*, lichens, mosses, ferns, and flowering plants, few of which are damaging, is dealt with in Chapter 21. Virus diseases form the sole subject of Chapter 12. Man, animals, and insects are discussed in Chapter 10, not with regard to the direct damage they do to trees, but because of the many ways in which their activities influence forest and tree diseases.

Diseases caused by fungi and bacteria, on the other hand, have not been grouped in any particular chapters. They are considered either under the host species, or, in the case of those with a widespread host range, in general chapters according to whether they occur chiefly (*a*) in nurseries, (*b*) on roots, or (*c*) on stems, shoots, or leaves.

The description of the fungus or bacterium involved in any particular disease is confined to those symptoms which can be observed with the naked eye or with a hand lens. For this reason notes on fungal fruit bodies are limited to their external appearance, and no details are given of spore shape or size. No cultural details of bacteria or fungi are provided. In most cases, therefore, the final identification of a fungus or bacterium will require the use of other books. Synonyms have been given only where the alternative name is likely to be equally or more familiar. Nomenclature, as in the case of host trees, is deliberately conservative. In nearly every case the name which is thought to be most generally familiar has been used. The reasons for this somewhat unorthodox practice have been explained in the Introduction. Authorities for names are normally given only once, at the first mention of the name of the fungus or bacterium in that part of the book where the disease it causes receives its main consideration. The Class and Order of the fungus concerned are given in the same place, as an aid to its location in other works of reference. In general fungi have been placed in their Orders in agreement with Ainsworth and Bisby (1954).

BACTERIA

Bacteria are microscopic, unicellular organisms, larger than viruses, but smaller than most fungal spores. They can be seen as individuals only under the higher powers of a microscope. They multiply by cell division and can increase with extreme rapidity. They absorb the food substances they require through their cell walls. Their pathogenic activity depends both on the substances they absorb and on those that they excrete, both of which disturb the normal biochemical processes of the host plant. Because of their unicellular nature, bacteria can move easily from one part of a plant to another, so that many bacterial diseases are markedly systemic. They are identified by their chemical reactions in culture, as well as by their size and shape and the presence or absence of the fine thread-like flagella which are sometimes attached to them. A detailed description of plant pathogenic bacteria is given by Dowson (1957).

There are not very many known bacterial diseases of trees and shrubs, but some are very important. In Britain, Bacterial canker of poplar and Watermark disease of Cricket bat willow are diseases of major significance.

FUNGI

Fungi, like bacteria, are devoid of chlorophyll, and obtain the carbohydrates they require from dead or living organic matter. The body of a fungus can consist of a single cell, but is much more often a mass of branching threads, the hyphae, known collectively as the mycelium. The hyphae are only one cell in thickness, but often have numerous cross walls, so that they become in fact chains of cells. Their growth is apical. In some fungi the hyphae can aggregate to form resting bodies or sclerotia, which may enable a fungus to survive periods of climatic stress. Hyphae can also aggregate into strands, often with a hard resistant covering. These are known as rhizomorphs and are frequently produced by the ubiquitous fungus *Armillaria mellea*. Their morphology and development have been studied by Townsend (1954).

In a few cases the mycelium of parasitic fungi is largely external to the host, for instance that of Oak mildew, *Microsphaera alphitoides*, is clearly visible as a white bloom on the leaves. More commonly it is internal, in which case it may be either inter- or intracellular. Intercellular mycelium often develops special processes, known as haustoria, which penetrate the cells of the host plants, and enable the fungus to extract nourishment from them. As in the case of bacteria, the damage caused by fungi depends on the substances they absorb from the host plant and on those they excrete into it.

Included among the fungi are the yeasts, existing as single cells, and especially important in the fermentation of sugar. The true yeasts are of no importance in forest pathology, but some fungi attacking trees can produce yeast-like forms which enable them to move more rapidly up and down the tree than would be possible merely by mycelial growth. One such fungus is *Ceratocystis ulmi*, the cause of the vascular disease of elm, usually known as Elm disease (Chapter 31).

The fungi reproduce by spores, small bodies of microscopic size, which are sometimes produced in extremely large quantities. The classification of the fungi is based very largely on the nature of the fruit body in or on which the spores are produced and on the form of the spores themselves. Spores may be produced following the sexual union of two cells, or asexually without any such preparatory union. Where sexual and asexual spores are produced by the same fungus, they are distinguished as the perfect and imperfect stages respectively. In many cases spores are disseminated by the wind, but there are many other methods by which they can move from place to place. They can be washed downwards, or splashed upwards from the ground on to the lower parts of a plant by rain. They can be carried on seeds or plant material which is being moved, or on other mobile living organisms, particularly insects. Where, in later chapters, individual diseases are described, the method of transmission is usually only mentioned if it is of an exceptional nature or has some particular bearing on the development of the disease.

Parasitic fungi can invade the host tissues in a variety of ways, but normally when a spore has alighted on the surface, it produces a germ tube which penetrates the host either directly or through a stoma, lenticel, or other aperture. The mycelium is soon produced by the growth and ramification of this germ tube. Successful germination and penetration require suitable external conditions, in particular a certain level of temperature and moisture, and a suitable substratum, which may involve a particular host plant, or a particular condition of the host tissues, or both. For instance, young newly developed needles are required for successful infection in the case of *Rhabdocline pseudotsugae* on Douglas fir. Some fungi, for example *Botrytis cinerea* (Chapter 15), are more successful if they can first colonize dead or moribund tissue and pass thence to the living parts of the plant. In many cases the control of a disease may depend on knowing the conditions under which infection is possible, so that protective measures may be correctly timed.

Classification of the fungi

The basis of separation of the four main classes of fungi is described below, but morphological details are given only to the small extent required for a proper understanding of the text. Points of difference between the orders, let alone families and genera, are beyond the scope of this book. Reference should be made to Gwynne-Vaughan and Barnes (1937), Bessey (1950), Gäumann (1949, 1952), Alexopoulos (1952), and F. Moreau (1952–3). Illustrations of the different forms of fungal fruit bodies are given by Ainsworth and Bisby (1954), J. S. Boyce (1948), and Snell and Dick (1957).

Phycomycetes

In contrast to the other classes the *Phycomycetes* in most cases have no cross walls (septa) in the hyphae. The more advanced forms have typically long, continuous, branched hyphae containing many nuclei. This group is primitive and some spore forms are motile in water. Only one Order, the *Peronosporales*, concerns us. In this Order asexual reproduction is by motile zoospores, but under some conditions the zoosporangium, in which these are borne, may germinate directly, thus functioning as a conidium. Low temperature and moisture favour the liberation of zoospores, high temperature and dryness the direct germination of the zoosporangium. Resting oospores are formed as a result of sexual fusion. The *Peronosporales* contains a number of soil fungi important in forestry, particularly those of the genera *Pythium* and *Phytophthora*.

Ascomycetes

In the Ascomycetes, the hyphae are septate, and the spores of the sexual stage are borne in asci, elongated sac-like bodies, containing a definite number of spores, usually eight, though two or other multiples of two occur in some species. These asci are usually produced in groups, within a protective wall of sterile hyphae, the peridium, but in some primitive forms the peridium is lacking. An ascus-containing fruit body with a peridium is usually known as an ascocarp. These are mainly of two types, apothecia, where the asci are borne on cup or saucer-like structures on the surface of the host, and perithecia, where the asci are borne in enclosed or semi-enclosed flask-like structures, which are sometimes superficial, sometimes in the host tissue, or sometimes embedded in a stroma formed by the fungus itself. In both apothecia and perithecia, the asci are mixed with long sterile cells, known as paraphyses. Ascospores are often ejected into the air from the ascus with considerable force, a useful aid in their dissemination.

Most of the Ascomycetes also produce asexual spores known as conidia. This is usually referred to as the conidial or imperfect stage, in contrast to the ascus or perfect stage. The conidia are borne on conidiophores, which vary widely in structure and are produced in a variety of ways. The asexual fruit bodies take many forms, for instance if the conidiophores are gathered into a fascicle, like a sheaf of corn, it is known as a coremium, if they are borne in a flask-like structure, resembling the perithecium of the ascus stage, it is known as a pycnidium. Sometimes an Ascomycete can have more than one conidial stage. In some groups this may be a form with very small, apparently functionless, spores, known as microconidia or spermatia.

No conidial stage has yet been found for some members of the Ascomycetes. Where both stages are known the conidial or imperfect stage tends to fructify before the perfect stage. In some cases the perfect stage may be rare or may occur only in certain places; in other areas or other countries the fungus may persist by the production of the conidial stage alone.

Some diseases had already become familiar under the name of the imperfect stage before the Ascomycete stage was discovered. In the text the most familiar name is normally given first, whether it be the perfect or the imperfect stage Where both are equally familiar, the perfect stage is given priority.

The Ascomycetes contain a very large number of important tree pathogens, particularly those attacking bark, twigs, and leaves.

Basidiomycetes

The Basidiomycetes, which also have septate hyphae, are characterized by the production of specialized cells, the basidia, on which the basidiospores, usually four in number, are formed. This is the sexual or perfect stage. The basidia, which are microscopic in size, may be scattered on the substratum, but are usually mixed with paraphyses in a definite layer, the hymenium. The hymenium is borne on or in the fruit body, which may be a very simple structure or highly elaborate, for example a toadstool. Some Basidiomycete fruit bodies are perennial, adding each year a fresh layer of hyphal tissue and a new hymenium. Some of the Basidiomycetes have a conidial (imperfect) stage; this is the case with *Fomes annosus*.

Most of the familiar toadstools belong to the Basidiomycetes. This Class also contains most of the wood-decaying fungi, both saprophytes and parasites, including the highly important fungi *F. annosus* and *Armillaria mellea*. Some of the wood-destroying fungi produce bracket-like fruit bodies on the trunks of attacked trees. These fruit bodies, which are sometimes very large and conspicuous, are the 'conks' of American literature. In the toadstools and bracket fungi the spore-bearing hymenium covers the gills or lines the pores on the underside of the fruit body.

The Basidiomycetes also include a very specialized Order, the Uredinales or Rust fungi, which must be considered in more detail. They are almost unique among the fungi, for many of them are heteroecious; that is to say they spend part of their life cycle on one host and part on another. They are all obligate parasites. Their life cycle is best explained by reference to an actual example. The life history of *Coleosporium tussilaginis*, attacking Two-needled pines on the one hand and Groundsel, *Senecio vulgaris*, and other herbaceous hosts on the other is set out diagrammatically in Fig. 30.

A fully developed Rust fungus has five spore forms:

Stage	*Fruit-body*	*Spore*
0	pycnia	pycniospore
I	aecia	aeciospore
II	uredia	uredospore
III	telia	teliospore
IV		sporidia

The numbers given in the first column are sometimes used as a convenient method for referring to the different stages.

Generally the pycnia and aecia are produced on one host, and the uredia and telia on the other. The pycnia are small bodies, producing minute unicellular spores, which have no infective function, but act as male fertilizing cells in the production of aecia. The aecia occur in clusters in or on the surface of the infected plant, and in close proximity to the pycnia. They are usually covered by a thin membrane, the peridium, which ruptures when they are ripe. The unicellular aeciospores are produced in chains; being orange or yellow in colour, they tend to make the aecia very conspicuous. They are borne on the first host, in our example Scots pine, but infect the second host, in this case *Senecio vulgaris*. Uredia appear on the second host, following infection by the aeciospores. They also are orange in colour, and often conspicuous because they are produced so thickly. They can also arise as a result of secondary infection by other uredospores; for this spore form can only infect the same host species or group of species as that on which it was borne. The uredial stage thus enables the fungus to build up on its second host, and is sometimes referred to as the summer stage.

Telia are also produced on the second host. They are usually darker, less conspicuous, and borne at a later date than the uredia. The teliospores, the equivalent of the normal basidia, often carry out the function of resting spores, enabling the fungus to survive the winter. They eventually germinate to produce four, or less often two, sporidia, which are small, unicellular, and comparatively short-lived spores, able to infect the first host, in our case

Scots pine, again, thus completing the cycle. The development cycle of a typical rust normally takes one year, the aecia being produced in the spring, the uredia in the summer, and the telia in the autumn.

Sometimes the Rust fungus is perennial in the host, producing fresh crops of fruit-bodies at the appropriate season each year. This is the case with the

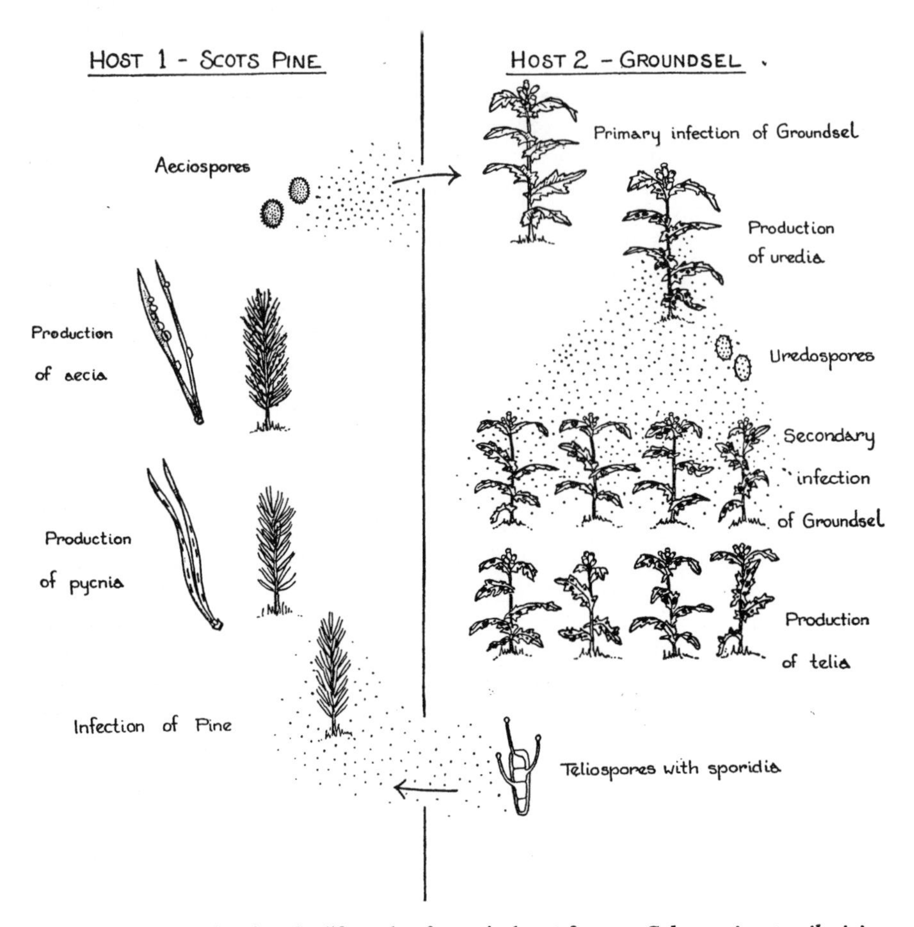

FIG. 30. Diagram showing the life cycle of a typical rust fungus, *Coleosporium tussilaginis*. (J. C.)

aecial stage of White pine blister rust, *Cronartium ribicola,* on Five-needled pines (Chapter 22). The existence of two hosts introduces the possibility of controlling the disease on one of them by the removal of the other. This has been done on a large scale in the case of White pine blister rust, where *Ribes* species are the uredial and telial hosts.

In some cases part of the life cycle is missing. In the genus *Gymnosporangium,* for instance, there is no uredial stage. Indeed some Rust fungi are not even heteroecious. Numerous exceptional cases will be met with when the diseases caused by Rust fungi are described later. Sometimes two or more species or strains may be separable on one host group, but indistinguishable on the other. *Coleosporium tussilaginis* is known to have a number of strains specific to different herbaceous uredial and telial hosts, but which cannot be distinguished on pine, and which have therefore been lumped as a single species.

Fungi Imperfecti

This is not a natural group, but comprises those fungi with septate mycelium which do not apparently possess a perfect stage, so that they cannot yet be assigned to the Ascomycetes or the Basidiomycetes. Some of them are probably forms that permanently reproduce themselves asexually. Others, grouped in Mycelia Sterilia, do not appear to produce spores at all, but exist only as mycelium. Ultimately many of the Fungi Imperfecti will be found to be imperfect stages of Ascomycetes, the group to which most of the linked ones already belong.

A number of the Fungi Imperfecti are associated with tree diseases, for instance *Verticillium* species with wilt of maple and other hardwoods, *Meria laricis* with Leaf-cast of larch, and *Cryptostroma corticale* with Sooty-bark disease of sycamore.

Parasites and saprophytes

It is not possible to deal here with the extremely complex biological activities which underlie the observable phenomena of infection and development by the pathogen, or of susceptibility and resistance in the host. These have been very fully covered by Gäumann (1950). Here they are dealt with, when the individual diseases are considered, only to the extent to which they throw light on the observable behaviour of the disease and on its control. For many diseases, of course, this fundamental information is still lacking.

Fungi and bacteria can be roughly classified into (*a*) saprophytes, which live on dead organic matter, (*b*) facultative parasites, which are able to attack living organisms, but can also live as saprophytes, and (*c*) obligate parasites, which can live only by attacking live tissue. Obligate parasitism would appear to be a disadvantage to the fungus, since it can only multiply on a suitable live host, whereas facultative parasites can build up on dead organic matter. Nevertheless, many of the obligate parasites are highly successful causes of disease; among tree disease fungi *Microsphaera alphitoides*, the cause of Oak mildew, *Keithia thujina* on *Thuja*, and *Rhabdocline pseudotsugae* on Douglas fir, as well as all the Rust fungi, are obligate parasites. Knowledge as to whether a fungus is an obligate parasite or not is often of value in devising control measures. If a fungus can exist only on the living host or in the form of spores, the locations where it will be found and where control can be applied are obviously limited.

There are cases when it is difficult to distinguish between parasite and saprophyte. A parasite is an organism living upon, and deriving food from, another live organism. Therefore fungi attacking the dead parts of a tree are not strictly parasites. The fact that those dead parts may still be attached, often very intimately, to the living tissue does not alter the position. Thus fungi decaying the dead heartwood of a tree are not parasites. They differ only in position from obviously saprophytic fungi attacking felled logs or debris on the forest floor. Some decay-causing fungi are facultative parasites, of which *Fomes annosus* is an obvious example, but others can live for many years in the heart of a decaying tree without damaging the living tissues.

12

VIRUS DISEASES

ALTHOUGH our knowledge of viruses has increased greatly in the last few years, it is still very difficult to define their real nature. They exist on the borders of life, where our present definitions of living and non-living are insufficient to define their position. They are very small, and cannot be seen even with the higher powers of an optical microscope; but it is now possible to photograph them by the use of an electron microscope, working with radiation of a shorter wave length than visible light. By this means much information on the shape and size of virus particles has been obtained. It appears that viruses are obligatory intracellular parasites, though the possibility of saprophytic bodies of a like nature existing in plant cells cannot be ruled out. Their pathological importance is based mainly on two factors, their ability to multiply within the plant, and their interference with the plant's physiological processes. This interference normally leads to visible external symptoms, by which the presence of the virus is first revealed.

There is at present no generally accepted system for naming plant viruses. It has been the practice merely to give descriptive names such as Apple mosaic, or Plum rough bark, and such names are apt to become confusing, when, as is often the case, the same virus is later discovered on another host, on which it may produce quite different symptoms. These descriptive names are based on the symptoms produced on a particular plant, and the identification of a virus rests mainly on its ability to produce these symptoms. Proof of the identity of a virus occurring on two different hosts depends on its transference from the second to the first host and its ability to produce typical symptoms on it.

Virus symptoms are very varied; perhaps the most typical are mottling of the leaves and distortion of the leaves and shoots. Increased knowledge of the sort of symptoms to be expected on different kinds of plants often leads to the discovery of new virus diseases. The almost complete lack of recorded virus diseases on conifers may be due to our inability to recognize certain coniferous disease symptoms as being of virus origin. It is possible for a virus to multiply in a resistant plant without producing external symptoms, or with the symptoms greatly reduced. The actual damage to such plants is negligible, but they can act as 'carriers', and be sources of infection to other plants more susceptible to the same virus.

Viruses are normally transmitted in the cell sap. The commonest means of transference from plant to plant is by the agency of sucking insects. Viruses can also be transmitted by grafting, so that naturally occurring root grafts may act as means of spread in some trees. Artificial grafting, which has considerable practical importance because of the number of trees and shrubs propagated by this method, is another means of transmission. Grafting has considerable value both as a means of diagnosis and in testing the host range and the host susceptibility of a particular virus. Ability to transmit a disease, for which there is no obvious bacterial or fungal cause, by grafting is considerable evidence in favour of its being due to a virus. The comparative difficulty of grafting conifers is likely to be a minor trouble in the investigation of possible virus diseases on them. Transmission by mere contact between healthy and infected tissue is possible with some viruses. They can also be transmitted on tools and on the hands of the workers using them.

The primary symptoms of virus infection are a clearing of the veins of infected leaves, and the appearance of small necrotic or chlorotic spots or rings at the points of infection. These are followed by one or more of the following symptoms: mosaic mottling of the leaves, chlorotic rings on the leaves, actual lesions on the leaves or young stems, distortion of the leaves or shoots, which may sometimes take the form of a reduction of the leaf tissue so that only the midribs remain, stunting of growth, dieback, and generalized chlorosis.

Virus infection does not usually result in the death of a plant, but often brings about a very marked reduction in growth. There is a tendency for virus complexes to build up by infection with different viruses from year to year, so that in agriculture and horticulture virus diseases are particularly serious in crops which are propagated vegetatively, such as potatoes or strawberries. Transmission by seed does occur with some viruses, but it can normally be assumed that a crop raised from seed will initially be more or less virus free. The perennial nature of trees makes virus infection in them particularly serious, because the disease will have such a long period of uninterrupted increase. Further information on the nature and behaviour of viruses is given by Bawden (1943) and K. M. Smith (1951).

Fortunately there are apparently few serious virus diseases of forest trees in the world, and none of any consequence in Britain. Of course, some of the diseases which still await explanation may eventually prove to be of virus origin.

VIRUS DISEASES OF BROADLEAVED TREES AND SHRUBS

The complexities of the situation as far as hardwoods are concerned are well illustrated by the successful inoculation of tobacco ringspot virus into a number of woody hosts by Wilkinson (1952). It produced mosaic and stunting on *Prunus avium* and *Ulmus americana*, mosaic and narrowing of the leaves on *Acer negundo*, and *Fraxinus pennsylvanica*, ringspots and line patterns on *Sambucus canadensis*, and mild mosaic on *A. ginnala*. *P. avium* and *A. negundo* tended to recover and eventually produced healthy leaves, but the other species remained systemically infected.

The list below sets out many of the virus diseases, or apparent virus diseases, which have been recorded on broadleaved trees and on the more important shrubs. It is not suggested that the list is complete. Firstly, it is probable that a great many virus infections, most of them probably quite unimportant, have not yet been recorded. Secondly, a number of possible virus infections, particularly variegations, have not yet been investigated. Thirdly, since records of virus diseases often appear mixed with those of diseases of a different origin in check lists and reports, many records may have escaped the author. The list as it stands gives some idea of the wide range of trees and shrubs already known or suspected to suffer from virus infections, most of which, fortunately, appear to be comparatively trivial in their effect.

The two tree viruses causing most economic loss are Spike disease of Sandal in India, and Phloem necrosis of Elm in North America. These are considered separately below.

Spike disease of Sandal

Spike disease of *Santalum album*, a tree of considerable economic importance, was probably the first recorded forest-tree virus disease. Despite the fact that both the tree and the disease are virtually confined to India, its position as the most damaging virus disease of a timber tree yet recorded justifies brief consideration of it here. Although it had been known for many years, it was not till early in the nineteen-twenties that the suggestion that a virus was the cause was first put forward (Coleman 1923). The principal symptoms are a reduction in the size of the leaves, which gives the 'spiked' appearance, shortening of the internodes, and death of the roots, leading eventually to death of the whole tree (Narasimhan 1928). The disease was later discovered to be transmissible by insects (Rangaswami and Sreenivasaya 1935), though it is still not known if this is the normal means of spread.

The question of resistance is complicated by the fact that *Santalum* itself is a parasite, and that its resistance, or at any rate its reaction, to the disease is influenced by the host or hosts on which it is growing (Rangaswami and Griffith 1939). On the other hand, there is experimental evidence that certain strains of *Santalum* are more resistant than others, even when the host is the same. Thus use of resistant strains of the tree, the cultivation of *Santalum* in association with resistance-inducing hosts, and the eradication of diseased trees by poisoning to lessen the spread of infection have all been suggested as possible control measures. Infected trees do not always show symptoms, so that the eradication of 'carriers' is likely to be difficult (Sreenivasaya 1930).

Recent reviews of the extensive research which has been done on the disease have been made by Sreenivasaya (1948), Muthanna (1955), and Iyengar (1958).

Phloem necrosis of elm

This virus disease extends over a large area in the central United States (Swingle, Whitten, and Young 1949). Phloem necrosis takes the form of a gradual decline in the health of the tree, but in large trees part of the crown may be affected before the rest. The foliage at the tips of the highest branches is first affected, the leaves droop because of the curvature of the petioles, and the leaf blades turn up at the margin, often becoming stiff and brittle (Swingle 1942). They then wither and fall, possibly partially as a result of the death of roots. The phloem of infected large roots turns yellow and then brown; this discoloration usually extends to the lower part of the trunk, but may also occur higher in the tree. Discoloured bark has a characteristic odour of wintergreen. Infected trees invariably die. Localized spread of the disease may occur by root grafting, but the chief vector is the leafhopper, *Scapnoideus luteolus*

Van Duz. (Baker 1949). The disease occurs on the native species *Ulmus americana* and *U. alata*. It is not yet known whether European species of elm are susceptible.

The only control, apart from the removal of affected trees, is the annual preventive spraying of elms with insecticides such as D.D.T. (Swingle, Whitten, and Young 1949). Two sprays are required to prevent feeding by the leafhopper, starting as soon as the leaves are fully formed, and spraying again one to two months later. Spraying the bark of elms in the spring, before the insects have hatched, is also of value. But these measures are very expensive and would, of course, have to be repeated annually.

Other virus diseases of broadleaved trees and shrubs

The list below, though incomplete, will serve to indicate the large number of genera already known to suffer virus infections, and the variety of symptoms produced. It will be noted that 'variegation' is one of the commonest effects. Not all variegations are due to viruses, but a good many, including some of those maintained by vegetative propagation because they are of horticultural value, are of this nature. The distribution of the virus diseases is not normally given in the list, for with such incomplete records the meagre information available might be very deceptive. Fruit-tree genera are omitted, unless they also include ornamental species. Tropical trees and shrubs have also been left out.

Abutilan

Mosaic (Smith, K. M. 1957): the resultant variegated plants of several species have a distinct horticultural value.

Acer

Mosaic (Atanasoff 1935; Brierley 1944): on *A. negundo*; said to be capable of doing considerable damage.

Aesculus

Infectious variegation (Atanasoff 1935; Blattný 1938*a*; Brierley 1944).

Berberis

Vein mosaic (Blattný 1933): on *B. vulgaris*, causing complete sterility on affected plants.

Mosaic with reddish blotches (Wilkinson 1953): on *B. thunbergii*.

Betula

Vein mosaic (Blattný 1938*a*).

Interveinal mosaic (Blattný 1938*a*).

Dieback (Hansbrough and Stout 1947; Berbee 1957): it is possible that the widespread deterioration of birch in the north-eastern United States may be a virus disease (Chapter 34).

Buddleia

Mosaic (Smith, K. M. 1952): produces narrow malformed leaves and reduced flowering.

Camellia

Yellow mottle leaf (Hildebrand 1954; Plakidas 1954): results in a slow deterioration of the plant.

Chimonanthus

Mosaic (Foister 1961).

Cornus

Mosaic (Atanasoff 1935): on *C. mas.*

Corylus

Mosaic (Atanasoff 1935): on *C. avellana.*

Line-pattern (Scaramuzzi and Ciferri 1957): causes elongated yellow or pale-green areas on the leaves between the veins; recorded in Italy, probably not serious.

Cydonia

Posnette (1957) has reported a number of virus diseases on quince used as root-stocks for pears.

Daphne

Mosaic (Brierley 1944; Smith, K. M. 1952; Beaumont 1956; Milbrath and Young 1956): on *D. mezereum* and *D. odora.* It does considerable damage to the former in Britain, resulting in a severe reduction of flowering or even in complete absence of flowers. It is unfortunately becoming increasingly common.

Eucalyptus

Mosaic (Fawcett 1942).

Euonymus

Infectious variegation (Atanasoff 1935; Brierley 1944): some variegations in this genus are not of virus origin.

Fraxinus

Infectious variegation (Atanasoff 1935; Brierley 1944): recorded in Britain, but not damaging.

Witches' broom (Plakidas 1949): on *F. berlandieriana* (Arizona ash).

Gleditschia

Mosaic (Atanasoff 1935).

Hydrangea

Ringspot (Brierley and Lorentz 1957*a*; Brierley 1957): causes lesions on and distortion of the leaves of florists' hydrangeas in the United States. Though most varieties are sufficiently resistant to escape serious damage, a few may be rendered unsaleable.

Ilex

Infectious variegation (Atanasoff 1935; Brierley 1944).

Jasminum

Infectious variegation (Atanasoff 1935; Brierley 1944).

Juglans

Brooming or Bunch disease (Dodge, B. O. 1947; Berry and Gravatt 1955): on various species. It is widespread in the eastern United States, but not apparently very serious.

Laburnum

Infectious variegation (Atanasoff 1935; Brierley 1944; Katwijk 1953): occurs on *L. anagyroides* and *L. vossii*. *L. alpinum* can be infected.

Ligustrum

Infectious variegation (Atanasoff 1935; Brierley 1944).

Lonicera

Infectious variegation (Brierley 1944).

Morus

Dwarfing or Curl (Zaprometov 1932; Atanasoff 1935).
Mosaic (Cook, M. T. 1931; Endo and Kurasawa 1937).

Myrica

Bayberry yellows (Raychaudhuri 1952).

Passiflora

Infectious variegation (Atanasoff 1935).

Paulownia

Witches' broom (Tokushige 1951, 1955): seriously damaging in Japan, frequently resulting in death, especially of young trees.

Populus

Mosaic (Atanasoff 1935; Blattný 1938*a*; Perisic 1951): on various species, resulting in premature leaf-fall.
Variegation: found in Britain on *P. candicans*, and occasionally as temporary infections on other varieties growing near. It has been sold horticulturally as *P. candicans* var. 'Aurora'.

Prunus

A number of virus diseases have been recorded on plum and cherry in Britain (Luckwill 1950; Posnette 1954; Wormald 1955; Smith, K. M. 1957). They cause stunting of growth, splitting of the bark, and a variety of leaf symptoms, and are responsible for much poor growth and considerable crop reduction. Some of them occur on wild and ornamental species of *Prunus* (Thomas and Rawlins 1939; Zeller and Milbrath 1942; Scaramuzzi and Corte 1957; Cockran and Pine 1958). Indeed, it has been suggested that flowering cherries may act as carriers for viruses affecting fruiting varieties (Reeves, Cheney, and Milbrath 1955). Elm mosaic virus is able to infect a wide range of *Prunus* species, but several of the Japanese varieties are immune (Callahan and Moore 1957).

Pyrus

Apples and pears are also subject to a number of virus diseases (Luckwill 1950; Wormald 1955; Smith, K. M. 1957; Posnette 1957). One of the most interesting, because its symptoms are somewhat unusual, is 'Rubbery wood', in which the main effect is an abnormal flexibility, and therefore weakness, of the branches. It is reasonable to suppose that this disease may occur also on ornamental varieties of *Pyrus*.

Rhamnus

Variegation (Atanasoff 1935; Brierley 1944).

Ribes

Cultivated red and blackcurrant are subject to a number of virus diseases, the most important of which is Reversion of blackcurrant (Wormald 1955; Smith, K. M. 1957). Some of them must certainly extend to flowering currants, but to what extent is not known.

Robinia

Brooming (Grant, Stout, and Readey 1942). The earliest symptom of this disease is a vein-mosaic, followed by reduction and distortion of the leaves and twisting of the petioles, accompanied by proliferation of buds and branches (Fig. 31). The disease, which is common in parts of the United States on *Robinia pseudacacia*, results in a marked reduction in growth when infection is severe, but is not normally fatal. It has also been reported on the Continent, but not in Britain.

Rubus

Cultivated raspberries are subject to a considerable number of virus diseases (Wormald 1955; Smith, K. M. 1957). Some of these certainly occur on the wild forms. On blackberry the most serious virus disease is Blackberry dwarf (Prentice 1950). This is known to occur on wild blackberries and raspberries (De Fluiter and Thung 1951).

Sambucus

Mosaic (Anon. 1933*a*; Atanasoff 1935; Brierley 1944).

Sorbus

Infectious variegation (Atanasoff 1935; Brierley 1944).

Syringa

Mosaic (Atanasoff 1935): causes premature defoliation, which results in winter injury and dieback.

Ring spot (Atanasoff 1935; Protsenko and Protsenko 1950; Nikolić 1951; Beale and Beale 1952): on *S. vulgaris* both in Europe and in America, causing moderate damage. It is very readily perpetuated, because lilac is so often propagated by grafting (Katwijk 1955).

Witches' broom (Lorentz and Brierley 1953): on *S. japonica*.

Smolák and Novák (1950) report a wide variety of virus symptoms on *S. vulgaris*, many of them resulting in wilting and death. K. M. Smith (1940) recorded a virus disease of lilac, with symptoms very like those of the incompatibility disease, known as Graft blight (Chapter 10).

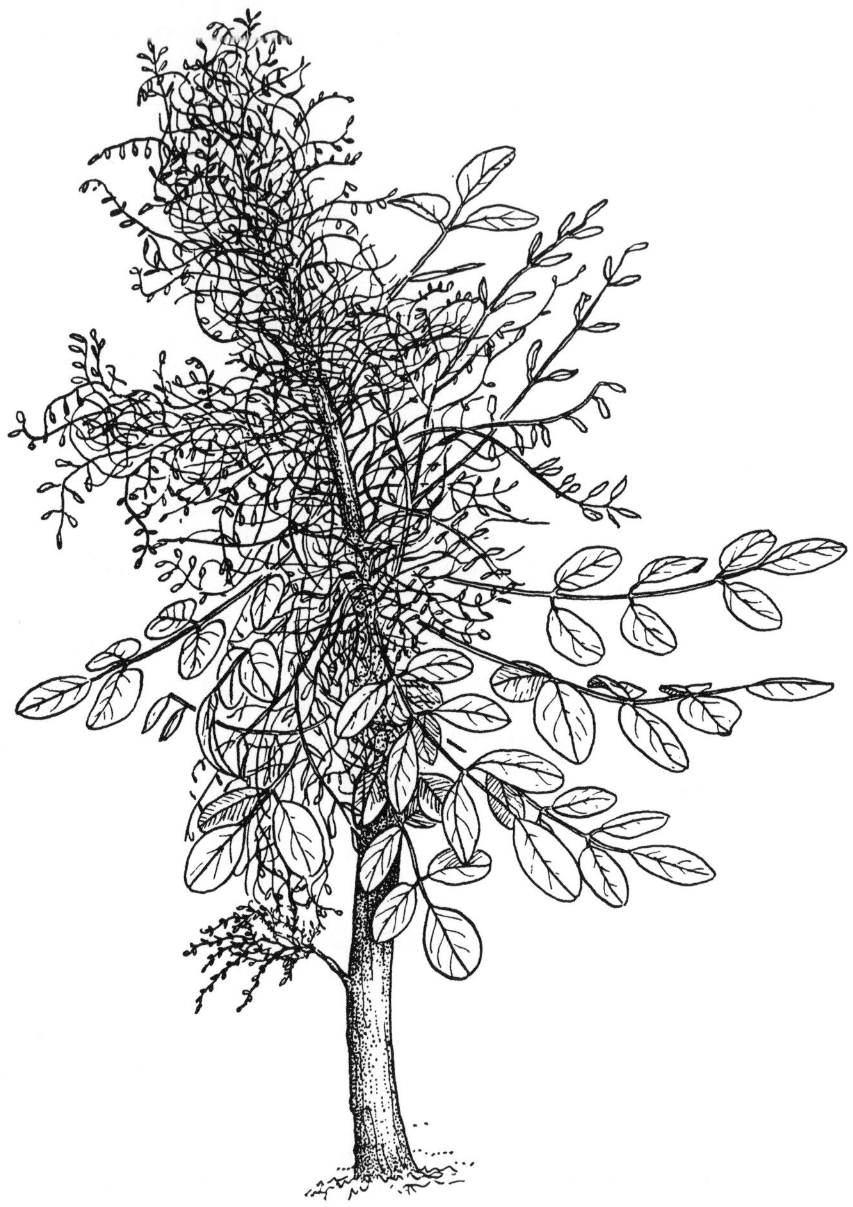

FIG. 31. Virus brooming of *Robinia* in the United States (×½). (J. C. after a photograph by T. J. Grant.)

Tilia

Little leaf (Blattný 1938*b*): causes stunting on *T. cordata* in Czechoslovakia.

Cowl- or Cup-leaf (Blattný 1938*b*; Klášterský 1951): distorts the leaves of *T. cordata* and *T. platyphyllos*, also in Czechoslovakia.

Ulmus

Mosaic (Atanasoff 1935; Blattný 1938*a*; Swingle, Tilford, and Irish 1943; Brierley 1944; Bretz 1950): causes a gradual decline in vigour. It can be transmitted by seed and by pollen (Callahan 1957).

Zonate canker (Swingle and Bretz 1950): interesting because twig cankering and girdling, which are not very usual virus symptoms, occur. Mosaic symptoms are also found on the foliage of infected trees, so that it may be related to some forms of mosaic.

Phloem necrosis (see above).

Fortunately all the more serious virus diseases of elm appear to be restricted to North America. There is no suggestion that the mosaics found in Central Europe are at all serious. Widespread reduction and distortion of foliage, strongly suggestive of a virus disease, was found on a single large elm at Oxford (Fig. 32). The tree destroyed.

Vaccinium

False blossom (Shear 1916): occurs only in North America, mainly on *V. macrocarpon*, but also on the European *V. oxycoccus*.

Ring spot (Hutchinson and Varney 1954).

Stunt (Wilcox, R. B. 1942; Smith, K. M. 1957): only known in the United States. It occurs on a wide range of *Vaccinium* species.

Mosaic (Varney 1957).

Shoestring (Varney 1957): causes red streaks on the shoots and narrowing and distortion of the leaves.

Vitis

Fruiting vines are subject to a number of virus diseases, which may well occur also on garden varieties, but have not so far attracted attention on them.

Wistaria

Mosaic (Brierley and Lorentz 1957*b*).

VIRUS DISEASES OF CONIFERS

There is only one record of a suspected virus disease on a conifer (Smolák 1948). Mosaic and chlorosis symptoms were observed on young *Picea excelsa*. Chlorotic and healthy foliage could be found on different branches of the same tree, and even on different sides of the same twig. Aphids, which were present, might have been responsible for transmitting the disease from tree to tree. The diagnosis of this disease as due to a virus is mainly based on the absence of any other feasible explanation, but aphid transmission has been carried out experimentally (Blattný 1956).

Occasionally, symptoms suggestive of virus disease have been seen on conifers in Britain, but never at a serious level, and completely without proof of virus origin. Unfamiliarity with the virus symptoms to be expected on a coniferous host may well be the reason why virus infections on conifers have not been recorded to the extent they have on broadleaved trees.

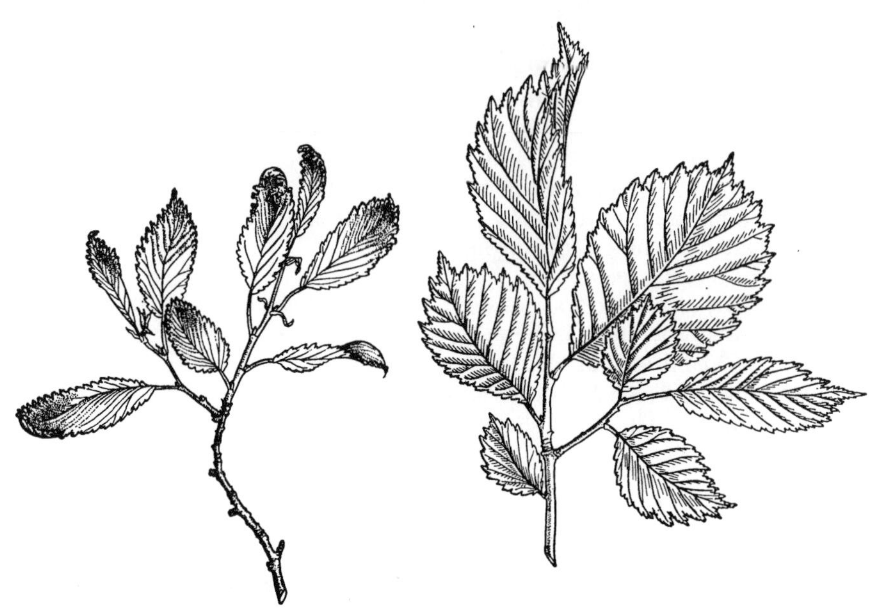

FIG. 32. Suspected virus disease on elm, University Parks, Oxford, August (×½). Left, affected twig; right, normal twig. (J. C.)

THE CONTROL OF VIRUS DISEASES

In agriculture and horticulture the control of virus disease is based mainly on the raising and maintenance of virus-free stocks. Such stocks are grown in isolation to lessen the chances of infection, if possible in areas where the insect vectors do not exist. If necessary such crops are protected by spraying against attack by insect vectors. Much work has also been done on the selection of virus-resistant varieties. It has been found possible in a few cases to kill the virus by heat without permanently damaging the host plant. Chemotherapy is a possible means of control, but little work has been done on this aspect. Destruction of diseased plants, and control of the insect vectors by spraying, are other lines of attack on virus diseases.

It will be seen that few of these measures are as easily applied in forestry as they are in horticulture or agriculture. This is another example, of course, of the usual difficulties of forest disease control (Chapters 9 and 36). Maintenance of virus-free stocks would be of little value, except for trees like poplar which are propagated vegetatively. Resistance to virus diseases could, of course, be made part of a forest-tree breeding programme, and probably could be tested on comparatively young trees. The use of heat appears to have little or no application, but the development of chemotherapy might be of value, at any rate for trees of individual importance. The removal of diseased plants would inevitably become a part of thinning practice, as is the case with other diseases, but the control of insect vectors, which involves preventive, rather than curative, spraying is hardly possible economically.

The appearance of a really damaging virus disease on trees would therefore be particularly serious, not only because the long life of the tree would provide it with such a lengthy period of development, but also because control would obviously be difficult.

13

INJURY TO TREES BY HIGHER PLANTS AND DISEASES OF FOREST WEEDS

PARASITIC PLANTS

Mistletoes

ALL the mistletoes are members of one specialized family of world-wide distribution, the Loranthaceae. They are parasitic on the aerial parts of trees and shrubs.

Viscum

The only mistletoe native to Britain is *Viscum album* L., a shrubby evergreen with thick, leathery, yellow-green leaves. It is usually dioecious, the female plants bearing whitish sticky berries in the branch axils. The seeds, which are distributed by birds, especially thrushes, feeding on the berries, germinate in cracks in the bark sending down sinker 'roots' as far as the cambium. Lateral 'roots' then develop in the bark and from these a series of further sinkers grow down to the wood surface. As new wood is laid on by the tree these sinkers increase in length by basal, not terminal, growth, so that their tips are gradually enveloped by the wood and present the erroneous appearance of having grown into it. With age the ramifications of the laterals and sinkers become more and more complex (Thoday 1951, 1958). Some of the earlier-formed sinkers die, becoming detached from the mistletoe plant and buried in the wood. Heavily infected wood is thus considerably weakened. The stem of the host plant tends to become swollen around the infected area.

Mistletoe is a semi-parasite. Since it has green leaves it can photosynthesize some of the organic food it requires, but for minerals and water it is entirely dependent on the host tree. There is evidence that it is xerophytic and that its water requirements are low (Cove, unpubl.). The degree of parasitism is difficult to gauge. Well-developed parasitic tissue in the host plant is occasionally produced with very little aerial growth, suggesting that it can become nearly a complete parasite. Some other members of the Loranthaceae have greatly reduced leaves and must be almost wholly parasitic. The degree of parasitism of *Viscum* is probably very variable, but it can certainly absorb carbohydrates as well as minerals and water from its host.

The effect on the host tree of a light infection is not detectable, but heavy infections result in reduced growth and chlorotic discoloration of the foliage. If only a limited number of trees are involved, however, the loss of increment may be offset by the market value of the mistletoe. Deliberate cultivation of mistletoe on old apple trees is quite frequent, especially in the western counties of England. The actual amount of damage to the tree cannot be judged by the visible growth of mistletoe on it. Péter-Contesse (1930) and Dallimore (1932) found that the 'root' development of the mistletoe and the degree of hypertrophy of the host tissue is often inversely proportional to the aerial growth of the parasite.

In Britain mistletoe occurs on a very wide range of broadleaved trees, being particularly common on sycamore, hawthorn, Hybrid black poplars (*P. euramericana*), apple, and lime (White, J. W. 1912; Somerville 1914; Nicholson 1932). The most recent and comprehensive account of *Viscum* in Britain has been given by Cove (1956). It is much less common on oak, and has very seldom been recorded on elm, walnut, or Grey poplar (*P. canescens*). It seems quite definite that mistletoe does not occur on beech. According to Boodle (1924), *Viscum* has been wrongly recorded on oak on the Continent, when in fact it was parasitizing its near relative *Loranthus europaeus* (see below) which does occur freely on *Quercus* species. Though actually on the *Loranthus*, the common mistletoe would appear to a casual observer to be growing on the oak. Indeed Nicholson (1932) states that *Viscum* can parasitize itself. When this occurs the second plant appears as a rather irregularly joined internodal branch on the first. There is disagreement between some lists of host susceptibility. For instance, Lombardy poplar (*Populus nigra 'italica'*) was said by Nicholson (1932) in his very comprehensive list to be commonly affected, whereas Somerville (1914) and others regarded it as rarely or never attacked.

On the Continent *V. album* occurs on conifers as well and Tubeuf (1907, 1908, 1910) separated it into three varieties, one on broadleaved trees, one on *Abies*, and the third on pine. The forms on conifers are regarded by some as two varieties of a separate species *V. austriacum* Wiesb. The variety on *Abies* appears to be restricted to that genus, occurring mainly on *A. alba*; but the pine variety is found occasionally on Norway spruce as well as on a considerable number of pine species, while it can be established by inoculation on larch and cedar. In Britain there is one record of mistletoe on larch (Nicholson 1932) and one on *A. alba* (Cove 1956); possibly these were introductions from the Continent. It has certainly not become established on conifers in Britain. Both Tubeuf and Nicholson found that the growth form and leaf colour of mistletoe varied on different host trees, and there may really be botanical varieties; one with much narrower leaves has been described on the Continent, but such differences certainly are less important than the three main host distinctions mentioned above. Cove (1956) has suggested the possibility of two sub-species with overlapping host ranges occurring in Britain, one on Ulmaceae and *Tilia*, but not on Rosaceae, the other on Rosaceae and *Tilia*, but not on Ulmaceae. Differences in host susceptibility in different regions also suggest the possibility of local varieties. A variety with a strong preference for poplar would account for the extraordinary prevalence of mistletoe on that host in parts of France.

It is fortunate that mistletoe is not firmly established on conifers in Britain, for although the variety on pine is not considered very damaging, that on *Abies* is said (Péter-Contesse 1937; Lanternier 1944; Plagnat 1950*a*) to do considerable damage to *A. alba* in parts of Europe, causing as much as 20 per cent. reduction in growth rate (Klepac 1955), weaknesses and defects in the wood, and sometimes death of the crown. In addition, cracks in the bark caused by the mistletoe may lead to subsequent infection by wood-rotting fungi. It is most serious in older stands where its effect on the growth rate is much more pronounced (Gäumann and Péter-Contesse 1951).

In Britain mistletoe occurs naturally only from Yorkshire southwards and is most prevalent in the south-western counties. It is particularly noticeable on old apple orchards in the Severn valley. Deliberately sown plants may occasionally be found in Scotland (Harrison 1932).

Loranthus

A large number of species of *Loranthus* occur on trees in the tropics, especially in India (Mathur 1949), but only *L. europaeus* L. occurs in Europe. It is somewhat similar to *Viscum album*, but has shorter, dark-green, deciduous leaves and yellow berries, which are not axillary like those of *Viscum*, but borne in a terminal raceme. It generally causes much more hypertrophy of the host tree than *V. album*, very large gnarled swellings forming at the points of attack. Attacked branches often die back to the point of infection. Severe injury to oak standards and coppice has occasionally been reported from Europe. It has a much more restricted host range than *V. album*, occurring chiefly on species of oak, but occasionally on chestnut (*Castanea*). It has been introduced into Britain, but is certainly not commonly cultivated, and does not seem to have spread naturally to any extent.

Elytranthe

Several species of this genus occur on *Nothofagus* in New Zealand, where apparently they are esteemed for the beauty of their flowers, an unusual phenomenon with mistletoes (Guthrie-Smith 1936). They have not been introduced into Britain, where their host genus is being increasingly planted.

Arceuthobium

The so-called Dwarf mistletoes which comprise this genus are all parasitic on conifers. They are serious only in the United States, where five species occur embracing a very wide coniferous host range (Gill 1935; Kuijt 1955). They are very small compared to *Viscum album*; the shoots, which are yellow, brown, or green in colour, vary from 1 to 20 cm. in length according to the species, and the leaves are reduced to opposite pairs of scales. Like *Viscum* they are dioecious, the female plants bearing olive green or blue berries which, when ripe, explode with sufficient force to propel the seed distances up to 30 feet.

Dwarf mistletoes cause extensive damage to the north-western coniferous forests in North America, being second only to decay fungi as a cause of loss. They bring about the formation of witches' brooms, and sometimes of cankers, they reduce growth and cause crown die-back, in the case of trunk infections they lower the amount of saleable timber, and in addition reduce the seed crop. There is a tendency in felling to leave mistletoe-infected seed trees because they have less timber value. Thus not only is less seed produced for regeneration, but a source of infection is left for the new crop. The wood in the witches' brooms tends to be abnormally resinous, so that when they are left on the ground after felling they tend to decay much more slowly than the rest of the brash, and thus extend the period of high fire hazard.

The sole European species, *A. oxycedri* (DC.) M. Bieb., occurs only in central and southern Europe (Turrill 1920), causing witches' brooms on *Juniperus communis*, *J. oxycedrus*, and other species of juniper (Heinricher 1914). It has also been reported on *Chamaecyparis thyoides* growing in Europe (Heinricher 1930). It is of no great importance since it does comparatively little damage and since *Juniperus* is not an important genus in Europe. *A. oxycedri* is in cultivation on juniper in a number of botanical gardens in Britain. Only the male plant was imported, so that increase can only be by cuttings from infected plants, and no risk of uncontrolled spread is involved.

The genus *Arceuthobium* is unique among mistletoes in its ability to stimulate witches' broom formation on the part of the host tree. Several others, notably *Viscum*, give a witches' broom appearance to an infected tree, but in their case the 'broom' consists of the twigs of the parasite and is not produced by the host.

Several fungi attack *Arceuthobium*. *Wallrothiella arceuthobii* (Pk.) Sacc. (Ascomycetes, Sphaeriales) infects the female flowers and stops seed formation. It may restrict the development of Dwarf mistletoe in some regions (Thomas, R. W. 1953). *Septogloeum gillii* Ellis (Fungi imperfecti, Melanconiales) attacks the stems of the mistletoes and causes considerable damage to them. It is considered to have possibilities as a biological control (Ellis 1946). There are no fungi able to exercise a similar influence on *Viscum album* in Europe.

Phoradendron

The larger mistletoes of the *Viscum* type are represented in North America by the genus *Phoradendron*, of which there are many species, mainly on broad-leaved trees, but also on conifers. They are similar to the European mistletoes in their general appearance and behaviour. At Christmas they are used in the same way and tend therefore to be regarded as a benefit rather than as a pest, despite the fact that heavy infections may cause appreciable damage. Many other species of this genus occur in South and Central America and elsewhere.

Control of mistletoes

The most obvious control for mistletoe would be the removal of all infected branches, but often this would involve severe mutilation of the tree. In America and on the Continent, where mistletoes occur on a forest scale, pruning would be too costly. In such cases the selective felling of badly infected trees and the encouragement of healthy ones can be practised with varying degrees of intensity (Plagnat 1950*b*; Hawksworth and Lusher 1956). Spread from tree to tree is usually slow and restricted. Hawksworth (1958) quotes a figure of 30 feet in twenty years for *Arceuthobium americanum*. This certainly assists eradicatory control. Flame guns have been used to destroy *Loranthus* on *Eucalyptus* in Australia. The mistletoe can be withered without seriously damaging the tree, but it is doubtful if the parasite would be completely killed, and in any case the method could be used only on low and easily accessible trees.

The best possibility of control appears to lie in chemotherapy. Considerable success, chiefly in the case of *Loranthus* on *Eucalyptus*, has attended the use of 2,4-D, either as a foliar spray on the mistletoe (Hartigan 1949) or by injection of the chemical into holes bored in the trunk (Greenham and Brown 1957). *Loranthus* has been controlled in India by the injection of copper sulphate, 2,4-D, or chlorinated phenoxy-acetic acids (Kadambi 1954; Muthanna 1956). In all cases the mistletoe was killed, while the host tree was virtually unharmed. Additional work is necessary before general recommendations can be made. Tree reactions, especially to 2,4-D, are very variable, and a safe dose for one tree species might be fatal to another.

Other parasitic flowering plants

Lathraea

Two species of *Lathraea* are parasitic on the roots of trees in Britain. The commoner, *L. squamaria* L. (Toothwort), produces a flowering shoot 8–30 cm. in height having a few scales at the base, and bearing a one-sided raceme of white bell-shaped flowers sometimes tinged with dull purple. It occurs quite commonly on the roots of a large number of broadleaved hosts, particularly on *Ulmus* and *Corylus* (Hartley and Ellis 1931). It is found chiefly in moist woodlands on good soils, and is locally common in some limestone areas in Britain. There is no evidence that it causes any appreciable damage to the host plants. On the Continent *L. squamaria* has been recorded on a number of coniferous genera (Heinricher 1906; Lilpop 1923), but in Britain it does not seem to occur on conifers.

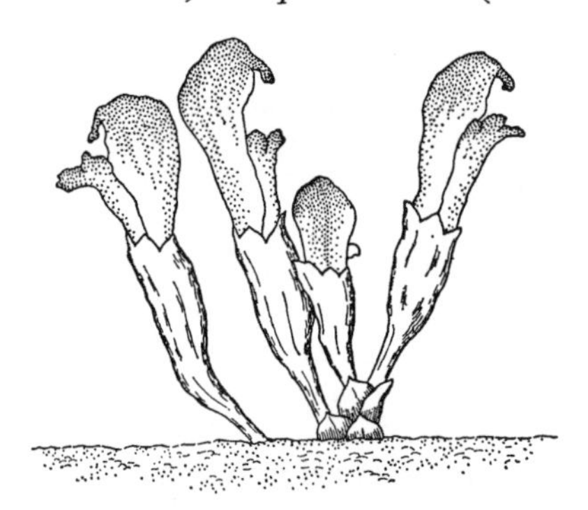

FIG. 33. *Lathraea clandestina* (× ½). (J. C.)

The other species, *L. clandestina* L., is confined to the roots of willows and poplars (Anon. 1949*a*). It has decorative flowers, purple in colour and somewhat reminiscent of crocuses (Fig. 33). Originally introduced into Britain as a garden plant, it has become naturalized in a few places. It occurs mainly in damp and moderately shady situations. In this case also there is no suggestion of damage to the host trees.

Orobanche

Two of the many species of broomrapes that occur in Britain are parasites on the roots of trees, but neither causes appreciable damage. *Orobanche rapum-genistae* Thuill. is found on the roots of shrubby members of the pea family, particularly *Cytisus* and *Ulex*. *O. hederae* Duby. occurs very locally on ivy. Both produce flower spikes 15–75 cm. in height with dull purplish-cream flowers. As with *Lathraea* the flower spike is the only part of the plant that appears above ground.

Cuscuta

The dodders are annual plants, which twist round the host and attach themselves by suckers, through which they extract their nourishment. Although they germinate on the ground, they soon become detached and are carried upwards by the growing host plant. They bear small dense heads of pinkish flowers at intervals along the otherwise bare stems. None of them is at all important on woody plants, but *C. europaea* L., the Large dodder, which is found mainly on nettles and hops, occasionally attacks woody hosts, such as *Corylus*, *Salix*, etc., especially on the Continent. *C. epithymum* (L.) Murr., much commoner in Britain, is found on *Ulex*, *Calluna*, and various other

FIG. 34. *Cuscuta epithymum* on *Ulex*, Guernsey, August (×1½). (J. C.)

herbaceous and woody hosts (Fig. 34). Dodders can cause considerable damage to some agricultural crops, but are of no importance in forestry.

Semiparasites

Several genera of the sub-family Rhinanthoideae of the family Scropulariaceae, although possessing green leaves, are parasitic on other green plants. These include *Bartsia*, *Rhinanthus*, and *Euphrasia*. Most are meadow plants, but *Melampyrum pratense* L. (Common cow wheat) and *M. sylvaticum* L., both of which have spikes of yellow flowers, occur in woodland, and are probably weakly parasitic on hardwoods.

The two ill-distinguished species of *Monotropa*, *M. hypopithys* L., and *M. hypophegea* Wallr., occur normally in woodland, producing spikes with scale-like leaves at the base and small bell-like flowers at the top, all creamy-white in colour and entirely without chlorophyll. They are, however, generally regarded as saprophytes. Certainly they do no detectable damage.

NON-PARASITIC PLANTS CAUSING MECHANICAL INJURY

Hedera

Ivy, *Hedera helix* L., is a common occupant of the forest floor, often covering large areas of ground, but climbing by means of specialized roots up any reasonably rough surface. Unlike our other woody climbers (see below) it seldom makes much progress on young trees and never damages them, unless, as Milne-Home (1952) assumes, it has a special significance as a competitor for nutrients and moisture. Admittedly ivy, by its ability to climb, remains

FIG. 35. The aerial roots of ivy; left, from a tree; right, from a wall. The development is the same on both substrates (× ½). (J. N.)

alive in the light when other vegetation is shaded out by the trees above. In this way it continues to compete longer than most other forest weeds.

The aerial roots formed along the climbing stems appear to act solely as a means of adhesion. On trees they penetrate only the fissures in the outer bark, and are not in contact with live tissue. It is probably significant that exactly similar roots are produced on surfaces such as walls from which they certainly cannot extract appreciable nutriment (Fig. 35). There is no evidence whatsoever of parasitism (Hadfield 1956; Brown, W. R. 1957). The ivy on a tree normally dies when its stem is severed at the base, clearly indicating that it is not getting appreciable amounts of water or nourishment from its aerial roots. Ivy was one of the many plants found by Willis and Burkhill (1884) growing in the crowns of old pollard willows, usually with no attachment to the ground, but feeding on the debris and rotten wood in the centre of the pollard, not on the live parts of the tree. Occasionally ivy may grow tightly round young trees and cause distortion or partial strangulation (Tubeuf 1915), but this does not often happen since its growth up the tree is usually nearly vertical. When the ivy develops its flowering branches in the crown of the tree, a type of growth that unlike the purely vegetative parts demands full sunlight, there will be competition for light between the foliage of the ivy and that of the tree, a struggle which the ivy often wins. So that while ivy allowed to develop unchecked in the crown of a tree is certainly harmful, ivy climbing up a tree is doing no serious damage, and its indiscriminate cutting from trees is generally waste of money. Stripping ivy from thin-barked trees exposed to the sun may be definitely harmful, leading to sunscorch (Chapter 5) of the hitherto shaded bark.

The diseases of ivy are dealt with in Chapter 35. None of them has any significant effect on its survival or spread.

Lonicera

The common honeysuckle, *Lonicera periclymenum* L., occurs very commonly in woodlands all over Britain. In reafforestation it can cause serious injury by twining round and constricting the stems of young trees. This eventually results in considerable swelling and deformation of the timber (Jacquiot and Viney 1954), with weak points at which the tree may subsequently break, and very occasionally death of the top of the tree.

During the weeding and cleaning stages honeysuckle should be rigorously cut, and any trees severely affected should be removed in the first thinning. When trees have closed canopy, honeysuckle causes no further trouble. The diseases of honeysuckle are dealt with in Chapter 35; none of them has any significant effect on its survival or spread.

Clematis

Clematis vitalba L., variously known as Old man's beard or Traveller's joy, is common on calcareous soils from Yorkshire southwards, and sometimes causes damage to trees by climbing up them and depriving them of light by covering their foliage with its own. In addition the sheer weight of the climber will often bend over a pliable tree. Occasionally, as trees grow taller, *Clematis* continues in the crown, connected to its own roots by long trailing stems, like the lianas of tropical forests. As with honeysuckle, cutting the *Clematis* away from the trees before canopy closure is the only method of control. The diseases of *Clematis* are dealt with in Chapter 35; none of them has any significant effect on its survival or spread.

DISEASES OF FOREST WEEDS

It is no part of this work to consider in detail the effect of weed plants on forest growth. In Chapter 6 the influence of weed growth on the water-supply of trees was discussed, and weeds such as *Lonicera* and *Clematis*, which have a damaging mechanical action on trees, are mentioned above. Here we are concerned only with the possibility of diseases exercising some control over weed species, and thus lessening their effect on trees; unfortunately this seldom happens.

Bracken

Bracken, *Pteridium aquilinum*, is one of the most frequent weed species in forest plantations. Though a considerable number of fungi have been recorded on it, only two appear to have been established as parasites. The first of these, *Corticium* (*Ceratobasidium*) *anceps* (Bres. & Syd.) Gregor. (Basidiomycetes, Aphyllophorales), kills the fronds and spreads a short distance down the leaf stalks. The dead leaf segments shrivel and fall off, so that in a severe attack little may be left except the bare axes of the fronds covered with the felt-like mycelium of the fungus (Gregor 1932*a*, 1935).

The second, *Phoma aquilina* Sacc. & Penz., together with a species of *Stagonospora* and *Ascochyta pteridis* Bres. (all Fungi Imperfecti, Sphaeropsidales), is associated with dieback and blackening of expanding shoots. Older fronds may become lop-sided or truncated by the death of the apex. The pathogenicity of the *Phoma* has been proved under greenhouse conditions; but the status of the two other fungi remains obscure. This disease has so far been found only in two places (Angus 1958). Gregor (1932*b*) associated a species of *Mycosphaerella* (Ascomycetes, Sphaeriales) and Angus (1958) *Didymella hyphenis* (Cke.) Sacc. (Ascomycetes, Sphaeriales) with severe dieback of bracken, but there is no definite proof of pathogenicity. *Cryptomycina pteridis* (Reb.) v. Höhn. (Ascomycetes, Phacidiales) causes leaf-roll on bracken. The fungus is systemic in the underground stems and buds. The disease forms slowly extending patches. Spore infections over longer distances are rare (Bache-Wiig 1940). It occurs in North America, Asia, and on the Continent, but not apparently in Britain.

Attempts to spread bracken fungi artificially have proved inconclusive (Anon. 1932*a*). Since the older diseases do not appear to have increased over the past twenty years, and since the more recent ones are very limited in extent, there is little possibility that they will have any appreciable effect on bracken, whether they be aided by man or left to spread naturally.

Fomes annosus (Chapter 16) has been found on bracken rhizomes (Rishbeth 1950). In his opinion the fungus was unlikely to grow on them vigorously enough for the bracken rhizomes to be a serious source of infection to trees.

Bracken is often severely cut back by spring frost, but it sends up recovery fronds later in the spring, and does not appear to be appreciably weakened, except possibly in a few sites where frosts are of almost annual occurrence.

Senecio

The rust fungus *Coleosporium tussilaginis* (Pers.) Lev. (Basidiomycetes, Uredinales) has a number of herbaceous hosts, of which *Senecio* is perhaps the commonest. It is more frequently seen on *Senecio* species than on pine (Chapter 22) and appears to do more damage to them. It is most often found on *S. vulgaris*, Groundsel (Fig. 57), a common nursery weed, but also on other species, such as *S. jacobaea*, Ragwort, and *S. sylvaticus*, Wood groundsel. Although the rust may cause a severe reduction in the shoot growth of affected plants, it seldom kills them. Generally it hastens flowering, a serious matter with a weed species which in any case flowers young and sets seed rapidly. The fungus, therefore, is of no assistance in the control of Groundsel.

Heather

Heather, *Calluna vulgaris*, which is a serious competitor with young trees over large areas in Britain, has one significant disease caused by a rhizomorph-forming fungus, *Marasmius androsaceus* Fr. (Basidiomycetes, Agaricales). The fruit bodies are tiny cream or pale-brown toadstools about 10 mm. across, with thin black horny stems up to 60 mm. long. The rhizomorphs are far thinner than those of *Armillaria mellea*, being almost like horsehair. The fungus infects the branches, causing dieback and sometimes the death of the whole plant. It occurs on old shoots much more than on young ones, so that it is seldom found doing significant damage on heather that has been recently cut or burned. It may in a few instances have a really damaging effect on the health of areas of heather with old long stems (Braid and Tervet 1937; Macdonald, J. A. 1948). The disease has mostly been reported from Scotland, where it appears to be commoner in the west than in the east. There is little hope, however, of using it as a means of heather control. *Sporonema (Clinterium) obturatum* (Fr.) Sacc. (Fungi imperfecti, Sphaeropsidales) has been associated with dieback of *Calluna* in Argyll, but there is no proof of its pathogenicity, and it is certainly of no significance.

A. mellea (Honey fungus) has occasionally been recorded on heather, but both Alcock and Wilson (1927) and Delevoy (1928) considered that it had spread from infected tree stumps. It seems unlikely that *Armillaria* could cause appreciable damage to heather in the absence of stumps to act as centres of infection. *Fomes annosus* has also been found on heather (Watt 1927), but it is certainly uncommon on that host.

The leaves and shoots of heather are often considerably withered in cold winters. This is probably more damaging, and therefore of more help to the forester, than the rather occasional and local damage by *Marasmius*.

Other shrubby weeds

The diseases of shrubs that are grown as ornamentals, but which may on occasion act as weed species, are dealt with in Chapter 35. These include *Cytisus*, *Hippophae*, *Ligustrum*, *Rhododendron*, *Sambucus*, *Symphoricarpos*, *Ulex*, and *Vaccinium*.

Fomes annosus has been recorded on *Ulex europaeus* (Gorse) (Rishbeth 1950), and in one instance young pines had been infected by the fungus spreading from a gorse stump. It may therefore occasionally play a significant role as a source of infection by this fungus.

14

FUNGI AND BACTERIA ON FOREST SEEDS

SEEDS, like other exposed objects, acquire on their surfaces fungal spores and bacteria which become significant only if they become active. Many of these organisms are harmless, but those that are pathogenic can act in a number of different ways. According to their nature and the time when they become active, they may attack either the seed in the dormant state, the radicle immediately after emergence, becoming then the first stage of Damping off, or the young plant, acting then as a means of disease transmission from a plant to its seed progeny. Of course, fungi already in the soil, as well as those on seeds, can cause damping off, and seed transmission is only one of the many ways in which a disease can pass from an infected tree to young seedlings. General reviews of this subject have been given by Baldwin (1942) and by Crocker and Barton (1953).

Several lists of fungi occurring on tree seeds have been published, mostly referring to conifers (Garbowski 1936; Zhuravlev 1940; Salisbury, P. J. 1953; Gibson 1957), or to conifers and hardwoods (Orlova 1954; Holmes and Buszewicz 1953–7); but lists for oak have been given by Potlaichuk (1953) and Urošević (1957). The most complete and up-to-date list is that of Noble, de Tempe, and Neergaard (1958). Many of the fungi in these lists are harmless, some of those of pathological significance are discussed below.

Damage on the tree

A number of fungi and bacteria attacking cones and catkins are mentioned in other chapters, for instance *Cronartium strobilinum* and *C. conigenum* on pine cones, *Thekopsora areolata* and *Chrysomyxa pyrolae* on spruce cones, *Taphrina amentorum* on alder catkins, *Sclerotinia fructigena* on hazel nuts, *S. betulae* on birch catkins, *Xanthomonas juglandis* on walnuts, and *Gloeosporium ulmicolum* on elm seed. With some of these it is the cone or catkin that is attacked rather than the seed, but in either case the amount of viable seed may be greatly reduced, and the seeds which survive may act as spore-carriers for the disease.

Damage in storage

While fungi and, to a much smaller extent, bacteria play a part in the deterioration of tree seed in storage, the primary troubles with conifer seed are normally physiological. The reserves in the seed are gradually used up in respiration, while slow degenerative changes take place in some of the cell contents. In addition, excessive moisture or temperature stimulates activity, leading to the formation of by-products which may be toxic under conditions where normal germination is impossible. Less commonly, low humidities or low temperatures directly damage the seed. Conifer seed is usually stored at low temperatures and at air humidities below 75 per cent. (Holmes and Buszewicz 1958), which is generally too low for fungal development.

Similar damage may happen to seeds of broadleaved trees, but in their case fungal invasion of injured seed is so rapid that such attacks may appear to be primary; indeed there is much evidence that fungi do attack uninjured hardwood seeds. Many broadleaved tree seeds require comparatively moist storage conditions, so that fungi are much better able to develop, than on coniferous seeds stored at lower humidities. Potlaichuk (1953) found *Sclerotinia pseudotuberosa* Rehm. (Ascomycetes, Helotiales), *Phomopsis quercella* Died. (Fungi imperfecti, Sphaeropsidales), *Gloeosporium quercinum* West. (Fungi Imperfecti, Melanconiales), which is usually associated with leaf spotting of oak (Chapter 27), and *Valsa intermedia* Nke. (Ascomycetes, Sphaeriales) definitely harming acorns. Strauch-Valeva (1954) recorded a number of fungi attacking acorns in storage, the most damaging of which was *Phomopsis quercella. S. pseudotuberosa* can be seriously damaging to stored acorns (Uroŝević 1957), but it can be controlled by fungicidal treatment applied as early as possible after collection (Bornebusch 1941). This same fungus (sometimes known as *Ciboria batschiana* (Zopf.) Buchw.), the conidial stage of which is *Rhacodiella castaneae* Peyr. (Fungi Imperfecti, Moniliales), is particularly serious on *Castanea*, because of the commercial value of the edible nuts. The nuts can be infected while they are still on the tree, but usually after they have fallen. They eventually become black and rotten. Spherical sclerotia appear, on which the cup-like apothecia are borne in the following spring. It is often difficult to detect the disease when the nuts are gathered, so that it can spread in storage and cause serious losses (Voglino 1931; Arnaud and Bathelet 1936). The fungus spreads by means of its superficial mycelium and probably also by conidia, which are borne directly on it. A species of *Phomopsis* (Fungi Imperfecti, Sphaeropsidales), variously described as *P. viterbensis* Camici and *P. endogena* (Speg.) Ciferri, has also been found causing mummification of chestnuts (Camici 1948; Ciferri 1950). Tocchetto (1954) attributed similar damage on *Castanea crenata* in Brazil to a species of *Dothiorella* (Fungi Imperfecti, Sphaeropsidales). It is doubtful if any of these fungi occur in Britain.

The occurrence of superficial moulds on chestnut fruits has also received considerable attention (Voglino 1928; Bertotti 1930). Several species of fungi are involved. These include *Penicillium crustaceum* (Fungi Imperfecti, Moniliales), *Trichothecium roseum* Link. (Fungi Imperfecti, Moniliales), and *Mucor* spp. (Phycomycetes, Mucorales). The methods of seed treatment for chestnut are discussed under Chestnut blight (*Endothia parasitica*) (Chapter 29), since the transmission of this disease on nuts, imported either for eating or seed purposes, is unfortunately a possibility. A very large number of fungi attack the husks of walnuts, often spreading to the kernels and causing serious losses (Batchelor 1923; Hamond 1931). Příhoda (1955) recorded heavy losses of stratified hornbeam fruits, owing to attack by species of *Verticillium* (Fungi Imperfecti, Moniliales).

There is less published evidence of damage to coniferous seed. Garbowski (1936) found that *Pyronema omphalodes* (Bull. ex Fr.) Fckl. (Ascomycetes, Pezizales) and a species of *Botrytis* (Fungi Imperfecti, Moniliales) seriously lowered the germination of Scots pine seed. Satô (1955) considered that some of the fungi which he found on seed of *Cryptomeria japonica* were pathogenic. P. J. Salisbury (1953) reported penetration of Douglas fir seed by fungi, but did not consider such infections necessarily damaging. Huss (1952) and Gibson (1957) thought that fungal infection resulted in the failure of weak seed or seed damaged during extraction, but that its effect on good seed was negligible. It was suggested that greater care should be taken to avoid damage during de-winging. No work has yet been done on the pathogenicity of the very large number of fungi isolated from conifer and hardwood seeds in Britain (Holmes and Buszewicz 1953–7). Pathologically the most interesting are *Verticillium* spp., *Botrytis cinerea*, *Fusarium* spp., and *Pestalozzia hartigii*, all of which are active causes of tree disease. These, however, were not generally the fungi most commonly isolated, and others may well prove more important as sources of loss in storage, and at the time of germination. There is no doubt, however, that many of the fungi that are found on seeds are normally harmless species, which only become pathogenic if the conditions are abnormally favourable to them.

Damage on germination

When we consider damage from fungi on the seed at the time of germination we are in fact involved with the very earliest stage of Damping off, which is dealt with fully under 'Nursery Diseases' (Chapter XV). There is no evidence as to how much of this damage is due to fungi initially on the seed and how much to fungi already in the soil, but Zhuravlev (1940) showed that the percentage emergence of coniferous seedlings decreased with increasing contamination of the seed with a species of *Fusarium*. There is no doubt, however, that the extent of the injury which occurs in the early stages of germination has been underestimated, and in addition often attributed to low germination capacity of the seed rather than to the activities of fungi. There is still doubt, however, as to the relative importance of fungal attack and other factors in pre-emergence losses. Hartley, Merrill, and Rhoads (1918) considered that fungi in the soil could attack the seed before germination as well as the radicles after emergence, and evidence of radicle attack is given by Rathbun-Gravatt (1931) and by P. L. Fisher (1941).

Seed-borne diseases other than Damping off

There is abundant evidence that a considerable number of agricultural diseases are seed-borne (Moore, W. C. 1954). A number of lists of pathogens found on seed have been issued (Orton 1931; Anon. 1953*a*; Noble, de Tempe, and Neergaard 1958); though it must be remembered that the presence of the spores of a parasite on a seed does not mean that it will infect the resultant plant, let alone that this is the normal method of transmission. In fact in forestry, leaving out of consideration attacks on seed or very young seedlings, there is no proven case of disease transmission on seed. *Endothia parasitica* (Chestnut blight) (Chapter 29) has been found to be able to infect seed (Collins 1915), and it may be that it was by this means that it has once or twice reached the western from the eastern States and possibly France from Italy; but it is certainly not one of the normal methods of transmission.

Indeed, if the seed itself is attacked it will very probably not germinate, so that the possibility of direct infection of the resulting seedlings does not arise. *Guignardia aesculi* has been found on the fruits of Horse chestnut (Orton 1931) and the bacterium *Xanthomonas juglandis* on walnuts (Wormald 1930), but there is no proof that these diseases are transmitted from one tree to another by this means. Far too often the presence of the spores of a parasite on seed has been taken as evidence of seed-borne transmission. The much quoted case of *Keithia thujina* on *Thuja* (Pethybridge 1919) rests entirely on the supposition that, since there is no evidence of plants being imported into Britain, the disease must have come on seed. Rayner (1929) has described the transmission of mycorrhiza-forming fungi in the seeds of *Vaccinium* and *Calluna*, but there is no suggestion that this is the normal method by which mycorrhizal fungi reach developing seedlings.

Control of fungi on seeds

Lack of evidence that forest diseases are commonly seed-borne makes any general surface-sterilization of seed, even when imported from a pathologically dangerous region such as North America or eastern Asia, an added expense and trouble which can hardly be justified. It is only in special cases, such as that of Chestnut blight, that measures need be taken.

Considering pre- and post-germination losses, however, and the trouble that fungi can cause under the artificial conditions of seed test germinations, further work on the surface sterilization of seed before germination or before sowing does seem entirely justified. The effect of seed treatment of this nature is likely to continue for a time after germination has taken place, indeed it is one of the recognized, though not always successful, methods of controlling Damping off (Chapter 15).

The extensive literature on the fungicidal treatment of agricultural seeds will not be reviewed here. Undoubtedly many agricultural methods would be equally applicable to tree seeds. Dusting with fungicides or immersion in fungicidal solutions are standard methods (Satô 1955), but fumigation, possibly combining insecticidal and fungicidal action, can also be used (Lanza 1950; Darpoux and Ridé 1952). Fungicides have sometimes proved phytotoxic, which has stimulated research into other methods of removing or destroying fungus spores on seeds. Washing with water (Leben, Scott, and Arny 1956; Ivanoff 1958), or the use of high-pressure water jets (Zhuravlev 1952), are methods that have been advocated. Heating the seed to a temperature lethal to fungi, but not to the seed itself, is another possibility (Miller and McWhorter 1948; Busnel, Darpoux, and Ridé 1951). It is obviously important that seed should be stored clean and without admixture of plant and soil debris (Uroševič 1957).

15

DISEASES OF GENERAL IMPORTANCE—NURSERY DISEASES

IT is in the nursery that the pathology of trees and shrubs comes nearest to its agricultural counterpart. Older trees, by their size, the sites where they are planted, and their length of occupation of those sites, provide very different conditions for the development and control of pathogens. In nurseries, where the plants are small and only occupy the ground for a year or two, tree crops are much more akin to agricultural ones. This naturally affects our approach to nursery diseases, particularly in matters of control. Spraying, for instance, is much easier in the nursery than in the forest, while infected ground can be freed of crops or even sterilized with much greater ease. These matters are considered more fully in Chapter 36, 'The Control of Fungal and Bacterial Diseases'.

Discoloration of foliage

Before dealing with diseases due to specific pathogens, one rather general pathological phenomenon demands consideration. This is discoloration of foliage, particularly on conifers, which is very common in tree nurseries. It is usually symptomatic of some physiological derangement rather than of fungal or bacterial diseases. Pronounced purpling of the foliage of pine, reddening of spruce needles, and very striking browning of *Thuja* can result from exposure to low temperatures. Excess water in the soil or drought can both produce discoloration, usually purpling, of the foliage of conifers. Nutrient deficiencies (Chapter 7) also produce a wide range of foliar symptoms both on conifers and on broadleaved trees. If the damage does not proceed beyond the discoloration stage, it is reversible and the plants recover their normal green colour when the cause is removed or alleviated. Under these circumstances there is no evidence of any predisposition to fungal attack. If the damage becomes sufficiently severe to result in the death of whole or parts of needles or leaves, it may lead to infection by fungi such as *Botrytis cinerea* or by more selective ones such as *Lophodermium pinastri* on pine.

Damping off and Root-rot

This complex of diseases includes fungal attack on the succulent parts of seedlings and transplants, underground or at ground-level, from the sowing of the seed, to final removal from the nursery.

Pathogens

A very large number of fungi, many of them common soil saprophytes, has been associated with damping off and root-rot. The most important, in Britain at any rate, are *Pythium* spp. (Phycomycetes, Peronosporales), in particular *P. debaryanum* Hesse, *P. ultimum* Trow, and *P. irregulare* Buis., and *Corticium solani* (Prill. & Delacr.) Bourd. & Galz. (Basidiomycetes, Aphyllophorales) (syn. *Pellicularia filamentosa* (Pat.) Rogers), usually referred to by the name applied to its mycelial state, *Rhizoctonia solani* Kuhn. Species of *Fusarium* (Fungi imperfecti, Moniliales) are also frequently cited as causes of Damping off, but in general they are less common and less virulent than *Pythium* and *Rhizoctonia*. Among other fungi which have been associated with Damping off are *Macrophomina phaseoli* (Maubl.) Ashby (Fungi Imperfecti, Sphaeropsidales), *Phytophthora* spp. (Phycomycetes, Peronosporales), *Moniliopsis klebahni* Burch. (*Mycelia sterilia*), *Botrytis cinerea* (see below), *Verticillium* spp. (Fungi Imperfecti, Moniliales), *Alternaria* spp. (see below), and *Phoma* spp. (Fungi Imperfecti, Sphaeropsidales), but this list is far from exhaustive. Numerous authors have published lists of fungi associated with or capable of causing Damping off (Hartley 1921; Rathbun 1922, 1923; Rathbun-Gravatt 1925, 1931; Fisher, P. L. 1941; Vaartaja and Cram 1956). Most of these lists must be assumed to have only local significance, for there are often wide differences between the fungi occurring in different nurseries and even between those in the same nursery at different times of year (Pomerleau 1942). Griffin (1957) considered that *Pythium* spp. are the main agents of Damping off in Britain, but it is seldom possible to generalize.

Pythium is a very large and difficult genus, which has been fully described by Middleton (1943). *Pythium* species, particularly as causes of Damping off and root rots, occur on many other plants as well as trees and shrubs. The fungus is microscopic in all its parts, except that the white mycelium may occasionally be visible to the naked eye on the roots or collar of attacked seedlings. For many years the species of *Pythium* commonly found in association with Damping off was almost automatically assigned to *P. debaryanum*. It is now realized that, in Britain at any rate, *P. ultimum* is much more important. Eliason (1928) and Warcup (1952) both noted differences in pathogenicity between different species and strains of *Pythium*. Damping off due to *P. irregulare* has been dealt with particularly by Roth and Riker (1943*a*, *b*, *c*.)

Corticium solani is usually found solely in its mycelial and sclerotial stage, *Rhizoctonia solani*. The small, irregular, brownish-black sclerotia occur quite commonly on the mycelium. The fructifications of the perfect stage, which take the form of thin, irregular, wrinkled, buff plates on the surface of dead tissue, are seldom seen. This fungus has a very wide host range both in Britain and abroad, but does not seem to be nearly so important as *Pythium* in British forest nurseries. It is a very variable fungus, some strains being much more virulent than others. Its virulence is general, rather than specific, any one strain attacking all susceptible hosts with about the same degree of severity (Wiant 1929; Itô and Kontani 1951; Rushdi and Jeffers 1952; Satô, Ôta, and Shôji 1955*b*). Occasionally it attacks the aerial parts of plants. Storey (1955) recorded it on the leaves and shoots of azaleas, as well as causing Damping off; Itô, Kontani, and Kondô (1955) associated it with 'Web-blight' on Japanese larch seedlings.

Several species of *Fusarium* cause Damping off, and in this role they may sometimes prove more important than other fungi (Tint 1945*a*, *b*; Igmándy,

Milinkó, and Szatala 1954). Generally, however, *Fusarium* is not so important as *Pythium* and *Rhizoctonia* in forest nurseries, though in Britain Damping off of many other crops has been attributed to it. Buckland (1948) has reported a *Fusarium* spreading from the soil and attacking the aerial parts of Douglas fir seedlings.

Phytophthora seems to be far more important in the forest than in the nursery (Chapters 16 and 29). *P. cactorum*, however, attacks the cotyledons and succulent stems of beech seedlings, causing a disease of the above-ground portions, which has often been referred to as a form of Damping off (Chapter 28). As a cause of true Damping off, however, this genus is of relatively small importance in Britain. Crandall and Hartley (1938) recorded *P. cactorum* on a wide range of broadleaved trees in nurseries in the United States and also connected it with Damping off of conifers. It is probably much more local in occurrence than the commoner Damping off fungi.

Symptoms and development

Damping off has three phases each of which may and usually does merge into the next. These are (*a*) pre-emergent attack on the germinating seed or developing seedlings, (*b*) post-emergent attack on the succulent roots and hypocotyls, and (*c*) attack on the root tips of older seedlings and transplants, usually resulting in restricted growth rather than death (Hartley 1921). On broadleaved species with soft stems Wright (1944) has distinguished a fourth type, known as 'sore shin', in which the fungus forms a lesion on the stem at ground level, usually without girdling it.

In pre-emergence Damping off the seeds, or more commonly the developing seedlings, are attacked and killed before they appear above ground. In this case, of course, there are no immediately visible symptoms; the emergent crop is merely sparse and possibly patchy. It is seldom possible to find the remains of attacked seedlings in the soil, since they tend to decompose rather quickly. A late stage of this type of damage sometimes occurs if the seed-bed cover has become caked. Clusters of seedlings push up a plate of soil leaving a humid air space underneath in which the cotyledons and hypocotyls are attacked. It is usually assumed that pre-emergence Damping off is due to the typical Damping off fungi such as *Pythium* and *Rhizoctonia.* These may be already present in the soil or may exist as spores on the seed before it is sown (Chapter 14). The fungi which commonly occur on seeds, however, are often not species associated with Damping off (Salisbury, P. J. 1953; Holmes and Buszewicz 1953–7), though species of *Fusarium*, a fungus known to be capable of causing Damping off, occur fairly commonly on seed. In any case it is probable that fungi not usually associated with the later stages of Damping off can attack germinating seed (Rathbun-Gravatt 1931). Garbowski (1936) found that *Pyronema omphalodes* (Bull. ex Fr.) Fckl. (Ascomycetes, Pezizales), normally regarded as a soil saprophyte, occurred commonly on Scots pine seed, and was able to cause heavy germination losses. There is still uncertainty on the relative importance of fungi and other factors in pre-emergence losses.

The erratic effects of seed treatment or soil sterilization on seedling emergence suggest that the importance of fungi varies greatly in different nurseries and in different seasons.

Seed of known low germination capacity is usually sown at an abnormally high density to save waste of space; but Gibson (1956*c*) has suggested that the dead seeds provide pathogenic fungi with an additional substrate on which they build up and attack the remaining viable seed with more than normal violence. According to Zhuravlev (1952), planting pre-germinated pine seed, a condition often following stratification, leads to increased pre-emergence Damping off.

In typical post-emergence Damping off, which may be referred to as classical

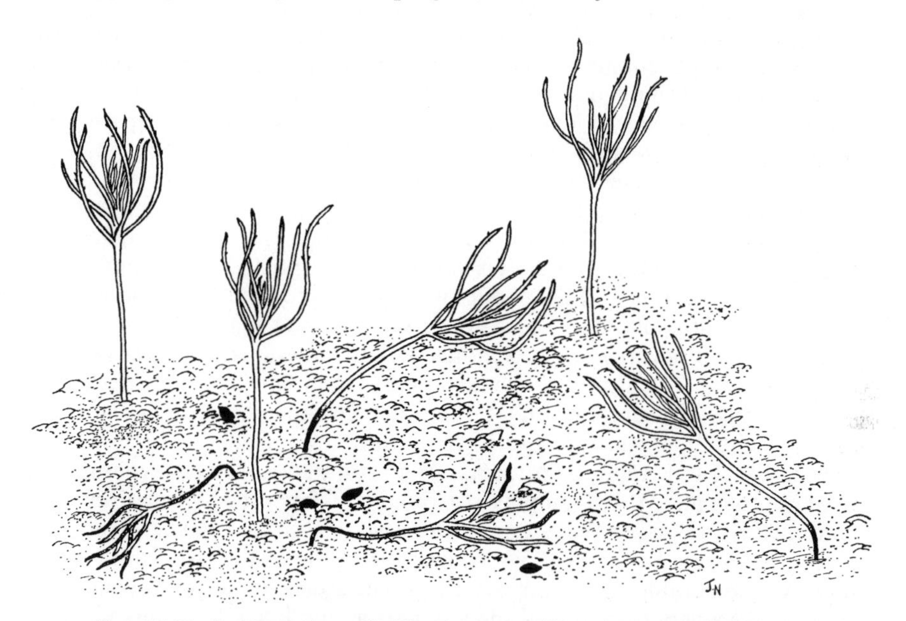

FIG. 36. Typical Damping off of Scots pine seedlings (nat. size). (J. N.)

Damping off, the seedlings collapse and wither, usually while they are still in the cotyledon stage (Fig. 36). This may be due to fungal attack on the hypocotyl, on the roots, or on both. In some cases the fungus appears to spread outwards from a centre so that the damage occurs in irregular patches. In the centres of these there may be few visible seedlings, most of the Damping off having been pre-emergent, or the dead seedlings having withered and disappeared (Plate VII. 2). Outside this empty centre is a ring of dead and withered seedlings, and outside that, one of seedlings currently attacked. But such regularity of attack does not always occur. In many cases the damage is much more scattered, healthy seedlings being intermixed with diseased ones. Because the attacked seedlings wither at ground-level and fall flat on the ground, the damage superficially resembles that caused by hot soil (Chapter 5). In practice, however, the conditions under which the two types of damage occur are so different that there is seldom any chance of confusion.

Fungal attack on the succulent root tips of older seedlings and transplants, which has been well described by Lindgren and Henry (1949), is not usually fatal. If soil moisture conditions are favourable, new roots develop, even if most of the original root system has been destroyed. If damage of this type is followed by drought or by excessively wet soil conditions, both of which discourage the formation of fresh roots, death of quite large seedlings or even of transplants may occur. In general the larger the plant the less likely is the whole of its root system to be attacked, and therefore the less likely death becomes. Rathbun (1922, 1923) found that the fungi which caused typical Damping off could also cause this type of injury, and it is logical to regard it merely as an extension of Damping off to plants too large and insufficiently succulent for all the below-ground portion to be attacked.

Nurseries in which conifers have been grown for a number of years often develop so-called 'conifer sickness', whereby growth of seedlings is poor and stunted. This trouble occurs particularly with Sitka spruce. It has been generally assumed to be due to an increasing concentration of pathogenic fungi in the soil. It was also thought that partial sterilization of the soil, which temporarily ameliorates the condition, acted by killing these fungi and allowing normal root development. Griffin (1956, 1957), working in a number of British nurseries, found that the death of root-tips was quite insufficient to account for the stunted growth. It is known that the poor growth is not entirely a matter of nutrition, and the possibility of toxin production by non-parasitic fungi in the rhizosphere has been suggested (Chapter 17).

Damping off occurs more commonly and more seriously in some nurseries than in others. It is also worse in some years than in others. The occurrence and virulence of Damping off of fungi are considerably influenced by the environment. It is generally agreed that Damping off does not occur in dry soil, and that wet, though not necessarily saturated, soils are favourable for its development. Most Damping off fungi are favoured by relatively high temperatures (Hartley 1921; Pomerleau 1942; Roth and Riker 1943*b*; Tint 1945*b*; Akai and Takeuchi 1954; Wright 1957*a*); nevertheless, if other conditions are favourable, Damping off fungi can attack under cool conditions, when the recovery reactions of the young trees are likely to be less vigorous. Shade, acting to some extent indirectly by its effect on soil moisture, but also directly as a matter of light intensity, increases Damping off (Hansen *et al.* 1923; Tint 1945*b*; Vaartaja 1952). Because of the effect of warmth, Damping off tends to develop more in high summer on late sowings than at the beginning of summer on seed sown early in the spring or in the autumn.

The acidity of the soil has a very marked effect on Damping off. The disease is nearly always reported to be more serious in alkaline or neutral soils than in acid ones (Roth, C. 1935; Jackson, L. W. R. 1940*b*; Roth and Riker 1943*b*; Tint 1945*a*; Griffin 1958). Damping off following the application of wood ashes to seed-beds was attributed to the resulting increase in soil pH (Anderson, M. L. 1930).

High humus content of the soil generally lessens Damping off. L. T. White (1956) found that the presence of humus lessened attack and he even suggested additions of forest litter to the soil as a means of prevention. Mikola (1952) considered that the beneficial effect was due rather to increased growth on the part of the seedlings than to any direct influence on the causal fungi. Vaartaja (1952), on the other hand, reported that Damping off increased with humus quality on some relatively poor soils under conditions of low light intensity.

The effect of humus is likely to be complex and possibly indirect. It may well decrease the soil pH, and very possibly alter the soil flora.

The activities of other components of the soil flora affect Damping off. In some cases the position may be further confused by Damping off fungi and nematodes both acting adversely on the same seedlings (Wright 1945). In fact without careful examination the activities of nematodes can hardly be distinguished from those of Damping off fungi. Effects involving the soil population are naturally complex and seldom direct. For instance, Gibson (1956*a*) found that treatment with ethyl mercury phosphate increased Damping off, probably by the destruction of organisms that were competing with the pathogenic fungi. Wright (1941) found that the addition of glucose to the soil lessened Damping off, possibly by stimulating organisms antagonistic to the Damping off fungi. Vaartaja (1954*c*) found that the addition of tetramethylthiuram disulphide to the soil increased the bacterial population and that some of the bacteria concerned were antagonistic to Damping off fungi. It follows that the control exercised by this substance may be partially indirect, due to its effect on the soil flora. The addition of antagonistic bacteria to the soil has even been suggested as a method of control (Krasilnikov and Raznitsina 1946). The practice of partial soil sterilization is briefly discussed below under 'Control'. There is no doubt that some of the beneficial effects which it produces are due to the encouragement of *Trichoderma viride*, relative to other soil fungi. *Trichoderma* is highly antagonistic to many fungi, including those causing Damping off (Weindling 1932; Weindling and Fawcett 1934; Allen and Haenseler 1935; Weindling and Emerson 1936), though its effects are not entirely consistent (Vaartaja 1957*b*).

Manuring is likely to produce complex and possibly unpredictable effects on Damping off, because of its influence on the soil flora and on the pH of the soil, as well as directly on the pathogenic fungi. Various suggestions have been made about the effects of manuring; for instance, W. G. Gray (1931) found that applications of basic slag gave very good control of Damping off. There has, however, been little attempt to investigate manurial effects, and there is no definite information on manures that would be beneficial or harmful in Damping off.

Some of the many other factors affecting the incidence of Damping off are discussed below in the section on 'Control'.

Distribution and damage

Damping off has been reported almost everywhere tree seedlings have been grown. It has been suggested that new nurseries on non-agricultural ground are usually free for the first few years, but this is not always the case. In Britain damping off is less serious now than a few years ago, possibly because the tendency to raise conifers in acid heathland or woodland nurseries has brought so many seed-beds on to acid soil, which is generally unfavourable to Damping off. It is still a major pathological problem, however, and a cheap method of control which could be generally applied regardless of local variations in soil and soil flora would be of great value. It is impossible to estimate the losses due to this disease complex, because there is so much confusion with other agencies of seedling damage such as frost or drought.

Damping off certainly occurs on seedlings in the forest as well as in nurseries. But its effect on natural regeneration has seldom been investigated. Wilde and White (1939) suggested that it might prevent the successful regeneration of pine species on heavy soils, and Duncan (1954) regarded it as the major cause of early losses of larch regeneration in Minnesota.

No conifer can be regarded as resistant to Damping off, but those that germinate quickly and grow fast are likely to be less affected than species of slow germination or slow early growth. Wright (1944) has given a long list of susceptible broadleaved species. He has also pointed out that the susceptibility of any particular host plant varies according to the Damping off pathogen concerned (Wright 1957). In considering control it is probably best to assume that all trees and shrubs are susceptible.

Control

Control can be exercised in a number of ways: (*a*) by adaptations of nursery practice, (*b*) by seed treatment before sowing, (*c*) by soil sterilization before sowing, or (*d*) by less drastic fungicidal treatment of the soil before or after sowing.

(*a*) **Modifications of nursery practice.** Though there are numerous cases where the amount of damping off is affected by the manner in which nursery operations are carried out, it is not often suggested that the disease can be completely and safely controlled by modifications in method. It is clear, for instance, that over-deep covering increases Damping off (Hansen *et al.* 1923), but shallow covering will not prevent it.

The type of covering material used is certainly important. If the local soil is used for cover, subsoil is safer than top soil (Igmándy, Milinkó, and Szatala 1954). Several authors favour sawdust or sand as a seed cover (Delevoy 1926, 1927; Buckland 1948; Dick 1950); but Cockerill (1957) found sawdust quite ineffective, and in any case its use presents certain difficulties.

Since most Damping off fungi become fully active only in warm weather, early sowing should enable seedlings to get past their most vulnerable stage before the attack has reached its height. Early sowing has frequently been suggested as a control measure (Delevoy 1926; Pomerleau 1942; Tint 1945*b*; Dick 1950).

Less information is available on the effect of method and density of sowing. Delevoy (1926) advocated drill sowing. Hansen *et al.* (1923) contended that dense sowing decreased Damping off. Gibson (1956*c*) got the opposite effect, which he attributed to dense sowing providing the attacking fungi with a greater amount of substrate per unit area. Similarly Satô, Ôta, and Shôji (1955*b*) found that the presence of weeds increased Damping off, probably because they also were attacked and served to increase the degree of fungal infestation. Possibly for the same reason Wycoff (1952) found that green manuring increased Damping off.

Since Damping off fungi are less active in acid soils, soil acidification has naturally been suggested as a control method for conifers. It has not, however, proved successful (Cockerill 1957). Johnson and Linton (1942), working in

Canada, pointed out that its effect was destroyed in wet years by rain and in dry years by irrigation water.

(*b*) **Seed treatment.** It has already been pointed out that some of the fungi found on seeds may attack the developing seedling, so that surface sterilization designed to destroy superficial spores could be of some value. In practice, however, the fungicidal effect of seed treatments should persist on and possibly around the seed in the soil, and thus protect it in the early stages of germination and development. In this role a wide range of products have proved more or less successful. In earlier experiments cuprous oxide and zinc oxide, particularly the former, were the best seed dusts (Ogilvie and Hickman 1938; Johnson and Linton 1942; Hamilton and Jackson 1951; Peace unpubl.). Recently attention has turned to more complex substances. Organic mercurials, tetramethylthiuram disulphide, and many other substances have given promising results (Riker *et al.* 1947; Holtzmann 1955; Molin 1955; Jacks 1956; Volger 1957; Peace unpubl.). The effect of such seed dressings can probably be increased by pelleting the seed, so that a greater quantity adheres (Berbee, Berbee, and Brener 1953; Vaartaja and Wilner 1956). Dressings must be applied to seed before it is stratified; after stratification some dusts may prove highly toxic, since they will come into direct contact with the emerging radicle (Cram and Vaartaja 1956). Seed disinfection is certainly one of the most promising means of control, especially if its influence could be extended to last until the seedlings had grown beyond serious danger.

(*c*) **Soil sterilization.** Partial sterilization of the soil has many other effects besides the control of Damping off. It kills many purely saprophytic soil inhabitants and may lead to rapid development of other more resistant ones, such as *Trichoderma viride*, in the clear field left to them. It may also provide a rapid increase of available nutrients for growing seedlings. It kills weed seedlings and weed seed and thus reduces weeding costs in seed-beds. It has already been pointed out that its undoubtedly favourable effect on 'conifer sick' soils is not primarily due to the destruction of root-attacking fungi. The general effects of soil sterilization have been discussed by Warcup (1951) and the whole subject reviewed by him (Warcup 1957).

In America a number of fumigants have been used successfully for soil sterilization. These have been reviewed by Parris (1958). Methyl bromide and ethylene dibromide have proved particularly successful (Lindgren and Henry 1949; Henry 1950; Wycoff 1955; Foster, Cairns, and Hopper 1956; Cockerill 1957). In Britain attention has been focused on chloropicrin and in particular on formaldehyde because of their greater ease of application (Edwards 1952; Faulkner and Aldhous 1956). Nearly all these authors have reported increased numbers of seedlings, which can presumably be attributed to a decrease in Damping off losses. But this increase does not invariably occur, and it is clear that the improved growth of seedlings in sterilized soil cannot be attributed primarily to lack of fungal attack on the roots.

Soil sterilization is an expensive and often troublesome process when used on a large scale in forest nurseries. If applied as a drench it may soak the beds at a time when they should be drying out for seed-bed preparation. It is certainly justified in some nurseries for its combined effect on seedling numbers, seedling size, and weed control. It could hardly ever, if at all, be justified solely for the control of Damping off.

(*d*) **Fungicidal treatment of the soil.** Much of the earlier work on the control of Damping off was based on the application of fungicides to the beds after the disease had appeared. However, by the time the first seedlings start to collapse many others will already be infected, so that, even if the disease is noticed at a very early stage and the fungicidal application made is successful, considerable losses will still occur. It was eventually realized that the disease was seldom detected early enough for this method of treatment to be satisfactory, and the present tendency is towards a heavy application at the time of sowing or periodic applications after sowing made without regard to the presence or absence of the disease. Such preventative treatments are certainly likely to be more effective than those made after the fungus has gained a hold.

In horticulture, particularly in glasshouses, Cheshunt compound (2 parts copper sulphate, 11 parts ammonium carbonate, used in solution) achieved a considerable reputation as a post-emergent treatment. It never became popular in forest nurseries, however, where treatment with a weak solution of potassium permanganate was often recommended (Delevoy 1926). This substance was sometimes quite ineffective (Gibson 1955*b*) and has now been superseded by more modern chemicals. A great deal of work has been done on the testing of these (Strong 1952; Berbee, Berbee, and Brener 1953; Gibson 1955*a*, 1956*b*; Vaartaja 1956*b*; Cram and Vaartaja 1957). A brief general review of recent work on post-emergence fungicides has been made by Vaartaja (1956*c*). Out of the many substances that have been tried, tetramethylthiuram disulphide, organic mercurials, and captan appear particularly promising. Rates of application cannot be given as they are likely to vary with the proprietary preparation actually used.

Considerable success has also been achieved with antibiotic substances (Gregory *et al.* 1952; Vaartaja 1957*a*), though Vörös (1954) found such substances ineffective in unsterilized soil, where they tended to decompose too rapidly. Frohberger (1956) has put forward the interesting suggestion that quinonoxim-benzoylhydrazone becomes systemic in the seedlings and thus provides more effective protection.

The position is made more complex by the fact that fungicides applied to the soil affect the saprophytic as well as the parasitic members of the soil population, and may thus produce unexpected secondary effects. Rushdi and Jeffers (1956) have pointed out that the nature of the soil has a profound effect on the efficacy of fungicides applied to it. The effect of fungicidal treatment is thus bound to vary from nursery to nursery and until more is known some failures will be experienced. Gibson (1955*a*, 1956*b*) particularly stressed the variability of results in different nurseries.

Obviously fungicides applied to the delicate tissues of germinating or developing seedlings must have a low level of phytotoxicity. On the other hand, they must be potent enough to destroy the Damping off fungi. This inevitably means that most fungicides used against the later stages of Damping off are very near to the phytotoxic level of the seedlings they are supposed to protect. Several authors have stressed the possibility of injury (Hartley 1915; Wiant 1929; Gibson 1955*a*; Cram and Vaartaja 1957).

The fact that after so long a period of research no firm recommendations can be made for the control of Damping off reflects the complexity of the whole problem. It is clear that treatments applied after the disease has been detected are likely to be too late. It is doubtful if soil sterilization can be justified solely as a preventative for Damping off. Its use is certainly economic in nurseries where it also improves growth and controls weeds. Seed treatments are worthy of trial, though it is doubtful whether their effect can be expected to last beyond the pre-emergence stage. Some of the newer chemicals appear very promising both for seed and soil treatment, provided the difficulty of their variable action in different soils and under different conditions can be overcome. Among them tetramethylthiuram disulphide shows exceptional promise, both as a seed and as a soil treatment.

Other root diseases

Rhizina inflata (Schäff.) Karst. (Ascomycetes, Pezizales) is chiefly important in Britain as the cause of Group dying of plantation conifers, and is dealt with fully in Chapter 16. It has occasionally been recorded killing young conifers in nurseries both in Britain and abroad, always in patches around the sites of bonfires (Davidson 1935).

Helicobasidium purpureum Pat. (Basidiomycetes, Auriculariales), the cause of Violet root rot, is described in Chapter 16. Its non-fruiting mycelium is called *Rhizoctonia crocorum* Fr. (syn. *R. violacea* Tul.) (Mycelia sterilia). It has occasionally been recorded in Britain and in Ireland attacking Scots pine, Sitka spruce, Norway spruce, Douglas fir, and oak seedlings in nurseries (Somerville 1909; Watson, H. 1928; McKay and Clear 1957). The brownish-violet, velvety fruit bodies are formed on stems of attacked plants, and on the surface of the ground around, with the violet mycelium covering the upper roots. Affected plants should be destroyed, and the soil sterilized before another crop is planted or sown.

Several species of *Rosellinia* (Ascomycetes, Sphaeriales) are concerned with root disease in trees and shrubs. *R. necatrix* Prill., White rot, is not a nursery disease and is dealt with in Chapter 16. *R. quercina* Hartig which does damage in nurseries occurs on oak and is described in Chapter 27. It is typical of these fungi that their mycelium envelops the upper roots of attacked trees. *R. aquila* (Fr.) De Not. has occasionally been reported on spruce seedlings in Britain (Chapter 23), the upper part of the roots and the base of the stem being covered with a web of grey mycelium (Wilson, M. 1922). In the United States *R. herpotrichioides* Hept. & Dav., a species which has not been recorded in Europe, attacks the lower needles of Douglas fir, causing considerable damage. Suppressed seedlings may be killed outright. In seed-beds, where the water table is high and the air immediately above the ground particularly humid, damage can be very extensive (Salisbury and Long 1955). Georgescu and Gasmet (1954) found *R. byssiseda* (Tode.) Schroet., a species which has rarely been recorded in Britain, matting and killing the needles of overdense spruce seedlings. It is characteristic of all the *Rosellinia* diseases that they are only locally and occasionally serious. None of them has achieved widespread or sustained importance.

Verticillium wilt

Verticillium wilt, caused by *V. albo-atrum* Reinke & Barth. (Fungi Imperfecti, Moniliales) and possibly by other species of *Verticillium*, is chiefly important on the genus *Acer*, and is therefore discussed fully in Chapter 30. However, it can attack a very wide range of woody species (Engelhard 1957). The fungus is chiefly important in the nursery. Since infection is from the soil, it may attack successive crops on infected ground.

Thelephora

There are a number of species of *Thelephora* (Basidiomycetes, Aphyllophorales) in Britain which grow over the surface of the soil. The commonest is *T. terrestris* (Ehrh.) Fr., which forms dark-brown, velvety fruit bodies on acid soils. It is not a parasite, but envelops or even covers small objects as it spreads (Fig. 37). It may cause the death of the lower needles of seedlings or even smother very small ones (Weir 1921). Seedlings enveloped by the fungus are held together very tightly. Efforts to separate seedlings in this condition for lining out may result in considerable mechanical injury, needles and the bark to which they are attached being stripped off the plants. Mainly for this reason, care should be taken to remove the fungus from seed-beds during weeding. Its most serious development is likely to be in heathland nurseries, where the soil is both acid and comparatively weed free, so that the fungus has time to develop between the infrequent cultivations. It usually dies out following transplanting, even on plants to which a substantial amount of the fungus was initially attached.

Grey mould

Pathogen

This disease is caused by the ubiquitous fungus *Botrytis cinerea* (Fr.) Pers. (Fungi Imperfecti, Moniliales), the perfect stage of which is *Sclerotinia fuckeliana* (de By.) Fckl. (Ascomycetes, Helotiales). It is possible that some strains of *B. cinerea* may belong to other species of *Sclerotinia* (Moore 1959). The fungus is known to be very variable both in appearance and in virulence (Paul, W. R. C. 1929; Groves 1946).

The fungus is usually seen as a sparse web of grey mycelium over the affected portions of plants. On this mycelium the conidiophores arise sometimes singly, sometimes in tufts, the conidia being borne in clusters at the tips. The conidiophores can easily be distinguished under a hand lens. Late in the year small black sclerotia are formed, on which the apothecia are produced in the spring. Under British conditions, however, the sclerotia usually give rise to conidiophores, not apothecia, on germination. The mycelium continues to grow at very low temperatures, so that although the sclerotia may aid overwintering, they are not necessary.

Symptoms and development

Botrytis is almost universally present on dead and dying vegetable matter, so that spores are always available when the conditions are suitable for infection. Though it can invade healthy tissues, it frequently first becomes established on the dead or moribund parts of a plant, and spreads thence into

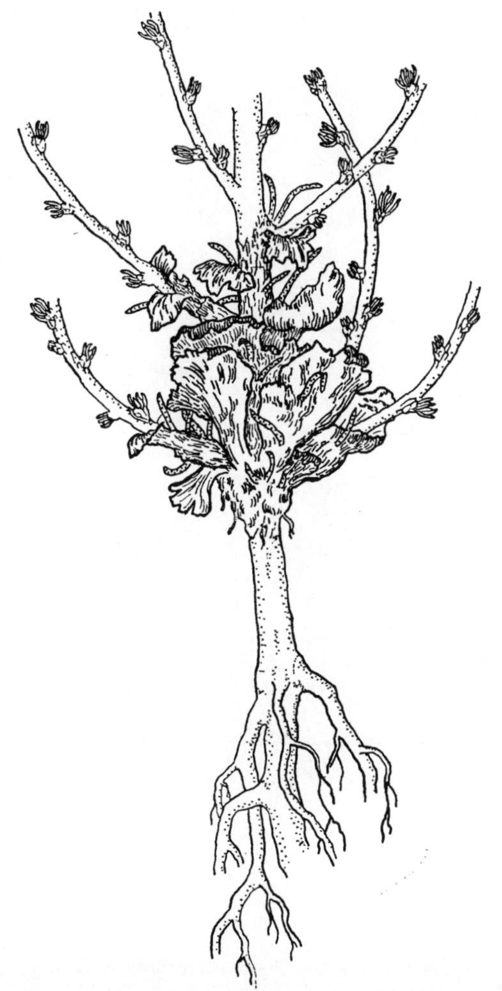

FIG. 37. *Thelephora* sp. from Japanese larch in an acid nursery, Wareham, Dorset, April ($\times\frac{3}{4}$). (J. N.)

adjacent healthy tissue (Brown, W. 1915, 1916, 1917). The presence of *Botrytis* is usually betrayed by the web of grey mycelium and the characteristic conidiophores (Fig. 38). When succulent shoots are attacked they quickly collapse and hang in a withered condition. Small lesions are sometimes formed on succulent stems, and are typically sunken. *Botrytis* is best able to attack soft succulent tissue, and flourishes particularly under conditions where the air is moist and stagnant. For this reason it is frequently found attacking the lower needles of coniferous seedlings in dense seed-beds, presumably having spread from initial infections on needles which were dying from lack of light. This type of damage is particularly common on Japanese

FIG. 38. *Botrytis cinerea* on a *Cupressus macrocarpa* seedling ($\times \frac{3}{4}$). (J. C.)

larch and *Pinus contorta*. Analagous damage has been recorded on broadleaved seedlings. In this case the fungus spread down the petioles of shaded leaves, forming lesions on the stems, and sometimes girdling them (Fig. 39). The most serious attacks in Britain have usually been in the autumn and early winter on the succulent tips of seedling conifers. In some cases the initial infection has been on frosted tips, but this is not always the case. The fungus continues to spread right up to the end of the year, unless the weather is exceptionally cold, and though few seedlings are killed outright, dieback and consequent bushiness may occur on many plants. There is no doubt that succulent shoots are particularly prone to attack when they have ceased growth, but are not yet hardened. Satô, Ôta, and Shôji (1955*a*) found that

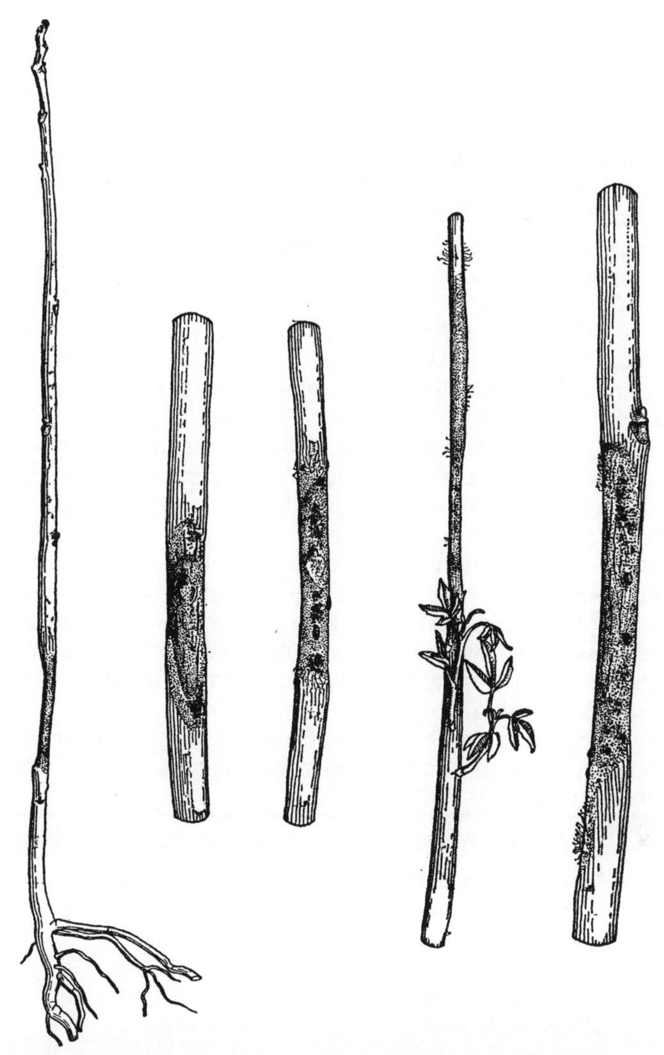

FIG. 39. *Botrytis cinerea* on *Ailanthus glandulosa* seedlings, October
From l.–r.: ×½, ×½, ×½, nat. size, ×2 (J. C.)

Botrytis damage to *Cryptomeria japonica*, a very susceptible species, could be lessened by treating the plants with maleic hydrazide, which brings about an early cessation of growth. The fungus has sometimes proved troublesome on coniferous cuttings in frames, presumably benefiting from the moist, still atmosphere. The fact that spraying with water gives a measure of control (Calavan *et al.* 1953) suggests that the fungus may not be a menace in mist propagation. *Botrytis* has often been reported causing Damping off, but in this role it is far less important than *Pythium* spp., *Rhizoctonia solani*, and other fungi.

Distribution and damage

The fungus appears to be universally present in all temperate regions. No part of Britain can be reckoned safe from attack, though the most serious outbreaks on conifers have mostly been in the lowlands of Scotland. It has an enormously wide host range on broadleaved trees and on conifers. It does not only occur as a nursery disease. On broadleaved trees, for instance, it is sometimes associated with quite severe dieback of shoots. This type of damage has been reported on *Aucuba*, *Cydonia*, *Daphne mezereum*, *Aesculus pavia*, *Jasminum*, *Syringa*, and *Cistus*. Small lesions on the branches have been reported on *Laburnum* and *Rhododendron*. Succulent seedlings may be girdled near the base. This type of damage has been found on *Catalpa bignonioides* and *Ailanthus glandulosa* (Fig. 39). In other cases flower buds are attacked. This has been recorded on *Magnolia*, *Gardenia*, *Lavandula*, *Syringa*, and *Stachyurus praecox*. The last named is an interesting example of damage caused very early in the year, rather than in the summer or autumn. All these types of damage may follow initial injury by other agencies. For instance, on *Cistus*, *Botrytis* almost invariably extends damage due to winter cold. This list of British records (Moore 1959; Peace unpubl.) is far from complete, and there is no doubt that the fungus could be found by more careful search on a much wider range of broadleaved trees and shrubs. Abroad, *Botrytis* has been widely associated with shoot dieback in *Eucalyptus*, usually on young trees or in the nursery. It seems to be the most widespread disease of that genus (Raggi 1947; Abrahão 1948; Nattrass 1949).

Nearly all conifers can probably be infected by *Botrytis*, but it is much more serious on some than on others. The list below gives some of the more susceptible species.

HIGHLY SUSCEPTIBLE	MODERATELY SUSCEPTIBLE
Sequoia sempervirens	*Picea sitchensis*
Sequoia gigantea	*Tsuga heterophylla*
Cupressus macrocarpa	*Larix leptolepis*
Cupressus sempervirens	*Pinus contorta*
Cryptomeria japonica	*Pseudotsuga taxifolia*
Pinus virginiana	*Abies* spp.

This list is far from complete, and it is possible that some other conifers ought to appear, at any rate in the second column. Even those deliberately omitted on account of their known resistance, for instance Scots pine, Norway spruce, and European larch, are occasionally attacked when conditions are particularly favourable to the fungus. It is interesting that the fungus has already been recorded damaging *Metasequoia glyptostroboides* (Palm, B. 1952).

In Britain serious damage by *Botrytis* only occurs in some years, and is usually confined to certain nurseries. Severe injury to more than 50 per cent. of the seedlings in one-year-old beds of Sitka spruce has been recorded. With most species damage is confined to seed-beds, particularly to the autumn and winter of the first year. Only with highly susceptible species, such as the *Sequoias*, does damage occur in transplant lines. Such species are almost invariably attacked wherever they are raised, and sometimes virtually the whole crop may be lost. *Botrytis* certainly ranks as one of the most serious nursery diseases in Britain. *Botrytis* is not common on plantation conifers, but appreciable damage to succulent shoots of Japanese larch 2 to 3 metres high has been observed.

Abroad, apart from its attack on *Eucalyptus*, the fungus has attracted very little attention on trees and shrubs. Most of the American records of damage are back in the past (Boyce, J. S. 1948). On the Continent little has been published. In Japan serious attacks have been reported on *Cryptomeria japonica* (Satô, Ôta, and Shôji 1955*a*) and on *Larix leptolepis* (Itô and Hosaka 1953).

Control

There is no possibility of finding or establishing nurseries free from this disease. *Botrytis cinerea* is too widespread. The possibilities of control by antagonistic organisms have been investigated, but do not appear feasible in our present state of knowledge (Wood 1951; Newhook 1957*b*). A great deal of work has been done on the control of *Botrytis* on horticultural crops, particularly winter lettuces and strawberries, by spraying. A large number of different sprays have proved successful, though results have not been consistent. This probably indicates that there is much to learn about times and methods of application. Spraying experiments on forest-tree seedlings have so far clearly indicated that the disease cannot be controlled in this way once it has become established. On the other hand, spraying shortly before the development of the disease has given good control, tetramethylthiuram disulphide, already used with success on herbaceous crops (Andrén 1946), proving the most satisfactory substance (Murray unpubl.). However, in view of the irregular occurrence of the disease, many years being virtually free from attack, regular preventative spraying would normally be somewhat wasteful. Highly susceptible species, such as the *Sequoias*, should always be sprayed; resistant species, such as Scots pine, never need treatment. The difficulty arises among the species of intermediate susceptibility, such as Sitka spruce or Japanese larch, on which severe and damaging attacks will sometimes develop in the absence of preventative spraying. It is obviously desirable to try to discover the conditions under which *Botrytis* attacks start to build up. It might then be possible to anticipate serious outbreaks, and to apply preventative sprays before disease had developed beyond control.

Certain obvious precautions can be taken in nurseries where the disease is known to occur seriously. A reduction in sowing density should be made for susceptible species, so that the plants are not so crowded. Protection against autumn frost should be applied so that the disease is not encouraged by the presence of frost-injured shoots.

Snow moulds

Pathogens

Two fungi, *Phacidium infestans* Karst. (Ascomycetes, Phacidiales) and *Herpotrichia nigra* Hartig (Ascomycetes, Sphaeriales), are the chief causes of this disease, but other species of *Phacidium* are occasionally involved. In North America *Neopeckia coulteri* (Pke.) Sacc. (Ascomycetes, Sphaeriales) causes similar damage. *Botrytis cinerea* has also been associated with snow moulding (Korstian 1923). The information below, however, refers only to the two first named fungi.

P. infestans, which is chiefly important on Scots pine, produces in the autumn its small, black, circular fruit bodies mainly on leaves browned the previous spring. Microsclerotia are produced on the mycelium in the spring and may also serve as sources of infection in the autumn. Its behaviour has been very fully described by Björkmann (1949). *H. nigra*, which attacks a wide range of conifers but is most important on spruce, forms greyish-brown mycelium on affected plants. It also produces small black perithecia on the underside of the needles. It has been fully described by Gäumann, Roth, and Anliker (1934). Both fungi spread under snow cover by means of profuse web-like mycelium. In the case of *Herpotrichia* this is brownish, while that of *Phacidium* is white.

Symptoms and development

Both fungi are only seriously damaging at high elevations or in high latitudes, where snow cover is deep and prolonged. The snow must be at least 40–60 cm. deep for *Phacidium infestans* to spread successfully (Mattsson Mårn and Nenzell 1941). Both fungi are capable of growing at freezing-point and even at a few degrees below, but most of the mycelial spread occurs, and most of the damage is therefore done, while the snow is melting in the spring. As the snow disappears the withered needles, matted together by the enveloping mycelium, can be seen.

Distribution and damage

Both fungi occur in Britain, but, owing to the absence of prolonged snow cover, they do not cause any appreciable damage. Abroad, *Phacidium infestans* is particularly serious in Norway, Sweden, and Finland on Scots pine. In North America this fungus attacks a wide range of conifers, but is particularly serious on spruce and on *Abies balsamea* (Faull 1930). *Herpotrichia nigra* has attracted most attention in the Alps, where it is important on Norway spruce. It can also attack a wide range of conifers.

The disease occurs in nurseries, in young plantations, and among natural regeneration, causing death of small plants and severe injury to the lower, snow-covered branches of larger ones. In some areas it is thought to be responsible for the absence of regeneration over quite large patches of ground (Mattsson Mårn 1944).

Control

Moderately successful control has been obtained by autumn spraying with lime sulphur (Faull 1930; Björkmann 1949), tar carbolineum (Nenzell 1943), pentachloronitrobenzene (Jamalainen 1956), and copper-containing sprays (Oechslin 1949). Control by dusting is also possible (Zobrist 1950; Meierhans 1951). There is obvious difficulty in finding fungicides that will remain effective throughout a whole winter under the snow.

It has been found that trees growing in the shelter of fallen logs (Sjöström 1946) or under the shade of larger trees are less affected, because the depth of snow is less, and the contact with the snow less intimate. Under heavy snow cover the needles and twigs are pressed together and the passage of the fungus from one part of the tree to another is thus facilitated. Shelter over plants is therefore a possible method of control, obviously more feasible in nurseries than in the forest (Korstian 1923; Oechslin 1957). Snow moulds are particularly serious in dense natural regeneration or in dense seed-beds. Wide spaced, rather than group, planting is therefore preferable in areas subject to the disease (Sjöström 1937; Björkmann 1949).

Rapidly growing trees will obviously be above the snow-level sooner; they will then be less subject to attack. The disease is therefore most serious on weakly trees, and on difficult sites where growth is naturally slow. Any silvicultural or genetical measures which hasten growth will help to lessen the effect of the disease (Stöltenberg 1934; Mattsson Mårn 1944).

Pestalozzia dieback

Two species of *Pestalozzia* (Fungi Imperfecti, Melanconiales), *P. hartigii* Tub. and *P. funerea* Desm., are found in nurseries associated with dieback of conifers. These fungi are inconspicuous, but their minute, black pustular fruit bodies can be detected on infected tissue.

The former has been recorded particularly in connexion with basal girdling of seedlings of spruce and Silver fir (Hartig 1894). In these cases the injury attributed to the fungus is indistinguishable from that caused by damaging by hot soil (Chapter 5), and it is probable that the fungus is only a weak secondary parasite on tissues initially injured by heat.

In recent years there have been a number of reports of basal swellings on seedlings, particularly of Silver fir and Japanese larch, mainly in nurseries in south Scotland. These swellings are clearly different from those which arise above the point of girdling in the case of hot soil damage. They are formed at soil-level, but taper at both ends to normal stem and root thickness (Fig. 40). The stems of affected plants are not girdled, though shoot growth is often very weak. Affected plants have been aptly termed 'carrot-rooted'. The swellings consist mainly of abnormally large starch-filled cells. *P. hartigii* was isolated from some of these seedlings, but there is as yet no proof of its pathogenicity. The primary association of *P. hartigii* with basal swellings on seedlings remains therefore a matter of considerable doubt. *P. hartigii* has also been isolated in Britain from seedlings of Norway and Sitka spruce which have died back without any basal swellings. In Canada it has been associated with shoot dieback of *Picea alba* in the nursery (Vaartaja 1956*a*).

P. funerea is usually associated with damage to the foliage and young shoots of conifers. It has attracted most attention on *Thuja* (Laing 1929; Boudru 1945) (Chapter 26), but has been cited as a cause of disease on a wide range of other conifers (Wenner 1914; Salisbury, P. J. 1957).

There is considerable doubt about the pathogenicity of both these fungi. Doyer (1925) was unable to infect conifer seedlings with either *P. hartigii* or

FIG. 40. Unexplained swellings on seedlings of *Abies grandis* (nat. size). (J. C. after a photograph by D. M. Henderson.)

P. funerea. C. Christensen (1932) was equally unsuccessful with a number of strains of *P. funerea*. Itô and Kontani (1954), on the other hand, made successful wound inoculations with several species of *Pestalozzia*, including *P. hartigii*. Batko (unpubl.) was able to infect young spruce with both species of *Pestalozzia*, but the fungus failed to develop beyond the point of inoculation on healthy seedlings. They are probably both weak wound-parasites, operating only under favourable conditions.

Other leaf and shoot diseases

Cylindrocladium scoparium Morg. (Fungi Imperfecti, Moniliales) causes a shoot wilt of cherries and roses in Britain. In the Argentine it has been associated with dieback of *Eucalyptus* (Chapter 35). It undoubtedly has a wide host range (Timonin and Self 1955; Terashita and Itô 1956), but in Britain it has not been recorded as a general nursery disease. Cox (1953*b*), however, found it causing serious damage to the roots, stems, and foliage of conifers in a nursery in Delaware, and recommended a control programme involving partial soil sterilization and subsequent spraying of the seedlings with Bordeaux mixture or tetramethylthiuram disulphide.

Ascochyta piniperda Lind. (Fungi Imperfecti, Sphaeropsidales), which is chiefly important as the cause of shoot blight of spruce (Chapter 23), has recently been recorded in several nurseries in north-east Scotland causing dieback of the shoot-tips of seedling *Pinus contorta*. Its consistent and early appearance on the affected shoots suggests that it is probably a primary parasite. It appears that the fungus on pine may be a different variety from that on spruce.

Diplodia pinea (Desm.) Kickx. (Fungi Imperfecti, Sphaeropsidales), chiefly important on pine abroad (Chapter 22), has been recorded in nurseries in the United States on Douglas fir, as well as on a wide range of pine species (Slagg and Wright 1943). It is certainly capable of infecting other conifers (Capretti 1956). The fungus occurs in Britain, but is of no importance.

Alternaria tenuis Auct. (Fungi Imperfecti, Moniliales), a very common saprophyte on dead plant material, has occasionally been associated with shoot dieback on young conifers (Borzini 1940; Carranza 1949). The most striking symptom was the curvature of the wilted shoot tips. It has also been reported as a cause of Damping off. This fungus has been found in Britain on a wide range of coniferous and broadleaved hosts, usually in the nursery or on small plants in the forest. It was usually associated with shoot dieback, but in no case was there proof that it was the cause of the damage.

Losses during storage and transport

Movement of young trees and shrubs from place to place involves exposure of the roots to the air. This may result in loss of water, sufficient to cause considerable damage to the roots or even to bring about death of the plant as a whole. After transplanting, water loss from the foliage of trees which have not yet had time to develop absorbing roots may also result in dieback or death. Heeling-in plants in waterlogged soils can also bring about serious root dieback, especially if they are left under such conditions for a long time. All these matters are dealt with in Chapter 6.

In Britain the ground is seldom frozen for long periods. This means that nursery plants can often be dug up shortly before they are required, so that prolonged retention of lifted seedlings and transplants is usually unnecessary, and the problem of damage in storage does not often arise. Occasionally, if trees are packed damp and stored under moderately warm conditions, fermentation of the dead material may raise the general temperature sufficiently to kill living tissues. In the worst cases these tissues in their turn ferment, causing further damage. This so-called 'heating' of closely packed young trees may occasionally cause severe losses during prolonged transit. It should not occur if the plants are reasonably free from dead matter and are not too wet when packed. The fungi which sometimes develop on plants damaged in this way are purely secondary.

In countries with more severe winters it is often desirable to lift plants in the autumn for use as soon as the ground becomes workable in the spring. In North America prolonged storage of nursery stock is a standard practice, but little information is available on the extent or causes of storage losses. G. Y. Young (1943) quotes cases of serious storage losses due primarily to frost. Species of *Fusarium* and *Alternaria* invaded the damaged plants in a secondary capacity. Wilner and Vaartaja (1958) found that keeping the roots moist was the most important factor in storage survival. Fungicidal treament was not beneficial.

It is possible that the use of polythene bags for plant transport may lead to pathological problems, especially if the trees are left in them too long or packed too wet. On the other hand, their use should greatly decrease excessive water loss.

It does not necessarily follow that storage conditions are worse than those affecting plants which have not yet been lifted or those which have already been planted out. The latter in particular may suffer from cold drying winds, to which the stored plants are, of course, not subject. A case has been recorded in which attack by *Botrytis cinerea* continued to develop on spruce seedlings still in the ground, but ceased to spread on those bundled and stored in basket-work crates.

16

DISEASES OF GENERAL IMPORTANCE ON ROOTS

THERE are several fungi which attack the roots of a wide range of species, and which cannot therefore properly be dealt with under any particular host genus. There are others which have a wide host range, but which are especially important on one particular genus. These latter are dealt with fully in the chapters devoted to the diseases of particular genera, and are only briefly referred to here. The symptoms of root disease have already been discussed in Chapter 2. The very important influence of water and air conditions in the soil on root health has been covered in Chapter 6. Nursery root diseases are described in Chapter 15. Included in the present chapter are some important pathogens which are technically root-rots since their primary attack is normally in the root system, but which subsequently attack tissues in the stem as well. These are more commonly thought of as butt or heart-rot fungi.

The fact that roots are normally hidden has inevitably resulted in there being much less information on their behaviour in health and in disease than is available for other parts of the tree. We know that under some circumstances a tree can survive, and even grow at a normal rate, after it has lost a considerable proportion of its root system. Sometimes under favourable conditions a tree is able to regenerate an almost completely new root system. Banking up soil over the roots, to promote fresh root growth, has been advocated as a method of combating root disease (Rhoads 1942). It is possible, of course, to refer to any death of roots as disease, though in fact a certain degree of root loss may be of no more significance than the shading out and gradual death of the lower branches. If minor root losses are accepted as normal, it is obviously difficult to say at what point normal loss ends and disease starts, especially as a degree of root damage which does not matter in the least under one set of climatic and edaphic conditions may become highly significant when those conditions alter, even though the change may be purely temporary.

The effect of root death from natural conditions, such as a rise in the water table or drought, on subsequent attack by root-invading fungi has probably been exaggerated. Nevertheless, it is certainly of considerable importance (Day, W. R. 1955). Natural conditions may, however, exercise a nearly controlling influence on the subsequent behaviour of the infected tree. They not only affect the rate of progress of the disease, but in particular they may determine whether the tree is killed, and, if that happens, the time at which death occurs. For instance, the death of trees with only partially functional root systems often results from drought, though the primary cause is the fungus which has destroyed many of the roots.

An important factor in root disease is the frequency and stability of root contacts, especially in dense forest, where the roots are inevitably interlaced. Direct transfer of a fungus from tree to tree is thus much easier than above ground. Root-grafts between trees of the same species are comparatively common (Dallimore 1917; Yli-Vakkuri 1954; Šafar 1955; Kuntz and Riker 1956; Molotkov 1956) and these also play a part in the transmission of disease, in particular of vascular diseases, where a direct union of the vessels is obviously more important than mere contact.

The whole subject of root disease has been dealt with very fully by Garrett (1956*a*). He has stressed particularly the importance of considering root pathogens in relation to other fungi, often saprophytes, occurring with them in the dead roots, stumps, or other substrates on which they are growing and also in the soil surrounding those substrates. The interaction between the pathogen and its usually saprophytic competitors may be extremely complex, but often throws light on otherwise inexplicable behaviour in root diseases. For instance, it is likely that soil-inhabiting fungi, particularly those of the rhizosphere, are mainly responsible for site differences in root-disease attack

ARMILLARIA ROOT ROT

Pathogen

Armillaria mellea (Vahl.) Quél. (Basidiomycetes, Agaricales), known in Britain as Honey fungus, has a world-wide distribution. The fungus and its effects have been briefly described in a Forestry Commission leaflet (Anon. 1958*c*). The fruit bodies, which are of the typical toadstool type, vary from 5 to 15 cm. in diameter, with stalks rather thinner than those of the Common mushroom, and 7·5 to 15 cm. in length. There is a pronounced annulus or ring round the stalk. When young the fructifications are always umbrella-shaped (Fig. 41), but as they age the edges of the cap turn upwards, so that older specimens are often concave on the upper surface (Fig. 42). They are very variable in colour, being orange-yellow, bright brown, or dark brown. The spore-bearing gills on the underside of the cap are yellowish. The fruit bodies are produced usually in clusters, and sometimes in great numbers, at the base of diseased trees, on stumps, on roots, and occasionally several metres up the stems of dead trees. They usually appear in September or October, and perish when the frosts come, if not sooner, collapsing into a black sodden mass which soon disappears. As a diagnostic character they are of limited value, not only because of their short season, but also because they are not always produced on diseased trees.

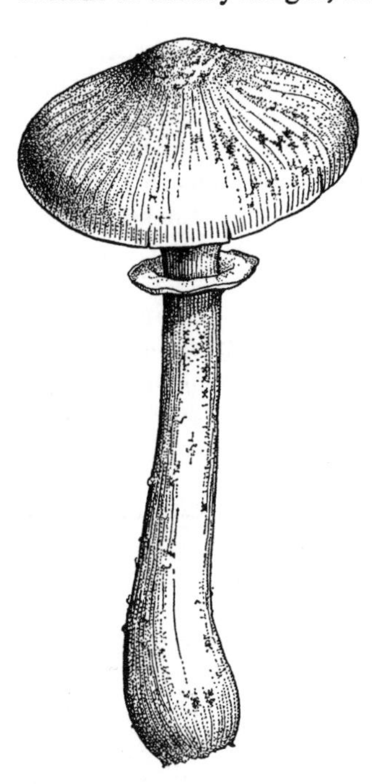

FIG. 41. Typical fruit-body of *Armillaria mellea*, November (×½). (J. C.)

The fungus also produces black bootlace-like rhizomorphs which frequently grow through the soil for as much as 5 metres, and even farther under the bark of dead trees or fallen logs, or along drain pipes (Fig. 43). The rhizomorphs give the fungus its American name of Shoestring fungus. Their development has been studied in detail by Garrett (1953) and Townsend (1954). They form the principal means of infection of living plants. When the fungus is well established in the host it forms characteristic sheets of creamy-white mycelium with dark streaks between the bark and the wood, particularly at

FIG. 42. A group of fruit-bodies of *Armillaria mellea*, November ($\times \frac{1}{2}$). (J. C.)

the base of the tree, and this is a very useful diagnostic feature. It often causes decay of the roots and this sometimes extends into the lower 50 cm. of the trunk and exceptionally farther. In the early stages the decayed parts are divided by numerous black zone lines (Campbell, A. H. 1934). Eventually the fungus reduces the wood to a stringy, sodden mass, which is whitish in colour in hardwoods, but brown in conifers.

The fungus is known to be very variable in culture. These variations exist not only between cultures from different hosts, but also between isolations from a single host; they may even arise in monospore cultures (MacLean 1950; Raabe, R. D. 1953).

Symptoms and development

Primary infection of living tissue is normally due to rhizomorphs. These penetrate the outer bark and develop mycelium, which, given a susceptible host, then proceeds to real parasitism (Day, W. R. 1927*a*; Thomas, H. E. 1934). Infection by root contact can take place, however, and in some places may be the normal method (Příhoda 1957*b*). Rhizomorph infection often takes place near the collar and a young tree may be girdled and quickly succumb without appreciable root death. In older trees, however, the roots are invaded more slowly, and root death may stop before the tree is girdled and while it still has enough unattacked roots to keep it alive, although decay

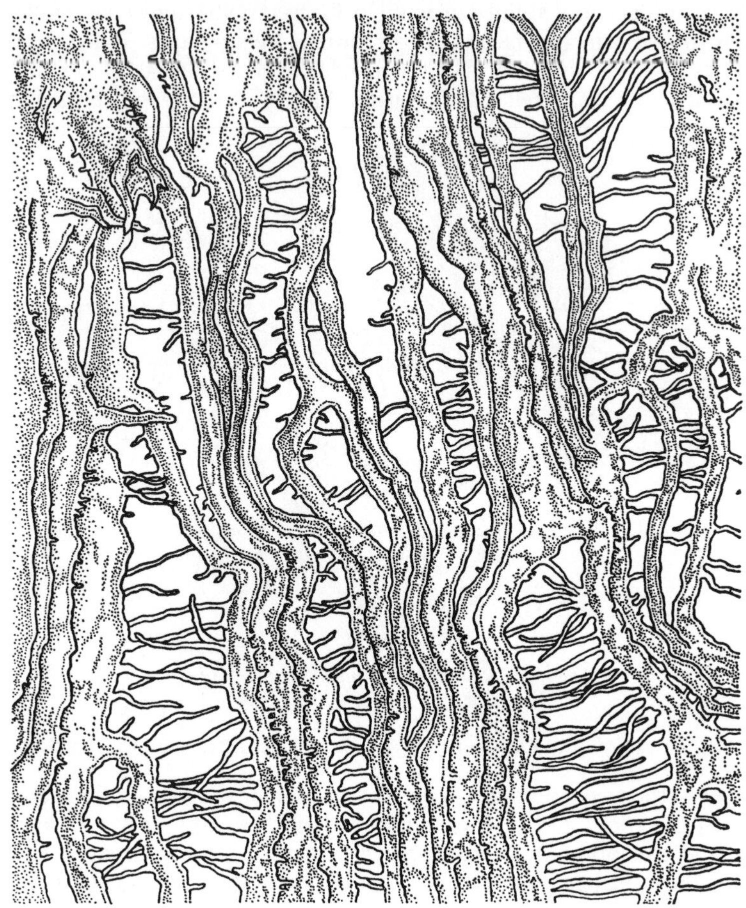

FIG. 43. Rhizomorphs of *Armillaria mellea* from under the bark of a dead tree ($\times \frac{3}{4}$). (J. N.)

may well continue. The ability to infect depends on the volume of the substrate from which the rhizomorph has arisen and on the distance of the substrate from the root (Garrett 1956*b*). The part that root contact plays in infection is not known, though apparent spread by such means has been observed in the forest. Marsh (1952) found that the pattern of spread in orchards was suggestive of infection by root contact. Since spread is so often from stumps, a large stump is more dangerous than a small one, and a stump near at hand more dangerous than one at a distance. Group attacks in the forest are almost invariably associated with the presence of large infected stumps.

Armillaria attack often results in the death of small trees or even of those of pole-size. The death of the tree is often hastened by adverse factors, such as drought, or waterlogging. In fact death may sometimes finally be due to the direct action of physical factors, *Armillaria* attack having left the tree with so few living roots that it can no longer withstand influences which would have had little effect on a tree with a complete root system. On the other hand, attacked trees may often survive, with little apparent damage, though there may be weakened root anchorage and decay of the bottom of the trunk. In general the chances of survival following attack by *Armillaria* increase with age. Fig. 44 is a section across the base of a *Thuja plicata* which

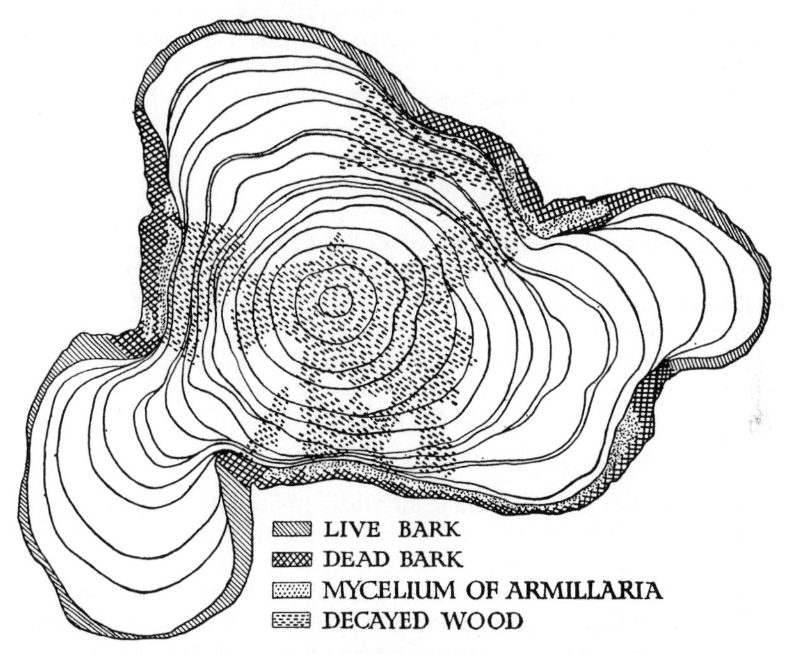

FIG. 44. Extent of *Armillaria mellea* infection at the base of an eighteen-year-old *Thuja plicata*. (A. C.)

was still growing vigorously at least ten years after *Armillaria* had attacked and killed several of the major roots. If the tree is seriously affected the attack eventually causes crown symptoms, such as yellowing of the foliage, reduced growth, abnormal production of cones, and eventually wilting of the young shoots, followed fairly rapidly by withering and death.

It is commonly believed that *Armillaria* attack is more serious on trees weakened by other causes or even that it is confined to such trees (Day, W. R. 1929*b*; Rayner 1930; Birch 1937). It was generally thought to be secondary to Oak mildew and caterpillar attack in the widespread death of oak which was experienced in various parts of Europe in the mid-twenties (Chapter 27), though Georgévitch (1926) considered it the primary factor. It was found by Buckland (1953) to be capable of invading Douglas fir of all vigour classes, but on the stronger trees only a few roots were killed and decay developed, while many of the weaker trees were killed outright. Although weakened trees may sometimes be more liable to attack than vigorous ones, it does appear that the main effect of external influences is to hasten death and thus destroy the possibility of a balance being established between the fungus and the host. Any action which will increase the vigour of attacked trees should increase their ability to cope with *Armillaria* infection.

Effective spread of the fungus always appears to be from some kind of organic substrate, a diseased tree, an infected stump, or even a portion of infected root. Hewitt (1936) has reported orchard infection from root debris washed down by floods from infected woodland in the hills above. Boughey (1938) found rhododendrons infected in a 'rootery', a kind of peat-filled rockery with old stumps used in the place of rocks. There is no information as to whether thinning-stumps are subject to spore infection by *Armillaria* as with *Fomes annosus*, though Käärik and Rennerfelt (1957) have found *Armillaria* present on pine and spruce stumps very soon after felling. On old hardwood areas, conifer thinning-stumps may be rapidly invaded from below. Here the fungus acts in a purely saprophytic role and no increase in parasitic activity in the stand appears to result. In Britain there is no evidence of the fungus becoming well established, or doing damage in pure conifer crops, on sites where hardwoods did not previously grow. *Armillaria* is essentially a fungus of areas with a hardwood forest history. Even where conifers have been grown for more than one rotation immediately adjoining hardwood forest, it has not become established in the same way as *F. annosus*. Indeed it appears that, where conifers completely replace hardwoods, *Armillaria* damage is absent or much diminished in the second rotation.

Distribution and damage

Armillaria mellea is found almost anywhere in both tropical and temperate regions, though locally, of course, there are areas which are free of the fungus because of the lack of suitable substrates. It is characteristic of forests, established woodlands, old gardens, and the like. It occurs almost universally in such places in Britain, though some areas are much more heavily infected than others. There are large areas of agricultural land and moorland which are virtually free, though even there it may be found in hedges. In Britain it is certainly less serious than *Fomes annosus*, but on the Continent it is often regarded as the more important of the two (Falck 1930).

The actual damage done by the fungus is surprisingly small considering its prevalence. It is very seldom that it kills enough trees in one place to cause irreparable injury. At its worst it may leave only a few scattered trees remaining over an area of several acres, but normally, even in a bad attack, it only makes a crop gappy and lessens the yield from the earlier thinnings. It is of limited importance as a cause of decay, because the rot only goes such a short distance up the stem. The fact that affected trees are less windfirm may sometimes be a considerable danger to the remainder of the stand as well as to themselves. It is certainly a mistake to underrate the importance of this fungus in the forest. In Britain, where so many old hardwood and scrub sites are now planted with conifers, it is of widespread occurrence and, apart from *Fomes* in some special localities, it is the most important cause of their death. In seed orchards, arboreta, and gardens, where each tree has an individual value, it can be very damaging. It is for this reason that it has attracted so much attention in fruit orchards in temperate regions and in plantation crops, such as tea, in the tropics.

Honey fungus is essentially a parasite of woody plants, probably mainly because they provide sufficient material for it to maintain a sustained and progressive attack. It is sometimes found on very small shrubs and even on herbaceous plants (Raabe, R. D. 1958). M. Wilson (1921*b*) and others have recorded it on potato and Alcock and Wilson (1927) on heather, but in these cases it had spread from nearby stumps, and the attacks would almost certainly have died out if the stumps had been removed.

Armillaria has been recorded on so many woody hosts that it is impracticable to give a list. It is very doubtful whether any woody species is entirely resistant. There are, of course, numerous records of occurrence, but not much information on relative susceptibility. In the opinion of the author and his colleagues Douglas fir is the only commonly planted conifer in Britain with a quite definite claim to resistance. Young trees are occasionally killed by the fungus, but older trees are seldom if ever rotted. European larch also has quite a good record, but is certainly not as resistant, either to killing or decay, as Douglas fir. The table below gives the position, as far as it is known, of some of the commoner British trees.

	Killing	**Decay**
RESISTANT	Douglas fir *Populus* vars.	Douglas fir Scots pine Corsican pine European larch Japanese larch
MODERATELY SUSCEPTIBLE	European larch Japanese larch *Abies* spp. Sitka spruce Norway spruce *Tsuga heterophylla* Ash Oak	*Abies* spp.
HIGHLY SUSCEPTIBLE	*Thuja plicata* Scots pine Corsican pine *Picea omorika* *Cedrus* spp. Walnut Cricket bat willow *Prunus* spp.	*Tsuga heterophylla* *Thuja plicata* Sitka spruce Norway spruce

Baxter (1933) is one of the few authors to have given a list of susceptibility. This was based on observations in Bagley Wood near Oxford, and in general supports the order given above.

Thuja plicata is peculiar in that the rot is nearly always on the outside of the stem in conjunction with the diseased roots, not central as it is, for instance, in the spruces. Poplar is seldom seriously attacked during the comparatively short rotations on which it is normally grown. It can safely be used to replant areas where Cricket bat willow, which is highly susceptible, has been attacked. Birch and oak probably provide most of the infected stumps which form centres of infection for subsequent coniferous plantings. Nevertheless, trees of these genera are not often killed by the fungus. In gardens, Honey fungus has attracted considerable attention by its attacks on rhododendrons, but its prevalence on this genus may well be due to the fact that rhododendrons are so commonly planted in cleared or partially cleared woodland, rather than to any particular susceptibility.

Vloten (1936) and R. D. Raabe (1955) have both reported wide variations in the pathogenicity of different strains of the fungus. Apparent variations in the resistance of particular host plants from place to place may well be caused by such differences in the specific pathogenicity of the fungus.

Control

Much work has been done on the control of *Armillaria*, but most of the methods suggested were devised for fruit orchards (Bliss 1944) and are therefore better adapted for use in arboreta and seed orchards than in the forest.

Trenching round an infected area is effective only if the soil is shallow over rock or chalk. In deeper soils the fungus can pass underneath the trench. In any case trenches soon become ineffective if they fill up with leaves and debris, while the severance of roots involved in their excavation is definitely disadvantageous, both by providing fresh material for the fungus, and by wounding roots which it may infect.

The removal of infected stumps is an essential part of control where trees of individual value have to be protected. The more thoroughly the stump and root system are removed, the less is the danger to the living trees around. Even quite small portions of infected roots can act as centres of spread. In the forest of course stump removal can hardly be advocated. At the moment there is no easy way of treating stumps to render them useless to the fungus as centres of spread. Orloś (1957) has suggested biological control by infecting stumps with harmless decay-causing fungi, such as *Polyporus borealis* Fr. and *Fomes marginatus* (Fr.) Gill. (Basidiomycetes, Aphyllophorales), which would compete with *Armillaria*. The practicability of such measures remains to be proved.

R. Leach (1937, 1939) elaborated a method for the control of *Armillaria* in tea plantations in cleared jungle areas. The jungle trees, the stumps of which normally served as sources of infection, were girdled several years before felling. It was suggested that while the trees were still standing their stumps and roots slowly became exhausted of food reserves, and after felling did not act as serious infection centres. This method might well be adapted for use under other conditions, for instance in clearing of existing woodland for special purposes in Britain; though further experiment is required before it can be regarded as generally applicable.

For reasonably permeable ground containing infected roots which cannot be entirely removed, or as a final measure of control in areas from which roots have been cleared, soil fumigation can be adopted. The most successful substance is carbon disulphide. This should be placed in staggered holes 20–30 cm. deep and 45 cm. apart at the rate of 60 c.c. per hole. Control in this way appears to be due in some measure to the strong development of the harmless competitor *Trichoderma viride* Pers. ex Fr. (Fungi Imperfecti, Moniliales) in the sterilized soil (Bliss 1951; Aytoun 1953; Darley and Wilbur 1954).

There is comparatively little that can be done for already infected trees. In warm climates exposure of infected roots to the summer sun, by removing the soil that covers them, has been advocated. This can be done fairly quickly and without damaging live roots by the use of a powerful water jet (Levitt, E. C. 1947). Many suggestions for combining root exposure with chemical treatment have been made. Pettinga (1950) suggested treating the exposed roots with 1 per cent. mercuric chloride, Gard (1925) and R. T. Jenkins (1952) with iron sulphate, and Depoerk (1946) with copper sulphate. In some cases admixture of these chemicals with the soil, when it is replaced around the roots, has been advocated.

In the forest, however, none of these methods is easily applicable, and little can be done beyond the removal of dying or dead trees, roots and all if they are small enough to be easily lifted, and the substitution, if the gaps are large enough to require it, of more resistant species, either hardwoods or Douglas fir.

Clitocybe root rot

Clitocybe tabescens (Scop. ex Fr.) Bres. (Basidiomycetes, Agaricales) differs from *Armillaria mellea* mainly in the absence of rhizomorphs and of the annulus on the stem of the fruit body. Rhoads (1945), after careful examination of the two fungi occurring naturally and in culture, considered that they were clearly different, but Heim and Jacques-Félix (1953) thought that they were not separable. A form of *Armillaria* with no annulus is said to occur in Britain (Rea 1922); this might equally be a form of *C. tabescens*. There are a few other British records. As a cause of disease *C. tabescens* has attracted attention mainly in the southern United States. Like normal *A. mellea* it has a very wide host range, but infection is by root contact between healthy and diseased roots, as with *Fomes*, not by rhizomorphs as with *Armillaria* (Rhoads 1950).

FOMES ROOT ROT

Pathogen

This disease is caused by *Fomes annosus* (Fr.) Cke. (Basidiomycetes, Aphyllophorales), sometimes known as *Polyporus annosus* Fr. or as *Trametes radiciperda* Hart. It has been briefly described in a Forestry Commission leaflet (Anon. 1957*b*). The fruit bodies are normally bracket-like and are produced on stumps, at the base of dead and occasionally live trees, on timber left lying on the forest floor, and on roots, nearly always under sheltered conditions (Fig. 45). They may occasionally arise as much as 2 metres above

FIG. 45. Fruit-bodies of *Fomes annosus* on Scots pine, November ($\times \frac{1}{2}$). (J. N.)

ground-level. They are frequently produced underground in tunnels formed by rodents, on the underside of wind-lifted root systems, or on cut roots along the sides of drains. The formation of these sporophores is very dependent on external conditions as well as on the stage of development of the fungus in the tree, so that their abundance cannot safely be used as a measure of the severity of attack. Once formed, they are rather resistant to changes in weather conditions and spores are produced at all times of the year. The fruit bodies, which are perennial, may be as much as 45 cm. across, but are more often in the range of 5 to 15 cm. The upper surface is at first reddish-brown, but becomes darker with age, and is furrowed with zones parallel to the margin, which is creamy-white in colour as long as the fruit body is growing. The lower, pore-bearing surface is also whitish, but sometimes turns biscuit-colour with age. The texture is tough and leathery. The pores are produced in layers, each representing a period, often annual, of active growth; as many as ten layers may be found. Characteristically the developing fruit body grows round and encloses any small objects, such as twigs or pine needles, which may lie in its path. Such objects may later decay or fall out, leaving small holes in the fruit body. Small pustules of creamy mycelium on which pores may appear, but which do not develop into proper fruit bodies, are often found, particularly on the surface of infected stumps and on roots (Fig. 46). Very thin layers of mycelium, less striking than the mycelial fans of *Armillaria mellea*, are sometimes formed under the bark. The fungus also produces microscopic conidia in small heads. This stage has been referred to as *Heterobasidion annosum* Bref. and as *Oedocephalum lineatum* Bakshi (Fungi Imperfecti, Moniliales). The conidial stage is produced freely in culture, or on infected wood incubated under still damp conditions. In Britain it has been found occurring in the forest in such situations as infected stumps covered with needle litter and where infected timber has been left lying on the ground. It is not thought to be important in the spread of the disease, but it forms a valuable aid to identification.

FIG. 46. Pustules of *Fomes annosus* on a root of Sitka spruce, June ($\times \frac{1}{6}$). (J. C.)

The fungus causes characteristic decay, both in the roots which it has killed and in the heartwood of living trees. Attacked timber at first assumes a lilac, pink, or purplish tint, but later turns red, red-brown, or purplish-brown. Small white pockets then appear in the red region; these are sometimes preceded by black specks (Plate VIII. 1). Later the tissue in the white patches crumbles away, leaving small cigar-shaped cavities. The final state of the decayed wood varies from fibrous to slimy according to the amount of moisture present. In the stem the final stage is nearly always fibrous. The decay often spreads far up the stem, sometimes advancing quite rapidly, so that several metres of the stem of a young tree may be rotted while it is still of pole size. The fungus seldom decays less than 1 metre, quite often goes up 4 or 5 metres, and exceptionally extends for 10 metres up the stem (Rennerfelt 1957).

There is little evidence of variation in the fungus. Etheridge (1955) compared a number of cultures from Europe with others from America and could find no significant differences. Roll-Hansen (1940) found no differences between the fungus on Norway spruce and on Scots pine in Norway, but minor differences in cultural behaviour have been recorded between isolates from different host genera (Glaser and Sosna 1956; Persson 1957).

Symptoms and development

The disease in its various aspects has been described by a large number of authors (Peace 1938; Rennerfelt 1946; Rishbeth 1950; Anon. 1957*b*). The fungus can act in two ways, either as a killing agent, usually on comparatively young trees, or as a cause of butt-rot (Rishbeth 1951*c*). In either case the fungus usually enters the tree by contact between infected and healthy roots. It was formerly thought that entry was through roots killed, or at least weakened, by other agencies, but Rishbeth (1950) found no evidence to support this view; indeed, roots killed by other agencies are usually quickly invaded, to the exclusion of *Fomes*, by harmless fungi. Rishbeth found that the principal means of infection is by contact between the live roots of healthy trees and the infected roots of stumps or of diseased trees (Rishbeth 1951*a*). He found also that stumps which were not diseased before felling could easily become infected by spores alighting on the cut surface. In areas where *Fomes* is active and plenty of fruit bodies are available, spores have been found in abundance in the air and on pine needles (Rishbeth and Meredith 1957; Rishbeth 1958); but even in isolated areas, *Fomes* spores are present in small quantities and the beginnings of spread into virgin plantations even from very limited or distant spore-sources have been found (Low and Gladman 1960). Thus it appears that the chief means by which the fungus invades previously *Fomes*-free plantations is by infection of the stumps provided by thinnings. When fruit bodies are produced on such stumps, the disease builds up further by the infection of the stumps of later thinnings by locally, as well as distantly, produced spores, and by underground root spread. Experience in East Anglia, where *Fomes* acts mainly as a killing agent on pine, has shown a very clear relation between the degree of thinning and the intensity of attack. Similar results have been obtained in other countries (Hepting and Downs 1944; Henriksen and Jørgensen 1954). Rishbeth (1957) pointed out that a number of other stump-producing operations, besides thinning and felling, may initiate or intensify the disease. Cutting of extraction racks is a notable example, and even the removal of large Christmas trees can provide stumps which may become infected.

The idea of *Fomes* infecting and spreading from stumps, which is now considered in Sweden to be one of the methods whereby the fungus extends (Rennerfelt 1949, 1952, 1957; Molin 1957) and which occurs in all parts of Britain, is not accepted in some other European countries, where there is a general opinion that *Fomes* is a widespread soil saprophyte, which attacks especially when the growing conditions of the tree are unsuitable, and in particular when the biocenosis or natural biological balance of the soil is absent or destroyed (Flury 1926; Weck 1952; Orloś and Dominik 1960). The suggested use of mixed plantings of hardwoods and conifers as a measure of control against *Fomes* attack is based largely on the idea that such conditions are more 'natural'. The evidence for the existence of *Fomes* as a soil-inhabiting fungus appears to be based mainly on its ability to grow in sterilized soil. It will not, however, grow in unsterilized soil in competition with other naturally occurring soil organisms (Treschow 1943; Rishbeth 1950; Braun 1958). It is significant that in Britain, where most of our plantations are on sites which have not previously carried conifers, the fungus has always been found occurring after thinning or where some other operation such as cleaning or Christmas tree removal has provided a means of entry. It has never been found before or at the time of first thinning in previously untouched first rotation crops.

The conception that the conditions under which the tree is growing influence its susceptibility to infection and the course of attack and its eventual outcome is much better supported. The disease in both its aspects is known to be worse on some sites than on others. Rishbeth (1950) found that it was worse in East Anglia on old arable land which had been limed, and on old alkaline pastures, than on acid heath. West-Nielsen and Oksbjerg (1953–4) found that liming the soil increased *Fomes* attack. Peace (1938) was able to show correlation between butt-rot and unsuitable site conditions for several tree species. Other workers have also given support to this general idea that conditions adverse to the tree increase the severity of attack (Priehäusser 1935, 1943; Løfting 1937; Day, W. R. 1948*b*; Populer 1956). Authors, too numerous to mention, have observed that attacks are worse on old plough-land, though the explanations put forward have been extremely varied and nearly all based on conjecture.

While it is certainly true that growing conditions exert an influence, nevertheless severe attacks of *Fomes*, particularly as a butt-rotting agent, have occurred in plantations which appeared to be growing under excellent conditions as judged by their general health and rapidity of growth. Site differences do not act entirely by their influence on the host tree; both Rishbeth (1951*b*) and Rennerfelt (1949) have stressed the importance of other micro-organisms both in the stumps and in the soil surrounding them. Rishbeth showed that the relatively severe attack on alkaline sites was due to the absence from them of the competing saprophyte, *Trichoderma viride*. Nissen (1956) has found that several other soil-inhabiting fungi are antagonistic to *Fomes*. Wood-destroying fungi, such as *Peniophora gigantea* (Fr.) Mass. (Basidiomycetes, Aphyllophorales), often act as harmless competitors within the stump (Rishbeth 1957). In Britain there are two major site types on which *Fomes* development is often very limited. These are old hardwood sites now planted with conifers and sites with wet acid peat soils. In both instances it seems possible that conditions in the rhizosphere restrict the passage of the fungus between hosts.

There are, of course, other methods of infection besides root contact. These mainly involve direct invasion of exposed living tissues by germinating spores. It appears that the growth of the fungus is more restricted when starting from a spore on living tissue than from one which has first built up a strong inoculum potential in a thinning stump, and is then better able to overcome the resistance to invasion of the host. Brashing wounds are occasionally infected, though not frequently enough to make them a major risk. Pruning wounds from the removal of multiple leaders may likewise become infected. Tushing and other inadvertent wounds may serve as means of entry (Rhoads and Wright 1946) and in Britain are the most important alternative mode of infection. Roots eaten by rodents have been suggested as a means of entry (Huet 1936). But the evidence for the infection of underground wounds by means other than contact is very incomplete, and there is none at all for direct infection of undamaged roots by spores washed down into the soil.

The fungus has long been known as a killing agent, particularly in pine. On the Continent 'La Maladie du Rond', which occurs in *Pinus pinaster* and other pines, is usually attributed to *Fomes annosus* (Anon. 1930; Huet 1936). In Britain killing on a significant scale is limited to certain soil conditions. It occurs where the soil is alkaline, and on any site where there is a previous history of agricultural tillage. Most pine in Britain is not on such sites and is thus not affected. Killing occurs particularly at two stages in the life of the crop, and continues for a varying length of time. The first stage appears at about four to eight years after planting, when the fungus is spreading from the stumps of the previous conifer crop, and the second usually three or four years after the first thinning. In a first-rotation crop the former does not occur, and post-thinning death will normally be the first appearance. On old woodland, on the other hand, the first stage is often still slowly operative when the second starts. It appears that trees become more resistant with age, so that in general the older the crop the less likely death becomes. After the pole-stage is passed deaths are comparatively rare. The entry of *Fomes* as a cause of decay normally involves the death of at least part of a root. Death of the tree is dependent upon the proportion of the root system killed and on the powers of recovery of the tree. Fig. 47 shows a European larch, all but one of the major roots of which were killed by *Fomes*, but which remained alive and growing slowly until it was blown down because of its impaired anchorage. Douglas fir often remains alive and growing well with more than half its roots *Fomes*-killed, and its root anchorage seriously damaged.

There is some evidence that trees are not only less likely to be killed when they are older, but are also less liable to infection and subsequent decay. This means that infection occurs more readily in the early part of the rotation, when it is most serious in terms of potential decay.

In most cases spread from infection centres is quite limited in a first rotation crop on ground not formerly woodland. There are exceptions to this, particularly plantations formed on old arable land. But there is abundant evidence, where butt-rot is concerned, that second-rotation crops are normally much worse infected with butt-rot than the crops they succeeded (Jørgensen, Lund, and Treschow 1939; Low and Gladman 1960). Infections as high as 80–90 per cent., though more often of the order of 30 per cent., can build up in the second rotation from infections as low as 10 per cent. in the first, and by the time of the first or second thinning the fungus is often higher up the stems than it was in the preceding mature crop. Various explanations have been put forward for this, by far the most likely of which is that the presence of even a small percentage of diseased stumps at the beginning of the rotation allows early infections at the time when the trees are most susceptible. Secondly, infection soon spreads through the root systems of stumps apparently uninfected at the time of clear felling so that a crop which showed only 5 per cent. infection at time of felling may have most of its stumps infected some years later. Certainly the high degree of freedom from *Fomes* found in most of the British plantings on virgin ground is no cause for complacency, since a slow but steady increase in *Fomes* infection is likely to take place if no remedial measures are taken.

Attack by *F. annosus* may sometimes lead to insect invasion and the death of trees that would otherwise probably have remained alive though attacked. Jørgensen and Petersen (1951) found this connexion with *Myelophilus piniperda* L. on Scots pine, and Francke-Grosmann (1948) and Kangas (1952) with *Dendroctonus micans* on Sitka spruce.

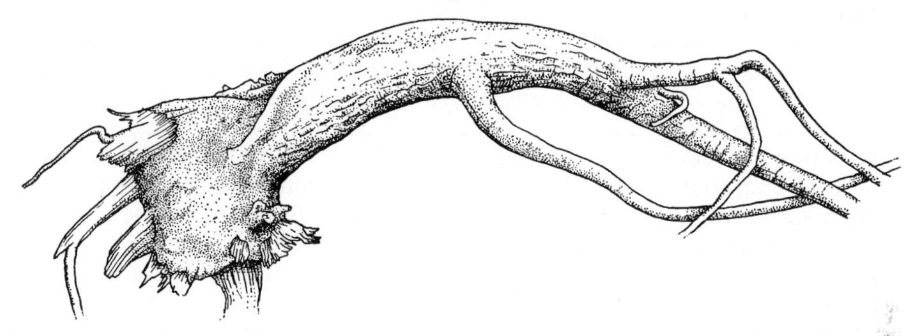

FIG. 47. The root system (viewed from below) of a European larch attacked by *Fomes annosus*. Only one root is still alive, and that has enlarged to support the whole tree (greatly reduced). (J. C.)

Distribution and damage

Fomes annosus has a very wide distribution in the northern temperate regions, but has attracted more attention in Europe than elsewhere. In North America it was hardly noticed before the nineteen-forties, and was at first regarded solely as a plantation fungus. It was soon realized, however, that it was an important pathogen in natural as well as planted crops (Miller, J. K. 1943; Hepting and Downs 1944; Wagener and Cave 1946). In Europe *Fomes* has probably attracted most attention in Sweden (Rennerfelt 1946, 1957), and it is certainly the most important forest pathogen in Britain (Peace 1938).

The loss caused by *Fomes* attack does not depend only on death and decay, though those two aspects, and especially the latter, are extremely important. There is also a substantial loss of increment associated with root-death (Henriksen and Jørgensen 1954). Swedish workers agree that this leads to greater economic loss in spruce than the deterioration of timber from decay (Petrini 1946; Arvidson 1954; Rattsjö and Rennerfelt 1955). Another important source of loss is windblow resulting from the reduced stability due to root-rot. No general estimate has been made of the loss caused by this fungus in Britain. In East Anglia deaths by *Fomes* have caused a 5 per cent. loss by volume in thinnings. Even now, in what may be regarded as the early stages of its development, since so many of the stands on virgin ground are hardly affected, it certainly causes greater loss than any other tree disease. In view of its wide host range and rapid progress in second-rotation crops, its future possibilities are very serious. It is fortunate that there is a practical method of partial control.

Our information on susceptibility and resistance is still far from complete. A comprehensive list of trees attacked in Denmark has been given by Jørgensen (Goor *et al.* 1954). This includes twenty-six species of conifers and twenty of hardwoods. An appreciably longer list of conifers could be made for Britain. M. Wilson (1927) gave a number of British records on hardwoods, but the fungus is really important only on conifers. Probably all conifers planted in Britain are to some degree susceptible. *Fomes* has been recorded on bracken and gorse (Rishbeth 1950) and on heather (Watt 1927), but it is certainly not common on such plants, and there is no evidence that they play any significant part in its spread.

It can be broadly stated that Scots pine, Corsican pine, *Pinus contorta*, and probably some other highly resinous pines are normally resistant to butt-rot, but can be killed by *Fomes* on alkaline soils or old plough-land. Whether this resistance to decay is in fact a matter of resin production remains to be discovered. Other conifers are liable to be killed by the fungus only when quite young, though exceptionally Sitka spruce, Douglas fir, and European larch have been known to die after the first thinning. The larches, the spruces, and *Thuja plicata* are highly susceptible to *Fomes* butt-rot. *Tsuga heterophylla* is outstandingly susceptible. On the other hand, Douglas fir and *Abies* spp., including *A. grandis* and *A. alba*, show appreciable resistance to decay. Douglas fir, despite the decay resistance of its stem, suffers considerable root-rot and is often rendered unstable by *Fomes* attack. The Silver firs have a strong reputation for resistance (Jørgensen, Lund, and Treschow 1939), but there are in fact quite a number of records of *Fomes* damage on *A. alba* and other species (Biraghi 1949*b*; West-Neilsen and Oksbjerg 1953–4; Sasaki and Yokota 1955; Bronchi 1956). Nevertheless, *Abies* species are the safest conifers to plant on areas known to be infected. Pines can also be used on infected sites, except those which are alkaline or were once under plough.

The fungus has been recorded on several occasions doing appreciable damage by killing hardwoods on alkaline or old agricultural sites. In most instances, however, this has occurred in mixture, when spread from infected coniferous stumps has been possible. Oak, beech, and birch have all suffered quite severe injury under these conditions. The use of hardwoods to plant up gaps caused by *Fomes* deaths has for this reason not been very successful.

There is little evidence of variation in resistance to *Fomes* within a tree species, though it is possible that this may exist. Treschow (1958) failed to find any resistance variation between different races of Norway spruce, but Mánka (1956) has advocated the use of native rather than exotic provenances of Norway spruce in Poland, on the ground that they are more resistant.

Control

Most of the methods of control suggested for *Armillaria mellea* (see above) should also be applicable to *Fomes*. But they all suffer from difficulties which limit their use to the protection of trees of especial value. It is disquieting that there is no ready answer to the problem of replanting a site known to be infected by *Fomes*. Leaving the site unplanted until *Fomes* dies out is not practicable, since stumps normally retain infection for periods of forty years and more. Experiments are now in progress in Britain to test the effects of treatments such as girdling, poisoning the trees before felling, poisoning the stumps after felling, or even removal of stumps on the development of *Fomes* in the next rotation. In view of the very rapid build-up of the fungus in second-rotation crops, described above, even expensive measures may prove to be economically justified.

It is quite clear, however, that prevention is better than cure, and the method of stump treatment introduced by Rishbeth (1952, 1957) on pines in East Anglia offers considerable hopes of success. If the cut surface of the stump is copiously painted with creosote of good quality immediately after felling, the successful entry and development of *Fomes* spores is prevented until such time as the stump has been invaded by other harmless saprophytic organisms. If the fungus is not able to colonize such stumps, it can invade *Fomes*-free areas only slowly, if at all. Such protection is not, of course, absolutely successful. A few stumps may still become infected. In addition there is the possibility of occasional direct infections, through brashing and tushing wounds, and of stump infection where extraction injuries expose unprotected tissue. But experiments in East Anglia and elsewhere have indicated that these risks are comparatively small, and that prompt treatment of thinning stumps, properly applied, gives good protection. There is every reason to believe that stump treatment will be as successful in preventing butt-rot in other conifers as it is in preventing death among pines. There is also the possibility of substituting for creosote other substances which will not be purely surface protectants but will affect the deeper tissues of the stump. This will lessen the danger arising from subsequent damage to the protected surface. There are also possibilities of differential toxicity, which might be used to favour other fungi at the expense of *Fomes*. The matter of alternative materials for stump treatment is now being explored.

This control method is obviously based on the explanation of infection and spread from stumps, which is well known in Britain, but has not yet been generally accepted. Foresters, especially on the Continent, who believe that maintaining the 'natural' biological balance is all-important, advocate mixed planting of hardwoods and conifers, or even restorative rotations of pure hardwoods, as the proper treatment for the disease (De Koning 1923; Priehäusser 1935; Rennerfelt 1946; Braun 1958). Even if such methods were silviculturally possible on typical British forest sites, they provide no quick solution for our immediate problem, and in most cases they are economically quite unacceptable.

Many other suggestions on *Fomes* control, such as liming, fertilizing, and the removal of litter, have been based on misconceptions concerning the behaviour of the fungus, and in particular on the belief that the disease is favoured by acidity. They will not be discussed here.

An unexpected source of infection has been described by E. Jørgensen (1955). Fence posts made from *Fomes*-rotted trees acted as sources of infection to the living roots of trees that came into contact with them. Fencing forms an obvious outlet for material too rotted for the saw-bench, but still strong enough cleft or in the round for fence posts; but obviously such material should not be used to fence coniferous plantations.

Other Fomes root diseases

Fomes lignosus Klotz. and *F. noxius* Corn. (Basidiomycetes, Aphyllophorales) are both tropical root parasites, particularly important on rubber, *Hevea braziliensis*. These diseases, especially the former, have been carefully studied because of their economic importance (Napper 1932*a*; Newsam 1954; Pichel 1956). *F. lignosus* produces rhizomorphs, so that its action is essentially different from that of *F. annosus*. The high value of the crop to be protected and low labour costs permit the removal of the stumps and major roots when the jungle is cleared, followed by periodic inspection of the upper roots of the planted crop, accompanied by the removal of rhizomorphs and of infected roots. It is possible to forgo the initial clearance, using the periodic inspection to detect rhizomorphs and follow them back to the source stumps, which are then removed (Napper 1932*b*; Young, H. E. 1952). Young has queried the desirability of jungle clearance by poisoning, as it does not prevent the colonization of the roots by *F. lignosus*, which can then spread to the planted crop. The behaviour of these tropical root parasites has been compared to that of *F. annosus* in Britain by Rishbeth (1955).

ROOT DISEASES CAUSED BY OTHER WOOD-DESTROYING FUNGI

A few other fungi, primarily of importance as agents of decay, are known to be capable also of parasitic infection of roots. Probably further investigation would disclose that this is true of others not suspected of ability to attack live roots. It is almost certain, however, that none are as important as *Fomes annosus*, *Armillaria mellea*, or *Polyporus schweinitzii*.

Polyporus schweinitzii

In Britain *Polyporus schweinitzii* Fr. (Basidiomycetes, Aphyllophorales) is second in importance to *Fomes annosus* as a cause of butt-rot of conifers (Peace 1938). Recent investigations have disclosed that it is commoner than was originally thought, decaying pole crops, particularly of Sitka spruce and Douglas fir. Attack on such young crops provides strong circumstantial evidence of pathogenicity. On the other hand, it has not been found in Britain on crops younger than the thinning stage, nor associated with death.

Wood rotted by this fungus is dry, brown, and friable, and tends to break into roughly rectangular blocks, the cracks between which are sometimes filled with yellowish-white mycelium (Plate VIII. 2). The fruit bodies, which are 10 to 40 cm. across, are dark rusty-brown, circular or bracket-like, and sometimes have a short stem. The underside carries yellowish-green pores. A conspicuous feature is the downy upper surface, which has earned the fungus the name 'Velvet top' in America. When active the fructifications have yellow rims, and often exude drops of moisture from the underside. They appear in the late summer and produce spores for a comparatively short time. In contrast to those of *Fomes*, they are soft and annual, not perennial. The fungus normally attacks conifers, but has been found as a cause of decay on a number of broadleaved species. It has fairly often been recorded attacking conifers under conditions where it is almost certain that it spread from the stumps of a previous broadleaved crop.

In the United States *P. schweinitzii* has been associated with large-scale root-decay and death of planted *Pinus strobus* and *P. resinosa* (York, Wean, and Childs 1936). It was proved by inoculation to be capable of parasitizing the roots of young *P. strobus*. The pathogenicity of the fungus increased with the alkalinity of the medium in which the trees were growing (Wean 1937). Berk (1948) proved by inoculation that it could attack elms.

In Britain it appears to be a potentially dangerous parasite, but less to be feared than either *Fomes* or *Armillaria*. Although it is almost certain that it spreads from the stumps of an old crop, there is no definite evidence that it colonizes freshly cut stumps in the way that *F. annosus* does. It remains to be seen, therefore, whether it can be controlled by stump treatment. In practice it is hoped that measures adopted against *F. annosus* will also serve to control the spread of *P. schweinitzii*.

Polyporus dryadeus

Polyporus dryadeus (Pers.) Fr. (Basidiomycetes, Aphyllophorales) is fairly common as a cause of decay in oaks in Britain. The bracket-like fruit bodies, which may be up to 30 cm. across, are produced rather infrequently at the base of infected trees during the summer and autumn. They are only of annual duration. They are yellow when young, but turn brown with age. Characteristically small drops of brown liquid are found in small depressions around the margins of actively growing fruit bodies. The rot, which never spreads very far up the trunk, reduces the wood to a soft, whitish, pulpy mass.

In America the fungus has been recorded by Long (1930) as a root parasite on a large number of species of *Quercus*, including *Q. rubra*, and on *Abies concolor*. Fergus (1956*b*), who observed the fungus on *Q. coccinea*, considered it to be of more importance as a killer of roots than as a cause of decay in the trunk.

In Britain it has not been associated definitely with death of oaks, though it is known to lead to windthrow by the destruction of the anchor roots. It has been found (Day unpubl.) on oaks which were dying back in the crown, but it would be rash to assume that it was the cause of the damage.

Ganoderma

Root parasitism has been attributed to a number of species of *Ganoderma* (Basidiomycetes, Aphyllophorales). *G. pseudoferreum* (Wakef.) Over & Steinm. attacks the roots of rubber, *Hevea braziliensis*, cacao, and other trees in the tropics. This fungus does not occur in Britain, but *G. applanatum* (Pers. ex Wallr.) Pat., which decays many broadleaved trees in Britain, usually entering through wounds, can also cause a root disease of tea and coffee bushes in the tropics. *G. lucidum* (Leyss.) Karst., which is common in Britain, decaying the base and upper roots of oak and other broadleaved trees, is regarded by Pirone (1957) as a serious parasite on shade trees in the United States. In South Africa, Lückhoff (1955) has recorded *G. colossus* (Fr.) Baker attacking the roots of twentyfive-year-old *Pinus hondurensis*, a local race of *P. caribaea*, as well as those of several species of *Eucalyptus* and *Callitris robusta*. This fungus, which is capable of killing all the species attacked, does not occur in Britain.

Stereum sanguinolentum

In Britain, *Stereum sanguinolentum* (Fr.) Fr. (Basidiomycetes, Aphyllophorales) invades the trunks of conifers, usually through wounds or dead branches, and causes decay. Entering in this way it is not usually regarded as a parasite. In this role it is one of the more important decay-causing fungi, chiefly because it occurs rather often in pole-size trees. The fruit body is a thin, cream to greyish-brown skin, usually reflexed at the margin, so that it forms a small bracket with the bulk of the fruit body adhering to the stem on which it is growing. The fructifications are borne in groups and vary in size from 1 to 8 cm. across and turn red when scratched. The decay, which starts as a reddish discoloration, develops into a white pocket-rot.

There is more evidence for its parasitism on roots. Hubert (1935*b*) found the fungus killing pole-stage Douglas fir and other conifers in North America. The attack was most noticeable on the upper roots and the collar. Lückhoff (1955) found the same fungus killing seventeen-year-old *Pinus taeda* in South Africa, and considered that the initial infection was through the smaller roots. So far there are no reports of this fungus being parasitic in Britain.

Poria weirii

Poria weirii Murr. (Basidiomycetes, Aphyllophorales) does not occur in Britain. In north-western America it is one of the most serious causes of decay in Douglas fir, *Tsuga heterophylla*, and other conifers. The same or a closely related fungus also does considerable damage by killing Douglas fir of any age from six years upwards; it spreads by root contact between diseased and healthy roots (Buckland, Molnar, and Wallis 1954). The local spread of the fungus has been successfully controlled by trenching (Wallis and Buckland 1955), but it may well be able to spread longer distances by means of spores, as does *Fomes annosus*.

Ustulina vulgaris

Ustulina vulgaris Tul. (Ascomycetes, Sphaeriales), together with the tropical form of the same species, *U. zonata* Lev., has a world-wide distribution on broadleaved trees and shrubs. Wilkins (1936) regarded *U. vulgaris* as an active parasite capable of killing lime trees (*Tilia*), and *U. zonata* was at one time considered to be a dangerous parasite on a number of tropical trees and shrubs. The recent tendency has been to regard both fungi as not more than weak parasites, secondary to other fungi or colonizing injuries caused by other agencies.

GROUP DYING OF CONIFERS

Pathogen

This disease has been fully described by Murray and Young (1961), with particular reference to its activities in pole-crops in Britain. The causal agent,

Rhizina inflata (Schäff.) Karst. (Ascomycetes, Pezizales) (syn. *R. undulata* Fr.), is a very distinctive fungus. The fruit bodies, which are produced on the forest floor, appear first as small brown buttons with a yellowish edge; when mature they are small, irregular, chestnut-brown hummocks, darkening with age, 2 to 6 cm. in diameter, often occurring in crust-like groups (Fig. 48). They are produced in the late summer, and normally decay during the winter, though it may sometimes be possible to detect their blackened remains on the ground the following year. The fructifications are produced freely in hot summers, but may be rare or completely lacking in cold, wet ones. The mycelium, which is aggregated into fine cream or yellow mycelial strands, is

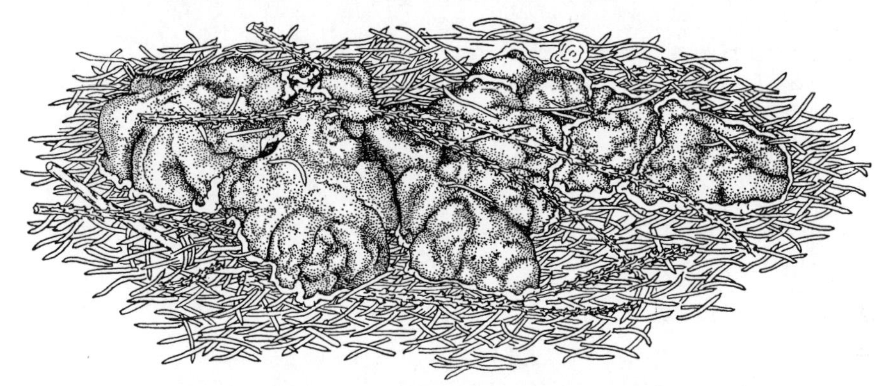

FIG. 48. Fruit bodies of *Rhizina inflata* growing on humus near Sitka spruce, October (×½). (J. N.)

attached to the underside of the fruit bodies, and grows through the soil. On the surface of attacked roots these strands form a network, which has characteristic branching angles where the strands cross one another or join (Fig. 49).

Symptoms and development

The fungus is a common saprophyte in conifer woods, occurring particularly on the sites of fires. It is often found in places where there is no evidence of the disease. It has occasionally been found killing young conifers in nurseries and in the forest, on or immediately around the sites of fires. In these cases death of the top follows the death of the roots, which are invested with mycelium.

The fungus is most important in Britain, however, as a cause of death of pole-crops. In this case small necrotic patches occur on the major roots of infected trees. Examination shows that the mycelium of the fungus has penetrated the roots at these points. The size and number of lesions gradually increase until the whole root is dead. Sometimes the fungus girdles the upper part of a root so that the terminal portion dies back. On all but the most favourable sites, death of a number of the major roots results in a reduction of growth of the tree and often in premature coning. Further loss of roots results in death. The fungus appears to spread radially from the sites of fires, normally those made by forest workers in the course of thinning operations to boil water for tea or to warm themselves during meal breaks. Thus the disease is not typically found in plantations until after some operation such as brashing or thinning has been carried out. Very often groups are associated with centres of activity such as extraction roads or stacking grounds, where fires are often lit. As long as a group is active, dead trees are found at the

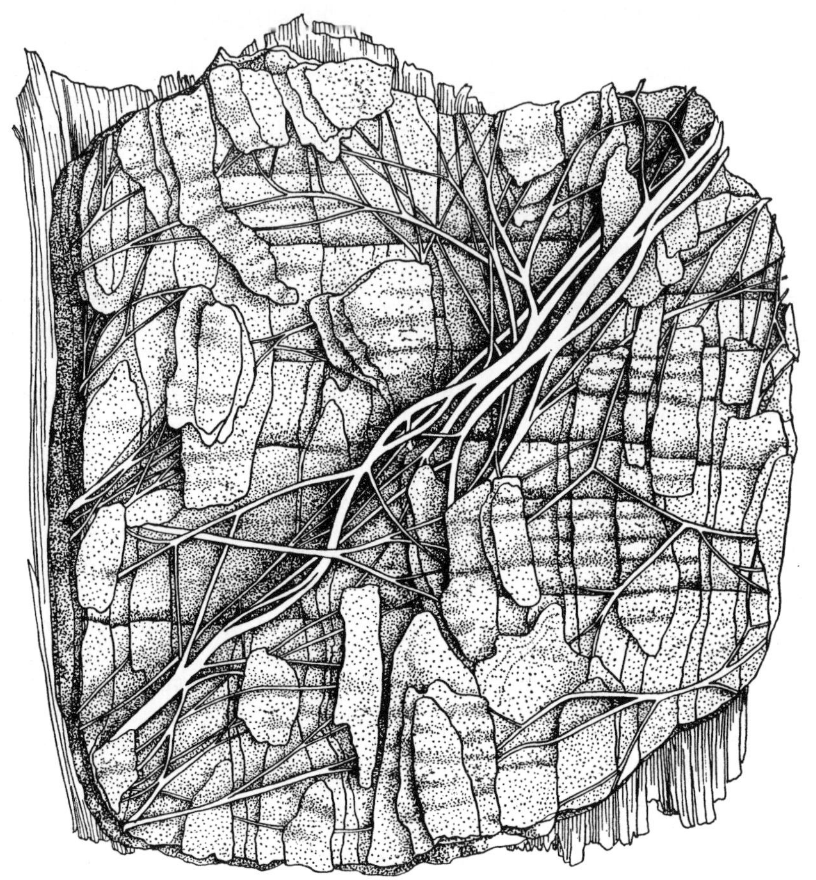

FIG. 49. Mycelium of *Rhizina inflata* on the surface of a Sitka spruce root (×2). (J. C.)

centre and trees with some dead roots on the periphery. The fruit bodies of the fungus are produced in an irregular discontinuous ring among the actively diseased trees. After a few years, spread of the fungus ceases, by which time the group may be as much as 0·1 hectare in extent, but seldom larger. Groups of greater size than this are usually due to the coalescence of two or more smaller groups.

Group dying has been found on crops as young as 15 years and as old as 60, but is commonest in the 20–30 year age range. It has occurred in first-rotation crops on previously non-woodland ground as well as on old woodland sites. It is still not known whether the establishment of *Rhizina* on fire sites rests on some question of nutrition, on the partial sterilization of the soil under the fire, permitting initial unrestricted colonization by the fungus, or on some other factor. There is no suggestion that *Rhizina* is secondary to any other pathogen or agency. Many of the affected crops were growing well on relatively favourable sites.

Distribution and damage

As a saprophyte the fungus appears to be very generally distributed in coniferous woods in Britain. As a cause of group death of pole-sized conifers it is largely confined to the western half of Britain. This may be due to the distribution of Sitka spruce, its commonest host, or to the fact that fires are more likely to be lit in the wetter regions of the west than in the east where the danger of their spreading is greater. Death of newly planted or nursery conifers, which in Britain at any rate is much less common, has been reported in the east. In this role it was first recorded by Brooks (1910), but as a cause of Group dying in older crops it was not noticed in Britain until 1936.

On present evidence it is difficult to explain the distribution of the disease. The fungus certainly occurs in many places where the disease has not developed. The disease does not develop in nearly all the places where the fungus and the fires which appear to be necessary for its pathogenic spread are present. Its development in some of the forests where it does occur has been on a much larger scale than in others. It has been found on a very wide range of sites and soils, and so far it has not been possible to associate its virulence with any climatic or edaphic factor.

The fungus has been recorded in America causing death of young conifers of various genera both in the nursery and under natural conditions (Davidson 1935; Zeller 1935). Wier (1915) successfully infected young conifers by watering the ground around them with a suspension of *Rhizina* spores. On the Continent *Rhizina* has not only been associated with death of young conifers (Hartig 1894), but has also been regarded as one of the possible causes of the death of groups of pine, often called 'La Maladie du Rond'. Here, however, there has been considerable confusion with the damage caused by *Fomes annosus* and *Armillaria mellea*, and the real importance of *Rhizina* still remains uncertain (Mangin 1912; Dufrénoy 1922*c*; Guyot, R. 1933). Recently, however, Gremmen (1958*a*) has found groups definitely associated with *R. inflata* in pole-crops of several species of pine in Holland. The groups originated from fire sites in the forest rides. Groups exactly similar to those in Britain and also associated with *R. inflata* have been recorded in Sitka spruce and *Pinus contorta* in Eire (McKay and Clear 1955). The dearth of recent references to the disease on the Continent and in America clearly indicates that it is not sufficiently serious there to attract continuous attention. On the other hand, if only superficially examined, it is very easily confused with groups killed by other fungi, such as *F. annosus* or *A. mellea*, or by lightning.

Very few cases have been recorded in Britain where the disease has attacked young or newly planted conifers, despite the fact that burning of lop and top is a normal practice when areas are cleared for replanting. Group dying is much more widespread, but the actual area of pole crops killed by the fungus is very small in relation to the total acreage planted. The gaps are particularly dangerous, however, because they form breaks in the canopy and lead to windfall, which may eventually devastate areas much greater than the original group. Wind damage is all the more likely because the death of some of the roots of the trees marginal to the group inevitably renders them somewhat unstable.

The majority of the groups have been found in Sitka spruce, but they have also been recorded fairly often in Corsican pine. Other conifers occasionally affected are European and Japanese larch, *P. contorta*, Norway spruce, and Scots pine. Group dying has never been found in Douglas fir, despite its occurrence in areas where the disease was serious on other species, and on present evidence it can be regarded as resistant to attack.

Control

The obvious means of control is prohibition of lighting of fires in or on the margins of plantations. Cases where brush has been burnt on rides have shown that the fungus can extend from such fire sites to the adjoining plantations. Special trenched fire sites would normally be safe provided the trenches were kept open. On present evidence restriction of fires can be limited to areas where Group dying, i.e. active attack by *Rhizina inflata*, is known to occur. In view of the scattered nature of the disease and the large number of forests where it has not developed, a general restriction is unnecessary. At present, also, there is no need to prohibit fires at the time of clearance for planting, except in the rare cases where death of newly planted stock, due to *Rhizina*, is known to occur.

The spread of the disease frequently halts at well-maintained drains or ditches. Many groups have one straight side along a drain, and others, pinched between two drains, have eventually become long and narrow in shape, spread between the drains being unrestricted. This suggests trenching as a possible method of control, provided the groups were detected at an early stage. Trenching a group that was nearing or had reached stability would, of course, be a waste of time.

ROOT DISEASES DUE TO OTHER SOIL-INHABITING FUNGI

Phytophthora root rot

A number of root diseases caused by species of *Phytophthora* (Phycomycetes, Peronosporales) are described in other chapters. They are:

On *Pinus*	Chapter 22	*P. cinnamomi*
On *Chamaecyparis*	Chapter 26	*P. lateralis* and *P. cinnamomi*
On *Fagus*	Chapter 27	*P. cambivora*, *P. syringae*, and *P. cinnamomi*
On *Castanea*	Chapter 29	*P. cambivora* and *P. cinnamomi*
On *Erica*	Chapter 35	*P. cactorum* and *P. cinnamomi*
On *Juglans*	Chapter 35	*P. cambivora*, *P. cactorum*, *P. cinnamomi* et al.
On *Rhododendron*	Chapter 35	*P. cinnamomi*

These are mostly typical root diseases, and the full description given for the so-called Ink disease of chestnut in Chapter 29 is applicable to the disease on most other trees. The pathological behaviour of *Phytophthora* has been reviewed by Hickman (1958).

Phytophthora species have been associated with root disease on a considerable number of other tree hosts, some of which were included in a general distribution and host list covering the whole genus by Tucker (1933). This has since been supplemented for *P. cinnamomi*, the most widespread species on tree hosts, by Thorn and Zentmyer (1954), and for all the species of *Phytophthora* occurring in Argentina by Frezzi (1950). A considerable number of the records refer to very young plants, but *Phytophthora* as a nursery disease is not sufficiently important to merit separate attention. There are many records on conifers, including several on various species of *Abies*, *Cedrus*, *Larix*, *Picea*, and *Pinus*. Vaartaja (1957*c*) successfully inoculated several species of pine, spruce, and larch with *P. cactorum*. The only serious diseases on conifers, however, are Little leaf of pine and Root rot of cypress, which are here dealt with separately.

There is a constant tendency to associate *Phytophthora* root attacks with difficult soil conditions, in particular excess soil water or large variations between summer and winter water content. This may be connected with the necessity of free water in the soil to transport the spores and for their germination. The fungus can survive in the soil for considerable periods (Hickman 1958).

Phymatotrichum root rot

Phymatotrichum omnivorum (Shear) Dug. (Fungi Imperfecti, Moniliales) attacks the roots of a very wide range of herbaceous and woody plants in the southern United States. The conidia, which are born on fluffy white mats of mycelium on the surface of the soil, seem to play little part in the spread of the fungus, which persists in the soil, often for several years, in the form of small brown sclerotia about the size of mustard seeds and as slender brown mycelial strands, which were originally regarded as a form of Mycelia Sterilia and called *Ozonium omnivorum* Shear.

The States where it mainly occurs, Oklahoma, Texas, Arkansas, and Arizona, are not heavily wooded, and in any case the fungus is largely confined to the moist lower-lying land, which is primarily agricultural. Its attack on trees and shrubs became important only in connexion with the large-scale shelterbelt plantings in the nineteen-thirties. The fungus is killed by low winter temperatures, its northern limit lying roughly along the line where a winter minimum of $-10°$ F. has been recorded at least once (Ezekial 1945). It might therefore be dangerous in the milder parts of the British Isles were it to reach this country.

There are wide differences in resistance between the many trees and shrubs which have been subjected to infection by this fungus (Peltier 1937; Wright and Wells 1948). Walnut, elm, poplar, and many other hardwoods, and a number of conifers, including several species of pine, are known to be susceptible. Several trees, including *Celtis occidentalis*, *Platanus occidentalis*, and species of juniper, are highly resistant.

Ammonium phosphate, ammonium sulphate, tetrachlorethane, pentachlorethane, and xylene have all been used with some success as soil fungicides

(Ezekial and Taubenhaus 1934; Anon. 1937), but the large area infected makes the use of resistant species a less expensive method of control.

Violet root rot

Helicobasidium purpureum Pat. (Basidiomycetes, Auriculariales) attacks the roots of a large variety of plants, including trees (Viennot-Bourgin 1949). It is widely distributed in England and Wales, but rare in Scotland (Moore, W. C. 1959). The brownish-violet velvety fructifications of the fungus are produced on the surface of the ground immediately around the stems of attacked plants. Buddin and Wakefield (1929) established the connexion between this fungus and the more commonly found violet mycelium, known as *Rhizoctonia crocorum* Fr., and at one time placed in Mycelia Sterilia. The fungus can grow through the soil, provided it has a food base from which to spread. Sclerotia are formed which enable the fungus to persist in the soil. They survive longer in alkaline than in acid soils (Valder 1958). The disease is unimportant on trees and shrubs in Britain, though it has been recorded on spruce in nurseries. In Eire it has been found attacking nursery plants of Douglas fir and *Pinus contorta* (McKay and Clear 1958). Magerstein (1928) reported serious damage to basket-willow plantations in Czechoslovakia, while Dana and Wolff (1931) recorded damage to several broadleaved species under forest conditions in Texas.

In Japan another species, *H. mompa* Tan., though sometimes a nursery disease, also attacks quite large trees of a very wide range of species, being particularly serious on mulberry and *Paulownia* (Itô 1949). It is rather similar to *H. purpureum*. The fruit body is a purplish-brown, velvety mat on the surface of the dead bark at the base of the tree. Purple rhizomorphs are formed on the surface of dead parts, and small purplish-brown sclerotia are also produced. Japanese experience suggests that this fungus is a virulent parasite. It is known to occur in Japan and Formosa, and probably extends to parts of China.

H. compactum Boed., which is a parasite on tea and various other plants in the tropics, has been recorded attacking *Pinus longifolia* in South Africa (Pole Evans 1934).

White root rot

A number of species of *Rosellinia* (Ascomycetes, Sphaeriales) cause purely nursery diseases, but *R. necatrix* Prill., the cause of White root rot, often occurs on older plants. In Britain it is largely confined to south-west England, but has occasionally been found in East Anglia in hot summers. It has been recorded in Britain on privet, *Jasminum officinale*, elm (only in the Scilly Isles), and a few other woody hosts (Moore, W. C. 1959). It does considerable damage to mulberries in Italy (Voglino 1929) and in Japan (Matuo and Sakurai 1954). Zambettakis (1955), who regards it only as a weak parasite, has dealt with it in considerable detail, giving a full list of the known hosts, which include *Acer*, *Castanea*, *Juglans*, *Quercus*, and *Salix* among woody genera, in addition to those already mentioned above.

Roots killed by this fungus are covered with a dense web of mycelium, which is white at first but turns greyish. Slender rhizomorphs and small black sclerotia are formed. Conidiophores may arise from the mycelial tufts, but they are not always present; the perithecia, which are produced below ground, have not been recorded in Britain.

Cylindrocarpon radicicola

Cylindrocarpon radicicola Wollenw. (Fungi Imperfecti, Moniliales), which has occasionally been accused of parasitism on roots, is very frequently isolated in cultures from diseased roots of trees, sometimes alone, sometimes in association with other fungi. There is very little evidence for its pathogenicity on any tree or shrub, though Govi (1952) inoculated it into adult *Prunus cerasifera* and produced small lesions. Whether it has any significance as a competitor with more dangerous fungi remains to be discovered.

ROOT DISEASES CAUSED BY BACTERIA

Crown gall

Bacterium tumifaciens E. F. Smith & Towns., now known as *Erwinia tumifaciens*, causes tumours on a very wide range of herbaceous and woody plants. These occur sometimes on the aerial portions, but typically at the collar or on the upper roots. Sometimes there is a single tumour, often a considerable number are produced. They vary in diameter from 5 mm. to 15 cm. Secondary tumours may be produced as a result of the translocation of growth substances produced by the primary tumour, without any transference of the actual bacterium. The bacterium probably requires wounds for infections; in fact Rack (1953) considers wound tissue a necessary predisposing stimulus without which the bacterium is ineffective.

The disease occurs in nature on a large range of different hardwoods, but is much less common on conifers, though Smith (1937, 1939) was able to produce small tumours on a considerable number of different conifers by inoculation. Lopatin (1936) found that comparatively few of the plants that he tested by inoculation were immune to *B. tumifaciens*; among the immune woody species were *Tilia cordata* and *Buxus sempervirens*.

Usually the tumours have remarkably little effect on the host plant; sometimes, however, they result in slightly stunted growth. But often a quite unexpected tumour will be found on an apparently perfectly healthy tree, when for some reason or other its roots are exposed. The disease is far too rare on trees to require any control measures.

Hairy root

Hairy root, which occurs particularly on apple in America, is due to *Bacterium* (*Erwinia*) *rhizogenes* Riker *et al.*, which is very closely related to *B. tumifaciens*. It often occurs on the same plants as Crown gall. Profuse masses of roots grow out from a tumour at the collar. The disease has been recorded on other broadleaved trees as well as apple. It has not been found in Britain, but there are a number of continental records.

17

MYCORRHIZA AND OTHER MICRO-ORGANISMS OF THE RHIZOSPHERE

ABOVE ground the contact between trees and micro-organisms is strictly limited to those actually established in or on tree organs. The roots of trees, on the other hand, are growing in a medium which normally supports a large population of micro-organisms, varying considerably from site to site both in species and in numbers. These may affect the tree in a variety of ways. They may be pathogens and attack the roots. They may form symbiotic associations with the roots; such are mycorrhiza. In other cases their effect on the tree may be indirect. They may break down the surface litter, rendering some of its components available for absorption by the tree. They may fix atmospheric nitrogen and thus increase the nitrogen content of the soil. They may compete with the tree roots for available minerals. Naturally there is competition and sometimes actual antagonism between the different components of the soil flora itself. This may lead to the suppression of a pathogen by a harmless organism, thus lessening the liability of root attack.

Obvious difficulties make it impossible to get a complete picture of the soil flora and its interrelations in any particular site. A vast number of investigations have thrown valuable light on various aspects, but it is still impossible to weld this information into a coherent whole. Thus we cannot yet fully understand the whole biological effect of any particular soil on the trees growing in it, though we may have clear information on certain aspects of that effect. Unfortunately, despite this lack of knowledge, a theory has been propounded that the balance of micro-organisms in the soil of the 'natural' forest is suitable for and probably beneficial to the growth of trees, and that any disturbance in this balance is harmful. It is inevitable that such a balance should exist, though it must obviously be an unstable one, constantly affected by changes in the environment. In forests where the trees are reasonably healthy it is obvious that the soil flora cannot be actively inimical to tree growth, but it is unreasonable to postulate that this balance, the so-called 'biocenosis' of the natural forest, is actively beneficial to trees, and that all departures from it, caused, for instance, by the forestry activities of man, are damaging to tree health. This is an assumption which our present knowledge of the effect of the soil flora on tree roots and tree growth is quite inadequate to support.

It is impossible here to give a full account even of our incomplete knowledge of the flora of forest soils. Much of the information available has no obvious or immediate connexion with tree disease. Only those aspects will be considered which appear to have a definite bearing on the maintenance of health or the development of disease in trees. Some of these are dealt with elsewhere. Root pathogens outside the nursery are covered by Chapter 16. Fungi attacking the roots of seedlings and transplants in the nursery, and in particular the disease of seedlings known as Damping off, are discussed in Chapter 15. Soil sterilization, in so far as it affects fungal attack on the roots of small trees, is also dealt with in that chapter. This leaves for consideration here, mycorrhiza, root nodules formed by fungi and bacteria on certain tree genera, and very briefly the fungi of the rhizosphere, which may be defined as the surface of the root and its immediate neighbourhood, with particular reference to their effect on root diseases.

MYCORRHIZA

Mycorrhizal fungi, which exist in intimate association with the active feeding roots of most trees and shrubs, have been studied for a very long period, during which time very different interpretations have been placed on their activities. Even now, when there is a much greater measure of agreement on their method of life and significance to the tree, there are still differences of opinion on the extent and importance of their effect on tree growth and health.

In view of the existence of several recent reviews on the subject (Harley 1956, 1959; Melin 1953; Garrett 1956), it is unnecessary to cover the whole field of mycorrhizal research here. Only a very small selection from the abundant literature on the subject will be quoted.

Mycorrhiza are usually divided into two groups, 'endotrophic', where the fungal hyphae grow between and mainly within the root cells, and 'ectotrophic', where the hyphae clothe the root with a mycelial mantle and penetrate only between the root cells. In a few instances intermediate forms are found. Usually only one of the two types is formed on any particular host or by any particular mycorrhizal fungus. Most of the important forest trees grown in Britain have ectotrophic mycorrhiza, but taking the plant kingdom as a whole endotrophic mycorrhiza are much more widespread. Among trees, Lawson cypress, *Sequoia*, *Thuja plicata*, *Robinia pseudacacia*, and *Ailanthus glandulosa* are some that have endotrophic mycorrhiza. There is no reason, however, to suppose that the modes of action of endotrophic and ectotrophic mycorrhiza are essentially different. In a short discussion of this kind there is no point in maintaining the distinction between them.

Root tips infected with ectotrophic mycorrhiza are thickened and repeatedly branching (Fig. 50). In pines they fork repeatedly, forming very typical coralloid clusters. This branching or forking may be due to auxins produced by the mycorrhizal fungi (Slankis 1949, 1955; Levisohn 1952), but it has also been suggested that this type of growth may result from mycorrhizal infection keeping alive root tips that would otherwise have died away.

The fungi concerned with ectotrophic mycorrhizal formation are mostly Basidiomycetes belonging to such genera as *Amanita*, *Boletus*, *Hebeloma* (Plate IX), *Cortinarius*, *Lactarius*, and *Russula*, and include many of the familiar forest toadstools (Laing 1932; Modess 1941; Melin 1948; Lobanov 1951). However, *Gyromitra esculenta* Pers., an Ascomycete, has been recorded forming mycorrhiza on *Populus tremula* (Sirén and Bergman 1951), while *Cenococcum graniforme* Ford. & Winge, which may also be an Ascomycete, can do so on a wide range of species (Lihnell 1942). Most of the fungi concerned with endotrophic mycorrhiza are still unidentified.

It is now generally assumed that mycorrhiza are formed only when there is an excess of simple carbohydrates present in the roots (Björkman 1942). The fact that higher light intensities, acting through increased assimilation by the tree, stimulate mycorrhiza formation is additional evidence for this (Björkman 1940; Wenger 1955). Mycorrhizal formation increases with increasing deficiency in plant nutrients in the soil, notably when there is a lack of nitrogen or phosphorus. It is for this reason that mycorrhiza are less well developed in, or occasionally even absent from, fertile soils (Mitchell, Finn,

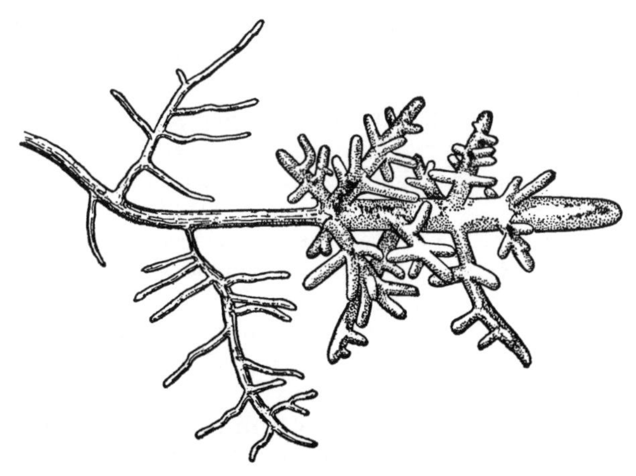

FIG. 50. Mycorrhizal roots on Deodar, October (×6). (J. C.)

and Rosendahl 1937; Björkman 1944). The fungi normally take over the absorbtive functions performed by the root hairs on normal roots and are particularly efficient in this capacity (McComb 1943; Harley and Brierley 1955; Melin and Nilsson 1955; Koch, J. 1957).

From the point of view of tree pathology it is idle to argue whether mycorrhiza are parasitic. Strictly, since they use carbohydrates provided by the tree, they must be so regarded. But the forester is concerned only with their effect, which is known to be sometimes beneficial, normally harmless, and probably never harmful (Levisohn 1958), though the possibility of damaging parasitism under some circumstances cannot entirely be ruled out on the evidence at present available.

A few fungi, notably *Rhizoctonia* spp. and *Mycelium radicis atrovirens* (both Mycelia sterilia) invade roots in much the same way as mycorrhiza, but are considered to be pathogens, rather than symbionts (Levisohn 1954). They are usually known as pseudomycorrhiza. Their pathological significance is unknown.

It is clear that the site has a big influence on mycorrhizal development (Levisohn 1957*a*). Other factors may affect their occurrence besides the nutrient status of the soil and the availability of surplus carbohydrates in the roots of the host. They are likely to be influenced by other components of the soil flora. Indeed it has been suggested that in some soils mycorrhizal development may be hindered or even prevented owing to antibiotic substances produced by otherwise harmless soil fungi (Brian, Hemming, and McGowan 1945).

The fact that they do perform absorbtive functions and thereby sometimes aid the tree has unfortunately led to a feeling that they exist for this purpose, and even to a belief that their presence under all circumstances is not only beneficial to the tree, but necessary for the maintenance of its good health. While this view is grossly exaggerated, there is nevertheless abundant evidence that under some circumstances, and particularly in nutrient deficient soils, they exercise a very beneficial influence on tree growth (Hatch 1936; Mitchell, Finn, and Rosendahl 1937; Miller, F. J. 1938; Rayner 1938, 1947; Rayner and Levisohn 1941; McComb 1943; Björkman 1944; McComb and Griffith 1946; Wright 1957*d*; Gilmour 1958).

This aspect of work on mycorrhiza has been fully reviewed by Levisohn (1958), who has pointed out that much of the evidence is circumstantial and that relatively few critical experiments have yielded incontrovertible proof of a beneficial effect on tree growth. There is always the possibility that some third factor may be encouraging good growth and good mycorrhizal development at the same time (Levisohn 1958).

Since mycorrhiza are particularly efficient nutrient absorbers, and since they tend to develop earlier in soils of relatively low nutrient status, their absence from such soils may well lead to a more pronounced development of deficiency symptoms in trees than would have occurred had mycorrhiza been present. It is obviously idle to try to define whether deficiency symptoms, which can be wholly or partly remedied by the introduction of mycorrhiza, are due to lack of nutrients or lack of mycorrhiza, since the two are so closely interrelated. It would certainly be a grave mistake to assume that the presence of mycorrhiza would enable healthy tree growth on all deficient soils.

Planted trees usually carry mycorrhiza acquired in the nursery. There is evidence that mycorrhizal fungi can be seed-borne (Garcia Salmeron and Breis 1958), but direct-sown seedlings may suffer from lack of mycorrhiza in certain soils until the necessary fungi arrive in the form of airborne spores, or unless artificial infection is provided. This renders the introduction of mycorrhiza into nurseries formed on agricultural soils a useful precaution even though in many cases the fertility of the soil will enable the trees to make healthy initial growth without mycorrhizal aid. In some cases manuring may give better results than the introduction of mycorrhiza (Latham, Doak, and Wright 1939). Introduction can be carried out by moving in transplants from an established nursery where mycorrhizal fungi are already present, but it could be argued that a forest of the species concerned is more likely to have all the desirable mycorrhiza and is therefore a better source of inoculum. Seed-beds should then be restricted to ground which has already carried a crop of such transplants. Harley (1957) takes the view that mycorrhizal fungi, since they are known to spread by means of spores, should eventually reach all afforested areas. The chance of a site suffering from a permanent lack of mycorrhiza is therefore remote.

The rapid and healthy growth of tree seedlings in partially sterilized soil from which mycorrhiza are lacking (Edwards 1952; Faulkner and Aldhous 1956; Wilde 1954), and the frequent occurrence of healthy trees with sparsely developed mycorrhiza (Wilde, Voight, and Persidsky 1956) make it quite clear that mycorrhizal infection is not always necessary for healthy tree growth. It is probably correct, however, to regard the presence of mycorrhiza as a normal condition of forest trees.

Disease symptoms have frequently been attributed to lack of mycorrhiza, but in several such cases there is some doubt whether this was the real trouble, or whether the symptoms were caused by soil deficiency. For instance, *Pinus radiata* in Australia, suffering primarily from zinc deficiency, was found to lack mycorrhiza. In this case there was some improvement in health following mycorrhizal inoculation (Kessel and Stoate 1936). Neilson Jones (1941, 1945) associated failure to form mycorrhiza with 'Fused-needle' disease of pines, but he also found that spring drought and possibly mineral deficiency were contributing causes. H. E. Young (1940) considered mycorrhizal deficiency one of a whole complex of factors contributing to the same disease in Australia. Redmond (1955) found that lack of mycorrhiza was one of the accompanying symptoms of a very complex dieback of birch, which he attributed mainly to the effect of hot soil on the fine roots (Chapter 34).

While chlorosis and stunting owing to lack of mycorrhiza have been frequently reported under nursery conditions (Mitchell, Finn, and Rosendahl 1937; Miller 1938; McComb 1943; Pryor 1956; Linnemann and Meyer 1958), information about the effect of lack of mycorrhiza in the forest is much less definite, though Hatch (1936), Björkman (1944), and Klotz (1956) have all suggested lack of mycorrhiza as one of the reasons for checked growth in conifers used to afforest non-woodland soils. It is easy, though dangerous, to extend this conception to a belief that mycorrhiza play an essential part in tree growth under all conditions, and that their absence or even their derangement may lead to other pathological developments, such as infection by root-rot fungi (Dominik 1946). We have much to learn, but on present evidence it would be wrong to attribute too great a pathological significance to the presence or absence of mycorrhiza.

ROOT NODULES

Root nodules, caused by bacteria or by organisms very near to bacteria, occur on several tree genera. On Leguminous trees they are caused by species of *Rhizobium* (Fred, Baldwin, and McCoy 1932) and on *Alnus*, *Elaeagnus*, *Myrica*, and *Hippophaë* by species of *Actinomyces* (Hawker and Fraymouth 1951; Bond, Fletcher, and Ferguson 1954; Bond, G. 1955, 1956).

The nodules, which are usually borne in profusion on the finer roots, are small and irregular in shape, and individually seldom larger than a pea, though they often occur in large clusters. The organisms causing them have the power of fixing atmospheric nitrogen. Under some circumstances this is beneficial to the host tree, and probably to other plants growing in mixture with them (Gants 1940; Virtanen 1957). Their presence is probably necessary for the healthy growth of some Leguminous species. However, they are nearly always present on trees of those genera which are subject to them, and there are few records of deficiency symptoms developing in their absence, although this presumably could happen on soils of low nitrogen content. Thus, though they are certainly abnormal growths and due to bacterial invasion, they appear to have very little pathological significance.

OTHER MICRO-ORGANISMS OF THE RHIZOSPHERE AND ROOT SURFACE

As yet we have only a very limited understanding of the complex interactions of the soil flora, but it is becoming increasingly evident that saprophytic organisms in the soil may have a profound effect on tree health, by their influence on parasitic root fungi or on mycorrhiza. This is particularly the case when the organisms are in the rhizosphere, that part of the soil which is in contact with the roots. It is now realized that many of the soil-inhabiting organisms produce antibiotic substances which may markedly influence the growth of other organisms near them. This subject has been fully reviewed by Stallings (1954).

Trichoderma viride Pers. & Fr. (Fungi Imperfecti, Moniliales) and a number of other soil fungi are antagonistic to the root-rot fungus *Fomes annosus* (Rishbeth 1951*b*; Nissen 1956), to *Phytophthora* (Kalser 1938), and to *Armillaria mellea*, as well as to the fungi causing Damping off (Chapter 15). The use of carbon disulphide injected into the soil as a method of controlling root attacks by *A. mellea* (Chapter 16) acts apparently through its differential effect on components of the soil flora, *Trichoderma viride*, which exerts an antagonistic influence on *Armillaria*, being encouraged (Bliss 1951; Aytown 1953; Darley and Wilbur 1954; Garrett 1957). *Alternaria tenuis* Auct. (Fungi Imperfecti, Moniliales) and some species of *Penicillium* (Fungi Imperfecti, Moniliales) are known to be antagonistic to mycorrhizal fungi (Levisohn 1957*b*; Brian, Hemming, and McGowan 1945). How frequent and how important this indirect effect of soil saprophytes may be awaits further elucidation.

Levisohn (1953, 1956) claims that mycorrhizal fungi can exert a beneficial influence before they have actually invaded the host. It has been established that toxins can be produced by the rhizosphere flora (Klotz and Sokoloff 1943; Kerr 1956). There is, therefore, a strong possibility that non-pathogenic organisms around the roots may exercise beneficial or harmful effects on them. It is possible that the production of toxins by the rhizosphere flora may partially explain the so-called 'conifer sickness', which results in poor, stunted seedling growth in nurseries where conifers have been grown for long periods, and which can normally be remedied by soil sterilization (Chapter 15).

18

DISEASES OF GENERAL IMPORTANCE ON STEMS, SHOOTS, AND LEAVES

A CONSIDERABLE number of pathogens found on the aerial parts of trees attack more than one genus. Several have been described in those chapters devoted to the genera on which they are most important; but all those which have any pathological significance are reviewed, at least briefly, below. Only *Nectria* canker requires at all lengthy treatment. Fungi causing decay in standing trees are considered in Chapter 19.

STEM DISEASES—BARK AND CAMBIUM

Nectria canker and dieback

Several species of *Nectria* (Ascomycetes, Hypocreales) have been associated with dieback and canker on hardwoods and much less commonly on conifers. Members of this genus also occur commonly as saprophytes on dead bark, and in many of the diseases they probably invade and extend injuries initiated by other agencies. The identification of *Nectria* species has always presented difficulties and, with the exception of *N. cinnabarina*, the attribution of a disease to any particular species of *Nectria* should always be accepted with caution. Booth (1959) has reviewed its taxonomy in Britain. He has unravelled much confusion, but only by the introduction of several new names.

Dieback and canker of beech, associated with *N. coccinea* (Pers.) Fr., *N. galligena* Bres., and *N. ditissima* Tul., are dealt with in Chapter 28, ash canker, associated with *N. galligena*, in Chapter 34, dieback of maple and elm, associated with *N. cinnabarina* (Tode.) Fr., in Chapters 30 and 31, *N. coccinea*, associated with canker on poplar, in Chapter 32, and *N. cucurbitula* (Tode.) Fr., connected with basal cankers on Sitka spruce, in Chapter 23.

N. cinnabarina is easily distinguished from the other species by the bright salmon pink pustules of the conidial stage, *Tubercularia vulgaris* Tode. (Fungi Imperfecti, Moniliales). These pustules, which give the fungus its name of Coral spot, are 1 to 2 mm. across and are produced in vast quantities on the dead bark of a wide range of hardwoods, in particular on *Ulmus* and *Acer*, and much less commonly on a few species of conifers. A full description of the fungus and a host list have been given by H. A. Jørgensen (1952). The fungus is a weak parasite often causing small cankers and minor dieback, fairly commonly on elms and maples, but rarely on other genera. It is usually held that it has strains of different virulence (Vloten 1943; Schipper and Heybroek 1957), but this is denied by H. A. Jørgensen (1952). Uri (1948), working mainly with maples, found that the most dangerous period for infection was at the close of the dormant season. Spring-pruning wounds were readily infected, and the fungus could be carried from one wound to another on pruning tools. The damage caused by this fungus in Britain, even on elms and maples, is very slight, and hardly justifies any special precautions beyond the removal of infected branches on ornamental trees.

The other species of *Nectria* commonly attacking hardwoods in Europe are *N. coccinea, N. ditissima,* and *N. galligena.* The last named is most important as the cause of canker on apples and pears. There is still some doubt about the conidial stages of these species of *Nectria,* but they are usually referred either to *Cylindrocarpon* or to *Fusarium* (both Fungi Imperfecti, Moniliales). Petch (1938) lists a number of other species of *Nectria* on hardwood trees and shrubs, but there is no evidence that these have any pathological importance. The commonly occurring species have been dealt with in some detail by Richter (1928) and Moritz (1930). They all appear to have strains of differing virulence, but there is little evidence of host specialization. Moritz (1930) and Ashcroft (1934) have both studied the development of *Nectria* cankers. They are both agreed that the fungus kills the host tissue ahead of its own development, so that the mycelium is never found in living cells. Grant and Spaulding (1939) found that *Nectria* cankers on hardwoods usually developed round buds, short spurs, or small branches.

Despite the very common occurrence of species of *Nectria* on hardwoods in Europe, they, and the cankers and dieback with which they are associated, have not attracted much attention, firstly because in many instances they are so obviously associated with wounds, and secondly because in the forest the percentage of trees infected is very low. Certainly, apart from the specific cases listed above and dealt with in other chapters, *Nectria* spp. demand little attention as causes of canker on the general range of hardwoods grown in Europe.

In eastern North America, on the other hand, *Nectria* canker of hardwoods has considerable pathological importance. The species most concerned is considered to be *N. galligena* (Lohman and Watson 1943). A large number of tree species are attacked, but the disease is much more serious on some, for instance *Betula lutea,* than on others (Grant and Childs 1940). The high susceptibility of some of the American hardwood species may explain the relatively greater importance of the disease there. The damage varies from small sunken patches in the bark to large conspicuous target cankers (Fig. 91), and is more serious because it weakens growth, damages timber, and may result in windbreak, than because it leads to serious dieback or death. As many as 50 per cent. of the trees in a stand may be infected (Welch 1934). The removal of cankered trees as a sanitary measure, without regard to sale value, has been put forward as the best method of control (Spaulding, Grant, and Ayers 1936).

Phomopsis pseudotsugae

Phomopsis pseudotsugae Wils. (Fungi Imperfecti, Sphaeropsidales) is chiefly important on Douglas fir, and is therefore described in Chapter 25. It has also been associated with cankers on the stem of Japanese larch (Chapter 24). Hahn (1957*a*, 1957*b*) has suggested that it should properly be called *Phacidiopycnis pseudotsugae* (Wils.) Hahn. He also regards it as identical with *Phomopsis strobi* Syd., which is of minor importance on *Pinus strobus* in the

United States and on *P. radiata* in New Zealand (Chapter 22). He has now found the perfect stage on *P. strobus*, and has named it *Phacidiella coniferarum* Hahn (Ascomycetes, Phacidiales). The black, discoid ascocarps, which are 0·25 to 1·0 mm. across, are produced together with the pycnidia on the bark.

The fungus has also been recorded on *Larix decidua*, *L. siberica*, *Cedrus atlantica*, *C. deodara*, *Sequoia gigantea*, *Picea excelsa*, *Pinus sylvestris*, *Abies grandis*, and *A. alba* (Wilson, M. 1925; Hahn 1930; Robak 1952*b*). It is known to be a weak parasite mostly associated with twig canker and dieback, and frequently occurring on twigs damaged by other agencies such as frost. Further observation will undoubtedly lead to its discovery on other conifers.

Cytospora dieback

Many species of *Cytospora* (Fungi Imperfecti, Sphaeropsidales) have perfect stages in the genus *Valsa* (Ascomycetes, Sphaeriales). Several species have been associated with dieback of trees. Among these are *C. kunzei* on spruce and Douglas fir (Chapters 23 and 25), *C. chrysosperma* on poplar (Chapter 32), and *C. abietis* on Silver fir (Chapter 26). *C. abietis* Sacc. has a wide host range among conifers, and has been recorded in Britain on Japanese and European larch, as well as on frosted Douglas fir. W. R. Day (1958) found *C. abietis* among other fungi associated with larch canker (Chapter 24). Many *Cytospora* species are only weak parasites, and it is typical that Dearness and Hansbrough (1934) should have found several species of this genus attacking a wide range of trees and shrubs following fire injury in British Columbia.

Physalospora dieback

Several fungi of the genus *Physalospora* (Ascomycetes, Sphaeriales) have been associated with dieback and canker of trees and shrubs. Spaulding (1958) gives an extensive host list for a number of *Physalospora* species, of which the most important in temperate regions is *P. obtusa* (Schw.) Cke. This fungus, the imperfect stage of which is *Sphaeropsis malorum* Peck. (Fungi imperfecti, Sphaeropsidales), is best known as a cause of leaf-spotting, dieback, and canker of apples and pears. It occurs on fruit trees in Britain, but is relatively unimportant. Both as a fruit-tree disease and on other hardwoods it appears to be important only in the United States (Marshall, R. P. 1939). On pine it produces symptoms like those caused by the closely related fungus *Diplodia pinea* (Jump 1937) (Chapter 22). It has also been recorded attacking initially weakened trees of *Abies concolor* in the United States (Fowler 1936*a*). Other species of *Physalospora* have been found attacking *Eucalyptus* and *Paulownia* (Chapter 35).

Most of the *Physalospora* species, including *P. obtusa*, are probably only weak parasites on trees already damaged by some other cause, but *P. miyabeana* on willow (Chapter 33) is certainly a vigorous parasite.

Dermatea livida

Dermatea livida (B. & Br.) Phillips (Ascomycetes, Helotiales) (syn. *Durella livida* (B. & Br.) Sacc., *Pezicula livida* (B. & Br.) Rehm.), the imperfect stage of which is *Myxosporium abietinum* Rostr. (Fungi Imperfecti, Melanconiales), is a very common saprophyte on conifers. The *Myxosporium* stage produces small pustular acervuli and the *Dermatea* stage small disk-like apothecia on the bark. M. J. F. Wilson (1928, 1931) considered that there was some evidence for weak parasitism, but there is little to support this, though the fungus is certainly a very rapid colonist of twigs and bark damaged by other agencies.

Botryosphaeria ribis

Botryosphaeria ribis Grossenb. & Dugg. (Ascomycetes, Sphaeriales), which occurs only in America, is chiefly important as the cause of a dieback of currants, but has an extensive host range on other broadleaved species. The dieback and canker disease it causes on *Cercis, Forsythia, Rhus,* and *Platanus* are dealt with in Chapter 35, on *Tilia* in Chapter 34, and on *Salix* in Chapter 33. Frezzi (1952) found it attacking Lawson cypress and *Eucalyptus,* among other trees in the Argentine. It has also been recorded in nature on elm (Luttrell 1950) and on *Liquidambar* (Toole 1957). C. O. Smith (1934), by artificial inoculation, found that it had a very wide host range indeed, though its pathogenicity varied from host to host.

Silver leaf

Stereum purpureum (Fr.) Fr. (Basidiomycetes, Aphyllophorales), the cause of Silver leaf in plums, is described under *Prunus* in Chapter 35. It has been recorded on a very large number of other woody plants, including some conifers (Moore, W. C. 1959). Some records probably refer to its saprophytic activities on dead wood, but even as a pathogen its range is wide. It usually causes progressive dieback, but the silvering of the foliage, which is so characteristic of the disease on plums, does not occur on all its other hosts.

Bleeding canker

In the United States *Phytophthora cactorum* (Phycomycetes, Peronosporales) has been cited as the cause of 'bleeding' cankers on a wide range of broadleaved trees (Howard, F. L. 1941). It is probably important there only on maples (Chapter 30). There are no records of *Phytophthora*-induced cankers in Britain, where species of this genus appear to act primarily as root-attacking fungi.

Crown gall

Bacterium tumifaciens E. F. Smith & Towns. causes tumour formation on many plants, including trees and shrubs. The tumours are usually below soil-level, so this disease is dealt with in Chapter 16, but infections do sometimes occur on the aerial parts, though they are rarely serious. Magerstein (1931), however, reported injurious attacks by this pathogen on willows in Czechoslovakia.

Fire blight

The bacterium *Erwinia amylovora* (Burr.) Winsl. *et al.* causes a very serious dieback and canker disease of pears, but also attacks a very wide range of Rosaceous hosts, including *Crataegus* (Chapter 35, where the disease is very briefly described), *Pyracantha, Cotoneaster,* &c. (Reid 1930; Shaw 1933).

It is possible to infect walnut by inoculation, but the bacterium does not occur naturally on this host (Smith, C. O. 1931). Until recently this disease was confined to the United States and New Zealand. It has now appeared in Britain (Crosse 1958, 1959).

STEM DISEASES—WILTS

Verticillium wilt

Verticillium albo-atrum Reinke & Berth. (Fungi imperfecti, Moniliales), and possibly several other species of *Verticillium*, cause a wilt disease of hardwoods. Among trees this disease is important only on the genus *Acer*, and is therefore dealt with in more detail in Chapter 30. The fungus normally enters the plant from the soil through the roots, but produces a vascular wilt rather than a root disease. *V. albo-atrum*, and the very closely related species *V. dahliae* Kleb., together have a very wide host range, including a large number of woody plants. Among these are elm (Chapter 31), ash, lime, numerous species and varieties of *Prunus*, and many species of *Rhus*. A complete host list of plants attacked by *V. albo-atrum* and *V. dahliae* has been compiled by Engelhard (1957). There is no evidence for the existence of strains specialized to different host plants, though there may sometimes be differences in pathogenicity between strains (Donandt 1932; Ende 1958).

DISEASES OF LEAVES AND SHOOTS

Grey mould

Botrytis cinerea Fr. (Fungi Imperfecti, Moniliales) is occasionally found on dying or dead succulent shoots on trees. The initial injury is usually due to some other agency such as frost, and the fungus is only secondary. Though quite unimportant in the forest, *Botrytis* causes a serious nursery disease, and is therefore dealt with fully in Chapter 15.

Pestalozzia

A number of species of *Pestalozzia* (*Pestalotia*) (Fungi Imperfecti, Melanconiales) have been recorded on the young shoots and foliage of trees. *P. guepini* Desm., which attacks *Camellia* and *Rhododendron*, has been dealt with in Chapter 35 and *P. hartigii* Tub., primarily a nursery fungus, in Chapter 15. *P. funerea* Desm., which has a fairly wide host range, has attracted most attention on *Thuja* (Chapter 26), but has also been associated with severe withering of the foliage of walnuts in Switzerland (Gäumann 1927). Servazzi (1934*a*, 1935*c*) recorded several species of *Pestalozzia* on various trees and shrubs in Italy. The comparative infrequency of records of damage by fungi of this genus, coupled with their wide and rather erratic host range, suggests that they are only weak parasites on leaves and shoots probably damaged by other causes.

Glomerella cingulata

Glomerella cingulata (Stonem.) Spauld. & Schrenk. (Ascomycetes, Sphaeriales) is best known as the cause of Bitter rot of apples. In Britain it has not been recorded damaging trees or shrubs, but abroad it has been associated with dieback of the shoots of *Aesculus*, *Camellia*, *Ligustrum* (privet), and *Paulownia* (Chapter 35).

Phyllactinia corylea

Phyllactinia corylea (Pers.) Karst. (*P. suffulta* (Rab.) Sacc.) (Ascomycetes, Erysiphales) causes a mildew on the leaves of hazel, ash, birch, hornbeam, alder, and some other hardwoods. Hammarlund (1925) considered that the fungus on each host was probably a separate strain. Blumer (1933) raised some of them to the status of species. The superficial mycelium on the leaves is less striking than that of Oak mildew. The small perithecia, at first yellow and later black, are borne mainly on the underside of the leaves. The fungus is seldom damaging.

19

DECAY AND STAIN IN LIVING TREES

Wood decay is certainly responsible for greater economic loss in forestry than all other diseases together. Not only standing trees but also felled logs and the products made from them are affected. In this book consideration of decay and stain is limited to their occurrence in living trees. Decay in felled timber and forest products in Britain has been adequately dealt with by Cartwright and Findlay (1959), and on a more popular basis by Findlay (1953). J. S. Boyce (1948) and Baxter (1952) devote considerable space to the deterioration of felled timber and forest products as well as to decay in standing trees. The writer (Peace 1938) listed the fungi recorded in a widespread investigation of butt-rot of conifers in Britain. Decay in living trees, the subject with which we are concerned here, has been very fully reviewed with a large bibliography by Wagener and Davidson (1954). Much of the available literature refers to north-western America, where the over-maturity of many of the forests renders decay of outstanding importance.

The fungi causing decay

A very large number of fungi can cause decay in trees and in timber. Most are Basidiomycetes, but a few, such as *Ustulina vulgaris* Tul. and *Daldinia concentrica* (Bolt.) Ces. & De Not. and *Xylaria* spp., are larger Ascomycetes. Savory (1954) has described the role that smaller Ascomycetes and Fungi Imperfecti play in wood deterioration. They produce a characteristic type of decay known as 'soft-rot', usually under very wet conditions, such as in the framework of cooling towers. Whether they have any importance in the living tree has still to be discovered.

The fungi decaying standing timber are generally different from those which attack felled timber and forest products, though a few species do both. Decay fungi which are present in standing trees do not die out immediately they are felled, indeed under suitable conditions they may continue to spread in the felled logs. In general, however, when a tree is felled and converted, the changes in the physical conditions, particularly of moisture, result in a change in the fungal flora. The number of fungi occurring in standing trees is still too large for separate consideration here, and reference should be made to Cartwright and Findlay (1959) for detailed descriptions. Some species of decay-causing fungi which are important on the Continent, and many of importance in North America, do not occur in Britain. Others, such as *Fomes annosus* and *Trametes pini* (Thore) Fr., have a very wide distribution, including Britain, the rest of Europe, and North America.

It is very important in considering wood-rotting fungi and their activities to realize that the heartwood of a tree is dead, and that a fungus attacking it is not therefore strictly a parasite. Only those fungi attacking live tissue, as well as rotting dead heartwood, can truly be regarded as parasites. A few decay-causing fungi are vigorous parasites. The most notable examples are *F. annosus* and *Armillaria mellea,* which are dealt with in Chapter 16. Many other wood-destroying fungi are thought to be parasitic, some on very good evidence. Others are only weakly parasitic, being able to spread a short distance through live tissue, or to attack the live parts of trees weakened by other causes.

A number of parasitic wood-rotting fungi, which are considered in other parts of the book are listed below. No attempt has been made to give a list of the much larger number of saprophytic fungi decaying different tree species.

Chapter 16	Diseases of General Importance on Roots. Most of these fungi are serious sources of decay in the stem, although they may enter through the roots.	*Fomes annosus* Fr. *Armillaria mellea* (Vahl.) Quél. *Polyporus schweinitzii* Fr. *Polyporus dryadeus* (Pers.) Fr. *Ganoderma* spp. *Poria weirii* Murr. *Ustulina vulgaris* Tul. *Stereum sanguinolentum* (Fr.) Fr.
Chapter 23	Diseases of Spruce.	*Pleurotus mitis* (Pers.) Berk.
Chapter 27	Diseases of Oak.	*Polyporus hispidus* (Bull.) Fr. *Stereum rugosum* (Fr.) Fr.
Chapter 28	Diseases of Beech.	*Fomes fomentarius* (L.) Kickx. *Polyporus adustus* (Willd.) Fr.
Chapter 32	Diseases of Poplar.	*Fomes fomentarius* *Fomes igniarius* (L.) Fr.
Chapter 34	Diseases of Other Forest Hardwoods.	*Polyporus betulinus* (Bull.) Fr. on birch. *Fomes igniarius* on birch.

It is very probable that other wood-destroying fungi, now thought to be purely saprophytic, will be found to be weak parasites, though their practical importance may still depend mainly on their wood-rotting activities.

The species of fungi decaying any particular tree or part of a tree depend on a number of circumstances. The tree can obviously be attacked only by those fungi which are in a position to infect either by mycelial growth or by spores. Which fungus is successful depends not only on the means of ingress available, but also on the internal condition of the tree, which may be suitable for some fungi and not for others. A tree, or even part of a tree, may be decayed by two or more fungi, though where decay fungi occur in mixture it is usual for one to predominate.

After decay is well established, most wood-rotting fungi produce fruit bodies on the outside of the tree. Some are very striking in appearance, taking the form of large brackets on the side or at the base of the tree (Plate VIII. 3); others are less conspicuous, consisting of sheets of spore-bearing tissue appressed to the bark surface. Some of these fruit bodies are ephemeral, others perennial and woody. Factors controlling their production are the species of fungus concerned, its mycelial growth within the tree, the age and internal condition of the tree, and climatic factors. Fructifications appearing on a rotted tree are not necessarily those of the fungus chiefly involved. Large, old trees, because of the advanced state of decay associated with them, are much more likely to show fruit bodies than small young trees. In the western United States and Canada, where much of the timber is still over 200 years old, fruit bodies, there known as 'conks', have proved a valuable, though occasionally deceptive, aid in estimating the decay condition of standing timber. In Britain this method is seldom of any value.

The dissemination of decay-causing fungi

The main means of spread of decay fungi is by spores. These are produced in enormous quantities, and a single sporophore may shed over 100 billion spores in a day (Buller 1909, p. 263; White, J. H. 1920). The period of discharge varies from a few days in the case of the more ephemeral species, such as *Armillaria mellea*, to the whole year, apart from very cold or very dry periods, in the case of species with long-lived sporophores, such as *Fomes annosus*, *F. igniarius* (Riley 1952), or *Trametes suaveolens* (L.) Fr. (Hirt 1932). The climate and other factors affect decay fungi in different ways, so that the spore discharge period of one species may be quite different from that of another (Orłoś 1958). A knowledge of conditions favouring spore discharge and of the periods when it is likely to occur is a necessary precursor to any practical study of a decay-causing fungus. The spores are carried long distances, for instance those of *F. annosus* are present over most of Britain and have been trapped 30 miles from land over the Irish Sea (Rishbeth 1958). The spore concentration in the air, however, and, therefore, the chances of successful infection, are naturally highest in the neighbourhood of active sporophores and decrease fairly rapidly with distance from them. The presence or absence of a decay fungus on a site is therefore one of the most important site factors influencing decay.

Many decay fungi produce only basidiospores, but a number, including *F. annosus*, have also a conidial stage. In the case of *Fomes* it appears that only the basidiospores are important in dissemination; but little is known about the importance of *Fomes* conidia or of those produced by other decay fungi.

Decay fungi which extend into the roots can spread directly, either by root contact or by root grafts. Those, like *A. mellea*, which produce rhizomorphs can spread through the soil without the aid of tree roots. Fortunately, for such fungi have an obvious pathological advantage, *A. mellea* is the only common pathogenic rhizomorph-producer in Britain.

Many decay fungi can be found rotting fallen logs and branches on the forest floor. Very occasionally they may infect a standing tree by direct contact between such material and the root or stem. In North America, where the amount of large material on the ground is much greater, sporophores produced on it may be of considerable importance as sources of spores. There is no evidence that decay fungi are able to live freely in the soil in the absence of woody material, or to infect directly from the soil.

Means of entry

Much more is known about the way fungi enter the wood of trees than about their behaviour in the tree. However, knowledge about the means of entry is particularly valuable, since at that stage there is most possibility of avoidance and control.

An intact, healthy tree has a complete covering of live tissue composed of the outer living part of the wood and the inner living part of the bark. Provided this remains intact, penetration is necessarily limited to fungi which are parasites. Some of these may have difficulty in gaining a hold unless a substantial amount of dead tissue is available to provide them with an initial 'build-up'. Some of the root fungi (Chapter 16) are able to attack more virulently when they can build up on stumps or on the roots of dead trees, and then enter living trees through root contacts. Most of the decay-causing fungi, however, require a wound in the tree, penetrating to the heartwood, or at least dead tissue extending to the heartwood, before they can attack successfully. As a working distinction, fungi entering through the crown are known as 'top-rots', and those entering through the roots as 'butt-rots'. This is not an entirely definite division, since wounds on or at the base of the trunk can provide entry for fungi of either type. Both *Fomes annosus*, normally regarded as a butt-rotting fungus, and *Stereum sanguinolentum*, normally considered a top-rotting one, can invade basal wounds, as well as wounds on the trunk make by brashing and in other ways.

There is a large literature on means of entry (Englerth 1942; Hornibrook 1950; Foster and Foster 1951; Foster, Craig, and Wallis 1954; Wagener and Davidson 1954), and many of them have been dealt with in other parts of this book. Fire scars, which are much more important in North America than in Britain, are considered in Chapter 5; breakages by ice and snow leading to decay are mentioned in the same chapter; frost cracks and drought cracks, which nearly always lead to decay, are discussed in Chapters 4 and 6; damage by man, particularly in the course of tree felling and haulage operations, by animals and by insects are dealt with in Chapter 10. While insects and animals are of little importance in this connexion, damage by man, which increases the more intensively the forest is managed, provides many channels for decay to enter. Even increment borings, blazes made to mark trees, or bruises caused by using the butt of an axe to sound for decay, can provide avenues of infection. The possibility that pruning wounds may allow decay to enter trees is discussed in Chapter 37. In this connexion it is very important to remember that decay can also enter through unpruned branches or through wounds resulting from natural pruning. Chapter 37 also deals with the entry of decay into stems arising from coppice stumps. In timber stands the singling of forked or multiple stems, common in species such as Lawson cypress, can lead to serious decay, especially if it is done after heartwood has started to form. The rate of healing and the use of protective treatments are obviously of importance in any consideration of wounds as a means of entry. These matters are also dealt with in Chapter 37. In North America, where the dwarf mistletoes on conifers are common and often very damaging, the wounds which they cause are commonly invaded by decay fungi. In Britain mistletoe is of little, if any, importance as a means of entry for decay.

The various ways in which decay may enter a tree are shown diagrammatically in Fig. 51. Two of them, which have not been fully covered elsewhere, remain for consideration. These are entry through dead limbs and entry through roots. The formation of protective layers at the bases of dead branches is described in Chapter 37, where it is pointed out that in

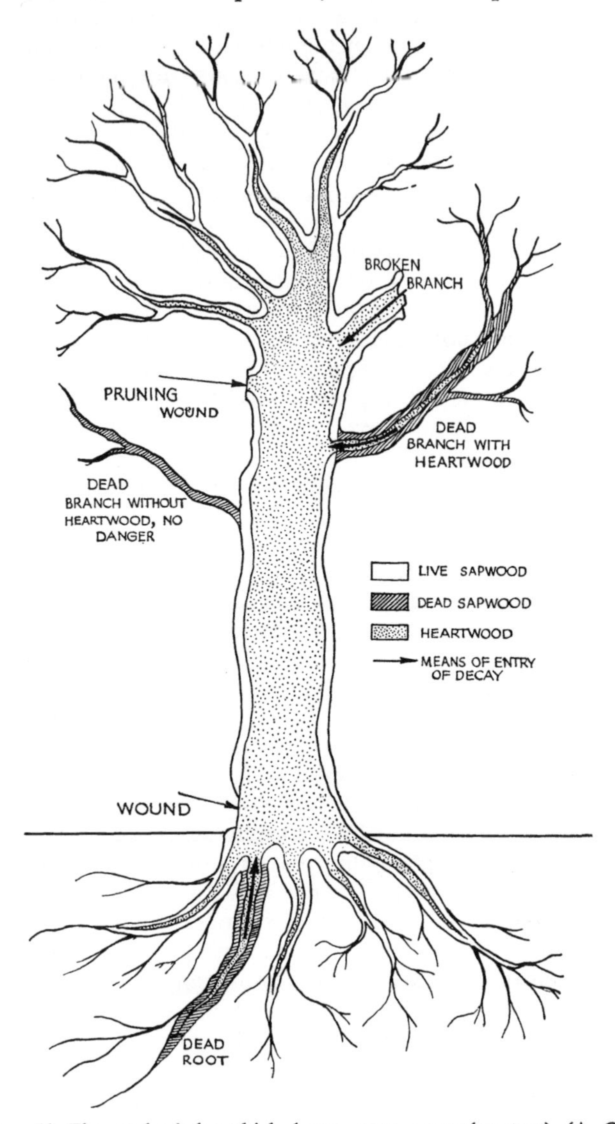

FIG. 51. The methods by which decay enters a tree (see text). (A. C.)

hardwoods these cover only living tissue, while in conifers the resin produced may extend some way into dead wood. This means that in hardwoods a dead branch containing heartwood, since it lacks any protective barrier across the heartwood portion, provides direct access for decay-causing fungi. In conifers, while a barrier may extend across small branches, large dead branches are still definite risks. Andrews and Gill (1941) found a clear relation between the number of large dead branches on *Pinus ponderosa* and the amount of decay by *Polyporus ellisianus* (Murr.) Sacc. & Trott. Liese (1936) pointed out that Scots pine under thirty years of age were hardly ever attacked by *Trametes pini* because they lacked branches containing heartwood. It is obvious that the older a tree gets the more large branches it will possess, and that eventually, however it is treated, some branches will die and become channels of entry for wood-destroying fungi. This is one way in which increasing age leads to increased decay. Equally it is obvious that the silvicultural treatment the tree receives will have a profound effect, both on the production of large branches and on the time they remain alive. A tree grown in the open will form many large branches, which, however, will remain alive for a long time since they have adequate light. Trees growing very close together may never form any heartwood-containing branches at all, since all their lower branches will be shaded out while still quite young. In a plantation properly thinned so as to maintain a reasonable growth rate, large heartwood-containing branches must eventually be formed, and the risk of fungal invasion will arise as soon as they die. Some treatments may accentuate this tendency. For instance, much of the oak over 100 years old in England became very wide spaced partly through deliberate policy, because at the outset short-boled, heavy-limbed oaks were favoured for shipbuilding, and later because many of the larger, straighter trees were removed for tan bark. These open-grown oak developed large limbs, which started to die from suppression when the canopy eventually did close (Plate X). At this stage the stocking was too low to allow of further thinning in order to keep the branches alive, and their death inevitably led to the entry of fungi, in particular *Stereum gausapatum* Fr., into the heartwood of the main stem (Day, W. R. 1934*b*). Some fungi, of which those mentioned above are examples, are better able to enter through branches which are more or less intact, though dead. Others such as *S. sanguinolentum* can enter more readily through broken dead branch stubs, as well as through wounds (McCallum 1928; Englerth 1942).

Heartwood-containing branches or branch stubs are generally required for successful entry, but Haddow (1938) has pointed out that *Trametes pini* can enter through twigs as small as 1·5 cm. in diameter. However, the general behaviour of top-rotting fungi, in particular their normal failure to invade unwounded young trees, suggests that small-branch infection is exceptional.

Apart from *Armillaria mellea* and *Fomes annosus*, described in Chapter 16, surprisingly little is known about entry through roots. Both fungi are true parasites and are therefore able to penetrate uninjured roots. The probability of entry through roots killed by other agencies, such as unsuitable soil water conditions, has often been suggested, even for such established pathogens as *F. annosus* (Day, W. R. 1934*b*, 1948; Hopffgarten 1933) or *Polyporus schweinitzii* (Hepting and Downs 1944). *Poria subacida* (Pk.) Sacc., a serious cause of decay in North America, is also said to enter through dead roots (McCallum 1928). W. R. Day (1934*b*) has described *Ustulina vulgaris* decaying roots initially killed by adverse soil conditions assisted by the fungus *Phytophthora cambivora*. The collection of direct observational evidence on this question is a matter of great difficulty. Very often the decay observed in a root has spread down into it from the stem, a fact that is seldom apparent once the root is thoroughly rotted. Most workers now think that the majority of the root-infecting fungi are probably parasites, able to enter uninjured roots.

The nature of decay and stain

The process of decay is the conversion by fungi of the cell walls of the wood into substances which those fungi can use as food. The nature of the decay depends mainly on the extent to which the different constituents of the cell wall are utilized. Fungi which attack only the cellulose cause the wood to darken in colour and to shrink. The shrinkage is usually accompanied by cracking into brittle, rectangular pieces. Such decays are generally termed 'Brown rots'. When the lignin also is broken down, the wood becomes paler in colour and is eventually reduced to a whitish, fibrous mass. These are the so-called 'White rots'. Good examples of White rots are those caused by *Stereum gausapatum* in oak, and by *Fomes annosus* in conifers. Typical Brown rots are those caused by *Polyporus sulphureus* in oak or by *P. schweinitzii* in conifers. In fact there are far more variations in the nature and symptoms of decay than the two broadly different types mentioned above. The nature of the decay depends not only on the fungi involved, but also on the species of host tree. Detailed descriptions of the various types of decay are given by Cartwright and Findlay (1959) and in more popular terms by Findlay (1953).

In most cases the timber is stained by the fungus before there is any real decay. This staining is reddish, brown, cream-coloured, or even purplish in colour, according to the host tree and fungus concerned. Most stains are characteristic to the expert eye, but they can be confused with non-fungal stains or with stains due to fungi incapable of causing decay. For this reason it is dangerous to base estimates of future decay condition of a plantation on the amount of stain found in the butts of thinnings.

Laboratory examination of a decayed stem typically discloses three zones. Farthest from the point of entry there is a narrow zone where microscopic examination reveals fungal mycelium which has not yet produced any visible effect on the wood. Inside this there is a wider zone where staining is visible but where the wood is still relatively undamaged. Finally there is the zone of decay where naturally the earlier rotted parts are more decomposed than those recently attacked. Eventually decayed parts may collapse altogether, leaving a hollow in the centre of the trunk.

Wood stained by the early stages of decay-causing fungi must inevitably be regarded with some suspicion, though provided such timber is used under dry conditions there is little chance of further fungal development. This sort of stain seldom occurs over a sufficient volume of timber to be utilized in the form of naturally coloured wood, but the early stages of attack by *Fistulina hepatica* (Huds.) Fr., known as 'Brown oak', are occasionally found on a sufficient scale to have a commercial value (Cartwright 1937).

After felling, timber is subject to invasion by a number of staining fungi, particularly species of *Ceratostomella* (*Ceratocystis*) (Ascomycetes, Sphaeriales) which discolour the wood without decaying it. They have been reviewed by Findlay (1959). Such stains occur occasionally in standing trees, sometimes being introduced by bark beetles. They can be quite serious on trees killed standing and not immediately harvested, but their effects are generally much less extensive than in felled timber.

Non-fungal stains

In some tree species there occur stains which appear to be of chemical origin. Some of these fade so much on exposure to the air that they become inconspicuous, but others persist and may seriously lower the value of the timber. None of these stains has been thoroughly investigated, so that the information about them is very incomplete. In most cases they are associated with certain soil types, and in some instances they occur more in one variety of a species than in another.

One of the best known stains of this type is 'Black heart' in ash, in which the heartwood is a very dark, greyish-brown in colour, rendering it entirely unfit for sports goods and other specialized uses. Little is known about this stain, but it is usually considered to develop only on wet sites (Bosshard 1955). Oberli (1937) has pointed out that the diameter of the stain is usually greater in the crown than at the butt. It often extends into the branches.

Dark heart of poplar, which is less serious since it often fades markedly after felling, has been studied by Meiden (1958). He found that the commoner, dark-brown type, which fades after felling, was associated with wounds, though no living organism appeared to be involved. The entry of oxygen may be the underlying cause. The more serious bluish-black stain of the heartwood of poplar was found only in a few trees on poorly drained soils where there was an accumulation of iron compounds. Earlier work on dark heart of poplar had suggested a bacterial association (Clausen and Kauffert 1952), and there is no doubt that bacteria are sometimes present in stained stems. There may here be some confusion between true staining and the water-soaked condition of the heartwood, known as 'wetwood', which is normally associated with bacteria (Chapter 20). Roth (1950), investigating staining of the heartwood of *Liriodendron* in America, inoculated trees with ten different species of bacteria which had been isolated from stained wood. None of these produced staining. In fact the stain appeared to result from wounding, regardless of whether the wounds were inoculated with bacteria or not.

Paclt (1953) attributed the formation of false heartwood in beech to wounds which admit air. Red discoloration of beech heartwood occurs commonly in Britain and on the Continent, and is usually attributed to the effect of frost (Custer 1934; Larsen, P. 1944). There are probably several different kinds of discoloration (Nečesaný 1956), not all of which are caused by frost. Raunecker (1956) inclines to the view that discoloration is due to the transmission of toxins from the roots. Beech staining, particularly in Britain, still awaits proper analysis and explanation. There is obviously room for more research work on all these non-fungal stains.

The progress of decay in trees

Progress in the individual tree

The rate of spread of decay in individual trees has frequently been measured by observing the extension both from inoculations and from injuries, such as fire scars, which could definitely be dated. The results of this work have been reviewed by Wagener and Davidson (1954) and are not dealt with in detail here. Very variable rates have been recorded, not only for different fungi and different host trees, but even for the same fungus in the same host species. For instance, Hirt and Eliason (1938), using inoculation methods, found that *Fomes pinicola* (Swartz) Cke. moved 100 cm. a year in *Picea rubra*, but only 3·3 cm. in *Tsuga canadensis*. Rennerfelt (1946) got an average movement of around 30 cm. per year from *F. annosus* inoculated into *P. excelsa*, but the maximum was over 100 cm. a year. Low and Gladman (1960) found that *F. annosus* invading a standing tree from a nearby infected stump had an average rate of growth of 55 cm. a year. Infection of extraction wounds in trees of the same age had in most cases made little progress during the same period, the highest figure being one of 18 cm. a year. Even wider variations have been recorded under natural conditions. Lagerberg (1919) found that decay in the snow-broken tops of spruces had progressed at rates varying from 0·5 to 112·5 cm. a year. J. S. Boyce (1920) found that the rate of spread of *Polyporus amarus* Hedge. in *Libocedrus* varied from less than 1 mm. to nearly 100 cm. a year. Toole and Furnival (1957), working on a large number of fungi decaying Red oak from fire scars, got annual extensions varying from 1·75 to 20 cm. Most of the figures given above certainly refer to spread along the grain.

Differences in rate of spread are probably also due to infection occurring at different times of year, though this affects initial development more than later progress. Spread is sometimes much more rapid initially than later (Lagerberg 1919; Hepting 1941). In fact some fungi may stop growing altogether after a rapid start.

The average rates of progress, as reported by a large number of workers, are not high. Figures seldom exceed 20 cm. a year in conifers or 10 cm. a year in hardwoods such as oak or ash (Wagener and Davidson 1954). It is easy to over-estimate the rate of spread, so that a crop may be considered to be on the verge of losing all its value by decay, whereas in fact a long period may elapse before its condition has seriously worsened. Nevertheless, the true rates are sufficiently high for quite serious damage to occur in a period of ten years or even less.

Typical butt- and top-rots start at the base or at the top of the utilizable stem respectively. Thus all their vertical spread is in one direction either up or down the tree. A fungus such as *Stereum sanguinolentum*, which habitually infects wounds on the stem, is potentially more serious since it can spread both upwards and downwards from the same point of entry.

Radial spread, as well as vertical, greatly affects the amount of timber spoilt, but a pipe-rot of rapid vertical extension, such as *S. gausapatum* on oak, can spoil a very large volume of timber by its long narrow 'pipes' of decay, even when the actual volume of rotted timber is extremely small.

Progress in stands

The progress of decay in natural forest or plantations could be assessed by periodic sampling. In fact practically no such work has been done, though estimates have been based on the sampling at one time of stands of the same species, but of different ages. This may give a rough idea of the increase of decay with age, a subject which is discussed below, but in view of the many factors other than age involved, the information it provides on the rate of progress of decay is unlikely to be accurate, unless the number of areas sampled is very large. Even then it will only give an average figure which could not be applied specifically to any one stand.

Samples from managed plantations will normally consist of thinnings, and may therefore be biased. However, a few comparisons of *Fomes annosus* rotting conifer plantations in Britain have suggested that thinnings do not usually show a widely different amount of decay from that which would be found in a truly random sample of the crop.

In really old forests, such as those in north-western America, death or windblow of the worst decayed trees may sometimes lead to an apparent fall in the amount of decay when the residual trees are sampled (Boyce and Wagg 1953). Even in forests managed on normal rotations, this situation may arise in the case of butt-rots, which damage the roots and impair the trees' anchorage, for windblow may selectively remove decayed trees, leaving the crop apparently less rotted than it was before.

Factors affecting decay

Internal factors

Remarkably little is yet known about the factors affecting the progress of decay in trees. It is certain, however, that a supply of oxygen is required for the growth of wood-destroying fungi. Thacker and Good (1952) take the view that sufficient oxygen is normally present in wood, but it seems possible that the carbon dioxide content may sometimes reach levels at which fungal growth is reduced or even stopped (Bavendamm 1928; Zycha 1937).

Moisture is obviously necessary for the growth of wood-destroying fungi. In most trees there is probably enough moisture to permit fungal growth, but Etheridge (1958*a*) found that the rate of decay of spruce by *Coniophora puteana* (Schum. ex Fr.) Karst. is sometimes depressed by low moisture content in the wood. Indeed he suggested (1957) that the distribution of fungi in the tree may depend to some extent on their moisture requirements, butt-rots operating at the base of the tree, where the moisture content of the wood is normally higher, because they require or at any rate tolerate more water than top-rots. This cannot be accepted as a general principle, however, because other workers have recorded different moisture distribution patterns in trees. Huckenpahler (1936), for instance, found that the moisture content in Shortleaf pine (*Pinus echinata*) increased with height, in which case Etheridge's theory on the distribution of butt- and top-rots would be quite inapplicable. Fielding (1952) states that moisture content may decrease or increase with height up the tree, that young trees are moister than old trees, and fast-growing trees are moister than slow. There are also substantial seasonal variations in moisture content, which have a general pattern of maximum content in the late spring and minimum in the summer, but which display wide variations for different species (Clark and Gibbs 1957), which may also affect the rate of spread of decay. These seasonal variations are much greater in deciduous broadleaved trees than in evergreen conifers.

Wood-destroying fungi have optimum temperatures for growth (Humphrey and Siggers 1933; Lindgren 1933; Cartwright and Findlay 1959). In temperate regions this optimum is always well above the minimum winter temperature. Temperature in the trunk fluctuates less than the air temperature outside (Reynolds 1939; Petrov 1955), and in some cases the activity of the decaying fungus itself may have an effect on the stem temperature (Rypáček, Tichý, and Hejtmánek 1951). Nevertheless stem temperatures never differ widely from those of the air outside, so that summer wood temperatures encourage decay and winter temperatures are low enough greatly to reduce or even completely to stop fungal growth. Boyce and Wagg (1953) consider that increased decay by *Trametes pini* on Douglas fir is dependent on a number of site factors all of which are associated with higher temperatures. Temperature also influences the rate of decay of slash, so that in warm regions much less unrotted material accumulates on the forest floor than in cold. In really cold climates slash may accumulate to a degree which is damaging to the crop. Saharov (1952) pointed out that stem temperatures vary on different site types and this is true also for different silvicultural treatments. Site and treatment may therefore influence the rate of decay in the tree in this way.

The anatomical characteristics of the wood are obviously likely to affect the rate of decay. In particular, rot often spreads more quickly in fast-grown conifers than in slow-grown ones (Lagerberg 1919; Spaulding and Hansbrough 1944). Gäumann (1948) found that decay in larch decreased with altitude, and this may possibly be associated with its slower growth at high elevations. In some cases the correlation between rate of growth and rate of decay merely means that the percentage of timber rotted is the same for slow-grown and fast-grown trees, though of course the actual volume of decay in the fast-grown trees is greater (Bier, Foster, and Salisbury 1946; Bier, Salisbury, and Waldie 1948). This relationship, however, does not invariably hold, even for conifers. For instance, Scheffer and Englerth (1952) found no difference in the rate of decay of fast- and slow-grown Douglas fir. It would be quite unjustified to encourage slow growth as a protection against decay or to condemn fast growth because of it. In hardwoods there seems to be no relationship between rate of growth and decay (Schmitz and Jackson 1927; Hepting 1935; Roth and Sleeth 1939).

Because some woods of high specific gravity, such as yew, have been found to be very durable, there has been a tendency to associate decay resistance with this property. There is certainly no general relationship; Southam and Ehrlich (1943), for instance, failed to find any correlation between rate of decay and specific gravity in wood of *Thuja plicata*.

The chemical nature of the wood, and of substances deposited in it, are known to have a marked effect on decay. Particular attention has been paid to *T. plicata*, the timber of which is resistant to decay by many fungi (Scheffer 1957; McLean and Gardner 1958), but the natural durability of other genera such as *Libocedrus* (Anderson, Zavarin, and Scheffer 1958) and *Eucalyptus* (Rudman and Da Costa 1958) has also received attention. The chemical constituents of the heartwood are largely responsible for differences in decay resistance between different species (Cartwright and Findlay 1959; Rennerfelt 1956).

Gradations in decay resistance associated with the concentration of chemical substances have been observed in the heartwood of a number of species. In general the older heartwood is more prone to decay, both in conifers (Cartwright 1942; Rennerfelt 1947) and in hardwoods (Scheffer, Englerth, and Duncan 1949; Scheffer and Hopp 1949). No doubt this has some bearing on the tendency for complete decay followed by hollowing to appear first in the centre of the tree. It is obviously partly responsible for the increased decay susceptibility of trees as they become old.

Site factors

The site has a very marked effect on the incidence and development of decay in trees. Numerous authors have recorded site-linked differences in decay (Peace 1938; Fenton 1943; Wagener and Davidson 1954). However, since the effect of the site on decay varies with the fungus, with the species of tree concerned, and with the silvicultural system used, as well as with age, it is impossible to generalize. If site quality is based on the rate of growth of a tree species, there is still no fixed relationship between site quality and decay. Etheridge (1958*b*) found that decay due to several different fungi in a complex of subalpine spruce species increased with improving site quality. Foster, Craig, and Wallis (1954), dealing with decay of Western hemlock in the interior of British Columbia, found that decay due to *Echinodontium tinctorium* E. & E. decreased, while that due to *Trametes pini* increased with improving site quality. Foster and Foster (1951), dealing with the same host species in the Queen Charlotte Islands, found that the overall decay caused by a number of fungi decreased with poorer site quality. Sometimes, however, there is no observable relationship. Englerth (1942) was unable to find any connexion between decay volume and site quality for Western hemlock in Oregon and Washington. Most of these observations apply to decay in mature forests. It is possible that young stands, such as we are mainly concerned with in Britain, may behave differently.

Numerous instances of more complex site relationships are known. For instance, the slight development of *Fomes annosus* in the early stages of plantations on land previously agricultural is largely conditioned by its absence from the site at the outset, though this may in a sense be regarded as a site factor. The very rapid development of *Fomes* at a later stage on such sites, however, is probably associated with a number of factors, including the comparatively high pH of the soil due to agricultural liming (Peace 1938; Low and Gladman 1960).

A relationship between site and decay may come about indirectly through the effect of site on silvicultural treatment or through some other site-linked factor. For instance, early thinning, resulting from rapid growth on a good site, will lead to the early provision of stumps, which may be invaded by and act as centres of spread for *F. annosus*. The steepness of the terrain may lead to greater extraction damage than normal, thus providing more means of entry for decay. The same may happen on a site which by reason of its vegetation is prone to scar-forming ground fires. Reference to the earlier section on 'Means of entry' will show that many of the means discussed can be affected by site.

In assessing the effect of site quality, it is important to distinguish between absolute and proportional decay losses. A certain volume of decay represents a higher percentage in a small tree than in a large one. Thus the same actual volume of decay on two contrasting sites will appear as a higher percentage, and truly do more damage to the crop on the poor site where the trees are small than on the good site where the trees are large. This point has been recognized in a number of investigations. Riley (1952), dealing with *F. igniarius* in aspen poplar, could establish no relation between decay and site on actual cubic volume, but when measurement was made in board feet, that is on a utilization basis, there was greater loss on the poor sites. Day and Peace (unpubl.) found that *Stereum gausapatum* in oak gave much higher percentage decay on the poorer sites where the height growth was less and the timber length consequently short. Thus, despite the fact that the actual volume of decay was roughly the same on the two sites, it would be best to dispose of the smaller, poor-site trees first.

Further examination of the literature would yield numerous other examples of relationships between site and decay. In most cases these have been recorded rather than explained. In view of the many interrelated factors involved which may act differently under different circumstances, and which are not necessarily the same in each instance, it is most unlikely that any generally applicable principles will eventually emerge. The problem of decay must be solved for each individual case, certainly on the assumption that other instances may give useful guidance, but never in the expectation that it will fit tidily into any overriding pattern of behaviour.

The effect of the host species on decay

Most decay-causing fungi have wide host ranges, though they may be much commoner on some genera than on others. A very few, for example *Polyporus betulinus* on birch and *Armillaria mucida* (Schrad.) Fr. on beech, are virtually confined to a single genus. Wide host ranges do not mean that decay-causing fungi normally attack all their possible hosts, or even the species and varieties within those genera, with equal frequency and vigour. Many are serious only on one or two genera, though recorded on others. Cartwright and Findlay (1959) give details of the known host ranges of most of the decay fungi attacking standing trees in Britain. Specific susceptibilities to *Fomes annosus*, and to some other fungi entering through the roots, are here listed in Chapter 16.

Specific differences in decay resistance have been clearly illustrated by the work of Scheffer, Englerth, and Duncan (1949), who tested the resistance of seven species of American oaks to a number of wood-rotting fungi. Scheffer and Hopp (1949) found clonal differences in decay resistance in *Robinia pseudacacia*. Clonal differences probably occur quite commonly, though their existence is likely to be masked by other decay-controlling factors, and in particular by the distribution of means of entry. The possibility that some decayed trees may be inherently susceptible raises doubts on the advisability of leaving such trees as sources of seed for natural regeneration. In really old stands many of the worst rotted trees will have blown or died before the final felling, resulting in some degree of natural selection before the seed trees are marked (Bier, Foster, and Salisbury 1946; Boyce and Wagg 1953). But in plantations run on normal, shorter rotations there is little such selection, and for genetic reasons it is inadvisable to leave obviously decayed trees.

The effect of age on decay

The fact that decay in a stand increases with age has been clearly demonstrated by many workers (Röhrig 1934; Peace 1938; Schober and Zycha 1948; Thomas and Podmore 1953; Foster, Craig, and Wallis 1954). If a decayed stand is allowed to remain long enough, the extension of decay eventually exceeds the increment, so that the volume of utilizable timber starts to fall (Day, W. R. 1929*a*; Foster and Foster 1951). Where the decay is in the butt, extension of the rot is at the expense of the most valuable timber and thus the value ratio between rot increment and timber increment will be more unfavourable than the volume ratio. Thus there comes a time, not always easy to determine, when the continued existence of the crop leads to a fall rather than to a gain in value.

The advance of rot with age shows both in the amount of decay in the individual tree and in the number of trees affected. The most important factor is merely that in old stands the decay has had longer to develop, but the question of opportunities for entry is also important. Parasites, such as *Fomes annosus* or *Polyporus schweinitzii*, which can attack uninjured trees and therefore affect comparatively young ones, have a great time advantage over other fungi which require easy access to the heartwood. But even with the parasites, old age may provide greater opportunities for access. For instance, repeated thinning provides *F. annosus* with a progressively larger number of stumps to act as bases for the invasion of the remaining trees. The possibilities of entry for saprophytic fungi obviously increase with the age of the tree. Apart from any special circumstances, the chances of wounding, branch breakage, and so on are basically related to age. An old tree has stood through more thinnings and therefore has been more liable to felling and extraction damage than a young one. A tree must reach a certain age before its branches have heartwood, and will be older still before they die and afford means of entry for decay fungi. Old trees with their large exposed crowns are more liable to wind and snow breakage than young ones, and with their reduced powers of root regeneration are also more liable to the death of major roots. The increased susceptibility of the heartwood as it becomes older has already been mentioned.

Increase in age involves not only a greater volume of decay but a larger number of fungal species involved. Thus some of the old natural coniferous stands in north-western America have a much richer flora of decay-causing fungi than younger, though silviculturally mature, stands in Europe. This is partially a reflection of the mixed nature of many of the American stands, which thus provide a wider range of host species, but it is also a reflection of age. No doubt as the average age of British forests becomes higher the variety of decay fungi attacking them will increase.

Age differences may sometimes lead to misunderstanding over fungal distribution and behaviour. For instance, *Trametes pini* is a serious source of decay on Douglas fir in western North America (Boyce and Wagg 1953), while in Europe, though it is locally serious on Scots pine, it has as yet attracted no attention on Douglas fir, the oldest examples of which in Europe are still very young compared with many of the natural stands in America.

The effect of decay on the tree

It is very doubtful whether decay fungi have any detectable effect on the growth or general health of the trees they are attacking, unless they are acting parasitically. Englerth (1942) found no connexion between lack of vigour and the presence of heartrot due to *Echinodontium tinctorium* in Western hemlock. Gosselin (1944) found the same with *Polyporus circinatus* Fr. in spruce, though Christensen (1940) regarded it as a primary cause of stagnation and senescence. In general most reports of decay fungi affecting growth rate or tree health can be associated with parasitism. This is the case with *Fomes annosus* (Chapter 16), where the loss of roots can affect growth rate to an extent sometimes more serious than the damage by decay itself.

Decay can manifest itself indirectly. Windbreak, which is much commoner in decayed than in sound trees, is a case in point. Spegazzini (1925) described a disease of *Populus alba*, associated with very complete decay of the heartwood by *Trametes trogii* Berk., in which the stems arched over until the crowns were resting on the ground. Similar collapse of larch stems has been seen in Britain.

Decay greatly influences tree safety. A sound tree may lose a limb or blow over, but the presence of decay greatly increases such possibilities. In law a sound tree is usually regarded as safe, but the presence of detectable decay may render an owner liable for any damage that may occur if the tree breaks or falls. Periodic inspection of roadside trees is therefore advisable. Adkin (1945, 1946, 1947) has dealt with the legal aspects of decay in roadside trees.

The detection and diagnosis of decay

It may sometimes be difficult to tell whether a tree is rotted at all, but it is certainly easier to do this than to discover the extent of the rot. For a whole stand the best estimation would be that based on a random sample. But in practice, if the remaining trees are to be left standing, the sampling must of necessity be done on some sort of silvicultural pattern, and may therefore become biased. Very little evidence has been collected, but it can generally be assumed that, though an estimate of decay based on thinnings will not be exactly the same as one based on a random sample, it will nevertheless be sufficiently close to be regarded as giving a reasonable indication of the decay condition of the stand.

Particularly in the case of top-rots it is sometimes possible to estimate the amount of decay by observing the number and condition of the possible means of entry, such as broken branches, wounds, etc. (Hepting, Garren, and Warlick 1940; Englerth 1942; Hornibrook 1950). In fact tables of so-called cull factors, to be used in the estimation of decay losses in standing timber, have been produced, and have proved useful, provided that their use is restricted to the tree species and region for which they were originally drawn up (Kimmey 1950; Kimmey and Hornibrook 1952). Resin exudation and swelling of the stem at the butt have been suggested by Rohmeder (1937) to denote the presence of *Fomes annosus* decay in spruce. It is certain that neither of these is a reliable indicator, for many decayed trees show no swelling, and resin-flow may have a variety of causes. Indeed many workers have found estimations based on external symptoms of little value (Haddow 1938; Bier and Foster 1946; Rennerfelt 1946). The presence of fruit bodies has also been used to estimate decay. Some fungi fruit much more readily than others, so that this method requires a knowledge of the behaviour of individual fungi, especially with regard to the relationship of fruit-body production to volume of rot. For instance, Horton and Hendee (1934) found that the amount of decay caused by *F. igniarius* in aspen poplar could be judged by the size and number of the fruit bodies. Riley and Bier (1936) confirmed this, pointing out that sporophores could be produced on quite small volumes of decayed wood, so that their number and distribution were the important factors. But cases such as this are exceptional. Bier, Foster, and Salisbury (1946) found that *Trametes pini* was the only fungus out of the large number they investigated on Sitka spruce for which fruit bodies could be used in estimating decay.

No systematic work has been done on these lines in Britain. It is clear, however, that some idea of the decay condition of trees, especially hardwoods, can be gained from an examination of possible means of entry. Indeed it is on such evidence that an opinion on the danger or safety of a roadside tree may have to be based. The presence of sporophores on a tree almost certainly indicates that, locally at any rate, decay is in an advanced stage. In the case of *F. annosus*, the presence of fruit bodies is of more value as an indication of the spore potential of the stand, and therefore of its importance as a source of infection to other areas, than as an indication of the amount of decay in the stand itself.

Other methods have been tried for the detection of decay in standing trees. Some operators claim to be able to detect decay from the sound made when the stem is hit with the butt of an axe (Boyce, J. S. 1930; Hornibrook 1950), but this method is obviously crude, and may lead to bark damage and subsequent fungal infection. The extraction of cores with an increment borer is more precise (Rennerfelt 1946; Hornibrook 1950), but unless several borings are made only a small portion of the trunk will be sampled. Under these conditions small amounts of rot may easily remain undetected. In any case increment borings may in themselves serve as means of entry for decay fungi or staining bacteria.

X-rays have been used successfully to estimate decay in standing trees and in standing telephone poles (Henriksen 1951; Truman 1953). This method is only able to detect decay which has reached a stage when it affects the density of the wood. It tends to be both expensive and, because of the apparatus involved, cumbersome. Improvements, such as the use of thallium or other radioactive substances as the source of radiation, and methods other than photography to measure its transmission, may well eventually make it more practicable for use in the field (Eslyn 1959). The distribution of radioactive isotopes in the wood has been used to detect decay in living Silver firs by Iizuka (1956). Waid and Woodman (1957) measured the interference in ultrasonic wave transmission through decayed and sound wood. However, other variations in wood structure, besides decay, cause interference and the interpretation of the results in terms of decay alone is fraught with difficulty.

It is unlikely that any method will be evolved that can give a complete picture of the rot distribution in whole stands such as could be produced by cutting up the felled stems. Nevertheless, even an incomplete picture, supplemented by information on the decay condition of thinnings, could be of great assistance in assessing the future possibilities of a stand and in getting advance knowledge of decay and its relationship with species and site.

The diagnosis of the fungi actually responsible for decay cannot safely be based on the sporophores produced. Some fungi fruit much more easily than others, so that the species fruiting, while responsible for some of the decay, is not necessarily the major agent. Expert examination of the decayed wood if the tree is felled, or less certainly inspection of an increment core, may betray the nature of the rot, especially if the range of possible causal fungi is a fairly narrow one. The safest method is culturing from decayed wood, but even with this method the decay must be properly sampled. If more than one fungus is present, a single random sample may disclose only a minor species.

Losses due to decay

Much of the work on decay in North America has been done primarily to provide information on potential decay losses. The overall picture for the whole of the United States based on such work indicates losses ranging from 5 to 20 per cent. according to species (Anon. 1947*b*). In America abundance of timber and very high extraction costs have involved abandoning the whole log, any badly or even moderately decayed stem being left on the felling site. The wastefulness of this was clearly illustrated by the fact that, under the stress of war-time shortages, the cull trees which had been left on some felled areas were reworked for timber. This attitude towards decay, whereby the whole log or at best the whole decayed length is discarded, has tended not only to make American estimates of cull high by European standards, but also to make its estimation much simpler.

In Europe, where utilization is much more intensive, it is much harder to calculate decay losses, for every effort is made to use the better parts of rotted logs. Some estimates of losses resulting from *Fomes annosus* have been made in Sweden and in Germany (Chapter 16). Similar detailed estimates, based on actual sawmill losses, have been worked out for *Acer saccharum* in the Lake States (Hesterberg 1957) and for oak in Yugoslavia (Marinković and Marinković 1957). Figures of percentage number of stems decayed give little indication of utilization loss, even if they are linked with a rough estimation of the extent of the decay in the tree. There is an obvious need for some method of relating reasonably simple field observations to actual sawmill losses. Until such a method has been evolved, decay losses, in Britain at any rate, will remain virtually unevaluated, though it is almost certain that they outweigh all other fungal losses on trees combined.

The avoidance of decay

Little work appears to have been done on the direct control of decay fungi by the use of systemic fungicides. Orłoś and Brennejzen (1957) injected zinc chloride, sodium fluoride, and copper sulphate into pine trees decayed by

Trametes pini and *Fomes annosus* and aspen poplars decayed by *F. igniarius*. This killed the mycelium without damaging the tree; but such methods, even if successful in stopping the spread of the fungus in the tree, could hardly be applied on a forest scale, though they might be applicable to park and street trees, or to the destruction of the fungus in stumps from which it might spread to surrounding trees. In general, control must be based on adjustments in silvicultural treatment, with particular emphasis on efforts to prevent the development of means of entry. Such efforts may justify the use of special measures, such as the creosote treatment of stump surfaces to prevent the spread of *F. annosus* (Chapter 16).

Thinning and felling policies are bound to have an effect on decay, and in some cases silviculturally correct treatments may be undesirable from a decay point of view. Heavy early thinnings, for instance, not only provide abundant stumps for colonization by root fungi such as *F. annosus* and *Polyporus schweinitzii*, but also lead to the development of large branches which may later form means of entry for top-rotting fungi into the remaining trees. Thinning or selection felling, on the other hand, may be adapted to include the removal of decayed trees, if they can be discovered. This may be fairly easy for old stands (Nordin 1956*a*), but is likely to be difficult for younger crops, unless some ready means of decay detection can be developed. It is possible, however, to remove large branched trees and favour small branched ones (Andrews and Gill 1941).

It is likely that uneven-aged forests are more subject to felling and extraction wounds than even-aged ones. An even-aged stand has a long initial period, extending at least as far as the second thinning, in which damage should be negligible because the trees to be felled and extracted are small and easily handled. In uneven-aged stands, on the other hand, damage to young trees by the felling and extraction of heavy timber is a constant hazard. The structure of the uneven-aged forest also allows some trees to develop heavy branches, again increasing the decay risk.

Pruning and its effects on decay are discussed in Chapter 37, which also deals with the prevention and treatment of decay in amenity trees. It seems clear that forest pruning, if properly carried out, need not substantially increase the decay risk.

Little information is available on the influence of sporophore production on the spread of decay fungi, though it is clear in the case of *F. annosus* that areas where the fungus is fruiting act as centres of spread (Low and Gladman 1960). It is difficult, therefore, to say whether the removal of sporophores or sporophore-bearing trees is a worth-while measure. J. S. Boyce (1927) regarded the retention of decayed seed trees as a potential danger to the succeeding crop. Aoshima (1951) urged the removal of dead or fallen birches as a sanitation measure against *Poria obliqua*. Felling of pines bearing sporophores of *T. pini* has been suggested and sometimes adopted as a control measure in Germany (Möller 1910; Röhrig 1934). Liese (1936) considered that this policy had successfully reduced the importance of the fungus; but it seems likely that a general fall in the average age of the pine forests—*T. pini* is essentially a decay of old trees—may have been the true cause. Under American conditions Haddow (1938) found that fallen logs were more important than standing trees as sporophore sources for this fungus; but Hepting and Roth (1950) pointed out that sporophore production on fallen timber for a number of decay fungi ceases after a period of ten to fifteen years, while standing trees can remain a danger for much longer periods than this. Under British conditions, of course, fruiting on fallen stems is a matter of less importance, for such stems do not occur frequently in the forest. Lack of spores may be a limiting factor in the spread of many of the British decay fungi, since old trees, suitable for sporophore production, are relatively rare in many areas.

On the Continent, planting of conifers in mixture with hardwoods has been generally advocated as a means of lessening decay, particularly by *F. annosus* (De Koning 1923; Priehäusser 1935; Rennerfelt 1946). Röhrig (1934) considered that pine in mixture with beech was less attacked by *T. pini* than pure pine. Several Continental workers, on the other hand, have failed to find this effect of mixed planting (Falck 1930; Schober and Zycha 1948), nor was the author able to confirm its existence in Britain (Peace 1938). On present knowledge there is no justification for advocating admixture with hardwoods as a control for decay in conifers.

Various methods of control applicable to *F. annosus* and *Armillaria mellea*, and almost certainly to several other root fungi, are discussed in Chapter 16.

20

ABNORMAL GROWTHS AND OTHER PATHOLOGICAL PHENOMENA

THE symptoms normally resulting from the effect of damaging non-living agencies, or from the attack of fungi and bacteria on trees, have already been discussed in Chapter 2. We are here concerned only with what may be termed abnormalities of growth, some of them stimulated by living or non-living pathogenic agencies, but many of them arising within the tree itself without apparent external stimulus. These abnormalities appear in most cases to arise by the stimulation of existing meristems, but this is not invariably the case, almost any living tissues in the tree being capable of abnormal growth activity under appropriate circumstances (Butler 1930). One aspect of such stimulation is the growth of healing tissues over wounds, but this is dealt with in Chapter 37, when pruning and tree surgery are under consideration. Many abnormalities in trees are the result of insect attack; these will not of course be considered here.

A very large number of the abnormalities, briefly described below, have been classified and illustrated by Klein (1913), who attributed varietal significance to many of them. The more permanent abnormalities in trees provide an interesting study for the arboriculturalist, but are of little importance to the forester, except in so far as they have pathological significance.

Witches' brooms

Witches' brooms, large bunches of closely set, frequently-branching twigs, sometimes with abnormal foliage, are one of the commonest and certainly one of the most striking abnormalities met with on trees. Many of them are of fungal origin, for instance *Melampsorella caryophyllacearum* causes very conspicuous brooms on Silver fir (Chapter 26), and *Taphrina turgida* does the same to birch (Chapter 34). Viruses can also cause witches' brooms; for instance, there are brooming diseases of *Robinia*, *Paulownia*, and walnut (Chapter 12). They are also produced as a result of attack by some of the Dwarf mistletoes in other countries, but not by our *Viscum album* (Chapter 13). Others arise as the result of attack by mites and insects. But still the origin of a considerable number of brooms, particularly on conifers, cannot be explained as the immediate result of attack by any kind of pathogen. General reviews of this subject have been made by Liernur (1927) and by Tubeuf (1933).

Witches' brooms apparently of non-pathological origin are quite often met with on conifers (Buckland and Kuijt 1957), but are not sufficiently frequent or damaging to be regarded as more than interesting curiosities. Sometimes they are very large. In one European larch, recorded in Norfolk, the entire crown, some 8 metres in diameter and 5 metres in depth, was a witches' broom

(Plate II. 4); but this is exceptional, most brooms being much smaller, restricted to the abnormal development of what would have been a single branch. Conifer witches' brooms usually have abnormally short needles, as well as the dense, restricted, much-branched growth characteristic of all brooms.

It is generally considered that witches' brooms not attributable to any pathogenic cause start as bud mutations. They can be propagated by cuttings or grafting and retain their dwarf form. Many of the dwarf conifers in commerce originated as witches' brooms. Occasionally they flower and bear cones, a proportion of the seed from which gives rise to dwarf forms (Liese 1933; Hintikka 1933).

Tubeuf (1933) regarded witches' brooms as in a sense parasitic on the trees that bear them. Schmitz (1916) investigated this aspect in connexion with the brooms caused by *Taphrina cerasi* on cherry (Chapter 35). He concluded that the brooms were probably self-supporting. It seems almost certain from their appearance that the leaves on many brooms must be functional, and many trees support large brooms without any apparent decline in vigour of the non-broomed parts. It would be rash, however, to suggest that they never draw from the tree's general food resources, and they must, of course, receive water and mineral salts from the tree's roots.

Something approaching witches' broom formation may be observed on trees subject to repeated injury by frost, drought, or any other factor that kills the tips of shoots without causing major dieback. The development of dormant buds or the initiation of new buds as a result of the twig dieback produces dense bushy growth which is occasionally confused with true witches' brooms. Blomfield (1924) considered that the twigs in witches' brooms on birch caused by *Taphrina turgida* did not ripen properly in the autumn and were therefore damaged by early frosts, thus making the brooms even bushier than they would otherwise have been.

This may be the appropriate place to refer to the production of dense conglomerations of cones. They apparently result from an abnormal production of female flower buds on short lengths of twig, but the basic stimulus is unknown. Recent examples have been recorded by Fergus (1956*a*) on *Pinus pungens* and by Looney and Duffield (1958) on Douglas fir.

Witches' brooms are never sufficiently damaging for control measures to be necessary. On ornamental trees they can usually be cut out, if they are considered unsightly.

Burrs, galls, and tumours

It is doubtful if any definitions can entirely separate and clarify the three terms normally used to describe abnormal swellings on trees. Burrs are usually roughened excrescences on the trunk or main limbs, usually relatively small in relation to the diameter of the stem on which they occur. Galls, on the other hand, are usually very large in relation to the stem, roots, or even leaves or flowers bearing them, and are not necessarily roughened, as is always the case with burrs. The term tumour is sometimes used to cover cases which do not fit either of the above categories, and also as a term embracing all kinds of swellings on any part of a tree.

These deformities are often caused by fungi, such as species of *Gymnosporangium* on juniper (Chapter 26) or *Exobasidium vaccinii* on azaleas and *Vaccinium* (Chapter 35). Swellings at the base of conifers, sometimes described as bottle-shaped, are occasionally produced as a result of internal decay by wood-destroying fungi. Tumours can also be caused by bacteria; *Bacterium tumifaciens* in particular causes tumours on the roots and stems of

FIG. 52. Galls on *Sequoia sempervirens*, possibly due to *Bacterium tumifaciens*, Garboldisham, Norfolk (nat. size). (J. N.)

a very wide range of host species (Fig. 52) (Chapter 16). Bacterial tumours on conifers, which are rare in Britain, have been studied in detail by Petri (1924*b*) and Dufrénoy (1925*b*). *B. pseudotsugae* sometimes causes numerous galls on twigs of Douglas fir in North America (Hansen and Smith 1937). Some Dwarf mistletoes cause tumours on their host plants; even *Viscum album*, the only species of mistletoe native to Britain, causes distinct swellings at the point of attachment to its host tree (Chapter 13). Honeysuckle sometimes stimulates the trees round which it twines to produce very large swellings (Chapter 13). Insects, mites, and nematodes are also common causes, particularly of the gall type of swelling. Lek (1929) has described so-called burr-knots on apples and other trees, which appear to be callus outgrowths associated with root initials, and which are indicative of easy vegetative propagation. But many tumours and burrs, particularly those occurring on the trunks of hardwoods and less commonly of conifers, still remain unexplained.

The origin and structure of plant tumours, a subject which cannot be dealt with here, has been considered in detail by M. J. Cook (1923) and Küster (1930). Briefly, tumours may consist simply of masses of proliferating tissue, but in some cases they possess a definite bud structure, and are in effect a mass of fused twigs each arising behind an individual bud (Fig. 53). Burrs formed

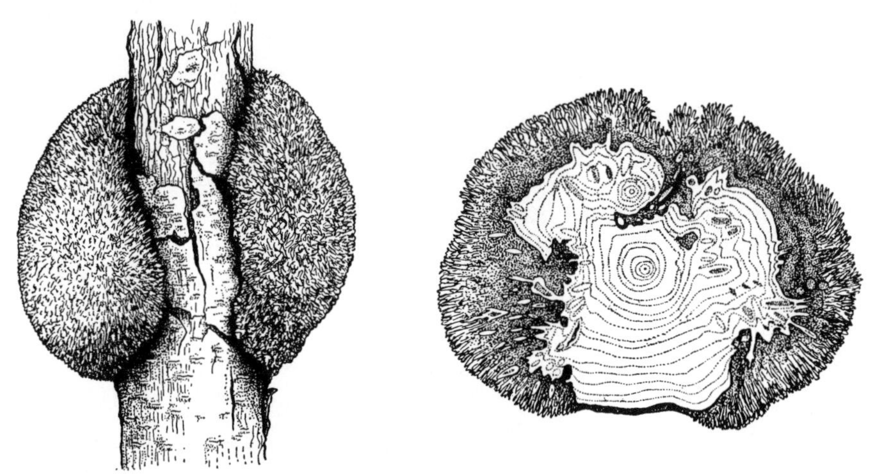

FIG. 53. Tumour of unknown origin on Sitka spruce, Slattadale Forest, Ross and Cromarty (greatly reduced). (J. C.)

in this way often possess an attractive grain and provide veneers of particular value for high-grade furniture, thus providing one of the few instances where a pathological condition actually increases the value of a tree.

Burrs which cannot be used for furniture veneers, because they are on unsuitable species or too discontinuous in structure, reduce the value of the logs on which they occur. Since we do not know how many of them originate, control measures cannot be suggested.

On a number of occasions, swellings, consisting mainly of large starch-filled cells, have been found, mainly in south Scotland, at the collar of conifer seedlings. These are considered in more detail in Chapter 15, partly because of a possible relationship with the common nursery fungus *Pestalozzia hartigii*. In some cases they appear to cause the death of the seedlings, in other cases recovery takes place. Dufrénoy (1922*d*) recorded similar swellings at the base of *Eucalyptus* and *Arbutus* seedlings, but did not regard them as pathological.

Fasciation

Fasciation involves a flattening of the normally cylindrical stem (Fig. 54). A fasciated stem is usually much heavier than a normal shoot. The flattened growth is due to the formation of a row of linked meristems, instead of a single one at the apex. It occurs both on conifers and on hardwoods, as well as on many other plants, but is commoner on conifers. In some cases the flattened stems may bifurcate, producing formations curiously reminiscent of fallow deer antlers. Fasciation is sometimes only of annual duration, some of the terminal buds resuming normal shoot growth again the following year, but it may continue for a longer period. In any case if a leader is affected, the damage prevents it from returning success fully to its dominant position. Fasciation normally affects stems of trees, but Owen (1955) has recorded it on the roots of *Ilex glabra*. It has been described in detail on *Pinus resinosa* by Kienholz (1932) and on *Betula* by Fowler (1936*b*). O. E. White (1948) has made a complete review of the whole subject.

It has been suggested that fasciation is due to wound stimulation, possibly as a result of insect attack, or to overnutrition of the affected plant. There seems little evidence to support either hypothesis. In some cases it is certainly a mutation, which can normally be propagated vegetatively, and which may come true from seed. The fact that it happens comparatively rarely hardly suggests an underlying cause of general occurrence. It is sufficiently uncommon in Britain to rank merely as a pathological curiosity, but Neves (1955) has reported fasciation on *P. pinaster* in Madeira so profuse on individual trees as to cause death. Such a condition suggests virus infection.

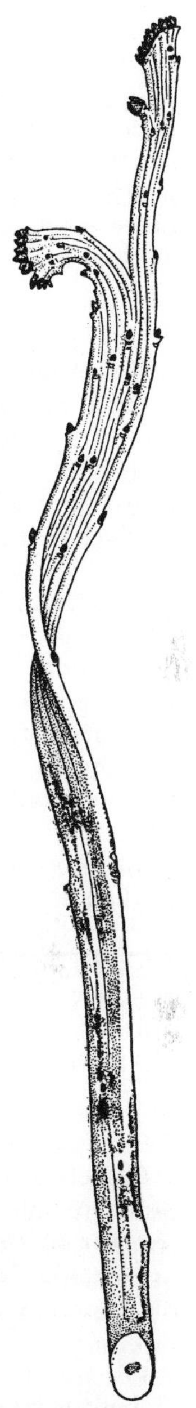

FIG. 54. Fasciation of ash (× ⅛) (J. C.)

Winged cork

The formation of raised ridges of bark, usually known as 'winged cork' on the twigs of certain species of elm, maple, and other broadleaved trees is sometimes thought, by those unfamiliar with it, to be a symptom of disease. It occurs most commonly on vigorous shoots and does not appear to have any pathological significance (Smithson 1952, 1954). It may be regarded as a somewhat inconstant botanical character. In Britain it is most commonly seen on *Acer campestre*, the Field maple, on *Ulmus* × *hollandica*, the Dutch elm, and on *Euonymus europaeus*, Spindle berry.

Lammas shoots

Renewed growth sometimes occurs towards the end of the summer, as a result of high rainfall accompanied by sufficient warmth, on shoots that have previously stopped growth, and in some cases actually formed buds. The resultant shoots are usually known as 'lammas shoots'. They sometimes have a pathological significance.

The phenomenon of Extra-seasonal growth of pine, wherein some of the lateral buds make short growths in the autumn and then, when growth starts again in the spring, rival the actual leading shoot and produce a tendency to forking, is dealt with in Chapter 22. Leibundgut (1955) has reported the same phenomenon in spruce. Production of lammas shoots in pine may also lead to double-branch whorls (Paul 1957).

In some cases the production of lammas shoots may lead to attack by fungi, since they are formed at a time when spores are likely to be present in abundance in the air. This is the case with Oak mildew, *Microsphaer alphitoides*, which attacks lammas shoots with particular severity (Chapter 27). In other cases Lammas shoots produced late in the year may be subject to autumn-frost damage.

Epicormic shoots

Epicormic or water shoots, arising on the stems of trees as a result of some external or internal stimulus, are a common phenomenon in forestry. They represent a normal reaction to certain types of dieback disease, occurring abundantly on the main stem in progressive canker and dieback of larch or below the cankers in Bacterial canker of poplar and Chestnut blight. Their production is also stimulated by the sudden removal of a considerable number of live branches, by sudden defoliation, for instance by caterpillars, and in some species by the sudden exposure of the stem to additional light. It is debatable whether they arise solely from dormant buds; they may also be adventitious (Stone, E. L. Jr. 1953*b*). They certainly occur more commonly on suppressed than on dominant trees (Seeholzer 1934; Rohmeder 1935). This observation has received experimental support from Skilling (1957), working on Sugar maple. They are certainly much commoner on hardwoods than on conifers. Their occurrence in a mixed hardwood stand has been studied in some detail by Jemison and Schumacher (1948). As they produce small knots in the wood, silvicultural measures are usually adapted as far as possible to avoid their production.

When they occur in quantity on diseased trees they can usually be taken as evidence that the tree is beyond reasonable chance of recovery. It may prove desirable to remove them from vigorous trees to enhance the timber quality. This can be done either by pruning or possibly by the application of herbicides (Aage 1941; Splettstösser 1957).

Branchiness

Branchiness in trees might at first seem to be hardly a subject for consideration here, and yet there are strains, provenances, or individuals in some tree species which possess a capacity for repeated forking, and consequent branchiness, so extreme as to be virtually pathological. Such trees are obviously undesirable in forestry and should be avoided in selection for tree breeding.

Beech is particularly apt to produce branchy types. Abnormal branchiness in conifers is less common, but Schlüter (1956) recorded excessive production of branches between the main whorls in pine, and Paul (1957) definite double branch whorls in *Pinus strobus* arising from lammas-shoot production and a resultant second ring of lateral buds. No doubt this kind of thing occurs also in other genera.

Lack of branches

Lack of branches, due to failure of the side-buds to develop or to flush, is occasionally met with in conifers, particularly spruce. In that genus it is sufficiently common to have acquired the name of 'Snake spruce' (Bennett 1936). It has also been found on Corsican pine, Scots pine, and Silver fir, and no doubt occurs on other conifers. Most affected trees appear to produce a few normal whorls of branches before they cease side-branch production, and those observed continued to make fairly normal height growth, though subnormal diameter growth, for a number of years. There seem to be no records of the eventual fate of such trees. They do not occur with sufficient frequency to have any practical importance.

Abnormal leaves

A number of trees possess varieties with abnormal leaves. Most of these probably originated as bud mutants and can be propagated by grafting or by cuttings. In some instances seed from abnormal trees produces a proportion of similarly abnormal seedlings. A number of forms of leaf-blade reduction, which include *Fraxinus excelsior monophylla*, *Sambucus nigra* var. *heterophylla*, in which the leaflets consist of little more than the mid-rib, and cut-leaved beech, have been considered by Saarnijoki (1955). The work of Hylander (1957*a*, 1957*b*) on *Alnus* and *Betula*, in which he describes as many as thirteen reduced-leaf forms of *A. incana* and twelve of *B. verrucosa*, clearly illustrates the complexity of this matter. A large number of forms with purple leaves also occur, of which the Copper beech is far the best known. Forms with golden or silvery-blue foliage occur in a number of coniferous genera. Variegations occur commonly with broadleaved species, as well as with some conifers. A number of these variegations and possibly some of the leaf reductions are of virus origin (Chapter 12), but the majority have arisen originally as mutations.

Many of the forms briefly mentioned above are very attractive in appearance, and others are cherished as interesting curiosities. In fact in this particular field it is impossible to say where normality ends and abnormality begins. These conditions are only damaging in so far as the reduction in leaf area, or the overlaying of the chlorophyll with some other pigment, results in reduced growth rate. In any case, since most of these trees are propagated solely for horticultural purposes, reduction in growth is of little importance.

Fluting

Protruding ridges at the base of the stem immediately above the major roots are quite a normal occurrence in trees, but similar ridges higher up the stem are abnormal and may detract considerably from the timber value. The production of such ridges is usually referred to as 'fluting'. In a few instances fluting may be the direct result of fungal attack, the depressions between the ridges being in fact elongated strips of dead cambium. This is the type of damage that occurs occasionally on Sitka spruce, and has been associated with

Nectria cucurbitula (Chapter 23). In many cases, however, fluting is only pathological in the sense that it is abnormal, occurring only on certain strains or on individual trees. The depressions are usually sited immediately above and below branch insertions. This type of abnormality is particularly common in Britain on birch. It has been studied in some detail on *Pinus taeda* by Newman (1956); but no explanation of its occurrence was put forward.

Spiral growth

Normally the fibres in the stem of a tree are arranged nearly vertically, but sometimes they run at an angle round the stem, producing what is usually known as spiral growth or grain (Plate XI. 1). Northcott (1957) believes that spiral growth is normal, and that trees with truly vertical grain should be regarded as abnormalities. If very slight divergences are defined as spirality, he may well be right. But divergences from the vertical sufficient to give a twisted appearance, to show up clearly when the wood is split, and seriously to affect its utilization, are certainly sufficiently uncommon to be regarded as abnormal. The deformity is often repeated in the bark, and is then visible on the outside of the tree. The twisting of young trees, and of the leading shoots of older trees, is probably a juvenile aspect of the same phenomenon. In the text below 'twisting' will be used to describe the phenomenon in young trees, 'spiral grain' in trees of utilizable size, and 'spiral growth' to cover both. These phenomena occur in both broadleaved trees and conifers. In Britain spiral grain is particularly common in Spanish chestnut, and twisting is sometimes rather serious in some seed origins of Japanese larch (Fig. 55). Both phenomena occur in a much wider range of species. There are a number of broadleaved trees where the tendency to twist has become so pronounced in certain individuals that they are propagated vegetatively as horticultural curiosities; instances of this are *Corylus avellana* 'contorta', the Twisted hazel, *Salix matsudana* 'tortuosa', and *Chamaecyparis lawsoniana* 'lycopodioides'.

Naturally spiral growth affects the utilization value of the wood. The stresses set up tend to produce cracking and warping when the tree is converted, and there is appreciable loss of strength (Rault and Marsh 1952; Banks 1953; Mayer-Wegelin 1956). There is a definite tendency for drought crack in Sitka spruce to occur in trees with spiral grain (Chapter 6).

It has been suggested that anti-clockwise growth is predominant in the northern hemisphere and clockwise in the southern, but there is little evidence to support this. Nearly all accounts, admittedly mostly referring to the northern hemisphere, find anti-clockwise the commoner direction, but A. T. Jones (1931) noted that in Massachusetts maples grew anti-clockwise and elms clockwise. Several workers have found that in conifers the twist starts clockwise, but changes over to the opposite direction after a variable period of time, sometimes very quickly (Champion 1925; Burger 1941; Kennedy and Elliott 1957). On this basis, older trees with clockwise spiralling would be those in which the change had not yet taken place. On the other hand, Kennedy and Elliott found that in alder the change with age was from anti-clockwise to clockwise, and this may hold for other hardwoods. Northcott (1957) also reported change of direction in a number of species, and went so far as to suggest that straight-grained trees were merely those which were passing through zero spiral on a change from one direction to the other. Lasschuit (1952) found that individuals of *Pinus merkusii* twisted in a clockwise direction in youth, but tended to straighten as they grew older. He was less certain than Champion of the direct connexion of twisted shoots with spiral grain, though he agreed with him in finding that the latter starts clockwise, and reverses later. In Britain few observations have been made, but spiral grain in older trees is certainly predominately anti-clockwise in most species. In some instances the spiral grain extends into the roots (Herrick 1932).

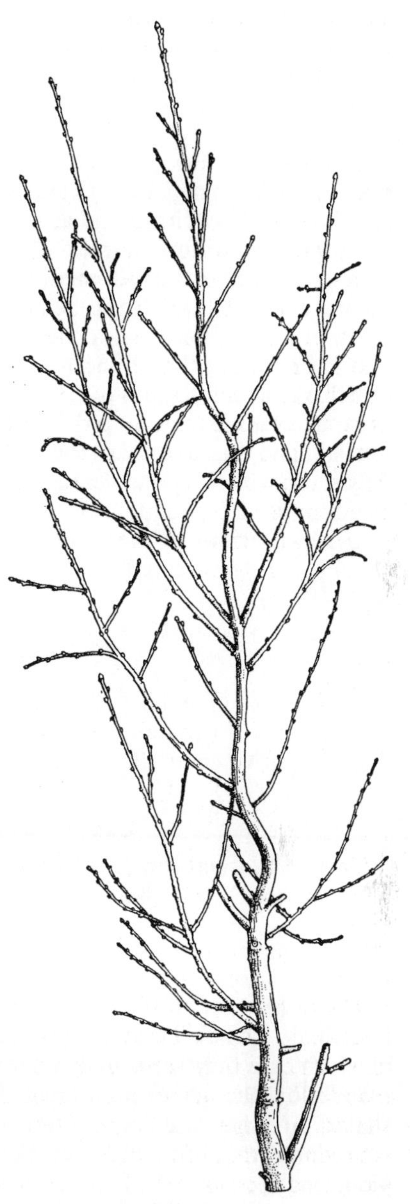

FIG. 55. Twisting of the stem in young Japanese larch. (J. C.)

The available evidence suggests that spiral growth cannot be directly associated with environmental factors. Rault and Marsh (1952), though they did not entirely rule out site influences, found that soil type, climate, and aspect had no significant effects on spiral grain in *P. longifolia*. They did conclude that both age and rate of growth exercised a marked effect. The degree of spirality was greatest in trees between thirty and sixty years of age, and increased in proportion to rate of growth, so that silvicultural restriction of diameter increment could be used as a control measure. Kennedy and Elliott (1957), on the other hand, found that in alder the effect of age out-weighed that of rate of growth, so that for trees of the same diameter the younger and therefore faster grown showed less spirality. This would suggest exactly the opposite silvicultural treatment to that recommended by Rault and Marsh. N. F. Howard (1932) attributed spiral grain to regular changes in wind direction and Brynski (1930) to unequal exposure of the crown, but these explanations have not been generally accepted. Champion (1925, 1930) in India, and Rault and Marsh (1952) in South Africa, found that it is definitely an hereditary defect in *P. longifolia*, and can be avoided by the use of seed from straight-grained trees. Seed from such a source produced straight trees even on sites where the previous crop had been almost entirely spiralled. Bolland (1957*b*) obtained evidence of the heritability of the direction and to a lesser extent of the degree of spiral growth in pine. Further support to a genetical basis has been given by Haskins and Moore (1933), who induced twisting in *Citrus* seedlings by irradiating the seed with X-rays. Nevertheless, there is still a general tendency to believe that environmental as well as genetical factors may be involved.

No information is available on the economic importance of this abnormality, either in Britain or elsewhere. Very high percentages of spiral growth may occur in some cases; Champion (1925) reported 100 per cent. in some stands of *P. longifolia* in India, and Koehler (1931) 99 per cent. in Alpine fir, *Abies lasiocarpa*, in Colorado. A very high percentage of twisting has been seen in a few stands of Japanese larch in Britain, but there is no information on the extent to which this later affected their utilization value. In general spiral grain in older trees in Britain is only an occasional nuisance.

It is important that obviously twisted trees should be removed as early as possible as thinnings and that none of them should be used as sources of seed. In some cases it may be possible to recognize potentially spiral trees in the nursery, and thus to discard them at an early stage (Kadambi and Dalral 1955). Any stand with a high percentage of spiral growth should be completely blacklisted for seed production.

General reviews of the subject have been made by Champion (1925), Rao (1954), and Northcott (1957). The anatomical aspect has been studied by Kohl (1933).

Timber stains

Dark heart of poplar, Red heart of beech, and related phenomena apparently not attributable to attack by fungal or bacterial pathogens are discussed in Chapter 19.

Compression and tension wood

These abnormal forms of wood produced mainly by the influence of the wind are briefly discussed in Chapter 5.

Shakes

The occurrence of internal cracks in the stem of a standing tree is usually known as 'shake'. These cracks may be radial, often starting from the pith, in which case they are usually referred to as star shakes or rift cracks, or they may follow the line of an annual ring, when they are known as ring or cup shakes. In either case they are serious defects when the log is converted. There is no doubt that some radial cracks are associated with drought; in fact they are drought cracks which have not extended to the surface or at any rate have not split the bark (Chapter 6). Ring shakes sometimes occur along the line of frost rings (Chapter 4). Shake occurs on some soils more than on others. For instance, a case is known where oak on sandy soil was much worse shaken than neighbouring plantations of the same species on clay. Most writers have assumed that the shearing stresses exerted in the stem during high winds are the final cause of rupture, but growth stresses in the wood itself have also been blamed (Koehler 1933; Boyd 1950). Lachaussée (1953) found that shake in oak was worst on acid soils, where the absorption of iron by the tree was easier. He attributed the shake to a weakening effect of the absorbed iron on the wood. The whole subject is badly in need of further investigation.

In Britain shake is common in oak, but certainly most serious in Spanish chestnut, where it is said to occur particularly in stems which have arisen from coppice stools. It occurs also in many other hardwoods. Much of the shake found in conifers, however, appears to be directly attributable to drought or frost.

Abnormal rings

A number of non-living agencies cause abnormal rings in the wood by their effect on the cambium during its developmental stages. Such is the case with frost (Chapter 4), drought (Chapter 6), and lightning (Chapter 5). Partial abnormal rings extending on either side of a wound are, of course, a common pathological symptom.

Wetwood and Slime-flux

The presence of excessive water in the trunks of trees, accompanied in most cases by high gas pressures, has been reported on numerous occasions and is now known as 'wetwood'. It has been found particularly in elm and poplar. Carter (1945) described its occurrence in elm in some detail and attributed it to the activities of a bacterium, *Erwinia numipressuralis* Cart. Trunk pressures in affected trees commonly ranged from 0·3 to 2·0 gm. per sq. cm., but exceptionally reached 4·0 gm. per sq. cm. Hartley and Davidson (1950) described the same phenomenon in a number of broadleaved species and a few conifers. It has been suggested that wetwood sometimes leads eventually to dieback. Carter (1945) found this to occur in elm and Hartley and Crandall (1935) in poplar and willow.

Under such conditions sap is apt to ooze from wounds and cracks. Such sap flow may be of comparatively short duration, in which case little harm is done; but if flow is prolonged, the exuding sap becomes infected by yeasts, fungi, and bacteria, and then ferments. Ogilvie (1924) described slime-fluxes on elm, Horse chestnut, apple, and willow in which fungi of the genera *Fusarium* and *Oospora* were involved, as well as yeasts and bacteria. The fermented sap tends to kill the tissues around the wound and the bark down which it flows, thus substantially enlarging the original injury (Guba 1934). Apart from this, slime-fluxes on ornamental trees are very unsightly and occasionally malodorous. They are usually extremely attractive to insects. In Britain slime-flux is most commonly met with on elm, sycamore, and Horse chestnut.

The control of slime-flux is not easy. Cleaning out the wounds and dressing with tar or a thick bituminous paint may stop the flow. In the case of very small wounds or cracks a peg or wedge of hardwood may be driven in, cut off flush and painted over (Le Sueur 1949). In America boring holes below the wound and inserting drainage pipes to reduce the pressure is advocated; but this is not an easy method and might well lead to further trouble when the pipes are finally withdrawn. These methods, of course, are only applicable to park and garden trees. In the forest slime-fluxes occur so seldom that they can be ignored.

21

THE SUPERFICIAL FLORA OF TREES

TREES, like practically any other object exposed in the open, eventually acquire a superficial flora mainly of simple forms of plant life. Many are saprophytic, but some, although growing primarily on the surface, may penetrate the host cells and extract nourishment from them; this is the case with the mildew fungi, e.g. *Microsphaera alphitoides* on oak. Such organisms are, of course, dealt with in this book in their appropriate places as actual causes of disease. Climbing plants such as ivy, although they are mainly supported in the crowns of trees, are rooted in the ground, and in so far as they cause damage are dealt with in Chapter 13. Those fungal and bacterial inhabitants of the soil, and in particular of the rhizosphere, which are not causes of disease, are mentioned briefly together with mycorrhizal associations, some of which are also superficial, in Chapter 17. Here we are concerned mainly with non-parasitic plants growing on the aerial parts of trees.

ALGAE

A number of algae can be found on trees, especially in damp shaded situations. Most are very simple forms, one of the commonest being *Pleurococcus*, which often gives a green bloom to the bark on the north side of trees. They are entirely superficial and have no effect whatsoever on the health of the tree. They may also occur in association with Sooty moulds (see below) on the leaves of trees (Tengwall 1924). However, a green alga, *Cephaleuros virescens* Kunze, is responsible for a quite serious disease of the leaves of tea in southern Asia, and has been recorded causing spots on the leaves of a number of ornamental shrubs, including *Magnolia grandiflora*, *Ligustrum*, and *Jasminum* in the sub-tropical climate of Florida. The algal filaments penetrate the leaf, growing mainly between the cells. Infected spots are characterized by hairlike algal outgrowths which form a velvety, reddish-brown or orange coating on the upper sides of the leaves (Wolf 1929–30). There are no records of any parasitic algae on trees in temperate regions. The ecology of algae on trees has been dealt with by Schmid (1927) and Prescott (1956).

LICHENS

Lichens are associations of fungi and algae. Their chief ecological characteristic is their ability to exist where nutriment is always, and water periodically, scarce. They are primary colonists on rocks, and occur frequently, and sometimes very abundantly, on the trunks and branches of trees, particularly in moist situations. They are mostly grey or grey-green in colour, and vary in form from a thin crust appressed to the bark, to branched forms attached only at the base and hanging in woolly masses from infested branches. They can be divided into three groups, 'crustose', growing closely appressed to the substratum, 'thallose', which are also prostrate, but leaflike and not so firmly attached, and 'fruticose', which are bushlike and erect or hanging.

Different species tend to grow on different parts of the tree according to variations in light and moisture conditions (Raup 1930; Hosokawa and Odani 1957). They require much more light than either algae or mosses, and are therefore often found in situations exposed to full sunlight. In addition, the water-holding capacity of the outer bark certainly affects the species that occur. Rough-barked trees are more favourable for lichens than smooth-barked ones; though a considerable number of crustose lichens can be found on young smooth-barked trees, and some species are more or less restricted to such bark.

The development of lichens varies with climate, the trees in some localities having much more growth on them than those in others. The most important factor concerned here is humidity; thus there is generally a greater development of lichens in the west of Britain than in the east. But there are also more local differences, dependent in some cases on areas where mist formation is frequent.

Most of the work done on lichens on trees has been either purely systematic or ecological with little regard to the tree, except in its function as a substratum. For instance, W. Watson (1936) described the lichen flora of beechwoods. The actual effect of lichen infestation on the tree itself has seldom been investigated. The best general account of the lichens is probably still that given by Lorrain Smith (1921, amplified 1933).

The most conspicuous lichens which occur on trees in Britain are *Hypogymnia physodes* (L.) Nyl., various species of *Parmelia*, all thallose varieties and *Evernia prunastri* (L.) Ach., *Platysma glaucum* (L.) Nyl. mainly in the west and north, *Ramalina* spp. and *Usnea* spp., which are fruticose (Fig. 56).

It is generally agreed that lichens are not normally harmful to trees, though J. F. V. Phillips (1929) considered the fungal component of an *Usnea* species to be parasitic on tissues outside, and sometimes inside, the cork cambium. They may, however, hasten the death of weakened branches or even whole trees by interfering with the normal functioning of the lenticels in the bark (Liou 1929). Particularly in the tropics they may cover living foliage and thus interfere with photosynthesis. Serious damage of this nature to spruce, due to profuse growth of *U. barbata* Weh., has been recorded in Switzerland (Rieben 1940). Lichens have frequently been observed to occur more abundantly on unhealthy trees than on healthy ones (Fig. 56) (Romele 1923). There appear to be two reasons for this; firstly, an unhealthy tree is likely to be growing slowly and will therefore shed its outer bark less frequently. This gives the lichens more time for development before they are sloughed off with the bark. Secondly, a tree with sparse unhealthy foliage allows more light to reach the lichens growing on it, and thus promotes their more luxuriant growth. Thus abundance of lichens can be used to some extent as a measure of the health and growth rate of a tree, always remembering that the basic level of development may be much higher in one locality than another. Münch (1936*a*) suggested that the profuse development of lichen on larches suffering from canker and dieback may be due to the high content of the food substances in the twigs killed rapidly by frost or by *Dasyscypha*, as compared with the low content in twigs that have died naturally.

Lichens are extremely susceptible to atmospheric impurities (Wheldon and Travis 1915; Jones, E. W. 1952), and thus form useful indicators of pollution

FIG. 56. Lichens of the genera *Usnea* and *Parmelia* on a slow-growing, unhealthy Douglas fir twig, Haldon, Devon, September (× ½). (J. C.)

in different areas. *U. comosa* (Ach.) Röhl. is piartcularly susceptible, while some *Parmelia* species are much more resistant. Some crustate lichens are more resistant than any thallose or fruticose species. Estimates of the degree of pollution can be based, therefore, not only on the amount of lichen development, but on the relative frequency of different species.

Normally there is no need to remove lichens from trees, but occasionally it may be desirable to do so on ornamentals. On deciduous species strong winter washes of the type used on fruit trees are likely to be successful. Copper-containing sprays have been used on *Citrus*, where, because of the evergreen foliage, spray damage has to be considered (Reichert and Palti 1947). J. F. V. Phillips (1929) suggested that lichen infestation might be lessened by avoiding heavy thinnings, which by letting in more light encourage lichen growth.

FUNGI

A number of fungi, generally known as the 'Sooty moulds', grow entirely superficially on the leaves and shoots of trees. Several belong to the genus *Capnodium* (Ascomycetes, Dothidiales), others to *Fumago* (Fungi Imperfecti, Moniliales). *Cladosporium herbarum* Fr. (Fungi Imperfecti, Moniliales), the perfect stage of which is *Mycosphaerella tulasnei* (Jancz.) Lindau (Ascomycetes, Sphaeriales), also occurs commonly as a normally harmless superficial mould. *Hormiscium pinophilum* (Nees ex Fr.) Lindau (Fungi Imperfecti, Moniliales) is found frequently on the surface of conifer needles; while *Pullularia pullulans* (De Bary) Berkh. (Fungi Imperfecti, Moniliales), which has some pathological significance, also appears occasionally as a Sooty mould. These fungi have been studied by Neger (1919).

Sooty moulds occur both on broadleaved trees and on conifers, frequently growing on the sugary exudations known as 'Honeydew' (Hey 1956), which are usually produced by aphids, but can occur without insect stimulation. Both Honeydew and Sooty moulds are particularly common on lime trees, *Tilia* spp. When present in quantity, Sooty moulds give leaves a distinctly sooty appearance, clearly visible to the naked eye, and they are therefore apt to attract more attention than their real importance justifies. They are not parasitic, but they are probably mildly harmful, in so far as they interfere with the reception of light by the leaf and with its gaseous exchange. Luttrell (1940*b*) attributed the poor growth of young oaks infested with *Morenoella quercina*, which is more or less a Sooty mould, to the reduction in light reaching the foliage (Chapter 27).

MOSSES

Mosses are generally shade-loving plants, and on trees are found mainly on the lower parts well away from the light. Like lichens their occurrence is affected both by the water-holding capacity and roughness of the bark. This partly explains their more frequent occurrence on broadleaved trees than on conifers. Mosses require rather more foothold than do lichens, and there is often a succession in time from relatively small, compact forms, such as *Ulota crispa* (Hedw.) Brid., *U. bruchii* Hornsch., and *Dicranoweisia cirrata* (Hedw.) Lindb., to taller species such as *Eurhynchium* (*Isothecium*) *myosuroides* Schp., and *E. myurum* Dixon, which are less tolerant of drought. Mosses are very dependent on water, so that they occur in moist sheltered positions, such as branch crotches and near the base of the trunk (Potzger 1939), and much more abundantly on trees growing in areas of high humidity. They are more restricted to the trunk and major branches than are lichens, but in areas of heavy rainfall they are sometimes found on quite small twigs.

Most of the work on the woodland mosses has been concerned with their systematy or with their ecology, the tree being regarded merely as one of their substrates (Watson, W. 1936; Hosokawa and Odani 1957). No studies seem to have been made of their effect on trees, but it is doubtful if they exert any detectable influence. Like lichens, their presence tends to indicate that their host tree is growing slowly. On fast-growing trees they are sloughed off with the old bark.

Arboreal mosses, like lichens, are particularly susceptible to smoke and fumes. Richards (1928) found that they had almost disappeared from the Middlesex area; though a number of species had been recorded there in the past before pollution became so serious. Even around Oxford some species have vanished in the last fifty to seventy years.

FLOWERING PLANTS AND FERNS

There is not a large flora of epiphytic plants in Britain. Some of the ferns, particularly Common polypody, *Polypodium vulgare*, occur commonly on tree trunks, often rooting in the dead remains of mosses or lichens, or in accumulations of rotted leaves in branch crotches; but no true flowering plants occur normally in such situations. Occasionally under special circumstances, especially in old pollarded trees, quite a rich epiphytic flora may spring up in the leaf-mould accumulations among the cut branch stumps in the crown. Eighty species were found in this position on old pollard willows near Cambridge by Willis and Birkhill (1884). More recently a stable flora of thirty-five species has been recorded in the same situation (Cannon and Cannon 1957). There is no reason, however, to suppose that plants of this nature have any measurable effect on the health of the trees on which they are growing.

22

DISEASES OF PINE

THE genus *Pinus* has provided the only native conifer of forest importance in Britain, Scots pine, *P. sylvestris*, though some of the strains grown have certainly been introduced from abroad. This and Corsican pine, *P. nigra calabrica*, are in the front rank of our forest trees. Several others, including *P. contorta*, an extremely variable species, *P. strobus* (Weymouth pine), the use of which is seriously limited by White pine blister rust, and *P. pinaster* are quite important in British forestry. Other forms of *P. nigra* have been used considerably, under the name Austrian pine, for shelterbelts, especially near the sea, while the Mountain pine, *P. montana*, is locally valuable for extremely exposed situations. The Californian *P. radiata* grows well in the milder parts of the country and is used mainly for shelter in the south-west of England. Many other pines, in particular Five-needled species, occur in gardens and arboreta.

Several species, for example *P. sylvestris* and *P. nigra*, are very variable in their reaction to climate and probably in their reaction to disease.

DAMAGE DUE TO NON-LIVING AND UNEXPLAINED AGENCIES

Most of the pines used in forestry in Britain are hardy, though quite severe damage has been recorded on Scots and Corsican pine in an exceptionally late June frost (Murray, J. S. 1955*b*). Scots pine, however, is extremely susceptible to 'blast' damage to the needles when growing in exposed situations (Peace 1953*a*). Near the sea this may be due to salt spray, but inland it appears to be the result of excessive transpiration (Chapter 6). Apart from this, Scots pine generally grows better in the east than in the west of Britain. It is easy to suggest that the higher rainfall in the west does not suit it, but very difficult to say why. There is evidence that there exist strains of Scots pine which would tolerate western conditions better; but the large bulk of work which has been done on Scots pine races on the Continent, and particularly in Sweden, where there is a big range of variation from north to south (Langlet 1934), is not easily applicable to Britain. In Britain the vigour of growth tends to improve with the use of more southerly provenances, but on the other hand provenances from appreciably south of British latitudes are seldom successful.

Scots pine is particularly sensitive to nutritional diseases. On limestone and chalk soils it often suffers from lime-induced chlorosis. Needle fusion in *P. contorta* and other pines in Britain has been attributed to inadequate development of mycorrhiza (Chapter 17), and the same symptoms on *P. radiata* and other pines in Australia to mineral deficiencies (Chapter 7).

Extra-seasonal growth

Jump (1938*a*) described a phenomenon in Red pine (*Pinus resinosa*), which he called 'Extra-seasonal growth'. One or more of the lateral buds in the terminal bud-cluster made a short shoot in the autumn. On the resumption of growth in the spring these shoots assumed a more vertical position than normal and were unusually vigorous. Eventually they gave the tree a forked appearance. Failure to make a good union, because of the narrow angle between the branch and the main stem, often led to cracking under the weight of snow, followed by breakage, or fungal infection. Jump, who regarded it as a serious trouble, later (1938*b*) attributed it to the stimulating action of growth substances produced by the fungus *Pullularia pullulans* (De Bary) Berkh. (Fungi Imperfecti, Moniliales) (see *Sclerophoma* leaf-cast below). This phenomenon has been observed on Corsican pine in Britain, though it was not kept under observation long enough for its final effect on the tree to be seen. It did not occur on a high proportion of trees, so that it would not in any case be serious. In Britain there is no evidence that it is associated with *P. pullulans* or any other fungus.

Pole blight

In the western United States and Canada *Pinus monticola* suffers from an unexplained disease, termed 'Pole blight'. This is characterized by long narrow streaks of dead cambium on the trunk, leading to resin flow, reduced growth, and needle fall (Buchanan, Harvey, and Welch 1951; Leaphart, Copeland, and Graham 1957). Several fungi have been isolated fairly consistently, including *Europhium trinacriforme* Parker (Fungi Imperfecti, Moniliales), but they are not considered to be the primary cause (Parker, A. K. 1953, 1957; Leaphart and Gill 1955). No association can be found with soil conditions (Ferrell 1955), except that there is a clear connexion with drought and high temperature, particularly when they occur together (Wellington 1954). Pole blight bears some resemblance to Resin bleeding of *P. contorta* (see below) and of Douglas fir (Chapter 25), two British diseases not yet explained.

Resin bleeding

Resin bleeding of *P. contorta* has been recorded in two forests in east Scotland. In both places it is extensive on pole-stage crops. Small dead patches, from which resin exudes copiously, occur on the branches and upper main stem. The attack varies from tree to tree, and there is no tendency for diseased trees to occur in groups. Although some cases result in dieback, most of the affected trees are still making normal growth. No fungal association has yet been discovered, nor any obvious connexion with site conditions.

NURSERY DISEASES

In the nursery pines of all species react strongly both to nutritional deficiencies and to disturbance of the water balance. Purple discoloration of the foliage of pine seedlings is a very common phenomenon usually attributable to one or other of these causes.

P. contorta in dense beds is very susceptible to infection by *Botrytis cinerea*, which attacks the lower needles rather than the tip (Chapter 15). *Lophodermium pinastri* occurs as often in the forest as in the nursery and is therefore dealt with under 'Diseases of Leaves and Shoots', where suggestions on nursery spraying, and on treatment of infected plants, are made. *Diplodia pinea*, also discussed later, occasionally occurs in nurseries.

Pine, like many other conifers, is attacked by the Snow moulds *Herpotrichia nigra* and *Phacidium infestans*, which are of no importance in Britain, but are discussed briefly in Chapter 15 because they are particularly found on small snow-covered plants in the nursery. On various pines in America, and on *P. montana* in the Carpathians, a third Snow mould, *Neopeckia coulteri* (Peck) Sacc. (Ascomycetes, Sphaeriales), causes similar damage, forming a brown felty mycelium over twigs or branches covered by snow during the winter (Savulescu and Rayss 1929). It does not occur in Britain.

Recently *Ascochyta piniperda* Lind. (Fungi Imperfecti, Sphaeropsidales), chiefly important on spruce (Chapter 23), has been found associated with collapse of the succulent shoots of *P. contorta* seedlings in north Scotland (Murray and Batko unpubl.). The damage was very similar to that usually associated with *Botrytis cinerea*.

Needle rust

Pathogen

Leaf rust, caused by the fungus *Coleosporium tussilaginis* (Pers.) Lév. (Basidiomycetes, Uredinales), is only of importance in the nursery. The name *C. tussilaginis* is now that of a group-species covering some ten varieties which were previously specifically separated on the basis of the herbaceous hosts on which the uredial and telial stages occurred. They are not distinguishable on pine, and have therefore been amalgamated (Wilson and Bisby 1954). Among these herbaceous hosts are: *Tussilago farfara* (Coltsfoot), *Campanula* spp., *Euphrasia* spp. (Eyebright), *Odontites verna* (Red bartsia), *Parentucellia viscosa* (Yellow bartsia), *Rhinanthus minor* (Yellow-rattle), *Melampyrum arvense* and *M. pratense* (Cow-wheats), *Petasites* spp. (Butterburs), *Senecio* spp. (including Groundsel and Ragwort), *Calendula officinalis*, and *Sonchus* spp. (Sow-thistles). In forest nurseries the most usual herbaceous host is Groundsel, *Senecio vulgaris*, a very common weed.

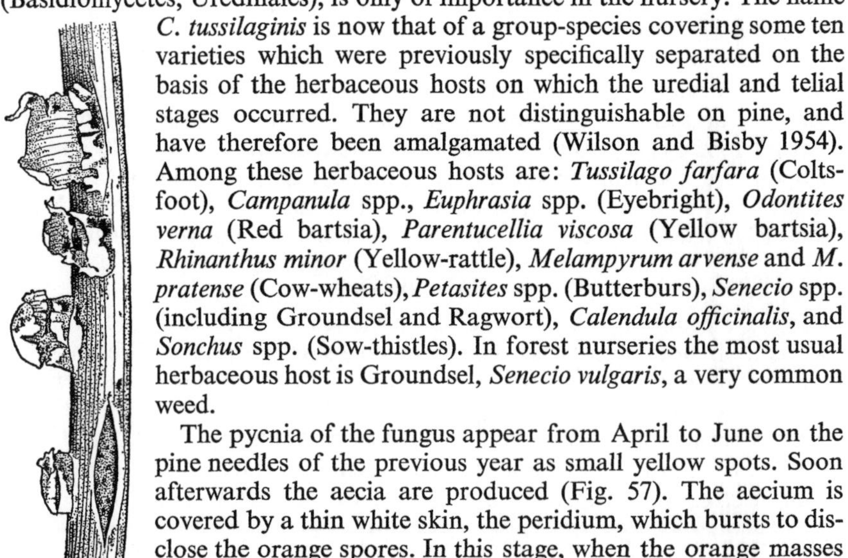

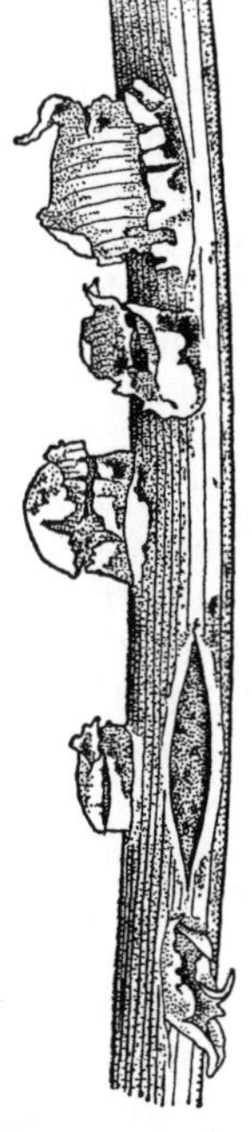

Fig. 57. Aecia of *Coleosporium tussilaginis* on a needle of Scots pine (×5). (J. C.)

The pycnia of the fungus appear from April to June on the pine needles of the previous year as small yellow spots. Soon afterwards the aecia are produced (Fig. 57). The aecium is covered by a thin white skin, the peridium, which bursts to disclose the orange spores. In this stage, when the orange masses of spores are contained by the tattered remnants of the white peridia, the fungus is very conspicuous. The aeciospores infect the herbaceous host, on which the uredia soon appear as orange spots on the leaves and sometimes on the stems (Fig. 58). The life cycle of the fungus is illustrated diagrammatically in Fig. 30. Under favourable weather conditions several generations of uredospores are produced in a season, so that a slight infection from pine can give rise to widespread disease of the herbaceous host. In the milder parts of the country, uredospores are produced all the year round, enabling the fungus to overwinter in the absence of pine. For instance, the fungus maintains itself on *Senecio smithii*, a garden

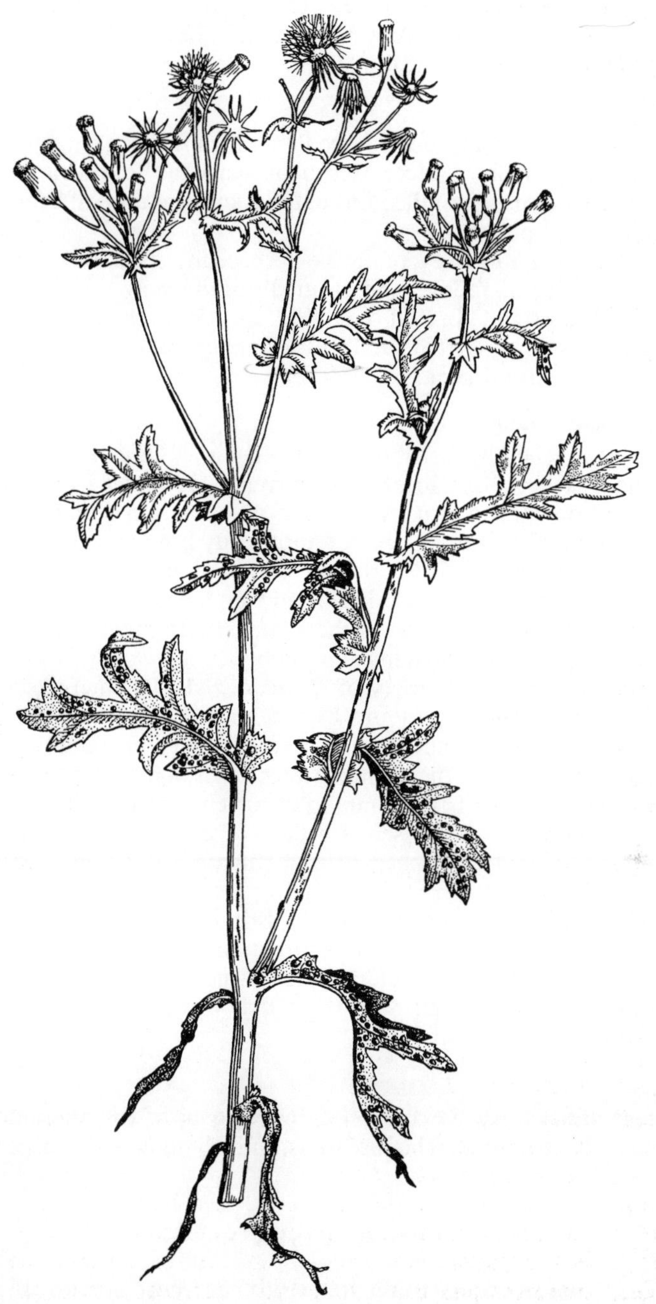

FIG. 58. Uredia of *Coleosporium tussilaginis* on *Senecio vulgaris*, September (× ½). (J. C.)

species, on the island of Coll off the west Scottish coast, where the nearest pines are ten miles away (Wilson, M. 1924). In the United States *C. solidaginis*, which should probably come under the same group-species, has been recorded on Aster in the spring before aecia had appeared on the pine needles (Weir and Hubert 1918).

Symptoms and development

Little damage results to infected needles, which usually remain green, and may sometimes produce a second crop of aecia the following year. Infrequently the needles may be killed but defoliation is seldom, if ever, complete even on individual plants.

Injury tends to be more severe on the herbaceous hosts, but although considerable withering of foliage and stunted growth may result, the plant is seldom killed outright. Indeed infection on Groundsel tends to hasten flowering and seeding, so that far from lessening the necessity for weeding, it makes it even more urgent.

Distribution and damage

The fungus is generally distributed over Britain, although some forms with rare herbaceous hosts are local in occurrence. It attacks only Two- and Three-needled pines, and is usually found on Scots pine though it has also been recorded on Corsican. It has not apparently been found on *P. contorta*.

A very large number of *Coleosporium* species on pines have been recorded in the United States (Hedgcock 1928), and in Japan. Some have the same alternate hosts as the British ones, others almost certainly belong to the same group-species. The most damaging is probably *C. solidaginis* (Schw.) Thuem., which is probably another variety of *C. tussilaginis*. Its chief alternate hosts are Golden rod (*Solidago* spp.) and Michaelmas daisy (Aster spp.). Baxter (1931) reports it as sometimes causing severe damage in stands of Red pine, *P. resinosa*. In the west of the United States this fungus, or a variety, occurs on *P. contorta*. In view of the common occurrence of the alternate hosts in gardens in Britain and the present freedom of our *P. contorta* from *Coleosporium* attack, it would obviously be unfortunate if this fungus reached this country.

The damage done by *Coleosporium* in Britain is very slight, though the conspicuous aecia attract considerable attention. It occurs on the Continent, where it is equally unimportant. Only in America have these fungi been regarded at all seriously, and even there they are not in the front rank of forest diseases.

Control

Persistent attacks can be controlled by removal of the alternate host, in practice usually groundsel. The disease is unlikely to occur, except sporadically, in a well-weeded nursery.

ROOT DISEASES

Armillaria mellea occurs quite frequently on most species of pine and occasionally causes severe losses in young plantations on old hardwood ground. The killing activities of *Fomes annosus*, in contrast to its much more important role as a cause of butt-rot, are mainly restricted to old agricultural and alkaline sites, particularly in Thetford Chase on the borders of Norfolk and Suffolk. Both the principal species of pine, as well as *P. contorta*, are affected. Several dead groups associated with *Rhizina inflata* have been recorded in Corsican pine. All these diseases are dealt with much more fully in Chapter 16.

Paxillus fairy rings

Pathogen

Paxillus (*Leucopaxillus*) *giganteus* (Sow.) Fr. (Basidiomycetes, Agaricales) spreading in the form of 'fairy rings' has been associated with progressive dieback of young Scots pine (Peace 1936*a*). The fungus produces large creamy-white toadstools, the caps of which are depressed in the centre and up to 30 cm. across.

Symptoms and development

It is almost certain that the fungus, which commonly forms 'fairy rings' in pastures, was present on the sites, which were open heath, before the pines were planted. The development of the rings is shown in Fig. 59. In the centre were healthy trees, which were planted after the fungus had moved outwards. Then there was a zone of dead trees; on the inside of this many of the pines were missing, and those that remained were markedly smaller than the rest of the crop. This suggested that they were killed soon after planting. Outside this was a zone where practically all the trees were unhealthy, with dead roots and short, yellowish needles, and many dying; here the trees on the inside tended to be in worse health than those on the outside. Each autumn a ring of fruit bodies was produced near the inside of this zone. Finally occurred the healthy trees, forming the bulk of the plantation. The rings varied in diameter from 20 to 100 metres, but were often extremely irregular in outline.

It appeared that the fungus spread radially outwards, killing the pines as it went. A thick mat of mycelium occurred in the soil under the ring of fruit bodies. Damage, which took the form of shedding of the older needles, and reduced growth, followed by progressive dieback, started well outside this zone of dense mycelium, however, and could not therefore be attributed to competition between the fungus and the trees for water, an explanation that has been put forward in other instances to explain the action of fairy rings on vegetation. There was no evidence of parasitism. Possibly the fungus produces a toxin which is active even in soil only recently penetrated by the advancing mycelium, and which adversely affects the roots of the pines before the main mycelial mat has reached them.

Distribution and damage

The disease was found only in the pine areas of East Anglia, and in one instance in east Scotland, where the fungus was identified as *Paxillus extenuatus* Fr. from a poor specimen, but may well have been *P. giganteus*. It has not been observed in recent years. While it was of considerable pathological interest, the total area affected was very small. The fungus occurs outside

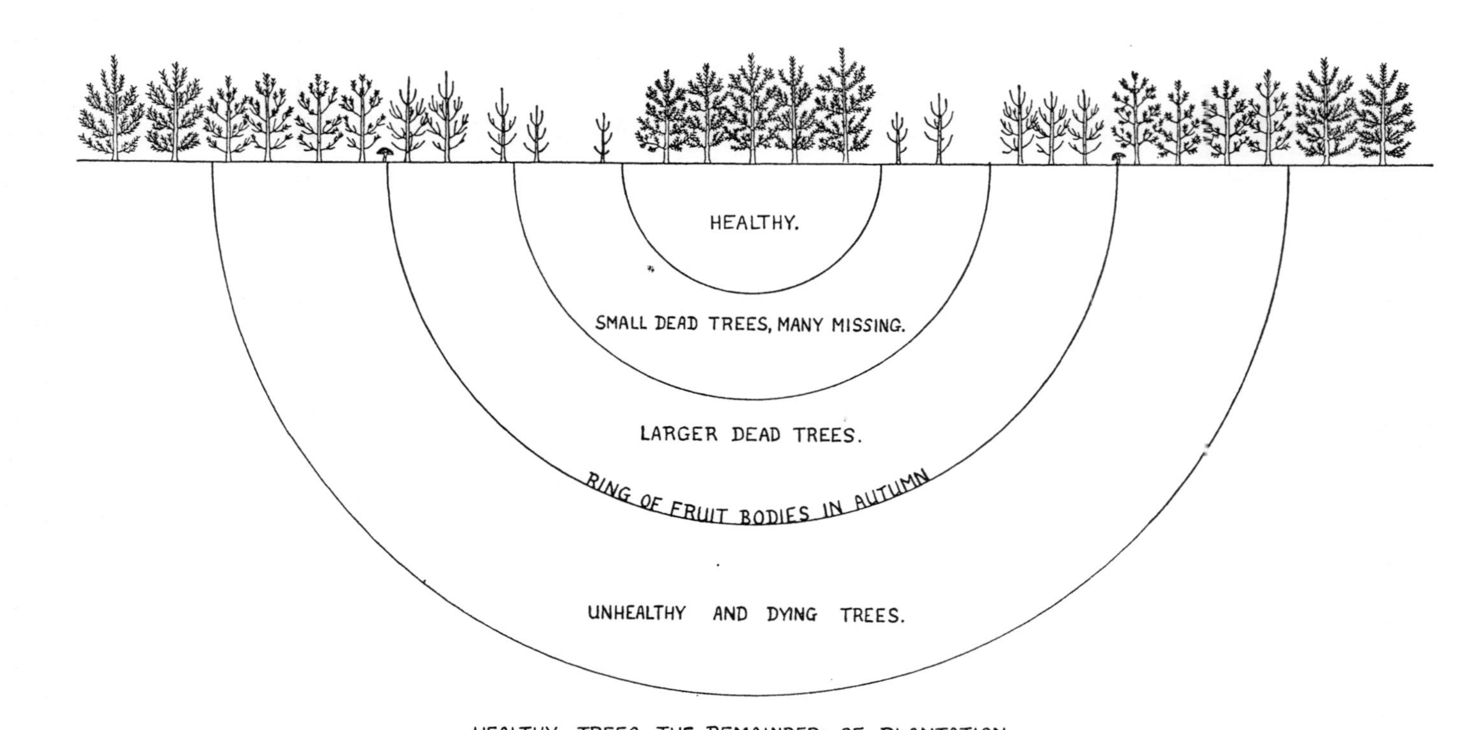

FIG. 59. Diagrammatic representation of an attack by *Paxillus giganteus* on Scots pine, showing the different zones. (J. N.)

Britain, but does not seem to have been recorded elsewhere as a cause of disease.

Control

All the cases in East Anglia occurred on shallow sand over chalk. They were easily controlled by trenching down to the chalk, into which the fungus was unable to penetrate, so that further extension of groups ceased.

Little leaf

Little leaf, a disease of Shortleaf pine (*P. echinata*) in the United States, which has been briefly described by Zak (1957), was for a long time considered to be a physiological disorder. A great many possibilities were explored and some factors found that obviously contributed to the disease (Copeland and McAlpine 1955). A definite relationship was found between the disease and poor internal drainage of the soil (Copeland 1949). It was not surprising therefore that *Phytophthora cinnamomi* Rands. (Phycomycetes, Peronosporales) was eventually isolated from the diseased roots (Campbell, W. A. 1948, 1951), for various species of this genus have been associated with root disease in broadleaved trees under conditions of bad drainage.

The disease is characterized by a gradual deterioration in the health of the tree, evidenced by reduction in growth and in the length of the needles, hence the name 'Little leaf'. It appears to be increasing (Roth, E. R. 1954; Campbell and Copeland 1954). The disease also occurs much less seriously on Loblolly pine, *P. taeda*. There is evidence that other pines may also be moderately susceptible (Zak and Campbell 1958). Nevertheless, the suggested treatment for infected areas is the substitution of *P. taeda* and other pine species for *P. echinata*. A similar disease, also associated with *Phytophthora cinnamomi*, has been reported on *P. radiata* shelterbelts in New Zealand (Newhook 1957). In view of the world-wide distribution and enormous host range of the genus *Phytophthora*, this disease may suggest a possible explanation of other cases of unexplained conifer deterioration.

STEM DISEASES—BARK AND CAMBIUM

White pine blister rust

Pathogen

This disease, which is one of the most important on forest trees, is caused by the Rust fungus, *Cronartium ribicola* Fisch. (Basidiomycetes, Uredinales). It can attack most if not all of the Five-needled pines and a large number of species of *Ribes* (currants and gooseberries). The most recent general accounts are given by Mielke (1943) and Hirt (1956). There is an enormous and varied literature, mostly American.

The fungus is a typical two-host rust, as described in Chapter 11. The sporidia, produced from the teliospores on currant leaves, germinate on pine needles during summer and autumn, the fungus entering directly through the epidermis, and causing slight spotting of the needles (Hirt 1938). Infection can take place on the needles of the current year, but is more frequent on older needles (Pierson and Buchanan 1938*b*; Hirt 1944). A year or two after infection, usually by September of the year following infection, the fungus, which has grown down the needle into the stem, becomes obvious by a yellow or orange discoloration of the bark, accompanied by slight swellings. Later, pycnia appear as small blisters on the bark (Hirt 1939). These burst,

FIG. 60. Effect of attack by *Cronartium ribicola* on a young *Pinus strobus*, June ($\times \frac{1}{2}$). (J. C.)

discharging drops of a honey-coloured, sweet-tasting liquid, which attracts insects. The infected patches gradually become more swollen, roughened and cracked, and bear annual crops of aecia in late April and May. These are prominent white blisters, which burst to disclose masses of orange spores (Fig. 60). The aeciospores can travel long distances, and infect young currant leaves. Within a few weeks of infection small, yellow, uredial pustules appear on the lower leaf surfaces. The uredospores infect other currants, but are incapable of infecting pines. Late in the summer the telia are produced; they are brown bristle-like structures, which sometimes occur in such numbers as to give a faintly felt-like appearance to the underside of the leaves.

Symptoms and development

Pines of any age can be infected, but of course only on needle-bearing shoots. Thus direct infection of the lower main stem of a large tree is impossible. Brashing or pruning wounds cannot lead to infection. The higher wind and lower humidity around the crown of a large, as compared with a small, tree make the infection of small trees much more probable than that of large ones. In addition, infections on the long branches of older trees are likely to die out before they reach the main stem (Lachmund 1934*b*; Slipp 1953), so that attack on such a branch will result only in its loss, not in damage to the crown as a whole by main-stem invasion. So although damage can occur in trees of any age (Buchanan 1938), it can be assumed that a tree of thirty years without main-stem infection is likely to survive for the rotation. In the hotter, drier parts of the United States the spread of the fungus is limited, principally by the lack of sufficient moisture for sporidial formation and germination (Arsdel, Riker, and Patton 1956). Unfortunately this limitation is quite unlikely to operate in Britain.

Infection often takes place in the nursery, but, unless planting out is delayed, such young trees are unlikely to show obvious symptoms until after they have reached the forest. Since the fungus usually reaches the main stem through branches, trunk cankers are often around branch origins, though the fungus can remain active in the main stem long after the branch has been killed. The scattered dead branches and the irregular, cracked swellings produced by the fungus make diagnosis fairly easy (Fig. 60), though on fast-growing stems the infected area may be sunken, because of the collapse of necrotic tissue and the healing growth of the unaffected parts of the stem (Plate XII). The soft spongy swollen bark appears to be particularly attractive to rodents, such as squirrels, mice, and voles, so that gnawing of the roughened areas is quite a common subsidiary symptom. Mielke (1935) suggested that rodents feeding in this way may significantly reduce the production of aeciospores.

Infections on the main stem are usually fatal, since sooner or later they girdle the tree. The chance of carrying on an infected crop to maturity can usually be judged with reasonable accuracy at the second or third thinning. If then enough uninfected trees remain to make a final crop, the plantation is worth retaining. If not, it may be better to utilize the trees at once, rather than salvage each infected tree as it dies.

Infection of *Ribes* takes place most readily on young developing leaves (Pierson and Buchanan 1938*a*), so that variations in climate or elevation, affecting the time of foliation of the currants, may have a considerable effect on the number of aeciospore infections and the subsequent development of the disease on the *Ribes* (Lachmund 1934*a*). The fungus is very easily detected on currants by virtue of the small orange uredial spots and later by the felty brown telia. Repeated annual attacks, causing premature defoliation, can appreciably weaken currant bushes.

Distribution and damage

It is supposed that *P. cembra* was the original host of this fungus both in Europe and in north-east Asia and Siberia. In each case the fungus persisted on the pine and on wild species of *Ribes*. In the relatively small areas, mainly in Switzerland, where *P. cembra* is indigenous, the pine showed considerable resistance and the disease attracted little attention, but the form of *P. cembra* growing in Asia was rather more susceptible and the disease was there more noticeable (Gäumann 1950). It was not until the latter part of the nineteenth century, quite a long time after the introduction of *P. strobus* into Europe, that the disease became epidemic. Large-scale plantings of *P. strobus*, which appeared to be a desirable addition to Europe's forest conifers, combined with widespread cultivation of the highly susceptible Black currant, *Ribes nigrum*, produced suitable conditions for an epidemic development of the disease. It seems likely that its spread on to *P. strobus* was westwards from the Siberian disease area, or possibly from a small relict area of *P. cembra* in the Carpathians (Rouppert 1935), rather than from the localized Swiss outbreak; and that it was the time required for this spread that delayed the appearance of the disease in western Europe. However, by 1900 it was epidemic in all the main Weymouth pine areas of western Europe, and sometime before 1906 it was imported into the United States on infected nursery stock. At that time forest nurseries were little developed in North America, and a sudden demand for planting stock was met by considerable imports of *P. strobus* and other conifers from established European nurseries. In 1921 another introduction into Vancouver, British Columbia, brought the fungus into the even more important stands of *P. monticola* in the west, and an epidemic which has cost the United States and Canada many millions of dollars was started.

In Britain the disease has appeared in nearly every place where Weymouth or other Five-needled pines have been planted. The presence of Black currants in many gardens has made this widespread infection extremely easy. Black currants will produce an economic crop in most parts of Britain, even at quite high elevations. They are often the only fruit to be grown around remote cottages and small farms in the Highlands of Scotland or in the Welsh mountains. Successful spread from currants to pines in the United States is said to be seldom effective over distances of more than 1·5 kilometres. However, distance of infection is known to be related to climate, and in different parts of the United States the maximum varies from over 3 kilometres to a few metres. Spaulding (1929) suggested that spread under European conditions might be more extensive, possibly a reflection of the extremely heavy spore production on Black currant as compared with that on some of the American *Ribes* species. It appears, however, in Britain that infection falls off rapidly beyond 0·8 kilometre so that pines should be virtually safe 2 to 3 kilometres from the nearest currants.

It seemed, therefore, that the almost complete cessation of planting of Weymouth pine in Britain was unnecessary, and that sites more than 2 kilometres from the nearest currants could be used with reasonable safety. In fact such sites were extremely hard to locate, nearly all the ground that was far enough from currants being at high elevations, often near the planting limit, and frequently on soils unsuitable for a rather exacting tree (Peace 1957*a*).

A great deal of information on the relative susceptibility of different species and varieties of *Ribes* is available (Schellenberg 1923; Kimmey 1938). From the British point of view it is probably sufficient to know that *Ribes nigrum*, the Black currant, is highly susceptible, whereas Red currants, which originated from several *Ribes* species, are slightly resistant, as is Gooseberry, though some varieties may show moderate infection. The Flowering currants, mostly varieties of *R. sanguineum*, which are often planted in gardens, are said to be susceptible, but they are usually much less severely attacked than *R. nigrum*. Female plants of the other British native species, *R. alpinum*, which is dioecious, are moderately susceptible, but male plants are resistant (Hahn 1939). In practice, because its susceptibility is so high and its planting so widespread, Black currant is the only species which need be regarded seriously as a source of infection in Britain. Red currant and Gooseberry are seldom seriously infected. Although all the cultivated species, as well as *R. alpinum*, occur wild, they are relatively uncommon and never form a dominant constituent of the vegetation; hence their influence, even that of the susceptible wild Black currant, is likely to be very small.

Among the Five-needled pines, the North American species *P. strobus*, *P. monticola*, *P. flexilis*, and *P. lambertiana* are all highly susceptible (Hirt 1940). All these species, if unattacked by the fungus, grow well under British conditions. *P. excelsa*, the Himalayan pine, and *P. cembra* are appreciably less susceptible (Boyce, J. S. 1926). *P. holfordiana*, a hybrid between *P. excelsa* and the Mexican *P. ayacahuite*, is also susceptible (Peace unpubl.). *P. peuce*, the Macedonian pine, is resistant though not immune (Tubeuf 1931). This species grows well in Britain (Macdonald *et al.* 1957), where the fungus does not appear to have been recorded on it, and in other parts of Europe; but pleas for its wider use (Tubeuf 1927) appear so far to have met with little response. The resistance possibilities of hybrids between *P. strobus* and *P. peuce* are being investigated, both in Britain and in America (Fowler and Heimburger 1958). Some hybrids between *P. excelsa* and *P. strobus* have proved resistant (Meyer 1957). Childs and Bedwell (1948), however, consider *P. excelsa* more resistant than *P. peuce*, putting it in their most resistant class together with *P. armandi* and *P. koraiensis*, neither of which is of any importance in Britain.

In America, White pine blister rust has undoubtedly cost more than any other tree disease. Enormous sums have been spent on the eradication of currants, and many promising stands of pine have been severely damaged. But it is by preventing the successful regeneration of these particularly valuable species that the fungus causes the greatest loss. In Britain *P. strobus* and the other Five-needled pines have never been widely planted, so that the actual damage to existing plantations is small; nor has much money been spent on control. The loss, in Britain, is that the fungus prevents the addition of a very valuable species to our list of timber-producing conifers.

Control

Control of the disease in the past has been based mainly on the fact that, although the aeciospores from the pine can spread over vast distances, infecting currants more than 150 kilometres away, the sporidia from currants seldom carry infection to pines more than 1·5 to 3 kilometres. Thus if all the *Ribes* in the immediate neighbourhood of susceptible pines are killed or removed, those pines will be safe from infection. *Ribes* eradication, particularly in and around *P. monticola* regeneration in the north-western States and British Columbia, has proved very costly. The various chemical and mechanical methods have been summarized by Offord *et al.* (1952). High local concentrations of *Ribes* have been sprayed from a helicopter. The hormone weedkillers, 2,4-D and 2,4,5-T are probably the most efficacious substances for use in this way (Offord, Quick, and Moss 1958). The economic aspects of this eradication programme, which has now been in progress for many years, were discussed by J. F. Martin (1938). Similar action in Europe, though suggested by Tubeuf (1935), has never proved possible because of the wide distribution and relatively high value of cultivated Black currants.

Silvicultural measures can influence the disease particularly when they affect the occurrence of wild currants. Moss and Wellner (1953) found that clear felling and burning, followed by complete replanting, gave much better results than attempts at group regeneration, because the second method actually encouraged the regeneration of *Ribes* as well as pine in the gaps. Maintenance of a closed canopy can be used to suppress the growth of *Ribes* in the forest. In cool, moist climates in America infection on pine is restricted if they are growing under a canopy of another species; but in Britain low light intensities would not allow Five-needled pines to be treated in this way. In fact in Britain, where the main source of infection is garden currants, silvicultural methods are of little value.

The eradication of the alternate host is naturally of first importance. Prevention of sporulation on the pine, unless carried out with great thoroughness over very large areas, would have little effect on the general progress of the disease. However, it is possible, mainly by pruning branches before infection has reached the main stem, greatly to reduce the incidence of the disease in a pine stand (Stewart, D. M. 1957), or to save trees which would otherwise die (Martin and Gravatt 1954). This might be important on ornamental trees or in small plantations of particular importance. Treatment of partly excised cankers with the antibiotic acti-dione (cycloheximide) has killed the rust mycelium and allowed healing (Moss 1958). So far this appears to be a very promising method of local control.

Considerable attention has been paid to the production of rust-resistant currants for commercial planting. Among Red currants the variety 'Viking' is virtually immune (Hahn 1943*a*). Work in Canada has produced a number of highly resistant Black currants (Hahn 1948). The most successful for fruiting appears to be the variety 'Consort'. In Britain, though these varieties might be used to replace susceptible currants in a few remote farmsteads and thus slightly increase the very limited disease-free areas available for Weymouth pine cultivation, any large-scale replacement is unthinkable, in view of the firm establishment of many of the existing commercial varieties.

Possibly the main hope for the future lies with resistant strains of pine. Resistant individuals have already been selected and propagated from both *P. strobus*, and *P. monticola* (Riker *et al.* 1943; Hirt 1948; Patton and Riker 1957). There appear to be possibilities of getting even higher resistance by a properly controlled breeding programme (Bingham, Squillace, and Patton 1956). There are nearly always unaffected trees in diseased stands of Five-needled pines in Britain, though these are more probably chance escapes than resistant individuals. It will be a long time before seed guaranteed to give resistant plants is available in quantity.

The fungus *Tuberculina maxima* Rostr. (Fungi Imperfecti, Moniliales) has been recorded attacking the aecia of the rust on pine, both in Europe (Tubeuf 1930) and in America by Hubert (1935*a*) and others, who did not regard it as a possible means of control. It has been recorded in Scotland.

Peridermium pini (Resin top) and cronartium flaccidum

Pathogens

Peridermium pini (Pers.) Lév. (Basidiomycetes, Uredinales) causes swellings on the branches and trunks of Scots pine. Conspicuous yellow aecia are produced on these (Plate XI, 2), the aeciospores directly infecting pines. It has sometimes been confused with the aecial stage of *Cronartium flaccidum* (Alb. & Schw.) Wint., formerly known as *C. asclepiadeum*, which it very closely resembles. *C. flaccidum* has its aecial stage on the branches of Scots pine, and its uredial and telial stages in Britain on *Tropaeolum* and Paeony. On the Continent the most important alternate host is *Vincetoxicum officinale*.

Symptoms and development

Infection by *P. pini* probably takes place through the needles, or through shoots of the current year (Bolland 1957*a*), and swellings bearing the aecia develop after an interval of two to three years on the branches at the point of infection (Rennerfelt1943). The disease then behaves very much as White pine blister rust, the fungus girdling the branches or spreading down to, and possibly girdling, the main stem. Attack is often in trees in the pole-stage or older, so that the top rather than the whole tree may be killed. Indeed, according to Mülder (1955) infection hardly occurs on trees under twenty years of age, and reaches its maximum in the 40–50-year-age class. This, coupled with the heavy resin content of the infected wood, has given the disease its Continental name of 'Resin top'.

Distribution and damage

Information on the distribution of this fungus may be confused by its similarity to *Cronartium flaccidum. P. pini* is almost certainly much the commoner of the two, as would be expected from its lack of dependence on an alternate host. It is locally damaging, though on a chronic rather than on an epidemic scale, in Germany, in parts of Scandinavia, and in eastern Europe (Rennerfelt 1943; Mülder 1951). In Britain, presuming no confusion with *C. flaccidum*, *P. pini* occurs occasionally up the eastern side of the country, but particularly in that part of Scotland which lies north-east of a line from Aberdeen to Inverness. In this area it is locally common enough to demand the removal of infected trees in thinning. It has been reported from Eire (Pethybridge 1911). It is commonly found on Scots pine, but has also been recorded on Corsican pine. It does not rank as a damaging disease in Britain, though when it does occur the conspicuous aecia often lead to its discovery.

It is almost certain that *C. flaccidum* is rare on its herbaceous as well as on its coniferous hosts, and that it is confined to the southern half of England. On the Continent it has also been confused with *P. pini*, so that its pathological significance there is difficult to assess. It is, however, of more consequence than in Britain.

Control

If it were more serious, selection for resistance would be a possible method of control. Mülder (1955) reports provenance differences in susceptibility, Scots pine from Hungary being completely resistant. Variations in resistance, said to be hereditary, have been reported by Klebahn (1924) and Liese (1936). Therefore it is a mistake to collect seed from infected trees. Diseased trees should be felled, since they act as foci of infection to those around them (Mülder 1951).

Other stem rusts

A considerable number of other species of *Cronartium* (Basidiomycetes, Uredinales) occur abroad on pine stems and branches, causing swellings, distortion, and often dieback. Conspicuous yellow or orange aecia are borne annually on infected pines. The uredial and telial stages occur on the leaves of a wide variety of woody and herbaceous plants. None of them, apart from those already mentioned, occur in Britain.

They are primarily important in North America, where the chief species are:

	AECIAL HOST	UREDIAL AND TELIAL HOST
C. cerebrum Hedg. & Long.	*P. echinata*, *P. virginiana*, and other pines	*Quercus* spp.
C. fusiforme (A. & K.) Hedg. & Hunt	*P. caribaea*, *P. taeda*, and other pines	*Quercus* spp.
C. harknessii (Moore) Meinecke (Woodgate rust)	*P. sylvestris*, *P. ponderosa*, *P. contorta*, and other pines	Various American *Scrophulariaceae*
C. stalactiforme A. & K.	*P. contorta*	The same plants as *C. harknessii*
C. filamentosum (Peck) Hedg.	*P. ponderosa* and *P. jeffreyi*	The same plants as *C. harknessii*
C. comandrae Peck	*P. ponderosa*, *P. sylvestris*, *P. contorta*, and many other pines	*Comandra* spp.
C. comptoniae Arth.	*P. banksiana*, *P. contorta*, and other pines	*Myrica* spp.
C. occidentale Hedg., Berth. & Hunt	*P. edulis* and other pines	*Ribes* spp.

Pine species planted in Britain have been included in the lists above, even if they are only of minor importance as aecial hosts in America.

The rusts of pine and oak have been considered in detail by Hedgcock and Siggers (1949). *Cronartium fusiforme*, the most damaging, is a threat to the cultivation of *P. taeda* and *P. caribaea* in certain areas. Siggers (1951, 1955) considers that infection may take place as early as immediately after germination of the seed, and found that control in the nursery by spraying is quite possible. In plantations the maintenance of dense stands has been put forward as a means of prevention (Siggers and Lindgren 1947; Lindgren 1948). Pruning of cankered branches has also been suggested against this disease (Lindgren 1950), as well as against *C. ribicola*. In *P. caribaea* there are genetic variations in resistance, which could be used in the development of resistant strains (Barker, Dorman, and Bauer 1957).

C. cerebrum was at one time considered synonymous with *C. quercuum* Miyabe, which occurs in eastern Asia on *P. sylvestris* and other pines (Ljubarsky 1952), but Hedgcock and Siggers think that the two are separate. A rust, *Uredo quercus* Brand., occurs on oak in Britain (Chapter 27), and may be the uredial stage of *C. quercuum*, but in the absence of any aecial stage it is impossible to be certain.

Only recently (Boyce, J. S. 1957*a*), *C. harknessii*, which had long been known on *P. contorta*, *P. ponderosa*, and other pines in the Rocky Mountains, has been recognized to be the same fungus as the so-called Woodgate rust, which had hitherto only been known by its aecial stage on Scots pine (York 1929). The name Woodgate rust was derived from the locality in New York State where the disease was first discovered, but it is also well established in Canada (McCallum 1929). In the Rocky Mountains the alternate hosts are a number of Scrophulariaceous plants, of which *Pedicularis* is the only genus occurring in Britain. However, direct aecial infection of pines occurs commonly, and on Scots pine in eastern America there is no suggestion that the intervention of an alternate host occurs. The disease causes death of young plants and malformation of older ones. Up to 18,000 of the rounded galls which the fungus produces can occur on a single tree (York 1929; True 1938). Since the Scots pines attacked in America are of European origin, there is no reason to suppose that the disease would be less damaging here than in America. In fact this rust would probably be the most dangerous of the group if imported into Europe.

C. stalactiforme is locally serious on *P. contorta* in the United States. It is closely related to, and has sometimes been compared with, *C. filamentosum*. It causes elongated cankers, which may be as much as 9 metres in length, on the main stem of quite large trees, but despite their size these cankers do not normally result in girdling. The diseased bark at the edge of the cankers is eaten by rodents, and in the past much of the damage basically due to the fungus has been blamed on these animals (Mielke 1956*a*).

C. filamentosum, which sometimes causes serious damage to *P. ponderosa* (Mielke 1952), has uredial and telial stages on the same hosts as *C. harknessii*, but there is a strong suspicion that this rust also can pass directly from pine to pine.

C. comandrae is locally damaging, particularly on *P. ponderosa* (Hedgcock and Long 1915). Recently severe attacks have developed on *P. contorta* (Mielke 1957*b*). However, the fungus is restricted by the local occurrence of its alternate host, the hemiparasitic plant *Comandra*, a genus which does not occur in Britain. The fungus might find our closely related Bastard toadflax,

Thesium humifusum, a suitable host; fortunately that has a very local distribution in Britain.

Cronartium comptoniae is also damaging on various Two- and Three-needled pines, especially in nurseries. Spaulding and Hansbrough (1932) have suggested that these should be sited at least 1½ kilometres away from any large concentrations of the alternate host *Myrica* spp. and surrounded by a *Myrica*-free belt at least a hundred metres in width. One of the alternate host species *M. gale*, Bog myrtle, is locally common in Britain, often in afforestation areas, so that the introduction of this rust could be dangerous to *P. contorta* and possibly to other pines.

C. occidentale is of less importance, because the pines which it attacks are of slight commercial value (Hedgcock, Bethel, and Hunt 1918). It is not easy to distinguish its uredial stage on *Ribes* from that of *C. ribicola*. Thus, when uredia are seen on *Ribes* in a new area, it is difficult to decide whether or not a new outbreak of White pine blister rust has started, and whether *Ribes* eradication should be undertaken (Kimmey 1946; Ford and Rawlins 1956).

Outside America much less is known about *Cronartium* rusts on pine. In India *C. himalayense* Bagc. produces typical Blister rust damage on *Pinus longifolia*, a sub-tropical species (Bagchee 1929, 1933). The alternate hosts are species of *Sweertia*, a genus of the Gentianaceae, which also occurs in America, but which occurs only very occasionally in gardens in Britain. This rust appears, therefore, to be no danger to Britain, but possibly to North America, if it can attack other pines (Spaulding 1952).

C. gentianeum Thüm. with its aecial stage on Scots pine and uredial and telial stages on *Gentiana asclepiadea*, the Willow gentian, causes blister rust on Scots pine in Switzerland (Widder 1948). This fungus is very closely related to, and may be only a race of, *Cronartium flaccidum* (Klebahn 1939). The Willow gentian is quite commonly grown in British gardens, but the fungus might be able to use other species of gentian as alternate hosts and so might establish itself in Britain.

Other diseases of bark and cambium

Several species of *Atropellis* (Ascomycetes, Helotiales) are associated with pine cankers in the United States (Lohman and Cash 1940). This genus is closely related to *Crumenula* (see below), but its effect on the tree is different, since it produces more definite cankers characterized by a greyish-green to blue-black discoloration of the wood beneath. With *A. pinicola* Zell. & Good. the cankers are mainly on the branches and attack occurs particularly on *P. monticola* and *P. contorta* (Zeller and Goodding 1930). *A. piniphila* (Weir) Lohm. & Cash is found chiefly on *P. contorta* and *P. ponderosa*, causing flattened cankers particularly at branch whorls. *A. tingens* Lohm. & Cash girdles the smaller branches and causes perennial target cankers on the larger branches and stems of a wide range of pine species, including *P. strobus* and *P. resinosa*, in the eastern United States. It has also been found on *P. sylvestris*, *P. pinaster*, and *P. nigra*, and for a short period was considered quite serious on the first named (Diller 1935, 1943). The literature on *Atropellis* species suggests that they do most damage in crowded stands and that their worst effects are on weakly trees. Nevertheless, the perennial nature of the cankers is a definite indication of parasitism. They do not occur in Britain.

Several species of *Tympanis* (Ascomycetes, Helotiales) occur on the bark of pines in Europe and in America (Groves and Leach 1949). A species near to *T. pithya* Karst. has been found on Corsican pine in Britain (Batko unpubl.). An undetermined species was described by Hansbrough (1936) as the cause of stem cankers on Red pine, *P. resinosa*. In this case infection took place through dead branches, and elongated cankers formed at the nodes. The small black apothecia, like flat pinheads, were borne abundantly on the dead bark. This disease occurred only on pines planted south of their optimum range and was worse on weak than on vigorous trees. Inoculation experiments proved its parasitism, but it is only a weak parasite. *Tympanis* species seem unlikely ever to assume major importance.

A number of fungi generally regarded as species of *Dasyscypha* (Ascomycetes, Helotiales), but possibly in some cases more properly placed in related genera, such as *Lachnellula*, occur on pines. They are found on Scots and Corsican, among other pines, in Britain and in other countries. They are often found on cankers initiated by other fungi (Bingham and Ehrlich 1943), and on trees weakened by fungal attack or difficult climatic conditions (Boudru 1947; Björkman 1957). Their nomenclature is so confused and their importance so slight that they need not be dealt with specifically.

Cucurbitaria pithyophila (Fr.) de Not. (Ascomycetes, Sphaeriales) has been reported occasionally on the Continent causing canker, swelling, and dieback on *Pinus sylvestris*, *P. strobus*, and *P. cembra*. It was regarded by M'Intosh (1915) as a cause of damage to *P. sylvestris* in Scotland, but has attracted little attention in Britain since. Conspicuous black encrusted fruit bodies, sometimes almost covering infected twigs, are produced. In America it was found in association with scale insects on *P. monticola* (Boyce, J. S. 1952), but most of the injury, which in any case was slight, was due to the insects rather than to the fungus. It may behave similarly in Britain and in the rest of Europe, where in any case it is not now regarded as a serious source of damage.

Caliciopsis pinea Peck (Ascomycetes, Dothideales) is apparently the cause of superficial cankers and roughening of the bark on *P. strobus* and other pine species in the United States (McCormack 1936). Its pathogenicity has been proved by inoculation (Ray 1936), but it is said to occur mainly on suppressed trees and shaded branches, and is of no great importance. It does not occur in Britain.

So-called 'Pitch canker' of pines in the south-east of the United States is caused by *Fusarium lateritium* var. *pini* (Nees.) Hepting (Fungi Imperfecti, Moniliales) (Snyder, Toole, and Hepting 1949). No perfect stage has yet been found on pine. On young trees the fungus causes girdling and dieback of the branches and leaders, but on larger stems sunken cankers are developed, characterized by profuse resin-flow. It has been suggested that inoculation with this fungus might be used to stimulate resin production in pines, such as *P. caribaea* and *P. virginiana*, from which resin is collected commercially (True and Snow 1949; Hepting 1954). The fungus has been established on Scots pine by inoculation (Hepting and Roth 1953).

F. lateritium occurs in Britain, where it is associated with a bud rot of apple trees and with canker and dieback of mulberry. It has been recorded on pine in Britain, but there is no evidence of pathogenicity.

DISEASES OF LEAVES AND SHOOTS

Pine twisting rust

Pathogen

Melampsora pinitorqua Rostr. (Basidiomycetes, Uredinales), which produces its aecial stage on Two-needled pines and its uredial and telial stages on aspen (*Populus tremula*), has now been merged with *M. tremulae* Tul. They are indistinguishable on aspen, but the latter produces its aecia harmlessly on the needles of larch (Wilson and Bisby 1954). This arrangement is unsatisfactory from a forester's point of view, not only because the effects on the coniferous hosts are so very different, but also because *M. tremulae* is generally distributed over Britain, while *M. pinitorqua* appears to be restricted to the southern half of England. The name *M. pinitorqua* is therefore retained in this book.

The fungus produces pycnia, which appear as small yellow flecks on the elongating pine shoots, and soon afterwards mealy masses of yellow aeciospores from aecia which have no peridia. The uredia are small yellow spots mainly on the underside of the aspen leaves, followed later by brown telia. A general account of the present status of the disease in Britain has been given by Peace (1957*a*).

Symptoms and development

Infection of the elongating pine shoots, and development of the aecia, cause a cessation of growth on the diseased side of the shoot. Since the opposite side continues to elongate, a bend develops. After the fungus has formed one bend, the shoot often produces a second at a higher level in order to resume vertical growth. S-bends are therefore a frequent symptom (Fig. 61). Often, however, the fungus girdles shoots at the point of attack. They then hang down and gradually wither, when the damage superficially resembles that due to Pine shoot moth (*Evetria buoliana*), from which, however, it can easily be distinguished because shoots attacked by the insect collapse at the base and are hollowed, while girdling by the fungus usually takes place an inch or two up the shoot, which remains solid.

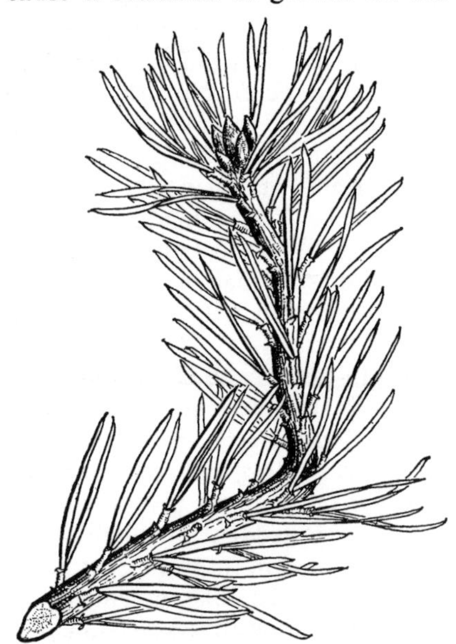

FIG. 61. Shoot distortion of Scots pine caused by *Melampsora pinitorqua* (×½). (J. C.)

On aspen the uredia, indistinguishable from those of *Melampsora tremulae*, are very conspicuous, but do little damage beyond causing slightly premature leaf-fall. They occur mainly in late summer and early autumn. The telia are formed in the autumn on the under surface of the leaves, which soon acquire a blackened appearance. The fungus overwinters on the fallen leaves.

Distribution and damage

The disease has been recorded only in the southern half of England. There are records from Yorkshire, Lincolnshire, Northamptonshire, Essex, Kent, Sussex, Hampshire, Worcestershire, and Dorset, but no doubt it occurs in other southern counties. The reasons for this restricted distribution, which limits the disease to areas where pine and aspen do not occur naturally together, are unknown. It is usually found where Scots pine has been planted on poor hardwood coppice areas containing aspen, which suckers freely among the pine. The fungus was formerly regarded as rare because not until the nineteen-thirties were such areas commonly planted with Scots pine (Peace 1944). If the fungus reached Scotland, where aspen and pine are often found together in semi-natural association, it might be much more damaging. It appears to be largely European in distribution, but has recently been recorded in British Columbia (Ziller, 1961).

The only species seriously affected in Britain is Scots pine. In Kent, under conditions where this species was heavily infected, a few aecia were found on *Pinus pinaster*, and one distorted shoot with no aecial development on Corsican pine. It has recently been reported on *P. pinaster* in the Landes in France, while in Italy it occurs on *P. pinea*, *P. pinaster*, and *P. sylvestris*, and rarely on *P. nigra* (Moriondo 1957; Biraghi 1954), being especially severe on *P. pinea*. It has also been reported in Cyprus on *P. halepensis*. In a test area, where *P. sylvestris* and *Populus tremula* are both infected, *P. radiata* and *P. resinosa* have remained undamaged for three years (Peace and Young unpubl.). Gavris (1939) found that strains of Scots pine raised from relatively heavy seed showed increased resistance, as compared with strains from light seed. No other work is known on the resistance of strains or provenances.

The most important poplar host is aspen, *P. tremula*, but *M. pinitorqua* also occurs on the White poplar, *P. alba*, and the Grey poplar, *P. canescens*. It is not found on Black or Balsam poplars. Exact knowledge of its host range on other poplars of the *Leuce* group, for instance the American aspens, is difficult to procure because of confusion with other species of *Melampsora* on poplar leaves, but it can occur on *P. tremuloides* and *P. tremula*×*tremuloides* hybrids (Rennerfelt 1953), though the latter are said to show resistance to attack (Anon. 1952*d*).

The damage to Scots pine can be quite serious locally, small trees occasionally suffering sufficient loss of shoots to cause death. Loss of leaders or distortion of leaders and consequent bushiness and forking are the usual results of attack.

Control

The disease is easily avoided if Corsican is substituted for Scots pine in the presence of aspen. A survey of an infected area in Hemsted Forest in Kent (Fig. 62) showed that severe attack on pine occurred only within 150 metres of the aspen, and no detectable infection was found more than 300 metres away. Rennerfelt (1953) found that infection fell to a low level 50 metres from the nearest aspen. Preliminary surveys to locate aspen are recommended for coppice areas in England, where it is intended to plant pine. Occasionally infection may depend on a single aspen tree which can be felled, although of course suckers will appear. More often, extensive patches of suckers are involved. Poisoning may be feasible, but early and frequent cutting of the aspen suckers may suffice to protect the pine until they close canopy. After that there is little risk of attack, because the pine will have suppressed the aspen.

The fungus can overwinter in twigs and buds of poplar (Moriondo 1954; Regler 1957). Thus the disease could maintain itself on aspen in the absence of pine, and an area could not be considered disease-free merely because pines were absent. All areas containing aspen must therefore be considered potentially dangerous, even if the pine to be planted is the first grown on the site.

Brunchorstia defoliation and dieback

Pathogen

The principal fungus associated with this disease, often referred to as *Brunchorstia destruens*, is now known as *Crumenula pinea* (Karst.) Ferd. & Jørg. (Ascomycetes, Helotiales). But its nomenclature and in particular its separation from other species of the genus *Crumenula* have always been a matter of argument (Ettlinger 1945), while recently Vloten and Gremmen (1953) have transferred it to the genus *Scleroderris*, under the name of *S. abietina* (Lgbg.) Gremmen. In Britain the perfect stage has not been found, only the imperfect stage *Brunchorstia pinea* (Karst.) v. Höhn. (Fungi Imperfecti, Sphaeropsidales), more familiar as *B. destruens*, occurs. Additional confusion also arose from the belief that *B. pinea* was the imperfect stage of *Cenangium ferruginosum* Fr. (*C. abietis* (Pers.) Duby.) (see below). In view of this confusion no attempt will be made to consider other species of *Crumenula* separately. It is impossible to say which fungus was involved when any particular attack was described. But most of the evidence suggests that only the one species, *C. pinea,* need be considered seriously as a parasite.

The small dark-brown pycnidia of *Brunchorstia,* from which the spores ooze in minute creamy tendrils during damp weather, are usually produced abundantly on infected pines, rarely at the base of the needles, more commonly on needle scars, and also on the twigs and on buds.

Symptoms and development

In Britain the disease affects mainly Corsican pine (*P. nigra calabrica*). It is usually considered secondary to adverse climatic and edaphic conditions. Corsican pine seems badly suited to high elevations, or areas of high rainfall and humidity. It is possible that the basic association of the disease is with low summer temperatures. Subjected to such conditions it normally deteriorates. This means that Corsican pine is generally successful only in the south and east of Britain, though there are exceptions, usually small in area, in the north and west. The distribution of diseased and healthy Corsican pine, as far as it has been recorded, is shown in Fig. 63. Curiously in some instances Corsican pine grows quite well for ten or fifteen years before needle loss and dieback become really apparent. This may be the stage when fungi, particularly *Brunchorstia*, start their attack.

The first symptom is usually loss of the older needles, so that a badly affected tree carries only two years' needles for a short period in the early

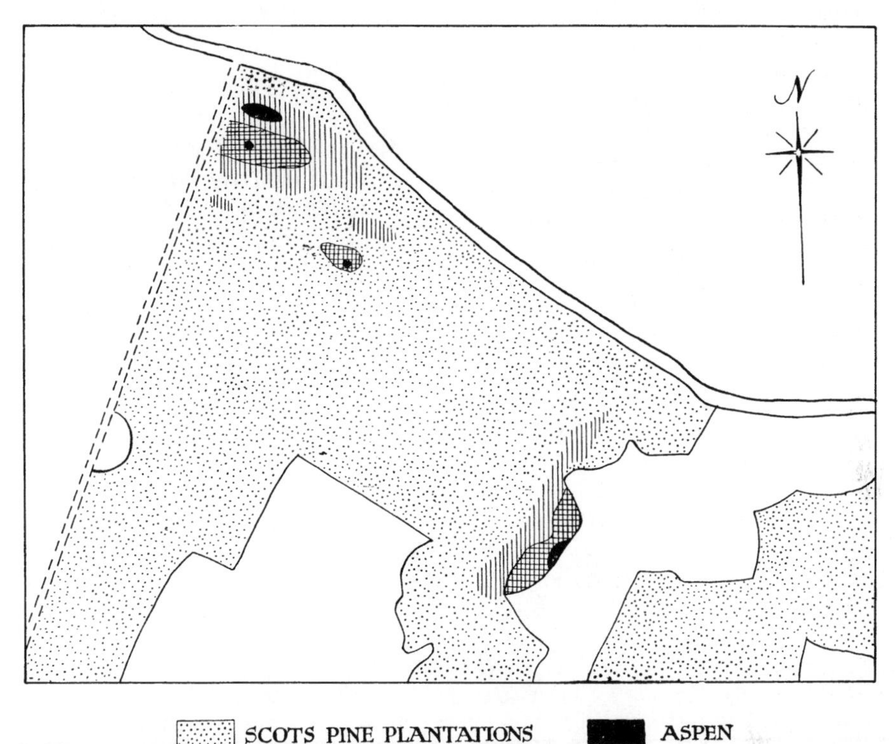

FIG. 62. Distribution of *Melampsora pinitorqua* on Scots pine at Hemsted Forest, Kent (scale *c.* 6 in. to 1 mile). (A. C.)

summer. This is followed or often accompanied by progressive dieback of the shoots, twigs, and branches. At this stage short tufted recovery-shoots are often produced at the twig axils or where the branches join the main stem. These in their turn die as the disease becomes more extensive (Fig. 64). Sometimes small cankers or areas of dead bark develop on the twigs or main stem.

Examination of diseased trees will usually, but not always, disclose the presence of the blackish brown pycnidia of *Brunchorstia*. Other fungi may also be found, in particular *Hendersonia acicola* and *Phoma acicola* on the needles, and *Phomopsis* spp. on the shoots. *Lophodermium pinastri* also sometimes occurs on the needles, but may be partially independent of the general deterioration of the trees. The occurrence of so many fungi in itself suggests that they are present in a secondary capacity, and that this may be the case also with *Brunchorstia*. *Brunchorstia* seldom, if ever, occurs in a pathogenic role on Corsican pine growing on the drier side of Britain. One instance has been recorded, admittedly in a fairly high rainfall area in Scotland, of an

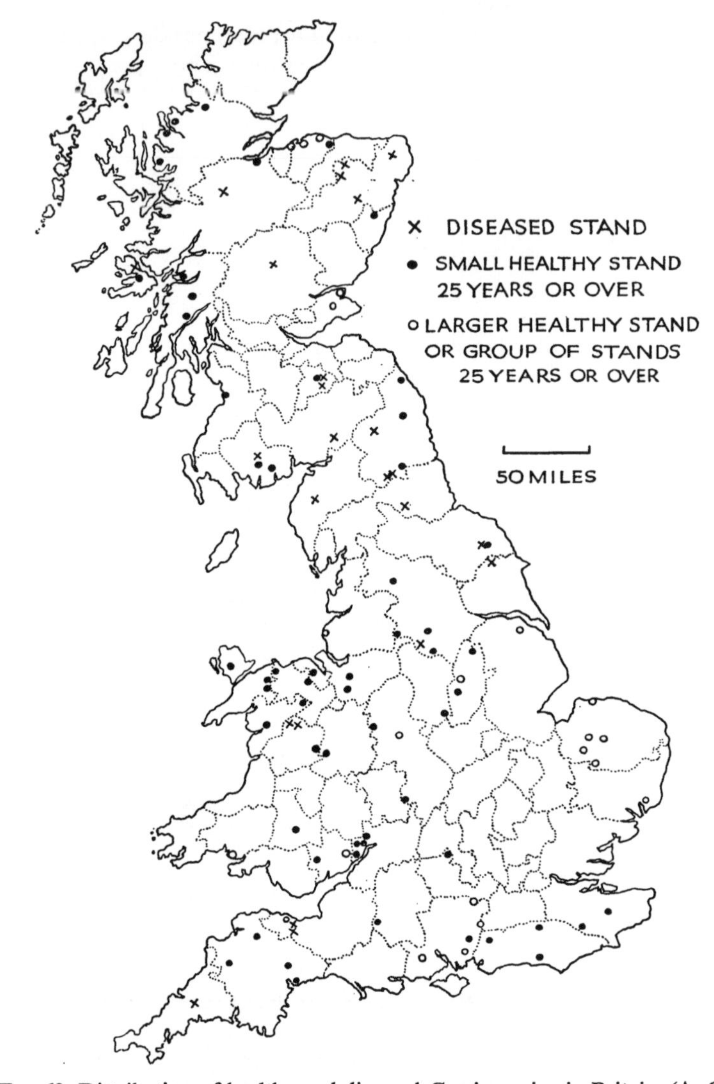

Fig. 63. Distribution of healthy and diseased Corsican pine in Britain. (A. C.)

attack by the fungus spreading through a plantation, where the trees still free of it were perfectly healthy and growing well. In this case the evidence strongly suggested direct fungal attack. Boudru (1947) in Belgium and Vloten (1946) in Holland regarded the fungus as secondary to frost; W. R. Day (1945) took the same view with regard to Corsican pine dieback in Britain, associating injury particularly with areas where extremely low night temperatures were to be expected (Chapter 4). In some areas, however, the topographical distribution of the disease does not support this view; attack seems more likely to be connected with high rainfall, high humidity, and exposure, resulting in low summer temperatures, possibly leading in some instances to subsequent

FIG. 64. Typical symptoms of Corsican pine dieback. (J. C.)

frost injury to imperfectly ripened shoots. On the Continent, however, the fungus has frequently been described as a primary parasite.

Distribution and damage

Brunchorstia pinea is very generally distributed, but does not occur in North America. It is probably present over the whole of Britain, but is important only on Corsican pine, though Waldie (1926) and Leven (1932) regarded it as more severe on Austrian pine (*P. nigra austriaca*) and even suggested that Austrian pine should be removed from the vicinity of Corsican pine as a protection. Jørstad (1925) considered it responsible for the virtual disappearance of Austrian pine from Scandinavia. On the other hand, in Britain it is far more important at present on Corsican than on Austrian pine. It has often been recorded on Scots pine both in Britain (Vloten 1929) and abroad (Guyot, A. L. 1934). The latter found it an active canker-causing parasite on trees growing under difficult conditions. In addition it occurs in Britain on *P. cembra*, *P. montana*, and *P. pinaster*. On the last named it has been found associated with symptoms exactly similar to those exhibited by Corsican pine growing on the same high-elevation site. This instance could be attributed to climatic unsuitability, but Martínez (1933) reports serious damage on *P. pinaster* in Spain at low elevations, where the pine should presumably grow well. Kujala (1948) associated it with considerable damage to *P. contorta* in Finland. It does not appear to have been recorded on this species in Britain. Indeed in one instance healthy *P. contorta* have been found growing in intimate mixture with badly diseased Corsican pine on a high-elevation site. Possibly there are strains of the fungus of varying pathogenicity on different pine species, or maybe the reaction of the host tree to the local climate is as important with *P. contorta* as it is with Corsican pine.

Control

The only action that can be taken against this disease is to restrict the planting of Corsican pine to warm, relatively dry sites at low elevations, or to places where the existence of healthy trees of the same species more than twenty years old indicates that local conditions are suitable for continued growth. Wettstein-Westersheim (1933) has indicated the possibility of selecting resistant pines, but in view of the overriding effect of climate this method seems to have only very limited possibilities.

Cenangium dieback

Cenangium ferruginosum Fr. (*C. abietis* (Pers.) Duby.) (Ascomycetes, Helotiales) is a very common saprophyte. Probably because it occurs so frequently on dead pine branches, it was for a long time regarded as the perfect stage of *Brunchorstia pinea*. In Britain it has never been accused of pathogenicity, though it has been recorded as a saprophyte on a number of pine species. There are numerous records on the Continent and in America. In both places it is sometimes regarded as a cause of disease, though it is recognized that it is usually saprophytic.

Both as a saprophyte and as a parasite it appears to have a wide host range, being found on several other conifers, including spruce, as well as on numerous pine species. Badoux (1922*b*) reported quite serious damage on *Pinus strobus* in Switzerland; Long (1924) and Molnar (1954) reported injury to shoots and branches of *P. ponderosa* in North America; Laubert (1926) described twig dieback on Scots and Austrian pine in Germany; Schoenwald (1931) recorded similar damage to Scots pine, but only on trees already weakened by root-attacking fungi.

Typically the disease is a twig dieback, infection taking place near the ends of the twigs and travelling down them, but the fungus can penetrate into older wood, and occasionally whole branches may die. These are usually the lower, weaker ones, so that the fungus has been described as having a pruning effect.

Diplodia dieback

Pathogens

Three closely related fungi, all Fungi Imperfecti, Sphaeropsidales, are concerned with this disease, which is of no importance in Britain. *Diplodia pinea* (Desm.) Kickx., which is sometimes called *Sphaeropsis ellisii* Sacc., is certainly the most important. The others are *S. malorum* Peck and *D. conigena* Desm. These fungi all form very small black pycnidia at the bases of infected needles and on twigs, where they are often arranged in rows, but are less conspicuous than on the needles.

Symptoms and development

The disease, which has been fully reviewed by Capretti (1956), results in stunting of the new growth and browning of the needles, and in severe attacks leads to extensive dieback. It is normally found on young pine in plantations, but has also been recorded on several species of pine in nurseries (Slagg and Wright 1943), causing a dieback or 'blight'. It is usually thought to attack only pines weakened by drought or growing on otherwise unsuitable sites (Birch 1936; Waterman 1943*a*, 1943*b*; Capretti 1956). It can also infect wounds caused by hail (Laughton, F. S. 1937; Rawlings 1955; Grayburn 1957), by frost (Jump 1937; Georgescu and Mocanu 1956), by insects (Haddow and Newman 1942), or by other agencies. Purnell (1957) found that wounding was a necessary pre-requisite for successful inoculation. Nevertheless, it has at times been accounted a serious disease on pine, especially in the Southern hemisphere.

Distribution and damage

The disease can attack a very wide range of pine species. A little information on varietal resistance is available, but the data are not considered conclusive, since in this disease the reaction of the host species to site conditions is so important. *P. radiata* is apparently much more susceptible than *P. caribaea* (Cardoso 1951) or *P. patula*. *P. nigra* has been quoted as particularly susceptible, and the fungus is frequently reported on Scots pine.

Diplodia pinea appears to be of world-wide distribution, having been associated with serious injury on *P. radiata* in New Zealand (Curtis, K. M. 1926), South Africa (Laughton, E. M. 1937), and Brazil (Cardoso 1951), on various exotic pines in Australia (Young, H. E. 1936) and the Argentine (Saravi Cisneros 1950), on *P. nigra* and other pines in the United States (Pirone 1948), on *P. sylvestris* and *P. nigra* in Romania (Prodan 1935), on *P. halepensis* in Portugal (Oliveira 1944), and on *P. sylvestris* and *P. nigra* in Spain (Martínez 1942), as well as in several other countries.

In Britain, though the fungus is not uncommon, it is not even a species normally associated with debilitated pines. It was reported as a widespread cause of disease by Bancroft (1911), since when it has attracted no attention at all as a pathogen.

Control

Quite obviously control, which in any case is not required in Britain, can best be effected by planting pine under conditions where they are likely to grow reasonably well. In areas where attack is severe, change of species may become advisable. *P. patula* is being used instead of *P. radiata* in some parts of South Africa.

Other dieback diseases

Several species of *Phomopsis* (Fungi Imperfecti, Sphaeropsidales), in particular *P. occulta* (Sacc.) Trav., which is considered by Wehmeyer (1933) to be an imperfect stage of *Diaporthe eres* Nits. (Ascomycetes, Sphaeriales), have been recorded on pines in Britain. There is no evidence for their pathogenicity in Britain, though Birch (1935) reported a dieback of young plants of *P. radiata* and other pines in New Zealand associated with *P. strobus* Syd. The disease, however, occurred only at high elevations, where the pine were subject to unseasonable frosts. This fungus on pines in New Zealand was probably behaving in a similar way to *Phomopsis pseudotsugae* on Douglas fir in Britain (Chapter 25), i.e. merely as a weak parasite colonizing and spreading on frosted plants.

P. strobus was regarded primarily as a North American species whence it is supposed to have been imported into New Zealand; but Hahn (1957*b*) now considers it identical with *P. pseudotsugae*. To separate the different species of *Phomopsis* and assign them to Ascomycetous perfect stages is still so difficult that distribution and host records must be treated with great caution.

Hypodermataceous leaf-casts

A very large number of species of the family Hypodermataceae (Ascomycetes, Phacidiales) occur on pine needles. These include the genera *Lophodermium*, *Hypoderma*, *Elytroderma*, *Hypodermella*, and *Naemacyclus*, which have been fully described by Darker (1932). They all produce small black fruit bodies, and although some species can be distinguished by the shape and size of these, in general their identification is a matter for the expert.

All these fungi are probably facultative parasites, which can readily adopt a saprophytic existence. They are thus able to colonize pine needles already damaged by other agencies, particularly by cold winter winds, or possibly by the more virulent members of their own group (Peace 1953*a*). For instance, *Lophodermium pinastri* behaving as a primary parasite can, given suitable conditions for infection, severely defoliate Scots and other pines. But it is frequently also a saprophyte, while the degree of pathogenicity of the other pine defoliators in this group is very uncertain. In addition, since they can infect needles damaged by other causes, it is very difficult to discover whether they normally possess a definite cycle of infection, development, and spore production. Such knowledge is a necessary precursor to any attempt at control by spraying in the nursery.

Lophodermium needle-cast

Pathogen

By far the commonest fungus on the needles of pines in Britain is *Lophodermium pinastri* (Schrad.) Chev. On the tree it is more often observed in the conidial stage *Leptostroma pinastri* Desm. (Fungi Imperfecti, Sphaeropsidales), the perfect stage being produced mainly on fallen needles. The pycnidia are minute black spots, but the apothecia, which are also black, are much more striking, being 1–2 mm. long and 0·5 mm. broad. This fungus is considered by some to be a member of the *Hysteriales*, but is for convenience considered here in the *Phacidiales* with most of the other needle-cast fungi.

Symptoms and development

The first signs of infection are small pale spots on the needles. These become yellow and then brown, spreading and merging, so that the needles take on a mottled appearance, gradually becoming completely brown. During this process narrow black bars, which consist of destroyed mesophyll cells impregnated with black gummy substances, and which probably indicate the limits of each infection, are produced across the needles (Jones, S. G. 1935). They are clearly visible to the naked eye, and indicate that the leaf was still alive, though not necessarily fully healthy, when infected. These rings may occur with other needle pathogens, so that caution should be exercised in using them as a definite diagnostic feature for *Lophodermium* alone. The leaves fall when completely withered. The withering is associated with excessive water loss, and may result partially from the fungus interfering with the normal closing of the stomata (Langner 1933).

In heavy infections almost complete loss of the previous year's needles takes place; when defoliation is only partial the needles tend to fall in patches rather than singly, so that one shoot may lose most of its needles and another remain unharmed, or even one length of a shoot becomes bare while the parts above or below it remain normal. In this it is in strong contrast to attack by *Sclerophoma pithyophila* where infected needles are intimately intermixed with healthy ones.

Infection appears to be solely due to ascospores. The normal cycle is that these germinate on the young needles in early or mid-summer, the pycnidia appearing towards the end of the summer, and leaf-fall taking place during the winter or in the early spring. The apothecia are produced on the fallen needles the spring or summer following infection. In practice, however, the disease does not necessarily follow this straightforward course. It is certain that apothecia can ripen and infection take place throughout the summer (Rack 1955), and that they are occasionally produced at other times of the year. The fungus can remain alive in fallen needles under favourable conditions for considerable periods. Infection can take place on mature as well as on developing live needles, and it is clear from the universal infection of the needles on cut branches that it can invade dead or dying needles of any age. Thus while the worst cases of defoliation usually appear in the late winter or spring, as would be expected from the normal cycle, browning and needle-cast apparently associated with this fungus can often be found at other times of year. These variations in the time of infection and in the rate of development obviously make control by spraying a difficult problem.

There is considerable disagreement about the climatic conditions favouring attack. In Britain *Lophodermium* appears to be more severe following wet summers, suggesting that the damp summer weather favours infection; this is also the opinion of Hagem (1926). Hesselink (1927), however, associated the disease mainly with winter and spring weather unsuitable for pine, considering the weakening of the host more important than the actual climatic conditions for infection. Olberg (1955) stated that mild weather in the autumn and early winter, which favoured the development of the mycelium of the fungus in the host, was the most important factor in producing epidemic attacks. Rawlings (1955), dealing with epidemic outbreaks on *P. radiata* in New Zealand, and Mánka (1956) in Poland also considered that mild winters favoured attack, but supported British experience by also associating heavy infections with wet summers. Unfortunately no prolonged series of annual observations on degree of attack in relation to climatic variations has ever been made.

Distribution and damage

The fungus has a very wide distribution in Europe and North America and occurs throughout Britain. It is particularly serious in the lowlands of central Europe.

It is most commonly met with on Scots pine, but has also been recorded in Britain on Corsican pine, on *P. contorta* on which it seems to be rather uncommon, on *P. radiata*, on Red pine (*P. resinosa*), and on Japanese red pine (*P. densiflora*). On the Continent and in America it has been found on a very large number of other pine species (Darker 1932). There seems little doubt that more careful observation, especially on fallen needles, would greatly increase the British list of pines known to be infected by *Lophodermium*.

There are considerable variations in attack on different provenances of Scots pine (Burger 1930–1; Rubner, K. 1937; Dengler 1955; Mánka 1956; Schütt 1957), but there is not sufficient agreement to enable definite advice to be given on the safest provenances for use where *Lophodermium* attacks are frequent. Marked differences have been observed in the progeny of individual mother trees. Breeding for *Lophodermium* resistance, which can be very easily tested by natural infection in the nursery, appears quite a possible line of attack on the disease.

The damage done by the fungus to plantations in Britain is not normally very serious. Trees will usually recover fairly quickly the following year even from the complete loss of one year's needles, unless they were previously weakened by some other cause. Only suppressed trees are likely to succumb to *Lophodermium* attack. The tops of dominant trees are very seldom affected, and the loss of the needles from the lower parts of the tree only hastens the death of the most shaded branches and possibly results in a small growth check on the remainder. No doubt if repeated annual defoliations took place, as with *Rhabdocline pseudotsugae* on Douglas fir, the position would be very different. So far, however, severe *Lophodermium* attacks have been only occasional, normally being separated by periods of several years, so that comparatively little harm has occurred. In the nursery, however, the disease is serious if it affects plants intended for export, either to the forest for planting or for sale. Defoliated plants do not survive the difficulties of establishment in the forest, and plants without the current year's needles, or possibly in the case of two-year seedlings with scarcely any needles at all, can hardly be regarded as saleable. If such plants are carefully lined out, or better still left *in situ* for another year, quite a high proportion, even of those worst affected, will recover. With plants left in position, however, there is a risk of reinfection from the fallen needles on the ground. The fallen needles should therefore either be raked away and burnt, or covered with a thick mulch of peat or leaf mould.

On the Continent the disease is generally regarded as much more serious than in Britain, and many papers describing outbreaks of the fungus and means of controlling them have been published (Hagem 1926; Englebrecht 1928; Orłoś and Brennejzen 1954; Olberg 1955; Rack 1955). In New Zealand *Lophodermium* has produced epidemic attacks on *P. radiata* (Rawlings, 1955). On the other hand, it has never been regarded seriously as a cause of disease in America and recently J. S. Boyce, Jr. (1951), has cast doubts on its pathogenicity there. Possibly America possesses a strain of lower virulence than Europe.

Control

Bordeaux mixture and lime-sulphur, particularly the former, have been very generally recommended on the Continent against this disease in the nursery. In recent spraying experiments against *Hypoderma lethale*, a related fungus, organic mercurials were very successful (Morris 1953). Schönhar (1956*b*) and Rack (1958) have tried other modern fungicides, such as organic zinc compounds, with success.

Some consider that spraying should start as soon as the needles are half developed, whereas Rack (1955) would delay the first spraying till the beginning of August. Several workers have suggested mid-July as the appropriate starting time. Generally, once spraying has started, repeated applications at three-weekly or monthly intervals are suggested until the end of September. It would obviously be safest to start when the needles were half-developed, but this might involve as many as five applications. In any case in most nurseries in Britain outbreaks are so comparatively infrequent that annual spraying might well cost more than the losses experienced in the occasional epidemic attacks. Fungicidal treatment can be recommended therefore only in nurseries where attacks are known from past experience to occur at least one year in three. To be effective spraying must be carried out immediately before and during the period when infection is taking place. Sprays applied after the symptoms of damage have appeared will have no effect.

It has already been suggested that plants left in position, subject to the removal of the infected needles on the ground or covering them with a mulch, will probably recover. The same can be said of plants lifted and carefully lined out. In this connexion it must be remembered that even if there are losses after lining out, they are not necessarily the result of the fungal defoliation; deaths after lining out occur frequently even with initially healthy plants.

W. Fischer (1957) considers pure plantations far more apt for the development of this disease than mixed ones. He particularly advocates an admixture of broadleaved trees, the leaves of which will cover the pine needles on the ground and prevent reinfection from them. The disease is not sufficiently serious in British plantations to make an admixture of other species with pine desirable. M. Leroy (1957) found that plants closely surrounded by vegetation were badly affected and therefore advocated clean weeding of young pine plantations.

There is clear evidence that the fungus can spread on to neighbouring pines from the needles of live branches cut and left lying following thinning or live pruning. On young trees, infection of this kind can be severe enough to result in death (Murray and Young 1956). They found that heavy infection extended only about 8 metres from the source of spores, suggesting that safety belt 20 metres in width, formed possibly by dragging back the branches and tops that distance into the thinned plantation, should be adequate to prevent severe attacks of this kind on neighbouring young plantations. Certainly this sort of precaution should be taken in any pine-raising nursery situated in the immediate neighbourhood of plantations of the same genus. Recent evidence suggests, however, that it is undesirable to raise pine in any nursery where existing pine trees occur immediately adjacent to the boundary. Under these conditions infected needles can blow directly on to the seed-beds (Murray, J. S. unpubl.). Pine raising should be concentrated in nurseries that are well away from standing pine trees, or where there is at least a belt of a few metres of trees of some other genus, before any pine are reached.

The possibility of increasing resistance to *Lophodermium* by selection of provenances or individual seed trees, and by breeding from resistant strains, has already been mentioned. This disease appears to offer one of the most promising fields for work of this kind.

Hypodermella and Hendersonia needle-casts

Pathogens

Hypodermella sulcigena (Rostr.) Tub., which at one time was known as *Hypoderma pinicola* Brunch., is probably not a very common fungus in Britain. It is often found associated with *Hendersonia acicola* Münch Tub. (Fungi Imperfecti, Sphaeropsidales), which has been regarded, possibly without definite proof, as its imperfect stage, and which is very common on pine needles in Britain. Darker (1932), however, considers it more likely that *Hendersonia*, when found in association with *Hypodermella*, is a separate fungus acting in a secondary capacity. *Hypodermella* produces long, narrow, black apothecia; the pycnidia of *Hendersonia acicola* appear as black pinheads clearly seen against the pale whitish-grey background of the diseased needles.

A further difficulty arises because Darker (1932) described a fungus on *P. sylvestris* in Scotland as a new species, *Hypodermella conjuncta* Dark., a name which is occasionally used almost as if it were synonymous with *H. sulcigena*. In practice if there are two species of *Hypodermella* involved they have seldom been separated in the field, and must therefore be dealt with together.

H. concolor (Dearn.) Dark. is associated with serious defoliation on

P. contorta and *P. banksiana* in the United States (Mielke 1956*b*). The loss of needles is sufficient to restrict growth, but not to kill the trees.

Symptoms and development

Both *Hypodermella* and *Hendersonia* cause a whitish-grey discoloration of the needles of pine in the summer, followed some time later by needle-cast. It is doubtful, however, whether it is really possible to separate the early stages from the symptoms of *Lophodermium* needle-cast on the basis of the pale discoloration of the needles alone without reference to the fruit bodies and spores. *Hypodermella* is difficult to detect, since the elongated reddish fruit bodies are deeply embedded in the needles (Wilson, M. 1920); but *Hendersonia*, which produces abundant black pycnidia, is easily noticed, and this may account for the many records of it in Britain.

Distribution and damage

Hypodermella has not been commonly recorded in Britain. It is known to occur on both Scots and Corsican pine. But *Hendersonia*, which has been described by Laing (1929), appears to be very generally distributed over Britain, occurring on Scots, Corsican, and Mountain pine. It is, however, generally less common than *Lophodermium pinastri* on Scots pine, but it may occur as often as *Lophodermium* on the other two pines mentioned.

It is impossible to assess the importance of *Hypodermella* or *Hendersonia* in Britain, since so few records distinguish clearly between them or even separate their effects from those due to *L. pinastri* or non-living agencies. On the Continent, *Hypodermella sulcigena* has occasionally attracted attention as a cause of defoliation on Scots and Mountain pine (Kalandra 1938; Terrier 1943), but it is generally regarded as far less serious than *Lophodermium.*

Control

Both these fungi have usually been recorded in the forest, rather than in the nursery, so that direct control by spraying is hardly feasible. Nothing is known of the effects of silvicultural treatment on them.

Other hypodermataceous needle-casts

Elytroderma (*Hypoderma*) *deformans* (Weir) Dark. causes severe defoliation of *P. ponderosa* locally in western North America. Infected needles are killed the first year and in the worst areas up to 90 per cent. death has occurred among pine of all ages as a result of repeated defoliation (Wagener, Childs, and Kimmey 1949; Lightle 1954; Childs 1955). It also attacks *P. jeffreyi* and some other western American pines, but significant damage has only been found on *P. ponderosa.* The fungus sometimes penetrates into the twigs and provokes the formation of witches' brooms. So far this potentially dangerous disease seems to be very local, at any rate as a cause of serious damage. It does not occur in Europe.

Various species of the closely related genus *Hypoderma* have been associated with needle-cast in pine. Three doubtfully separate species, *H. brachysporum* (Rostr.) Rub., *H. desmazieri* Duby, and *H. strobicola* Tub., occur on pine needles in Britain (Wilson, M. 1920). Darker (1932) reduced these to one species under the name *H. desmazieri*, and this treatment is best for practical pathological purposes. *H. desmazieri* produces oval black apothecia about twice as wide as they are long, and occurs mainly on *Pinus strobus* and other Five-needled pines, but it has also been recorded in America on *P. banksiana*, *P. resinosa*, and *P. sylvestris*, and in Britain on more than one occasion on *P. radiata* (Batko unpubl.). So far it has not been recorded on Scots pine in Britain. It is probably one of the least important of this group of leaf casts.

Hypoderma lethale Dearn., which is common on *Pinus rigida* and other pines in the United States, has been successfully controlled by spraying with organic mercurials (Morris 1953); but it can generally be assumed for all *Hypodermataceous* leaf-casts that spraying is only possible in the nursery, and there only if the life-cycle of the fungus is sufficiently understood for the period of infection to be known and spraying restricted to it.

Naemacyclus niveus (Pers. ex Fr.) Sacc. has been associated with defoliation of various species of pines, both on the Continent (Příhoda 1950*b*) and in the United States (Darker 1932). The tiny fruit-bodies, which are borne on the needles, are brown at first, but fade to the same colour as the leaf and are, therefore, not at all conspicuous. In Britain it occurs on fallen needles of Scots pine, but has not been associated with defoliation.

Sclerophoma, phoma, and pullularia

Sclerophoma pithyophila v. Höhn. (Fungi Imperfecti, Sphaeropsidales) is locally common on pine needles. It has been associated particularly with damage to needles of the current year on Scots pine, where the Pine gall midge, *Cecidomyia baeri* Prell., appears to be the primary agent. The affected pairs of needles collapse at the base, which is where the Gall midge burrows and also where the fungus first develops. The needles then hang downwards for some time before they fall. On severely affected shoots all the needles may be attacked, and complete defoliation takes place, but frequently only some of the needles drop, and these are normally scattered among the healthy ones, so that when they fall, they tend to remain hanging on the healthy needles below them. So far this form of damage has only been found in Britain on Scots pine five to ten years planted, varying in height from 1 to 5 metres. The needles collapse in the latter part of the summer, and the small, black, protuberant fruit bodies, clearly distinguishable from the much larger elongated fructifications of *Lophodermium*, are produced in the autumn and winter. The relationship between insect and fungus and the real importance of the latter remains to be elucidated.

Isolations from diseased material have repeatedly yielded the ubiquitous fungus *Pullularia* (*Dematium*) *pullulans* (De Bary) Berkh. (Fungi Imperfecti, Moniliales) (Batko, Murray, and Peace 1958), which had previously been connected with *S. pithyophila* by Robak (1952*a*). Fructifications of *Sclerophoma* are produced on sterilized pine needles placed in *Pullularia* cultures. *P. pullulans* is generally regarded as a widespread saprophyte, occurring frequently on various conifers as well as on many other hosts. It is known to be somewhat variable, and it is quite possible that only one form of it is connected with *Sclerophoma*. It has also been suggested that *Phoma acicola* Sacc. (Fungi Imperfecti, Sphaeropsidales) is yet another form of this fungus.

Records of *S. pithyophila* have so far been rather local. Wilson (unpubl.) found it in Berkshire and Kent, the writer and his colleagues (Batko, Murray, and Peace 1958) have found it in Norfolk, where it is extremely common, Suffolk, Dorset, Wiltshire, Staffordshire, and Northumberland, generally on Scots, but in one case on Corsican pine.

Phoma acicola has been recorded much more widely both on Scots and Corsican pines. It is common on the needles of Corsican pines which are dying back owing to unsuitable climatic conditions, aided by *Brunchorstia destruens* (see above).

In Russia *Sclerophoma pithyophila* has been associated with dieback of pine and the formation of witches' brooms, distinctly different symptoms from those found in Britain (Nazarova 1936). Severe dieback associated with this fungus has also been reported on *P. contorta* in Finland (Kujala 1948). Gibson (personal communication) has found *Sclerophoma* on the dead shoots of *P. radiata* and *P. patula* in Kenya, but there was no evidence of pathogenicity. Generally this fungus has attracted little attention abroad.

Patton and Riker (1954) ascribed loss of current year's needles on *P. resinosa* to drought; though *Pullularia pullulans* was consistently isolated, they were unable to get any evidence of its pathogenicity and regarded it as purely secondary. Jump (1938*b*) (see above) associated *P. pullulans* with extra-seasonal growth and subsequent forking of *P. resinosa.*

All these associations of *Sclerophoma* and *Pullularia* with disease in pine are distinctly different from those described above for Britain. But Jahnel and Junghans (1957) associated *S. pithyophila* with damage by the Gall midge, *Cecidomyia baeri,* and subsequent needle collapse of Scots pine in Germany, though they described the needles as dying from the tip, not from the base as in Britain. Almost exactly similar symptoms to the British ones have been described by Haddow (1941) on *P. resinosa* in Canada. In this case *Pullularia pullulans* followed and increased initial damage by a Cecidomyid gall midge; apparently no fruit bodies of *Sclerophoma* were observed. Further work is required on this interesting and apparently widespread insect-fungus complex.

Other needle casts

Scirrhia acicola (Dearn.) Siggers (Ascomycetes, Dothideales), the imperfect stage of which is known as *Septoria acicola* (Thüm.) Sacc. (Fungi Imperfecti, Sphaeropsidales) causes a leaf-blight of Longleaf pine (*P. palustris*) in the southern United States (Wolf and Barbour 1941; Siggers 1944). It has also been associated with active disease on *P. ponderosa* (Luttrell 1949*b*), on *P. taeda* (Boyce, J. S., Jnr. 1952) and on the Five-needled pines *P. strobus* and *P. monticola* (Boyce, J. S., Jnr. 1959). It is of particular interest because controlled burning of the ground cover has been used as a method of control. The dead needles, which form the main source of reinfection, are burned, but the live pines, provided the burning is not done under too dry conditions, remain virtually unharmed. It has not been recorded in Britain.

Dothistroma pini Hulb. (Fungi Imperfecti, Sphaeropsidales) has been associated with defoliation of Austrian and Ponderosa pines in the United States (Hulbary 1941; Thomas and Lindberg 1954). Chlorotic spots appear on the needles in the autumn and winter. These gradually spread and turn brown, the distal end of the needle withering while the base still remains green. In the early spring, black stromata, in which the conidia are borne, erupt through the epidermis. A very similar fungus has been found in association with dying needles of Corsican pine, *P. radiata*, and *P. ponderosa* in Dorset in Britain (Murray and Batko unpubl.). In the one locality involved, damage was severe on *P. ponderosa* and there was serious browning on Corsican pine.

Rhizosphaera kalkhoffii Bub. (Fungi Imperfecti, Sphaeropsidales), which is of limited importance on spruce (Chapter 23), has been recorded on the needles of *P. nigra* and *P. montana* in Britain, but was not doing significant damage.

DISEASES OF CONES

In the United States, cones of *P. caribaea* and *P. palustris* are attacked by the rust *Cronartium strobilinum* (Arth.) Hedg. & Hahn (Basidiomycetes, Uredinales), while those of *P. leiophylla* are attacked by *C. conigenum* (Pat.) Hedg. & Hunt (Hedgcock, Hahn, and Hunt 1922). The infected cones become swollen and distorted, bearing masses of yellow aeciospores on the scales, and no seed is produced. The seed crop can be reduced by as much as 60 per cent. (Foster 1956). The alternate hosts are species of oak. These two rusts do not occur in Europe.

Sphaeropsis necatrix Petr. & Adam (Fungi Imperfecti, Sphaeropsidales) attacks the cones of *P. pinea* in Italy. The disease in its more severe form causes premature desiccation of the cones and considerable seed losses (Verona 1950). It has not been recorded in Britain.

23

DISEASES OF SPRUCE

It is surprising, considering the size of the genus, how few spruces are of forest importance in Britain. Sitka spruce (*Picea sitchensis*) is our most commonly planted forest conifer, and Norway spruce (*P. excelsa*) occupies a position scarcely less important, not only because of its value as a timber tree, but because of its preponderance in the Christmas-tree trade. The only other species which have been used on a forest scale, the Serbian spruce (*P. omorika*) and the White spruce (*P. alba*), appear unlikely ever to play an important part in British forestry.

Many other spruces are planted in arboreta and gardens. These include the Blue spruce, *P. pungens*, especially in its more brightly coloured forms, the Oriental spruce, *P. orientalis*, and a large and confused group of Asiatic species including *P. asperata* and *P. jezoensis*.

The predominance of *P. pungens* in pathological records of the genus is rather puzzling. As many pathogens have been recorded on this species as on the much more commonly occurring *P. excelsa* and *P. sitchensis*. This may in part be due to its ornamental nature. It is a tree of great beauty from which a dead branch or loss of needles would obviously detract. Possibly this has led to the earlier and commoner discovery of diseases on *P. pungens* than on other species.

DAMAGE DUE TO NON-LIVING AGENCIES

Most spruces are susceptible to spring frost. On sites where spring frosts occur regularly, repeated trimming by frost and the abundant production of recovery shoots by the trees often produce spruces of almost cypress-like outline, much older than their height and size suggest. Norway spruce provides one of the best examples of a species with early and late flushing individuals, the latter being distinctly more frost hardy. These are not strictly races, since they normally occur mixed, though in certain very frosty places stands are said to have become purely late flushing by a process of natural selection. Even then it is likely that early flushing trees would occur in the seed progeny.

Spruces are usually shallow-rooted, and this may be one of the reasons why they, and particularly Norway spruce, are very subject to drought damage. Indeed, Norway spruce appears to be peculiarly liable to difficulties with water-supply. A disease of this species which occurs very commonly in Britain, and is usually referred to as 'top-dying', appears to be due to failure of sufficient water to reach the crown, sometimes because the soil water content is insufficient, but more often because the conducting system of the tree is unable to meet the water requirements of the crown (Chapter 6).

Sitka spruce, on the other hand, is more liable than any other conifer to drought-crack in hot, dry years.

NURSERY DISEASES

Spruce is subject to a number of nursery diseases, several of which also attack other conifers and are therefore dealt with in Chapter 15. These include dieback associated with *Pestalozzia hartigii* and *P. funerea*, Grey mould caused by *Botrytis cinerea*, which is particularly serious on Sitka spruce, and the two Snow moulds, so called because the fungus spreads over the plants when they are covered with snow, *Herpotrichia nigra* and *Phacidium infestans*, neither of which is of any importance in Britain.

Phomopsis occulta Trav. (Fungi Imperfecti, Sphaeropsidales), considered by Wehmeyer (1933) to be an imperfect stage of *Diaporthe eres* Nits. (Ascomycetes, Sphaeriales), has been fairly frequently found associated with a seedling blight of both Norway and Sitka spruce. The top of the seedling dies and the collar is ringed with white mycelium. Later, small, black, pustular fruit-bodies appear on the stems of the dead seedlings. The damage is very similar to that caused by *Pestalozzia hartigii*, so that there may be confusion between the two diseases. Indeed, they sometimes occur together. Fortunately *P. occulta* is not very common, though it can damage spruce seedbeds severely. It has been recorded causing similar damage to *P. pungens* and other spruces in the United States (White, R. P. 1929).

Rosellinia aquila (Fr.) De Not. (Ascomycetes, Sphaeriales) has occasionally been reported causing severe damage to spruce in the nursery (Wilson, M. 1922). The lower parts of the stem and the upper parts of the root become covered with a greyish mycelium, which envelops and binds together the lower needles. Black spherical perithecia, 0·7–1 mm. in diameter, are produced in clusters just above soil-level on the web of mycelium. Most of the records have been on Norway spruce, but it has also been found on *P. omorika*. *Rosellinia* is not serious, being less common than *Pestalozzia hartigii* or *Phomopsis occulta*, which are associated with much the same kind of injury.

Ascochyta piniperda Lind. (Fungi Imperfecti, Sphaeropsidales) occurs in plantations, as well as in the nursery, and is therefore dealt with later in this chapter.

ROOT DISEASES

Armillaria mellea occurs fairly frequently on most species of *Picea* on infected sites. There is some evidence that *P. omorika* is particularly susceptible. Possibly the frequency of root grafts in shallow rooting spruces may contribute to the transmission of root diseases from tree to tree. *Fomes annosus*, mainly as a cause of decay rather than death, attacks the two principal spruces very severely. The other species of *Picea* are probably equally susceptible. Group dying associated with the fungus *Rhizina inflata* (Chapter 16) occurs far more commonly among Sitka spruce than other conifers. Individual groups spread more rapidly and attain a greater size in Sitka spruce. All the evidence so far suggests that the species is particularly susceptible to *R. inflata*.

On the Continent the general debility of Norway spruce in some places and Sitka spruce in others has been attributed mainly to the combined action of fungi and insects, chiefly bark beetles. The fungi most commonly mentioned are *F. annosus* and *A. mellea*. Nothing quite comparable seems yet to have occurred in Britain.

STEM DISEASES—BARK AND CAMBIUM

Stem canker

Pathogen

Nectria cucurbitula (Tode.) Fr. (Ascomycetes, Hypocreales), the conidial stage of which is *Cylindrocarpon cylindroides* Fr. (Fungi Imperfecti, Moniliales), and which is a common saprophyte on dead conifer branches, is the fungus commonly associated with injury to spruce stems in Britain (Laing 1947). It has been associated with similar injury on the Continent also (Zycha 1955*a*). Since it has been particularly associated with spruce canker, it is dealt with here, rather than with other members of the genus *Nectria* in Chapter 18. The typical red fruit-bodies of the fungus normally occur abundantly on the dead bark. It must be admitted that there is very little certainty about its pathogenicity. Zycha considered frost and deer damage might both be predisposing factors. W. R. Day (1950*c*) attributed cankers on pruned stems of Norway spruce entirely to frost. Banerjee (1956) suggested that two other fungi, *Pleurotus mitis* (Pers.) Berk. (Basidiomycetes, Agaricales) and *Stereum sanguinolentum* (A. & S.) Fr. (Basidiomycetes, Aphyllophorales), are the active causes of the disease, and that the *Nectria*, which cannot always be found, is purely secondary. *S. sanguinolentum* is known to infect wounds on spruce resulting from careless brashing. *P. mitis*, which produces small pure white toadstools 1–2 cm. in diameter on the dead bark, has been described as the cause of a very similar disease on spruce and *Abies alba* in Switzerland (Gäumann and Jaag 1937). In their case successful inoculation experiments were carried out, so that there seems no doubt that *P. mitis* can be a parasite on spruce. Further investigation would be required to elucidate the relative importance of the three fungi in the causation of this disease in Britain.

Symptoms and development

Two types of damage, merging into each other, characterize this disease. Firstly, small, dead, slightly sunken patches are found around dead branches or brashing wounds; secondly, long strips of dead tissue occur normally covered by dead but intact bark, and extending often from ground level up the whole of the brashed length. The intervening wood continues to grow, so that eventually these necrotic areas appear as long vertical hollows on a heavily ridged stem (Laing 1947). Similar symptoms were found in the disease described by Gäumann and Jaag (1937), but in their case the cankers extended into the crown. European records of cases where the disease is attributed to *Nectria* describe more varied symptoms, the fungus sometimes attacking and girdling much smaller branches, and therefore causing crown dieback.

In Britain the principal damage is the deformation of the stem, which results in a considerable reduction in utilization value, but occasionally the disease causes the death of the tree.

Distribution and damage

Long cankers of the type described above have been found fairly widely on Sitka spruce, mostly in Scotland, but also in England and Wales. Typically, affected trees are scattered in the stand. In fact the disease would often pass unnoticed were it not for the conspicuous fluting. At present the amount of damage is small.

In Britain the disease occurs chiefly on Sitka spruce; but on the Continent it has long been known on Norway spruce, and has been recorded on *P. alba*. In Europe *Nectria cucurbitula* has been found also on various species of *Abies*, on *Pinus strobus*, and on other conifers, but is mainly important on spruce.

Control

Cankered trees should be removed in thinning, for the fluting is likely to become more exaggerated with time, and decay fungi will enter through the dead areas.

Other diseases of bark and cambium

In America the most prevalent stem disease of spruce is a canker caused by *Cytospora kunzei* Sacc. (Fungi Imperfecti, Sphaeropsidales), the perfect stage of which is *Valsa kunzei* Fries. (Ascomycetes, Sphaeriales). The most obvious symptom of this disease is the dying of individual branches, usually near the base of the crown, but sometimes higher up. Resin exudes from the base of infected branches and tends to hide the inconspicuous cankers. In dry weather the fruit-bodies of the fungus are inconspicuous, but are easily detected when wet, since they exude yellow spore-tendrils (Gilgut 1937; Strong 1953).

The fungus is serious on ornamental spruces, particularly on varieties of *P. pungens*, where the death of even a few branches detracts from the ornamental effect. The fungus attacks also *P. excelsa*, on which it tends to produce trunk, rather than branch cankers.

In the United States other varieties of *C. kunzei* attack other coniferous genera such as *Abies*, *Larix*, and *Pseudotsuga*, but only on *Picea* is the disease serious (Waterman 1955). *C. kunzei* has been recorded on dead twigs of spruce and other conifers in Britain and on the Continent, but has never been found there as a parasite.

Several closely related *Ascomycetes* of the group *Helotiales* have been associated with minor bark diseases of spruce. These include *Durella livida* (B. & Br.) Sacc., more familiar as *Dermatea livida* (B. & Br.) Phillips, the perfect stage of *Myxosporium abietinum* Rostr., which is dealt with in Chapter 18, *Pezicula* (*Dermatea*) *eucrita* Karst., and *Cenangium abietis* (Pers.) Rehm., which has been reported by Badoux (1922*a*) causing serious damage to Norway spruce in Switzerland, but which is much commoner on pine. All these *Helotiales* occur in Britain, but are of no pathological importance on spruce.

Similarly, two species of *Dasyscypha* (Ascomycetes, Helotiales), *D.* (*Trichoscyphella*) *resinaria* (Cooke & Phill.) Rehm., and *D. calyciformis* (Willd.) Rehm., have been accused on comparatively little evidence of causing lesions on spruce stems.

DISEASES OF LEAVES AND SHOOTS

Chrysomyxa needle rusts

Pathogens

A number of species of *Chrysomyxa* (Basidiomycetes, Uredinales) attack spruce needles, but only two, *C. abietis* Unger. and *C. rhododendri* de Bary, are found in Britain. The first occurs only on spruces, but the second has *Rhododendron* species as alternate hosts. In parts of the Continent, *C. ledi* (A. & S.) de Bary, the alternate host of which is the small Ericaceous shrub *Ledum palustre*, is of considerable importance. *C. ledi* does not occur in Britain, where in any case the alternate host is rare.

At least seven species of *Chrysomyxa* occur on spruce needles in America. One of these is *C. ledi*. The others, except *C. weirii* Jacks, which occurs like our *C. abietis* on spruce alone, have various alternate hosts.

Symptoms and development

The development of the two *Chrysomyxa* diseases which occur in Britain is very different. In *C. abietis* the young needles are infected in the early summer by the sporidia produced from the teliospores, borne on the infected needles of the previous year. As the year advances the needles become mottled with yellow patches, and by the end of the winter they are bright orange-brown in colour. Heavily infected trees are very conspicuous. The golden-yellow telia appear as elongated pustules on the underside of the needles in the early summer, when infection of the next crop of needles takes place (Fig. 65). Infection of older needles does not occur. Successful infection appears to depend on coincidence between the production of the sporidia and a short period when the young developing needles are in a suitable condition for infection. In normal years in Britain the new needles have passed the vulnerable stage before the telia are ripe. Only when the spring is cold and shoot development delayed may the two coincide and heavy infection take place (Delforge 1930; Murray, J. S. 1953). As soon as the sporidia have been produced the needles are shed.

The fungus *Darluca filum* Cast. (Fungi Imperfecti, Sphaeropsidales) sometimes parasitizes the fructifications of *C. abietis* and checks the production of sporidia. It is very doubtful, however, whether it has any significant effect on the development of the disease.

With *C. rhododendri*, on the other hand, the development in the spruce needles is more rapid. After infection by the sporidia, produced on rhododendron leaves and young shoots in the early summer, the young needles soon develop conspicuous yellow bands and later become completely discoloured. The effect of the two fungi on the needles is very similar. The aecia, appearing as small, columnar, white pustules, are borne on the lower surfaces of the spruce needles in late summer. Aeciospores from the spruce then infect rhododendron leaves. Infected spruce needles are thus shed only a few months after they develop. The fungus overwinters as mycelium in the rhododendron leaves until the following spring, when the orange uredia are produced on the leaves and young shoots. Occasionally uredia may be produced in the

FIG. 65. Telia of *Chrysomyxa abietis* on needles of Norway spruce, Newcastleton Forest, Dumfriesshire, July (nat. size, ×2). (J. C.)

autumn. The uredospores spread infection among rhododendrons. Later, dark-brown telia are formed, the sporidia from which infect the young spruce needles. The fungus can survive on rhododendron in the absence of spruce, but not on spruce in the absence of rhododendron.

Distribution and damage

Chrysomyxa abietis occurs mainly in Scotland. It is most commonly met with on Norway spruce, but has been recorded on Sitka spruce, *P. pungens*, *P. rubra*, and *P. engelmannii*. On the Continent it appears to be much less important than *C. rhododendri* and *C. ledi*.

There is some evidence of provenance differences in susceptibility to *C. abietis*. Possibly these are linked with variation in time of flushing. With *C. rhododendri*, large differences in susceptibility are often noticed between adjacent trees (Dufrénoy 1932). It is almost certain that they also are linked with differences in dates of shoot development.

C. rhododendri is also chiefly a disease of Norway spruce, but it has recently been recorded on Sitka spruce as well. It also occurs more commonly in Scotland than in England, but has been recorded as far south as Cornwall. The chief *Rhododendron* hosts in Europe are *R. hirsutum* and *R. ferrugineum*, but it also occurs on a number of other species and has recently been recorded in Britain on *R. ponticum* (Wilson and Bisby 1954). If this species proves very susceptible (present indications are that it is not), the disease could become more important. Most other rhododendrons, including *R. hirsutum* and *R. ferrugineum*, are found only in large gardens, but *R. ponticum* has become generally established on acid soils, and is often found growing near spruce plantations. In the Alps, where spruce and the wild rhododendron hosts often grow in close proximity, the disease is occasionally very severe, though there are big fluctuations in intensity of attack from year to year (Fischer, E. 1933).

An importation of *C. rhododendri* into the United States on rhododendron plants (Gould, Eglitis, and Doughty 1955) clearly illustrates one of the difficulties of plant quarantine. At the time of importation the fungus existed as mycelium in the leaves, and would not be detected by the most careful visual examination. By the time the conspicuous uredia appeared in the spring, the plants had been widely distributed.

In north-east Europe the closely related *C. ledi* replaces *C. rhododendri* (Melehov 1946). *C. ledi* does not occur in Britain. The most important host, *Ledum palustre* is possibly native over a small area in Perthshire and occurs as an escape in a few other places. It is occasionally planted in gardens, as are a few other *Ledum* species.

In America the two most damaging species appear to be *C. cassandrae* (Pk. & Clint.) Tranz., the alternate host of which is *Chamaedaphne calyculata*, a small Ericaceous shrub very occasionally planted in British gardens (Lindgren 1932), and *C. ledicola* (Pk.) Lagerb. which, like *C. ledi*, has species of *Ledum* as alternate hosts (Nordin 1956*b*). In view of the scarcity of the alternate hosts in Britain, the introduction of these rusts into this country would probably not be serious.

The injury to spruce caused by *Chrysomyxa* rusts is in all cases limited by the irregularity of their occurrence in epidemic form. If defoliation took place every year, they could be extremely serious; behaving as they do, the occasional loss of one year's needles results, at the worst, only in a temporary check in growth.

Control

No feasible control method has been suggested for *C. abietis*. In the case of *C. rhododendri* and other species with two hosts, removal of the alternate host, or separation of the two host plants, are obvious methods. The range of infection of the spores produced on rhododendron is probably limited to a few hundred yards across open ground, and would be less through a thick plantation. This being so, spruce could probably be kept uninfected, if they were separated from rhododendrons, by a substantial belt of some other tree genus.

Other needle and shoot rusts

A number of other rust fungi occur on spruce in America. *Pucciniastrum americanum* (Farl.) Arth. and *P. arcticum* (Lagerb.) Tranz. both have *Rubus* species as alternate hosts and attack spruce needles. *Peridermium coloradense* (Diet.) Arth. & Kern. causes witches' brooms on spruce with conspicuous, deciduous, yellow needles, similar to the brooms caused by *Melampsorella caryophyllacearum* on *Abies* in Britain. These rusts are not normally considered serious, but *P. coloradense*, the alternate host of which is not known, has recently been reported (Bourchier 1953) causing loss of increment and occasionally death of infected trees in Alberta.

Thekopsora areolata, normally attacking spruce cones (see below), can also infect one-year-old shoots of spruce, in the same manner as *Melampsora pinitorqua* does on pine, but this form of injury has not been reported on spruce in Britain.

Lophodermium needle-cast

Three species of *Lophodermium* (Ascomycetes, Phacidiales) have been associated with needle-cast of spruce. The best known is *L. macrosporum* (Hart.) Rehm., sometimes known as *Hypodermella macrospora* Lagerb., which with *L. piceae* (Fckl.) v. Hohn. occurs in Europe. In America *L. filiforme* Dark. causes similar injury, but *L. piceae*, which also occurs there, is not considered important. It has, however, been suggested by Terrier (1953) that *L. macrosporum* and *L. filiforme* are the same.

All are characterized by black apothecia on the underside of the needles. In *L. macrosporum* these are much longer than in *L. piceae*. With *L. piceae* dark bars are formed across attacked needles, similar to those produced in pine needles by *L. pinastri*, but these do not occur with *L. macrosporum*.

It seems almost certain that these two fungi have frequently been confused. It is impossible, therefore, to say with certainty whether their effect on the tree is different. Reports on *L. macrosporum*, which has in the past been regarded as far the more virulent, have shown great variation, leaf-fall taking from one to three years, and apothecia being produced at various times of year, sometimes on needles still adhering to the tree and sometimes on those that have fallen. This wide range of behaviour is suggestive of more than one species being described, and until their life histories and differences have been separately investigated it would be a mistake to try to give definite information on the time, method, and duration of attack. It is probable that these fungi are often secondary on needles that are dying from other causes.

Both species occur on *P. excelsa*, *P. sitchensis*, and *P. pungens*. They have been found on other spruces and may well occur on most members of the genus. *L. piceae* is the commoner in Britain, being quite frequently recorded, but seldom associated with appreciable defoliation. Some Continental reports mention quite severe defoliation, specially by *L. macrosporum*. Nevertheless, *Lophodermium* needle-cast of spruce does not rank as a really serious disease.

Rhizosphaera needle-cast

Possibly the commonest fungus on spruce needles is *Rhizosphaera kalkhoffii* Bubák. (Fungi Imperfecti, Sphaeropsidales). This fungus produces small, black, spherical pycnidia, which emerge through the stomata on both the upper and lower surfaces of the needle. As they emerge they carry up small masses of white waxy material from the mouths of the stomata, and these can be seen with a lens, still adhering to the pycnidia. *Rhizosphaera* is therefore quite easily distinguished from *Lophodermium*. It was described by Wilson and Waldie (1926*a*) as a serious defoliator of *P. pungens* in Britain and also as occurring on a number of other *Picea* species, including *P. alba*. More recently it has been found on *P. omorika*. It occurs widely on the Continent, and has also been described as a cause of disease on *P. pungens* in the United States (Waterman 1947).

In Britain it is the commonest fungus on spruce needles damaged by other causes. J. S. Murray (1957) has described it occurring constantly on Norway spruce affected by Top-dying, a disease primarily due to water shortage in the crown (Chapter 6). Went (1947) in Holland found *Rhizosphaera* associated with injury to Norway spruce attributed primarily to drought. In these cases it may play quite an important role; the defoliation, which it causes, weakening beyond possibility of recovery trees or parts of trees which might otherwise have survived.

On a few occasions in Britain the fungus has been found associated with very severe defoliation of young plantations of Sitka spruce. In these cases it was very noticeable that only certain trees were affected. Others near them remained quite healthy, and the damage occurred only during one season, there being no reappearance in following years. On affected trees the damage was almost universal, nearly all the shoots of the current year being affected. Neither the uniform severity of the disease on affected trees, nor its complete failure to reoccur are suggestive of primary fungal attack, and in addition *Rhizosphaera* could not be isolated from needles in the early stages of discoloration. Thus, while the origin of this locally severe disease remains obscure, it does provide evidence that *Rhizosphaera* is a very rapid and efficient colonist of spruce needles damaged by other causes. Even on *P. pungens* its pathogenicity must remain doubtful until successful inoculations have been made. Inoculations so far carried out have always failed.

Wilson and Waldie (1926*a*) reported the fungus on a wide range of *Picea* species, and also on species of *Abies*, *Pseudotsuga*, and *Pinus*. No further records on genera other than *Picea* appear to have been made, and the fungus can be regarded as important only on spruce.

Ascochyta shoot blight

Ascochyta piniperda Lind. (Fungi Imperfecti, Sphaeropsidales), sometimes known as *Septoria parasitica* Hart., causes a shoot blight of spruce, both in plantations and in nurseries, in Europe and North America. The first indication of this disease is a brown discoloration in early summer of the needles at the base or sometimes in the middle of the current year's shoot. These needles fall, and this is usually followed by the loss of the remaining needles, though occasionally a few may remain attached near the tip. The shoot withers and hangs down, and it is the drooping defoliated shoots which are so characteristic of the disease. The fungus may also spread a short distance back into two-year-old wood, and thus infect side shoots at the base. Damage due to this fungus is sometimes confused with frost damage, but with the latter the tip of the shoot or the whole of the shoot is affected, whereas with *Ascochyta* needles at the base or middle of the shoot die first.

Minute black pycnidia are produced in very variable quantities on the dead parts. Sometimes they are obvious, but at other times a careful search among the scales at the base of the shoot is required to find them (Laing 1929; Lagerberg 1933). It has been suggested (Rúžička 1938*b*) that provenance may have a major effect on susceptibility, but this has not been generally reported. The disease has been recorded on Norway and Sitka spruce, and also on *P. pungens*.

In recent years drooping and partially defoliated shoots, suggestive of this disease, have been recorded on a number of occasions, chiefly on Sitka spruce. The damage, which was mainly on trees 1 to 5 metres high, was sufficient to attract attention without doing the trees serious injury. No pathogen has been isolated, but it is possible that *Ascochyta* was responsible. If so, *Ascochyta* would rank as a common, but relatively harmless, parasite. Otherwise it must be regarded as of comparatively rare occurrence on spruce in Britain. It has recently attracted attention in north Scotland as the cause of shoot dieback of nursery plants of *Pinus contorta* (Chapter 22).

Bud blight

Cucurbitaria piceae Borthw. (Ascomycetes, Sphaeriales) causes death of buds, particularly on *P. pungens*. Infection is followed by a marked swelling of the buds, and in addition there is often distortion of the shoots. Most of the infected buds die; any shoots that are produced from them are thin with short needles, and tend to die prematurely. The fungus forms a crust-like stroma completely enveloping the bud; on this numerous black perithecia are eventually produced. Infected trees have a curiously sparse and irregular appearance (Müller, E. 1950). Locally damage may be quite severe.

The fungus was originally recorded in Scotland by Borthwick (1909) on Norway and Sitka spruce and on *P. pungens*. More recently it has been found in Denmark on Norway spruce and on *P. alba*, and in Switzerland and Germany on *P. pungens*. It seems almost impossible that a disease producing such striking symptoms could be overlooked, so that the complete lack of records in Britain, since Borthwick first described it, suggests that the fungus, at any rate as an active pathogen, has died out.

Other needle and shoot diseases

A number of species of *Phoma* (Fungi Imperfecti, Sphaeropsidales) have been associated with injury to the needles of spruce in Holland (Went 1947), in Britain, and in America (Spaulding 1912). There is no suggestion that the damage caused is serious, and probably they are often secondary.

Diedickia piceae Bonar (Fungi Imperfecti, Sphaeropsidales) causes yellowing and browning of the needles of Sitka spruce in California (Bonar 1942). There is some leaf-fall, but the disease does not appear to be serious.

DISEASES OF CONES

Cone rusts

A number of rust fungi (Basidiomycetes, Uredinales) attack the cones of spruce. The best known of these is *Thekopsora areolata* (Fr.) Magn., sometimes called *Pucciniastrum padi* Diet. In the spring the sporidia infect the female flowers of the spruce. As the cones develop, whitish pycnidia are formed mainly on the inside of the scales, followed later in the summer by numerous spherical brown aecia. The scales of attacked cones remain open in wet weather, when those of normal cones are shut. Infected cones produced impaired seeds or may be completely sterile. The uredia appear as small brown spots on the under-surface of the leaves of Bird cherries, including *Prunus serotina* and *P. virginiana*, as well as our native *P. padus*. The dark-brown telia appear later on the upper surface of the leaves and the spores from them infect the spruce cones. The disease is not very common and does little damage in Britain, where it has been recorded only on Norway spruce. It is common on the Continent, and has been reported by Roll Hansen (1947) in Norway to be capable of attacking the young shoots of Norway spruce, causing distortion and dieback in the same way as *Melampsora pinitorqua* does on pine (Chapter 22).

Chrysomyxa pyrolae Rostr. has spruce and *Pyrola* species as alternate hosts. In Britain it occurs occasionally on species of *Pyrola*, small herbaceous plants of limited distribution, on which it produces the uredial and telial stages, but it has not been found on *Picea*. In Scandinavia, where the aecial stage on spruce cones does occur, it sometimes causes appreciable loss of cones and consequent interference with regeneration. It also occurs on a number of other spruce species in America, but not on Sitka spruce; but it is not very damaging.

Recently a new cone rust on Sitka spruce in British Columbia has been reported by Ziller (1954). This closely resembles *C. pyrolae*, but has been named *C. monesis* Zill. The alternate host is the small herbaceous plant *Moneses uniflora*, which is closely related to *Pyrola* but which is much rarer in Britain, occurring only in north and west Scotland. Should this rust, which in any case does not seem very damaging in British Columbia, ever reach Britain, the rarity of the alternate host would probably limit both its spread and the severity of its attack.

24

DISEASES OF LARCH

THE two species of larch most prominent in British silviculture are *Larix decidua*, the European larch, and *L. leptolepis*, the Japanese larch. Of recent years the hybrid between these two, *L. eurolepis*, has come into prominence. Many of the plants of this hybrid now being used were raised from second-generation seed, and cannot therefore be regarded as a pure strain. One may expect, therefore, that their disease resistance will be very variable. European larch has a wide range of variation, and different provenances are known to give very different disease reactions.

The American larches have not been much planted in Britain; but the Tamarack, *L. laricina*, sometimes does well, and the Western larch, *L. occidentalis*, though usually a failure, has occasionally succeeded. Most of the Asian and Siberian larches, other than *L. leptolepis*, appear unsuited to our climate, the short shoots coming into leaf as early as February if the weather is warm, and ceasing growth in the middle of the summer, before they have had the full benefit of the hotter weather.

DAMAGE DUE TO NON-LIVING AGENCIES

It follows that those larches which come into leaf excessively early, for instance *L. sibirica* and *L. gmelini*, will be frost susceptible. The fact that the terminal buds often flush much later than the short shoots does not save the trees from serious damage. Even the more commonly planted species are frequently damaged by spring, and in the case of Japanese larch by autumn, frosts. The importance of this susceptibility to frost in the causation of larch canker and dieback is dealt with later in this chapter, but it is also an important cause of damage in its own right.

Japanese larch, particularly during its early years, is very subject to drought injury. This may occur at any time during the first fifteen to twenty years of its life, particularly on shallow soils over rock, or where there is competition for water with a heavy grass mat. The initial difficulties with the grass mat, or other water-demanding vegetation, will of course be overcome as the trees root deeper and the competition is suppressed. The apparent absence of drought damage in older plantations on rocky sites may merely indicate that trees on the shallow patches of soil have already succumbed.

Both the principal species of larch are commonly affected by chlorosis and later dieback if planted on highly calcareous soils. Neither of them, despite their deciduous nature, is at all tolerant of industrial pollution.

NURSERY DISEASES

Swelling at the base of the stem accompanied by death of roots, and sometimes by eventual death of the whole plant, has been reported more commonly on Japanese larch than on other conifers. The cause is still unknown (Chapter 15). Leaf-cast, *Meria laricis*, is primarily a nursery disease, but is dealt with under 'Leaf Diseases' below.

ROOT DISEASES

Japanese larch is among the conifers most susceptible to *Armillaria mellea*, and all larch species seem highly susceptible to *Fomes annosus*, as a cause of decay rather than death, particularly under dry conditions. Both European and Japanese larch are subject to decay by *Polyporus schweinitzii*. *Rhizina inflata*, associated with Group dying, has been found in a few instances on both European and Japanese larch, but the disease is much less important on larch than on Sitka spruce (Chapter 16).

STEM DISEASES—BARK AND CAMBIUM

Canker and dieback

Pathogens

It is generally agreed that this disease, which has been briefly described in a Forestry Commission leaflet (Anon. 1948*a*), is due to a complex interaction of frost and the fungus *Dasyscypha* (*Trichoscyphella*) *willkommii* (Hart.) Rehm. (Ascomycetes, Helotiales). The role of these two factors will be considered later. When Hiley (1919) made his study of larch canker, the first substantial work on the disease, only one species of *Dasyscypha* was considered to occur on larch, and he accepted the name *D. calycina* Fckl. for this. He regarded *D. willkommii* as a synonym, but pointed out that the fungus was smaller in all its parts when growing saprophytically and not on a canker. It was left to Hahn and Ayers (1934*a*) to suggest that two species were involved, one *D. willkommii*, a parasite, and the other, *D. calycina*, a saprophyte confined to dead twigs and the dead parts of cankers. Morphologically the two species, if such they be, are very close indeed; their ranges of measurement on any character tend to overlap. For this reason Robak (1951*b*) suggested that they should be regarded as varieties of a single species. The most recent work on these fungi has been carried out by Manners (1953), who accepted the division into two species, but proposed certain changes in nomenclature, consequent on the transference of both species to the genus *Trichoscyphella*. It appears that this change is not the final elucidation of a very difficult group, and, in view of the likelihood of further changes, the names used by Hahn and Ayers are retained here.

The apothecia of both fungi are similar, being saucer-shaped and short-stalked (Figs. 66 and 67). The rim and under-surface are white, and the concave upper-surface salmon orange, that of *D. calycina* usually being brighter. In *D. willkommii* the apothecia are usually from 1 to 4 mm. in diameter but occasionally larger, in *D. calycina* they are 0·5 to 3 mm. In both species larger fruit-bodies tend to have a broader white rim. Apart from this the rim is generally broader in *D. willkommii* than in *D. calycina*. However, it is almost impossible to tell the two species apart with the naked eye. If

D. calycina is growing on a food-rich substrate, for instance on a tree that has been killed rapidly by girdling, the average size of the apothecia tends to be greater, and the general appearance is more suggestive of *D. willkommii*. The

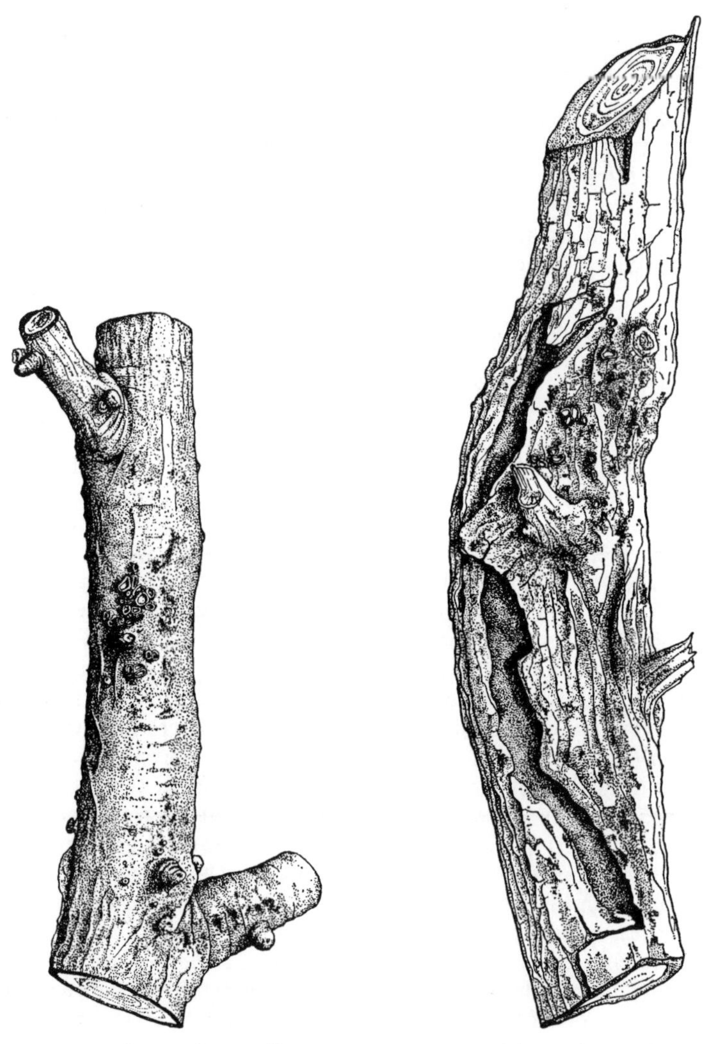

FIG. 66. Developing canker on European larch, with fruit-bodies of *Dasyscypha willkommii*, Lael, Ross and Cromarty, August (nat. size). (J. C.)

FIG. 67. Older canker on European larch, with fruit-bodies of *Dasyscypha willkommii*, Drummond Hill, Perthshire, August (×¾). (J. C.)

asci are borne on the upper surface of the apothecium and under suitable conditions of high humidity the spores are ejected into the air. The apothecia tend to become flatter as they develop, but even when fully expanded they will partially close in dry weather, opening to their full extent only when moist.

Whether two species are involved appears to be largely a matter of opinion. It is now, however, generally conceded that there are certainly two strains, one of which is a parasite and the other a saprophyte, and it is simplest to use the names *D. willkommii* and *D. calycina* for them.

Other fungi have been found in cankers. W. R. Day (1958*a*) recorded mycelium of *Cytospora abietis* Sacc., *C. curreyi* Sacc., and *Coniothyrium fuckelii* Sacc. (all Fungi Imperfecti, Sphaeropsidales), sometimes with and sometimes without *D. willkommii*, in the living tissue on the margins of cankers. Inoculations suggested that they were able to extend existing wounds, which suggests that the assumption that *D. willkommii* is the only fungus involved in the disease requires reinvestigation.

Symptoms and development

The symptoms of the disease can only be understood if we first consider its development in relation to the fungal and other factors responsible. The disease generally takes two forms, dieback of twigs on the one hand, and canker of twigs, branches, and even the main stem on the other. There has been a tendency in the past to separate these two types of damage and occasionally to refer to them as two separate diseases. In reality, as has already been pointed out (Chapter 2), they differ only in form, not in cause. Dieback occurs when the cambium is killed round the whole circumference of the twig, canker when part of the cambium is left alive and attempts at healing can take place. In the case of larch canker it is noticeable that, as the tree becomes progressively weakened, dieback occurs on larger and larger branches, indicating that in its debilitated condition the tree is unable to stop the extension of, and final girdling by, the canker.

The disease is most common on trees between ten and twenty-five years of age, but occurs over a much wider age range. Typically a diseased tree displays abundant small cankers, some of them healed, on the twigs and smaller branches. Cankers which have only just started to develop show as small slightly sunken areas (Fig. 66). There is also considerable dieback, particularly of smaller twigs, and occasionally there are cankers on the larger branches and main stem. The larger cankers are accompanied by cracking of the bark and distortion of the stem (Fig. 67). Regular 'target' cankers are comparatively seldom produced. Some of the cankers heal and show only as rough patches on the bark. Many, but not all, bear fruit bodies of *Dasyscyoha.* As the dieback progresses, epicormic shoots arise on the upper parts of the main stem, giving the young tree a curious appearance, sparce on the outside, but bushy at the centre of the crown. Recovery of such trees seldom occurs, but milder symptoms quite often disappear as the tree heals the cankers and replaces the dead twigs.

Hiley (1919) noted that many cankers have a dead twig or side branch in the centre. On this basis he put forward the hypothesis that infection took place by the spread of the fungus down the dead twig into the live stem. In this theory he was deceived by the abundant occurrence of the saprophyte, *D. calycina*, on dead twigs; in fact this fungus is unable to invade live tissue. The dead twig, however, may provide 'wounds', through which invasion by *D. willkommii* would be possible.

The theory that frost was the main causal agent of Larch canker was first put forward by W. R. Day (1931), who found abnormal tissue, typical of that known to be associated with frost injury, in the cankers. He also pointed out that in the spring the reawakening of cambial activity is a local process, and that the cambium at the base of a twig was known to become active before that on the rest of the stem. This would account for the fact, already noticed by Hiley, that most cankers had a twig or small branch in the centre. Day further noted that, once canker development has begun, the cambium is stimulated to earlier activity in that region than elsewhere, and therefore becomes liable to repetitive damage by frost in subsequent years. In addition it has been shown by the author (unpubl.) that the bark exerts a considerable protective influence during short periods of low temperature, such as are normal in damaging spring and autumn frosts, so that the exposed edges of the cambium in the canker are, for that reason also, liable to further injury. Day suggested, therefore, that frost was the real cause of Larch canker, and that *Dasyscypha* was, in the main, merely a colonist on the injured tissues. This view of the disease received general support from Münch (1936*a*) and Langner (1936), both of whose papers have been the subject of critical review and comment by W. R. Day (1937).

More recently the idea that frost was the real causal agent has been revived by Latour (1950), who considered, mainly on histological evidence, that *D. willkommii* was only capable of colonizing dead tissue.

Wound inoculations with the fungus were successfully carried out at a very early date by Hartig and others. These were repeated by Hahn in America and by Peace in Britain, and the results published by Hahn and Ayers (1943). These inoculations indicated clearly that *D. willkommii* was a parasite, that *D. calycina* was not, and that the former could successfully initiate cankers on *L. leptolepis*, *L. occidentalis*, *L. laricina*, and *L. gmelini*, as well as on *L. decidua*. The field inoculations carried out by Hahn were open to slight objection in that the trees used may have been subject to frost injury, while the greenhouse and frame inoculations carried out by Hahn and by Peace suffered from the fact that the conditions, while frost-free, were very artificial. In Hahn's inoculations on vigorous trees the development of the cankers was restricted, and extension soon ceased, but it is possible that he did not keep them under observation for a sufficiently long period. More recently, inoculations have been carried out by Manners (1953, 1957). Some of these were done on vigorous trees growing outdoors near Penzance in Cornwall, where it seems certain that frost injury could not occur. Manners reported continued development in some of his cankers.

Manners (1957) also found anatomical similarities between natural larch cankers, cankers caused by inoculation with the fungus, and cankers caused by local, artificial freezing of a portion of the stem. The most important difference was the lack of resin in frost-induced cankers. True frost rings extending round the stem away from the actual wound area were lacking in cankers due to inoculation with the fungus or to local freezing. Thus the frequent presence of such rings in cankered stems can certainly be taken as an indication that frost is involved, but the other anatomical abnormalities which occur may equally well be fungal in origin.

There appears, therefore, to be definite evidence that the fungus *D. willkommii* is a wound parasite, capable under some conditions of initiating and maintaining cankers. On the other hand, there is abundant evidence that frost is not only able to cause canker and dieback on larch, but actually does so in the field. The frequent occurrence of the disease in places known to be subject to radiation frosts in the spring is clear evidence of this. Instances have been observed, however, where the disease, starting in the bottom of a frosty valley, has spread up the slopes. This phenomenon can be explained on purely climatic grounds, for it is possible that when the larches on the slopes reach the thicket stage, they trap the cold air moving down the hillside among their branches, and are thus subjected to lower temperatures than earlier in their life, when their small size offered no impedance to air flow. On the other hand, it seems equally likely that *Dasyscypha*, having built up a substantial source of spores by infection of the frosted larches at the bottom of the valley, spreads up the slopes on to trees less injured or even uninjured by frost. Even if we assume, as inoculations suggest, that the fungus is a wound parasite, it is likely that insect-punctures, minor frost injuries, and even wounds made by bud-pecking or insect-hunting birds, would provide plenty of entries. The fact that the disease continues to progress in frost-free years provides additional evidence that the fungus is capable of acting as an independent agent. Thus, while further work is still required on the relationship between frost and fungus in the initiation and development of canker and dieback, both agencies clearly play an important part.

Distribution and damage

Very considerable areas of European larch, both in Britain and in other parts of northern Europe, have been virtually destroyed by this disease. Over still larger areas it has caused less devastating damage, producing slower growth and cankers on the main stem that may eventually lower the value of the timber.

The fungus *Dasyscypha willkommii*, and the disease with which it and frost injury are associated, occur in all parts of the British Isles and over most of Europe where larch is grown. The fungus was imported into the United States of American on larch plants and became locally established (Spaulding and Siggers 1927). Attempts were made to eradicate the fungus, but in 1953, when a re-survey was carried out, it was found still in existence on a relatively small number of trees (Fowler 1953; Fowler and Aldrich 1953). In Britain dieback and canker are much more liable to occur to a damaging extent in frosty sites, such as valley bottoms, shelves on hill-sides, and the like, but some signs of the disease are likely to be found in almost any plantation of European larch.

The disease is mainly one of European larch, though there are wide differences in susceptibility between different provenances. The fungus has been found associated with damage on Japanese larch on several occasions. Canker and dieback have also been recorded occasionally on all the species of larch grown in Britain, but some of the damage may, of course, have been due solely to frost injury. Spaulding and Siggers (1927) found the disease occurring naturally on *L. laricina* in America, while Hahn and Ayers (1943) were able successfully to inoculate, not only this species, but *L. occidentalis*, the

Western American larch, as well as *L. leptolepis* and *L. gmelini* from Asia. However, their inoculations gave no reliable information on the relative susceptibility of these species, and there is little evidence from field observations, except that Japanese larch is obviously much less susceptible than European. Whether this is due to resistance to frost or resistance to the fungus on the part of Japanese larch is not known. Schober (1958), who considered Japanese larch immune, found that *L. gmelini* and its varieties were generally resistant; *L. sibirica* was particularly susceptible.

Perhaps the most important point about Larch canker and dieback is the wide variation between the susceptibility of different European larch provenances. At the end of the last century Michie (1882) asked a number of nurserymen their views on the use of home-collected as against Alpine seed. The verdict was almost unanimously in favour of home-collected, mainly because the resulting plants were noticeably more frost resistant. Unfortunately little further notice was taken of this distinction, and continental seed continued to be imported in quantity. Throughout the nineteen-thirties it was increasingly realized that most of the severely attacked stands were of High Alpine, and many of the healthiest of Scottish, origin.

In the severe spring frosts of May 1935 (Day and Peace 1946) several instances were noted in which larch of Scottish origin was markedly less damaged than that of Swiss or Silesian origin. This matter was put on an experimental basis by W. R. Day (1951), who found a very big variation in frost susceptibility between larch of various European origins in artificial freezing experiments, markedly the best being one seed lot of Sudeten larch. The one Scottish provenance included in his experiments showed only average frost resistance. He later found that there were considerable variations between the frost resistance of different Scottish provenances (Day, W. R. 1958*b*). This experiment suggested that there may be strains in the Alps quite suitable for growing in Britain. But, apart from a strong suggestion that, allowing for elevation, the more northerly latitudes were best, the results gave no clear guide as to means by which these suitable races could be detected. The observations and experiments noted above, of course, refer solely to frost injury. Considering the disease as a whole, recent analysis of some of the older European larch provenance experiments (Edwards 1957) has given further evidence that Alpine origins are worse attacked than Scottish, while some Sudeten origins were also resistant. More recent international provenance experiments, in which a larger number of origins are included, are still too young for the disease to have reached its full development, but such results as have been recorded give general support to the superiority of Scottish and some strains of Sudeten larch. These experiments also tend to show that larch of Polish origin is resistant to canker and dieback, a result that was suspected already from the healthy growth of most of the small number of plantations of this strain that exist in Britain. Schober (1958) also found that provenances from Poland showed low susceptibility, but his results with Sudeten larch were inconclusive, since it was resistant on one site but susceptible on others.

It might be assumed from the observations above that differences in susceptibility among seed origins of European larch were purely a question of frost susceptibility, though of course many of the field observations refer to the disease as a whole, rather than directly to frost. Manners (1957), however, working on a small number of origins, found that larch from Munsterthal in Switzerland was more affected by *Dasyscypha* in the absence of frost than Scottish or Polish larch.

The undoubted disease resistance of most origins of larch raised from seed collected in Scotland is not entirely easy to explain. J. Macdonald (1949) examined many of the older records of larch planting in Scotland, and there is clear evidence that as early as 1848 canker and dieback were quite widespread in larch plantings. At that time much of the seed used was still imported from Europe. Thirty years later Michie (1882) reported the extensive use of Alpine seed.

The explanation appears to lie in the elimination of unsuitable provenances and individuals by the disease itself. The fairly common Scottish method of planting larch in mixture with Scots pine, which originated because of the high price of the larch plants, but still persists as an accepted silvicultural practice, may have tended to make the elimination of diseased larch more rapid. Healthy larch would outgrow the pine and survive, but diseased ones would be suppressed, and might well be forgotten when the plantation had become one of pure Scots pine. The stands from which seed is now being collected may be assumed, therefore, to be the disease-resistant survivors of the various origins originally introduced into Scotland. Certainly they provide the best and most certain sources for future seed collection.

Control

In view of our ignorance of the relative importance of frost and *Dasyscypha* in the disease, it is impossible to give definite recommendations for control. It has often been suggested in the past that the disease can be reduced by heavy thinning. In some instances this does seem to have a markedly beneficial effect. Apart from any thinning treatment, larch seem to suffer progressively less from the disease once they are more than twenty to twenty-five years old. This may mean that they are getting above the cold air layer and are less damaged by frost, or that conditions for the development of the fungus are not so favourable on taller trees. Thus the normal improvement with age may be attributed wrongly to thinning treatment. Thinning will also tend to remove the most unhealthy trees, which may lessen the quantity of *Dasyscypha* spores available to infect the remaining trees as well as giving an immediate improvement in the appearance of the plantation. Finally, larch is now known to respond to heavy thinning, so that opening up a plantation may improve the health of the trees by making them better able to grow away from the disease (Hopp 1957).

Nevertheless thinning, with particular attention to the removal of the worst diseased trees and the encouragement of the most vigorous, is a wholly proper operation in a slightly or moderately attacked stand. Thinning is unlikely to produce a satisfactory recovery from severe attacks, however, and in such cases it may be necessary either to clear-fell and replant, or if a scattering of healthy trees exists, to thin heavily and underplant with a shade-bearing species.

On the Continent the use of mixtures has also been recommended as a means of restricting the disease; though Hopp (1957) found that a high proportion of Norway spruce in mixture with larch increased the susceptibility of the latter. Here again the underlying reasons are probably complex. The second species may act as a nurse to the larch, protecting it to some extent from frost injury. The separation of the larch one from the other may lessen the ability of *Dasyscypha* spores to spread from tree to tree. The presence of a second species makes the removal of diseased larch an easier operation. All these factors may combine to improve the health of the remaining larch. Certainly examples of healthy European larch raised in mixture with other species are frequently found.

For new planting, however, the best means of avoiding the disease is by the use of species other than European larch on really frosty sites and by the selection of a good strain of larch for other places. Most Scottish origins may be expected to remain reasonably canker-free, except on seriously frosty sites.

Twig dieback and basal canker of Japanese larch

Pathogen

Here again we have a complex problem in which frost and a fungus may both play a part. The fungus in this instance is *Phomopsis* (*Phacidiopycnis*) *pseudotsugae* Wilson (Fungi Imperfecti, Sphaeropsidales), which occurs more commonly on Douglas fir, and is therefore described in Chapter 25.

Symptoms and development

The first report of this fungus on Japanese larch was made by M. Wilson (1921*a*). Later (1925) he recorded further occurrences, always on cankers on the main stem. He noted that the cankers on larch were often considerably larger than those associated with the same fungus on Douglas fir, being as much as 25 cm. long and 12·5 cm. wide. Twig dieback was attributed to this fungus by Robak (1946).

Frost damage on Japanese larch is known to be of quite common occurrence, and can be serious (Day and Peace 1946), but it was not singled out for special attention until W. R. Day (1950*a*) described two separate forms of frost injury, dieback of the crown and canker of the main stem. He pointed out differences between the symptoms on Japanese larch and those associated with canker and dieback on European larch. The cankers on the former developed on the main stem after canopy had been formed. Apart from this the dieback took place from the ends of twigs and usually in the top of the tree, in contrast to European larch, where dieback is more at the base of the tree, and takes place as a result of girdling of the branches by injuries, which occur at some distance from the tip. These differences were attributed to autumn frost being the main cause on Japanese larch, as contrasted with spring frost in the case of European. There was supporting evidence, such as the occurrence of multiple buds, for frost being involved in this injury; *Phomopsis pseudotsugae*, if it occurred, was not recorded. Day thus described as due to frost the two types of injury with which we are here concerned.

The supposed relationship with *Phomopsis* was put on a more definite basis by Vloten (1952), who found that infection by the fungus followed brashing in a larch plantation. Infection only took place if the operation was carried out during the winter months, the less the activity of the tree the greater the possibility of infection. He also found that the amount of infection depended partly on the method of brashing, saw brashing resulting in rather less infection than when the branches were knocked off. Bruising the bark with a hammer, introduced in the experiments as a control, almost invariably led to the entry of the fungus. It thus seemed possible that crushing of the bark beneath the brashed stem, as it was knocked downwards, accounted for the more serious effect on stick-brashed trees, as compared with saw-brashed ones. The source of inoculum was thought to be *Phomopsis* fruiting abundantly on dead twigs in the crowns. The fungus, having infected the main stems, spread only during the winter of entry, thereafter the cankers healed.

The occurrence in Britain of main-stem cankers associated with *Phomopsis* has been recorded above. The fungus has not often been noticed on twigs, but it has been found on a few occasions in places where frost injury to the twigs seemed unlikely. In a frosty site it would probably be impossible to separate the two agencies, especially since *Phomopsis* would probably colonize frost-killed twigs.

Recent occurrences of severe main stem cankering on Japanese larch cannot as yet be attributed with certainty either to the fungus or to frost. In some instances very extensive cankers on the lower main stem of Japanese larches have only been disclosed when the bark covering them started to crack, exposing large dead areas, partially healed, and very often with a brashing wound in the centre (Plate XIII). Normally all the damage took place at one time, immediately following brashing. It is tempting, in view of Vloten's work, to attribute these cankers to the fungus. Unfortunately, because of their delayed discovery, all traces of the fungus, even if it was present immediately following injury, would have disappeared. The sites where this type of damage has occurred do not appear to be particularly frosty, nor has there been subsequent extension of damage, such as normally occurs with frost cankers on conifers. It is possible to make a case for frost based on the stimulation of the cambium to early activity as a result of the brashing wounds, and on alteration of the local climate following the removal of the lower branches. It is equally possible, however, that *P. pseudotsugae* is the real cause. Earlier discovery of cankers of this type should settle the question.

Distribution and damage

Twig dieback is so comparatively unimportant, and so easily confused with frost injury, that no information on its distribution exists. Basal canker has been found in a number of localities scattered over Britain. However, its frequency of occurrence, in relation to the area of Japanese larch brashed each winter, is very small indeed. There must be some other limiting factor on infection by the fungus besides the season of brashing. This can hardly be the distribution of the fungus, which is of general occurrence. Without this unknown limiting factor, the disease could be very serious. At present it has no effect on the cultivation of Japanese larch.

Somewhat similar injuries have been recorded on other conifers, which, in view of the rather wide host range of *P. pseudotsugae* and other species of the genus, may well be found to be associated with *Phomopsis* species.

Control

The present frequency of the disease does not justify a restriction on winter brashing. If it were to become commoner, summer brashing between the months of April and August would certainly be desirable in the neighbourhood of plantations already infected. On the basis of Vloten's work the disease also provides a slight argument in favour of the use of saws for brashing larch, as against rods or billhooks. The latter, although theoretically a cutting instrument, often break out larch branches instead of cutting them, and in any case is more likely to bruise the bark below the branch than a saw.

DISEASES OF LEAVES AND SHOOTS

Needle-cast

Pathogen

Leaf-cast of larch, caused by the fungus *Meria laricis* Vuill. (Fungi Imperfecti, Moniliales) is purely a needle disease; the young shoots are not attacked. The fungus has been fully described by Peace and Holmes (1933) and more briefly in a Forestry Commission leaflet (Anon. 1955*a*). It produces small clusters of conidia on the lower, and occasionally on the upper, surface of larch needles. These spore clusters emerge through the stomata, and therefore occur in parallel rows along the needles. Through a high-power lens they can be seen as white dots, but confusion with the white, waxy surfaces of the stomata is very possible, and the only safe diagnosis involves microscopical examination. This can be done either by selective staining, when the spore clusters will stand out as dark spots, or by scraping off the fruit-bodies and observing under the microscope the characteristic oval spores, constricted in the centre. Biggs (1958) found four strains of the fungus differing in cultural characteristics, stable in culture, but unstable in the host. They do not seem to have any pathological significance.

Symptoms and development

The first sign of the presence of the fungus is a yellowing and wilting of the needles at the base of the shoot. This is usually seen in early May. The attacked needles soon go brown and wither (Fig. 68). Damage due to *Meria* is occasionally confused with frost injury or in the autumn with injury due to *Botrytis*. In fact they are easy to distinguish. *Meria* damage takes some time to develop after infection so that needles at the tip, though possibly infected, are always green, whereas frost and *Botrytis* both quickly wither the terminal needles. *Meria* normally infects the middle or end of the needle and spreads down it, so that the bases of *Meria*-attacked needles may still be green, while the tips are withered and brown. Needles damaged by frost or *Botrytis* usually quickly and completely wither. Finally *Meria* does not infect the shoot, so that after the needles are killed, an abscission layer is formed, and they are eventually shed. With frost and *Botrytis* the shoot is killed as well as the needles, which remain attached in a withered condition well into the winter.

Damp weather is required for successful infection, so that the heaviest attacks are to be expected in wet seasons. The fungus over-winters on fallen needles and, in nurseries, on the terminal tufts of needles which remain attached to the young plants. For this reason the disease is usually more serious on plants which have remained two seasons in the same place in the

FIG. 68. *Meria laricis* on a European larch transplant, Millbuie Forest, Ross and Cromarty, August ($\times \frac{3}{4}$). (J. C.)

nursery. If the weather is dry in the spring, infection may be delayed, so that the first-produced needles escape infection. In that case the disease will first appear in the middle, rather than at the base of the shoot (Fig. 68). There is no evidence that the initial health of the plant has any effect on *Meria* attack. A weak plant is likely to suffer more from defoliation than a strong one, but this is a question of recovery rather than of susceptibility.

Distribution and damage

The disease appears to be generally distributed over Britain, both in nurseries and in plantations. It is therefore unlikely that a new nursery would remain free from the disease for any length of time, even if all the larch in it was raised from seed. It has been recorded in various parts of Europe, though seldom as a seriously damaging disease. In the United States it has so far only been found in Idaho on *L. occidentalis* (Shearer and Mielke 1958).

It was originally regarded as a disease of European larch. Peace and Holmes (1933) found it naturally only on *L. decidua*, and on one needle of a Hybrid larch. In their investigations it was never found on Japanese larch. Inoculation experiments, using methods which gave a heavy infection of *L. decidua*, produced equally heavy infection on *L. occidentalis*, one infected needle on *L. leptolepis*, and no infection at all on *L. gmelini* and *L. sibirica*. In Britain therefore *Meria* came to be regarded purely as a disease of European larch.

However, Robak (1946) recorded the fungus in Norway, not only on European larch, but also on *L. sibirica* and *L. leptolepis*, damage to the former being appreciable. A year later in Sweden, Sylvén (1947), reporting on an international provenance trial which was subject to severe and persistent attack by *Meria laricis*, mentioned that both the origins of *L. leptolepis* used in the experiment were completely resistant to the disease, and that many resistant individuals occurred among *L. decidua*; these he regarded as hybrids between the two species. Schönhar (1958) found *L. decidua* and *L. occidentalis* susceptible, but regarded Japanese and Hybrid larch as highly resistant.

In 1954 *Meria laricis* was found on Japanese larch in four nurseries in Britain, two in south Scotland, one in Wales, and one in Cornwall (Batko 1956). In two of these nurseries Hybrid larch was also infected. There was no common source of Japanese or Hybrid larch plants to these nurseries. It was therefore impossible to explain the widespread appearance of the disease on varieties hitherto regarded as resistant as being due to the distribution of a new strain of the fungus from a common centre.

Since 1954 *Meria* has been found frequently, though not invariably, on Japanese and Hybrid, as well as on European, larch in nurseries. It has also been found on a number of occasions causing easily visible and occasionally severe injury on young plantation Japanese larch. In nurseries, however, the attack is always less severe on Japanese and Hybrid than on European larch. There is normally little difference between individual plants of the same species.

Biggs (1958) found that following simultaneous infection the symptoms appear first on European, then on Hybrid, and finally on Japanese larch. The same successional appearance on the three larches has also been noted under natural conditions in nurseries. Cross-inoculation experiments do not indicate any difference in pathogenicity in the fungus isolated from different species. The fungus from one host can infect both the others. There is thus no easy explanation for the apparent absence of *Meria* from Japanese and Hybrid larch in the past and its recent limited occurrence thereon.

Leaf-cast is most important as a nursery disease, and in seasons favourable to the disease very serious defoliation may result. With small nursery stock a few of the weakest plants may die as a result, but the main effect is a reduction in growth, with a consequent higher percentage of cull plants. The fungus is more serious, therefore, in a poor nursery, where in any case the plants are small, than in a fertile one, where the main effect will be a reduction in the size of already comparatively large plants.

However, the fungus occurs commonly on plantation European larch, and sometimes on Japanese and Hybrid larch as well. The defoliation caused in plantations is seldom significant, though severe attacks have been recorded, but the number of spores produced makes an infected plantation a dangerous neighbour to a larch-producing nursery.

In view of the comparative ease of control, the disease on European larch, though a nuisance, has not so far proved really serious. There is certainly no reason at the moment to regard *Meria* in any way as a threat to the continued cultivation of any of the larches.

Control

In the nursery the disease can be kept under control by the use of suitable sulphur sprays. The range of other materials tried has been very small, but copper sprays appear to be ineffective (Peace 1936*b*). Schönhar (1958) got good control with several proprietary fungicides. Two recognizable stages in the growth of larch buds in the spring should be noted in connexion with spraying. In January or February the buds begin to swell and show green at the tips, but no further development takes place for several weeks, after which the needles begin to elongate. So long as the buds are in the first stage, a comparatively strong solution (referred to as the 'winter strength') can be applied, but after the needles have begun to elongate more dilute sprays ('summer strength') should be used. Liver of sulphur at 14 lb. per 100 gallons of water for winter strength and 7 lb. per 100 gallons for summer strength, both applied at the rate of 6 gallons per 100 square yards, is a quite satisfactory material to use. But some of the proprietary polysulphides are probably less liable to damage the foliage, in which case the winter strength should be that recommended by the makers for a winter fruit-tree wash, and the summer strength that for use on potatoes or tomatoes. Colloidal sulphur is equally useful, and in that case the summer strength should be equal to, or slightly higher than, that recommended for horticultural crops, and the winter strength three times the summer one.

The first spraying should be given at the end of February in southern and western counties, and a week to a fortnight later in the north. This will be long before the active appearance of the disease, and is designed to deal with the fungus on fallen needles and on those still attached to the plant before new infection has taken place. It should be followed by summer strength sprayings from the end of March and thereafter at intervals of two to three weeks. In settled dry weather they can be discontinued, and in any case they should cease entirely early in August. To avoid scorching, spraying should never be done immediately after cover has been removed from the beds, nor following a heavy weeding. The plants should be given a week to harden after their sudden exposure. Sunny days should also be avoided.

The survival of *Meria* on needles attached to the plants, and particularly on fallen needles on the ground, makes it evident that plants remaining in the same place for more than one year are very liable to heavy reinfection in the second summer. Indeed serious attacks are commonest on two-year seed-beds and on transplant lines left two years in the same place. Every effort should be made, therefore, to raise seedlings large enough to transplant at one year and to plant in the forest at the end of two, thus avoiding two years in the same place in the nursery. This suggests that it would be good practice to restrict the raising of larch, as far as possible, to nurseries sufficiently fertile to carry out this programme.

On the other hand, the more larch that is being raised in a nursery, the more frequently it will occupy the same pieces of ground, and the greater therefore the possibility of the carry-over of infection. A balance has to be struck between the desirability of restricting larch to those nurseries where its satisfactory growth permits yearly transplanting, and the undesirability of having so much larch in the one nursery that a proper rotation of the crop becomes impossible.

Other needle diseases

A number of other fungi have been associated with leaf-cast of larch. *Hypodermella laricis* Tub. (Ascomycetes, Phacidiales) occurs both in America and Europe, attacking shoots as well as needles. The damage is striking because the dead shoots with the withered needles remain attached to the tree up to and sometimes during the winter; but the actual injury to the tree is not usually regarded as serious. However, death of natural reproduction has been recorded (Shearer and Mielke 1958). This disease is clearly distinguishable from *Meria*, firstly because the attacked needles remain attached, and secondly because the black apothecia, which are clearly visible to the naked eye, appear in a single line on the upper side of the needle. Although it occurs in some parts of Europe from which our larch originated, the fungus does not seem to have reached Britain. The full host range has not been investigated, but in Europe it occurs on *L. decidua*, and in America on both *L. laricina* and *L. occidentalis*.

On the Continent *Mycosphaerella* (*Sphaerella*) *laricina* Hartig (Ascomycetes, Sphaeriales) is also associated with quite serious defoliation of European and Japanese larch. A closely related species, *M. larici-leptolepis* Itô, Satô, and Ôta, caused similar damage on *L. leptolepis* in Japan (Itô, Satô, and Ôta 1957), but no species of *Mycosphaerella* has been reported on larch in North America. The needles go brown in patches during the summer, and, while they are still attached to the tree, conidia are borne in small black pustules. The perithecia, also small and black, are produced on the fallen needles in the following spring. Unlike *Hypodermella laricis* this fungus does not attack the shoots. It also appears to be absent from Britain.

Interesting suggestions have been made on the effect of mixed plantings on this disease. When larch is mixed with spruce, the larch needles are said to hang on the lower branches of the spruce throughout the winter, and are therefore in a better position to reinfect the larch when the perithecia are produced on them in the spring. In pure stands the needles fall to the ground, but can still act as sources of infection. On the other hand, when larch is mixed with beech, the leaves of the latter, falling later than the larch needles, cover them on the ground and prevent the spring reinfection (Hartig 1900).

M. laricina is often confused with *Lophodermium laricinum* Duby (Ascomycetes, Phacidiales). The conidial stage of this fungus, *Leptostroma laricinum* Fckl. (Fungi Imperfecti, Sphaeropsidales), is said to produce its small shining black pycnidia while the needles are still on the tree; but they also occur on fallen needles. They are followed the next spring by elongated black apothecia. This fungus has been recorded in Britain on fallen needles of *L. decidua*, but never as a cause of disease. It is present on the Continent and in North America, but it is doubtful whether it is anywhere of economic importance.

All three of these fungi can easily be distinguished from *Meria laricis* by their black fruit-bodies, visible to the naked eye. It is doubtful whether any of them constitutes a threat to larch in Britain; but it would be a pity if either of the first two were imported, in view of their occasional virulence abroad.

Needle rusts

A number of rust fungi (Basidiomycetes, Uredinales) have larch needles as their aecidial host. The British species are listed below, with their hosts as far as they are known:

	AECIAL STAGES ON	UREDIAL AND TELIAL STAGES ON
Melampsora larici-caprearum Kleb.	*L. decidua* and *L. leptolepis*	*Salix* spp.
Melampsora epitea Thuem.	*L. decidua*	*Salix* spp.
Melampsora larici-pentandrae Kleb.	*L. decidua*	*Salix pentandra*
Melampsora tremulae Tul.	*L. decidua*	*Populus* spp.
Melampsora larici-populina Kleb.	*L. decidua* and *L. leptolepis*	*Populus* spp.
Melampsoridium betulinum (Desm.) Kleb.	*L. decidua* and *L. leptolepis*	*Betula* spp.

In all cases the fungi are much more important on their broadleaved hosts, and are considered, therefore, in more detail in other chapters. The damage to larch is quite negligible, indeed it is often hard to find the fungus on the needles even in the neighbourhood of heavy infection on the alternate host. It is suspected with some species of *Melampsora*, and certainly with *Melampsoridium betulinum*, that the fungus can over-winter on the broadleaved host without recourse to larch. However, there is evidence that neighbouring larch plantations can seriously increase the severity of *Melampsora* attack on poplars (Chapter 32). The list of aecial hosts given above should be regarded as tentative. No doubt in due course some of the other *Melampsora* species will be found on *L. leptolepis*, and possibly on other larch species, in Britain.

The aecia of the *Melampsora* species on larch are indistinguishable by eye. They appear as small orange spots on the needles and are easily overlooked. The aecia of *Melampsoridium*, on the other hand, have each orange spore-mass surrounded by a conspicuous white peridium, and are therefore much more noticeable.

Melampsoridium hiratsukanum Hiratsuka has been found in Britain producing its uredial and telial stages on *Alnus* (Chapter 34), but the aecial stage, which is known to occur on *L. decidua* and *L. leptolepis* in Japan, has not been recorded here.

The two larch rusts which occur in America are quite unimportant on their coniferous hosts. *Melampsora bigelowii* Thuem. has its aecial stage on *L. decidua*, as well as on American *Larix* species, and its uredial and telial stages on *Salix* species. *M. medusae* Thuem. has its aecial stage on *L. laricina* and *L. occidentalis* and its other stages on *Populus* species.

25

DISEASES OF DOUGLAS FIR

ONLY one species of Douglas fir, *Pseudotsuga taxifolia* Brit., is of any importance in Britain. *P. taxifolia* has at various times been divided to make two or even more species and numerous varieties, but the distinctions on which these have been based break down if a large number of trees are examined. It is best to regard *P. taxifolia* as a complex of variables between two extreme types, approximating to the normal conceptions of *P. taxifolia*, the Coastal Douglas fir, and *P. glauca* Schneid., the Colorado Douglas fir (Peace 1948). In British plantations, Douglas fir from the interior of British Columbia has often been separated as Intermediate Douglas fir, under the name *P. taxifolia caesia* Asch. & Graebn. Some authors include northern Idaho and Montana in the area occupied by the Intermediate variety, and the map given by Vloten (1932) shows this distribution. These three regional divisions, though they are untenable on taxonomic grounds, have considerable importance when considering variations in resistance to needle-cast fungi, particularly to *Rhabdocline pseudotsugae*. In the author's opinion it is best to regard Douglas fir coming from the seaward side of the coastal mountain ranges in British Columbia, Oregon, Washington, and California as the Coastal variety, trees from the interior of British Columbia as the Intermediate variety, and trees from the Rocky Mountain and other interior mountain ranges in the United States, as far south as Mexico, as the Colorado variety.

DAMAGE DUE TO NON-LIVING AGENCIES

Douglas fir is particularly susceptible to losses following transplanting in the nursery and particularly following planting in the forest. These are liable to occur if the soil into which the trees have been moved is wet and cold, so that the initiation of new root growth is delayed.

NURSERY DISEASES

Douglas fir is moderately susceptible to, and rather frequently attacked by, Grey mould, *Botrytis cinerea* (Chapter 15). *Diplodia pinea* (Desm.) Kickx. (Fungi Imperfecti, Sphaeropsidales), chiefly important as a cause of dieback in pines (Chapter 22), has also been recorded causing a 'blight' of Douglas fir seedlings in the United States (Slagg and Wright 1943). Withering of the needles of seedlings has also been attributed to *Rosellinia herpotrichoides* Hept. and Dav. (Ascomycetes, Sphaeriales) (Salisbury and Long 1955) (Chapter 15). *Phomopsis pseudotsugae*, which frequently occurs in nurseries, is also commonly found in plantations of Douglas fir, and is therefore dealt with later.

ROOT DISEASES

Douglas fir is of particular interest because it shows more than average resistance to *Armillaria mellea* and in some ways to *Fomes annosus*, the two most important plantation root diseases in Britain. Though resistant to stem decay by *Fomes*, Douglas fir is susceptible to root-rot caused by that fungus, and the resultant instability leads to windthrow. In addition Douglas fir shows more than average susceptibility to *Polyporus schweinitzii*, another important root pathogen. Group dying, caused by *Rhizina inflata*, has never been found in Douglas fir (Chapter 16).

DAMAGE BY FLOWERING PLANTS

Douglas fir is the tree most seriously affected by the Dwarf mistletoes, *Arceuthobium* species, in North America (Chapter 13). Fortunately these parasites are not present in Britain.

STEM DISEASES—BARK AND CAMBIUM

Phomopsis dieback

Pathogen

The fungus chiefly concerned with this disease is *Phomopsis pseudotsugae* Wilson (Fungi Imperfecti, Sphaeropsidales), recently transferred by Hahn (1957*a*) to another genus as *Phacidiopycnis pseudotsugae* (Wils.) Hahn. Robak (1952*b*) found that this fungus was identical with *Discula pinicola* (Naum.) Petr., which causes staining in the felled wood of spruce and other conifers. It was thought until recently that *Phomopsis pseudotsugae* was probably the imperfect stage of a *Diaporthe* (Ascomycetes, Sphaeriales), but Hahn (1957*a*) has finally found the perfect stage on living and dead trunks and branches of *Pinus strobus* in North America, and named it *Phacidiella coniferarum* Hahn (Ascomycetes, Phacidiales).

The fructifications of both stages are borne on dead twigs or on dead patches of bark on live twigs. Those of the perfect stage, when they occur, are saucer-like, black, and borne in clusters on the dead twigs, often mixed with those of the imperfect stage. They are 0·25–1 mm. in diameter. The black pycnidia of the *Phomopsis* stage project through elongated or oval slits in the bark, sometimes singly, sometimes two or three together. The individual clusters of pycnidia are usually thickly distributed on the twigs, often only 2–3 mm. apart.

A number of other species of *Phomopsis* have been recorded in Britain on Douglas fir. *P. conorum* (Sacc.) Died., which macroscopically is easily confused with *P. pseudotsugae*, but which is easily separated by examination of the spores, is a saprophyte on dead twigs, but according to Hahn (1928) may occasionally be weakly parasitic. Confusion is also possible with *P. occulta* Trav., the perfect stage of which is *Diaporthe conorum* (Desm.) Niessl., which Wehmeyer (1933) regarded as a form of the complex genus *D. eres* Nits. According to Hahn (1930) *P. conorum* frequently occurs on dead twigs of Douglas fir.

Neither of the species mentioned above appears to be of any pathological importance. In the United States, however, *P. lokoyae* Hahn has been associated with active cankers on young Douglas fir (Hahn 1933).

Symptoms and development

The symptoms associated with *P. pseudotsugae* have been described by M. Wilson (1925) and in a Forestry Commission leaflet (Anon. 1948*c*). Three

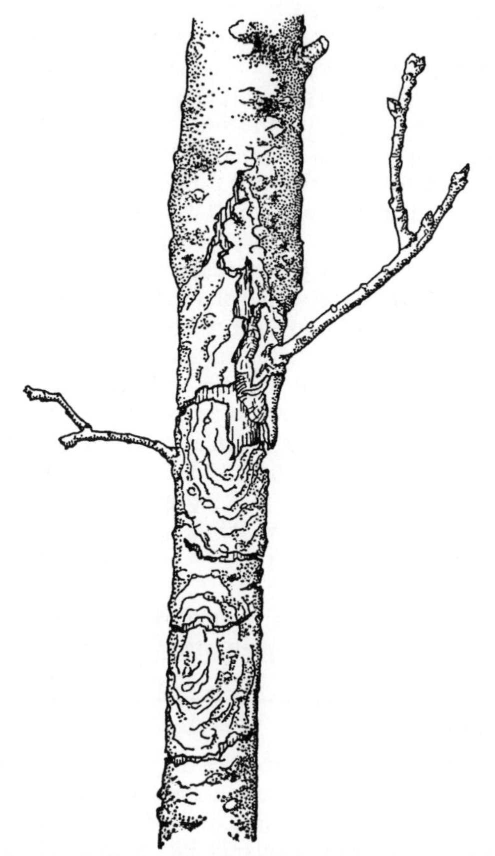

FIG. 69. Lesion on Douglas fir associated with *Phomopsis pseudotsugae* ($\times \frac{2}{3}$). (J. N.)

types of damage can be distinguished. Firstly, leading and lateral shoots may die back for as much as 30 cm., the foliage yellowing, browning, and falling fairly rapidly. Diameter growth usually continues immediately below the dead portion, so that an abrupt reduction in twig diameter from live to dead tissue is characteristic. Secondly, infection of wounds or side twigs on young stems may result in girdling. Again the difference in stem diameter is striking; in particular the diameter of the live twig above the canker (Fig. 69), until it eventually dies from lack of root-absorbed nutrients, is markedly greater than that of the girdled length below it. Finally, on larger stems, the fungus may be found fruiting on elongated sunken cankers, which may be up to 15 cm.

long and 7·5 cm. wide. In all cases the marked contrast in level between the dead part and the live tissue immediately adjoining it must be regarded as due more to the ability of the host tree to continue growing right up to the edge of the dead parts than to any characteristic of the fungus. It is therefore not a safe diagnostic feature. Nor can the presence of *Phomopsis* fruit-bodies be taken as evidence that the fungus is the cause of the death of the bark on which it is fruiting, for it is a common and conspicuous saprophyte. Only the culturing of the fungus from the early stages of damage can definitely establish its status as a parasite in any particular case.

There is no doubt that it frequently follows other primary causes of damage, often no doubt as a weak parasite. In particular, it is associated with frost injury, but there is no doubt that the sunken cankers and girdling and, in particular, the dead tops of lesser diameter than the live tissue below them, can be caused by frost, unassisted by the fungus. It has been mentioned above that Douglas fir is particularly liable to root death when moved into cold wet soil. Root injury of this nature, if not fatal, often leads to dieback of the leading shoots, and may easily be confused with *Phomopsis* injury, though owing to the weak growth of the whole plant the contrast in this case between the diameter of live and dead tissue is not so striking.

When the fungus is truly parasitic, its spread is restricted to the winter months (Wilson, M. 1925), but it appears unlikely that it is always limited to a single winter, though this was found to be the case by Vloten (1952) in his infection experiments on Japanese larch. There is no doubt that the fungus can be an active wound parasite on Douglas fir and other conifers. Wound inoculations by Hahn and Wilson in Scotland (Wilson, M. 1937*b*) and by Hahn (1957*b*) in America make this abundantly clear. It is very difficult, however, in the case of this fungus to tell when it is acting in a primary and when in a secondary capacity.

The damage caused by *P. lokoyae* in north-western America resembles almost exactly that described for *P. pseudotsugae* (Boyce, J. S. 1933; Thomas, G. P. 1950).

Distribution and damage

P. pseudotsugae has now been recorded in many parts of Europe. It has also been described under that name on Douglas fir in the eastern United States; but as *P. strobi*, now considered identical, it has a much wider range in North America and in New Zealand, but on pine (Chapter 22), not on Douglas fir. *P. lokoyae* appears to replace *P. pseudotsugae* on *Pseudotsuga* in its native range.

There is no evidence for any variation in resistance to the fungus between Douglas fir of different provenances or varieties. Indeed such variations would hardly be expected for a fungus with so wide a coniferous host range outside the genus *Pseudotsuga*. Apart from its occurrence on *Larix leptolepis*, where it has some pathological importance, and as *Phomopsis strobi* on *Pinus strobus* and other pines, it has also been recorded on *L. decidua*, and on species of *Abies*, *Cedrus*, *Sequoia*, and *Tsuga* (Hahn 1957*b*). It is unlikely that this represents its full host range, though on many species it may be only a very weak parasite.

In view of its frequent occurrence in association with damage due primarily to other agencies, it is impossible to evaluate its real importance as a cause of disease. There is certainly no evidence that it has initiated any epidemic attacks either in Europe or in America. It is regarded in Britain with much less apprehension now than it was about 1930. *P. lokoyae* probably occupies a similar position in the native Douglas fir stands, for it has attracted only occasional notice.

Control

There is little need for any direct control against *Phomopsis*. Measures suggested in the past (Wilson, M. 1925; Anon. 1948*c*) for the removal and burning of infected plants are unnecessary, since the fungus is so commonly present on dead Douglas fir twigs. Most of the precautions that should be taken, such as the avoidance of frosty sites and of cold wet soils, relate more to other primary damaging factors than to *Phomopsis*. The avoidance of wounds, both in the nursery and in plantations, is an obvious precaution, for *Phomopsis* may well invade and enlarge them.

Resin bleeding and similar diseases

In several places in Britain, Douglas fir has suffered from a disease characterized by profuse resin exudation arising from lesions in the main stem and at the base of the branches. The lesions occur particularly at the base of the live crown. They may appear first as dead, flat patches on the bark, or as cracks (Fig. 70). The latter are particularly frequent on the basal 10 centimetres of the branches, and could be caused by the weight of snow on them. The bleeding is accompanied by a gradual decline in the health and vigour of the tree, the lesions often extending and the outcome in some cases being death. Somewhat similar bleeding has been found on other conifers, notably Sitka spruce and *Pinus contorta*, but the effect on the health of the tree is much more pronounced in the case of Douglas fir.

Several fungi have been consistently isolated from the lesions, in particular *Pycnidiella* (*Zythia*) *resinae* Fr. (Fungi Imperfecti, Sphaeropsidales), which is also found in its perfect stage, *Biatorella resinae* (Fr.) Mudd. (Ascomycetes, Helotiales), *Myxosporium abietinum* Rostr. (Fungi Imperfecti, Melanconiales), which is a common saprophyte on conifers (Chapter 18), and *Dasyscypha oblongospora* Hahn & Ayers (Ascomycetes, Helotiales). There is no evidence that any of these is parasitic. N. O. Howard (1929) associated a species of *Dasyscypha* with lesions and profuse resin bleeding on Douglas fir, but in his case, injury was confined to the lower 3 to 5 metres. Several species of *Dasyscypha* have since been associated with lesions on Douglas fir in North America. These include *D. oblongospora* (Hahn and Ayers 1934*a*), *D. ellisiana* (Rehm.) Sacc. (Hahn and Ayers 1934*b*), and *D. pseudotsugae* Hahn (Hahn 1940).

In the United States Pitch girdle, a disease of pole-stage Douglas fir, which has some resemblance to British Resin bleeding, is now known to be caused by *Cytospora kunzei* Weh. (Fungi Imperfecti, Sphaeropsidales). This fungus is of more importance on spruce and is therefore described in greater detail in

Chapter 23. On Douglas fir the fungus causes more definite cankers than Resin-bleeding, but the exudation does extend for 6 to 7 metres up the tree. Roeser (1929), who originally described the disease, associated it with overcrowding, but it was left to Wright (1957*b*) to prove a definite association with *C. kunzei* and to suggest that difficult soil conditions leading to periodic drought were a primary factor of considerable importance. The disease still appears to be of very local occurrence.

In Britain Resin bleeding first attracted attention in 1950, and although it is now known to occur at several places in the southern half of England, it

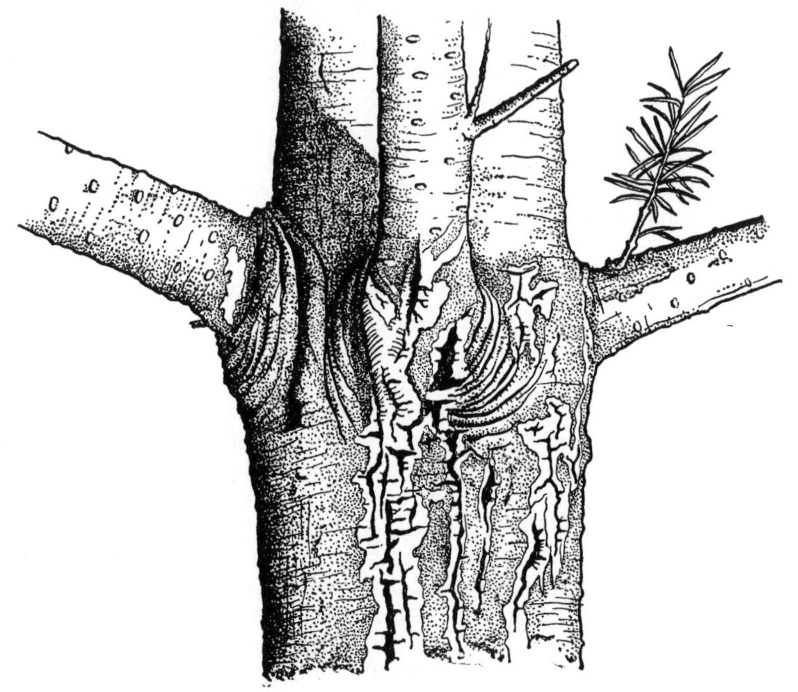

FIG. 70. Typical cracking associated with 'Resin bleeding' of Douglas fir, Shere Forest, Surrey, December (× ⅓). (J. C.)

has done relatively little damage. It remains an open question as to whether it is due to a fungal pathogen, to natural causes (drought has been suggested as a possibility), or to both.

Other diseases of bark and cambium

Young Douglas firs, three to fifteen years old, in California are liable to develop twig and stem galls, which may be as much as 10 cm. in diameter. These are caused by *Bacterium pseudotsugae* Hans. & Smith (Hansen and Smith 1937). The disease occurs mostly on marginal sites, where the timber value of the tree attacked is likely to be very slight, but where its ornamental value is often considerable.

DISEASES OF LEAVES AND SHOOTS

Rhabdocline needle-cast

Pathogen

This disease is caused by *Rhabdocline pseudotsugae* Sydow (Ascomycetes, Phacidiales). Elongated, bright orange-brown apothecia up to 3 mm. in length burst through the epidermis in early summer. They are normally borne on the underside of the needles. The fungus has been described by Vloten (1932) and more briefly in a Forestry Commission leaflet (Anon. 1956*d*). There is no certain knowledge of an imperfect stage, but *Rhabdogloeum pseudotsugae* Syd. (Fungi Imperfecti, Melanconiales) has been suggested. Ellis and Gill (1945) found *R. hypophyllum* Ellis & Gill growing in constant association with *Rhabdocline* in native stands of Douglas fir, and considered that it might be the imperfect stage. There is, however, no proof of this, and neither species of *Rhabdogloeum* has been found in Europe. The pathogenicity of *Rhabdocline* has been proved by inoculation (Vloten 1932). It is an obligate parasite and has not been successfully cultured.

In Britain the fungus, as far as it has been examined, is very constant in appearance, but variations, some of which are extremely marked, have been found in North America. For instance a variant with notably longer fruit bodies has been observed (Wilson, M. unpubl.; Peace unpubl.); Childs (unpubl.) recorded a form on Mount Hebo in Oregon and elsewhere which had brighter orange fruit-bodies, borne more on the upper than on the underside of the needles, and spores that were never constricted in the middle. It is quite probable that only one clone of a very variable fungus has reached Britain.

Symptoms and development

Infection normally takes place in early summer on the young developing needles of the current year by spores produced on the needles of the previous year, and carried by air currents. The infected needles remain green until the autumn, when yellow blotches appear. During the winter these blotches enlarge and deepen in colour, until by the spring they are purplish-brown. Finally in late spring or early summer the epidermis bursts, disclosing the elongated orange fruit-bodies (Fig. 71). Those parts of the needle not occupied by fructifications usually remain green, giving infected needles a strikingly mottled appearance.

The mottling caused by *Rhabdocline* is quite distinct from that due to the aphid *Adelges cooleyi* Börn., the Douglas fir Chermes. The chlorotic patches resulting from attack by this insect are smaller, and develop during the growing season. They are often, but not always, accompanied by distortion of the needle. The insect-caused patches tend to disappear during the winter, which is the period when those due to *Rhabdocline* develop most rapidly.

Needles severely infected by *Rhabdocline* fall very shortly after spore discharge, moderately infected ones later in the summer, while lightly infected ones may stay on the tree and produce fruit-bodies in the second, or exceptionally even in the third spring. In general, however, a severe attack involves the loss of most of the previous year's needles, so that a badly affected tree will only carry one year's foliage, except for a very short period in the early summer after the new shoots have developed, and before the older needles have been shed. A. B. Brown (1930) has attributed needlefall to a marked drop in the moisture content of infected needles, but Vloten (1932) considered that the toxic action of the fungus also played a part. In severely attacked trees

FIG. 71. Fruit-bodies of *Rhabdocline pseudotsugae* on Douglas fir, Santon Downham, Norfolk, May (nat. size, ×2). (J. C.)

this annual loss of needles usually continues until the tree dies, a process which takes at least five years and often much longer.

In western North America the author (unpubl.) noted a much greater irregularity in behaviour. The degree of attack sometimes varied markedly from year to year, so that a tree might be found with alternate years' needles missing. In addition the retention of infected needles was more frequent. Fruit-bodies on two-year-old needles were found more often than in Britain and they occasionally occurred on three- and four-year-old needles. It is possible that this variability of symptoms in America is associated with the variability of the fungus mentioned above.

In Britain the disease is usually found in plantations rather than in the nursery. In fact it is comparatively rare in nurseries, though it has been recorded in them. In North America damage in nurseries is apparently more serious. There is little information on its occurrence on trees over thirty years of age, partially because of difficulties of observation. Really susceptible individuals are likely to succumb before they reach thirty years, but it is probable that light attacks will continue on more resistant trees regardless of age.

It was suggested by Liese (1932) that late-flushing confers resistance, and that this might account for the very clear differences in susceptibility between different provenances. Other workers, however, do not favour this explanation (Rohde 1934; Lyr 1955). It has been pointed out by Vloten (1932) that viable ascospores are in fact still being shed when the late flushing trees come into leaf. The author, who made a large number of observations on diseased Douglas fir in north-western America, failed to find any connexion between date of flushing and degree of infection (Peace unpubl.). Recent statistical examination of some 300 trees in a provenance experiment at Glentress in Scotland also failed to disclose any relationship between *Rhabdocline* attack and date of flushing (Peace unpubl.). The real basis of varietal resistance remains unexplained, though Vloten (1932), Meyer (1951*b*), and Lyr (1955) have suggested structural and chemical differences that might have some effect on the penetration and development of the fungus.

Distribution and damage

The fungus is native to North America where it probably extends over the whole range of Douglas fir. It has also been found on plantation Douglas fir in the eastern United States.

It was first found in Europe in the mid-nineteen-twenties (Wilson and Wilson 1926; Day, W. R. 1927*c*; Liese 1931). It now occurs over most of central and northern Europe, and certainly over the whole of Britain, where it is not confined to the west like the other important needle-cast fungus *Phaeocryptopus gäumannii* (Fig. 72). It can be presumed that it was originally imported into Europe on infected nursery stock.

On severely attacked trees subject to regular annual defoliation the fungus is definitely a killing agency. Such trees show severe reductions in increment before they die. Less severely attacked trees also suffer incremental losses (Liese 1932), but no estimates have been made as to how seriously this affects timber production.

The damage done by the fungus in Europe is greatly mitigated by the preference which it shows for certain varieties of Douglas fir. It is generally agreed that Colorado Douglas fir is the variety most severely attacked (Gerlings 1939; Schober 1954). Indeed in Britain this variety has almost been wiped out by the disease. Most authors consider Coastal Douglas fir by far the most resistant of the three regional varieties normally distinguished, though it is not immune (Liese 1931, 1932, 1936*c*; Rohde 1932; Gerlings 1939; Lyr 1955). Opinions on the Intermediate variety are more divided. Liese and Lyr ranked Intermediate with Colorado Douglas fir as highly susceptible. In Lyr's case this may well have been because his so-called Intermediate provenances were

FIG. 72. Map showing the distribution of *Rhabdocline pseudotsugae* and *Phaeocryptopus gäumannii* on Douglas fir in Britain. (A. C.)

all from Idaho and Montana, and therefore nearer to the Colorado variety than British Columbian origins. Schober (1954) found variable susceptibility in the Intermediate variety, some provenances being as resistant as the Coastal variety. Rohde (1936) found in one of his trials that Intermediate Douglas fir was more resistant than Coastal.

In general agreement with the rather variable results summarized above, British experience, both in the forest generally and in provenance experiments, has shown that the Colorado variety is almost invariably heavily attacked with only very occasional individuals showing resistance, that in the Intermediate variety, reactions, both between and within provenances, vary from high susceptibility to marked resistance, and that the Coastal variety is normally resistant, though individuals may develop slight to moderate attacks. In some of these provenance experiments, some or even all of the Intermediate provenances from the interior of British Columbia are no worse damaged than the Coastal ones.

It is obviously unsafe, therefore, to plant Colorado Douglas fir. There is danger also in planting the Intermediate variety, though it may grow relatively unharmed. Damage on known Coastal Douglas fir has never yet reached dangerous proportions. It must be admitted, however, that quite serious attacks have been found in Britain on plantations of unknown origin, which on general appearance seemed most likely to be of Coastal origin. In Britain restriction of planting to Coastal Douglas fir is no disadvantage since, apart from its disease resistance, it has silvicultural advantages. On the Continent, Intermediate Douglas fir was favoured for some sites on account of its greater winter hardiness, so that there the chance of damaging attack by *Rhabdocline* is greater. In Britain we can regard the disease with some complacency, though there are the uncomfortable possibilities that the fungus might produce a mutant capable of severe attack on the Coastal variety, or that such a variant might reach Europe from America. In this connexion it is interesting and encouraging that provenance experiments with Douglas fir in North America show roughly the same varietal reactions as in Britain.

Control

Rhabdocline was probably one of the first of the diseases of North-western American conifers to be imported into Europe, and its rather sudden impact on established European forestry provoked a spate of ideas on control and containment. A good deal of the existing European legislation on the import of conifers, including genera other than *Pseudotsuga*, originated mainly from efforts to stop the spread of *Rhabdocline*. In Germany a long controversy raged on the respective merits of felling diseased trees, or of leaving them, so that such resistant clones as did exist should be subject to the heaviest possible infection (Geyr 1932). Also in Germany legislation was enacted forbidding the cultivation of the Colorado and Intermediate varieties (Anon. 1936*b*), a very serious step in view of the suitability of the latter for some Continental conditions.

In view of the frequently observed individual variations in susceptibility to attack, there are obvious possibilities in the selection and breeding of resistant strains (Meyer 1951*a*). There is one complicating factor, in that such work ought also to take into account susceptibility to *Phaeocryptopus gäumannii* (see below).

The planting of Douglas fir in mixture with other species has been advocated as a control measure, but it is doubtful if this has any appreciable effect (Liese 1932). Once a tree has become infected the disease is likely to be self-perpetuating on it. Spraying with Bordeaux mixture has been suggested for use in nurseries. It would, of course, be impracticable on a plantation scale. Meyer (1951*b*) suggests two applications, one in the middle and one at the end of May, while H. Fischer (1938) puts forward a more ambitious programme involving sprays at ten-day intervals from the beginning of May until the end of June.

In Britain no control measures appear either feasible or justifiable. The disease prevents the growing of Colorado Douglas fir and makes the cultivation of the Intermediate variety very risky, but so far at any rate the effect on the important Coastal variety is not serious enough to cause concern.

Phaeocryptopus needle-cast

Pathogen

Phaeocryptopus gäumannii (Rohde) Petr. is the fungus responsible for this disease. There is some doubt about the systematic position of this fungus, but it is generally placed in the *Erysiphales* (Ascomycetes) (Petrak 1938*a*). It was originally described as *Adelopus gäumannii* Rohde and indeed was at first confused with *A. balsamicola* (Peck.) Theiss., a harmless and common saprophyte occurring mainly on dead needles of *Abies* species. No imperfect stage has been recorded. The disease is briefly described in a Forestry Commission leaflet (Anon. 1956*d*), and the early work on it was reviewed in *Forestry Abstracts* (Anon. 1939*c*).

The fructifications of the fungus can be found all the year round on infected needles protruding through the stomata. They appear as tiny black pin-heads under a lens (Fig. 73). They are scarcely visible to the naked eye unless present in great quantity, when they show as two sooty bars down either side of the underside of the needle. The main production of spores is in the early summer (Krampe and Rehm 1952), but the perithecia remain viable long after they have reached maturity, so that they can be found alive all through the year (Rohde 1937). It is possible that some spore production occurs whenever conditions are favourable. The fungus is not an obligate parasite and has been cultured.

Symptoms and development

The only reliable symptom of this disease is the presence on the needles of the fruit-bodies and they, of course, are detectable only by very close examination. Fructifications first appear on the needles in the early summer of their second year, when they are just over a year old. These fruit-bodies are presumed to result from infection taking place roughly a year earlier, soon after flushing. According to Krampe and Rehm (1952) the fruit-bodies are at first concentrated near the tips of the needles; but by the third year the bases of the needles are also covered, and yellowish discoloration has spread from the tips over the whole extent of the needles, which are subsequently shed.

These authors, however, recorded that some trees retained their needles, despite the fact that they were as heavily infected as the defoliated trees, while others showed a more definite resistance in that very few fruit-bodies were produced on their needles. Gaisberg (1937), who observed that the needles of a particular year might show heavier infection than those of adjacent years, found fruit-bodies on needles seven years old, which is well beyond the usual age to which healthy needles are retained. There is no good evidence as to whether the development of the fungus in the needle always follows from an

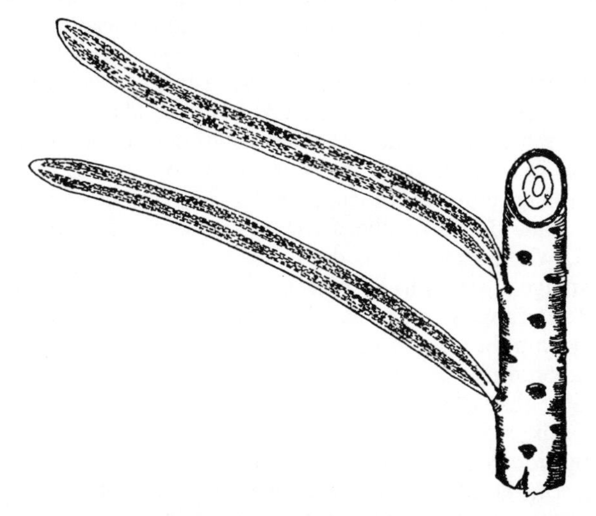

FIG. 73. Fruit-bodies of *Phaeocryptopus gäumannii* on needles of Douglas fir, Coed-y-Brenin Forest, Merioneth, June (×3). (J. N.)

initial infection in the first year. Rohde (1937), who stressed the great irregularity of fruit-body production and needle-fall with this disease as compared with that caused by *Rhabdocline pseudotsugae*, took the view that infection of older needles was possible, and indeed that in many trees it was more important than infection of the young needles. In the author's opinion the fact that all the needles of any one year on any particular tree show approximately the same number of fruit-bodies does suggest that infection took place over a relatively short period, and this would most likely be when the needles were developing, or at any rate when they were not fully hardened. Mańka (1956) also takes this view. Generally the fungus infects initially healthy needles, but Ferré (1955) reported an attack on needles damaged by cold dry winds the preceding winter.

Observational studies by the author (Peace unpubl.) have confirmed the views set out above on the complex behaviour of the fungus. On some trees there is a significant correlation between the amount of needle shedding and the severity of *Phaeocryptopus* infection as measured by fruit-body production, but on others this correlation is entirely lacking. This also applies to the relationship between needle discoloration and *Phaeocryptopus* to the extent that heavy fruit-body production is often, but not always, associated with pronounced yellowing of the needles. Some degree of needle discoloration appears, however, to accompany every *Phaeocryptopus* attack. Thus many infected trees or plantations, which appear in the absence of comparable uninfected trees to be perfectly normal in colour, will be found to be very slightly more yellow-green and less deep green in colour than absolutely healthy trees. Careful observers may be able to detect, even without comparison, that *Phaeocryptopus*-infected plantations, which have not really been damaged by the disease, are nevertheless slightly 'off colour'.

This is primarily a plantation disease. No serious attacks have been observed in nurseries, though the fungus has been recorded on seedlings (Krampe and Rehm 1952).

It is almost impossible to know how seriously *Phaeocryptopus* attack will affect an individual tree or a plantation as a whole. The effect may vary from loss of all but the current year's needles resulting in death after a few years, to an influence so slight that it is only detectable by direct colour comparison with healthy foliage from another source. This extreme variability in symptoms might throw some doubt on the pathogenicity of the fungus were it not for the fact that, where defoliation does occur, no other fungus which could cause the damage appears to be present. The occurrence of fruit-bodies and mycelium of the fungus in live needles, and inoculation experiments carried out by Hahn (unpubl.), leave no doubt that the fungus is a parasite, but on some trees it certainly comes very near to being a non-damaging one.

It is probable that the differences in the effects of attack on the tree have a genetical basis, since trees showing very different reactions can often be found side by side, and on any one tree the reaction is fairly constant from year to year. On the other hand, a whole plantation or even a whole forest often shows, within certain limits of individual tree variation, a fairly uniform degree of attack throughout. This could be explained either genetically, on the basis of provenance differences in reaction to attack, or environmentally, on the assumption that some climatic or other site factor was affecting the development of the disease.

We know that infection is favoured by high humidity and high rainfall especially in the early summer (Merkle 1951; Durrieu 1957). This relationship could explain the distribution of the fungus in Britain, where it is restricted to the higher-rainfall areas (Fig. 72). But rainfall and humidity probably only influence infection. There is no evidence that they affect the discoloration and shedding of needles, which may or may not follow initial infection.

Distribution and damage

The first record of the fungus was in Switzerland in 1925, and ten years later it was causing considerable concern in central Europe. It was first found in Britain in 1928 and described under the name *Adelopus balsamicola* (Wilson and Waldie 1928). Alerted by the agitation in Europe, American workers soon found it on planted Douglas fir in the eastern United States (McCormick 1939), though as soon as a search was made it was realized that it was indigenous on Douglas fir in its native range (Boyce, J. S. 1940). Its distribution is now known to extend from central Europe up through Germany into southern Scandinavia. It occurs in Eire (Liese 1939), as well as in Britain, where, however, it is absent from the east and south-east of England (Fig. 72), probably because the climatic conditions are unsuitable.

This is a classic case of a disease, of no importance in its native haunts, which has become damaging when transported to other areas. J. S. Boyce (1940) reviews a number of possible explanations of this phenomenon, and is inclined to accept change of climate as the most likely, holding that the dry spring and early summer experienced over most of the native range of Douglas fir would render infection difficult, and thus prevent the disease from reaching epidemic proportions. Another possibility is that the fungus in Europe, and presumably in eastern North America, is a mutation of greater pathogenicity than the original strain.

Very little evidence is available on varietal resistance or susceptibility to this fungus. There is a general opinion on the Continent that Colorado Douglas fir is more susceptible (Mańka 1956), but with this variety there is a strong possibility of confusion with attack by *Rhabdocline*. Indeed in Britain it is impossible to get any information on this point, for *Rhabdocline* has attacked nearly all trees of Colorado Douglas fir with such thoroughness that no foothold is left for *Phaeocryptopus* to occupy. There is, however, evidence from provenance experiments in Britain that seed origins from the interior of British Columbia (the so-called Intermediate variety) are more susceptible than those of Coastal origins (Peace unpubl.). Unfortunately, the seed origins of the few British plantations which have been severely damaged by this fungus are not known.

Since the effect of the fungus on the tree is so very varied, reports and opinions on the damage caused are somewhat contradictory. In parts of central Europe it was at one time thought that the disease would mean the end of Douglas fir cultivation. The fungus, sometimes assisted in the later stages by *Armillaria mellea* on the roots (Gaisberg 1937; Liese 1939), was frequently regarded as the primary cause of a killing disease. Where death did not ensue, reduced increment, resulting from the loss of needles, was thought to be quite serious (Zimmerle 1952). During the last twenty years the disease has attracted less attention in Europe, though recent publications (Merkle 1951; Krampe and Rehm 1952; Ferré 1955; Durrieu 1957) make it clear that it is still, locally and occasionally, sufficiently serious to attract attention and cause concern. In Britain the fungus is almost universally present on Douglas fir in the west and north (Fig. 72). In many forests, especially in west Wales, it is very hard to find a Douglas fir which is not at least slightly infected. Such a widespread attack by a fungus which is potentially pathogenic cannot be entirely disregarded. On the other hand, the area seriously damaged by the fungus is very small indeed in relation to the acreage planted with Douglas fir, and there is no evidence of loss of increment, except in the relatively small number of trees which have been really severely defoliated (Peace unpubl.).

Control

In Britain the comparatively slight damage caused by this fungus would not justify any substantial expenditure on control. In any case no feasible methods, other than selection and breeding for resistance, have as yet been suggested. The wide variability in resistance already reported (Meyer 1951*a*; Krampe and Rehm 1952; Ferré 1955) indicates that this is a very promising line of approach, if ever the severity of attack justifies it.

Other needle diseases

Two species of rust fungi (Basidiomycetes, Uredinales) form their aecial stage on the needles of Douglas fir in the United States. They are *Melampsora albertensis* Arth. and *M. occidentalis* Jacks. (Ziller 1955). Both are damaging to the alternate hosts, which are *Populus* spp., but neither does appreciable injury to Douglas fir. Indeed *M. occidentalis* has only recently been found on the latter.

26

DISEASES OF OTHER CONIFERS

THERE is very little information available on the diseases of some of the minor coniferous genera such as *Cunninghamia*, *Sciadopitys*, and *Podocarpus*. If genera are omitted below, it can be assumed that no specific diseases of importance have been found on them, and that they are not known to be particularly prone or resistant to any of the diseases of wide host range. *Armillaria mellea* has been recorded in Britain on *Athrotaxis* and *Metasequoia* among the less common conifers. Indeed it was the first pathogenic fungus recorded on *Metasequoia*, at any rate in Britain.

DISEASES OF *ABIES*

Although many species of *Abies* are found in parks and gardens in Britain, only two, *A. grandis* and *A. nobilis*, are used on a moderate scale in forestry. The European Silver fir, *A. alba*, would also be used were it not so severely attacked by the aphid *Adelges nüsslini* C.B.

Damage due to non-living agencies

Silver firs are commonly damaged by spring frost, and are particularly liable to stem crack (Chapter 6) in hot dry summers.

Crown dieback from undetermined causes is liable to occur in all *Abies* species when the trees are fifty years of age or more. Sometimes they continue to live with the tops of their crowns dead for many years, in other cases the trouble is progressive and finally fatal. It seems to occur more frequently and more seriously on the eastern side of Britain, and may therefore be associated with lower rainfall or lower humidity. The extent to which fungi and insects are involved has never been investigated. Fortunately trees usually reach utilizable size before they start to deteriorate.

Nursery diseases

Records of Damping off of *Abies* are surprisingly frequent in Britain, considering the comparatively small nursery area devoted to the genus. It may very well be particularly prone to attack by soil fungi, but there is no doubt that the rather sparse array of large seedlings, which is typical of most *Abies* seed-beds in this country, does provide conditions under which even mild attacks of Damping off are very noticeable. There is a possibility that soil fungi may be responsible for some of the pre-germination losses associated with the sparseness of the beds.

Seedlings of *Abies* have been found with unexplained swellings at the base on several occasions. Since these swellings have occurred on other conifers they are dealt with in more detail in Chapter 15.

Root diseases

It is generally considered that the Silver firs possess appreciable resistance to attack by *Fomes annosus* (Chapter 16). A good deal of observational evidence supports this contention.

Stem diseases—bark and cambium

A considerable number of fungi have been recorded on the stems of Silver firs, but few are damaging, and some do not occur in Britain.

Nectria cucurbitula (Tode) Fr. (Ascomycetes, Hypocreales), which has been associated with serious cankering on the trunks of spruce (Chapter 23), has also been recorded as a cause of damage on several species of *Abies*. In the case of *Abies*, however, the damage is confined to young wood one to three years old, and takes the form of small cankers and dieback. Robak (1951*a*) considered that the fungus entered the twigs through small frost injuries. It has been found on *A. cephalonica* in Argyll and on *A. concolor* in Gloucestershire, associated with severe twig canker and dieback of the type described by Robak. There was no definite evidence of any predisposing factor such as frost in these cases.

Cytospora abietis Sacc. (Fungi Imperfecti, Sphaeropsidales) has been found causing elliptical cankers, accompanied by resin flow and dieback, on *A. concolor* and *A. magnifica* in the south-western United States (Wright 1942). It was found that the fungus only behaved epidemically when the trees were weakened by drought or some other environmental factor. The fungus has been recorded in Britain on dead wood of a number of coniferous genera, but has never been associated with any damage to live tissues.

Dasyscypha calyciformis (Willd.) Rehm. (Ascomycetes, Helotiales) has been reported on the Continent attacking the stems of *Abies*. It is normally a saprophyte, and that is probably its role on *Abies* in Britain.

Cucurbitaria pithyophila (Fr.) De Not. (Ascomycetes, Sphaeriales), occasionally associated with damage to pine (Chapter 22), also occurs on Silver firs. Rough blackish fungal outgrowths arise, particularly at the nodes of the branches, giving them a warted appearance. These outgrowths are covered with the fruit-bodies of the fungus embedded in a black stroma. The black crust of the fungus has been found covering and protecting colonies of the damaging insect *Adelges* (Franz 1955). The fungus has been recorded injuring *A. nordmanniana* and *A. nobilis*, as well as *A. alba* (Mehlisch 1938). Though it occurs on various conifers in Britain, it has not been associated here with injury to Silver firs.

Aleurodiscus amorphus (Pers.) Rab. (Basidiomycetes, Aphyllophorales) has been recorded on a wide range of conifers. It occurs most commonly as a saprophyte, but has been associated in the United States with elongated cankers on Silver firs. These are found on the main stem and may occasionally result in the death of saplings. The pale saucer-shaped fructifications of the fungus, 3 to 15 mm. in diameter, and somewhat reminiscent of *Dasyscypha* fruit-bodies, eventually develop abundantly on the cankers (Hansbrough 1934). It has been recorded on *A. grandis*, *A. nobilis*, and other species and is probably capable of weakly attacking any Silver fir under suitable conditions. It has been recorded as a saprophyte in Britain, but has not been associated with damage to Silver fir or any other conifer.

Several species of the genus *Phomopsis* (Fungi Imperfecti, Sphaeropsidales) occur on Silver firs. In most cases their pathogenicity is very doubtful. *Phomopsis abietina* (Hart.) Wils. & Hahn (*Phoma abietina* Hart.), which is the species most commonly associated with injury, appears to be restricted to the Continent (Hahn 1930). This fungus is said to cause pale, sunken patches on the bark, sometimes girdling even quite large limbs and causing dieback. In Britain a *Phomopsis*, identified as *P. pseudotsugae* Wils. (Chapter 25) has been associated in one instance with dieback of *A. grandis* (Wilson, M. 1925). *P. conorum* (Sacc.) Died. (*P. pitya* Lind.) has also been found on cankered *A. grandis* in Britain (Batko unpubl.), but may not have been the primary cause of the damage. In America *P. boycei* Hahn causes cankers on young trees of *A. grandis*, and in some seasons this leads to appreciable branch dieback.

C. M. Christensen (1937) reported a species of *Cephalosporium* (Fungi Imperfecti, Moniliales) associated with inconspicuous cankers on *A. balsamea* in the United States. The slightly sunken, oval cankers, which exuded resin, occurred mainly on suppressed trees and the damage was not serious. No member of this genus has been reported attacking Silver firs in any other country.

Gäumann and Jaag (1937) reported *Pleurotus mitis* (Pers.) Berk. (Basidiomycetes, Agaricales) causing elongated cankers on the main stems of *Abies alba* and Norway spruce in Switzerland (Chapter 23). In Britain it has been associated by Banerjee (1956) with similar injury on Sitka spruce, but does not seem to have been reported as a potential parasite on any species of *Abies*.

Scleroderris abieticola Zell. & Good. (Ascomycetes, Helotiales) is associated with short-lived cankers on *A. grandis* and *A. amabilis* in the north-western United States (Zeller and Gooding 1930). Canker development takes place only during the winter and ceases as soon as cambial growth starts in the spring. During the summer the cankers heal over and the dead bark is shed. Damage is restricted to twigs and small branches and is not regarded as serious. Small black apothecia, 0·5–1·2 mm. in diameter, are produced during the summer on the dead bark.

Rust fungi

Silver firs are attacked by an enormous number of species of rust fungi (Basidiomycetes, Uredinales) belonging to the genera *Melampsora*, *Pucciniastrum*, *Calyptospora*, *Hyalospora*, *Melampsorella*, *Uredinopsis*, and *Milesia*. The last two genera are peculiar in that the alternate hosts are all species of ferns. In all cases the pycnia and aecia occur on needles of *Abies*, while the uredia and telia are found on various herbaceous and fern hosts. In a few instances the alternate hosts are unknown. Most of these Silver fir rusts are so difficult to identify on *Abies* that differences in the pycnia and in the haustoria have been suggested as possible aids to separation and identification (Hunter 1927, 1948).

The following species have been recorded in Britain:

	AECIAL HOST IN BRITAIN	UREDIAL AND TELIAL HOST IN BRITAIN
Calyptospora goeppertiana Kühn. (Syn. *Pucciniastrum goeppertianum* (Kühn.) Kleb.)	*A. alba*, *A. nordmanniana*	*Vaccinium vitis-idaea* (telial)
Melampsorella caryophyllacearum Schröt. (Syn. *Pucciniastrum caryophyllacearum* (DC.) Fisch., *Melampsorella cerastii* (Pers.) Schröt.)	*A. alba*, *A. nordmanniana*, etc.	*Cerastium* spp. *Stellaria* spp.
Melampsorella symphyti Bub.	Not recorded on *Abies*	*Symphytum* spp.
Milesia blechni (Syd.) Arth.	*A. alba*	*Blechnum spicant*
Milesia carpatica (Wrobl.) Faull	Unknown	*Dryopteris felix-mas*
Milesia kriegeriana (Magn.) Arth.	*A. alba*, *A. cephalonica*, *A. nordmanniana*, and *A. grandis*	*Dryopteris* spp.
Milesia murariae (Syd.) Faull	Unknown	*Asplenium ruta-muraria*
Milesia polypodii White	*A. alba* and *A. concolor* by inoculation; unknown in nature	*Polypodium vulgare*
Milesia vogesiaca (Syd.) Faull	*A. alba* by inoculation; unknown in nature	*Polystichum setiferum*
Milesia scolopendrii (Fckl.) Arth.	*A. alba* and *A. concolor* by inoculation; unknown in nature	*Scolopendrium vulgare*
Milesia whitei Faull	Unknown	*Polystichum setiferum*
Pucciniastrum circaeae (Thuem.) Speg.	Not recorded on *Abies*	*Circaea* spp.
Pucciniastrum epilobii Otth.	*A. grandis*	*Epilobium* spp.
Uredinopsis filicina Magn.	Unknown	*Thelypteris phegopteris*

Few of these are of any practical importance, in fact it will be noted that many of them have not been found naturally on *Abies* in Britain, although that genus is likely to provide the alternate host. In North America there are over twenty rusts actually recorded on *Abies*, and in Japan at least fifteen (Kamei 1932).

Calyptospora goeppertiana produced two rows of white cylindrical aecia on the underside of the needles in the summer. The aeciospores infect *Vaccinium*, causing pale pink or whitish swellings on the shoots and witches' broom formation. The mycelium is perennial in *Vaccinium* and no uredial stage is produced (Faull 1939). This rust is very rare in Britain, indeed there is some doubt about the existing records (Wilson and Bisby 1954).

Melampsorella caryophyllacearum is certainly the most striking of the Silver fir rusts in Britain. The sporidia infect the young shoots of *Abies* in early summer. A swelling develops at the point of infection from which a witches' broom, often of considerable size, usually arises (Plate XIV. 1). Sometimes, however, only a canker is formed. The witches' brooms may persist and enlarge for many years. The shoot growth in them is restricted in length, but is much branched and very erect; the needles on the brooms are unusually short, yellow in colour, and deciduous. When in leaf the brooms are very conspicuous. Pycnia and aecia appear on these abnormal needles, the aeciospores infecting the herbaceous alternate hosts. The needles are shed at the end of the summer (Richardson, A. D. 1923). The fungus can survive on the alternate host in the absence of *Abies*.

In Britain this fungus has only been recorded on *A. alba*, *A. nordmanniana*, and *A. numidica*. On the Continent, where it is certainly more prevalent, and in America (Mielke 1957*c*), it has been found on a number of other *Abies* species, but not apparently on *A. grandis* or *A. nobilis*. Control measures are not required in Britain. Removal of the brooms has been occasionally advocated on the Continent.

The various species of *Milesia*, which are virtually indistinguishable on *Abies*, and some of which have never been recorded on it, are rare and have no practical importance in Britain (Hunter 1936; Wilson and Bisby 1954). The North American species have a limited pathological importance (Faull 1934).

Pucciniastrum epilobii has been fully studied by Faull (1938). It is of interest because it occurs on our most commonly planted Silver fir, *A. grandis*. Nevertheless, it is still rare in Britain and is of no pathological importance here, although damage has been reported from the Continent (Guinier 1931).

Uredinopsis, the other genus with Silver firs and ferns as alternate hosts, is represented only by one species in Britain and that is not known on *Abies*. Other species of this genus are of minor pathological importance on *Abies* in North America and in Japan (Kamei 1940).

Other diseases of needles and shoots

Apart from the rusts there are not many needle fungi on *Abies*. None is of importance in Britain.

Rehmiellopsis dieback

The commonest disease of Silver firs in Britain is probably that caused by *Rehmiellopsis bohemica* Bub. & Kab. (Ascomycetes, Sphaeriales). The small black pycnidia of the imperfect stage, *Phoma bohemica* Bub. & Kab. (Fungi Imperfecti, Sphaeropsidales), appear first on the upper side of the needles, followed by the scarcely distinguishable perithecia.

The young needles are infected soon after they expand. At first they are reddened, but later they turn dark brown and shrivel. They often remain attached to the shoots for as long as a year after they have been killed. In severe attacks the young shoots tend to die back (Wilson and Macdonald

1924). There has certainly been some confusion between the damage caused by this fungus and that attributable to the aphid *Adelges nüsslini*, for the fungus and the insect often occur on the same tree. In Britain *R. bohemica* attacks chiefly trees under fifty years of age (Wilson and Macdonald 1924), but in America a closely related species, *R. balsameae*, may be found on trees of all ages (Waterman 1945).

The fungus is most serious on *A. alba*, but it has also been recorded on *A. nobilis*, *A. pinsapo*, *A. cephalonica*, *A. pindrow*, *A. balsamea*, and *A. magnifica*; fortunately it occurs very rarely on *A. grandis* and is never associated with serious damage to it. *R. bohemica* is commoner in Scotland than farther south, though it has been recorded in England. It would probably be more important in Britain if *A. alba* were more widely planted. It has been reported from many parts of the Continent, and is sometimes quoted as the most important disease of European Silver fir, though there have been very few reports of it causing serious damage there.

In America a different species, *R. balsameae* Waterman, causes very similar injury. It occurs particularly on *A. balsamea*, but has also been found on *A. cephalonica* and *A. concolor*. A number of other firs including *A. nobilis* have been found by experiment to be susceptible to *R. balsameae* (Waterman 1945). The two fungi are certainly very similar, and it may well be that the occasional records of *R. bohemica* in North America should, in fact, refer to *R. balsameae*.

Hypodermataceous needle-casts

Various members of the *Hypodermataceae* (Ascomycetes, Phacidiales) occur on the needles of *Abies*, but most have been recorded in America and none in Britain.

In North America the most serious of these is *Bifusella faullii* Dark., which infects the needles during their first season, but does not discolour them until the following spring, and may not cause them to fall for two or three years. Another species of *Bifusella*, three species of *Hypodermella*, and one of *Hypoderma* are also associated with *Abies* needle-cast in North America, but none is important (Darker 1932).

In Europe, but not in Britain, *Lophodermium nervisequum* (DC.) Rehm. (*Hypodermella nervisequa* (DC.) Lagerb.) is locally serious as a cause of defoliation, mainly on *A. alba*. Discoloration does not appear until the leaves are two years old, i.e. at the beginning of their third year. The elongated black apothecia are produced the following spring, by which time many of the needles have fallen from the trees.

Other needle diseases

Acanthostigma parasiticum (Hart.) Sacc. (Ascomycetes, Sphaeriales), at one time known as *Trichosphaeria parasitica* Hart., spreads in the form of colourless, superficial mycelium on the underside of the shoots. The needles are eventually killed, but remain attached to the shoots by the mycelium. Usually only the lower shaded branches are affected. This disease is of some importance on the Continent, but has rarely been recorded in Britain (Watson 1933; Anon. 1941).

Rhizosphaera kalkhoffii Bub. (Fungi Imperfecti, Sphaeropsidales) has been found on the needles of *A. nobilis* and *A. alba* in Britain and probably occurs on other species, but its pathogenicity on *Abies* is even more doubtful than on *Picea* (Chapter 23).

Dimeriella terrieri Petr. (Ascomycetes, Erysiphales) has been recorded on live needles of *A. alba* in Switzerland (Terrier 1947). Minute black fruit-bodies occur in groups on patches of superficial reddish mycelium. The fungus occurs only on the lower branches under conditions of high humidity, and, despite some development of internal hyphae, does little or no damage to the live needles on which it is found. It has not been reported elsewhere.

DISEASES OF *ARAUCARIA*

No diseases of any significance seem to have been found on *Araucaria* in Britain, despite the fact that *A. imbricata*, commonly known as the Monkey puzzle, has been very widely planted as an ornamental. *Armillaria mellea* has been found on the roots of windblown trees. It is worth recording that *A. imbricata* is surprisingly resistant to exposure and is sometimes found growing quite well in very bleak situations. A few diseases have been recorded in Italy and in the southern hemisphere, where several species of *Araucaria* occur naturally; these diseases are briefly mentioned below.

Nursery diseases

Most of the recorded diseases on *Araucaria* have been found in the nursery or on very young plants. Damping off, particularly associated with the fungus *Rhizoctonia crocorum* (Pers.) DC., the sterile form of *Helicobasidium purpureum* (Tul.) Pat. (Basidiomycetes, Auriculariales), has been found by H. E. Young (1948) on *A. cunninghamii* in Queensland. He also associated *Botryodiplodia theobromae* Pat. (Fungi Imperfecti, Sphaeropsidales) with death of seedlings in Australia (Young, H. E. 1936). This could be the same fungus as the *Diplodia* found on dying seedlings of the same species of *Araucaria* by Simmonds (1933), who also recorded a species of *Fusicoccum* (Fungi Imperfecti, Sphaeropsidales) on the same plants. *F. araucariae* Vogl., with the perfect stage *Cryptosporella araucariae* Vogl. (Ascomycetes, Sphaeriales), was associated by Voglino (1932) with a collar rot of rather larger plants of *A. imbricata* in a nursery in Italy. The leaves turned yellow, swellings were formed at the collar, and the bark died at this point. Servazzi (1938*c*) found a somewhat similar fungus on potted *A. excelsa* suffering from a progressive dieback, and described it under the name *Cryptospora longispora* Serv. The lower twigs were attacked first, but the fungus was capable of killing six-year-old plants.

Diseases of leaves and shoots

In Italy *Didymella araucariae* Vogl. (Ascomycetes, Sphaeriales) was found producing elongated black pustules up to 2 mm. long on both surfaces of the leaves of *A. imbricata*. The affected leaves turned pale green or brown and were shed prematurely (Voglino 1933).

DISEASES OF *CEDRUS*

The three important species of cedar, *Cedrus atlantica*, *C. deodara*, and *C. libani*, have all been planted fairly widely in Britain as ornamentals. They are not very easily distinguished botanically, and they do not seem to differ in their pathological behaviour. *Deodar* has occasionally been used as a forest tree.

Root diseases

Apart from the occasional occurrence of the root fungi *Armillaria mellea* and *Fomes annosus*, only one fungal disease, *Phomopsis pseudotsugae*, has been recorded on *Cedrus* in Britain and remarkably few abroad. Nevertheless, reports of premature needle-fall, and sometimes of death of whole trees following this, are of comparatively frequent occurrence. The needle-fall, which sometimes happens quite early in the summer, is not accompanied by any fungal invasion of the needles, but is sometimes associated with drought. The death of large trees is sometimes finally due to attack by root fungi, but there may be other predisposing factors. It seems that in Britain cedars are particularly prone to suffer severely from alterations in their environment.

Stem diseases—bark and cambium

Phomopsis pseudotsugae Wils. (Fungi Imperfecti, Sphaeropsidales) has been found by M. Wilson (1930*b*) attacking *C. atlantica* and *C. libani*. In view of its wide coniferous range it may be assumed to be capable of attacking *C. deodara* also. The disease on cedar takes the form solely of girdling and dieback, no sunken cankers are formed as is frequently the case on Douglas fir (Chapter 25). Probably the fungus on cedar may sometimes follow frost or drought damage, as it does on Douglas fir.

Diseases of needles and shoots

According to Darker (1932), the only *Hypodermataceous* needle fungus on *Cedrus* is *Lophodermium cedrinum* Maire (Ascomycetes, Phacidiales). This has been recorded on *C. atlantica* in North Africa. It is possible that Konev's (1951) record of *L. pinastri* (Schrad.) Chev. on *C. libani* in the Baikal region of the U.S.S.R. may in fact refer to *L. cedrinum*. There the defoliation it causes is said to be the main factor in the death of both young and old trees.

DISEASES OF *CHAMAECYPARIS* AND *CUPRESSUS*

Pathologically it is almost impossible to deal with these genera separately for most of the fungi which attack them cover both.

In Britain Lawson cypress, *Chamaecyparis lawsoniana*, especially in its many foliage varieties, is greatly prized as an ornamental. It is quite frequently used for hedges and holds a minor place as a forest tree. *Cupressus macrocarpa* has been planted fairly commonly near the coast and, far more widely than it deserves, as a hedging plant. Although other species appear in gardens and arboreta, none has forest possibilities, except the intergeneric hybrid *Cupresso-cyparis leylandii*.

As will be seen below, none of the more serious diseases of cypresses occur in Britain.

Damage due to non-living agencies

C. macrocarpa combines a very high degree of resistance to sea-wind with a notably high susceptibility to winter cold. In practice it can only safely be grown in a comparatively narrow coastal belt. Within this, however, its resistance to sea-wind gives it a particularly high value for shelter. *C. leylandii* has so far displayed a high degree of resistance to frost and to exposure.

Nursery diseases

As one-year seedlings *C. macrocarpa* and *C. sempervirens* are particularly susceptible to attack by *Botrytis cinerea* Pers. ex Fr. (Fungi Imperfecti, Moniliales). Lawson cypress is not so severely attacked.

Species of *Pestalozzia* (Fungi Imperfecti, Melanconiales) (Chapter 15) have been recorded several times on cypresses, mainly abroad.

Phomopsis juniperovora Hahn (Fungi Imperfecti, Sphaeropsidales) has been associated with dieback of cypresses, as well as junipers, in nurseries in various parts of the world, particularly in the United States. R. P. White (1929) found that on *Chamaecyparis* it followed frost injury, but Foster (1956) considered it a primary parasite on *Cupressus arizonica*, as did Bottomley (1919) on *C. torulosa*, *C. arizonica*, and *C. macrocarpa* in South Africa. It is not known to occur anywhere in Europe.

Keithia chamaecyparissi Adams (Ascomycetes, Phacidiales) (see below) attacks seedlings in the nursery as well as older trees.

Phytophthora root rot (see below) does serious damage to Lawson cypress in nurseries in North-western America, as well as in older plantings.

Root diseases

Lawson cypress is above average in its susceptibility to *Armillaria mellea* and is susceptible to *Fomes annosus* (Chapter 16).

Phytophthora root rot

In North-western America, particularly in Oregon, Lawson cypress, there known as Port Orford cedar, is attacked by two species of the genus *Phytophthora* (Phycomycetes, Peronosporales), *P. lateralis* Tuck. & Milb., and the ubiquitous *P. cinnamomi* Rands.

The fungi first attack the roots and thence spread to the trunk. The first symptom observed is usually a progressive discoloration of the foliage, which eventually withers completely (Tucker and Milbrath 1942). The symptoms caused by the two fungi are similar, and it is therefore difficult to evaluate their relative importance, though Torgeson (1951) considers that in Oregon *P. lateralis* is more widely distributed. *P. lateralis* has also been found to cause infections on the foliage, lower branches, and even the main stem. These are clearly distinguishable from root infections by the fact that the dieback they cause is localized (Trione and Roth 1957). As is usual with *Phytophthora* root rots, the disease is said to be favoured by high water-levels in the soil (Torgeson 1954).

The disease is serious only on *Chamaecyparis lawsoniana* and its varieties; some varieties of *C. obtusa* are also susceptible but *C. pisifera* is resistant

(Tucker and Milbrath 1942). *P. cinnamomi* has also been recorded on *Taxus* (Torgeson 1951), but attempts to infect other conifers, such as *Thuja* and *Juniperus*, with *P. lateralis* have fortunately proved unsuccessful (Milbrath and Young 1949). At first the disease was thought to be limited to nurseries and ornamental plantings, but recently it has been found active in the relatively restricted native stands of Lawson cypress near the south-west Oregon coast (Trione 1959). In this area the disease is found mainly along roads and near houses. If the wide extension of the disease, since it was first recorded in this area in 1952, represents genuine spread, it has been more rapid than one would normally expect of a soil fungus. Its location does suggest that soil disturbance may have encouraged the pathological development of latent infections already present. Both species of *Phytophthora* have been found in nurseries and on planted trees of Lawson cypress in British Columbia (Salisbury, P. J. 1955).

P. cinnamomi has been recorded by Buddenhagen (1955) on Lawson cypress in Holland, otherwise this potentially dangerous disease of cypresses has not been found in Europe, although one of the pathogens, *P. cinnamomi*, is known to occur widely there. *P. lateralis* has not been found in Europe.

The only methods of control suggested are the use of uncontaminated ground for raising cypress, and the selection of disease-free stocks when vegetative methods are employed for the propagation of ornamental varieties (Milbrath and Young 1949).

Stem diseases—bark and cambium

Coryneum and monochaetia canker

The two closely related fungi, *Coryneum cardinale* Wag. and *Monochaetia unicornis* (Cooke & Ell.) Sacc. (both Fungi Imperfecti, Melanconiales), cause very similar diseases on cypresses. The perfect stage of the *Coryneum* is a species of *Leptosphaeria* (Ascomycetes, Sphaeriales) (Hansen 1956), but it is rarely found in nature. That of *M. unicornis* is a member of the Sphaeriales (Ascomycetes) (Nattrass 1950), but has not yet been described. In the case of *M. unicornis* Rudd Jones (1954*a*) has detected strains of varying virulence, and he suggests mutation from a strain parasitic only on a native host *Juniperus procera* as a possible origin for the disease on cypress.

The first evidence of infection by *Coryneum* is a browning of the live bark around the point of entry. This is commonly at a crotch or around the base of a branchlet, though infection may occur through wounds. The infected bark swells slightly and resin exudation occurs. Later the bark dies and canker formation commences. The cankers are elongated, being normally three to four times as long as they are wide. They soon girdle small branches, causing sporadic dieback. Girdling of larger limbs or the main stem is often a long process, and considerable deformation may take place around the canker. Trees of any age, size, or vigour are liable to attack, though normal open-branched trees seem more prone to infection than those which have become densely branched owing to pruning. The tiny, black, pustular fruit-bodies of the fungus are irregularly scattered over the surface of the dead bark (Wagener 1939, 1948).

The symptoms of the disease caused by *Monochaetia unicornis* in East Africa are almost exactly the same (Nattrass and Ciccarone 1947; Rudd Jones 1953), but young trees are found to be most susceptible and older trees are infected only in their rapidly growing parts. The fruit-bodies are superficially similar to those of *Coryneum.*

In California the *Coryneum* disease is serious only on *C. macrocarpa* and has led to the consideration of alternative species for use in shelterbelts. Fortunately the very restricted natural stands on the Monterey Peninsula are not infected. Other cypresses of little importance in Britain, such as *C. sempervirens*, are also susceptible, but Lawson cypress was found to be resistant (Wagener 1948). This is in agreement with the results obtained in Italy by Grasso (1952). Nothing is known of the resistance of the hybrid *C. leylandii*, or of its hardier parent *C. nootkatensis*, to attack by *Coryneum cardinale*.

In East Africa *C. macrocarpa* is the species chiefly damaged by *Monochaetia*. The disease was at one time regarded as a threat to its continued cultivation. *C. lusitanica* is less susceptible, but is also attacked, especially in the neighbourhood of diseased *C. macrocarpa* (Rudd Jones 1953). Neither of these species of *Cupressus* is native to East Africa, and the disease has appeared in plantations only over the last twenty years. By inoculation Rudd Jones (1954*b*) found that a large number of species of *Cupressus* and *Chamaecyparis*, including *C. lawsoniana* and *C. leylandii*, were susceptible to *Monochaetia* and none were resistant. Among related conifers tested, *Juniperus procera* and *Thuja orientalis* were susceptible.

In New Zealand both *C. macrocarpa* and *C. lawsoniana* are seriously attacked by *Monochaetia unicornis* (Fuller and Newhook 1954). In Tasmania this fungus is also serious chiefly on *C. macrocarpa*, while *C. torulosa* and *C. lusitanica* var. *benthamii* are regarded as resistant (Anon. 1957*c*). An undetermined species of *Monochaetia* has also been recorded on Lawson cypress on one occasion in the north-western United States (Hotson and Stuntz 1934), and *C. cardinale* on *C. macrocarpa* in the Argentine (Saravi Cisneros 1953).

In Europe *C. cardinale* was first reported by Barthelet and Vinot (1944), causing cankers 20 to 30 cm. long on branches of *C. macrocarpa* in southern France. This fungus also occurs in Italy, mainly on *C. macrocarpa*, but also on *C. sempervirens*, *C. lusitanica*, and *C. arizonica* (Grasso 1951, 1955, 1957). It appears to be spreading there, but is not yet causing serious damage. Neither fungus has been found in Britain.

The most significant difference between the two fungi is that *Coryneum cardinale*, which occurs in California, in the Mediterranean region of Europe and in the Argentine, does not attack Lawson cypress, whereas *Monochaetia unicornis*, which occurs in East Africa, New Zealand, and possibly in America, does.

Neither spraying nor cutting out cankers has proved of any real value for control on infected trees in California, though spraying with Bordeaux mixture was found to reduce the number of fresh infections. It is considered there that the best means of checking the disease is the removal of infected trees (Wagener 1948). In New Zealand, however, spraying in the autumn and early spring with Bordeaux mixture and also the removal of cankered limbs is advocated

(Fuller and Newhook 1954). Individuals of *C. macrocarpa* appear to be resistant to *Coryneum* so that there may be some possibilities in selection and breeding for resistance (Wagener 1948).

Other diseases of bark and cambium

A number of species of the rust genus *Gymnosporangium* (Basidiomycetes, Uredinales) occur on cypresses, causing swellings on the twigs, branches, and trunks and sometimes producing witches' brooms. Those attacking cypresses have attracted attention only in North America, and none occurs in Britain. The alternate hosts are mostly members of the *Rosaceae*, but the most damaging species in the United States, *G. ellisii* (Berk.) Farl., which attacks *Chamaecyparis thyoides*, has its aecial stage on species of *Myrica*. *G. biseptatum* Ell. is also damaging to *C. thyoides* in North America. Cypresses may die from the effect of the numerous branch burls. The aecial host of this fungus is *Ameanchier*. In both cases the telia, as with other species of *Gymnosporangium* (see under Diseases of *Juniperus* below) are borne on the coniferous host and there are no uredia. The production of witches' brooms on the cypress may be so profuse that death results (Dodge, B. O. 1934).

Diseases of needles and shoots

Keithia (*Didymascella*) *chamaecyparisii* Adams (Ascomycetes, Phacidiales) occurs on *Chamaecyparis thyoides* in the United States. The attack is usually light, but it can cause severe damage to the foliage of young trees, eventually killing the twigs. Conspicuous apothecia are produced on the needles (Adams 1918; Korstian and Brush 1931). This fungus has not been recorded in Britain.

Two species of *Lophodermium* (Ascomycetes, Phacidiales) occur on cypresses. *L. chamaecyparisii* Shir. & Hara causes defoliation of *C. obtusa* in Japan (Nishikado 1944). *L. juniperinum* (Fr.) De Not. occurs on *C. thyoides* in the United States; there is disagreement as to whether it is more or less damaging than *Keithia* (Adams 1918; Korstian and Brush 1931). *L. juniperinum* occurs on junipers in Britain and other parts of Europe, but has not been recorded here on cypresses.

DISEASES OF *CRYPTOMERIA*

Cryptomeria japonica, the only species in this genus and an important timber tree in Japan, occupies a minor place as a forest tree in Britain where it is very free from disease. In Japan several fungi have been found on it (Anon. 1950*b*), but only in nurseries are any of them damaging.

Nursery diseases

In Britain *Cryptomeria* is one of the conifers particularly prone to attack by *Botrytis cinerea* (Chapter 15). This is the case also in Japan, where damage often takes place under the winter snow cover (Itô and Hosaka 1951).

Guignardia cryptomeriae Sawada (Ascomycetes, Sphaeriales) attacks the shoots of *Cryptomeria* in Japan. Infection is normally by wounds, probably due to stabbing by the leaf apices as the twigs thresh about in the wind (Kobayashi 1957).

In Japan a very large number of fungi have been recorded on needles of *Cryptomeria* seedlings suffering from the so-called 'Needle blight' (Itô, Shibukawa, and Kobayashi 1952; Itô, Shibukawa, and Terashita 1954). The most important of these is considered to be *Cercospora cryptomeriae* Shir. (Fungi Imperfecti, Moniliales). This disease, which causes withering of the needles and then of the shoots and results in the death of seedlings, can be controlled by copper sprays applied during the rainy season (Nohara and Zinno 1955). A few individual trees are resistant to this fungus (Chiba 1955*a*).

DISEASES OF *GINKGO*

The Maidenhair tree, *Ginkgo biloba*, is considered by Pirone (1948) to be unusually free from fungal diseases. No diseases seem to have been recorded on it in Britain. *Glomerella cingulata* (Stonem.) Spauld. & Schrenk. (Ascomycetes, Sphaeriales) has been recorded damaging the leaves in the United States.

Two fungi of cosmopolitan distribution in sub-tropical regions have been specifically reported damaging *Ginkgo*. *Corticium* (*Pellicularia*) *kologera* (Cke.) v. Höhn. (Basidiomycetes, Aphyllophorales), known as 'Thread blight', covers foliage with superficial, though damaging, mycelium. It has been found attacking *Ginkgo* in Japan (Itô 1958). *Macrophomina phaseoli* (Maubl.) Ashby (Fungi Imperfecti, Sphaeropsidales) attacks the roots of a very wide range of plants. In China it invades and enlarges basal wounds, initially due to hot soil, on young *Ginkgo* seedlings. It can be avoided by shading or mulching (Fang *et al.* 1956).

DISEASES OF *JUNIPERUS*

Juniperus communis is one of our three native conifers. Many other species have been introduced into Britain, one of the commonest of which is *J. sabina*. None of them is of any forest importance. *J. communis* and *J. oxycedrus* are hosts for *Arceuthobium oxycedri* (DC.) Bieb., the only European representative of the Dwarf mistletoes, which are so damaging on conifers in North America (Chapter 13).

Nursery diseases

Only a small number of junipers are raised in nurseries in Britain and they have no diseases of importance, except possibly *Botrytis cinerea*. However, in America *Phomopsis juniperovora* Hahn (Fungi Imperfecti, Sphaeropsidales), earlier described as a species of *Phoma*, causes twig dieback of junipers, chiefly in the nursery, but occasionally on larger trees. The tips of the shoots turn brown, progressive dieback follows, and in severe attacks the entire plant is killed. The fungus often spreads down laterals into the main stem (Hahn, Hartley, and Pierce 1917). It has a wide host range on species of juniper, and can infect a number of other coniferous genera, particularly cypresses (Hahn 1943*b*). In the United States it is serious chiefly on *J. virginiana*. It has not been recorded in Europe. The disease can be controlled by regular spraying during the late summer with a number of proprietary fungicides (Caroselli

1957). Nitrogen manuring to improve growth can result in increased damage by this fungus (Davis and Latham 1939).

In mountainous parts of the Continent the 'snow mould' disease caused by *Herpotrichia nigra* Hart. (Ascomycetes, Sphaeriales) is often particularly serious on juniper. In Britain, where snow seldom lies for long, it is of no importance.

Stem diseases—bark and cambium

Monochaetia unicornis (Fungi Imperfecti, Melanconiales), chiefly important as a cause of canker and dieback on cypresses in East Africa, New Zealand, and elsewhere (see above), occurs naturally on *J. procera* in East Africa, and may have spread from the juniper to introduced cypresses (Rudd Jones 1953). Wagener (1948) found *J. chinensis* and *J. virginiana* slightly susceptible to *Coryneum cardinale* (Fungi Imperfecti, Melanconiales), a fungus which causes canker and dieback on cypresses in California and elsewhere (see above). In neither case is the disease serious on juniper.

Gymnosporangium rusts

Junipers are attacked by a very large number of species of the genus *Gymnosporangium* (Basidiomycetes, Uredinales), which injure the stems, young shoots, and needles. Most have Rosaceous trees and shrubs as their alternate hosts. They fall into three regional groups, European, North American, and Asian. Only three species, *G. aurantiacum* Chev., *G. juniperi*, and *G. clavariaeforme* are cosmopolitan. The rest are each restricted to one of the three regions (Crowell 1940). Four species occur in Britain:

	AECIAL HOST IN BRITAIN	TELIAL HOST IN BRITAIN
G. clavariaeforme (Jacq.) DC.	*Crataegus* spp., *Pyrus communis*	*J. communis*
G. confusum Plowr.	*Crataegus* spp., *Cydonia* spp., *Mespilus germanica*	*J. sabina*
G. fuscum DC. (*G. sabinae* (Dicks.) Wint.)	*Pyrus communis*	*J. sabina*
G. juniperi Link.	*Sorbus aucuparia*	*J. communis*

In all these species the production of telia on juniper is accompanied by pronounced swellings of the shoots. The teliospores are extruded in large, erumpent, gelatinous, yellow masses. There are no uredia. The effect of these fungi on the aecial hosts is dealt with under the appropriate genera in Chapter 35.

G. clavariaeforme is the commonest of the British species. *G. confusum* is comparatively rare (Moore, M. H. 1945), while the other two occupy an intermediate position, *G. juniperi* being much commoner in Scotland than in England. None of them is seriously damaging to juniper.

All the British species have wider host ranges on juniper abroad. For instance, *G. fuscum* is locally a serious disease of cultivated pears in southern Europe, where the juniperous hosts are *J. sabina*, *J. excelsa*, and other species

(Deckenbach 1927). The Mediterranean species, which include all the British ones with the exception of *G. juniperi*, have been exhaustively investigated by Bernaux (1956).

In North America the genus *Gymnosporangium* is represented by a very large number of species, mostly with the telial stage on juniper, though some infect cypresses (see above). Japan also has a large number of *Gymnosporangium* species on juniper (Hiratsuka 1937).

The British species of *Gymnosporangium* all cause stem swellings on juniper, but in North America and elsewhere different species infect leaves, young shoots, or older stems, causing various forms of hypertrophy, including leaf galls, stem swellings, and witches' brooms. There is a very large literature, not reviewed here, concerning North American species of *Gymnosporangium* on their aecial hosts, with particular reference to their occurrence and control on apple trees. So many species are concerned that consideration of their different preferences for various *Juniperus* species would be a very complex matter.

The most important species in North America, to which continent they are confined, are *G. juniperi-virginianae* Schw., *G. globosum* Farl., and *G. clavipes* Cke. & Pk. (Thomas and Mills 1929). *G. juniperi-virginianae* attacks the leaves and occasionally the fruits of apple, and causes galls on the young shoots of juniper (Crowell 1934). It has occasionally been found on imported apples (Moore, W. C. 1959), but has never become established in Britain. *G. globosum* causes very small galls on the foliage of juniper and has its aecial stage chiefly on *Crataegus*, but also on a large number of other Rosaceous hosts (MacLachlan 1935*a*, 1936). *G. clavipes* causes swellings on the branches of junipers and has its aecial stage on a wide range of Rosaceous trees and shrubs (Crowell 1935*a*).

Varying clonal resistance of juniper to *Gymnosporangium* has been observed (Thomas, H. E. 1933; Berg 1940). There is also evidence of strains of varying pathogenicity in some species of *Gymnosporangium* (Fischer, E. 1930; Crowell 1934).

Removal of junipers has often been advocated to control *G. juniperi-virginianae* on apples. It is known that the teliospores can travel seven or eight miles, but in practice the eradication of junipers within a one-mile radius of the orchards gives a high degree of protection (MacLachlan 1935*b*). Compulsory eradication of junipers was once tried in Austria to protect pears from *G. fuscum*. Where both hosts are grown together, as in gardens, control of *G. juniperi-virginianae* can be obtained by spraying in spring with Bordeaux mixture (Marshall, R. P. 1941) or cyclo-heximide (Strong and Klomparens 1955).

Other diseases of needles and shoots

Lophodermium juniperinum (Fr.) De Not. (Ascomycetes, Phacidiales) has been recorded on a wide range of juniper species in Europe and North America, producing elongated, shining black fruit-bodies on the needles (Darker 1932). It has been recorded frequently in Scotland and in north England on *J. communis* and is said to have been locally common at times and to have caused appreciable damage. But neither in Britain nor elsewhere has it recently attracted attention as a cause of injury.

Camarosporium juniperinum Georg. & Bad. (Fungi Imperfecti, Sphaeropsidales) causes needle-fall of *J. communis* in Romania, producing pycnidia, up to 0·5 mm. across, in pairs on the dead needles (Georgescu and Badea 1935). A species of *Camarosporium* has been recorded on dying *J. virginiana* in Yorkshire (Batko unpubl.), but there was no proof of its pathogenicity.

Hendersonia foliicola Berk. (Fungi Imperfecti, Sphaeropsidales) has been recorded on dying needles of *J. communis* on the Continent. This species has not been definitely recorded in Britain, but an unidentified *Hendersonia* was found on the needles of dying *J. virginiana* in Yorkshire (Batko unpubl.). Again there was no proof of pathogenicity.

Keithia (*Didymascella*) *tetraspora* (Phil. & Keith) Sacc. (Ascomycetes, Phacidiales) was found many years ago on *J. communis* in Scotland (Phillips, W. 1880); there are no recent records. It has also been recorded on the Continent (Grasso and Capretti 1955). Although, like other species of *Keithia*, it is almost certainly a parasite, it does not seem to have caused appreciable damage anywhere. The fungus attacks the needles, forming conspicuous brown apothecia on them. *Coccodothis sphaeroidea* Thies. & Syd. (Ascomycetes, Dothidiales), wrongly described by Miller (1935) as *Keithia juniperi* Mill., is superficially similar, and occasionally does similar damage to the foliage of *J. virginiana* in the United States (Miller, J. K. 1935).

Pithya cupressi (Batsch.) Fckl. (Ascomycetes, Pezizales) has been found on dying juniper foliage in Britain and on the Continent (Brouwer 1945), but there seems to be no proof of its pathogenicity.

DISEASES OF *LIBOCEDRUS*

Libocedrus decurrens, a native of North America, is the only species widely planted in Britain, but it has not been used for forest purposes. There are very few specific pathological records and none in Britain, although it is occasionally attacked by *Armillaria mellea.*

Lophodermium juniperinum (Fr.) De Not. (Ascomycetes, Phacidiales) has been found on the needles in California (Darker 1932). *Gymnosporangium libocedri* (P. Henn.) Kern. (Basidiomycetes, Uredinales) produces swellings and witches' brooms in America (Boyce, J. S. 1918). Neither fungus is seriously damaging to *Libocedrus*, but the latter was at one time regarded as a threat to pear, which, together with *Amelanchier* and *Crataegus*, is an aecial host (Jackson, H. S. 1914).

DISEASES OF *METASEQUOIA*

Metasequoia glyptostroboides, which was discovered in China as recently as 1945, has not been long enough in cultivation to have acquired many diseases. In the nursery, Damping off has been recorded in the United States (Damon and Snell 1948) and *Botrytis cinerea* in Sweden (Palm, B. 1952). In Britain a young tree some 2 metres in height has been attacked and killed by *Armillaria mellea.*

DISEASES OF *PSEUDOLARIX*

Pseudolarix amabilis, the one species in this genus, is only of very minor ornamental importance in Britain. The sole pathological record of interest is that of *Dasyscypha* (*Trichoscyphella*) *willkommii* (Hart.) Rehm. (Ascomycetes, Helotiales) in the United States (Miller and Aldrich 1936). This fungus, which is associated with dieback and canker in larch, is dealt with in Chapter 24.

DISEASES OF *SEQUOIA*

Both species of this genus, *Sequoia sempervirens*, the Redwood, and *S. gigantea*, Wellingtonia, which is now placed in the genus *Sequoiadendron*, grow well in Britain. *S. sempervirens* is much the less hardy of the two and is intolerant of exposure. In most places the foliage is browned, probably by dry cold winds, in all but the mildest winters.

Nursery diseases

Both species are extremely susceptible to *Botrytis cinerea* (Chapter 15). In fact it is almost impossible to raise *Sequoia* without some degree of *Botrytis* attack. *Phomopsis pseudotsugae* Wils. (Fungi Imperfecti, Sphaeropsidales) (Chapter 25) has been recorded both in Britain and abroad on *S. gigantea*, but does not seem to have done much damage.

Tumours

Galls have been recorded on the branches of *S. sempervirens* in France (Dufrénoy 1922*b*), at Eastbourne (Bull 1951), and in Norfolk (Peace unpubl.). Examination of these galls, which vary in size but may be as much as 20 cm. across, has always disclosed the presence of bacteria. E. L. Martin (1957) in Germany definitely associated galls on *S. giganteum* with *Bacterium tumifaciens*. Dufrénoy associated them with wounds, but this was not the case in England, the galls being so abundant as to suggest systemic infection of the trees on which they occurred. In one case infection was so heavy that the weight of the galls broke branches.

Smith (1942) was able to produce galls on both species of *Sequoia* by inoculation with *Bacterium tumifaciens* Sm. & Towns., the cause of Crown gall, but there is no proof that this is the organism concerned in natural infections.

Needle diseases

Mycosphaerella sequoiae Bonar (Ascomycetes, Sphaeriales) causes considerable browning of the needles of *S. sempervirens* in California (Bonar 1942). Infection starts at the tip of the needle and spreads to the base. The damage on severely affected trees is very conspicuous.

DISEASES OF *TAXUS*

Yew, *Taxus baccata*, is one of our three native conifers. Apart from its occurrence wild, chiefly on chalk and limestone soils, it has been widely planted in parks and gardens, particularly as a hedging plant.

Pirone (1948) has found foliage yellowing and dieback in yew planted on badly drained, acid sites, for which he recommended drainage improvement and liming. In Britain planted yew appears to tolerate a much wider range of soils, including certain acid sands, than those on which it occurs naturally.

Root diseases

Yew appears to tolerate infection by fungi, such as *Armillaria mellea*, and to continue normal growth despite the death of some of its roots. Old yews almost invariably support a number of decay-causing fungi, some of which are probably parasitic. Damaging infection of the roots by *Phytophthora cinnamomi* Rands, which is more important on cypresses (see above), has been recorded in Holland (Buddenhagen 1955) and in the United States (Torgeson 1954).

Diseases of needles and shoots

The commonest fungus on the needles of yew is *Sphaerulina taxi* (Cke.) Massee (Ascomycetes, Sphaeriales), the imperfect stage of which is *Cytospora taxifolia* (Fungi Imperfecti, Sphaeropsidales). Infected needles turn brown, and heavy attacks are accompanied by shoot dieback. It has been suggested that this is due to the fungus entering the stem from infected needles and girdling it (Callen 1938). The upper surface of the needles becomes studded with the minute perithecia. A number of cases of damage to hedges have been attributed to this fungus. It has also been recorded on *T. brevifolia* in the United States.

Diplodia taxi De Not. (Fungi Imperfecti, Sphaeropsidales) is also said to attack living needles of yew. These turn brown, but do not fall for some time. The minute black pycnidia are scattered over the lower surface of the needles.

Physalospora gregaria Sacc. (Ascomycetes, Sphaeriales), the imperfect stage of which is *Phyllostictina hysterella* (Sacc.) Petr., attacks both the needles and twigs of yew (Callen 1938). It is certainly not common, and its pathological status is very doubtful.

All three fungi occur on the needles and shoots of yew in Britain, but no information is available on their real importance here.

DISEASES OF *THUJA*

The North American *Thuja plicata* is the only species of forest importance in Britain, though several other species and their varieties have been fairly widely planted in parks and gardens.

Damage due to non-living agencies

In the autumn many of the older needles turn brown and later fall. While still attached they tend to give the tree a diseased appearance, so that this normal condition occasionally leads to reports of disease.

Root diseases

T. plicata is markedly susceptible to *Armillaria mellea*, as a cause of decay as well as of death, and to *Fomes annosus* decay.

Nursery diseases

The foliage of young *T. plicata* turns a bronze colour in cold winters. This is often assumed to be a disease and has frequently been confused with damage due to the fungus *Keithia thujina* (see below). In fact it is easily distinguished by the even nature of the discoloration and the absence of fruit-bodies. It is reversible and in the spring the plants recover their green colour.

K. thujina, which is the only important nursery disease of *Thuja*, is considered separately below.

Keithia needle blight

Pathogen

This disease, caused by *Keithia thujina* Dur. (Ascomycetes, Phacidiales), has been described in a Forestry Commission Leaflet (Anon. 1958*a*). No imperfect stage has yet been found. It appears to be an obligate parasite, since it has proved impossible to grow it in culture. The fungus, together with other species of the genus *Keithia*, was transferred by Maire (1927) to *Didymascella*, on the grounds that *Keithia* had already been used for a genus of *Labiates*. In practice, however, the possibilities of confusion are negligible, and in Britain the change would be very inconvenient, because, lacking any English name, the disease has become widely known simply as 'Keithia'. In view of the difficulty of having to say that the *Keithia* disease of *Thuja* is caused by *Didymascella thujina*, it seems best to continue the use of *Keithia* to embrace this and the other species of the genus.

The apothecia of the fungus appear as bright brown cushion-like structures, visible to the naked eye, usually on the upper side of the needles. Generally one to three appear on each infected needle. With age the apothecia turn almost black, and after spore discharge is completed they shrivel or fall out, leaving cavities (Fig. 74). Apothecia can ripen and discharge spores at any time from early May to November. This long period of activity is one of the reasons why control is so difficult. No conidial stage has been found. Pawsey (1957) has found that the fungus can overwinter, either as immature apothecia which ripen in the spring, or in the form of ascospores adhering to the needles. The relative importance of these two methods of overwintering is still a matter of investigation. In America it has been noted that snow cover during the winter increases infection in the spring (Boyce, J. S. 1948). Possibly the snow helps to maintain the viability of overwintering ascospores.

Symptoms and development

The fungus first causes browning of individual needles; shoots with many infected needles on them soon die so that the disease tends to result in progressive dieback. The mechanism of shoot death is uncertain, there being no evidence that the mycelium spreads into the shoot from the needles. Often, however, the attack is less severe, resulting in a characteristic scattering of dead needles on the shoots. The fungus can attack trees of any age. There is, however, a clear indication that resistance increases with age (Søegaard 1956), and cuttings from older trees show markedly higher resistance than seedlings. The disease is most serious in nurseries, occasionally staging autumn attacks on one-year seed-beds, but developing far more profusely on two-year beds and on transplant lines. Even on small plants attack is always initially most severe on the lower twigs; this characteristic is very noticeable on plantation trees. In severe attacks death eventually takes place, though this seldom happens with plants more than four years old.

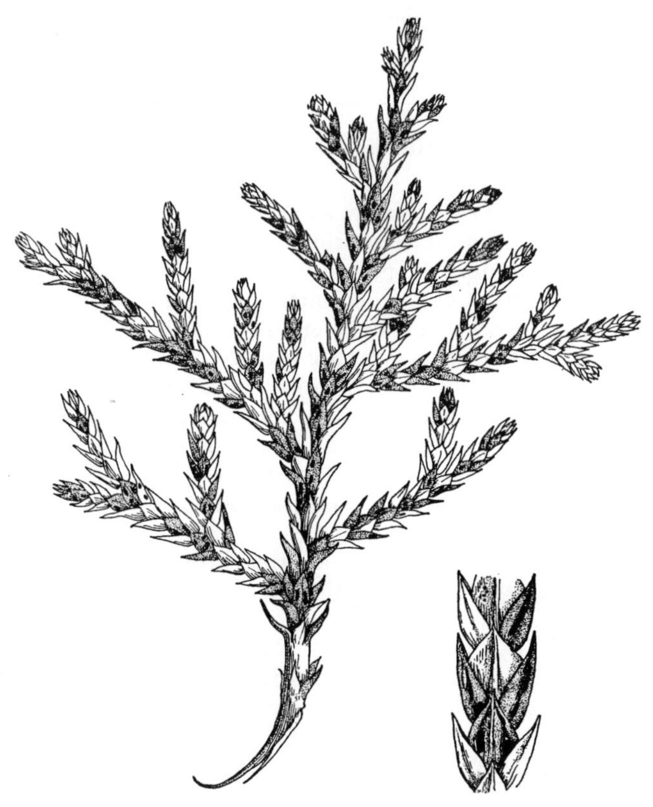

FIG. 74. *Keithia thujina* on *Thuja plicata*, Alice Holt Research Station, Hampshire, July (×1½, ×3). (J. C.)

Distribution and damage

The disease is known to occur on native *T. plicata* in North-west America, where it is more noticeable on forest trees than it is in Britain (Peace unpubl.). However, it has attracted very little attention there since Weir (1916) first described it. It must have been imported thence to Europe, and was first recorded in the British Isles by Pethybridge (1919) in Ireland. He suggested it might have reached there on seed; there is, however, no real evidence that seed transmission does occur. It was later described in Scotland by Alcock (1928). The fungus is now widely, but not absolutely generally, distributed in Britain; so far as is known there are no large areas still free. It has now been recorded in a number of other European countries and probably occurs in most places where *Thuja* is grown.

In plantations it is seldom damaging, though it occasionally checks the growth of trees in damp sheltered situations, or in the early thicket stages, by damaging or killing all the lower branches. It may also continue active for a time on trees already infected when they were planted out, thus adding to the difficulties of establishing already weakened plants. In nurseries in Britain the serious consequences of the disease are only mitigated by the comparatively small importance of *Thuja* as a forest species. In severe attacks it can kill over 80 per cent. of the plants, and cases where none of the surviving plants are worth retaining are comparatively frequent.

The disease appears to be confined to *T. plicata* and *T. occidentalis*. In most attacks, infection appears alarmingly uniform and there is no evidence of individual resistance, but two apparently resistant individuals of *T. plicata* have been found in an infected hedge. Søegaard (1956) found that the Japanese *T. standishii* was immune to the fungus. This immunity is dominant in *T. standishii* × *plicata* hybrids. The large-scale production of such hybrids might therefore be one means of avoiding the disease, assuming they are of equal forest value to *T. plicata*.

Control

So far this disease has proved very difficult to control by spraying. Early reports of success with several applications of Bordeaux mixture do not appear to have been soundly based. Incomplete experiments in Britain suggest that control by spraying may well be feasible, though possibly rather expensive, once a programme can be worked out to cover the extremely long period of spore production.

The fact that some nurseries are still free from the fungus suggested the possibility of concentrating the raising of *Thuja* in a limited number of such nurseries; a solution which would be acceptable to a state forest service with a choice of nurseries, but hardly of value to a landowner or commercial nurseryman. Out of eleven nurseries selected for experiment because they were remote from any existing *Thuja* and free of the fungus, nine have become infected after periods varying from nought to six years, and this despite the fact that in most cases no *Thuja* plants, only seed, were brought in from outside. Two of the nurseries under test still remain free after known periods of seven and nine years healthy *Thuja* production. They are both nurseries into which there is virtually no import of plants of any species (Peace 1958*b*).

A more practical possibility, where a number of nurseries is available, would be to rotate the cultivation of *Thuja*, raising only one crop in each nursery. Each year *Thuja* would be sown in a fresh nursery, the plants remaining there for three to four years until they were large enough for planting. Using six or seven nurseries, selected as having no *Thuja* in the immediate vicinity, each would be free of *Thuja* plants for at least two years before it had to be used again for that species. This should entirely prevent build-up of infection from one crop to the next. There is evidence that it takes two or three years to build up a serious attack from an initial light infection, so a damaging attack on any particular crop could only take place if it became infected in the first year. This method of *Keithia* avoidance is now under test.

It is obviously a grave mistake to raise *Thuja* in the presence of hedges or specimen trees of the same genus, since these often carry a light infection sufficient to start epidemic attacks on the young trees near them. Dense seedbeds are certainly worse attacked than sparse ones, probably because damp, crowded conditions favour infection (Alcock 1928). Sowing and transplanting *Thuja* in mixture with other species have been suggested as one means of lessening attack. There is some evidence that it does this, but it certainly does not prevent infection.

Other diseases of needles and shoots

Cercospora thujina Plak. (Fungi Imperfecti, Moniliales) causes a serious leaf and shoot dieback of *T. orientalis* in the United States. Chupp (1954) considers that this fungus should be placed in the genus *Heterosporium*. The fungus causes the needles and twigs to turn brown, and may eventually kill whole branches or even the entire tree. There is also formation of bark cankers on small twigs (Plakidas 1945). The disease can be controlled by copper sprays.

Very similar damage is caused to *T. orientalis* in Oregon by *Coryneum berckmanii* Milb. (Fungi Imperfecti, Melanconiales). Small black fruit-bodies are produced on the needles or on the needle-bearing stems (Milbrath 1940).

T. orientalis was found by Rudd Jones (1954*b*) to be susceptible on inoculation to *Monochaetia unicornis*. Wagener (1948) found *T. plicata* slightly susceptible, but *T. orientalis* and *T. occidentalis* resistant to natural infection by *C. cardinale*. Both fungi cause serious canker diseases of cypresses in different parts of the world (see above), but neither has developed to any extent in nature on *Thuja*. Boudru (1945) recorded *C. thujinum* Dearn., which has not been found in Britain, on plants of *T. plicata* severely attacked by *Keithia thujina* in Belgium. The minute black fruit-bodies are produced on the underside of the needles. In the presence of the more damaging fungus it was impossible to decide whether it was a parasite.

Fungi of the genus *Pestalozzia* (Fungi Imperfecti, Melanconiales), probably in most cases *P. funerea*, have frequently been found in Europe and America on plants of *Thuja* with dying twigs and foliage, in some cases in association with *Keithia* (Laing 1929; Boudru 1945). A *Pestalozzia* is not, however, as has been suggested, the conidial stage of *Keithia*. In Britain *Pestalozzia*, in some cases certainly *P. funerea*, has been found on a number of occasions apparently acting as a parasite. Laing (1929) carried out successful inoculation experiments with it on *T. plicata*. However, this fungus is neither so active nor so important a parasite as *K. thujina*.

DISEASES OF *TSUGA*

The Western hemlock, *Tsuga heterophylla*, is the only species of forest importance in Britain. Several others, particularly the Eastern hemlock, *T. canadensis*, have been fairly widely planted in parks and arboreta. Apart from root fungi and decay, *Tsuga* appears to be virtually free from dangerous diseases.

Root diseases

There is evidence that *T. heterophylla* is particularly susceptible to attack by *Fomes annosus*. An extensive literature on wood-rotting fungi attacking *T. heterophylla* in its native North-west America suggests that it is generally prone to decay.

Stem diseases—bark and cambium

Denyer (1952) reported a species of *Cephalosporium* (Fungi Imperfecti, Moniliales) causing resinous, sunken cankers on suppressed *T. heterophylla* in British Columbia. More recently a *Cephalosporium* has been associated with defoliation and dieback of *T. canadensis* in Ohio (Strobel and Vermillion 1956).

Phomopsis pseudotsugae Wils. (Fungi Imperfecti, Sphaeropsidales) has been commonly reported as occurring on dead shoots of *T. heterophylla* in Britain and on the Continent (Spaulding 1956). It is doubtful if these records are really critical and some other species of *Phomopsis* may well be involved; nor is there any proof that the fungus was acting as a parasite.

Diseases of needles, shoots, and cones

Keithia (*Didymascella*) *tsugae* (Farl.) Dur. (Ascomycetes, Phacidiales) is locally prevalent on *T. canadensis* in the eastern United States, but it is not regarded as a serious disease. The effect on the tree is similar to that of *K. thujina* on *Thuja* (see above). The sole European record is of a single tree near Peebles, in south Scotland (Wilson, M. 1937*a*), where *T. canadensis* was lightly infected in the neighbourhood of other unattacked *Tsuga* species, including *T. heterophylla*. This, and the fact that the fungus has not been recorded on *T. heterophylla* in North America, supports the hope that our only important forest species of *Tsuga* may be resistant or even immune to this fungus.

In America a number of rust fungi (Basidiomycetes, Uredinales) attack the needles and cones of species of *Tsuga*. The most damaging of these is *Melampsora farlowii* (Arth.) Davis, which produces reddish, waxy, erumpent telia on the needles, young shoots and cones of *T. canadensis* (Hepting and Toole 1939). The other stages are unknown. It is said to be controllable by weekly spraying in May with lime sulphur (Boyce, J. S. 1948).

M. abietis-canadensis (Farl.) Ludw., which forms its aecia on *T. canadensis*, has its uredial and telial stages on poplar. Infected *Tsuga* needles are distorted, but so few are infected that little damage is done. Young shoots and cones are also attacked, the latter being made conspicuous by the masses of powdery yellow spores. Some loss of viable seed may result, since infected cones wither and fall.

Two other rusts have their aecia on *T. canadensis* and *T. caroliniana* in North America. They are *Pucciniastrum myrtilli* (Schum.) Arth., the uredial and telial stages of which occur on various Ericaceous shrubs, and *P. hydrangeae* (B. & C.) Arth., the uredial and telial stages of which occur on *Hydrangea arborescens*. Neither is pathologically significant on *Tsuga*.

Two rust species produce their aecia on *T. heterophylla* and *T. mertensiana*; they are *Caeoma dubium* Ludw., and *Uraecium holwayi* Arth. No alternate stages are known. Neither species is of any pathological importance.

Death of needles and small twigs of *T. canadensis* due to attack by *Rosellinia herpotrichoides* Hept. & Dav. (Ascomycetes, Sphaeriales) has been recorded in North Carolina (Hepting and Davidson 1937). The symptoms are generally similar to those caused on other conifers, including *T. heterophylla*, by the snow mould, *Herpotrichia nigra* (Chapter 15). The needle-bearing parts of the twigs become covered with a greyish-brown mycelial mat, in which perithecia are produced on the lower twig surfaces and leaf bases. The disease tends to occur mostly on the lower branches of small trees, but can cause up to 80 per cent. defoliation on heavily attacked individuals.

Acanthostigma (*Trichosphaeria*) *parasiticum* (Hart.) Sacc. (Ascomycetes, Sphaeriales), which occurs much more commonly on Silver fir (see above), has also been recorded on the needles of *T. canadensis* on the Continent. The sole record in Britain was on *Abies alba*.

27

DISEASES OF OAK

Two species of oak, *Quercus robur* and *Q. sessiliflora*, are native to Britain. In forest practice they have seldom been considered separately, and this lack of distinction has extended to their pathology, so that there is scarcely any information available on their relative susceptibility to disease. The only other oak having any real importance in forestry at the moment is the American Red oak, *Q. rubra*, though the Turkey oak, *Q. cerris*, has been widely planted, occasionally in plantations. A very large number of other exotic oaks, of which the best known is probably the evergreen oak, *Q. ilex*, can be grown in Britain.

DAMAGE DUE TO NON-LIVING AGENCIES

Most of the deciduous oak species are often damaged by spring frosts, while the evergreen ones frequently suffer leaf injury in cold winters. Frost crack occurs rather commonly in oak.

Stagheadedness is a common condition with old oaks, especially on heavy soils. It is often associated with variations in the water table. Oaks are among the longest-lived trees in Britain, and thus, since they occupy the same site for so long, the likelihood of the site altering during their lifetime is greater than for shorter-lived trees. Despite their liability to dieback, however, they are very tenacious of life, and many oaks, already several centuries old, have continued to live in a state of decrepitude for more than another hundred years.

Shakes (Chapter 20) occur rather commonly in the timber of oak on some soils, causing considerable losses on conversion.

NURSERY DISEASES

Rosellinia root rot

Pathogen

Rosellinia quercina Hartig (Ascomycetes, Sphaeriales), the cause of this disease, is a soil-inhabiting fungus. On attacked plants a great deal of the mycelium is external, forming over the roots and collar a felt, which is at first white and later brownish-black. The mycelium is frequently aggregated into strands, and in this form it often grows through and over the soil. During the summer microscopic conidia arise from this mycelium, the spores from which probably play a part in spreading the disease. In autumn perithecia are formed in the mycelium around the oak stems, just above soil-level, and sometimes on the surface of the soil around infected plants. They are small, black, globose bodies 0·8 to 1·0 mm. in diameter. The ascospores are fully formed by

September, but are not released until the following spring. In addition small black sclerotia are formed among the mycelium. The fungus and the disease have been fully described by Waldie (1930).

Symptoms and development

The disease normally affects oak seedlings, not more than three years old. The first indication of its presence is the withering foliage of the attacked plants, which are usually found in patches. On lifting, the seedlings show the characteristic web of dark mycelium over the upper roots. Infection is probably by small side roots, but the mode of attack requires further investigation. Direct penetration of living tissue by the fungus is said to be possible (Woeste 1956). One instance has been recorded where infection appeared to have taken place through the severed ends of undercut roots; but there is certainly no general association between the disease and undercutting.

The fungus certainly spreads over the surface and through the soil from plant to plant. The part played by the conidia, and later by the ascospores, has not been ascertained. The sclerotia can persist in the soil for several years. Dry soil conditions usually stop the spread of the fungus, though seedlings already infected die all the faster, since their water-supply is restricted.

It has been suggested on several occasions that the disease can be carried from place to place on acorns, and there is some indirect evidence for this supposition.

Distribution and damage

This disease has been known for a long time in Europe, where it has also been recorded on Turkey oak and possibly on other hardwoods, but it has never been regarded as a serious factor in the raising of oak. In Britain its occurrence is very sporadic.

Despite its ability to persist in the soil, repeated attacks in the same nursery do not necessarily occur. Thus, when the disease is found, it is often unexpected and appears to bear no relation to the previous history of the nursery. This implies the existence of some controlling factor and it has been suggested that bad drainage may favour the disease. However, attacks have occasionally occurred in well-drained, friable soils, so unsuitable water conditions cannot be the sole explanation. Sometimes the disease has appeared in nurseries on old agricultural ground, only a few years after they were started, and in one instance on the first crop of oak. It is normally restricted to nurseries, but has also been recorded on oak sowings in the forest. The disease can result in nearly complete loss of oak seed-beds, but more often less than 20 per cent. are killed. If *Rosellinia* occurred more commonly it would be quite a serious disease.

Control

Waldie (1930) reported successful control by removal of all diseased seedlings and sterilization of the infected patches with a flame gun, followed by a 5 per cent. solution of carbolic acid. Other methods of soil sterilization would probably be equally effective. The infected ground should not be used again for fresh crops of oak for three or four years.

Seedlings from infected beds, even if they appear healthy, should not be lined out in other parts of the same nursery, and should never be sent to other nurseries. There is no reason why they should not be used for direct planting in the forest or lined out on the same ground, after the infected patches have been sterilized.

Oak mildew

Pathogen

The most important leaf disease of oak in Britain, Mildew, is caused by the fungus *Microsphaera alphitoides* Griff. & Maubl. (Ascomycetes, Erysiphales). It has been described in a Forestry Commission Leaflet (Anon. 1956*a*), and more fully by Woodward, Waldie, and Steven (1929). It was formerly known by the rather ambiguous name *M. quercina* (Schw.) Burr. It has been suggested that Oak mildew in North America is caused by a different species of *Microsphaera*. The conspicuous white mycelium is largely superficial on both surfaces of the leaves and on young shoots, the only penetration is by the haustoria. From mid-summer onwards, conidia are produced in chains, giving the leaf surfaces a powdery appearance. The conidia cause fresh infections during the remainder of the summer.

The perfect stage of the fungus, a cleistocarp with relatively long appendages, is comparatively rare, especially in Britain. The cleistocarps remain on the surfaces of fallen leaves producing ascospores in the spring. However, ascospores have seldom been observed to germinate, so that the perfect stage is probably of little importance as a means of overwintering. The usual method is by resting mycelium in between the scales and leaf primordia of the bud. It is not known how many buds are infected in the autumn, but the percentage of plants on which overwintering is successful, and from which infected shoots emerge in the spring, is very small. On the Continent overwintering in the form of chlamydospores on the surface of fallen leaves has been reported, but this method is not known to occur in Britain.

Symptoms and development

The disease appears in the spring, usually about mid-May, on the leaves and shoots arising from buds which contained resting mycelium during the winter. Cinnamon-coloured spots first appear, but these soon turn white and the superficial white mycelium then spreads over the leaf surfaces. As the attack develops, the leaves become distorted and the young leaves may be reduced in size (Plate XIV. 2). In a severe attack growth of the shoot may cease prematurely. Later the leaves wither and fall. Shoots defoliated late in the season may be cut back by early frosts, as they do not ripen properly. In the latter half of June secondary infections appear on plants on which the fungus failed to overwinter, and on neighbouring one-year-old seed-beds. The fungus attacks succulent growing tissue, so that after mid-summer damage is confined to shoots that are still growing on young vigorous plants, to lammas-shoots and to recovery shoots resulting from injury by this disease, or by some other damaging agency, earlier in the season. The effect of the disease is mainly a reduction in growth, leading to plants which are weaker and bushier than normal.

Towards the end of the summer *Microsphaera* is itself sometimes parasitized by another fungus, *Cicinnobolus cesatii* De Bary (Fungi Imperfecti, Sphaeropsidales), which has a dark mycelium, and so makes the leaves appear darker. Possibly this fungus sometimes lessens the severity of later attacks.

Distribution and damage

The fungus appears to be universally distributed on oak in Britain, and it has been recorded from most other countries where oak occurs. There is a suggestion that it spread over Europe from the west at the beginning of the century. It is certainly not mentioned in the earlier textbooks.

It has been recorded on a very large number of oak species, including *Q. robur*, *Q. sessiliflora*, which is considered to be more resistant than *Q. robur*, *Q. rubra*, *Q. lusitanica*, *Q. coccinea*, *Q. mirbeckii*, and *Q. cerris*, which, however, is said to be resistant in Britain. Information on the relative susceptibility of different varieties is very inconsistent. There are provenance and probably individual differences in *Q. robur* in susceptibility to the fungus, but they are too small to be of practical value (Cieslar 1923; Rack 1957*a*). Klika (1922) found that *Q. robur pectinata*, a variety with pinnatifid leaves, was highly susceptible, whereas the fastigiate form, *Q. robur fastigiata*, was particularly resistant. *Q. rubra* is sometimes regarded as resistant to the fungus (Klika 1922), and at one time it was suggested as a substitute for native oaks in regions of France badly afflicted with Oak mildew (Pardé 1928). The fungus has occasionally been recorded on beech.

As a nursery disease on oak it is often responsible for considerably increasing the number of culls and greatly reducing the quality of the remaining plants. It appears to do little damage to vigorously growing coppice oak, though the attacks are often rather spectacular. It seldom causes appreciable injury to young plantation oaks, once they have started to grow vigorously, though it occurs quite commonly on them. Newly planted oak may be delayed quite seriously by the disease (Guillebaud 1930), and spraying, though probably not economic in a plantation, resulted in greatly improved growth. Oak planted among existing oak coppice are particularly liable to infection.

It seldom attacks older trees to any appreciable extent if they are growing normally; but if they have been defoliated by caterpillars in the spring or cut back by frost, heavy attacks may develop on the recovery shoots, which arise at a time when abundant conidia from the normal attack on younger trees and coppice are available to infect them. Thus trees, already weakened by frost or insect defoliation, are often subjected to a second partial loss of leaves due to mildew. Since nearly complete defoliation by Oak roller moth and Winter moth is of very frequent occurrence in some areas of England, the addition of mildew as a secondary defoliator is quite serious. These two agencies acting together probably play a large part in the stagheadedness of oak, which is very noticeable in some parts of England. It must be borne in mind, however, that older oak are very sensitive to soil water conditions and some of the dieback, especially on heavier soils, is due primarily to this cause.

Armillaria mellea sometimes attacks oaks weakened by defoliation, adding a fourth factor to those already mentioned as associated with Oak crown dieback. This complex of factors has been described in Britain (Day, W. R. 1927*b*; Robinson 1927; Osmaston 1927), in France (Demorlaine 1927), in

Germany (Falck 1924), and in Jugoslavia, where it was especially serious (Langhoffer 1929). There is no doubt that Oak dieback, with mildew as one contributory factor, is of very general occurrence, but its effects have not been so serious as was at one time anticipated. Possibly, during the nineteen-twenties when it was regarded very seriously, the disease complex was favoured by climatic circumstances.

Control

The fungus is comparatively easily controlled by spraying. Its sensitivity to sulphur is well illustrated by its absence from industrial districts (Köck 1935). The best substance for spraying appears to be colloidal sulphur at rates around 3 lb. per 100 gallons of water per acre. It is essential that spraying should start at the first appearance of the primary infections and be continued thereafter at intervals of two to three weeks. If after three or four applications the disease appears to be completely under control, sprayings can be discontinued until its reappearance. Copper sprays and a number of other substances have been used successfully in Germany (Müller-Kögler 1954; Rack 1957*b*).

In the forest spraying would not be economic, and in any case the damage done by the fungus is usually comparatively slight. If oak has been planted among oak coppice, it may prove desirable to cut the coppice early in the summer to lessen infection on the planted trees.

Other nursery diseases

Coniothyrium quercinum Sacc. (Fungi Imperfecti, Sphaeropsidales) forms small, intensely black pycnidia on oak twigs and acorns. It has been recorded on the latter in Britain, but in Czechoslovakia it has been associated with a nursery disease of oak seedlings (Jančařík and Urošević 1958).

ROOT DISEASES

Armillaria mellea frequently attacks oak, and was reckoned as one of the factors responsible for a general decline of oak in parts of Europe in the nineteen-twenties (see Oak mildew, under Leaf and Shoot Diseases). *Fomes annosus* can attack and kill oak, but usually only in places where the fungus is already established on conifers. Red oak may be attacked in this way when it has been planted among conifers with the idea of improving soil conditions. *Polyporus dryadeus*, common as a cause of decay in oak, may sometimes be actively parasitic on it (Chapter 16).

STEM DISEASES—WILTS

Oak wilt

Pathogen

This disease, which is restricted so far to eastern North America, is caused by the fungus *Ceratocystis* (*Endoconidiophora*) *fagacearum* (Bretz) Hunt. (Ascomycetes, Sphaeriales). Before the perfect stage was discovered it was known by the name of the conidial stage, *Chalara quercina* Henry (Fungi

Imperfecti, Moniliales). The fungus can live in the sapwood vessels of the host and probably produces conidia there, which may aid the spread of the fungus within the tree. The perfect stage is produced, together with enormous quantities of conidia, on mycelial mats formed under the bark of recently killed trees. The fungus is heterothallic and perithecia are produced on the mats only following cross-fertilization by conidia of another mating type. Normally only a single type occurs on an infected tree and cross-fertilization is usually effected by insects bringing conidia of the opposite mating type from another tree. The perithecia are black flask-shaped structures borne inconspicuously in the mycelial mat (Bretz 1952*a*; Hepting, Toole, and Boyce 1952).

A very large literature, most of which is highly specialized, has grown up around this disease, but summaries have been given by Hepting (1955*a*), by Barnett and True (1955), and by Fowler (1958).

Symptoms and development

The disease takes the form of a progressive wilting and dieback. Dark streaks usually appear in the sapwood of infected trees. The mycelial mats mentioned above are a valuable diagnostic feature, but are not produced in the early stages of the disease. The wilt is due to blockage of the vessels and possibly to toxins produced by the fungus (White, J. G. 1955; Boyer 1958).

Over long distances the fungus is carried from tree to tree by insects, particularly by Nitidulid beetles, which heavily colonize the sporulating mycelial mats (Himelick and Curl 1958). Birds and squirrels may also act as carriers (Himelick and Curl 1955). Above ground, infection appears to take place only through wounds which, because of the fermenting sap, prove attractive to insects, some of which may have been previously feeding on a mycelium mat of *Ceratocystis*. Root grafts, which occur very commonly between neighbouring oaks of the same or closely related species, form a rapid means of transmission over small areas (Kuntz and Riker 1950*a*), and are the main means of local spread. Such spread may, therefore, be confined to species between which root grafting is possible.

Trees of the Red oak group (*Q. rubra*, *Q. palustris*, etc.) generally die in the year of infection, but those of the White oak group (*Q. alba*, *Q. stellata*, etc.) may survive for several years (Jones, T. W. 1958).

Distribution and damage

The disease is centred in the upper Mississippi valley in the United States, but has been recorded in some eighteen states, ranging from Nebraska and Kansas in the west to Maryland and Pennsylvania in the east, and from the Great Lakes in the north to Arkansas, Tennessee, and North Carolina in the south (Bretz and Jones 1958). Aerial surveys have been extensively used to check the distribution of this disease.

It has probably been present in parts of this area for thirty or forty years, slowly building up. Even now, although considerable damage has been done, the disease is regarded seriously, mainly because it appears as a threat to the future of oak in the United States. Careful inquiry by the European Plant Protection Organization reveals that the disease is still absent from Europe.

Bretz (1952*b*; 1955) reported that all species of oak and related genera in the family *Fagaceae*, subjected to natural or artificial inoculation with the oak wilt fungus, have proved susceptible. These include *Q. robur*, *Q. ilex*, and *Q. cerris*, as well as the American Chestnut, *Castanea dentata*. The fungus is also capable of producing active disease symptoms in apple, and can persist and spread without doing appreciable damage in several other woody genera (Bart 1957). The big variation in susceptibility of the American oaks in nature, however, does mean that further information on the behaviour of the disease under natural conditions is required before its potential danger to European oaks can be assessed.

Control

Efforts are being made in the United States to lessen the incidence of the disease by cutting and, if possible, burning infected trees. If the timber is retained, it is sprayed with an insecticide to reduce the chances of insect spread. Poisoning of infected trees has also been tried, but has not proved wholly successful. Neighbouring oaks are kept under careful observation, as the fungus may have spread to them by root grafts. Indeed, in woodland the poisoning of all the oaks within a 50-foot radius of an infected tree has been recommended. It is generally accepted that control must be limited to keeping losses to an acceptable minimum; complete eradication is regarded as impossible.

The disease is unlikely to be carried directly on acorns (Bretz and Buchanan 1957), though it could be carried by infected insects in or on them, nor is it very likely to be transported on young plants. Oak timber appears to be a greater danger, since the fungus can remain alive in oak logs for at least three months, if they are not too dry (Spilker and Young 1955). It does not persist in seasoned timber (Englerth, Boyce, and Roth 1956). Fortunately the trade in oak logs from America to Europe is very small. Under present British regulations the import of oak plants only is controlled.

Unfortunately the disease is obvious only on susceptible species. On resistant species it may well be confused with any of the minor diebacks, sometimes physiological in cause, that afflict oak trees. It cannot be assumed therefore that the disease, if it were to reach this country, would immediately disclose its presence by the severity of its attack. However, it might be found that it did not seriously damage European oaks under natural conditions. It is certainly a disease that should be kept out of Britain.

STEM DISEASES—BARK AND CAMBIUM

Diaporthe dieback

Pathogens

Two species of *Diaporthe* (Ascomycetes, Sphaeriales), *D. taleola* (Fr.) Sacc. and *D. leiphaemia* (Fr.) Sacc., have been associated with dead patches on the bark of young oak stems.

Small dark brown oval or circular stromata are formed in the bark. These are faintly visible on the bark surface, which later splits to reveal them. They then bear microscopic conidia and later the perithecia, the necks of which appear as small projections on the dead bark. The two species are very difficult to distinguish in the field, but *D. taleola* has a very distinct black border round the stromata which can be seen if the outer bark is pulled away.

There is still considerable confusion over the identification of the conidial stages of these fungi (Wehmeyer 1933). They have been placed in *Cytospora*, *Myxosporium*, *Fusicoccum*, and *Phomopsis* among other genera. Here belongs the disease that is occasionally attributed to *Myxosporium lanceola* (Fungi Imperfecti, Melanconiales), a supposed conidial stage of *D. leiphaemia*. Occasional reports of oak cankers associated with *Phomopsis* species (Fungi Imperfecti, Sphaeropsidales) may be based on a similar relationship.

Symptoms and development

The disease shows first as slightly sunken areas on the smooth bark of young oaks, or on the smooth-barked limbs of older trees. These lesions are purplish or greyish in colour and of irregular outline, though normally vertically elongated. Later they darken and may spread to encircle the stem. Alternatively, gradual healing may take place. The extension of these necrotic areas is normally accompanied by cracking of the bark. No very marked cankering results, because the fungus usually attacks weak trees, the healing responses of which are rather feeble. This is usually a disease of weakened, young trees, occurring particularly on the suppressed trees in dense stands and on trees suffering from drought. However, it has also been associated with branch and twig dieback of older oak (Moreillon, H. 1918).

Observations on 11–15-year-old infected stands in Lincolnshire indicated that the progress of the disease is very slow, indeed many of the stronger trees appeared to be recovering. In this case the fungus involved was *D. taleola*; but *D. leiphaemia* has been associated with drought damage to oak in Norfolk. *D. taleola*, however, is certainly the more important of these two fungi, and is the one normally associated with this disease on the Continent.

Distribution and damage

Probably both fungi are generally distributed on sickly and even dead young oaks. It is doubtful whether they can appreciably damage healthy trees. They are likely therefore to attract attention only when they infect oaks already weakened by overcrowding, drought, or some other factor.

Control

No control appears to be necessary. Affected trees are likely to be removed in the ordinary course of thinning.

Chestnut blight fungus on oak

The fungus causing Chestnut blight, *Endothia parasitica* (Murr.) A. & A. (Ascomycetes, Sphaeriales), has been recorded on a number of oak species (Chapter 29). In the United States only Post oak, *Q. stellata*, appears to be infected (Clapper, Gravatt, and Stout 1946), but in Italy it has been recorded on *Q. pubescens*, *Q. sessiliflora*, and *Q. ilex* (Biraghi 1951; Gravatt 1952), while in Jugoslavia it has been found on supposed *Q. robur* × *sessiliflora* hybrids (Krstić 1952). Except on *Q. pubescens*, which is attacked in the same manner as *Castanea*, the fungus produced only very slow-spreading or even stable cankers. There is no need, therefore, to anticipate a threat to the cultivation of oak in Europe. Young oaks, and particularly unbarked oak timber, may well act as carriers of the disease to new areas. This, rather than the damage to oak itself, is probably the chief danger in the susceptibility of oak to this fungus.

Other diseases of bark and cambium

A very large number of fungi have been associated with cankers and bark lesions on oak in various countries. Conversely cankers on oak, the causes of which have never been discovered, have been frequently recorded both in Britain and abroad. It is possible to mention only a few of the more important fungi. A great many of those recorded appear to have no actual or potential pathological significance.

Dothidea noxia Ruhl. (Ascomycetes, Dothideales) has been associated with branch canker and dieback of the tips of oak twigs in Germany (Ruhland 1904; Bavendamm 1935), but not in Britain. It causes a yellow or red discoloration of the bark, on which small pycnidial pustules appear during the spring and early summer. The pycnidia eventually fall out leaving tiny scars. This is the conidial stage *Fusicoccum noxium* (Fungi Imperfecti, Sphaeropsidales). After the death of the twig, groups of perithecia appear. The earlier records were on *Q. robur*, but later ones mainly on *Q. rubra*. The disease is not serious.

Roland (1945) associated *Diplodia quercina* West. (Fungi Imperfecti, Sphaeropsidales) with a twig dieback of oak in Belgium. The fungus was acting as a wound parasite and was unable to spread far, so that each affected twig or small branch was separately infected. He associated the disease with poor growing conditions, particularly lack of water. A closely related species *D. quercus* Fckl. has been recorded on twigs of *Q. robur* in Britain, but not as a cause of disease. It appears that *Diplodia* is one of the many fungi only able to invade weakened oaks.

Colpoma quercinum (Fr.) Wallr. (Ascomycetes, Phacidiales) has been cited (Twyman 1946*b*) as the cause of dieback in young oak trees, coppiced oak, and the smaller terminal branches of older trees. It has also been described under the name *Clithris quercina* (Pers.) Rehm. Pycnidia, producing apparently functionless spores, are formed before the apothecia, which appear as small longitudinal slits in the bark. The apothecia usually occur on dead twigs, but may sometimes be found on the diseased portion of living branches. Black zone-lines delimit the areas of diseased wood in the twigs. The fungus is widely distributed in Britain, but it is certainly not a serious cause of disease and may well, like other dieback fungi on oak, be dependent on previous weakening of the tree by other causes.

Various species of the genus *Nectria* (Ascomycetes, Hypocreales) (Chapter 18) have been connected with canker on oak, sometimes as supposed primary parasites, sometimes in a secondary capacity. Probably the most important instance of oak canker, associated with *Nectria*, has been described recently in France (Rol 1951; Barriéty *et al.* 1951; Moreau and Moreau 1952). This disease is attributed primarily to *Phytophthora cinnamomi* Rands.

(Phycomycetes, Peronosporales), one of the two species of *Phytophthora* (the other is *P. cambivora*) associated with the so-called Ink disease of *Castanea* and other hardwoods (Chapter 29). Because the same parasite is involved, this canker on oak has unfortunately been called the Ink disease of oak. Though *Phytophthora* initiates the cankers, they are invaded and possibly extended by a number of species of *Nectria*, among which a form of *N. rubi* Osterw. is the most important. The disease is so far localized in the Pays Basques, affecting chiefly *Q. rubra*, but also *Q. robur* and *Q. toza*. Though the disease is of little importance as a killing agency, the basal cankers spoil the most valuable length of timber. *Phytophthora* cankers on oak have also been recorded in the United States (Miller, P. A. 1941).

In America large target cankers on older oaks and girdling of younger trees are frequently associated with the fungus *Strumella coryneoidea* Sacc. & Wint. (Fungi Imperfecti, Moniliales) (Sleeth and Lorenz 1945; Hansborough 1951). The fungus apparently invades the main stem through small branches that it has killed. Small nodules, apparently sterile fruit-bodies, soon appear on the diseased tissue, but the fruit-bodies proper, which are dark-brown rounded pustules about 1·5 mm. in diameter, are only occasionally produced on the dead bark of cankers on living trees, though they occur abundantly after trees have been girdled and killed.

A number of wood-decaying *Basidiomycetes* have been associated with canker formation on oak. Among these are *Polyporus hispidus* (Bull.) Fr. (Toole 1955), *Poria spiculosa* Campbell & Davidson (Toole 1954), *Stereum rugosum* (Pers.) Fr. (Schönhar 1951), and *S. gausapatum* Fr., which is also responsible for the most serious decay of oak in Britain (Chapter 19). When fungi of this nature are associated with cankers, they are probably acting as weak parasites around the bases of dead branches, or in wounds, i.e. at positions where there is already some exposed dead tissue. In the case of *S. rugosum*, however, successful inoculations were carried out on healthy oaks (Banerjee 1956).

DISEASES OF LEAVES AND SHOOTS

Oak mildew

Though Oak mildew plays quite an important role in the forest, it is far more serious as a nursery disease, and has already been dealt with under that heading.

Other leaf diseases

A large number of fungi have been associated with leaf disease on oak, but none approaches oak mildew in importance. *Sclerotinia candolleana* (Lév.) Fckl. (Ascomycetes, Helotiales) occasionally appears epidemically on the leaves of *Q. robur* and *Q. sessiliflora* in Britain (Wilson and Waldie 1926*b*). Infected leaves show yellow spots, which later become light-brown areas of dead tissue, usually 5 to 6 mm. across, but sometimes spreading irregularly over the greater part of the leaf. Under moist conditions numerous black sclerotia, about 3 mm. across, are produced during the winter on fallen leaves. The apothecia of the perfect stage develops from these sclerotia in the spring and early summer of the following year. A thread-like stalk grows out, bearing a small flat reddish cup, also about 3 mm. across, in which the asci are borne. The ascospores infect the young leaves; shoots are not infected. Little is known about the occurrence of the fungus in Britain; it is probably fairly generally distributed, but very seldom does appreciable damage.

Taphrina coerulescens (Desm. & Mont.) Tul. (Ascomycetes, Taphrinales) causes leaf blisters accompanied by some distortion of the foliage on oak. The blisters vary in size from 1·5 mm. to 12 mm. in diameter, and are sometimes pink or purplish in colour. The spore-bearing surface on the underside of the blisters is at first silver-grey, but later becomes velvety-brown (Weber 1941). In America it does enough damage to necessitate control by spraying (Goode 1953), strong winter sprays being used to kill overwintering spores. Though it has been recorded on a number of occasions in Britain, it has never reached epidemic proportions.

Several species of *Phyllosticta* (Fungi Imperfecti, Sphaeropsidales) have been recorded abroad as causes of leaf spots on oak. *P. maculiformis* Sacc., the perfect stage of which is *Mycosphaerella maculiformis* (Pers.) Schroet. (Ascomycetes, Sphaeriales), has been recorded in Britain, but was not doing serious damage.

The damage caused by *Gloeosporium* (Fungi Imperfecti, Melanconiales) is more serious because the young shoots are often attacked, as well as the leaves. In the United States *G. nervisequum* (Fckl.) Sacc., the perfect stage of which is *Gnomonia veneta* (Sacc. & Speg.) Kleb. (Ascomycetes, Sphaeriales), was regarded as the sole fungus responsible for this anthracnose disease both on oak and on plane (Waterman 1951), though it is much more important on the latter (Chapter 35). Westerdijk and Luijk (1924) considered that in Europe two species were involved, the one on oak having consistently larger spores. However, isolations of the fungus from plane were able to infect oak, so that diagnosis cannot necessarily be based on the host genus alone. This work, including the cross inoculations, has more recently been repeated in the United States (Schuldt 1955). The form on oak has been separated in Europe as *G. quercinum* West. It has been recorded on a wide range of oak species in Germany and the Netherlands, but is not generally regarded as very serious, the damage often being restricted to brown spots on the leaves. The fungus appears to be quite common on oak in Britain, but has only once been associated with serious damage. On that occasion the brown spots on the leaves were so numerous as to give the tree a withered appearance, but there was no leaf-fall (Fig. 75).

The fungus *Morenoella quercina* Ell. & Mart. (Ascomycetes, Hemisphaeriales), which causes leaf-spots on oak in the south-eastern United States, is of particular interest, because the very conspicuous black mycelium is at first largely superficial, but nevertheless extracts some food substances from the leaves. At a later stage there is slight mycelial penetration, but the fungus never causes premature defoliation. It is thought that the reduced growth of heavily infected plants is due largely to loss of light. It occurs on *Q. rubra*, *Q. coccinea*, and various other species, causing minor damage to small plants, but none at all to older trees (Luttrell 1940*b*).

Only one rust fungus, *Uredo quercus* Brond. (Basidiomycetes, Uredinales), has been recorded on oak in Britain. Here and on the Continent the uredial stage only has been found. This shows as small orange pustules on the leaves. It occurs particularly on coppice shoots, but is quite rare and completely unimportant. It is possibly the same fungus as the Asian *Cronartium quercuum* Miyabe, but until the telial and particularly the aecial stages have been found in Britain, this cannot be certain. The aecia of *C. quercuum* occur on the branches of pines and the telia follow the uredia on the leaves of oak. *C. quercuum* was at one time thought to be the same as the American *C. cerebrum*, but Hedgcock and Siggers (1949) considered that they were different. Meinecke

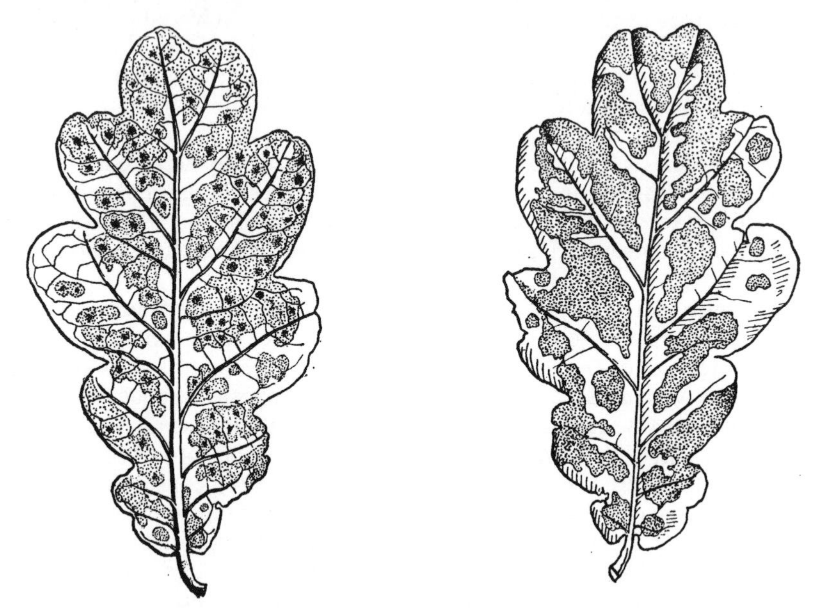

FIG. 75. *Gloeosporium quercinum* on oak, Farnborough, Hampshire, November ($\times \frac{2}{3}$). The fruit-bodies are on the lower surface of the leaf (left). (J. C.)

(1916) found that the American species could overwinter on evergreen oaks, and was thus able to persist in the absence of pine. It must be presumed that in Britain *Uredo quercus* can also overwinter on oak. Another rust fungus, *C. fusiforme*, also occurs on oak in North America. It displays strong preferences for certain oak species, in particular *Q. nigra* and *Q. phellos*, while others such as *Q. virginiana* are virtually immune (Lamb and Sleeth 1940). But this fungus and *C. comandrae* are both much more important on their alternate host, pine, than on oak, and are therefore dealt with in Chapter 22.

DISEASES OF ACORNS

Several fungi occur on acorns, they are considered in more detail in Chapter 14.

28

DISEASES OF BEECH

ONLY one species of beech, *Fagus sylvatica*, is important in Britain. It is widely used for forest and ornamental purposes, for shelter and for hedges. There are various ornamental varieties of it, notably the Copper beech, but there is no reason to suppose that their disease reactions are different from those of ordinary beech.

DAMAGE DUE TO NON-LIVING AGENCIES

Beech is frequently damaged by spring frost. There are considerable differences in susceptibility between individual trees, associated with variations in time of flushing, but no beech can be regarded as really resistant to late frost.

Sunscorch occurs commonly on initially shaded beech stems exposed to full sunlight by the removal of neighbouring trees (Chapter 5).

Beech can tolerate soils of high lime content, and is generally considered the best final forest crop for most limestone and chalk soils. Nevertheless, it can develop lime-induced chlorosis on such sites (Day, W. R. 1946).

There is no doubt that natural agencies play a big part in the Bark disease of beech, but, since the fungus *Nectria* has been so consistently associated with this disease, Bark disease is dealt with later in this chapter.

NURSERY DISEASES

Seedling blight

Pathogen

This disease, which is sometimes loosely referred to as 'Damping off', is caused by *Phytophthora cactorum* (L. & C.) Schroet. (Phycomycetes, Peronosporales). This *Phytophthora* on beech was once regarded as a separate species, and was named *P. fagi*; it also appears in some older papers under the name *P. omnivora.* As with all species of *Phytophthora*, the fruit-bodies are microscopic; so the disease is normally detected by its symptoms on affected plants.

Symptoms and development

The fungus attacks young succulent beech seedlings both in the nursery and among natural regeneration in the forest. Infection can take place at any time from germination onwards, in fact it has been suggested, without supporting evidence, that the fungus can be seed-borne. Normally early infection comes from the soil, where the fungus can persist for at least four years (Hartig 1894). Sometimes in early attacks the seedlings do not emerge, but more characteristically the cotyledons are first infected, and show dark-brown patches when the seedlings appear above ground. The patches spread to the stem, which also becomes brown and collapses. There may be further spread from seedling to seedling after emergence, if the weather conditions are suitably moist, but this is generally much less important than the initial attack. The disease is often associated with wet weather in the spring (Liese 1926; Manshard 1927). Once the seedlings start to become woody, attacks cease.

Distribution and damage

This disease has attracted attention particularly in Germany and Denmark, but even there, where beech is a very important forest tree, the fungus is not often seriously damaging. In Britain it occurs infrequently, but always attracts attention because of its conspicuous symptoms. Some nurseries may be free of the fungus, but the information available is insufficient for this to be stated with certainty.

Control

Pre-emergent infection could only be controlled by seed or soil treatments, and no work on these lines has been carried out. According to Manshard (1927), further spread after emergence can be controlled by spraying with Bordeaux mixture. Undoubtedly the most important precaution is to avoid raising beech seedlings on infected ground for at least five years, unless soil sterilization is carried out.

Other nursery diseases

Rosellinia quercina Hartig (Ascomycetes, Sphaeriales), which is primarily a disease of oak and is therefore dealt with in Chapter 27, also occurs occasionally on beech in nurseries.

ROOT DISEASES

Ink disease

The fungus *Phytophthora cambivora* (Petri) Buism. (Phycomycetes, Peronosporales) occurs on beech as well as on Spanish chestnut (Day, W. R. 1938, 1939). Day also recorded *P. syringae* Kleb. on the roots of beech. Day (1939) and Oyler and Bewley (1937) successfully inoculated beech with *P. cinnamomi* Rands; in Day's work *P. cinnamomi* appeared less virulent on this host than *P. cambivora* and he did not find the former on beech in nature.

The symptoms of *Phytophthora* attack on beech are similar to those on chestnut, but the disease is far less common and far less serious on beech (Chapter 29). W. R. Day (1946) states that this disease does not occur to any extent on the shallow calcareous soils where beech is particularly apt to develop lime-induced chlorosis and 'bark dieback' (see below). As with chestnut, the disease occurs particularly on water-retentive soils.

Other root diseases

The upper roots of beech, like its stem and branches, are particularly subject to invasion by weakly parasitic and saprophytic fungi, many of them decay-causing. These fungi may gain entry through wounds, or because of the general ill health of the tree due to environmental conditions or old age. Thus the appearance of fruit-bodies of wood-destroying fungi at the base, as well as higher up the stems of beech trees, is a very common phenomenon.

When conifer stumps are available to act as centres of spread, young or even middle-aged beech can be killed by *Fomes annosus*. Losses as high as 50 per cent. have been recorded among young beech (Chapter 16).

STEM DISEASES—BARK AND CAMBIUM

Bark disease

Pathogens

Although there is considerable evidence that non-living agencies play a large and possibly primary role in the initiation and development of this disease, it has been consistently associated with species of *Nectria* (Ascomycetes, Hypocreales). *N. coccinea* (Pers.) Fr. is most commonly recorded, but in Denmark *N. galligena* Bres. appears to be concerned. *N. ditissima* Tul. has been associated only with cankers on younger beech (see below), though in a taxonomically difficult genus like *Nectria* there may well have been errors in identification, so specific records should not be taken too seriously. All these species of *Nectria* produce red pin-head perithecia, which darken and finally blacken with age.

Two other fungi have been specifically associated with bark disease of beech. The first, *Phaeobulgaria inquinans* (Fr.) Nannf. (Ascomycetes, Helotiales), often known in the past as *Bulgaria inquinans* or wrongly as *B. polymorpha*, was thought by Tabor and Barratt (1917) to be responsible for localized bark dieback in old beeches. Pycnidia, which exude black spore-tendrils, and later characteristic large, flat, black, rubbery apothecia, 1 to 3 cm. across, with raised edges, are produced on the dead bark. The fungus is a very common saprophyte on the bark of felled or dead oak and beech, and the evidence for its pathogenicity is very doubtful. Infection experiments carried out by Tabor and Barratt were successful only on old trees. The second fungus, *Melogramma spiniferum* (Wallr.) De Not. (Ascomycetes, Sphaeriales), was associated with bark disease by Boodle and Dallimore (1911). Its small black fruit-bodies were found mostly near the base of the tree and even on the upper roots. It is a common saprophyte on dead beech bark, and again there seems little evidence for its pathogenicity.

The dead bark and wood of beech is very readily invaded by fungi, many of them decay-causing. Once the bark has been killed, therefore, colonization, often by a succession of fungi, takes place very quickly. Before long, fruit-bodies of Basidiomycetes, such as *Stereum purpureum*, *Polyporus adustus*, or *Armillaria mucida*, are produced. These secondary fungi may play a very minor part in the disease, but they are more important because they destroy the heartwood, thus reducing the value of the tree.

Symptoms and development

The symptoms of this disease have been described by several authors (Ehrlich 1934; Thomsen, Buchwald, and Hauberg 1949; Zycha 1951). They vary in detail, possibly because several agencies, living and non-living, are involved, and the cause and development of the disease in different areas may not be the same. Even in Britain the symptoms are very variable. Generally, however, the disease first shows as small black patches on the bark. Initially these may be only a few centimetres in diameter. Often the sap exudes, forming a black crust on the bark as it dries; sometimes well-developed slime-fluxes occur (Chapter 20). As the lesions enlarge they become a pale reddish-brown, appearing as lighter patches against the healthy bark. The fruit-bodies of *Nectria* usually appear fairly soon on the dead bark, and *Nectria* can be cultured from the discoloured tissue at an early stage. Frequently the bark is cracked or merely somewhat roughened in appearance (Plate XV). Occasionally healing takes place and the dead bark is pushed off by the ridges of wound tissue. There is seldom any real canker development. More commonly the tree makes only feeble efforts to heal the wounds, and eventually the dead bark falls off to expose the wood, which by that time is often rotten, as a result of invasion by wood-decaying fungi.

The actual extent of the dead bark is usually small compared with the total surface area of the tree, and girdling by the disease itself is comparatively rare. The final outcome of the disease depends much more on the effects of secondary invasion by wood-destroying fungi, which inevitably weaken the tree at the lesion, where it is apt to break off. This phenomenon, which occurs quite commonly, accounts for the name 'Snap disease' used in some parts of Britain (Prior 1913).

In the United States and Canada, on *Fagus grandifolia*, and in Denmark the disease has been associated with attacks by the Woolly aphid, *Cryptococcus fagi* Bär., which sometimes develops in such numbers that the trunks of infested trees have a white felty appearance. It was found by Ehrlich (1934), in the United States, that *Nectria coccinea* invaded the bark through the punctures made by the insects. The same train of events was found in Denmark (Thomsen, Buchwald, and Hauberg 1949), but there *N. galligena* was the fungus involved. In Britain, although *Cryptococcus* occurs abundantly at times, no connexion between the insect and the disease can be found. Very heavy attacks by the insect may have a slightly weakening effect on the tree, but certainly do not cause bark disease (Boodle and Dallimore 1911) or lead to fungal infection.

Evidence on the influence of non-living agencies is still very incomplete and somewhat contradictory. Schindler (1951) has drawn attention to the large number of explanations that have been put forward. Szántó (1948) attributed the disease in Hungary to infection by *Nectria* of cracks formed in the bark as a result of a cool moist spring following a warm dry winter. Leibundgut and Frick (1943) associated the disease with winter cold, and found by inoculation that *N. coccinea*, the species with which they were concerned, was only a weak parasite. In Britain there is some evidence associating the initiation of lesions with both winter cold and summer drought. This supports the views of Zycha (1951), who considered the disease to be largely dependent on these influences. The effect of drought on bark diseases is still largely unexplained, but recent work by Bier (1959) on infection of willows by *Cryptodiaporthe salicis* suggests that lowering of the moisture content of bark may assist fungal infection.

If drought were a factor it would explain the observed occurrence of the disease on shallow soils in Britain. W. R. Day (1946) called attention to the prevalence of this disease on shallow chalk soils, where lime-induced chlorosis of beech is also a common phenomenon. The disease in Britain tends to occur particularly on the more difficult soils; for instance on those that are very water retentive or where there is a large variation between winter and summer water content, as well as on shallow soils, where summer drought is likely to weaken the tree. It is also found mostly on old trees, particularly on those over 100 years of age. The older the tree the less important do site factors become, and on trees over 130 years of age the disease can occur on quite favourable sites. In a few cases the disease has developed following sunscorch of the bark (Chapter 5), but this is certainly not a common method of initiation.

Distribution and damage

In Europe the disease is found wherever beech is grown; but in America reports always carry the suggestion that the disease and the accompanying insect are spreading (Spaulding 1948; Brower 1949). *Cryptococcus fagi* is a European insect, introduced into North America. There may in fact be differences in the nature and development of the disease in the two continents. The American disease may be mainly due to a pathogenic combination of *Nectria coccinea* and *C. fagi*, which is building up and extending. The European disease may generally be due to climatic and soil factors, with *Nectria* playing a purely minor role, in which case one would expect, and in fact does find, local variations of attack both in place and time.

The damage done is considerable, since infection and deterioration of the wood is extremely rapid, once the protection of live bark has been removed. Much of the older beech in the country, particularly that being retained for amenity and shelter in parks and policies, is being, or will soon be, spoilt by this disease. On many sites beech cannot safely be left after it has reached 100 years of age. In the eastern United States and Canada the disease has done very serious damage to *F. grandifolia*. In some places up to 50 per cent. of the crop has been killed.

Control

Where *Cryptococcus fagi* is acting as a means of entry, its control by spraying is an obvious method of lessening the subsequent attacks of the fungus. For this purpose lime sulphur has proved effective (Brower 1949). No doubt this was used to effect dual control of insect and fungus. Any insecticide suitable for Woolly aphids should control *Cryptococcus*; tar oil winter wash has proved very successful. On individual trees painting affected areas with paraffin is a possibility. But in Britain such measures would control only the insect, which is seldom damaging. In the absence of the insect-fungus relationship, they would have little or no effect on the disease proper. Where this combination is not involved the disease can only be avoided by more care in the choice of sites for beech; but knowledge on this aspect of the disease is still far from complete. The after effects can be mitigated by the prompt felling and utilization of attacked trees. They should not be left standing to decay.

Nectria canker

Pathogen

The fungus associated with this disease, which is entirely separate from Bark disease (see above), is *N. ditissima* Tul. (Ascomycetes, Hypocreales). Like the other species of *Nectria* on beech, it produces red, pin-head perithecia, which darken with age.

Symptoms and development

Markedly sunken cankers on branches and stems of young beech are characteristic of this disease (Fig. 76). Often several cankers occur on the same stem, which may become extremely distorted. Often the stem is girdled, resulting in branch and crown dieback. The disease occurs mainly on twigs and branches varying in diameter from 1 to 20 cm., on trees under thirty years of age. It sometimes occurs on the twigs and smaller branches of older trees.

The fungus has been observed colonizing wounds made by weevils, but it is not suggested that this is the normal means of entry. However, it may well be a wound parasite. The cankers sometimes heal quickly, lasting only for a single year; but more commonly they remain active for four or five years, and continued activity for eleven years has been recorded. They finally heal, unless they have girdled the stem.

No association of the disease with frost or drought has been found. In Westbury Forest, where it was carefully studied, the worst diseased areas were on the higher ground and on the deeper clay soils. No explanation of this relationship was discovered (Murray and Young unpubl.).

Distribution and damage

Outside Britain this disease has received little attention, though it was mentioned by Langner (1936) in Germany. Welch (1934) lists *Fagus* as one of the genera in which young trees are damaged by cankers associated with *N. ditissima* and other species of *Nectria*. The general question of *Nectria* cankers on hardwoods is dealt with in Chapter 18.

In Britain the disease is usually of minor importance. Occasional infected trees can be found here and there in young plantations in many forests, particularly in south-east England. Only in Westbury Forest on the South Downs near Petersfield has the disease done sustained damage; though it has appeared alarming over small areas, and for short periods in a few other forests, where, however, its attacks gradually died out. So far no explanation has been found to account for its widespread occurrence on occasional trees and its failure in most places to develop as an epidemic disease. In Westbury Forest 80 per cent. of the beech plantations were infected, 20 per cent. seriously in that more than half the trees were cankered.

Control

On the basis of our present knowledge, no suggestions can be made about control, except that cankered trees should as far as possible be removed in the first thinning. This may well result in the virtual extinction of the disease in lightly infected areas; but in severe attacks it may be found that only the worst trees can be removed without unduly opening the canopy.

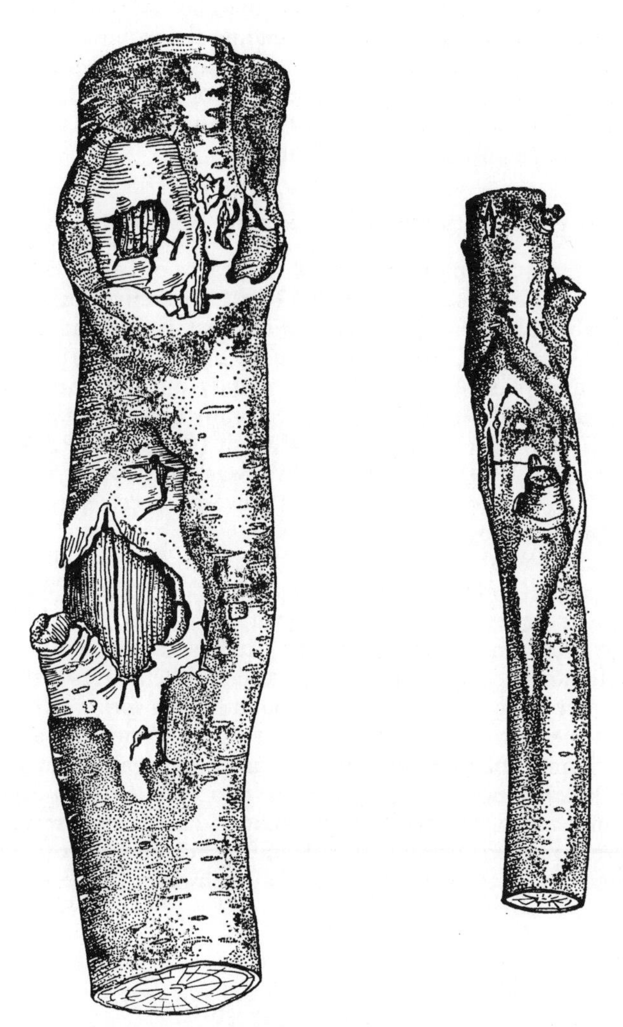

FIG. 76. Canker on beech associated with *Nectria ditissima*, Westbury Forest, Hampshire (nat. size). (J. C.)

Other diseases of bark and cambium

Endothia parasitica (Murr.) And. & And. (Ascomycetes, Sphaeriales), the cause of Chestnut blight (Chapter 29), has been successfully inoculated into beech by Bazzigher (1953); but it has not been found on beech in nature.

Pirone (1948) associated *Phytophthora cactorum* (Leb. & Cohn.) Schroet. with a cambial disease of beech and maple in the United States. Sap oozes from fissures overlying dead areas of cambium, and there is a reddish-brown discoloration of the affected inner bark, cambium, and sapwood. The disease is more important on *Acer* than on *Fagus*.

Some of the decay fungi, which invade exposed beech wood so readily, are probably able to act as weak parasites, extending the wounds by which they entered the tree. This is certainly sometimes the case with *Fomes fomentarius* (L.) Kickx. (Basidiomycetes, Aphyllophorales). For instance, Lohwag (1932) ascribed the death of longitudinal strips of cambium on beech trunks, resulting in pronounced 'furrowing', to this fungus.

DISEASES OF LEAVES AND SHOOTS

There are no really common diseases of the leaves of beech. Two mildews occur on beech, *Phyllactinia corylea* (Pers.) Karst. (Ascomycetes, Erysiphales) (*P. suffulta* (Rab.) Sacc.), which has a wide host range and is discussed more fully in Chapter 18, and *Microsphaera alphitoides* Griff. & Maubl. (Ascomycetes, Erysiphales), which occurs mainly on oak and is therefore dealt with in Chapter 27. *M. alphitoides* is certainly not important on beech, though it has several times been recorded on it (Cotton 1919; Woodward, Waldie, and Steven 1929). Cotton's record was on coppice beech in the immediate vicinity of heavily infected oak; however, beech normally remain healthy, even near mildewed oak.

Discula quercina (West.) v. Arx (Fungi Imperfecti, Melanconiales), usually known in the past as *Gloeosporium fagicolum* Pass., or *G. fagi* West., causes irregular brown patches and spots on beech leaves, particularly along the midrib. Under wet conditions the fungus extends rapidly until the whole leaf withers. Minute black pycnidia are produced on fallen leaves in the autumn, followed by rather larger black perithecia the following spring (Ritschl 1937). This fungus has not attracted much attention on the Continent, and it is not common in Britain, though it occasionally causes noticeable damage.

Dieback of beech shoots and twigs, one to two years old and occasionally older, sometimes associated with small cankers, has been observed several times in different parts of Britain. It is definitely not due to frost, but so far no causal fungus has been isolated. It is usually observed in the late summer.

29

DISEASES OF SPANISH CHESTNUT

ONLY one species of *Castanea*, *C. sativa*, the Spanish or Sweet chestnut, is common in Britain. It is not a native, but has been present in the country for a long time and is naturalized in the south of England. It is probably near its northern climatic limits in Britain, and growth is only really satisfactory in the warmer parts of the country. Its value as a timber tree is lessened by a tendency to 'shake', but poles, owing to early formation of heartwood, are unusually durable. It is the most valuable coppice species in Britain, and has been widely planted in south-east England to provide material for hop-poles and cleft chestnut fencing. In Britain chestnut is primarily a timber-producing crop, the nuts failing to reach full size in our short season. Over large areas in southern Europe chestnuts are grown primarily for their edible fruits.

The American chestnut, *C. dentata*, has scarcely been planted in Britain. The Asian species, *C. mollissima* and *C. crenata*, however, have been tried on a number of sites because of their resistance to Chestnut blight (see below), but generally they have not proved winter hardy. Little is known about the behaviour in Britain of the other two Asian tree species, *C. henryi* and *C. seguinii*, which in any case are not so resistant to blight (Peace 1953*b*). It is very doubtful, however, if any of the Asian chestnuts would prove a satisfactory timber tree in Britain.

DAMAGE DUE TO NON-LIVING AGENCIES

It has already been pointed out that in Britain the Asian chestnuts are frequently damaged in the winter. All *Castanea* species appear to be moderately susceptible to spring frost. Butin and Schuepp (1959) regard frost as a primary factor in the deterioration of chestnut north of the Alps in Switzerland.

The deleterious effect of a markedly variable water table on the roots is reflected in the susceptibility to *Phytophthora* root rot (Ink disease) of chestnuts growing under such conditions (see below).

ROOT DISEASES

Ink disease

Pathogens

Two species of *Phytophthora* (Phycomycetes, Peronosporales) are associated with the so-called 'Ink disease' of chestnut—*P.* (*Blepharospora*) *cambivora* (Petri) Buism., which was originally regarded as the sole cause of the disease, and *P. cinnamomi* Rands now realized to play an equal if not greater part in it. These fungi, which are capable of living for considerable periods in the soil (De Bruyn 1922), produce microscopic fruit-bodies on very inconspicuous mycelium, so that in practice their presence is usually deduced from the symptoms they cause. In addition they are difficult to isolate, partly because of the restricted zone occupied by their mycelium, and partly because they are so quickly replaced by other fungi, both in the tree and in culture (Dufrénoy 1925*a*). Thus general observations on the disease have been based mainly on the external symptoms and have not distinguished the species of *Phytophthora* involved. Urquijo Landaluze (1947) considered *P. cinnamomi* the species most commonly involved. It has a much wider distribution and host range than *P. cambivora*. Grente (1952), however, found no appreciable difference in virulence between the two species on *Castanea*. Far wider cultural surveys of the disease than have yet been made would be required to elucidate the relative importance of the two fungi.

Symptoms and development

The fungus gains entry through the fine roots, and in successful attacks eventually reaches the collar (Grente and Solignat 1952). The surface of infected roots has a slightly darkened, watery appearance and a sour smell. This darkening often extends in tongues up the main stem above dead main roots (Petri 1924*a*) (Fig. 77). To the naked eye there is no sign at all of mycelium, which clearly distinguishes this disease from that due to *Armillaria mellea*, which sometimes occurs on *Castanea* both as a primary parasite and as a secondary one following *Phytophthora*.

Later, when the roots are dead, the tannin in them reacts with the iron in the soil to produce the inky discoloration which has given the disease its name. The presence of the 'ink' is a quite unreliable symptom, for it can be produced by leaving freshly dead chestnut timber or bark in contact with moist soil, whatever the cause of death.

When most of the major roots are dead, leaves in the crown wilt and wither. Sometimes they first turn yellow, but in cases of rapid collapse this stage is omitted. With trees the whole crown often wilts simultaneously, but the first crown symptom may be branch dieback resulting from progressive death of the roots. With coppice stools a few stems may wilt while the remainder are still green. This happens because different coppice stems depend on different portions of the root system, and so in a sense act as separate plants. The fungus can be transmitted, possibly only for short distances, through the air. Movements of plants with infected soil adhering to their roots may account for its more extended spread (Petri 1923), but local spread is probably mainly by mycelium and by spores washed through the soil.

Earlier investigators regarded these fungi as active primary parasites, and though differences in degree of attack from place to place were noted, these were usually regarded as indicating variations in susceptibility of the host tree. It was often suggested, for instance, that planted chestnuts were more susceptible than self-sown ones. However, Petri (1916–17) considered that an impermeable subsoil and high clay content led to infection, while Morquer (1923) found the disease most prevalent on compact clay soils. This connexion between the moisture content of the soil and the disease was strongly supported by W. R. Day (1938), who found that in south and west England it was almost restricted to water-retentive soils. It is particularly prevalent on the

Hastings Beds in south-east England, for this formation often produces soils which are extremely wet in winter but very dry in summer. It is uncertain whether the important factor is the long period during which the soil is saturated, or the extreme variation in water content. The latter certainly affects the final stages of the disease, for the restricted root systems formed on sites with high winter water tables, having suffered damage from the fungus, will often prove inadequate under dry summer conditions. It remains to be settled whether the effect of soil water is to favour growth, sporulation, and

FIG. 77. Typical symptoms of 'Ink disease' at the base of a Spanish chestnut. Bark removed to show tongues of darkened tissue. (J. C.)

infection by the fungus, or to cause the death of rootlets which the fungus can then invade.

The disease is associated with high soil moisture content, not with infertility. However, the idea put forward by Grente and Solignat (1952), that the disease occurs particularly on fertile soils, is not generally supported by British experience. Most, but not all, of the most fertile chestnut sites in Britain are relatively free from Ink disease.

Distribution and damage

Phytophthora cambivora is virtually restricted to Europe, though it has been recorded in the United States. It has seldom been found on genera other than *Castanea*, though Pirone (1948) associated it with basal canker on maple (Chapter 30). When it was considered the sole pathogen involved in Ink disease, the suggestion was made that it had spread to Europe from the Azores. There is evidence, however, that the disease has been present in Spain since 1726 (Artaza 1949). Since the much more generally distributed

P. cinnamomi is also involved, and the disease has very definite site relationships, it is just as likely that epidemic development resulted from the extension of chestnut cultivation to unsuitable soils where the fungi were already present. Possibly, both unsuitable sites and the spread of the fungus may be involved. The second fungus, *P. cinnamomi*, has an almost world-wide distribution in temperate and tropical regions, and has a very wide host range on conifers, broadleaved trees, shrubs, herbaceous plants, and annuals (Thorn and Zentmyer 1954).

The disease caused by these two fungi on *Castanea* has attracted attention mainly in the chestnut-growing areas of Europe, particularly in Italy, France, and the Iberian Peninsula. *P. cinnamomi* has caused damage to chestnut in the United States (Crandall, Gravatt, and Ryan 1945), but its importance there has, of course, been quite eclipsed by that of *Endothia parasitica* (see below). In Britain the disease is sometimes responsible for the deterioration of quite large areas of chestnut coppice, and in places it has necessitated a change to other species. On chestnut grown for timber it occurs sporadically, occasionally causing appreciable losses, but very seldom destroying whole patches, as it does in coppice. In many places the disease still seems to be extending, but whether this is due to spread of the fungus, or to progressive deterioration of the soil conditions, leading to fungal attack, remains to be discovered.

It is generally agreed that while the European chestnut is susceptible, the Asian species, in particular *C. crenata*, are resistant (Dufrénoy 1933; Couderc 1936; Crandall, Gravatt, and Ryan 1945). W. R. Day (1939), however, found *C. crenata* susceptible to inoculation by both fungi, an indication that they may possess strains of varying virulence. He pointed out, however, that *C. crenata* might well behave differently under conditions of natural infection. Hybrids have been produced in Portugal between *C. sativa* and *C. crenata*, with the object of combining the good growth qualities of the first with the resistance to *Phytophthora* of the second (Fernandez 1949).

It is difficult to estimate the damage done by the disease. Undoubtedly it has caused considerable reductions in the acreage of chestnut cultivation, particularly in Spain (Urquijo Landaluze 1936; Artaza 1949). In parts of north-west Spain up to 75 per cent. of the chestnuts have been killed. But in other places reduction in the area under chestnut may also be due to a decline in the use of nuts as food. In Italy it is now rather difficult to separate the damage done by Ink disease from that due to Chestnut blight, and much of the damage earlier attributed to the former may in fact have been due to the latter. Marcelin (1951) has suggested that in France site deterioration is the main reason for the abandonment of chestnut groves. In England Ink disease slightly restricts the area available for chestnut coppice cultivation, but, although it may affect individual forests, mainly on water-retentive soils, its effect on chestnut cultivation in general is not serious.

Control

Various chemical methods designed to kill the fungus in the soil or on the roots have been advocated on the Continent. They have been summarized by W. R. Day (1939). Petri (1928) and Urquijo Landaluze (1936) considered that the removal of the soil from the upper roots and exposure of the latter to the winter cold was sufficient to kill the fungus. Root exposure can be combined with chemical treatment of the surface of diseased roots and of the soil around them, particularly with copper salts (Urquijo Landaluze 1951; Fernandez 1951). This method is now widely used in Spain and Portugal.

The replacement of susceptible *C. sativa* by the resistant *C. crenata* has frequently been suggested, but has not yet been carried out on any large scale. Indeed Dufrénoy (1930), while recommending this course, admits that *C. crenata* is a more demanding species than the European chestnut.

In Britain such replacement is ruled out by the frost susceptibility of *C. crenata* in our climate. The chemical treatments advocated on the Continent, and even there never generally adopted, could be used in Britain only on individual trees of particular value, and they, by the very nature of their growing conditions, seldom develop the disease. Ink disease can generally be avoided if chestnut is not planted on soils with a high water content, on badly drained soils, or where there is considerable variation between the summer and winter water table. These restrictions still leave large areas where chestnut can be cultivated successfully, and to which the disease is unlikely to extend.

Other root diseases

Armillaria mellea is frequently found on *Castanea*, often on trees or stools where the primary cause of damage was adverse soil-water conditions, the fungus *Phytophthora*, or both. Considering the large number of chestnuts weakened or killed by these agencies, *Armillaria* is perhaps not so common on *Castanea* as might be expected.

STEM DISEASES—WILTS

The Oak wilt fungus on chestnut

Bretz (1952*b*) proved by inoculation that the fungus causing Oak wilt, *Ceratocystis* (*Endoconidiophora*) *fagacearum* (Ascomycetes, Sphaeriales) (Chapter 27), can attack several species of chestnut as well as oak. It is unlikely that it occurs commonly on chestnut in nature, but it has been recorded in the field on *C. mollissima* (Bretz and Long 1950). *Castanea* species must therefore be regarded as potential carriers of the fungus.

STEM DISEASES—BARK AND CAMBIUM

Chestnut blight

Pathogen

Chestnut blight is caused by the fungus *Endothia parasitica* (Murr.) And. & And. (Ascomycetes, Sphaeriales). The pin-head pycnidia, from which long tendrils of orange spores ooze in wet weather, are very conspicuous, as are the perithecia, which are larger, though less numerous, than the pycnidia. The perithecia are produced later than the pycnidia and may remain functional for two years. The spores from the pycnidia are washed down the stem by rain, causing fresh infections, and since they are sticky can be carried long distances by birds and insects (Heald and Studhalter 1914). The ascospores, on the other hand, are shot into the air, and are mainly wind disseminated (Heald, Gardner, and Studhalter 1915).

This fungus has been confused with *Endothia fluens* (Sow.) S. & S., which also occurs on *Castanea*, and which is at the worst a weak parasite on trees growing under poor conditions (Baldassini 1948). Orsenigo (1951), in fact, considered them inseparable morphologically, and merged them under the name *E. radicalis* (Schw.) De Not. Kobayashi and Itô (1956) considered that these fungi were separable on morphological grounds, as did Biraghi (1953*b*), who refuted the suggestion that *E. parasitica* may be a virulent mutant of *E. fluens* arising locally, rather than an Asian species which has spread to America and Europe. Isolations of *E. parasitica* made in Italy agree exactly with those made in America (Picco 1948; Baldacci and Orsenigo 1952). Apart from one old record, almost certainly of *E. fluens*, from the New Forest (Wilson and Bisby 1940), the genus *Endothia* is absent from Britain. The two fungi are clearly distinguishable on pathological grounds by their varying virulence and by the fact that *E. parasitica* forms mycelial mats under the bark, while *E. fluens* does not. It is possible that confusion of the two species accounts for some of the doubtful records of supposed *E. parasitica*, for instance that in Belgium (Quanjer 1924).

Symptoms and development

The symptoms have been briefly described and illustrated in colour in a Forestry Commission Booklet (Anon. 1958*b*), and more fully by Darpoux, Ridé, and Bondoux (1957). Infections, which are mainly through small wounds, quickly cause bright-brown patches on young smooth-barked branches. These form a sharp contrast to the olive-green of the normal bark. On older stems infections are not so obvious, but they always produce some discoloration. If the bark and cambium are killed quickly a sunken lesion results; if the progress of the fungus is less rapid new layers of bark are formed under the diseased patch, and there is a certain amount of swelling accompanied by cracking of the outer bark (Fig. 78). The pin-head pycnidia quickly develop in enormous numbers on the infected bark (Fig. 78). The formation of mycelial fans in the inner bark is another characteristic symptom. These are pale fawn in colour and can be found by cutting away the outer bark (Fig. 79). Death of the stem usually takes place before there has been much formation of healing tissue, so that canker development in the usual sense hardly occurs. Thus while the term 'blight' is a misnomer, because the disease does not often cause twig dieback, the term 'canker', which has sometimes been applied to the disease, is also incorrect. When the fungus fails to encircle the stem and healing does take place it usually happens from beneath, the infected bark being sloughed off by the new bark growing underneath it.

The disease is accompanied by wilting and dieback of the whole crown, or of branches, resulting from girdling by the fungus. Vigorous recovery shoots arise immediately below the lesions. These symptoms clearly distinguish this disease from the so-called 'Ink disease'. With Chestnut blight, a tree can die back a branch at a time and coppice one stem at a time, this dieback being accompanied by the formation of epicormic shoots; with Ink disease the whole tree usually dies at once and even with coppice a whole section of the stool dies simultaneously. Recovery shoots are seldom produced in the case of Ink disease, because the upper roots are dead as well as the stem, and there is no live tissue from which they can arise.

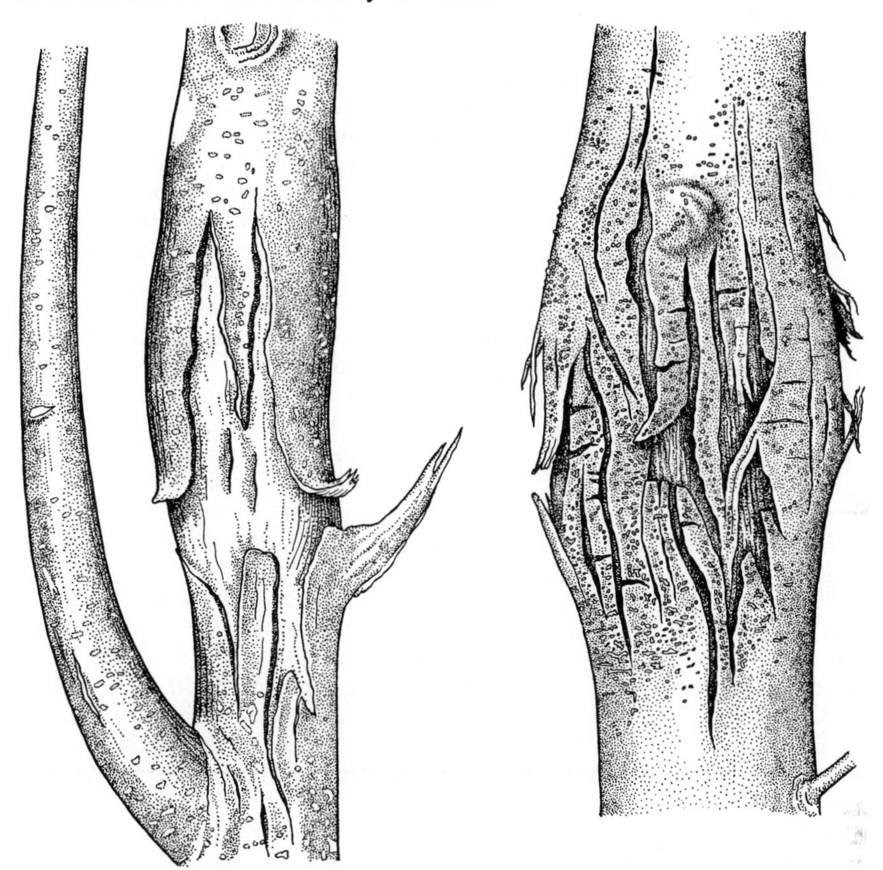

FIG. 78. *Endothia parasitica* on chestnut. Left, a fairly advanced infection; right, a lesion bearing the fruit-bodies of the fungus. (J. C. after a painting by Y. Guerrini.)

Endothia has generally been regarded as a fungus capable of attacking chestnut whatever its growing conditions. Indeed the fungus appears to grow more vigorously on healthy trees. This is certainly borne out in the United States, where it has swept through the chestnut area almost, if not quite, unaffected by elevation, soil, and climate. Gäumann (1951) has suggested that site deterioration may have affected the development of the disease in the Tessin canton of Switzerland, but offers no considered evidence in support of this supposition. Bazzigher (1953) has found that the fungus spreads more rapidly in the hottest months of the year. This might affect the progress of the disease if it ever reached Britain, which has a cooler summer climate than the countries where infection is present.

It was noted quite early in the history of the disease in America that infected trees which had been killed back by the fungus produced fresh shoots from the base, and that these sometimes remained healthy for considerable periods (Graves 1926). Healed cankers were observed on such shoots by Kelley (1948). It is almost certain that the escape of these recovery shoots is partially due to lack of spores, once the disease has killed the majority of the chestnuts and passed on, but the occurrence of healing cankers on a supposedly susceptible host clearly carries other implications. G. A. Zimmerman (1936) claimed to be able to induce immunity by the injection of a vaccine, but this work was never followed up. Recently much more definite evidence of recovery has been found in Italy on the coppice shoots arising from the stumps of several of the stands which were first attacked by the disease and felled as a result (Biraghi 1953*a*). There is clear evidence of a very general healing of lesions despite the fact that spores were present at the outset in large numbers and many infections were initiated. Healing is normally from beneath, apparently indicating that the fungus never killed the inner bark or the cambium (Fig. 80). In many cases the fungus has died out and all that now shows is a slight swelling and roughening of the surface bark. Inoculation experiments with the fungus isolated from healing lesions have proved that it has not lost its virulence and one is driven to assume that the tree has acquired some degree of resistance. This unexpected result makes the future of the disease in Europe very uncertain.

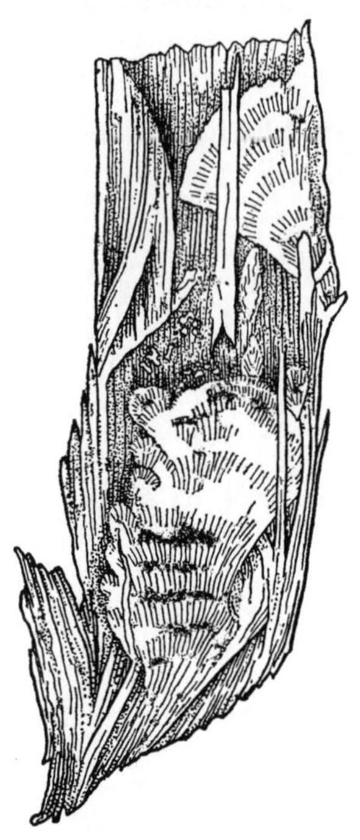

FIG. 79. Mycelial fan of *Endothia parasitica* on chestnut, revealed by stripping off the bark. (J. C. after a painting by Y. Guerrini.)

Distribution and damage

It is now considered that the disease was introduced into the New York Zoological Park in 1904, probably from eastern Asia on imported Asian chestnut plants. It occurs in Japan, China, and Korea (Kobayashi and Itô 1956), but has attracted much less attention than in America or Europe because of the relatively high resistance of the *Castanea* species there. It is now present over the whole of the natural range of *C. dentata* in the eastern United States, and has destroyed chestnut as a commercial crop in five-sixths of it. A comprehensive account of the disease in America has been given by Beattie and Diller (1954). It was discovered in Italy in 1938, but was probably present there some time earlier. It is now widely distributed in Italy, though there are still areas which are free (Baldacci and Orsenigo 1952; Biraghi 1955). It is present in the western part of Yugoslavia, and has spread over most of the chestnut areas in the Swiss canton of Tessin, which lies wholly south of the main ridge of the Alps. It has recently been recorded in India and is said to be quite severe on *C. sativa* in the Kumaon Hills (Gupta 1950). Still more recently it has been found in numerous localities in two Departments in southern France (Darpoux, Ridé, and Bondoux 1957).

In Spain, the only other European country involved, the fungus was found on inactive cankers on imported trees of *C. crenata*. One supposition is that

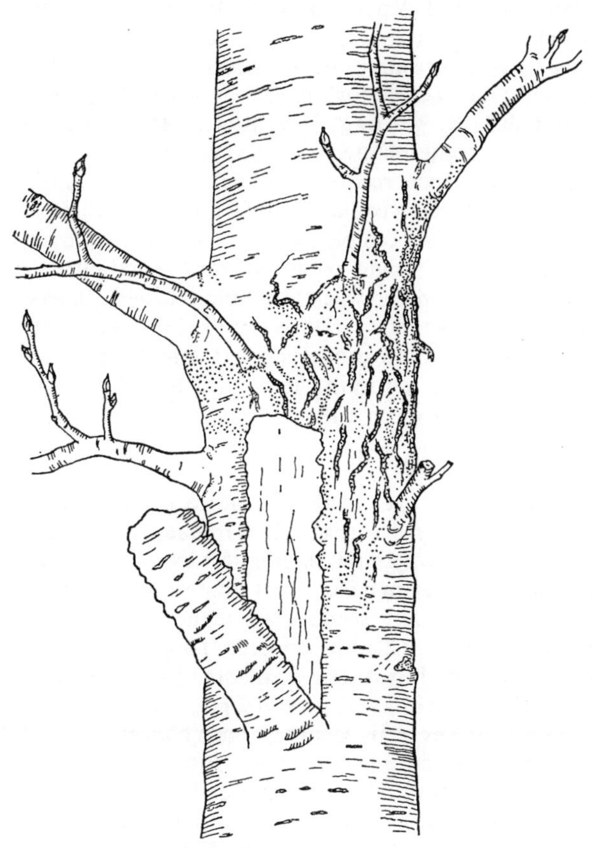

FIG. 80. Chestnut stem showing recovery from *Endothia parasitica*, Masone, Italy. Under the old lesion the bark is perfectly healthy. (J. N.)

it was introduced on the trees from Japan (Artaza 1949). European chestnuts in the vicinity of the infected Japanese ones remained free, although inoculations carried out in Italy with the fungus isolated in Spain indicated that it was a virulent strain (Biraghi 1948). The infected trees were removed and the disease now appears to have died out in Spain.

It is clear that *C. dentata*, the American chestnut, is highly susceptible to the fungus, though Diller (1957) states that a considerable number of relict old chestnut trees are still growing in areas devastated by the disease in the past. It is hoped that some of these may prove to be naturally resistant. *C. sativa*, the European chestnut, apart from its recent development of apparently acquired resistance, may be slightly more resistant. Graves (1950) has reviewed the resistance of all the principal species and hybrids. *C. mollissima* is the most resistant, *C. crenata* is next, and while the other Asian species are more resistant than *C. dentata* or *C. sativa*, they are not good enough to be considered for breeding purposes or to replace the more susceptible species. In America the planting of the more resistant Asian species and of some hybrids between them and American chestnut is now recommended (Anon. 1954; Nienstaedt and Graves 1955). In Britain, however, *C. mollissima* is even less hardy than *C. crenata*, and the use of hybrids, which might prove hardier, would only be possible for forest planting if they could be produced from seed rather than by grafting. The latter method would be practicable only if the trees were being planted at wide spacing for nuts.

The fungus has also been recorded on oak (Chapter 27), but, except on *Q. pubescens*, forms inactive cankers only and causes little or no damage. Oak is important only as a possible carrier of the disease from one chestnut area to another, since the fungus could be transported either on young oak plants or unbarked oak timber without being detected. It has been successfully inoculated on to beech by Bazzigher (1953), but the results did not suggest that the fungus would prove very virulent on this host. Should the fungus ever attack beech naturally, it would probably produce the same superficial cankers as it does on oak.

Control

No serious attempt was ever made to eradicate the disease in the United States. It had spread too far before the danger was realized. The position was exactly the same in Italy, where the felling of infected trees before the fungus reaches the base is recommended. This allows free coppicing from the uninjured root-stock, but any sprouts which become infected are also cut back. In this way reasonably healthy coppice crops can be grown, which may later be converted to high forest. It is also sometimes possible successfully to excise individual cankers (Biraghi, personal communication 1957).

Efforts at control, however, have largely been concentrated on the selection and breeding of resistant strains. The work done in America has been summarized by Graves (1950), and by Beattie and Diller (1954). Similar work has been started in Italy and Switzerland, but has not been in progress long enough to produce tangible results. The hybrids produced in southern Europe may prove of value in Britain, should the disease ever reach this country.

If the disease did reach Britain, attempts at eradication would naturally be made, but in view of the heavy spore production and easy transmission of the fungus they might well prove unsuccessful. Work by Orsenigo and Boncompagni (1950) indicated that chemical control of the fungus is difficult. Quarantine restriction is an obvious measure to prevent the entry of the fungus into Britain. Plants are already prohibited. Nuts, however, are imported into Britain, from Italy and other Mediterranean countries, not only on a large scale for eating, in which case the risk of infecting chestnut trees is slight, but on a much smaller scale for forest purposes. Genetically this may be ill advised, in view of the very poor timber quality of many of the trees from which the nuts are collected. It is considered that washing and drying before dispatch would free the nuts of most adhering spores and germinate and subsequently desiccate any that remain. Actual infection of the fruit is possible (Collins 1915; Rumbold 1915), but such damaged fruits are likely to be removed in the washing process. Various methods of surface sterilization of nuts have been investigated in France (Busnel, Darpoux, and Ridé 1951; Darpoux and Ridé 1952), and the use of formaldehyde washing has been made compulsory for chestnuts imported into that country.

The amount of unbarked chestnut timber imported into Britain from infected areas is very small, but the possibility of entry on unbarked oak is rather greater. The import of vegetable crates made of chestnut wood with bark still adhering is another risk. However, sterilization by heat is successful (Verneau 1953), and most of this material is treated with steam in the process of manufacture. No restrictions are at present in force on nuts or unbarked timber.

Cryptodiaporthe and diplodina canker and dieback

Pathogens

Cryptodiaporthe castanea (Tul.) Wehm. (Ascomycetes, Sphaeriales), the conidial stage of which is *Fusicoccum castaneum* Sacc. (Fungi Imperfecti, Sphaeropsidales), and *Diplodina castaneae* Prill. & Del., which has also been described under the name *Cytodiplospora castaneae* Oud. (both Fungi Imperfecti, Sphaeropsidales), are associated with very similar damage to chestnut. In all their stages they produce small black pustular fruit-bodies on the dead bark.

There has been considerable confusion between the two, and Grove (1935) suggested that they were all stages of the same fungus. However, they are now usually considered to be separate, although they sometimes occur together on the same stem. The latter view was taken by C. and M. Moreau (1953), who found them both on diseased chestnut shoots in the Forest of Marly, and considered *Diplodina* to be the primary parasite in this instance.

Symptoms and development

Both diseases appear first as slightly sunken, bright brown areas on the normal purplish bark of the twigs and young stems. In some cases the lesions encircle the stem and cause dieback; in other cases healing occurs and simple open cankers result. The lesions normally become much darker, possibly even blackish, with age. The fungi spread more easily up and down than around the stem, so that the lesions and cankers tend to be elongated. General accounts of the *Diplodina* disease have been given by W. R. Day (1930) and of the *Cryptodiaporthe* disease by Défago (1937).

In some cases the fungi are definitely wound parasites. They can, for instance, enlarge wounds on young chestnuts made by careless transplanting (Dufrénoy 1926). They occur frequently as saprophytes on chestnut shoots killed or damaged by frost and other agencies (Butin and Schuepp 1959). The fact that both fungi occur only very locally and usually die out after a few years suggests that they normally require some predisposing factor. In one recent outbreak of *Cryptodiaporthe* in Britain, nearly all the cankers started at the same time and after its initial spread the fungus made little further progress.

Distribution and damage

Cryptodiaporthe has attracted some attention in North America because it can attack the Asian chestnut species, which are resistant to the far more serious *Endothia parasitica* (Fowler 1938). Both fungi occur widely in Europe and attacks of *Diplodina* have attracted considerable attention in France under the name 'Javart' disease and in Switzerland (Butin and Schuepp 1959). In Britain both have been recorded as causes of disease, but they are not serious.

Control

Where outbreaks of any intensity occur in coppice, it is advisable to cut and burn, which will not only lessen the possibilities of further spread, but also encourages the stools to make healthy regrowth in place of the useless cankered stems.

Melanconis canker and dieback

Melanconis modonia Tul. (Ascomycetes, Sphaeriales) (syn. *M. perniciosa* Bri. & Farn.), the imperfect stage of which is *Coryneum kunzei* var. *Castaneae* Sacc. (Fungi Imperfecti, Melanconiales), is associated with elongated lesions and cankers on the young stems of chestnut. The fruit-bodies show as small erumpent black pustules, which can only be distinguished from those of *Cryptodiaporthe* and *Diplodina*, with which it sometimes occurs mixed, by laboratory examination. What little evidence there is suggests that *Melanconis* is the least important. These diseases, especially in their early stages, can easily be confused with *Endothia parasitica*; but the absence of mycelial mats under the bark, and of orange spore-tendrils, serve to distinguish them from *Endothia*.

The long lesions caused by *Melanconis* are brownish, and eventually produce cankers. Numerous adventitious shoots arise immediately below them (Dufrénoy 1926; Verplancke 1930). The fungus may also be found on the collar and possibly the upper roots, at any rate on stems already dead. Because of its occurrence in such positions it was at one time considered a possible cause of 'Ink disease', now known to be due to species of *Phytophthora* (Dufrénoy 1922*a*).

Melanconis has been recorded on Japanese chestnut, *C. crenata* (Dufrénoy and Gaudineau 1924), as well as much more commonly on the European chestnut. The fungus occurs commonly in Britain, sometimes in association with *Cryptodiaporthe*, but it has never been regarded as a cause of disease. Only on the Continent, where the cutting and burning of infected branches has been suggested as a means of control, is it regarded as a pathogen (Bazzigher 1956). Such action is unlikely to be necessary in Britain.

DISEASES OF LEAVES AND SHOOTS

Mycosphaerella maculiformis (Pers.) Schroet. (Ascomycetes, Sphaeriales) is the cause of a leaf-spot on *Castanea*. It occurs on the leaves of other hardwoods, and Klebahn (1934) separated the fungus on *Castanea* as *M. castanicola* Kleb. It may have more than one conidial stage, for *Phyllosticta maculiformis* Sacc. (Fungi Imperfecti, Sphaeropsidales) and *Septoria* (*Cylindrosporium*) *castanicola* Desm. (Fungi Imperfecti, Sphaeropsidales) both occur as constant associates with it. Klebahn was able to produce cultural evidence for the connexion with *S. castanicola*. The pycnidia of both the latter appear as minute black spots on the leaves, sometimes while they are still attached to the tree. The perithecia of *Mycosphaerella*, which are larger, are borne on fallen leaves the following spring (Klebahn 1934).

The disease causes very small, yellow and brown spots on the leaves. These spots eventually join up and the whole leaf withers. In a severe attack all the leaves may be brown by August. It has been suggested that the disease is favoured by cold wet weather during the summer (Cambonie 1932; Biraghi 1949*a*). It has occasionally become locally epidemic in various parts of Europe (Bazzigher 1956), but in Britain, although the fungus occurs very commonly, probably wherever chestnut is grown, it has never been associated with extensive damage.

It has usually been reported on *C. sativa*; the Japanese *C. crenata* is said to be resistant (Biraghi 1949*a*).

Very large spots, up to 1 cm. in diameter, on chestnut leaves are formed by *Monochaetia desmazieri* Sacc. (Fungi Imperfecti, Melanconiales). The spots wither completely and sometimes the dead tissue falls out leaving large holes in the leaves. A heavy attack leads to premature leaf-fall, but this occurs so late in the season that little damage is done to the trees. The disease has been reported in Japan (Nisikado and Watanabe 1953) and in the United States (Hedgcock 1929). It has not been found in Britain.

Oak mildew, *Microsphaera alphitoides* Griff. & Maubl. (Ascomycetes, Erysiphales), has been recorded on chestnut as well as on oak and beech, but is never damaging on *Castanea*.

DISEASES ON CHESTNUT FRUITS

A number of fungi are associated with damage to chestnut fruits. These have received particular notice, because over large areas in southern Europe chestnuts are grown chiefly for food purposes. These fungi are discussed more fully in Chapter 14, 'Fungi and Bacteria on Forest Seeds'.

In the United States *Glomerella cingulata* (Stonem.) Spauld. & Schrenk. (Ascomycetes, Sphaeriales), a fungus with a very wide host range, particularly associated with decay of apple fruits, has been found damaging the burrs and nuts of *Castanea mollissima* (Fowler and Berry 1958). Infection took place on the tree and the first visible symptom, browning of the burrs, started several weeks before the seed was mature. On some trees over 40 per cent. of the nuts were decayed, but other trees proved highly resistant.

30

DISEASES OF MAPLE AND SYCAMORE

IN Britain many species and varieties of the genus *Acer* are grown, and several are capable of reaching timber size, but only *A. pseudoplatanus*, Sycamore, and *A. platanoides*, Norway maple, are of forest importance. The Field maple, *A. campestre*, a common, but never dominant, constituent of semi-natural mixed hardwood woodland and hedgerows, is of little practical importance. None of the *Acer* species, other than *A. pseudoplatanus*, has been much studied from a pathological point of view, though many fungi have been recorded on them. Most of the information available, therefore, refers to Sycamore, though much of it may be applicable to *A. platanoides* and the other tree maples.

DAMAGE DUE TO NON-LIVING AGENCIES

Sycamore is one of the hardwoods very subject to dieback as a result of changes in the root environment. It is particularly sensitive to increases in soil water content. Fungi, of which a number occur on Sycamore twigs and stems, may assist in the later stages of dieback. Sycamore is extremely resistant to wind blast, and its use for shelter is limited mainly by its moderately high soil fertility requirements. It is also particularly resistant to industrial fumes. In both these cases its survival appears to rest more on its ability to suffer and recover from moderate leaf injury than on any real resistance to damage.

Being comparatively thin-barked, Sycamore and other maples are subject to sunscorch of the stem. The Japanese maple, *A. palmatum*, many varieties of which are cultivated in gardens, is also subject to sunscorch of the foliage, so that it cannot safely be planted in entirely unshaded positions.

ROOT DISEASES

Maples are occasionally attacked by *Armillaria mellea* (Chapter 16), though they are not particularly susceptible to it.

A basal canker of maple species, associated with the fungus *Phytophthora cambivora* (Petri) Buis. (Phycomycetes, Peronosporales), has been described in the United States by Pirone (1948). This is primarily a root disease, although one of the symptoms is the death of bark at the base of the stem just above ground-level. *Phytophthora* has not yet been recorded causing root disease of *Acer* in Britain, although it has on other genera, particularly *Castanea* (Chapter 29).

STEM DISEASES—WILTS

Verticillium wilt

Pathogens

Verticillium albo-atrum Reinke Berth. (Fungi Imperfecti, Moniliales), the hardly separable species *V. dahliae* Kleb., which has recently been merged with *V. albo-atrum* by Ende (1958), and possibly other species of *Verticillium*, cause a wilt in Sycamore and other maples. The same fungi also attack elms and some other hardwoods, but only on maples is the damage normally of any importance. Species of *Verticillium* are responsible for wilts on a large number of agricultural and horticultural crops including potatoes, tomatoes, raspberries, hops, etc. The species are microscopic and cannot be identified or recognized with the naked eye.

Symptoms and development

Externally the disease shows as a sudden wilting, followed by withering of the leaves and dieback of the shoots. Internally, affected shoots show brown or greenish-brown streaks, mainly in the outer rings of the sapwood. In cross-section these are more clearly defined than the diffuse markings associated with watermark disease in willow (Chapter 33), but much less definite than those of Elm disease (Chapter 31). These symptoms have been very well illustrated by Goidànich (1934). The actual wilting appears to be caused by toxins produced by the fungus (Green, R. J., Jnr. 1954; Caroselli 1955*a*).

Distribution and damage

The disease appears to be generally distributed, but is only occasionally serious, and then only in nurseries, for resistance seems to increase markedly with age. It is not known where or when fruiting occurs on infected trees, but the fungus can persist for at least two years as a saprophyte in the soil (Caroselli 1954). Given suitable conditions, probably high soil temperature accompanied by a reasonable degree of soil moisture (McKeen 1943), infection can occur through the roots; though it is not known whether invasion of intact healthy roots is possible. Infection through wounds is also said to occur, and the rather common occurrence of the disease following grafting suggests that grafting wounds may be a means of entry. While moist soil is necessary for infection, the amount of wilting and the degree of dieback, once infection has taken place, appear to be worse in dry soil (Caroselli 1955*b*).

Little information is available on the resistance of different varieties in Britain, but Sycamore is less affected than some other maple species. There is a marked variation in resistance between species and probably between varieties. It is possible, for instance, for grafted scions to wilt, while shoots arising from the rootstock remain healthy. In America *Acer platanoides* was second only to *A. saccharinum* in susceptibility (Dochinger 1956). A full list of the species and varieties known to have been attacked is given by Engelhard (1957).

Control

In the nursery trees may recover if cut back. Manuring with ammonium sulphate and other substances has been suggested as a means of encouraging recovery (Caroselli 1954, 1956), but the information on this aspect of control is rather contradictory. Soil infected with *Verticillium* can be fumigated to kill the fungus before replanting (Wilhelm and Ferguson 1953). The most promising substances appear to be chloropicrin, allyl bromide, and 55 per cent. chlorobromopropene, at 25, 30, and 45 c.c. per m.^{2} respectively, injected at points 30 cm. apart and to a depth of 15 cm. in the case of chloropicrin, which diffuses rather rapidly upward, and 7·5 cm. for the other two substances. In a mixed nursery infected ground should not be used for maples or Sycamore. Unfortunately, in view of the occurrence of the fungus on so many host plants, it cannot be assumed that a period under agricultural crops will clear the ground. Attack may follow the use of ground previously occupied by a crop such as potatoes (Meer 1926).

Any diseased plants which have been grubbed up should be burnt, while those retained in the hope of recovery should not be moved to another part of the nursery, lest they carry infection with them. Tools used for cutting down or pruning infected plants should be sterilized before being used on healthy trees.

Other wilt diseases

A wilt caused by the fungus *Ceratocystis* (*Endoconidiophora*) *virescens* (Davidson) Hunt (Ascomycetes, Sphaeriales) has been reported causing the death of large Sugar maples (*Acer saccharum*) in the United States (Hepting 1944). The internal symptoms are generally similar to those of *Verticillium* wilt, though the stain in the wood is grey to reddish; but the external symptoms are quite different, the foliage getting progressively smaller and paler green each year, and death following within two to four years of their first appearance.

There is some danger inherent in the fact that the fungus causing the disease is a common sapwood stain of hardwood logs in North America, and is therefore likely to be freely transported from place to place. On the other hand, the disease does not appear to have developed seriously since it was first reported in 1944. The host range is not known, but trees of *A. rubrum* mixed with infected Sugar maples have remained healthy.

STEM DISEASES—BARK AND CAMBIUM

Nectria dieback

Nectria cinnabarina (Tode). Fr. (Chapter 18) is associated with dieback of twigs and occasionally larger branches in Sycamore and maples, but is seldom serious. There is variation in susceptibility between different species of *Acer* (Uri 1948). Spierenberg (1937) suggested that cankers caused by this fungus on maples should be treated with a very strong fungicide, the dead and dying tissue then cut out, and the wound finally treated with Bordeaux mixture. *N. coccinea* (Pers.) Fr. and other species of *Nectria* have also been found associated with canker and dieback of maples.

Sooty bark disease

Pathogen

The fungus associated with this disease, *Cryptostroma corticale* (Ell. & Ev.) Gregory and Waller (Fungi Imperfecti, Moniliales), has already been fully described (Gregory and Waller 1951; Anon. 1952*a*). It forms hard black stromatic layers in the tissue of the bark. These layers eventually separate leaving cavities, the two sides of which are kept apart by black columns of fungal tissue up to 1 mm. long by 0·33 mm. in diameter. These columns usually persist as minute spikes on the exposed surface after the outer bark has been shed. When the bark is shed, the spore mass is revealed. The spores arise from the ends of single hyphae conidiophores, which are packed closely together to form a bluish-grey layer on top overlying the hard black stroma. This bluish-grey layer is exposed after the spores have been shed, but eventually wears away, showing the black stroma underneath. The minute spores are brownish-black in mass, and are borne in chains. The number of spores produced by the fungus is enormous, ranging from approximately 30 million to 170 million per square centimetre (Gregory and Waller 1951). The herbage at the base of infected trees is often blackened by the spores falling on it. It has been suggested by C. and M. Moreau (1954) that the perfect stage of *Cryptostroma* is the very common saprophyte on Sycamore, *Eutypa acharii* (Fries.) Tul. (Ascomycetes, Sphaeriales), but there is no good evidence to support this contention. In Britain *Eutypa* appears to have a far wider distribution than *Cryptostroma*.

Symptoms and development

The most obvious symptom of the disease is a blistering of the bark, the outer layer of which subsequently peels off, exposing the spore mass (Fig. 81). This may happen over the whole trunk of a dead tree, while on live trees it may be limited to quite small areas on the trunk, or to individual twigs or branches. On some trees small lesions may lie in a vertical line up one side of the trunk. If the tree is badly affected the foliage wilts over the whole or part of the crown and sometimes this is the first sign of the disease.

If an infected twig or branch is cut across, a greenish-brown to yellow stain, which is usually darker at the edges, is found. The stain is in a single column up the stem, not in streaks, as is the case with *Verticillium* wilt. Where the stain touches the inner bark, the spore-producing lesions are usually found. This stain extends far up and down the tree, but fades rapidly after the tree dies. Although it always occurs in trees attacked by *Cryptostroma*, in which case the fungus can be cultured from it, staining cannot be taken to indicate with certainty the presence of Sooty bark disease, because similar discolorations, due to other causes, are often found in Sycamore wood.

At the moment there is no actual proof that the fungus is the cause of the disease. Such inoculation experiments as have been tried (Robertson 1955) have not been fully successful, since the fungus did not spread appreciably into the living wood beyond the cut stubs to which the inoculum was applied. On the other hand, the fungus has been found consistently associated with wilting, and often death, of Sycamores that were undoubtedly in perfectly good health prior to attack, and the degree of dieback has been observed to be roughly proportional to the bark area colonized by *Cryptostroma*. Thus the circumstantial evidence in favour of its pathogenicity is very strong. Possibly the proper conditions for successful development of the fungus have not been achieved in the inoculation experiments. The number of successful

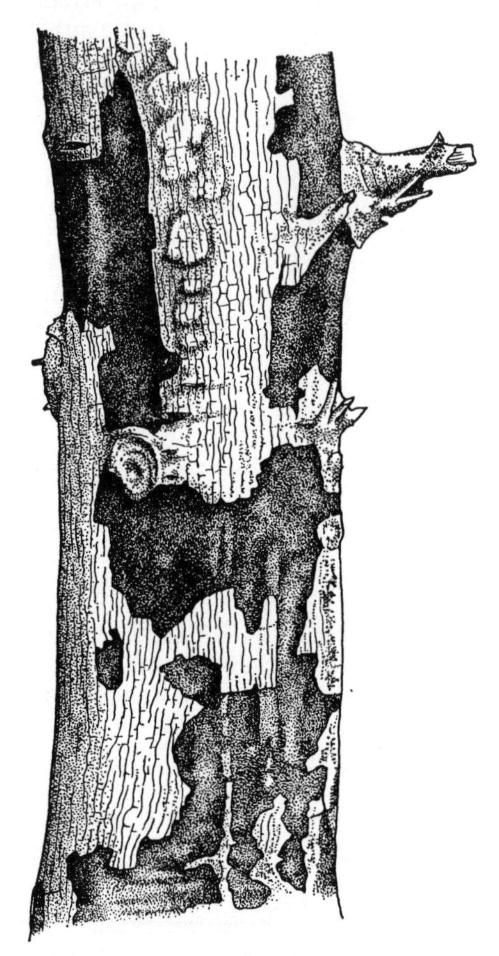

FIG. 81. *Cryptostroma corticale* on sycamore, Wanstead, Essex ($\times \frac{1}{12}$). (J. C.)

infections in the field is very low in proportion to the enormous spore production, and it would appear that special conditions may be required, both for successful infection and for subsequent spread. Townrow (1954) found that the germination of the spores and the growth of the fungus are favoured by relatively high temperatures; its growth rate at 10° C. is only one-third of that at 25° C. This could mean that infection and active spread were only possible in hot weather.

Distribution and damage

Active cases of this disease have not been recorded outside the Greater London area, and the majority of the damage has been in north-east London. It is tempting to suggest some connexion with town conditions, but it is difficult to believe that a tree, generally so well suited to town life as Sycamore, could be so weakened by smoke and fumes that the fungus could attack it. Outside London, *Cryptostroma corticale* has been recorded as far afield as Somerset and Norfolk, but nowhere commonly, and always as a saprophyte on Sycamore killed by other causes, or even on felled Sycamore poles.

In Britain the disease has usually been found on Sycamore, but it has also been recorded on *A. campestre*. Other species of *Acer* may well be susceptible, for it so happens that only Sycamore and occasional Field and Norway maples were growing in the areas where the disease was active. In France it has also been recorded on *A. platanoides*, and in one place on *A. negundo*.

The disease has been reported so far only from Britain and France (Moreau and Moreau 1954), but the fungus has been known for some time in the United States and Canada, under its former name of *Coniosporium corticale* Ell. & Ev., as a saprophyte on maple logs.

At the height of its development in 1949 and 1950, the disease appeared to be a real danger to the cultivation of Sycamore, but its activity reached a peak in 1950 and has declined steadily since (Peace 1955), with a slight recrudescence in 1956, according to a recent survey by Dowden (unpubl.). This is suggestive of activity being encouraged by the hot summers of 1947, 1949, and 1955. The amount of damage so far caused by the disease is slight in relation to our Sycamore population, and the recrudescence, reported in 1956, does not compare in virulence with the behaviour of the disease in the London area in the years 1948–50. The disease has now nearly, if not quite, disappeared.

Control

Trees badly affected by the disease should be felled and barked, the bark being burnt on the site. Though the fungus might remain alive in barked logs, it appears unable to fructify in the absence of bark. Where, as quite often occurs, only a branch or even a dead branch stub is affected, this should be pruned off. The disease is very difficult to detect, except on a severely attacked tree, so that any attempt at eradication, even over a limited area, would prove an impossible task.

Other diseases of bark and cambium

Scattered dead patches on the bark of Sycamore, suggestive of fungal attack but certainly not due to *Cryptostroma*, and leading eventually to the death of the affected trees, have been reported on two occasions from Yorkshire. The cause of the disease was not ascertained.

A somewhat similar disease, in which the main symptoms were the oozing of sap from vertical fissures overlying dead areas of cambium, and a reddish-brown discoloration of the affected inner bark, cambium, and sapwood, has been reported on maples in the United States (Caroselli and Howard 1940; Pirone 1948) under the name of Bleeding canker. The disease led to progressive reduction in the number and size of the leaves or in acute cases to crown dieback. It was attributed to the fungus *Phytophthora cactorum* (Leb. & Cohn.) Schroet. (Phycomycetes, Peronosporales).

Other causes of canker on maple in North America are *Eutypella parasitica* Dav. & Lor. (Ascomycetes, Sphaeriales) (Davidson and Lorenz 1938) and *Hypoxylon blakei* Berk. and Curtis (Ascomycetes, Sphaeriales) (Bier 1939*b*). Cankers caused by the former occasionally girdle small trees under 4 inches in diameter, but on larger trees they persist, and may eventually reach a very large size. They are characterized by the bark remaining attached, despite the production of broad, slightly raised rings of callus tissue. *Hypoxylon* forms irregular cankers, accompanied by cracking of the bark, and is also capable of girdling young trees. Neither of these canker diseases has been recorded in Europe.

A number of species of the bark-inhabiting genera, *Diaporthe* Nits. and *Cryptodiaporthe* Petrak (Ascomycetes, Sphaeriales) (Wehmeyer 1933), have been reported on *Acer* species. It is possible in view of the behaviour of other members of the genus that these may occasionally act as weak parasites on Sycamore and maples damaged by other agencies. Indeed *Septomyxa tulasnei* (Sacc.) v. Höhn (Fungi Imperfecti, Melanconiales), which is supposed to be the conidial form of *C. hystrix* (Tode) Petr., has been recorded in Britain on three occasions on dying shoots of Sycamore, though its pathological significance was not investigated (Batko unpubl.). No information is available on the symptoms caused or on the host range of these minor bark fungi.

DISEASES OF LEAVES AND SHOOTS

Tar spot

Pathogens

Tar spot of Sycamore and other maples is usually caused by the fungus *Rhytisma acerinum* (Pers.) Fr. (Ascomycetes, Phacidiales) which produces in summer black stromatic spots on the leaves. This disease has been fully described by S. G. Jones (1925). Small pimple-like spermagonia, which exude large quantities of apparently functionless spermatia, are formed on the underside of these spots from June until August. In the spring inconspicuous apothecia sunken in the stromata are produced on fallen leaves, and the ascospores from these infect the new developing leaves.

K. Müller (1912) and v. Tubeuf (1913) found that there were biological races of *R. acerinum*, adapted to different maple species, so that transmission of the disease from one host species to another was restricted. They separated the fungus on Sycamore as *R. pseudoplatani* Müller, but this distinction has not been consistently maintained, and most of the records of *R. acerinum* in Britain could be referred to *R. pseudoplatani* of Müller, though the disease is occasionally found on *A. campestre* (Jones, E. W. 1944). A separate but similar species, *R. punctatum* (Pers.) Fr., has been recorded on Sycamore in Britain and rarely on Field maple (Jones, E. W. 1944), but is certainly less common than *R. acerinum*. No information is available on the relative susceptibility of different *Acer* species.

Symptoms and development

Although infection takes place when the leaves are young, external symptoms take quite a long time to appear. Yellowish spots appear on the leaves in late June or in July, but in practice the disease is seldom noticed until the development of the black spots, usually in August. These spots vary in size, but are usually between 5 and 15 mm. in diameter. After the appearance of the spots the leaves gradually wither and fall, so that leaf-fall may be advanced by a month or six weeks. The stromatic spots remain clearly visible on the fallen leaves under infected trees throughout the winter.

The blotches caused on Sycamore by *R. punctatum* are said to consist of a large number of small distinct black patches, about 1 mm. in diameter, crowded together on a yellow-green ground.

Distribution and damage

The fungus appears to be present over most of the British Isles. It may be absent at first from very isolated plantings of Sycamore (Maxwell 1933), and from some islands. It does not appear, for instance, to be present in Skye (Batko unpubl.). It is often absent from areas of heavy industrial pollution. It is likely that the fumes exert a fungicidal action. Sycamore is notably a tree occurring in and around industrial towns, so that it is often found in places where the fungus may be unable to attack it.

Except on young plants, the disease appears to have little effect on the general vigour of infected trees, and the fungus is not often found in nurseries, so that its ability to reduce the growth of small plants is of little practical importance.

Control

The best control in nurseries, or for individual trees, is the removal of infected leaves from the ground beneath the trees during the winter. Apart from nurseries, the only reason for control would appear to be the disfiguring appearance of the disease during the late summer on sycamores or maples planted for ornament.

Other leaf and shoot diseases

The fungus *Uncinula aceris* (DC.) Sacc. (Ascomycetes, Erysiphales) occurs occasionally as a cause of mildew on Sycamore and maples. In Britain it is commonest on *A. campestre* (Jones, E. W. 1944). As is usual with mildews, the superficial myclium gives infected leaves a white mealy appearance. The disease is not common, and is of no practical importance. Little is known of the relative resistance of different maple species, but in Britain the fungus is much more common on *A. campestre* than on *A. pseudoplatanus* (Jones, E. W. 1944), while in Switzerland *A. platanoides* is said to be resistant (Klika 1922).

Two species of *Cristulariella*, *C. depraedans* (Cke.) v. Höhn. and *C. pyramidalis* Waterman and Marshall (Fungi Imperfecti, Moniliales), have been recorded causing greyish leaf spots and premature defoliation on maple and Sycamore in North America (Bowen 1930; Waterman and Marshall 1947). The fruit-bodies are somewhat similar to those of *Botrytis* (Chapter 15) and the fungus was at one time placed in that genus. Only *C. depraedans* occurs in Britain. It appears to be uncommon, but Wilson (unpubl.) recorded a serious outbreak on Sycamore near Dunkeld in 1935.

Three species of *Phyllosticta* (Fungi Imperfecti, Sphaeropsidales) have been recorded on *Acer* species in Britain. Leaf spots and twig dieback have also been associated with three species of *Phleospora* (Fungi Imperfecti, Sphaeropsidales), and also with *Leptothyrium platanoidis* Pass. (Fungi Imperfecti, Sphaeropsidales) (Grove 1935). No adequate descriptions exist of the symptoms caused by these fungi, some of which are almost certainly different stages of the same organism. Certainly none of them is a serious pathogen.

Diplodina acerum Sacc. & Br. (Fungi Imperfecti, Sphaeropsidales) has been isolated from irregular dead patches on Norway maple leaves from Scotland (Batko unpubl.). Very locally damage was quite severe.

Ribaldi (1948) found *Macrophoma negundinis* Rib. (Fungi Imperfecti, Sphaeropsidales) causing pale spots with a dark rim on the living leaves of *A. negundo* in Italy. The disease was purely local.

Plakidas (1942) associated *Venturia acerina* Plak. (Ascomycetes, Sphaeriales) with leaf spots on *A. rubrum*. The spots, which were dark reddish-brown above and grey beneath, were surrounded by a pale-green halo, and varied in size from 0·5 to 20 mm. This fungus has not been recorded in Britain.

Ark (1939) in the United States and Ogawa (1937) in Japan both recorded bacterial blights of maple. In Japan the bacterium, named as *Phytomonas* (*Pseudomonas*) *acernea* Ogawa, caused blackening and withering of the shoots of *A. trifidum*. In the United States the disease, which had similar symptoms, occurred on *A. macrophyllum*, but other species of *Acer* were successfully inoculated under humid conditions. Ark named his bacterium *P. aceris* Ark. There are no records of bacterial leaf-spots of maple in Britain.

The absence of further reports on the leaf and shoot diseases mentioned above strongly suggests that they have never become established in an epidemic form.

31

DISEASES OF ELM

ELM is one of the most important hedgerow and roadside trees in Britain. In the southern half of England, particularly the south-east, the elm population is made up of a considerable number of species and their hybrids. In the north of England and in Scotland, *Ulmus glabra*, the Wych elm, is the only common elm. *U. stricta* and its varieties had become popular for street and park planting, because of their upright habit and short branches, but since about 1930 the use of elms has been severely restricted on account of the danger of Elm disease (see below). There is quite a good market for elm timber, and this, together with its proven ability to grow well in hedgerows and copses, makes it a very useful tree.

DAMAGE DUE TO NON-LIVING AGENCIES

The native British elms are notably hardy both to winter and to spring and autumn frost, but the Asian elms, such as *U. pumila* and *U. parvifolia*, which have the advantage of being resistant to *Ceratocystis ulmi*, the cause of Elm disease, appear to be susceptible to winter cold, for they often die back considerably during the dormant season. Elms are very resistant to toxic fumes and smoke.

In parts of the north of England and Scotland, middle-aged and old Wych elms frequently suffer from a slow, chronic dieback, the symptoms of which are sparse, rather yellow foliage, accompanied by the death of twigs and sometimes of small branches. Infrequently the disease develops further and the trees become stagheaded. So far no pathological agency has been proved responsible, nor is the disease obviously associated with any site factor, apart from its tendency to occur in more northerly latitudes.

STEM DISEASES—WILTS

Elm disease

Pathogen

This disease, which transcends in importance any other disease of elm, is caused by the fungus *Ceratocystis* (*Ceratostomella*) *ulmi* (Schwarz) Moreau (Ascomycetes, Sphaeriales). It is often referred to as Dutch Elm disease, because so much of the early work on it was done in Holland. The fungus was recently transferred to the genus *Ceratocystis* (Moreau, C. 1952), and, since this change has been generally accepted, the new name is used here. It has also been known as *Ophiostoma ulmi* (Schwarz) Nannf. The conidial stage is *Graphium ulmi* Schwarz (Fungi Imperfecti, Moniliales). The fungus and the disease are described in a Forestry Commission Leaflet (Anon. 1958*e*) and also in two American publications (Walter, May, and Collins 1943; McCallum and Stewart 1958).

The fungus exists in the tree mainly in the form of small yeast-like bodies, which multiply rapidly by budding and which are carried along in the vessels. It produces a toxin which not only poisons affected branches, but also stimulates the production of tyloses, which partially block the vessels (Feldman, Caroselli, and Howard 1950). The resultant wilting is thus a combination of toxicity and water-supply restriction. Though the fungus forms some mycelium in the wood, it has considerable difficulty in growing outwards from one annual ring to the next, a fact which has a profound effect on the behaviour of the disease in the tree.

The delicate fructifications are produced only in damp sheltered situations, such as cracks in the bark. They occur particularly on the wood under dead bark which is just starting to loosen, and in the larval tunnels of Elm bark beetles. The black, flask-shaped perithecia, which extrude their spores at the tip in mucilaginous drops, are less than 0·5 mm. in height, and the thin, black, sheaf-like coremia, which also bear their spores in terminal mucilaginous drops, are about 1 mm. high. In Britain J. M. Walter (1939*b*) found that coremia usually appeared before perithecia on newly-dead material, and that the former were produced in suitable places all the year round, except during intensely cold periods; perithecial production on the other hand ceased completely for four to five months during the winter. In culture spores are also borne in clusters at the ends of hyphae. This is usually referred to as the *Cephalosporium* stage. This stage has been observed in sections of infected wood (Wollenweber 1927), but nothing is known of its importance in nature.

In culture the fungus is very variable in appearance (Walter, J. M. 1937; Walter and May 1937). It is also variable in pathogenicity, since some strains are virtually non-pathogenic (Tyler and Parker 1945; Frederick and Howard 1951). However, most strains that are capable of producing symptoms appear to have the same high level of virulence.

Symptoms and development

The toxins produced by the fungus, together with the water-restricting effect of the tyloses which it stimulates, combine to cause yellowing and wilting of the foliage on affected branches. This is the first external sign of infection. In Britain such symptoms may appear as early as June, but more often show from July until leaf-fall. After the foliage on young succulent shoots has withered, the tips of the shoots curl over, forming little 'shepherds' crooks', which persist and are of some value in the preliminary diagnosis of the disease. If twigs from the affected parts of the tree are cut across, they show dark spots in the current annual ring, often in sufficient quantity to form a definite dark ring. If the bark and outer wood are removed, these markings appear as discontinuous streaks in the spring wood. They will not necessarily be found in all the affected twigs and branches, but their total absence is a sure sign that the damage observed is not due to Elm disease. In a severe attack the infected branch or branches are killed before the end of the summer, but often, despite considerable yellowing or even wilting of foliage, only a few twigs die. The death of an elm in a single season as a result of infection by *Ceratocystis* is a very rare occurrence in Britain.

The markings occur only in the annual rings of those years when attack took place, and can thus be used to tell when any particular tree was first

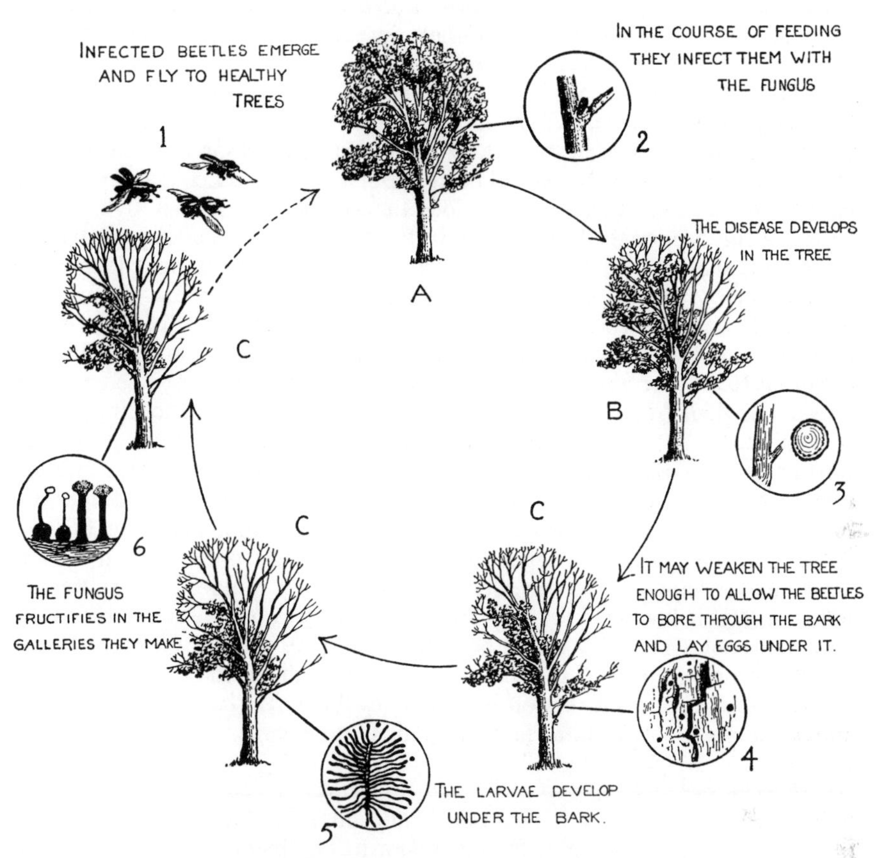

FIG. 82. Diagram showing the life cycle of *Ceratocystis ulmi*, the cause of Elm disease, and its connexion with Elm bark beetles. (J. C.)

affected. The fungus has great difficulty in growing out from one annual ring to the next, except possibly in the roots where the ring structure is less distinct. For this reason a tree attacked one season may escape the next, unless it is freshly infected. In fact recoveries from the disease are very common in nature, thousands of elms are still growing in Britain which have had the disease actively one or several times. The whole subject of recovery has been fully investigated by the author (Peace 1960).

In Britain the fungus is normally carried from tree to tree by Elm bark beetles, *Scolytus scolytus* F. and *S. multistriatus* Ratz. Both these insects breed freely in the bark of elms weakened or freshly killed by any cause, *S. scolytus* in the trunk and major limbs and *S. multistriatus* in smaller limbs, where the bark is not so thick. The beetles and their biology have been described by

Munro (1926) and R. C. Fisher (1931, 1937). Their relationship to the Elm disease has been dealt with in many papers, but particularly by Fransen (1939). In America transmission is mainly by *S. multistriatus*, which was inadvertently imported, as was the fungus, from Europe, and by a native bark beetle, *Hylurgopinus rufipes* Eichh., the host range of which extends beyond the genus *Ulmus*. The relationship of these insects to the fungus has been described by Parker *et al.* (1947).

In Britain both the *Scolytus* beetles breed freely in elms substantially weakened by the disease, and the fungus fructifies in the larval galleries. When the beetles finally emerge they carry on and in them spores of the fungus. Owing to variations in the period of development, emergence may take place anytime from May to October provided the weather is warm enough. Before breeding the beetles feed for a time in the tops of elms, cutting grooves or sometimes shallow tunnels in the thin bark of young shoots, particularly in the crotches of twigs. At this stage they attack healthy shoots, though not necessarily entirely healthy trees. It has been suggested that feeding beetles fly high, and that therefore infection is likely to be commoner in tall trees (Fransen 1939). Certainly elm hedgerows and small trees are not so commonly infected as their frequency of occurrence would warrant. The feeding beetles deposit spores in the cuts they have made, which are often deep enough to touch the developing vessels. In this way hitherto healthy trees are infected for the first time and trees which have recovered from previous attacks are reinfected. This cycle of infection is illustrated diagrammatically in Fig. 82.

Without the aid of the beetles the fungus can spread only through the air. This seldom happens, and in any case wounds are required for successful infection (Smucker 1935, 1937). Transmission by root grafts or through a common root system has often been recorded (Verrall and Graham 1935); but it can only happen if the fungus penetrates right to the base of the tree. Nevertheless, root transmission is sometimes the means by which the fungus spreads along a row of elms in a hedge or in an avenue.

Distribution and damage

The disease was first discovered in France in 1918 (Guyot, M. 1921), and within a few years in most of the neighbouring continental countries. In view of the comparatively wide range of the early discoveries, it is very probable that its introduction was earlier than 1918. Whence it came is still a matter for surmise. The most likely explanation is that the fungus is endemic somewhere in Asia on one of the resistant, but not immune, elm species which are native there, and that it was imported by some means into Europe, where it found far more susceptible host elms, and possibly conditions more favourable for its development. Accounts of the early history of the disease in Europe have been given by the author (Peace 1960) and by Clinton and McCormick (1936). The disease is now known to occur over most of Europe, with the exception of Norway, Greece, and Turkey (Anon. 1947*a*).

In 1927 it was first found in Britain at Totteridge in Hertfordshire, but there is evidence from ring discoloration in earlier years to suggest that the disease was in Britain at least as early as 1920 and very probably before that. In 1928 the disease was discovered to be widespread in southern England, which alone is evidence that it must have become established before 1927. The disease is now generally present all over England south of a line from Chester to Hull, and over the whole of Wales, wherever elm occurs. Thus all the major areas of elm in Britain are affected. In the northern half of England, elm occurs much more sporadically and the disease is much less common. It has been found a few miles south of Carlisle, and in the extreme north of Northumberland. In the latter area there was at one time quite a substantial outbreak, which extended over the border into Scotland. Northwards from this outbreak the disease can be found here and there as far as the outskirts of Edinburgh. One case has been located just north of the Firth of Forth; but apart from this the disease appears to be absent north of the Firth of Forth and west of a line from Edinburgh via Peebles to Hawick. Further information on the history and distribution of the disease in Britain has been given by the author (Peace 1932, 1960). It is now known to occur in Ireland (Heybroek unpubl.).

It was first recorded in the United States in 1930, probably having been introduced on elm burl-logs imported for furniture veneers (Beattie 1933). It appeared in Canada in 1944. Since then it has spread over nearly all the north-eastern States as far west as Illinois and as far south as Tennessee, with an isolated outbreak in Colorado (Swingle, Whitten, and Young 1949; Holmes, F. W. 1958). In Canada the outbreak, which originated separately from that in the United States, is limited to the provinces of Quebec and Ontario (McCallum and Stewart 1958; Holmes, F. W. 1958).

All the species and hybrids of elm native or commonly planted in Britain are more or less susceptible. Wych elm, *U. glabra*, is possibly the most susceptible, while the Cornish elm, *U. stricta*, and its varieties, and the Huntingdon elm, *U. hollandica vegeta*, are certainly more resistant, though they are often quite severely affected. None of the British species is quite so susceptible as the American elm, *U. americana*. A number of Asian species, in particular *U. pumila* and *U. pumila pinnato-ramosa*, show a high degree of resistance (Buisman 1935), but they do not thrive in Britain, never making more than small trees. Comparative inoculations on different species have been described by a number of workers (Westerdijk, Ledeboer, and Went 1931; Wollenweber 1931*b*; Walter, J. M. 1939*a*; Smucker 1941), while the whole position has been reviewed by the author (Peace 1960).

At one time it was hoped that individual elms, which remained unattacked in heavily infected areas, might represent clones of naturally high resistance. In most cases further testing proved them to be susceptible (Peace 1960). However, large-scale selection work in Holland provided a basis of resistant clones on which further breeding work and selection could be based. The first product of this work was a highly resistant clone, *U. 'Christine Buisman'*, which unfortunately proved abnormally susceptible to *Nectria cinnabarina*; a later clone, *U. 'Bea Schwarz'*, was also highly resistant, but of poor habit. The Dutch have now produced a number of clones, mostly of good habit, which are more susceptible than the earlier introductions, but probably sufficiently resistant for practical purposes (Went 1954; Heybroek 1957).

The course of the damage caused by the disease in Britain has been studied by the author (Peace 1960). It reached a peak in 1936 and 1937, since when it has slowly and irregularly declined, not only in the number of trees actively attacked, but more particularly in the average severity of attack. At present man is removing far more elms for timber and to clear ground for road widening or building than the disease is killing. The combined losses are to some extent made good by the emergence of sucker elms from hedges and copses. Without allowing for this, the total loss of elms in the southern half of England, as a result of the disease, probably lies somewhere between 10 and 20 per cent. The decline in virulence of the disease has not been fully explained. There is considerable evidence against both of the most obvious explanations; one, that the remaining elms are more resistant, and the other, that the virulence of the fungus has declined. When the disease was at its height, elm timber was hard to sell, so that dead and severely damaged elms remained standing, acting as sources of infected beetles. During the Second World War and since, elm has proved very saleable, so that much less breeding material has been available to produce infected beetles. This may well be one of the main reasons for the decline in the disease. It is impossible, however, to say with certainty that the decline will continue, so that the future outlook remains rather uncertain.

A similar reduction in the level of the disease seems to have taken place in many European countries, though there is less exact information available than for Britain. In some countries the damage has been higher than in Britain. Holland, for instance, has lost two-thirds of her elms (Went 1954), but that included the felling of all seriously diseased elms. This may have resulted in a higher loss than would a policy of *laissez-faire*.

In America there has also been confusion between direct losses due to the disease and the effect of felling diseased elms in an attempt at eradication. This policy was abandoned in 1940 and the disease is now taking a more or less unchecked course, apart from local efforts at sanitation. It is still virulent in areas where it has only recently become established, but there is no information available on its behaviour in the areas which it first invaded.

Most of the timber from diseased trees can be utilized, so that the damage done by the disease is more a question of amenity and the expenditure on efforts at control than any direct timber loss. In Britain the disease has certainly been far less damaging than was expected when it was at the height of its virulence in the nineteen-thirties.

Control

Since the wood of diseased or newly dead trees is fully utilizable, economic justification for any substantial expenditure on control is very difficult. Mainly for this reason the idea of eradication by felling was quickly abandoned in Britain. In 1929 the discovery that trees could recover after an attack made a policy of leaving elms alone, at any rate until they became unsightly or were past all hope of recovery, even more attractive, and this policy has since been followed consistently. In Holland the felling of diseased trees was only intended to prolong the existence of the remaining standing elms until such time as they could be replaced with resistant clones. To what extent this has been successful is a matter for conjecture. In the United States eradication was the original aim of the felling measures. There is now disagreement as to their effect. Zentmyer, Horsfall, and Wallace (1946) consider that little if any good was done, but Marsden (1953), while agreeing that their abandonment as a national policy was inevitable, feels that felling is still a worth-while policy for any town which wishes to protect its elms.

There is clear evidence that intense local infections take place around accumulations of material suitable for beetle breeding, such as unbarked elm logs, standing dead elms, or elm firewood (Parker *et al.* 1947; Liming, Rex, and Layton 1951). There is therefore very good reason for the removal or destruction of such material. Barking and burning the bark is probably the best method. Elm firewood from infected trees should never be stored either in the open or under cover in the neighbourhood of standing elms. The beetles can emerge even from a closed shed. It is therefore desirable to fell and bark dead trees, and also, where possible, to remove dead limbs which may also serve as breeding material for the beetles. This is a very different matter from the felling and destruction of all diseased elms, a practice which is definitely not advocated. The disease can be arrested by pruning, provided the infection is still in an early stage. The removal of wood must go well beyond the dark streaks (Zentmyer, Horsfall, and Wallace 1946).

The question of the selection and breeding of resistant elms has already been dealt with. In view of the present behaviour of the disease in Britain, limited planting of elm is quite permissible. The emergence of suckers from hedges, so that they can develop into trees, should certainly be encouraged, if only because the initial cost is so low. Some of these new trees will probably become infected, a few may suffer serious damage, but on present evidence the total risk is low. Planting in parks and along streets is a different matter. One severely diseased elm can make quite a serious gap in an avenue. It is for such situations that the Dutch resistant clones would be valuable, and a number of them are now being tested in Britain to see how they behave under conditions of natural infection.

Since the fungus is carried from tree to tree by feeding Elm bark beetles, prevention of beetle feeding should prevent infection. Practicable methods of doing this by spraying with insecticides, particularly D.D.T., have been developed in the United States and are being used there on a limited scale (Welch and Matthysse 1955; McCallum and Stewart 1958; Whitten and Swingle 1958). One method involving two sprayings with D.D.T., one in the spring and the second in the summer, was tried out in two localities in Britain (Peace 1954). The results were generally, but not consistently, successful, but the cost was around £1 per tree per annum for trees 6 to 10 metres in height. Since spraying is only protective, it would have to be repeated every year, making the cost so high as to restrict such a method to trees of exceptionally high amenity or sentimental value.

A good deal of work has been done in America on the possibility of controlling the disease by the use of systemic fungicides, either injected into the trunk or watered on the ground over the roots (Zentmyer, Horsfall, and Wallace 1946; Dimond *et al.* 1949; Potter 1956). A number of substances have given promising results in small trees, but in larger trees they have not been so successful. This type of treatment, like spraying with insecticides, is primarily a protective measure, so that the cost, which is probably of the same order as that for spraying, would involve an annual charge of about £1 per tree.

In Britain, therefore, felling of dead and unsightly elms, accompanied where possible by removal of dead limbs that might serve as breeding material for bark beetles, are the only control measures worth taking.

Cephalosporium wilt

This American disease is caused by a fungus now usually called *Dothiorella ulmi* Verrall & May (Fungi Imperfecti, Sphaeropsidales) which, however, produces in culture heads of spores typical of a *Cephalosporium* (Fungi Imperfecti, Moniliales), hence the earlier and perhaps better known name for the disease. The internal symptoms are very similar to those caused by *Ceratocystis ulmi*, but the markings in the wood are more diffuse and watery. It causes wilting of the foliage and dieback, but unlike *Ceratocystis*, *Dothiorella* sometimes leads to the formation of sunken cankers on small branches. The small black pycnidia develop on these cankers and on recently killed twigs and branches during the summer.

In the nineteen-thirties the disease attracted considerable attention in the United States (Goss and Frink 1934; Creager 1935; May and Gravatt 1937; McKenzie and Johnson 1939). At that time, though it never approached *Ceratocystis* elm disease in importance, it was regarded as much more serious than *Verticillium* wilt (Johnson 1937). Of recent years it has attracted no attention in the United States. It has not been recorded in Europe.

Verticillium wilt

Verticillium wilt, caused by *V. albo-atrum* Reinke & Berth (Fungi Imperfecti, Moniliales), affects many other hardwoods as well as elm. It is particularly common on maples and is therefore dealt with more fully in Chapter 30. Its wider aspects are covered briefly in Chapter 18, 'Diseases of General Importance on Stems, Shoots, and Leaves'.

In the case of elms, the disease is commonest in the nursery. It causes wilting, followed by dieback, the external symptoms being accompanied by rather faint, brown streaking in the wood, much less definite than the markings normally caused by *Ceratocystis*. The disease on elms has attracted some attention both on the Continent (Meer 1926; Wollenweber 1929; Goidànich 1935) and the United States, where it has caused appreciable damage to quite large trees (May and Gravatt 1937). Even there, however, it is far less important than *Ceratocystis* elm disease, and in some regions, at any rate, than *Cephalosporium* wilt.

In Britain *Verticillium* has been recorded on elm on only a few occasions, and then associated only with very slight damage. It is quite often found in maples, doing little or no damage. It is possible that this may also be the case with elm.

Other wilt diseases

A water-soaked condition of the heartwood in elm, known as wetwood, has been described by Carter (1945) in the United States, and attributed to to a bacterium, *Erwinia nimipressuralis*. This phenomenon, which may be connected with the observed proneness of elms to slime flux, is discussed more fully in Chapter 20.

STEM DISEASES—BARK AND CAMBIUM

Nectria dieback

Nectria cinnabarina (Tode.) Fr. (Ascomycetes, Hypocreales), the conidial stage of which is *Tubercularia vulgaris* Tode. (Fungi Imperfecti, Moniliales), is described in Chapter 18, since it occurs on other hardwoods, particularly maples, as well as on elm. It is, however, a common saprophyte on dead elm twigs, being very conspicuous on account of the bright, salmon-pink, pin-head pustules of the conidial stage. It also causes sporadic twig or even branch dieback of elm, in this role being fairly common, but never serious in Britain. Where branches are not girdled cankers are formed. When the leaves on individual twigs or small branches wilt or turn yellow rather suddenly, *Nectria* is always a likely cause.

It is nowhere serious as a cause of disease on elms, but unfortunately the first of the Dutch selections resistant to *Ceratocystis ulmi*, *Ulmus 'Christine Buisman'*, proved abnormally susceptible to *N. cinnabarina* and has been more or less discarded on this account (Went 1954). The injury to this variety is sufficient to cause obvious disfigurement. Testing for resistance to *N. cinnabarina* is now a part of the elm-breeding programme in Holland. The Siberian elm, *U. pumila*, and its variety *pinnato-ramosa*, both resistant to *C. ulmi*, are also abnormally susceptible to *N. cinnabarina* (Buisman 1935; Jacques 1944), though there is a possibility in this case that invasion by the fungus may follow initial frost injury. Carter (1947) has attributed canker and dieback of *U. pumila* to a new and different species of *Tubercularia*, *T. ulmea Carter*, but the general symptoms are the same.

Other diseases of bark and cambium

Three species of *Sphaeropsis* (*Botryodiplodia*) (Fungi Imperfecti, Sphaeropsidales) were associated with cankers on elms in the United States by Buisman (1931). One of these, *S. ulmicola* E. & E., was also recorded by H. A. Harris (1932) causing sunken dark brown cankers and considerable dieback of twigs and smaller branches on elms in Illinois. Bunches of recovery shoots produced from below the cankers gave a false impression of witches' brooms. Another, *S. malorum* Peck., the perfect stage of which is *Physalospora cydoniae* Arn. (Ascomycetes, Sphaeriales), and which causes leaf spot and canker on apples and other fruit trees, has been recorded causing dieback of elms in Italy (Melloni 1936). This fungus is not uncommon on fruit trees in Britain, but has not been recorded on elm.

A somewhat similar disease, also in America, is caused by one or more species of *Coniothyrium* (Fungi Imperfecti, Sphaeropsidales). In this case the fungus apparently enters through the young shoots in the spring and passes down them to form sunken brown cankers on larger twigs. The progressive dieback thus caused can be quite serious (Harris, H. A. 1932; Pierce 1934). Grove (1935) records two species of *Coniothyrium* on elm in Britain, but neither of them as a cause of disease. *Sphaeropsis* and *Coniothyrium* on elm have attracted little recent attention. They are probably of little importance.

Zonate, bleeding cankers on the trunks and main branches of *U. americana* in America are caused by *Phytophthora inflata* Caro. & Tuck. (Phycomycetes, Peronosporales) (Caroselli and Tucker 1949). Cankers of this kind have not been observed on elm in Britain, nor is the causal organism known to occur.

Diaporthe eres Nits. (Ascomycetes, Sphaeriales), a fungus with an enormously wide host range, occurs commonly as a saprophyte on dead elm twigs, blackening the bark. There is usually a distinct zone-line in the wood under the infected area. The imperfect stage is known as *Phoma oblonga* Desm. or *Phomopsis oblonga* (Desm.) v. Höhn. (Fungi Imperfecti, Sphaeropsidales). In Britain it has never been considered a parasite. In America, although Richmond (1932) associated it with cankers on elm, there seems no good evidence that it was the primary cause. Gram and Rostrup (1924) thought that *P. oblonga* preceded *Nectria cinnabarina* on elms dying back in Denmark. Possibly this fungus may be a weak parasite under certain conditions.

Cytospora ambiens Sacc. (Fungi Imperfecti, Sphaeropsidales), the perfect stage of which is *Valsa ambiens* Sacc. (Ascomycetes, Sphaeriales), is a common saprophyte in Britain on hardwood twigs, including those of elm. The spores extrude in yellowish tendrils from pycnidia in flat, blackish stromata which are about 1 mm. in diameter. In the United States (Harris, H. A. 1932) and in Italy (Melloni 1936) it has been associated with canker and dieback of elms. Possibly, in keeping with the usual behaviour of fungi of this genus, it was present in a secondary capacity.

Cytosporina ludibunda Sacc. (Fungi Imperfecti, Sphaeropsidales) is another fungus occurring in Britain on elm as well as other species of hardwoods, and generally regarded as a saprophyte. It also produces yellowish-pink tendrils of spores from pycnidia embedded in black stromata. In the United States, Carter (1935, 1936*b*) associated it with active cankers on *U. americana* and made successful inoculations.

Second only to *Ceratocystis* disease in importance is the virus disease of elms known as '*Phloem necrosis*' and described in Chapter 12. Fortunately it is so far confined to North America.

DISEASES OF LEAVES AND SHOOTS

The most conspicuous leaf-spotting fungus on elm in Britain is *Dothidella ulmi* (Fr.) Wint. (Ascomycetes, Dothideales), sometimes known as *Systremma ulmi* (Sch.) Theiss. & Syd. This fungus causes the so-called Tar-spot disease, producing conspicuous black stromata on fading leaves in the autumn. Pycnidia of *Piggotia astroidea* Berk. & Br. (Fungi Imperfecti, Sphaeropsidales), which is probably the imperfect stage, are produced on the leaves at that time, but the perithecia of the perfect stage are formed on fallen leaves the following spring. It does no real damage to the tree. It occurs on the Continent, but not in America.

There has been some confusion between this disease and that caused by *Mycosphaerella ulmi* Kleb. (Ascomycetes, Sphaeriales). The conidial stage of this fungus is *Septogloeum* (*Phleospora*) *ulmi* (Fr.) Wallr. (Fungi Imperfecti, Melanconiales). It causes numerous reddish-brown spots on the leaves, which tend to fall early. The conidia show as minute whitish patches on the underside of these brown spots (Miles 1921). This disease is common in Britain (Moore 1959) and sometimes quite conspicuous, but also does no real damage. In America Carter (1939) was able to control this disease by the use of sulphur or Bordeaux mixture.

Two species of *Gloeosporium* (Fungi Imperfecti, Melanconiales), *G. ulmicolum* Miles and *G. inconspicuum* Cav., also cause brown spots on elm leaves. The former particularly damages tissue along the veins and on the margins of the leaves; the lesions caused by the latter are more irregular. Only *G. inconspicuum* has been recorded in Britain. It also occurs on the Continent. Grove (1935) suggests that it may be another conidial form of *Mycosphaerella ulmi*, but this connexion has not been properly established and for the moment it is better to regard them as separate. In Britain, at any rate, the disease is quite unimportant. In the United States, where both species of *Gloeosporium* are occasionally damaging, successful control has been achieved by spraying with a number of substances, including Bordeaux mixture (Trumbower 1934). *G. ulmicolum* has been recorded on the fruits of elm as well as on the leaves, both in America (Ames 1952) and in Romania (Georgescu and Petrescu 1954). Though the destruction of seed is probably slight, infection of this kind might prove a ready method of transmission.

In North America the most serious leaf disease of elms is caused by *Gnomonia ulmea* (Schw.) Thüm. (Ascomycetes, Sphaeriales). This fungus and the disease that it causes have been very fully described by Miles (1921), by Pomerleau (1938), and briefly by Marshall (1940). The imperfect stage, which is the chief means of spread, is also a *Gloeosporium*, *G. ulmeum* Miles. The disease often, though not always, appears in early spring, soon after the leaves have unfolded. Whitish or yellowish blotches are formed on the upper surfaces of the leaves. The black pustules of the imperfect stage soon develop in the centres of these. The fungus can also attack petioles and shoots. When it starts early in the season, which it does more often than the other *Gloeosporium* leaf-fungi on elm, it can do considerable damage, particularly in wet summers. Most species of elm grown in America appear to be susceptible, but it is most important on *U. americana*. In the nursery it can be controlled by spraying or dusting with Bordeaux mixture, with sulphur or with various proprietary substances (Trumbower 1934; Carter 1939; Carter and Hoffman 1953). In Japan a closely allied fungus, *Gnomonia oharana* N. & M., causes an exactly similar disease of various species of elm (Nisikado and Matsumoto 1929). Neither fungus is known in Europe.

Taphrina ulmi (Fckl.) Johans. (Ascomycetes, Exoascales) causes green blisters, later turning brown, on the leaves of elm. The fruit-bodies show faintly as a scarcely detectable bloom on the underside of these blisters. Eventually the necrotic areas fall out, so that the leaves become ragged and perforated, rather as if they had been attacked by leaf-feeding insects (Bond, T. E. T. 1956). The disease has attracted little attention, possibly because in its early stages it is so inconspicuous and in its later ones confused with insect damage. It is known to occur quite widely in southern England, but has not been recorded in Scotland. It is found in other parts of Europe and in North America, but nowhere does it do any serious damage.

32

DISEASES OF POPLAR

MANY species and varieties of poplar are present in Britain, but only one, *Populus tremula*, is definitely native. The Black poplar, *P. nigra*, is possibly native. *P. canescens*, which is sometimes thought to be a series of hybrids between *P. tremula* and the White poplar, *P. alba*, is very widespread in a semi-natural condition. The Lombardy poplar, *P. nigra 'italica'*, has been very widely planted for amenity. The Hybrid black poplars, now grouped under *P. euramericana*, and in particular *P. euramericana 'serotina'* and *P. euramericana 'robusta'*, have been widely planted for timber, and it seems probable that these and other clones of this group will continue to be planted on suitable sites in the southern half of England. The Balsam poplars, in particular *P. trichocarpa*, seem better suited than the Black hybrids to the climate of northern Britain. Unfortunately most of the trees of this species planted in the past seem to have arisen from a single introduction of a clone highly susceptible to Bacterial canker. The possibility of substituting other clones with the same climatic needs, but of higher disease resistance, is discussed below.

The breeding of poplars has progressed farther than that of any other timber tree. The ease of vegetative propagation of most species, other than the aspens, makes the perpetuation of first-generation hybrids and special selections a very simple matter. As a result there are now more clones available than can reasonably be tested in any one country even for disease resistance, let alone for silvicultural behaviour and timber yield. This has led to considerable international co-operation in work on poplars and their diseases.

Poplars on suitable sites are capable of extremely rapid growth, and their timber can be utilized for match and plywood veneering at ages as low as twenty-five years in Britain and fifteen years in hotter climates such as that of Italy. This rapid growth, coupled with their ability to grow well as isolated trees, provides their chief attraction to the forester. Thus any diseases that seriously interfere with their growth rate or damage the wood so that it cannot be used for veneering are particularly serious. However, because of the short rotations on which poplars are usually grown, decay is of little importance, since there is not time for it to develop to a damaging extent.

Recent general accounts of poplar diseases have been given by Ciferri and Baldacci (1953), by Butin (1957*c*), and by the author (Peace 1952*b*; Anon. 1958*d*).

DAMAGE DUE TO NON-LIVING AGENCIES

In Britain the Hybrid black poplars (*P. euramericana*) are on the northern limit of their range, so that the farther north they are planted, the higher must be the quality of the sites if the trees are to achieve a satisfactory growth rate.

In northern England and in Scotland, Hybrid black poplars often display sparse foliage and sporadic twig dieback, apparently due to climatic unsuitability rather than to any particular pathogen. The Balsam poplars, in particular *P. trichocarpa*, and the aspen, *P. tremula*, appear better suited to the cooler parts of Britain.

Frost is not usually a serious problem with poplars, though they are often planted in low-lying, frosty sites. Poplar nurseries should not be sited in frost hollows, because spring-frost damage to young growth in the cold-air zone near the ground (Chapter 4) can result in serious forking. Some Asian Balsam poplars, such as *P. koreana*, come into leaf excessively early in the spring, so that the leaves are damaged by cold, dry winds in March and early April. Frost damage to poplar has been very fully discussed by Joachim (1957*b*).

In any case most poplars, with the exception of the aspens, are very site-demanding, so that slow growth, often accompanied by premature leaf-fall and by twig dieback, occurs also on sites where the soil is too wet, too dry, or insufficiently fertile. Even the less-demanding aspen, which does not grow well anywhere in Britain, so that it is of little forestry importance, does much better on good soils.

It has been suggested that some of the older clones, which have been propagated vegetatively for very long periods, and which thus represent in a sense a single old individual, have become senescent. This idea has sometimes been put forward to explain the rather widespread deterioration that has affected Lombardy poplars in various places and at various times. This matter is further discussed in Chapter 3, but the continued healthy growth of so many other Lombardy poplars does not support senescence as an explanation.

DISEASES DUE TO HIGHER PLANTS

In some countries, but not in Britain, mistletoe, *Viscum album*, is very common on Hybrid black poplars (Chapter 13). In north-eastern France spectacular infestations are often seen on commercial poplar plantings. Poplar is also a host for the attractive but harmless parasite, *Lathraea clandestina*.

NURSERY DISEASES

Poplars grow so rapidly in the nursery, sometimes reaching a height of 5 metres in two years, that there is really no clear demarcation between the nursery and immediate post-nursery period. No diseases are confined to the nursery, though many occur and some do considerable damage there.

ROOT DISEASES

Poplars are not particularly susceptible to root diseases. They are much less often attacked by *Armillaria mellea* than willows. The short rotation on which they are grown and their vigour when planted on suitable sites both militate against damaging attack by root fungi.

STEM DISEASES—BARK AND CAMBIUM

Bacterial canker

Pathogen

The bacterium concerned in this disease has been generally considered to be a form of *Pseudomonas syringae* Van Hall., which is pathogenic on a wide range of hosts including *Prunus*, *Syringa*, and *Forsythia* (Chapter 35). The form on poplar is now known as *P. syringae* f. sp. *populea* Sabet. In many earlier papers it appears under the name *P. rimaefaciens* Koning. The bacterium has been described by Sabet and Dowson (1952). Ridé (1958) has recently attributed the disease to another bacterium, which he named *Agrobacterium populi* Ridé. General accounts of the disease have been given by a number of authors (Koning 1938; Peace 1952*b*; Sabet and Dowson 1952; Dowson 1957; Anon. 1958*d*).

Symptoms and development

The first sign of the disease is usually an exudation of dull cream-coloured slime from cracks in the wood of one year or sometimes of older shoots (Fig. 83). The slime contains a number of bacteria, so that the difficulty of deciding which is the chief pathogen is not surprising. The cracks often form near dead buds, infection presumably being through the bud scales. Leaf infection is also possible, attacked leaves becoming partly, and in time, wholly blackened. Wounds, in particular insect punctures, probably provide other means of entry. Shoots of the current year frequently die back but this is often the result of the girdling of the older twigs on which they are borne. Small twigs are quickly girdled, and branches also may be girdled after a few years. On twigs and branches that are not girdled, large, very irregular, erumpent cankers are formed (Fig. 84). Such cankers often appear on the larger branches and main stems of poplars in positions where external infection seems unlikely. This, and the presence of continuous orange staining in the outer wood of affected trees, strongly suggest that the bacterium becomes systemic in the tree. If this were the case, the canker on the older branches and main stem might develop from internal rather than external infection. This is borne out by the appearance of cracks and bacterial exudation on inoculated plants some distance from the point of inoculation.

On the individual tree the slime is probably splashed about and washed down the twigs by rain. Transfer from tree to tree, and particularly from one area to another, appears to entail a vector, probably an insect, but there is no definite information on this matter. The spread of the disease has been observed up the west coast of Scotland, where it has steadily progressed from one area of susceptible poplar to the next, each infection involving a jump of several miles.

Where artificial inoculations are made, the bacterium cannot produce lesions in the absence of the slime. Pure cultures of *P. syringae*, reunited with slime sterilized by filtering, are able to produce lesions, however (Sabet 1953*a*). Thus according to Sabet it is the chemical nature of the slime, not the other bacteria present in it, which is important in infection. Ridé (1957), on the other hand, claims that inoculations with pure cultures of *Agrobacterium* *populi*, which he considers to be the cause of the disease, can result in canker development. According to him lesions develop the same year if inoculations are made in spring or early summer. Successful inoculations are also easily made by placing bacterial slime in small wounds on one-year-old wood during the spring or early summer. On susceptible varieties, this normally results in the formation of a lesion by the following spring. Inoculations at other times

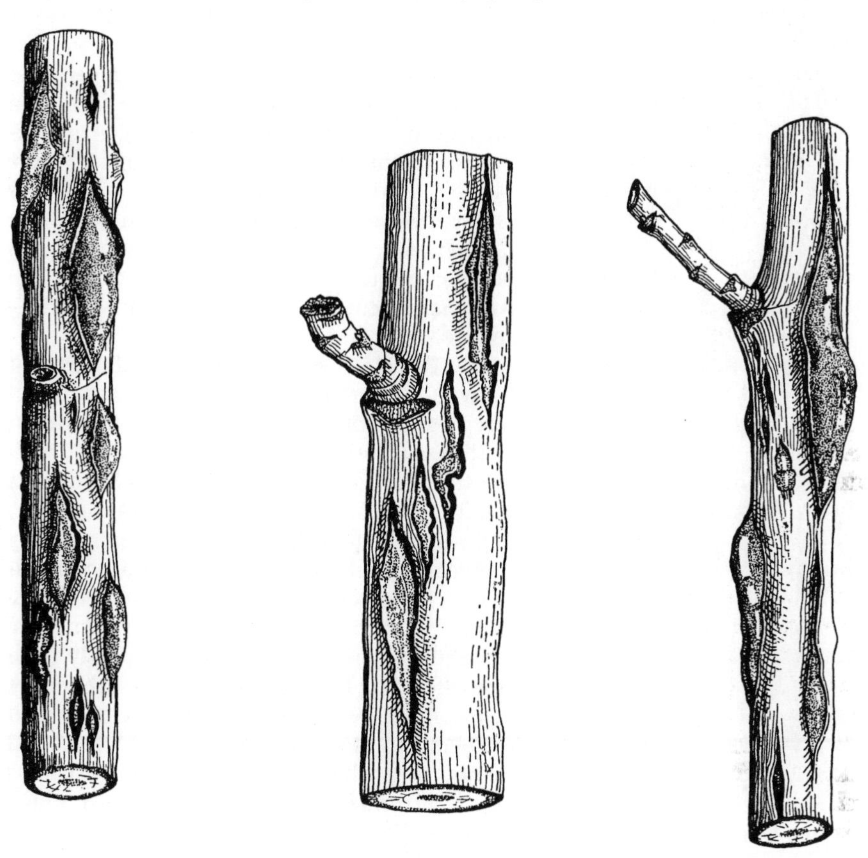

FIG. 83. Early stages of Bacterial canker on *Populus trichocarpa*, showing cracks in the bark and exudation of bacterial slime, Brandon, Suffolk, June ($\times \frac{2}{3}$). (J. C.)

of year are usually unsuccessful. Inoculation not only provides a ready means of testing resistance or susceptibility, but also throws some light on the rather variable behaviour of the disease (Fig. 85). The normal reaction of a highly resistant clone is complete healing of the wound without canker development. Highly susceptible clones are often girdled before the end of the second summer or may develop highly erumpent cankers. Between these two extremes there occur small lesions or mere roughened healing callus over the wound. The subsequent behaviour of this intermediate class is very variable. On some clones the effects get worse with time; on others, including a proportion of those where the initial lesion is quite extensive, the wounds recover. The reasons for this curiously variable behaviour are obscure, but it is hoped that a close appraisal of the results of inoculations over a period of years will disclose the existence of a number of definable 'behaviour patterns', which can be correlated with the reaction of each clone in the field. Work of this kind is now in progress both in Britain and in Holland.

FIG. 84. Erumpent canker on ×*Populus* 'generosa', Brandon, Suffolk, January (×⅜). (J. C.)

Usually the disease is serious only on trees ten years or more in age, but younger trees are sometimes attacked, and the bacterium has been recorded in nurseries causing dieback of the current year's shoots. Severe attacks cripple the trees, and may kill very susceptible varieties. Even slight attacks cause a marked reduction in the growth rate, a serious matter with poplars, while cankers on the main stem spoil the timber for veneering purposes.

W. R. Day (1948*a*) has suggested that frost plays a part in the initiation of Bacterial canker. Frost alone can cause cankers on poplar, but the bacterial disease is not associated with frosty sites, so that frost is unlikely to be a factor of general importance. Sabet (1953*b*) was unable to establish experimentally any connexion with frost. There is no good evidence for association with other adverse factors, though poplars growing under unsuitable conditions do not recover so vigorously from the effects of the disease, and may therefore appear to be worse attacked. Under bad growing conditions damage due to Bacterial canker is probably extended by fungi such as *Dothichiza* and *Cytospora chrysosperma*, which readily attack weakened trees (see below).

Distribution and damage

As far as is known Bacterial canker is limited to northern and central Europe. It is particularly serious in northern France, in Britain, and in north-west Germany, but is known to extend as far as Poland and Russia, and has recently been reported as widespread in Hungary (Györfi 1957). It was serious in Holland and Belgium, but phytosanitary measures and the consistent use of resistant clones for planting have greatly lessened its importance there. It does not seem to occur in southern Europe, though reports of the much less important, and quite different, bacterial necrosis common in Italy have often been assumed to refer to Bacterial canker. The disease is now present in most parts of Britain except parts of north-west Scotland. Within its general distribution healthy individuals of susceptible varieties can often be found. These usually become infected before they reach merchantable size, but a

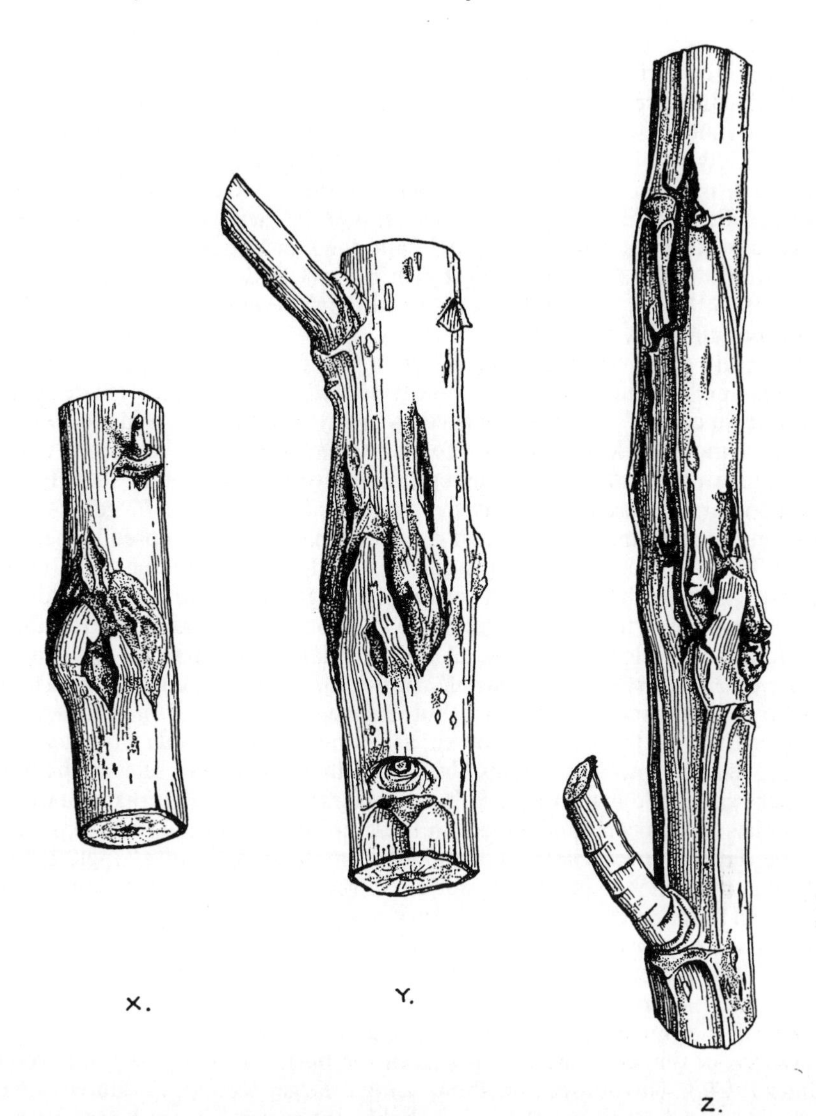

FIG. 85. Inoculations on poplar with bacterial slime, showing different reactions (nat. size). Left to right: uninfected healing on *P.* 'serotina' × *nigra*; formation of a small lesion on a clone of *P.* 'regenerata'; formation of a large lesion on × *P.* 'generosa'. All drawn in June from inoculations made just over a year earlier, Mundford, Norfolk. (J. C.)

few instances are known of undoubtedly susceptible *P. trichocarpa* reaching timber size unattacked.

The losses caused by the disease are likely to fall steadily as the use of resistant clones becomes general. Even now it is important mainly because it limits the clones that can safely be planted, and because the introduction of new clones of silvicultural promise must be delayed while their resistance to Bacterial canker is tested.

Much information is available on varietal and clonal resistance and susceptibility to this disease. This is based on field observations as well as on the results of inoculations. Much of the data from inoculations is biased by being based mainly on observations made after one year, which, as has already been explained, may give somewhat erroneous results for clones of intermediate susceptibility (Ende 1957*a*). Nevertheless, there is generally good agreement between observations made in the field and the inoculation results. Lists have been published by Peace (1952*b*), by Brink and Ende (1951), and by Ende (1953, 1955, 1957*a*, 1957*b*). Most of the author's recent work on this subject is still unpublished.

It is quite clear, however, that all varieties of the Black poplar, *P. nigra*, are resistant. This includes the Lombardy poplar, *P. nigra* 'italica' (Plate XVI). The resistance of *P. nigra* is inherited by many of its progeny (Ende 1957*a*), but much more work is required before any clear picture can be formed of the inheritance of resistance and susceptibility. Resistance among the Hybrid black poplars, *P. euramericana*, is very variable. Some, such as *P.* 'gelrica', are consistently resistant, others, such as *P.* 'brabantica' or *P.* 'pseudoeugenei' (Plate XVI), are consistently susceptible; while the remainder, such as *P.* 'robusta', occupy an intermediate position. *P.* 'regenerata' and *P.* 'marilandica', both names covering a variable group of clones, show different reactions according to the clone concerned. Among poplars of the section *Leuce*, many clones of *P. tremula* are susceptible, but *P. canescens* is generally resistant. *P. alba* also appears to be resistant, but its fastigiate variety, *P. alba* 'bolleana', is susceptible. So far the fast growing aspen hybrid, *P. tremula* × *tremuloides*, has proved susceptible in Britain, but a resistant progeny of selected parents might be raised. So many of the Balsam poplars are susceptible that the whole group has often been condemned on these grounds. As stated above, nearly all the older *P. trichocarpa* in Britain appear to stem from a single clonal introduction which is highly susceptible to the disease. Resistant clones have now been found in this species and also among its hybrids with *P. tacamahaca*, the Balsam poplar of eastern North America.

The Dutch lists show wide variations in susceptibility from one year to another. These are probably mainly due to the effect of climate on the early stages of the inoculations, which are made in Holland by inserting small quantities of diluted slime into pricks in the bark which are left uncovered (Ende 1957*a*). The author's method, using a larger wound, undiluted slime, and a protective covering, though probably not so critical, appears to be less easily affected by external conditions, so that results from year to year are more consistent. Other apparent discrepancies have arisen because of clonal variations within a single variety. For instance, *P.* 'regenerata' may appear as resistant or susceptible according to the clone tested. A few discrepancies may be due to errors in identification, for the taxonomy of poplars is a very difficult matter. It is quite clear that the question of resistance must be dealt with on a clonal basis.

Control

The best control is the use of resistant clones. This can be encouraged by some form of varietal registration, such as is now in force in Holland, Belgium, and some other countries. In Britain cuttings of selected resistant clones are made readily available and the import of susceptible varieties is discouraged by quarantine restrictions. Resistant clones can almost certainly be selected safely on the basis of inoculation experiments, repeated and observed for several years.

The removal of diseased trees, although enforceable by law in some countries (Anon. 1951), is of doubtful value, since really resistant clones are not appreciably affected by canker even in close proximity to diseased trees. Most cankered trees will disappear during normal felling and replacement during the next twenty years, since very few susceptible trees are now being planted. It has already been pointed out that the disease is not appreciably affected by growing conditions, so that it cannot be controlled by choice of site or special cultural practice.

Systemic antibiotics have been used with success against *P. syringae* on *Prunus* (Dye 1956) (Chapter 35). On poplar, however, their use could be justified only on susceptible varieties in arboreta. No work has been done on the chemical treatment of this disease on poplar.

Dothichiza bark necrosis and dieback

Pathogen

The fungus concerned in this disease is *Dothichiza* (*Chondroplea*) *populea* Sacc. & Briard (Fungi Imperfecti, Sphaeropsidales). The perfect stage was once thought to be *Cenangium populneum* (Pers.) Rehm. (Ascomycetes, Helotiales), solely because the two fungi were often found growing together on dead poplar bark. The perfect stage is now known to be *Cryptodiaporthe populea* (Sacc.) Butin (Ascomycetes, Sphaeriales) (Butin 1957*b*; Gremmen 1958*b*). In the spring the pycnidia of the *Dothichiza* stage appear, sometimes irregularly scattered, sometimes in concentric ovals, on the dead bark (Fig. 86). They are rather variable in size, averaging 1–2 mm. in diameter, black in colour and bursting through the outer bark. The spores exude in damp weather in cream-coloured droplets. This fungus can be confused with *C. chrysosperma*, but the pycnidia of *D. populea* are larger and are less densely scattered on the bark. Under the microscope the spores of *D. populea* are oval and distinctly larger than the sausage-shaped spores of *C. chrysosperma*, which exude in golden tendrils. The perfect (*Cryptodiaporthe*) stage is not found frequently enough to be of any importance, and has not yet been recorded in Britain.

Full descriptions of the disease have been given by Goidànich (1940), by Houtzagers *et al.* (1941), and by Schmidle (1953*b*).

Symptoms and development

The fungus produces dead patches on smooth-barked stems. The disease begins as a discoloration of the cortex under the outer bark. This is not easy to detect. The patches are often at the base of twigs, or at the junction of one- and two-year-old growth. They also occur round wounds on stems of any age.

The lesions are usually sunken, and often girdle small stems. The fungus is also said to produce a toxin, so that some of the dieback may be due to poisoning rather than to girdling (Butin 1956*a*, 1958; Braun and Hubbes

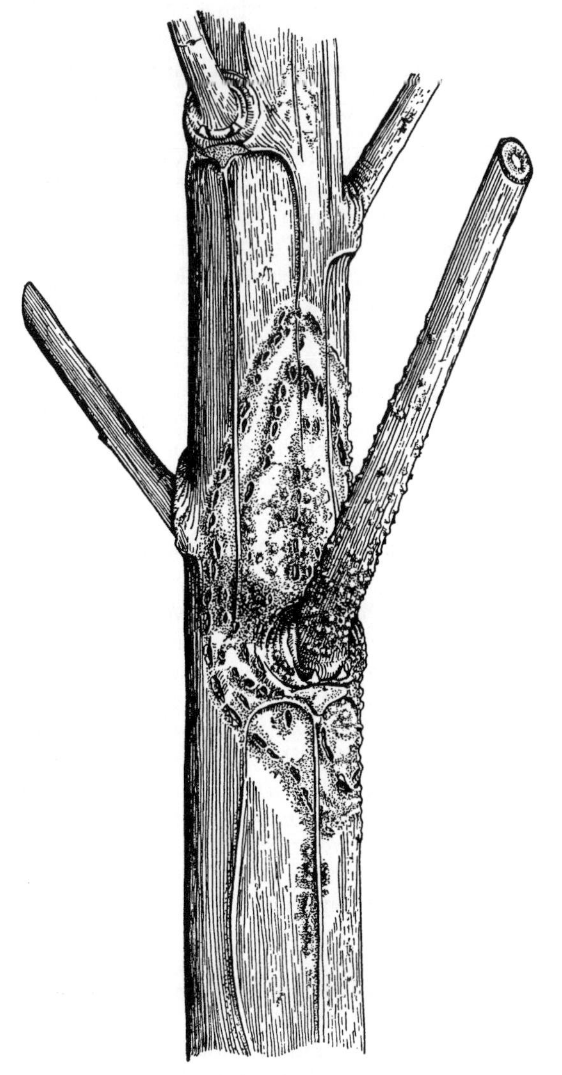

FIG. 86. *Dothichiza populea* on a stem of hybrid black poplar, Owston, Yorkshire, May (nat. size). (J. C.)

1957). There is comparatively little canker formation. If the tree is vigorous, it is able to heal over the lesion in a single season; if it is weak, it will lack sufficient vigour to produce callus tissue. Thus highly erumpent cankers of the kind associated with Bacterial canker of poplar are never produced.

Infection takes place most readily in the winter, when the resistance of the

tree is lowest (Vloten 1938). According to Schmidle (1953*b*), *Dothichiza* is less able than *Cytospora chrysosperma* to infect poplars during the summer months. Schönhar (1957*b*), however, considers that summer infection is not only possible but important, and therefore recommends summer spraying as a control. Braun and Hubbes (1957) consider that winter infection depends mainly on the inability of poplars to form wound tissue at temperatures below 12° C. Butin (1956*b*) found connexions between temperature, wound tissue formation, and infection by the fungus. He also found that the water content of the shoots was at its highest in May and June, and associated winter infection with low turgor (Butin 1957*c*). Taris (1957) gives November to March as the normal period during which the fungus enters the trees. Development of the lesions, on the other hand, which normally starts in February, may continue as late as August.

It has been proved by inoculation that the fungus cannot invade uninjured tissue (Taris 1957), but the entrance wound may be very small. Indeed Gremmen (1958*b*) found that the scars left by the bud scales, as well as leaf scars, were means of entry. Schmidle (1953*b*) has also confirmed infection through leaf scars. Hedgcock (1927) many years ago found that the fungus could enter through the leaves themselves, though Schönhar (1953) could not get infection in this way, and it is unlikely to be an important means of entry. Infection of the bud scars, as described by Gremmen, is one reason why lesions are so frequently found at the joint of one- and two-year-old wood. Other means of entry are pruning wounds, especially if they have not started to heal before the winter stops callus growth, and dead side twigs which often result from overcrowding in the nursery.

It is generally agreed that any adverse circumstances which weaken the tree increase the possibility of infection (Goidànich 1940; Schmidle 1953*b*). Such factors include drought (Schönhar 1953), waterlogging, transplanting, and frost injury (Franken 1956). Schmidle (1953*b*) found that the inoculation of healthy plants, though sometimes successful, resulted in very small lesions. Since small, transplanted poplars are more liable to be weakened by adverse conditions than large, established ones, *Dothichiza* occurs oftener in nurseries and unestablished plantations than on older trees. Spaulding (1958) notes that in the United States the disease is limited to planted, as opposed to natural, stands. It is also more likely to be noticed on young trees, where it may girdle the main stem, than on older trees where usually it causes only dieback of twigs and small branches. Nevertheless, it has been found damaging large trees, particularly on the Continent (Schönhar 1952). Serious damage has been done to twenty-year-old *P.* 'marilandica' in Holland, following initial heavy defoliation by *Melampsora* rust (Meiden and Vloten 1958). On some trees damage was sufficiently severe to result in death. In this case premature defoliation appears to have provided leaf scars particularly suitable for infection. In the nursery leaf scars resulting from *Melampsora* attack may also be of importance as a means of entry for *Dothichiza*.

Distribution and damage

This fungus was first recorded in Europe about 1906 and in the United States ten years later. In both continents it may have been present for some years before it was recorded. In both cases there is considerable circumstantial evidence that it has spread since its first discovery, and the literature suggests that it is still increasing both in distribution and in importance, particularly in Europe, though no doubt this is in some measure due to increased interest in poplar cultivation. It has now spread over much of the eastern United States and Canada, and over most of Europe. It has been found in Turkey. There are reports of its presence in the Argentine.

In most European countries, with the exception of Britain where it takes second place to Bacterial canker, it is now regarded as the most serious disease of poplar. In Germany it is sufficiently serious for legislation to have been introduced requiring the destruction of affected trees. It is impossible, however, to estimate the damage caused. There are instances, such as the outbreak on *P.* 'marilandica' in Holland, of serious injury to established trees. It is chiefly important, however, because it makes it so very difficult to raise or plant poplars under any but ideal conditions.

In Britain so far it has done relatively little damage, though severe attacks have sometimes developed on dry sites, on over-sized transplants, and on plants which had been crowded in the nursery. Nevertheless, *Dothichiza* is unlikely to cause much trouble in Britain, provided nursery practice and silviculture of poplars are correctly carried out.

Several authors have published lists of relative susceptibility and resistance (Vloten 1938; Schönhar 1953, 1957*a*; Waterman 1957). The lists are often in disagreement, and the earlier lists of an author may not agree with his later ones. No doubt in a disease which is so directly dependent, both for infection and for subsequent spread, on the health of the tree and therefore indirectly on the tree's environment, differences in varietal susceptibility may occur because varieties react differently to climatic and edaphic factors. Thus there is no general resistance relationship between the fungus and the host. Butin (1958) found that varieties normally regarded as resistant were as much damaged by the toxin produced by the fungus as susceptible ones. Resistance to the disease must, therefore, rest considerably on resistance to infection. The matter is further complicated because Hubbes (1959) has reported differences in virulence between different strains of the fungus.

Among the more commonly planted Black hybrids, *P.* 'robusta', *P.* 'serotina', and *P.* 'marilandica' are now considered moderately susceptible, but *P.* 'gelrica' and the so-called *P.* 'serotina erecta' are resistant. Several continental authors have listed *P.* 'eugenei' as resistant, but in America it proved particularly susceptible (Detmers 1923), and recent experiments in Britain suggest that it is susceptible there also. Possibly clonal differences within the variety *P.* 'eugenei' are involved. Many of the Balsam poplars are susceptible, though *P. candicans* is resistant. It is said to be very serious on the Balsam × Black hybrid, *P.* 'berolinensis', in Denmark. Poplars of the *Leuce* section seem to be particularly resistant (Bavendamm 1936; Schmidle 1953*b*). In the United States the disease is particularly serious on *P. nigra* 'italica'. Practically no information on varietal resistance or susceptibility has been collected in Britain. On the basis of our present knowledge it would be rash to assume that any variety or even any clone was resistant under all circumstances.

Control

The best control of *Dothichiza* lies in correct cultural practices. At all times plants should be kept growing vigorously. Overcrowding in the nursery should be avoided. Pruning of nursery plants, especially those likely to be planted out, should be done before the middle of August, so that some healing tissue can form before the cessation of growth. In Britain the use of large plants with two-year-old tops has often led to attacks by *Dothichiza*. The custom of using large plants, with two- or even three-year-old tops, is probably one of the reasons why the disease is more serious in parts of the Continent. Large plants are more liable to attack since they provide more avenues for infection and suffer a greater check on transplanting. The cutting and burning of diseased material in nurseries is often advocated, but is of doubtful value in view of the very general distribution of the fungus.

Summer spraying has been recommended by a number of authors, though it is difficult to see why it should have much effect since it is generally agreed that the winter is the most important period for infection. 0·5–1 per cent. copper oxychloride and 0·7 per cent. copper oxide have been recommended (Wöstmann and Goossen 1956; Schönhar 1957*b*); Bordeaux mixture has also been suggested (Schmidle 1953*b*).

Cytospora bark necrosis

Pathogens

Three species of *Cytospora* (Fungi Imperfecti, Sphaeropsidales) occur on poplars. These are *C. chrysosperma* (Pers.) Fr., *C. nivea* Sacc., and *C. ambiens* Sacc. All three have species of *Valsa* (Ascomycetes, Sphaeriales) for their perfect stages, *V. sordida* Nits., *V. nivea* Fr., and *V. ambiens* Sacc. respectively. All three fungi are very widely distributed, occurring on other broadleaved trees as well as poplar. Only *C. chrysosperma* is thought to have any pathological significance.

The three species are not very easy to distinguish, except under the microscope. They form small, stromatic, black pycnidia, which burst through the surface of the bark in the spring, and which vary in diameter from 0·5 to 2 mm. This fungus can be confused with *Dothichiza populea*, but the pycnidia of *C. chrysosperma* are smaller and are more thickly scattered on the bark. Under the microscope the sausage-shaped spores of *C. chrysosperma* are distinctly smaller than the oval spores of *D. populea*. Conspicuous tendrils of spores exude from the fruit-bodies (Fig. 87). These are white and later yellowish for *C. ambiens*, yellow to orange for *C. chrysosperma* and reddish for *C. nivea*. The black, flask-shaped perithecia of the *Valsa* stages are less common, so that the disease is usually referred to as *Cytospora* canker, bark necrosis, or dieback.

There is general agreement that *C. ambiens* is merely a saprophyte occuring on dead twigs of poplar and of many other trees. *C. nivea* is seldom of pathological importance. One or other of these species, usually *C. chrysosperma*, almost invariably occurs on dead poplar branches or twigs, whatever the cause of death. Several general accounts of this disease have been given (Schreiner 1931*a*, 1931*b*; Müller-Stoll and Hartmann 1949; Müller, R. 1953; Schmidle 1953*a*; Magnani 1958*a*).

Symptoms and development

C. chrysosperma, and to a lesser extent *C. nivea*, are capable only of infecting and spreading from wounds on trees weakened by other causes (Müller, R. 1953). They produce in the bark gradually extending lesions, which may girdle small stems and cause dieback. With a weak parasite there is likely to be rapid healing without canker formation if the general health of an attacked tree improves; on the other hand, if the tree remains feeble, the formation of healing tissue is restricted, and no canker is formed. Thus the name 'canker'

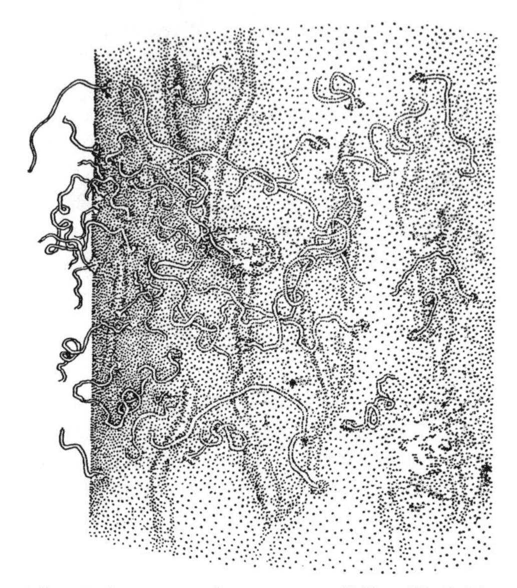

FIG. 87. Spore tendrils of *Cytospora chrysosperma*, Selby, Yorkshire, September (×2). (J. N.)

applied to this disease is a misnomer. The weakening factors allowing infection by *Cytospora* can be drought (Müller-Stoll and Hartmann 1950), difficult soil conditions (Wright 1957*c*), winter injury (Kuntz and Riker 1949), fire (Povah 1921), poor planting, excessive competition with vegetation, or overcrowding in the nursery. Butin (1955) has found a relationship between the percentage water loss from the tissues and infection by *C. chrysosperma*. Schmidle (1953*a*), who did large-scale inoculation experiments with this fungus, obtained only 4 per cent. infection on healthy plants. Schreiner (1931*b*) found that *C. chrysosperma* caused serious damage to cuttings in storage during the winter. The possibility of damage of this kind depends partly on the storage temperature, for Taris (1957) found that while temperatures of 6°–10° C. were favourable for infection, spread of mycelium in the host tissues required temperatures between 10° and 15° C. For this reason infection normally takes place during the winter or early spring, but spread of the fungus continues well into the summer. However, according to Schmidle (1953*b*) *Cytospora* is better able to infect in the summer than *Dothichiza*.

Most authors consider *C. chrysosperma* a much weaker parasite than *D. populea*, but R. Müller (1953) found the former the more virulent.

Distribution and damage

All three fungi have a wide distribution, occurring in North America and in Europe, including Britain. *C. chrysosperma* is regarded more seriously in the United States than in Europe (Schreiner 1931*b*; Kuntz and Riker 1949). It is particularly common there on *P. nigra* 'italica'. In Europe it is nearly always thought to be much less important than *D. populea*, though several workers have reported quite severe attacks (Müller-Stoll and Hartmann 1950; Schmidle 1953*a*). In Britain *C. chrysosperma* is extremely common, but has never been associated with serious dieback. It occurs occasionally on trees weakened by other causes, but is usually found on wood that was dead or nearly dead before infection.

Very little is known about varietal resistance. Müller-Stoll and Hartmann (1950) recorded varietal differences, but found that they were not consistent from year to year. The varying effects of growing conditions on the health of different varieties may be more important in the incidence of the disease than varietal resistance to the fungus. However, Bloomberg (1959) has associated resistance with the anatomical structure of the host, in particular with those features associated with greater water storage and translocation.

C. nivea has attracted little attention as a parasite. In the United States it is sometimes bracketed with *C. chrysosperma* as a cause of disease (Schreiner 1931*a*; Baxter 1952), and in Sweden, where it commonly occurs as a saprophyte on dead wood of *P. tremula*, it has seriously damaged *P. tremula* × *tremuloides* (Johnsson 1957). The interesting suggestion has been made that it may be able to enter through the large lenticels of the hybrid (Persson 1955).

Control

It is generally agreed that direct control measures against *Cytospora* are not justified. Indirect control, using every possible means to stimulate rapid and healthy growth, is the best method. In nurseries care should be taken to see that plants are not wounded, especially during the winter months (Müller, R. 1953). Cuttings should be stored at low temperatures, though it may not be possible to keep them below 2° C. as advocated by Schreiner (1931*b*).

Septoria canker

Pathogen

Several species of *Septoria* (Fungi Imperfecti, Sphaeropsidales) cause leaf-spotting on poplars, but only *S. musiva* Peck, the perfect stage of which is *Mycosphaerella populorum* Thomp. (Ascomycetes, Sphaeriales), can also cause dieback and canker. The pycnidia are borne on the leaves while they are still attached to the tree or on the bark, and the spores exude in minute pink tendrils. Later, in the autumn, dark, globose spermagonia are produced on leaves still on the tree and on the fallen leaves. The perithecia appear on the fallen leaves the following spring. Thus only the conidial stage is produced on the bark. General accounts of the disease have been given by Bier (1939*a*) and Waterman (1954).

Symptoms and development

Infection of the stem takes place through the petioles of infected leaves, but also directly through wounds and lenticels. The fungus kills twigs of the current year and then travels down them to older branches, where cankers are formed. In the early stages the cankers are sunken, like those caused by *Dothichiza populea*, but later they become erumpent, like those of Bacterial canker. *Cytospora chrysosperma* and other fungi invade the cankers in their later stages and may be responsible for enlarging them. Stems less than 2 cm. in diameter may be girdled by *Septoria* in one year. On resistant varieties the spread of the lesions is much slower, and they may heal completely after a single season. Attacks are favoured by wet weather in the spring and early summer and also by acid soils (Sarasola and Magi 1951).

Distribution and damage

This fungus is confined to North America and the Argentine. As a cause of leaf-spot it occurs on several species native to North America, including *P. deltoides* and *P. tacamahaca*, doing practically no damage. Damaging canker has been limited largely to hybrids, either imported or in some cases raised in North America. Susceptibility is not, however, found only in hybrids; for instance *P. koreana* and *P. laurifolia*, both Asian Balsam poplars, are highly susceptible. Nor have all poplars imported into America proved susceptible; for instance the Lombardy poplar, *P. nigra* 'italica' from Europe, is resistant (Sarasola 1945). Poplars of the *P.* 'berolinensis' group, possibly hybrids between *P. nigra* 'italica' and the Balsam poplar *P. laurifolia*, are particularly susceptible. The disease has not been recorded on poplars of the *Leuce* group. Hybrids differ greatly in susceptibility (Waterman 1954), and a number are sufficiently resistant to be safely planted where the fungus is present.

This disease is a good example of the serious development, as a cause of dieback and canker on imported and locally raised hybrids, of a fungus occurring as a harmless leaf-spot on native species. It would be a great pity if it ever reached Europe, for varietal susceptibility to it is quite different from that to Bacterial canker or *Dothichiza*. Thus its presence would still further restrict the number of clones that can safely be grown.

Control

The best method of control is obviously the planting of resistant clones. Sanitation measures are of little value in view of the common occurrence of the fungus as a more or less harmless cause of leaf-spot.

Hypoxylon canker

Pathogen

This disease is caused by *Hypoxylon pruinatum* (Klot.) Cke. (Ascomycetes, Sphaeriales). A conidial stage has been found both in culture and in the forest, but has not yet been named (Ponomareff 1938). The disease has been fully described by Bier (1940) and by Gruenhagen (1945), and briefly by R. L. Anderson (1956). The conidiophores are formed under the bark. Eventually they burst the surface bark, disclosing the grey spores. The open cracks soon dry out, leaving small, black, hyphal pillars projecting from the blackened underbark. These may still be visible when the fruit bodies of the perfect stage are formed. The cankers are usually three or more years of age before this occurs. Stromata, very variable in shape and size, but usually several millimetres across, are then formed on the dead bark. They are covered initially by a greyish bloom, but eventually become black. The perithecia of the perfect stage are sunk in these stromata.

Symptoms and development

Young cankers appear on the smooth bark as slightly sunken, yellowish-orange areas. Old cankers, which may be over a metre in length, are rough and blackened at the centre. On most cankers the fungus spreads so rapidly that callus formation is prevented. Stripping the bark reveals white mycelial fans in the cambial zone. In many cases the fungus quickly girdles the stem, so that branches and tops showing withered foliage are characteristic of the disease. In other cases decay-causing fungi invade the cankers and weaken the stems, so that they break at canker level.

The fungus appears to be a wound parasite; inoculations on unwounded bark have been quite unsuccessful (Bier 1940; Gruenhagen 1945). Insect wounds are certainly one important means of entry (Graham and Harrison 1954). The disease occurs mainly on trees up to forty years of age, probably because only smooth bark is infected.

The incidence of the disease is not directly connected with site quality or tree vigour. It does, however, do more damage on poor sites because the slower-growing trees remain longer in a condition suitable for infection and have less powers of recovery. The incidence of the disease decreases as stand density increases. Riley (1948) found that the incidence of attack increased with the degree of thinning. It is also lower in pure aspen stands than in mixed ones and increases as the proportion of other species in the stand increases (Anderson, R. L. 1952). The reason for this rather unusual behaviour is not known, but it is probably connected with the means of infection.

Distribution and damage

The fungus is confined to North America, where it occurs widely on aspen, both *P. tremuloides* and *P. grandidentata*. It occurs very occasionally on *P. tacamahaca*, and has been recorded on *P. alba* 'bolleana' and on *P. adenopoda*, the Chinese aspen (Anderson, R. L. 1956). On the American aspens it is locally very serious.

Control

There is no direct control, but cankered trees should as far as possible be removed in thinning. In regeneration the losses from the disease are greatly lessened if a well-stocked stand is raised. Every effort should therefore be made to get and maintain stands of reasonably high density. This will not only lessen the incidence of the disease, but will allow more latitude in the removal of infected trees.

Nectria canker

Species of *Nectria* (Ascomycetes, Hypocreales) (Chapter 18) have often been associated with canker on poplars both in Europe and in America. *N. coccinea* (Pers.) Fr. is most common, but *N. galligena* Bres. has been reported on a number of occasions. In many cases *Nectria* is secondary to other causes, and has probably attained more prominence than it deserves because of its conspicuous, red, pin-head fructifications. It has sometimes been assumed to be responsible for serious outbreaks of canker, which were probably due to other fungal or bacterial agencies. In the United States *N. galligena* was blamed for the production on aspen, *P. tremuloides*, of cankers which might have been due to *Hypoxylon* (Povah 1935). In Germany damage due to Bacterial canker was attributed to *N. coccinea*, and legislative action was taken against the latter (Anon. 1933*c*; Richter 1933; Scheffer Boichorst 1934). Koning (1938), however, who found it as a secondary agent in Bacterial cankers, proved by inoculations that it was capable of forming lesions unaided.

Nectria cankers are much more regular and less erumpent in appearance than those of bacterial origin; sometimes they are of typical target form. Varieties of *P. nigra* seem to be particularly susceptible to *N. coccinea*, whereas most of the Black hybrids (*P. euramericana*) are resistant (Koning 1939; Houtzagers *et al.* 1941). *Nectria* cankers on poplar are seldom serious, though the fungus may be important in enlarging cankers caused primarily by other agencies.

Other fungal diseases of bark and cambium

Botryodiplodia penzigii Petr. & Syd. (Fungi Imperfecti, Sphaeropsidales) was frequently isolated by Koning (1938) in Holland from cankers of the same non-erumpent type as those associated with *Nectria*. Its pathogenicity was proved by inoculation. It has not been recorded in Britain. As with *Nectria*, inoculations showed that varieties of *P. nigra* were particularly susceptible, and Black hybrid poplars mostly resistant (Houtzagers *et al.* 1941).

Ungulina inzengae De Not. (Basidiomycetes, Aphyllophorales), which is closely related to the well-known decay-fungus *Fomes fomentarius*, has been associated with a gradual dieback of old poplars in France (Heim and Lami 1950). It is only a slow parasite, but has nevertheless been held responsible for the death of considerable numbers of poplars along the River Seine in Paris.

Domański (1954) has found *F. igniarius* (L.) Fr. acting in the same way as a slow, killing parasite on poplars in Poland. *F. igniarius* occurs in Britain mainly as a decay fungus in old willows, though it has been recorded on poplar. In North America it causes widespread and serious rotting of aspen, though it is not there considered to be a parasite.

Stereum purpureum (Fr.) Fr. (Basidiomycetes, Aphyllophorales), cause of the well-known Silver-leaf of plums and other fruit trees, has a very wide host range which includes poplar. It occurs particularly on the latter after severe lopping, infecting the branch stubs. In this case it is certainly not an active parasite, but may be dangerous as a source of infection to nearby fruit trees (Putterill 1923). Poplars are often used to shelter orchards, and make a very rapid screen, but in this position they should be pruned frequently, to avoid the removal of large branches. If this is done, infection by *S. purpureum* is unlikely.

Macrophoma tumifaciens Shear. (Fungi Imperfecti, Sphaeropsidales) causes globose galls up to 4 cm. in diameter, as well as roughening of the bark on *P. trichocarpa* and *P. tremuloides* in America. The galls sometimes girdle small stems and result in dieback (Kauffert 1937).

In Austria *Diplodia gongrogena* Temme (Fungi Imperfecti, Sphaeropsidales) is said to cause excrescences on the bark of *P. nigra*, *P. alba*, and *P. tremula* (Bittmann 1933), but its pathogenicity has not been proved (Houtzagers *et al.* 1941).

The role of *Fusarium* (Fungi Imperfecti, Moniliales) in the pathology of poplars is very hard to determine. Various species have been isolated in Britain from diseased poplar stems. A *Fusarium* regularly develops in Britain on the small lesions caused by *Pseudomonas syringae* on aspens. This happens on cankers resulting both from natural infections or from inoculation. There is no evidence in any of these cases that the fungus is pathogenic.

Magnani (1954) isolated two species of *Fusarium* from poplars showing dieback in nurseries; but he failed to establish their pathogenicity and considered that they were secondary to adverse soil conditions. However, Gallucci and Perotti (1959) have redescribed the disease and consider *Fusarium* a pathogen. Taris (1957) has recorded *F. avenaceum* (Fr.) Sacc. as the cause of small erumpent cankers on the stems of nursery plants. Among a small number of clones tested, *P.* 'robusta' was particularly susceptible. The disease was worse on poorly cultivated sites. Bier (unpubl.) has found *F. lateritium* Nees causing canker on *P. trichocarpa*. Infection and spread of the fungus were associated with low turgor.

Other supposedly bacterial diseases of bark and cambium

In southern Europe, particularly in Italy, bacteria have been associated with bark lesions on Hybrid black poplars. These lesions are usually only a few centimetres across and remain active for only one year, though the larger ones may take several years to heal over (Ciferri 1951; Vivani 1957). Bacterial slime oozes from cracks in the centre of the lesions, but it is still not certain that the disease is caused by bacteria. It occurs on trees of any age from two years upwards, but is most serious on stems over 10 cm. in diameter. In such trees the lesions damage the utilizable part of the wood and allow the entry of staining fungi and bacteria. For this reason attacked trees are useless for veneering or any other purpose requiring unblemished wood. There has in the past been confusion, caused mainly by the use of the name Bacterial canker, between this disease and the much more serious canker discussed above, which does not seem to occur in Italy.

Similar damage has been observed on *P.* 'robusta' in southern France. In parts of Germany, where the disease is known as 'Braunfleckenrind', it is sufficiently common to discourage the use of *P.* 'robusta'. Joachim (1958*c*) believes that the disease is widespread in Europe. Schönhar (1956*a*) has isolated *Dothichiza populea* more often than bacteria from this type of lesion and Taris (1957) frequently found this fungus associated with a bark disease known as 'Taches brunes' which affects young plantations in northern France, but there is as yet no proof that it is a primary agent in these diseases. Drought or some other physiological disturbance may be involved.

Small, erumpent tumours up to 15 cm. in diameter, which occur on poplars in Italy, Cyprus, and probably in other Mediterranean countries, may be due to *Bacterium tumefaciens* (Servazzi 1938*a*) (Chapter 20). Wetwood of poplar, which may be due to bacterial infection of the sap, is also described in Chapter 20.

DISEASES OF LEAVES AND SHOOTS

Leaf rusts

Pathogens

A number of species of *Melampsora* (Basidiomycetes, Uredinales) are concerned in this disease, but since they produce similar symptoms on poplar leaves, it is best to consider them together. Like most rusts they have two host plants, forming uredia and telia on poplar and aecia on other hosts. Their

SPECIES	AECIAL HOST	UREDIAL AND TELIAL HOST	DISTRIBUTION
M. abietis-canadensis (Farl.) Ludw.	*Tsuga* spp.	Aspen, White, Balsam, and *Leucoides* poplars	North America
M. aecidioides (DC.) Schroet.	Not known	White poplars	North America, South America, Near East, India, Europe (not Britain)
M. albertensis Arth.	*Pseudotsuga taxifolia*	Aspen, Balsam, and *Leucoides* poplars	North America, Argentine
M. allii-populina Kleb.	*Allium* spp.	Black and Balsam poplars	Europe (inc. Britain), Near East, North Africa, Argentine
M. larici-populina Kleb.	*Larix* spp.	Black and Balsam poplars	Europe (inc. Britain), Near East, Argentine
M. larici-tremulae Kleb.	*Larix* spp.	Aspen and White poplars	Europe (inc. Britain)
M. magnusiana Wagn.	*Chilidonium majus*, *Corydalis* spp., etc.	Aspen and White poplars	Europe (not Britain), Japan
M. medusae Thüm.	*Larix* spp.	Aspen, Black, and Balsam poplars	North America, France
M. occidentalis Jacks.	Unknown	Black and Balsam poplars	North America
M. pinitorqua Rostr.	*Pinus* spp.	Aspen and White poplars	Europe (inc. Britain), Near East, Canada (B.C.)
M. rostrupii Wagn.	*Mercurialis perennis*	Aspen and White poplars; possibly rarely on Black and Balsam poplars	Europe (inc. Britain)

classification is to some extent dependent on the hosts on which they occur. Information on poplar hosts is very incomplete, and there has been some confusion between the different species of poplar, particularly in America.

The species of *Melampsora* known to occur on poplar are set out above. The recent merging by Wilson and Bisby (1954) of *M. pinitorqua*, *M. rostrupii*, and *M. larici-tremulae* under the one species *M. tremulae* Tul., which was made on the grounds that they were indistinguishable on the poplar host, is not here accepted (see Chapter 22, where *M. pinitorqua* is dealt with in more detail).

All species form orange or yellow uredia, mainly on the under surfaces of the leaves. Uredia may also occur on the upper surfaces and sometimes on

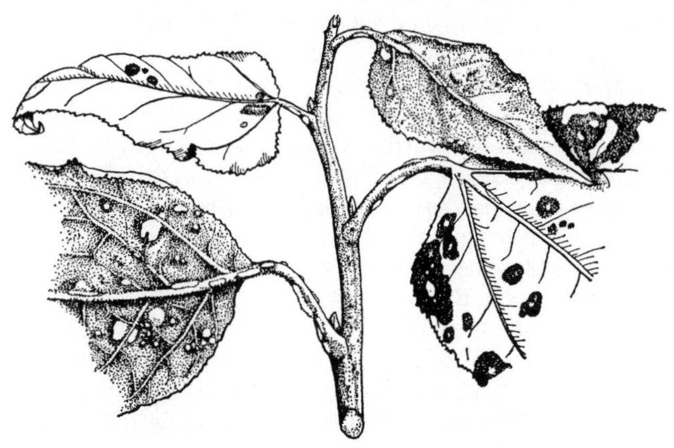

FIG. 88. *Melampsora* sp. fruiting on the leaves and petioles of *Populus grandidentata*, Alice Holt Research Station, Hampshire, September (× ½). (J. C.)

the petioles and on young succulent shoots. Shoot and petiole infection is particularly common on *P. grandidentata* (Fig. 88). The uredospores infect poplar leaves and are mainly responsible for the rapid build-up of the disease. The uredia appear from mid-summer onwards; in the autumn brown telia are formed. The sporidia from the teliospores infect the alternate host and lead to the formation in the late spring of white and yellow aecia, which are not always particularly conspicuous. Those on larch, for instance, are easily overlooked.

It is extremely difficult, and in some cases impossible, to distinguish the different *Melampsora* species on poplar. One guide, of course, is the poplar group which is attacked. Some species can be distinguished by microscopic characters. *M. larici-populina* and *M. allii-populina* can be separated in this way (Gremmen 1954). On larch the aeciospores of *M. larici-tremulae* are light yellow, while those of *M. larici-populina* are orange.

Symptoms and development

The orange uredia are produced in enormous quantities so that a heavy attack on poplar is very striking. As a result of infection the leaves wither prematurely, usually hanging for some weeks before they fall. This premature defoliation causes only a slight loss of growth, since it does not usually happen until late August or even early September. However, the premature loss of the leaves seems to interfere with the proper 'ripening' of the succulent shoots, so that severe *Melampsora* attack is often followed by autumn-frost damage resulting in shoot dieback. This is often much more serious than the initial reduction in growth.

It is primarily a nursery disease, but severe defoliation can occur on older trees. A severe outbreak in twenty-year-old *P*. 'marilandica' has recently been described (Meiden and Vloten 1958). In this case the severity of the *Melampsora* attack was due to the planting of Japanese larch in and near the poplar plantations. Not only was there a marked reduction in growth, but the poplars were so weakened by the rust attack that they became infected with *Dothichiza populea*, which caused severe branch dieback.

Overwintering is normally in the telial stage, the sporidia produced on the teliospores in the spring infecting the alternate host. Several species, however, are known to overwinter without the intervention of the alternate host. This is particularly the case in warm climates, and certainly happens in the Argentine. Moriondo (1954) has found *M. pinitorqua* mycelium in the cambial region of twigs of *P. canescens* and considers that it can overwinter in this way. *M. aecidioides* overwinters in the form of uredospores between the bud scales (Spaulding 1958). There is no definite evidence as to whether an alternate host is always necessary in Britain. Certainly attacks often develop when larch and *Allium* are absent from the immediate neighbourhood. There is evidence, however, that the presence of larch in the immediate vicinity of the poplar increases the severity of attack. In any case uredospores are produced so abundantly that a small initial infection on poplar leaves can quickly build up into a severe outbreak.

Distribution and damage

Admittedly incomplete information on the distribution of the different species has already been given. In Britain the most important species is *M. larici-populina*, which can cause severe defoliation, particularly in nuseries. The species attacking poplars of the *Leuce* group seldom attack severely in Britain, though *M. pinitorqua* can be seriously damaging on Scots pine (Chapter 22). It is possible that *Melampsora* on poplars may act as a weakening agent, leading to attack by other agencies more often than is at present realized.

The collection of information on varietal susceptibility is badly hampered by confusion between the different species of *Melampsora*. In addition there are clonal differences in susceptibility within species and varieties. Nagel (1949), for instance, found differences in defoliation ranging from 6 to 98 per cent. in different clones of the very variable species *P. deltoides*. Such differences may occur even between individuals supposedly of one clone. For instance, *P*. 'robusta', which is sometimes regarded as a single clone, is in reality a collection of very similar clones, which display, among other differences,variation in susceptibility to *Melampsora*. In addition there is evidence that the fungi themselves are variable, some species containing races of differing virulence (Vloten 1949).

Tables of relative susceptibility made in different countries by different workers are thus often in disagreement. For instance, differences between the lists given by the author (Peace 1952*b*) and by Leontovyč (1958) are probably due to the fact that the former refers mainly to *M. larici-populina* and the latter to *M. allii-populina*. Host lists without reference to susceptibility are, within their limitations, more reliable. These have been given by Wilson and Bisby (1954) and Vloten (1941) for the species of *Melampsora* occurring in Britain. In some cases information can be related to a single species of *Melampsora*. For instance, there are considerable differences among various aspen hybrids in susceptibility to *M. rostrupii* (Anon. 1955*c*).

In Britain some poplar varieties are frequently and severely attacked, probably mainly by *M. larici-populina*. These include *P.* 'generosa', most members of the *P.* 'berolinensis' group, and many, but not all, clones of *P. trichocarpa*. All these and some others may suffer severely as a result of the disease and subsequent frost injury. *P.* 'generosa', which has thick, soft shoots, is frequently damaged in this way. Several of the Black hybrids, in particular some clones of *P.* 'robusta', often suffer severe defoliation, but the resultant dieback is not so serious as with the susceptible Balsams and Balsam hybrids.

Control

Control of the disease must rest principally on the selection and use of resistant clones. Spraying with Bordeaux mixture several times during the summer has been found to lessen the severity of attack. However, even in the nursery, poplars are so large that spraying is expensive. While no doubt more effective sprays could be found, it is doubtful whether the improved growth resulting from the absence of rust would justify their cost.

Leaf blister

Very striking blisters are produced on the leaves of poplars by *Taphrina aurea* (Pers.) Fr. (Ascomycetes, Exoascales) (syn. *T. populina* Fr.). In early summer the leaves become distorted. The concave, spore-bearing surface, which is on the underside of the leaf, is bright yellow, while the convex upper side of the blister remains green. The yellow patches are very conspicuous and sometimes lead to confusion between this disease and that caused by the rust fungus *Melampsora*, the symptoms of which are really quite different. *Taphrina* does not usually cause leaf-fall, so that its effects are not serious. The fungus attacks most varieties of poplar, but is commonest on varieties of *P. euramericana*. Within this group some clones are obviously more attacked than others, though few comparative records are available. Servazzi (1935*a*) suggested that there are races of the fungus specific to different poplars. In Britain the disease is very common, but there appears to be no record on *P. tremula*. It has, however, been found on *P. alba*.

Two other species of *Taphrina*, *T. johansonii* Sadeb. and *T. rhizophora* Johans., cause golden-yellow swellings on the female catkins. The former has a wide host range, but is usually found on *P. tremula*. The latter is supposed to be confined to *P. alba*. It is doubtful if they are really separate species. Both have been recorded in Britain, but they are uncommon and unimportant.

Marssonina leaf blight

Pathogens

Several species of *Marssonina* (Fungi Imperfecti, Melanconiales) have been associated with leaf infections on poplar. Most commonly cited names are *M. populi* (Lib.) Magn., *M. castagnei* (Desm. & Mont.) Magn., and *M. rhabdospora* (Ell. & Ev.) Magn. Klebahn (1918) provisionally created several species of *M. populi* according to the host attacked, giving them names such as *M. populi-albae*. He regarded the perfect stages as species of *Pseudopeziza* (Ascomycetes, Helotiales), using *P. populorum* Kleb. as the basic species, but again splitting off some as *P. populi-albae*, etc. Much more recently G. E. Thompson (1954) has given *Pleuroceras populi* Thompson (Ascomycetes, Sphaeriales) as the perfect stage of *M. rhabdospora*. None of the perfect stages has apparently been recorded in Britain. Whether there are really several species of *Marssonina* on poplar still remains to be proved. From a pathological point of view it is better to consider them as a single species, embodying strains of varying pathogenicity on different hosts.

Symptoms and development

The lesions caused by this fungus usually appear after mid-summer as irregular, brown blotches, sometimes with darker margins, usually merging to cover the whole leaf. The conidia, exuding in white tendrils on the upper side of the leaves, can be detected with a lens. The fungus causes premature leaf-fall, and is sometimes associated with shoot dieback. Pirone (1948) considers that it actually invades the twigs, but it is possible that its action is more complex than this, premature defoliation interfering with the ripening of the twigs, so that they are damaged by autumn frost, and thus exposed to colonization by other fungi. This is known to occur on poplar in the case of defoliation by *Melampsora* rusts. The disease is favoured by high rainfall in the spring and early summer (Mielke 1957*a*). Perithecia are produced in the spring on fallen leaves.

Distribution and damage

If we regard this fungus as a single species, it is widely distributed both in North America and in Europe. In the United States and Canada it has been the cause of widespread defoliation of *P. tremuloides* (Nordin 1953; Mielke 1957*a*). It attacks other poplar species there also. In Europe it has not generally been regarded as a very serious disease, though very widely distributed.

In Britain it is the most probable primary cause of an increasingly common disease of Lombardy poplars (*P. nigra* 'italica'). In this disease, which has done quite severe damage in a few places in southern England, the leaves on the lower shoots wither prematurely and this is followed by shoot dieback. Eventually only the tip of the crown is left alive, and in a few cases trees have been completely killed. In Holland and elsewhere it also attacks *P. nigra* 'italica' (Poeteren 1938), often destroying its value as shelter for orchards, because of the severe defoliation. It also occurs there on other poplar species.

In Britain it has been recorded on many other poplars besides *P. nigra* 'italica'; but there are not sufficient data available to provide any definite information on varietal susceptibility, a matter that in any case would be complicated by the probable occurrence of several strains (or species) of the fungus. Mielke (1957*a*) noted that some clones of *P. tremuloides* were immune to attack, so that groups of healthy undefoliated trees, each group of a common clonal origin, could be detected even in photographs.

Control

Several control measures have been suggested (Poeteren 1938). These include sweeping up dead leaves under the trees during the winter to prevent the production of the fruit-bodies of the perfect stage, cutting off and burning dead branch tips, and spraying affected trees with Bordeaux mixture just after the leaves have unfolded and again when they are fully grown. These sprayings are, of course, designed to prevent infection and are therefore applied well before any symptoms of attack are likely to appear. On orchard shelterbelts extension of spraying against Apple scab to the poplars has been suggested.

Spring defoliation

Pathogens

Two fungi, apparently separate species though much confused, are involved in this disease. Both are species of *Venturia* (Ascomycetes, Sphaeriales). Their synonyms, which are very involved, have been given in full by Ciferri and Baldacci (1953). Only the commoner names, some of which are still in use, are given below.

Venturia populina (Vuill.) Fabr. Syn: *Endostigme populina* (Vuill.) Cif.
Didymosphaeria populina Vuill.
Conidial stage: (Fungi Imperfecti, Moniliales)
Pollaccia elegans Serv.
Fusicladium tremulae Frank.
Fusicladium radiosum (Lib.) Lind.

Venturia tremulae Aderh. Syn: *Didymosphaeria populina* Vuill.
Conidial stage: (Fungi Imperfecti, Moniliales)
Pollaccia radiosa (Lib.) Bald. & Cif.
Fusicladium tremulae Frank.

It will be seen from the list above that originally the two species were not separated. They can, however, be separated taxonomically, apart from the fact that they seem to have different host ranges (Servazzi 1938*b*). *V. populina* attacks particularly poplars of the *Aigeiros* (Black) section, and *V. tremulae* those of the *Leuce* (White and Aspen) section.

Symptoms and development

Both species cause the same general symptoms. Angular black spots develop on the young leaves in the spring or early summer. These gradually increase in size until the whole leaf is killed. The fungus then spreads down the petiole to the shoot, which also turns black. The tips of the withered twigs are characteristically hooked. The perithecia of the perfect stage are inconspicuous and sometimes are not produced. Conspicuous olive-green mats of conidia are formed on the dead leaves and shoots (Servazzi 1935*b*; Gremmen 1956). Wet weather in the spring greatly increases the severity of attack. The disease continues to develop during the summer, each fresh batch of recovery shoots being attacked in its turn, but in hot dry summers the disease may become completely quiescent (Servazzi 1935*b*; Goidànich 1938).

Hartig (1894) and Prillieux (1897) described one or other of these fungi under the name of *Didymosphaeria populina*, as a cause of severe twig dieback of *P. nigra* 'italica' trees in Germany and France respectively. Their description agrees rather closely with the symptoms of Lombardy poplar dieback now found in Britain (see under *Marssonina* Leaf blight), the shoots near the base of the tree dying first. *Venturia*, however, has not yet been isolated from any of the British cases, whereas *Marssonina* has been found on several occasions.

Distribution and damage

V. tremulae occurs over most of Europe, including Scandinavia, but its presence in Britain rests on two records on *P. tremula*, one in Scotland (Foister 1961) and one in Dorset (Sutton unpubl.). On the Continent it is locally severely damaging. *V. populina* is commoner in southern Europe and is particularly serious in Italy, where it certainly outweighs in importance any other leaf and shoot disease of poplar. *V. tremulae* occurs in North America, *V. populina* is probably also present (Hatfield 1946), but does not seem to have attracted much attention.

In Italy many Black hybrids (*P. euramericana*) are attacked, though some of the Italian selections are highly resistant. *P. nigra* and *P. nigra* 'italica' are also sometimes attacked, but *Venturia* is seldom found there on *P. alba* or *P. tremula* (Servazzi 1935*b*). This suggests that *V. tremulae* is uncommon in Italy. In Holland *V. tremulae* has attacked *P. tremula* hybrids and *P. alba*× *tremula* hybrids; both displayed clonal differences in susceptibility (Gremmen 1956). In Germany it has been recorded on *P. tremula* and *P. canescens*. Hybrids between *P. tremula* and *P. tremuloides* are said generally to be resistant (Andersson and Strand 1951).

Venturia populina has been reported in Germany and Denmark causing severe dieback of *P.* 'berolinensis', which is a hybrid between the Black and Balsam groups (Gremmen 1956). Donaubauer (1957) reports serious damage by *V. populina* on *P.* 'marilandica' in Austria. Much more information must be collected on varietal resistance to these fungi before the position can be properly clarified.

Control

Probably the only effective control measure is the removal of the dead twigs during the winter to prevent the overwintering of the fungus as mycelium and the formation of the conidial mats in the spring. If the disease ever becomes more serious in northern Europe, or even in Britain, selection of resistant clones is a possibility.

Septoria leaf spot

Pathogens

Several species of *Septoria* (Fungi Imperfecti, Sphaeropsidales) cause leaf-spots on poplar. The most important are *S. populi* Desm. in Europe and the Argentine, *S. populicola* Peck confined to North America, and *S. musiva* Peck in North America and in the Argentine. The last named is far more important as a cause of canker than as a cause of leaf-spot (see above). All three have species of *Mycosphaerella* (Ascomycetes, Sphaeriales) as their perfect stages. These are *M. populi* (Auersw.) Kleb., *M. populicola* Thomp., and *M. populorum* Thomp. respectively. *Cercospora populina* Ell. & Ev. (Fungi Imperfecti, Moniliales), which causes leaf-spotting of poplars in Japan and North America, also has a *Mycosphaerella*, *M. togashiana* Itô & Kob., as the perfect stage (Itô and Kobayashi 1953). The two American species of *Septoria* have been described by G. E. Thompson (1941) and by Sarasola (1944). The conidia are extruded in small tendrils, while the leaves are still on the tree. These are followed by dark, globose spermogonia in the autumn. The black globose perithecia are borne on the fallen leaves the following spring.

Symptoms and development

All these fungi form small black, grey, or brown spots, usually paler in the centre, on living leaves, particularly on the lower part of the tree. The colour of the spots varies according to the host, as well as to the fungus species. The size of the individual spots varies from 1 mm. to 15 mm. across, the average size being greatest for *S. musiva* and smallest for *S. populi*, those of *S. populicola* and *C. populina* being intermediate. The spots sometimes coalesce to form much larger blotches.

Distribution and damage

The distribution has already been given above. Only exceptionally are any of these fungi seriously damaging, but according to Sarasola and Magi (1951) premature leaf-fall due to *S. musiva* may weaken the trees. *S. populi* is not uncommon in Britain, but has never been associated with serious damage. It is not sufficiently important for any detailed information to have been collected on its host range, though it has been found on *P. nigra* and *P.* 'serotina'. *S. musiva* has a very wide host range including Black, Balsam, and *Leuce* poplars. *S. populicola*, on the other hand, has been recorded only on Balsam poplars (Thompson, G. E. 1941).

Control

Apart from *S. musiva*, which has been considered above, these diseases do not seem ever to be sufficiently damaging to justify any attempt at control.

Septotinia leaf blotch

Pathogens

Septotinia populiperda Wat. & Cash (Ascomycetes, Helotiales) has been recorded in North America and on the Continent, but not in Britain. Waterman and Cash (1950), who first described it, considered that the conidial stage, which they called *Septotus populiperda* Wat. & Cash, was identical with

Septogloeum populiperdum Moesz. & Smarods (Fungi Imperfecti, Melanconiales) described on poplar in Latvia. Ende (1952) considered that the imperfect stage of this fungus was a *Septogloeum*, which might well be identical with *S. populiperdum* Moesz. & Smarods. There is one unconfirmed record of an unidentified *Septogloeum* species on *P. euramericana* 'pseudoeugenei' in Britain.

Another species of *Septogloeum*, *S. rhopaloideum* Dearn. & Bisby, has also been recorded on poplar leaves. Ciferri (1951) considered that this was probably synonymous with *Pollaccia elegans*, the conidial stage of *Venturia populina* (see above). G. E. Thompson (1954), on the other hand, considered that the perfect stage of *S. rhopaloideum*, which is associated in the United States with pale spots on the leaves of *P. tremuloides*, was a *Guignardia* species, which he named *G. populi*.

Symptoms and development

The blotches on the leaves, starting in the spring as small brown spots, develop rapidly. Before long, white spore-bearing masses of the *Septogloeum* stage develop on both surfaces of the leaves; on the upper surface they tend to develop in concentric circles. The blotches are always associated with wounds, usually insect injuries, and it has been proved by inoculations that the fungus is always a wound parasite (Ende 1954; Schmidle 1955). The affected leaves, which are usually on the lower parts of the trees, are shed prematurely.

Small, black sclerotia are produced on the fallen leaves and from them the saucer-like apothecia are produced in the spring. These are brownish in colour and 2 to 7 mm. in diameter with thin stalks, 1 to 2 cm. long and 1 mm. in diameter. They are very characteristic, but are not of course produced while the leaves are still on the tree, and so may be difficult to associate with the disease.

Distribution and damage

The disease does not appear to be very important in the United States. In Europe it has been recorded in Holland, where it has done appreciable damage in nurseries, in Jugoslavia and in Germany, where it caused some alarm. It is doubtful whether it occurs in Britain.

S. populiperda can attack a wide range of species, and inoculation experiments indicate varietal differences in susceptibility (Waterman and Cash 1950; Ende 1954). So far it seems to have occurred only on poplars of the Black and Balsam groups and their hybrids.

Control

Practically nothing is known about the control of this disease; in any case, it is seldom sufficiently serious to necessitate any control measures.

Other leaf and shoot diseases

At least ten species of *Phyllosticta* (Fungi Imperfecti, Sphaeropsidales) have been associated with leaf-spots on poplar. Several of these supposed species are probably not separate; indeed Servazzi (1934*b*), who studied this disease in Italy, was prepared to include four under the name *P. populina* Sacc. This species has been recorded in Britain on a wide range of poplar varieties, causing irregular brown spots on the leaves, but never doing serious damage. *P. populi-nigrae* Allesch. has also been recorded in Britain.

Mildew, *Uncinula salicis* (DC.) Wint. (Ascomycetes, Erysiphales), occurs widely, but seldom seriously on poplar leaves. It has been recorded on poplar in Britain, but is even less common than on willow.

The other diseases listed below do not, as far as is known, occur in Britain.

Sphaceloma (*Hadrotrichum*) *populi* (Sacc.) Jenk. (Fungi Imperfecti, Melanconiales) has been associated at intervals over a long period with the withering of poplar foliage in southern Europe and in South America (Fresa 1936; Jenkins and Bitancourt 1939; Ciferri and Baldacci 1953). Its real importance has never been adequately assessed.

Ink spot disease, probably caused by the fungus *Sclerotinia bifrons* Whetz. (syn. *Ciborina whetzelii* Seav.) (Ascomycetes, Helotiales), is very prevalent in North America. It particularly attacks the aspen poplar *P. tremuloides*, but also *P. grandidentata* and other species (Pomerleau 1940). The first sign of the disease is the appearance of spots with brown margins on the leaves. These spots spread and often coalesce. Small black sclerotia are formed on them and eventually fall to the ground. Apothecia are formed on the sclerotia on the fallen leaves in the spring and the spores are carried to the young foliage by the wind. Very wet conditions are required for successful infection, so that only in exceptional seasons is the disease serious. In severe outbreaks, which usually occur in dense young stands, some trees may be killed, especially if defoliation continues for more than one season. Cash and Waterman (1957) have recorded *Plagiostoma populi* Cash & Waterman (Ascomycetes, Sphaeriales) causing brown discolorations late in the season on aspen leaves that had escaped attack by *Sclerotinia bifrons*, and thus adding to the total damage.

A disease of young poplar shoots caused by the fungus *Pestalozzia populi-nigrae* Sawada & Itô (Fungi Imperfecti, Melanconiales) has been reported from Japan (Itô 1950). It occurs in nature on *P. nigra*, *P. nigra* 'italica', and *P. deltoides monolifera*, all exotic varieties in Japan. On inoculation *P. simonii* from China proved moderately susceptible and *P. maximowiczii*, from Japan itself, resistant. The lesions first appear as well-defined brown areas on the shoots of the current year, and a conspicuous symptom is the presence by mid-June of one or more dead leaves on the affected shoots. The fungus spreads rapidly on the shoots, often girdling them. Later the affected areas become paler and small erumpent fruit bodies appear on them. At a later stage cankers may form on twigs that have not been girdled. *Pestalozzia populi-nigrae* has been recorded on the bark of felled poplars in Italy (Gambogi and Verona 1958). *Pestalozzia hartigii* Tub. has been found on a number of occasions on poplar in Britain, but has not been associated with any specific disease.

Corticium salmonicolor Berk. & Br. (Basidiomycetes, Aphyllophorales), which is widely distributed on many host trees in the tropics as the cause of 'Pink disease', has been associated with shoot dieback of poplars in Brazil (Rombouts 1936). The disease gets its name from the profuse, superficial, pink incrustations which the fungus forms on the twigs of affected trees. It is unlikely to occur in the temperate regions where poplars are mainly grown.

Physalospora populina Maubl. (Ascomycetes, Sphaeriales) was associated with shoot dieback of poplars in a nursery in Italy by Suliotis (1936). It is probably only a weak parasite.

Recently Dance (1957) has found a species of *Gloeosporium* (Fungi Imperfecti, Melanconiales) consistently associated with leaf-withering and shoot-dieback of *P. alba* × *grandidentata* in Canada. The disease has not been found on either of the parents. *G. tremulae* Pass. and *G. populi-albae* Desm. occur as unimportant leaf-spots of poplars in Europe, but have not been definitely recorded in Britain.

Linospora tetraspora Thomp. (Ascomycetes, Sphaeriales) has caused severe damage to the leaves of young trees and of the lower branches of old trees of *P. tacamahaca* in Canada (Thompson 1939). It causes lesions with irregular and diffuse margins. *L. populina* (Pers.) Schroet has been recorded in Britain, but only on dead poplar leaves.

33

DISEASES OF WILLOW

THERE are at least twenty species of *Salix* native to Britain, and a considerable number of hybrids between them. Many are shrubs, some even prostrate creepers; only *S. alba*, the White willow, *S. fragilis*, the Crack willow, and some of their hybrids and varieties can be regarded as producers of timber. The Cricket bat willow, *S. alba* var. *caerulea*, ranks high among these, mainly because of the particular suitability of its timber for the manufacture of cricket bats.

Three species, and a number of their hybrids, are grown in special beds and are cut annually to provide rods for basket-making. *S. triandra*, the Almond-leaved willow, and *S. viminalis*, the Common osier, are the most important of these; *S. purpurea* is used to a lesser extent. *S. alba vitellina* is grown by nurserymen to provide rods for tying bundles of trees. The Weeping willow, *S. babylonica*, is extensively planted as an ornamental.

The Sallows, *S. caprea*, *S. atrocinerea*, *S. cinerea*, and *S. aurita*, occur commonly in mixed broadleaved woodland, especially on damp soils, but none of them reaches any great size.

DAMAGE DUE TO NON-LIVING AGENCIES

Willow is considered a fairly hardy tree, but Mooi (1948) attributed serious cankering and dieback of *S. alba caerulea* in Holland to frost, and W. R. Day (1948) considered some of the abnormalities and stains of the wood, which greatly lower the value of the timber for cricket bats, to be due to frost injury.

Most willows can grow in wet soils, provided the soil water is not completely stagnant; some can grow healthily, but rather slowly, on dry sites. They are moderately adaptable to site changes, but crown dieback is quite common where sites have become markedly drier or wetter. Despite the dieback, however, they have considerable tenacity of life, often lingering on in extremely wet situations with the whole of their root system apparently immersed.

ROOT DISEASES

Armillaria mellea is occasionally troublesome in the cultivation of Cricket bat willow. If willow is being replanted, infected stumps should be removed, lest the fungus spreads from them to the new trees.

The parasitic flowering plant *Lathraea clandestina*, which has rather decorative purple flowers faintly suggestive of crocus, was introduced into Britain and is now found occasionally on roots of *Salix* and *Populus* (Anon. 1949*a*). There is no evidence that it does any damage to the host trees (Chapter 13).

Crown gall (*Bacterium tumifaciens*) (Chapter 16) occurs to a damaging extent on basket-willow stools in parts of the Continent, but has not been observed in Britain.

STEM DISEASES—WILTS

Watermark disease

Pathogen

The most serious disease of willow in Britain is the Watermark disease, so called because of the diffuse watery brown markings in infected wood. In this country the disease is caused by *Erwinia* (*Bacterium*) *salicis* (Day) Chester (Day, W. R. 1924; Dowson 1937), but a very similar disease in Holland has been attributed to a different bacterium, *Pseudomonas saliciperda* Lind. (Lindeijer 1932). Both Day and Dowson carried out successful inoculation experiments with *E. salicis*, as did Lindeijer with *P. saliciperda*. Three other bacteria, besides *E. salicis*, were isolated by G. Metcalfe (1940) in the later stages of disease development, but he did not establish their pathological significance.

A disease associated with water-soaked and eventually brown-stained wood, in which bacteria were abundantly present, was reported in the United States by Hartley and Crandall (1935), who remarked on its resemblance to Watermark disease in Europe. Its failure to attract further attention in the United States suggests that it has not developed as seriously as Watermark disease has in Britain.

Symptoms and development

A brief description of the disease is given in Forestry Commission Leaflet No. 20 (Anon. 1949*b*). The first signs of the disease may be seen as early as the end of April or the beginning of May. The leaves on affected shoots or branches wither and turn reddish-brown. These primary symptoms may appear up to the end of July. Later the affected parts may die back completely. Epicormic shoots often appear on the lower parts of affected branches, but these usually wither as well. Exudations of a thin, nearly colourless, sticky liquid takes place through cracks and insect wounds on the affected branches. Sometimes branches, withered by the disease one year, may come into leaf again the next, but generally the disease is progressive, the tree becoming more and more 'stagheaded'. Total recovery, as evidenced by the absence of any further dieback, does, however, occasionally occur. The wood of all affected twigs and branches shows a diffuse watery brown stain, initially darker towards the outside of the annual ring, which gives the disease its name of Watermark. This stain often extends down into the trunk and even into the roots. However, bacteria may be present in quantity in the wood before any clearly visible staining develops. The watermark gradually darkens, even inside the tree, so that in a willow which has been diseased for several years the inner (older) parts of the marking will be quite dark brown in colour. On exposure to air this darkening is much more rapid, and the cut surface of a water-marked twig darkens in a few hours. In dead wood the water-mark fades, and eventually becomes impossible to detect; thus living wood from the affected parts of the tree is required for diagnosis.

When the external symptoms appear in spring or in early summer, the stain is found in the annual ring of the previous year, presumably indicating that infection took place then. But later appearances of external symptoms are often accompanied by markings in the current ring (Metcalfe, G. 1940). The bacterium appears to spread with difficulty radially, but with ease up and down the tree, so that the occurrence of active symptoms in one year does not necessarily mean that they will appear in the next. The tree may lay down healthy wood over the watermarked wood. Symptoms very often do reappear, however. This is sometimes due to bacteria moving out from one annual ring to the next. They can do this through insect wounds, but what other routes they can use are not known. Reappearance of the disease on previously infected trees may sometimes, on the other hand, be due to new infections. The relative importance of internal spread and fresh infections in continuing the disease in infected trees has never been investigated.

The means of infection and of spread from tree to tree are still unknown. It is very probable that insects are involved. A great deal of work was done by Callan on insect vectors of the disease. *Salix* has a very large and varied insect population. Several species appear possible carriers, among these are Willow gall midge, Willow sawfly, Willow wood wasp, Goat moth, and a number of weevils, one of which, *Cryptorrhynchus lapathi* L., was considered by Lindeijer (1932) to be the chief vector of the disease in Holland. Callan (1939) was unable to confirm this, nor could he prove with certainty that any other insect carried the disease. E. Gray (1940), who investigated the Willow wood wasp, *Xiphydria prolongata* Geoffr., failed to obtain any definite proof that it was a vector. Carriage by birds is another possibility. Various birds, especially tits, often peck away the bark when seeking insect larvae, in so doing they might well transfer the bacterium on their beaks from one tree to another.

The affected vessels are substantially blocked by tyloses, the production of which is stimulated by the bacteria. It is not known whether the wilting of the foliage is primarily due to interference with the water-supply or to a toxin produced by the bacteria. The extreme rapidity of the wilting in some instances suggests that toxins may be important. A number of degenerative changes occur in diseased wood (Metcalfe, G. 1941), but these do not appear substantially to affect its gross mechanical properties. Such changes and the formation of tyloses, rather than the actual presence of the bacteria, give rise to the discolorations. The changes include dissolution of the middle lamellae, causing tangential cracks in the wood, which probably aid the bacterium to invade fresh parts of the stem.

The dead and dying branches are often invaded by the fungus *Cytospora* (see below), but there is no evidence that it acts as a parasite. Trees very seldom die completely from Watermark disease alone, but damage often reduces the growth rate considerably, and watermarked timber is of no value for cricket bats.

There is no evidence that growing conditions appreciably affect this disease; it attacks trees in good and bad sites alike. The symptoms may be more striking on vigorously growing trees, though the final damage to weakly growing ones may be greater, owing to their poorer powers of recovery.

Willows are usually planted as unrooted sets. In the case of Cricket bat willow these are raised on stools planted for the purpose in special nursery beds. Stools infected with Watermark have been recorded on a number of occasions; these were probably raised from cuttings taken from infected trees. The sets taken from such stools may sometimes, though not invariably, carry infection. It is not known how important infected sets are in disseminating the disease, but the comparative rarity of the disease in young trees suggests that it does not often happen.

Distribution and damage

The disease is particularly active in the principal Cricket bat willow growing districts, Essex, east Hertfordshire, Suffolk, and south Cambridgeshire. It has also been found as far afield as Gloucestershire, Wiltshire, and Yorkshire and is common in Bedfordshire on *S. alba*. It has already been mentioned that an almost exactly similar disease, attributed, however, to a different bacterial pathogen, occurs in Holland. A disease causing watermark symptoms has recently been recorded in Austria.

S. alba caerulea is the variety most commonly affected. Among the other tree willows, *S. alba* and hybrids between it and *S. fragilis* are commonly attacked (Dowson and Callan 1937), but *S. fragilis* itself and the sallows are only rarely infected. As far as is known the disease does not occur on basket-willows in Britain, but in Holland *P. saliciperda* has been recorded on *S. purpurea*.

The actual damage done by dieback and consequent reduction in growth rate is quite severe, but the most serious aspect of the disease is that once a tree has shown even the slightest sign of infection it is valueless for bats. No buyer will take a watermarked tree, since he cannot tell whether the markings extend into the trunk and thus render it useless for bat-making. Diseased trees, therefore, however slightly infected, must be sold in other much less remunerative markets.

Control

At present the only method of control seriously advocated is the removal of diseased trees. Pruning out diseased branches is seldom, if ever, successful, because the bacterial infection extends so far beyond the external symptoms. Diseased trees should be felled as soon as they are detected, the branches and twigs burnt, and the main stem removed. Though unsuitable for making bats, watermarked wood can be used for a variety of other purposes, such as chip basket, match, and match-box veneer, toys, and pulp.

On present evidence there is no great risk in moving infected timber, but it should not be left lying near living willows, lest insects should emerge from it and carry the bacterium to healthy trees. Poisoning of infected trees is a possibility which is now being tried. There is no evidence that infection can take place through roots, which in any case usually decay rapidly after death. Thus fairly rapid replanting of an infected site is permissible.

An Order for the destruction of diseased willows has been in force in Essex since 1933. Similar Orders have been issued more recently for Suffolk, the eastern part of Hertfordshire, and a small area in north-east Middlesex. Inspection under these Orders is carried out by the Local Authorities, who may require any owner to fell and destroy infected trees. Generally owners are permitted to sell the timber, but the branches and twigs must be burnt. Unfortunately no comparative records have been kept, but there is a strong impression that the operation of the Orders has lessened the occurrence of the disease in the area involved.

The Orders were designed to cover that part of the country where Bat-willow growing is concentrated, and where the financial interest of the growers would encourage full co-operation. The disease exists unchecked, particularly on *S. alba*, in bordering counties, and reinvasion of the protected area is, of course, a constant menace. But the difficulties of control in areas where willows are not a source of considerable financial return to their owners probably makes restriction of the Orders to the commercial willow-growing areas inevitable.

STEM DISEASES—BARK AND CAMBIUM

Willow is attacked by a very large number of fungi affecting the twigs, shoots, and leaves. These have been variously described under names such as 'Black canker' and 'Twig blight'. They are essentially similar, in that serious attack is mainly on the younger shoots, and they will therefore be considered later under the heading of 'Leaf and Shoot Diseases'. Here only the few fungi which normally attack older stems are dealt with.

Nectria galligena Bres. (Ascomycetes, Hypocreales) (Chapter 18) has been described in Holland by Mooi (1948) infecting frost injuries around branch stubs, and forming eventually large perennial cankers on *S. alba caerulea*. In Britain *Nectria* species do not appear to be important on willow.

One or more species of *Cytospora* (Fungi Imperfecti, Sphaeropsidales) occur almost invariably on the younger branches and larger twigs of willows dying back from Watermark disease, or from other causes. They are characterized by small pustular stromata in the dead bark, from which in wet weather long tendrils of spores extrude. It is often assumed that the fungus concerned is *C. chrysosperma* (Pers.) Fr., an important fungus on poplar (Chapter 32), but it may be one of several other species of *Cytospora* which have been recorded on *Salix*. In America it has been suggested that *C. chrysosperma* is a weak parasite on willow, but only on trees enfeebled by other causes. There is no suggestion that any species of *Cytospora* is parasitic on willows in Britain.

In the United States *Botryosphaeria ribis* Gross & Dugg. (Ascomycetes, Sphaeriales) causes cankers on the trunk and larger branches of various species of willow (Wolf and Wolf 1939). It has been reported also on *S. caprea* in the Argentine (Marchionatto 1952). Fortunately the fungus does not occur in Europe. The first symptoms of the disease show on occasional twigs and smaller branches as elongated depressed lesions, which often spread to girdle them. Later numerous, minute, circular, depressed lesions, 6 to 12 mm. in diameter, are formed on the trunk and larger branches. These may coalesce to give larger cankers, more than 5 cm. in length, which may eventually girdle the stem. Eventually dark stromata are visible, extruding from cracks in the bark, in which the fruit-bodies are formed. Damage can be very extensive.

DISEASES OF LEAVES AND SHOOTS

Twig blights and cankers

The very large number of fungi which occur on the young shoots of willow has led to great confusion. Symptoms caused by one fungus have been attributed to another, while in other cases the symptoms have been recorded without any indication of the causal agent. In some instances two or more fungi apparently act together, either simultaneously or successively. Inevitably, therefore, the fungi on young willow shoots must be dealt with as a group.

Pathogens

The fungi in the following list have been associated with injury to young twigs of willow, many of them occur also on the leaves. The list does not include the much greater number of fungi which have merely been recorded on willow twigs without any suggestion of pathogenicity.

Two of the many Rust fungi which attack willows also infect the young shoots as well as the leaves, but they are easily separated from other shoot diseases by their conspicuous orange-yellow fructifications, and are therefore considered with the remainder of the Rusts.

Fungi attacking young willow shoots

ASCOMYCETES

- Helotiales
  - *Scleroderris fuliginosa* Karst.
- Phacidiales
  - *Cryptomyces maximus* (Fr.) Rehm.
- Sphaeriales
  - *Physalospora* (*Glomerella*) *miyabeana* Fukushi; the imperfect stage is probably a *Gloeosporium.*
  - *Cryptodiaporthe salicina* (Curr.) Wehm.; perfect stage of *Discella carbonacea.*
  - *Venturia chlorospora* (Ces.) Karst.; possibly the perfect stage of *Fusicladium saliciperdum.*

FUNGI IMPERFECTI

- Sphaeropsidales
  - *Dothichiza populea* Sacc. & Br.; in Britain only on poplar.
  - *Discella carbonacea* (Fr.) Berk. & Br.; imperfect stage of *Cryptodiaporthe salicina.*
- Melanconiales
  - *Marssonina kriegeriana* (Bres.) Magn.
  - *Marssonina salicicola* (Bres.) Magn.
  - *Myxosporium scutellatum* v. Höhn.
  - *Myxosporium salicinum* Sacc. et Roum.; not in Britain.

Moniliales

Fusicladium (*Pollaccia*) *saliciperdum* (Allesch. & Tub.) Tub.; possibly the imperfect stage of *Venturia chlorospora*.

These fungi are dealt with separately below, but are finally considered together with regard to damage and control. Several are described from the practical angle in Bulletin No. 29 of the Ministry of Agriculture and Fisheries (Anon. 1931).

Scleroderris fuliginosa

This fungus was recorded as the second stage in a disease succession on *S. fragilis*, following *Cryptomyces maximus* (Alcock and Maxwell 1925; Alcock 1926). It entered the lesions caused by the latter and produced much more extensive dieback. It was described as producing clustered pycnidia, up to 0·25 mm. in diameter, over the surface of the flat black cushions formed by *C. maximus*. Later the apothecia, which are like tiny saucers with greyish centres, are produced.

It has been recorded in Europe on a considerable number of *Salix* species, causing brown lesions on the shoots; but there is little information available on the damage it causes. It does not appear to be a serious pathogen.

Cryptomyces maximus

Alcock and Maxwell (1925) and Alcock (1926) considered this the primary parasite in the disease complex which they described on *S. fragilis* in Scotland. It has also been recorded on *S. viminalis* in Scotland (Batko unpubl.). It has been recorded in various European countries on other species of willow, and has been found on Cricket bat willow stool-beds in the south of England.

The fungus produces long black cushions, 15 to 20 cm. in length, on the bark of young twigs; these cushions, which swell when wet, have a curious blistered appearance at the edges, but the top is flat and glistening (Fig. 89). The whole surface of the cushion forms a layer of apothecia. Later they often peel off in wet weather, leaving extensive scars; but the twigs sometimes remain alive, despite the fact that the fungal cushions are produced over the whole circumference.

This fungus also is very generally distributed, but not often associated with serious damage.

FIG. 89. *Cryptomyces maximus* on willow, Peebles, November (×1½). (J. C.)

Physalospora miyabeana

This, together with *Fusicladium saliciperdum*, is the fungus most commonly concerned in damage to willow twigs. It is usually called 'Black canker'. The obvious signs of damage by this fungus are blackened, withered leaves, and small lesions on the shoots, particularly around the bases of diseased leaves

(Nattrass 1928). The first symptoms are usually observed about the end of May. On the leaves, infection starts as reddish-brown or black areas along the edges or at the tips. Sometimes only the apical part is affected, in which case that withers and hangs down, but often the fungus affects the whole leaf and petiole, passing thence into the shoot. Lesions on the shoots are at first small discoloured areas, 2 to 3 cm. long. They become more and more sunken and eventually develop into cankers. Young succulent shoots infected at the tips often wither and bend over. Minute pink conidial pustules appear on the lesions in the summer; the conidia from these cause many secondary infections. Later in the summer small black perithecia develop, sunk in the dead bark on the same lesions. Fruit-bodies do not normally appear on the leaves.

Most species and varieties of willow appear to be to some extent susceptible. Among the basket-willows, *S. viminalis* is not often seriously affected, and the leaves, but not the shoots, of *S. purpurea* normally remain uninfected, possibly because they are covered with fine hairs which prevent the adherence of the water droplets necessary to give suitable conditions for infection. The disease is said to be particularly serious on *S. alba vitellina*, *S. triandra*, and the so-called *S. americana*, another basket-willow, which is probably a hybrid between *S. purpurea* and *S. triandra*. There is little information on its relative importance on different tree willows, though it occurs on many of them. However, there is no evidence that it causes serious injury to any tree willows, except possibly in nurseries. It is of no importance on tree willows in Britain.

The disease is of world-wide distribution, though there is evidence that it was introduced into North America from Europe. It is probably present throughout Britain, but may be more important in the west than in the east.

Physalospora is often found associated with *Fusicladium*, and it is very difficult to evaluate the relative importance of the two. Nattrass considered, as a result of observations and inoculation experiments on basket-willows, that *Physalospora* was a primary parasite and *Fusicladium* a secondary one. Further inoculation experiments were carried out by Dennis (1931) in Britain, and by Rupert and Leach (1942) in the United States; they all supported Nattrass' conclusions. However, some workers (see below) have produced evidence that *Fusicladium* is a primary parasite.

Cryptodiaporthe salicina

This fungus has been recorded on cankers on willow in various parts of the British Isles, more commonly in the conidial stage *Discella carbonacea*. It has once been found on *S. viminalis* in association with *Cryptomyces maximus* (Batko unpubl.). It has not been proved to be a pathogen. In Holland Broekhuizen (1934) described it on *S. viminalis* causing rapidly spreading brownish-green lesions, with darker-coloured protuberances containing the pycnidia of the *Discella* stage. Mooi (1948), also in Holland, found it on *S. alba* and *S. alba caerulea* only as a weak parasite following injury by winter cold.

Bier (1959) found this fungus in British Columbia causing canker and dieback on the young shoots of two native willows, *S. hookeriana* and *S. scouleriana*. He found a definite relation between the turgidity of the bark and the ability of the fungus to spread in it. High turgidities inhibited canker development, but in the winter the moisture content of the bark fell low enough to allow the fungus to extend.

Dothichiza populea

This fungus, which is the cause of a serious disease on young poplars (Chapter 32), has recently been recorded on Cricket bat willow in Belgium. Although normally a weak parasite on poplar, it can be very serious on trees weakened by other causes, such as dry weather following transplanting. It could be serious if it behaved in the same way on willow.

Marssonina

Nattrass (1930) described an anthracnose of the basket-willow *S. purpurea* occurring in Britain, caused by *Marssonina salicicola.* The fungus causes elliptical or circular lesions 2 to 4 mm. in diameter on the young shoots. They may occur singly or may coalesce, giving rise to larger elongated cankers of indefinite outline. When young, the lesions are black, the fruit-bodies appearing as one or more bladder-like protuberances in the centre. Eventually the epidermis splits, and a whitish spot is formed, which gradually increases in size, exposing the sub-epidermal tissue. Finally the lesions become slightly sunken, with a raised light-coloured centre and a dark brown or black rim. Though the lesions are usually on the shoots, they occasionally occur on the leaves. In a wet season, when the fungus develops freely, the disease can be of considerable economic importance. Not only do the lesions reduce growth, but they appear as unsightly scars on the peeled rods.

Nattrass also described *M. kriegeriana* as causing severe leaf-spotting of willows in Egypt; it also occurs in Britain. With this fungus the majority of the lesions are on the leaves, though early in the season a few may occur on the stems and flowers. The spots are reddish-brown with a light-grey centre, or dark brown with a black margin. In the summer they occur mainly on the upper surfaces of the leaves, where dew, which renders infection possible, is deposited. Earlier in the year, when rain is the main infection-assisting agent, they occur also on the underside of the foliage. No doubt their position in other countries would depend on climatic conditions at the time of infection. The white acervuli are visible to the naked eye as minute dots scattered over the surface of the larger lesions, or singly in the centre of the smaller ones. Shoots attacked by the fungus bear black lesions, similar to those caused by *M. salicicola.*

B. J. Murray (1926) reported *M. salicicola* causing severe injury to *S. babylonica* in New Zealand. The reduction of the long shoots was so severe that the infected trees almost lost their weeping habit. Lesions occurred also on the leaves and catkins.

Jauch (1952) described both species of *Marssonina* in the important willow-growing areas of the Paraná Delta in the Argentine. *M. kriegeriana* occurred as a severe leaf-spot on hybrid willows, and *M. salicicola* in the same role on *S. babylonica* and *S. alba caerulea.* On the latter, however, it also caused stem lesions and even cankers.

These fungi are almost world-wide in their distribution and it is curious that they have so seldom been recorded as causes of damage in Britain.

Possibly they may be more damaging than is supposed, their effects having been attributed on superficial inspection to better-known fungi.

Myxosporium

Myxosporium scutellatum was considered by Alcock and Maxwell (1925; Alcock 1926) to form the third stage in a disease succession on *S. fragilis*, following *Cryptomyces maximus* and *Scleroderris fuliginosa*. They describe it as enlarging the long dead areas of bark caused by the other fungi. On the dead patches are seen innumerable small disk-like markings of pin-head size, which are the immature fruit bodies. The cup-like fructifications are produced eventually from these spots.

Though admitting that it was generally only a saprophyte, they considered it capable of causing considerable damage to willows weakened by the other two fungi. It has been recorded more recently on diseased twigs, but always in association with other fungi, which were probably the primary causes of injury. It is very doubtful if it has any pathological significance.

M. salicinum, a species not found in Britain, has been recorded in Holland (Tuinzing 1946) as a cause of canker on basket-willows.

Fusicladium saliciperdum

This fungus is nearly always referred to by the name of its conidial stage *Fusicladium saliciperdum*. The perfect stage has usually been thought to be *Venturia chlorospora*, but Arx (1957*b*) has recently thrown doubt on this. Its popular name is Willow scab. It attacks the twigs and leaves, causing small black lesions (Fig. 90). There can be considerable loss of foliage, often accompanied by dieback of the branches. In the United States, where the disease is much more serious than in Europe, defoliation for several years in succession often results in the death of the tree. The conidia appear as an olive-brown mat on the lesions on the lower surface of infected leaves, particularly along the mid-rib and veins (Alcock 1924; Clinton and McCormick 1929; Kochman 1929). The perithecia, which have not been found in Britain, are seldom formed on the tree, but tend to occur on fallen twigs and leaves. Overwintering is apparently by perennial pustules on the shoots, which bear conidia in the spring.

Some workers have carried out inoculation experiments which appeared to prove that, while *Physalospora miyabeana* was an active parasite of willow, *F. saliciperdum* was a saprophyte or at the most a weak parasite following the other fungus. But, conversely, Clinton and McCormick (1929) did successful inoculation experiments with *Fusicladium*, which were repeated by Brooks and Walker (1935) who remarked particularly on the rapidity with which the damage to the leaves extended, and who concluded that the fungus was definitely an active parasite.

Fusicladium, like *Physalospora*, was apparently introduced into America from Europe, and appears there to be far the more serious of the two diseases. It has spread over a much larger area than *Physalospora*, so that there is not the same possibility of confusion between the two diseases as in Europe. In America it causes far more catastrophic damage than any of this group of fungi causes on willows in Britain.

Information on resistance and susceptibility is very incomplete, and somewhat contradictory, but *S. pentandra*, *S. triandra*, and *S. viminalis* seem to be fairly resistant, while *S. alba* and its variety *vitellina*, and *S. babylonica* are susceptible. *S. fragilis* is listed as susceptible by several authors, but Arx (1957*b*) considers it resistant. Thus in Britain, at any rate, *Fusicladium* appears to be largely a disease of the tree willows. On the Continent the American osier, the so-called *S. americana*, is particularly badly affected.

FIG. 90. *Fusicladium saliciperdum* on *Salix babylonica*, Tilford, Surrey, October (× ½). (J. C.)

General consideration of twig blights and cankers

In Britain there has been so much confusion between the diseases associated with these various fungi that it is hardly possible to consider them separately. The lesions on the twigs and the sporadic dieback which they produce are of comparatively common occurrence and occasionally very disfiguring. Once a tree is infected the disease tends to occur year after year, often making it unsightly in the early summer. On ornamental willows, particularly *S. babylonica*, this type of damage can be irritating, but only in very severe attacks is there likely to be any reduction in growth or real weakening of the tree. These diseases are not, therefore, a very serious factor in the cultivation of willows for timber.

In basket-willow beds, however, these diseases are much more serious.

Dieback of the tips results not only in a serious reduction in length, but also causes side shoots to develop, so that the rods are useless for basket-making. The value is also reduced by the shoot lesions, which appear as disfiguring marks and points of weakness on the peeled rods. These diseases restrict the cultivation of some types of *viminalis* × *purpurea* hybrids, but serious attacks on other types of basket-willow seem to be rare, and in fact the rusts, especially *Melampsora amygdalinae* (see below), which causes much the same kind of damage, are now regarded as the more serious.

Control of twig blights and cankers

On large trees control is impossible; spraying is too expensive, and attempts to cut out the infected twigs result in more serious mutilation of the tree than the effect of the disease. But on basket-willows spraying with copper fungicides should be carried out at two- to three-week intervals, starting just before the stools break, and continuing until mid-summer or until the onset of hot dry weather, which so lessens the chances of infection that further spraying is unnecessary. Diercks (1957) successfully controlled *Physalospora miyabeana* by the use of various proprietary sprays from the beginning of July to the end of September. In addition care should be taken to see that stools of basket-willows are cut low, so that the fungus cannot overwinter on the snags. Rejected rods should not be left lying in the vicinity of the beds, lest they serve as sources of infection, and care should be taken to see that new beds are raised from healthy cuttings.

Schmidt (1938) suggested that attacks of *Physalospora* on *S. babylonica* in nurseries could be greatly reduced by planting alternate rows of willow and poplar. This idea should be equally applicable to any of these diseases.

Rust fungi

A considerable number of rust fungi, all of the genus *Melampsora* (Basidiomycetes, Uredinales), occurs on the leaves of willow. Those which have been recorded in Britain are set out in the table below.

	AECIA ON	UREDIA AND TELIA ON
Melampsora allii-fragilis Kleb.	*Allium* spp.	*S. fragilis*, *S. pentandra*
M. allii-salicis-albae Kleb.	*Allium* spp.	*S. alba* and var. *vitellina*
M. amygdalinae Kleb.	*Salix* spp.	*Salix* spp.
**M. epitea* Thuem.	*Larix* spp. and *Saxifraga* spp.	*Salix* spp.
**M. euonymi-caprearum* Kleb.	*Euonymus europaeus*	*S. atrocinerea*, *S. cinerea*, *S. caprea*
M. larici-caprearum Kleb.	*Larix* spp.	*S. atrocinerea*, *S. aurita*, *S. caprea*, and other spp.
M. larici-pentandrae Kleb.	*Larix* spp.	*S. pentandra*
**M. repentis* Plowr.	Various orchids	*S. aurita*, *S. repens*

**M. ribesii-purpureae* Kleb.	*Ribes* spp. (British records doubtful)	*S. purpurea*
M. ribesii-viminalis Kleb.	*Ribes* spp. (not recorded in Britain)	*S. viminalis*

Hylander, Jørstad, and Nannfeldt (1953) group the four species marked with asterisks under *M. epitea*. On the other hand, Henderson (1957), who has studied *M. epitea* in the narrow sense on mountain willows and saxifrages, divides it into two species.

Other species with different alternate hosts occur on the Continent and in America, in addition to some of those listed above. Some of the British species are comparatively rare, the majority are of slight importance. In practice the different species are scarcely ever distinguished on willow, so that their relative importance is not really known.

All produce orange-yellow uredia on willow leaves during the summer and later brown telia. Considerable defoliation may result from the attack on the leaves. More serious damage, particularly in basket-willow beds, results from attack by *M. allii-salicis-albae* and *M. amygdalinae*, which also attack the young shoots, producing small black cankers somewhat like those of *Physalospora miyabeana*. These appear later as scars on the peeled rods and spoil them for basket-making. Such lesions also cause a considerable reduction in shoot growth. They may also girdle and kill young stems in the spring. Orange-yellow uredia of these fungi are produced on the stem lesions as well as on the leaves.

It seems probable (Scaramella 1931) that many of these rusts can overwinter without the intervention of the alternate host by means of a perennial mycelium in the twigs. This is certainly the case with *M. allii-salicis-albae* (Weir and Hubert 1918). But in *M. amygdalinae*, which produces all its stages on *Salix*, aecia are produced on the elongating shoots and the earliest formed leaves in the spring. The spores from these cause further infections on stems and leaves, on which uredia and later telia are borne. Overwintering in this case can be either as the telial stage on fallen leaves, or by the production of fresh uredia in the spring from lesions which have survived the winter. In willow beds, where the rods are cut annually, the former method is probably the more important (Ogilvie 1932).

Our knowledge of the host ranges of these species of *Melampsora* is very far from complete. The most serious, because they attack basket-willow species of commercial importance, are *M. amygdalinae* on *S. triandra* and *M. epitea* on *S. viminalis*. The incidence of attack differs enormously from year to year. Cool moist weather conditions are required for the germination of the uredospores, so that periods of hot weather greatly reduce the severity of the disease (Ogilvie and Hutchinson 1933).

In basket-willow beds the disease can be controlled to some extent by late cutting of the rods in the spring, or, if cutting has to be done earlier, by removal of the first crop of new young sprouts when they are a few inches high. In practice this can be done by grazing the beds with cattle or by the application of a tar-oil winter wash, which kills back the young shoots. By these means initial infection by the teliospores and subsequent production of aecia are checked, and the disease makes a much slower start (Ogilvie and Hutchinson 1933). The control of the disease in its later stages by more orthodox sprays is not easy, because fresh unprotected growth is continually and quickly being produced; but a proprietary form of cuprous oxide has given promising results (Stott unpubl.).

Other leaf diseases

A few other fungi affect the leaves of willows, but do not produce lesions on the young shoots. None is of economic importance. Mildew, *Uncinula salicis* (DC.) Wint. (Ascomycetes, Erysiphales), occurs in Britain, but does no appreciable damage. *Rhytisma salicinum* (Pers.) Fr. (Ascomycetes, Phacidiales) causes thickened black spots on the leaves of willow like those caused by the much better known *R. acerinum* on Sycamore (Chapter 30). *R. symetricum* J. Müller causes similar symptoms, but is even less common in Britain.

Occasionally *Septoria salicicola* Sacc. (Fungi Imperfecti, Sphaeropsidales) causes very small spots scarcely 1 mm. in diameter and white with a brownish edge; it is quite harmless. *Septogloeum salicinum*(Peck) Sacc. (Fungi Imperfecti, Melanconiales) causes a somewhat similar spotting. *Gloeosporium salicis* Westend. (Fungi Imperfecti, Melanconiales) produces numerous blackish spots, much smaller than those of *Rhytisma* and not erumpent; occasionally they cover nearly the whole leaf surface. This appears to be a different fungus from the much more important *Physalospora miyabeana*, which also has a *Gloeosporium* as its conidial stage. All these fungi occur in Britain, but none of them have ever proved of practical importance.

Several fungi have been recorded abroad, but not in Britain, causing appreciable damage to the leaves of willow. *Gloeosporium capreae* Allesch. (Fungi Imperfecti, Melanconiales), known for a long time as an unimportant parasite of willow in Europe, but not in Britain, was thought by B. J. Murray (1926) to occur commonly on *S. fragilis* and less frequently on *S. babylonica* in New Zealand, causing appreciable defoliation. Jenkins and Grodsinsky (1943) transferred Murray's fungus to the genus *Sphaceloma*, under the name of *S. murrayae* Jenk. & Grod. Under this name it was later recorded in the United States, in the Argentine, and in Europe (Jenkins, A. E. 1944). In New Zealand *Macrophoma salicis* Dearn. & Barth. (Fungi Imperfecti, Sphaeropsidales), previously recorded in America as a saprophyte, has been found causing severe defoliation on *S. fragilis* (Murray, B. J. 1926). In the United States, *Cylindrosporium salicinum* (Pk.) Dearn. (Fungi Imperfecti, Melanconiales) produces numerous small brown spots, resulting in early leaf-fall.

34

DISEASES OF OTHER FOREST HARDWOODS

THIS chapter covers those hardwoods which have not already been dealt with separately and which are of more importance in the forest than in parks, in gardens, or along roads. Only a few genera are involved, namely: *Alnus*, *Betula*, *Carpinus*, *Corylus*, *Fraxinus*, *Nothofagus*, and *Tilia*.

DISEASES OF ALDER

Only *Alnus glutinosa*, the Common alder, is native to Britain, but *A. incana*, the Grey alder, has been widely planted. *A. rubra*, the Oregon alder, and *A. cordata*, a southern European tree, have been used on a very limited scale as forest trees, and a few other species are planted occasionally in gardens.

Diseases due to non-living agencies

A. glutinosa, together with some species of willow, can probably grow under wetter conditions than any other tree common in Britain. For this reason it is quite frequently found growing in places where, even for it, the soil water-level is critical. A rise in the water table will completely flood the root zone, whereas a fall leaves the shallow root system in the dry surface soil and produces drought symptoms. Under these conditions crown dieback frequently occurs. This is very similar to the dieback disease, particularly of *A. rubra*, described below, which, however, often occurs on quite favourable sites. On wet sites the two diseases probably merge into a single complex.

Root diseases

Armillaria mellea (Chapter 16) occurs quite commonly on alder, and may sometimes assist in the final stages of Alder dieback, an obscure disease, which is discussed below.

The symbiotic root nodules of alder are dealt with in Chapter 17.

Stem diseases—bark and cambium

Dieback

Dieback of alder has been reported from several countries and attributed to various fungi. On the Continent *A. glutinosa* has been most affected, but in Britain *A. rubra* has suffered most. It is possible that a number of diseases are concerned, but they are more easily considered together.

The disease has been described on *A. glutinosa* in Germany by Münch (1936b), who found the fungus *Valsa oxystoma* Rehm. (Ascomycetes, Sphaeriales) generally present. He was unable, however, to get successful inoculations with it. He describes the disease as a general dieback starting in the crown, when the trees are about twelve years old, and ending in their death about eight years later. *V. oxystoma* had previously been associated with alder dieback on the Continent by Appel (1904) and others, but again with no proof of its pathogenicity. The fungus forms small black stromata under the dead bark, which ruptures, giving the appearance of small tubercules on the surface. The perithecia are formed in the stromata after they have ruptured the bark. *V. oxystoma* has been recorded in Britain. Though Appel (1904) regarded *V. oxystoma* as a parasite, he considered that it was only damaging on alder suffering from malnutrition or drought.

More recently the disease was described in Holland by Truter (1947), who recorded a very large number of fungi, including *V. oxystoma*, all of which she regarded as saprophytes on dead bark, or at the most very weak parasites. She considered that physiological factors must be mainly responsible, but was unable to determine them.

Among the fungi she recorded was *Cryptospora suffusa* (Fr.) Tul. (Ascomycetes, Sphaeriales), which has been cited as the cause of dieback disease in Denmark (Juel, C. 1940). The conidial stage of this fungus, *Cryptosporium neesii* Corda (Fungi Imperfecti, Melanconiales), has been recorded in Britain on alder showing dieback, but there is no proof of its pathogenicity.

Melanconis thelebola (Fr.) Sacc. (Ascomycetes, Sphaeriales) has been found in Britain on *A. cordata* showing dieback symptoms (Batko unpubl.). A species of *Myxosporium* (Fungi Imperfecti, Melanconiales) has been found on *A. rubra* in Scotland. In neither case, however, is there proof of pathogenicity.

The large number of fungi associated with this disease suggests a physiological basis. However, in Britain as in Holland, the disease has been recorded on such a wide range of sites that it is impossible even to suggest any common detrimental factor. In Britain the disease is serious only on *A. rubra* (Oregon alder), other species not being affected to any serious extent. On this alder, dieback occurs so frequently once the tree is twelve to fifteen years old that its cultivation is scarcely worth while. The course of the disease agrees very well with that described by Münch. Although quite vigorous recovery shoots may arise from below the dead portions, they usually die in their turn.

Other diseases of bark and cambium

Didymosphaeria oregonensis Goodding (Ascomycetes, Sphaeriales) causes canker on several species of *Alnus* in North-west America (Goodding 1931). It does not appear to have been recorded in Europe.

Diseases of leaves and shoots

Three members of the genus *Taphrina* (Ascomycetes, Exoascales) are found on alder in Britain (Henderson 1954). *T. tosquinetii* (West.) Magn., by far the commonest species (Bond, T. E. T. 1956), severely distorts both the leaves and young shoots of *A. glutinosa*. The disease is perennial in the twigs. Affected shoots have a thick and wrinkled cortex, so that when young they appear almost fasciated. The unfolding leaves are reddish and very crumpled, and become extremely distorted as they develop. Occasionally individual infections on the leaves appear as small blisters. The asci are borne on the leaf surfaces, giving them a whitish bloom. The fungus occurs particularly on coppice shoots. In Britain it appears to attack only *A. glutinosa*. Even when mixed with infected plants of the Common alder, *A. incana* has remained healthy.

T. sadebeckii Johans. causes yellow blisters on the leaves of *A. glutinosa*. *T. amentorum* (Sadeb.) Rostr. infects and distorts the female catkins of both *A. glutinosa* and *A. incana*. It has been more commonly referred to under the name *T. alni-incanae* (Kühn.) Magn. Some of the scales of affected catkins become greatly enlarged, projecting as curled reddish tongues which later become covered with a white glistening layer, the asci of the fungus. Only the first of these three species of *Taphrina* can be regarded as damaging in this country, and even it is not common. Other species of *Taphrina* attacking alder in North America have been described by Ray (1939).

Microsphaera alni (Wallr.) Wint. (Ascomycetes, Erysiphales) occurs commonly on the Continent on several species of *Alnus*. If it does occur on alder in Britain, it is certainly very uncommon on it. Another mildew, *Phyllactinia corylea* (Pers.) Karts. (Ascomycetes, Erysiphales), occurs occasionally on alder causing a white bloom on the leaves, but it is never serious (Chapter 18).

Two fungi have been associated with leaf spots on *Alnus*. *Sphaerulina alni* Smith. (Ascomycetes, Sphaeriales) has been recorded in Scotland causing spotting followed by leaf scorch. Commoner is *Gnomoniella tubiformis* Sacc. (Ascomycetes, Sphaeriales). The conidial stage of this fungus, *Leptothyrium alneum* Sacc. (Fungi Imperfecti, Sphaeropsidales) (syn. *Septoria alnicola* Cke.), shows as minute shining black dots on infected leaves. The perithecia of the ascigerous form are produced on fallen leaves. Neither of these leaf spots appreciably damages alder in Britain.

One rust fungus, *Melampsoridium hiratusukanum* Hiratsuka (Basidiomycetes, Uredinales), until recently known as *M. alni* (Thuem.) Diet., has been recorded on both *A. glutinosa* and *A. incana* (Wilson, M. 1924), the uredia appearing as small orange spots on the leaves. The telia, which are also borne on the alder leaves, are very inconspicuous. The aecidial stage, which has not been recorded in Britain, occurs on larch. The fungus must therefore be able to overwinter on *Alnus*. It is comparatively rare and innocuous.

DISEASES OF ASH

The only species of any forest importance in Britain is the Common ash *Fraxinus excelsior*, though several other species, notably *F. ornus*, the Manna ash, and *F. americana*, the White ash, are found in parks and gardens.

Diseases due to non-living agencies

Ash, despite its late leafing, is very subject to spring-frost injury, and on the Continent some attention has already been paid to the breeding and selection of strains that flush even later than normal, and are therefore more frost hardy.

Frost is one of the factors responsible for the regrettably frequent forking of young ash. In some plantations this occurs on so many trees that the number of straight unforked stems falls below an economic level. Unpublished investigations by Peace and Town suggested that three factors, frost, the Ash bud moth, *Prays curtisellus* Don., and failure of the bud to break, or in some cases to break normally and vigorously, were of about equal importance in the failure of terminal shoots to develop properly. The relative importance of these factors varies considerably from year to year. The failure of ash buds to break, and the growth of the terminal shoot being so weak that it is quickly surpassed by side shoots, are two phenomena which have never been properly explained. Hemberg (1949) reported very high concentrations of growth-inhibiting substances in terminal buds of ash, and it seems possible that the complete or partial failure of the terminal shoot is due to an abnormal retention of these substances. Apparently terminal bud failure does not occur year after year on the same tree, so the chance of selecting strains that do not behave in this way is small.

Older hedgerow ash, especially on clay soils, often show extensive crown dieback. This type of injury, not so common in plantations, appears to be due to the inability of older trees to tolerate the major variations in soil water content characteristic of heavy soils. Ash throughout its life is a tree particularly prone to go into check; this often happens following planting on a difficult site. Checking almost invariably occurs when ash is planted on a site which is too dry or too infertile for the rather high demands for water and nutriment of the species. In such cases early growth is often rapid, but ceases almost completely before the trees reach timber size. The production of satisfactory ash is virtually limited to high-quality sites.

Ash is resistant to industrial fumes. It has been suggested that *F. excelsior monophylla*, in which the normal pinnate leaf is reduced to a single blade, is even better able to tolerate town conditions.

On some sites ash develops dark heartwood which is not of fungal origin. This apparently chemical staining does not affect the mechanical properties of the timber (Kühne 1954); but it materially lessens its sale value for such uses as sports goods, for which the best prices are paid. The staining occurs on certain sites that may otherwise seem entirely suited to the growth of ash, but no proper investigation of the site factors involved appears to have been made (see Chapter 20).

Root diseases

Though ash is not very commonly attacked by *Armillaria mellea*, severe outbreaks have occasionally been recorded, especially on sites where the trees were not growing vigorously.

Stem diseases—bark and cambium

Canker

Ash is frequently cankered, the damage varying from large erumpent black cankers associated with extensive branch dieback to insignificant warts on the bark surface. Even the latter are important, for they interfere with the cleaving of ash, which is a necessary part of its utilization for some purposes. The agencies chiefly concerned with ash canker appear to be frost, *Nectria galligena* Bres. (Ascomycetes, Hypocreales), and the bacterium *Pseudomonas savastanoi* f. sp. *fraxini* (Brown) Dowson. The fungus and the bacterium are probably the most important. The bacterium is typically responsible for the erumpent black cankers, *Nectria*, and sometimes frost, for the more regular target cankers (Fig. 91). The cause or causes of the small superficial cankers, which appear to occur more on some soils than on others, remains obscure.

The bacterium differs from *P. savastanoi* proper, which is a parasite causing knots or tubercles on the young twigs of olive trees, in its host range only; the variety on ash will not attack olive, while the variety on olive only produces tubercles, not erumpent cankers, on ash (D'Oliveira 1939). It has been suggested that the bacterium enters the shoot through hail wounds and leaf scars. It is not known how it is carried from tree to tree. The disease first shows as small cracks in the bark of young shoots. Later, cankers develop; probably, as is the case with bacterial canker of poplar, the bacterium becomes systemic in the tree, so that cankers can be initiated on older stems internally without wound infection. Bacterial canker often occurs on sites obviously suitable for ash and relatively frost free. The bacterium has been successfully inoculated into *F. americana* (Brown, N. A. 1932; Škorić 1938; D'Oliveira 1939); but it has not yet been found occurring naturally outside Europe.

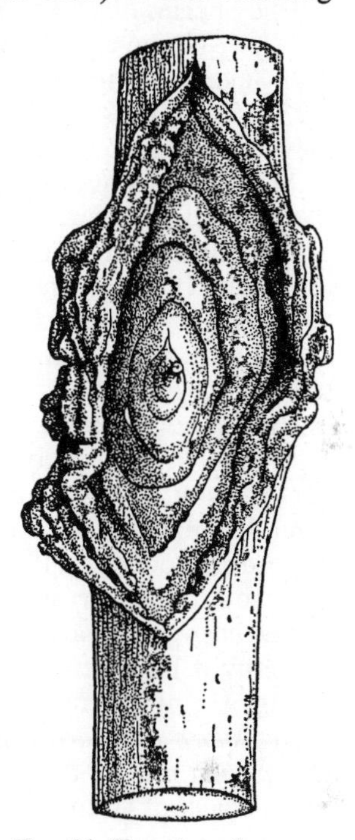

FIG. 91. Target canker associated with *Nectria galligena* on ash (× ⅜). (J. C.)

In a recent investigation in Switzerland, Riggenbach (1956) has suggested that the cankers are the result of a complex of pathogens, including not only the bacterium, but also *Fusarium lateritium* Nees. (Fungi Imperfecti, Moniliales), *Pleospora herbarum* (Pers.) Rabh. (Ascomycetes, Sphaeriales), and *Plenodomus rabenhorstii* Preuss. (Fungi Imperfecti, Sphaeropsidales). Inoculations made with this complex caused much larger cankers than those produced by the bacterium alone. All three fungi have wide host ranges both in Britain and abroad; they have not previously been associated with disease on ash.

Nectria galligena Bres., the conidial stage of which is *Cylindrocarpon mali* (All.) Wr. (Fungi Imperfecti, Moniliales), is the species of *Nectria* most commonly associated with canker on ash (Tubeuf 1936; D'Oliveira 1939) (Chapter 18). It is perhaps better known as the cause of canker on apple and pear trees. It is not clear whether it can initiate cankers on ash. It frequently invades and subsequently extends those caused by frost and may do the same in those caused by the bacterium, though Vliet (1931) regarded the activities of the two pathogens as entirely separate.

Ash of all ages from about ten years upward may be attacked, but there is no information on the relative importance of the various agencies on different ages of trees. Although quite common, canker has not yet proved sufficiently important for the full-scale investigation which would be required to elucidate the whole problem to be launched.

The only control that can be suggested on the basis of present meagre knowledge is removal of the worst cankered trees in the course of thinning.

Other diseases of bark and cambium

The most important causes of canker have been considered together above, because it is uncertain whether their activities are separable. Certain less important fungi may also at times take part in this complex, but, since they have been recorded individually, they are dealt with separately here.

Two species of *Phomopsis* (Fungi Imperfecti, Sphaeropsidales), *P. scobina* (Cke.) v. Höhn. and *P. controversa* (Sacc.) Trav., occur on ash stems. It has been suggested that these two fungi are identical but Macdonald and Russell (1937), who investigated their infection of ash in Scotland, considered that they were separate species. Wehmeyer (1933), however, considered them both to be conidial stages of a form of the species-complex *Diaporthe eres* Nits. (Ascomycetes, Sphaeriales) occurring on ash. *P. scobina* is the more active parasite of the two, forming brown sunken lesions up to 25 cm. long on which the small black pycnidia are borne in profusion. Sometimes quite large branches are girdled. Buds or small twigs are often killed and the recovery shoots may give rise to a witches' broom effect. This disease does not appear to be of any real significance.

Hysterographium fraxini (Pers. ex Fr.) de N. (Ascomycetes, Hysteriales) has been recorded in various parts of Europe causing depressed lesions in the bark of young stems and branches. The apothecia take the form of small black blisters with a furrowed opening, and are scattered on the dead bark. According to Zogg (1943) the fungus is normally a saprophyte which can become actively parasitic under conditions adverse to the host tree. In Britain it has been only rarely recorded, normally as a saprophyte (Bisby 1944). Recently it has been found in association with long lesions on the stems of young ash, but there is no proof that it was responsible for them.

Cytospora annularis Ell. et Ev. (Fungi Imperfecti, Sphaeropsidales) (syn. *Cytophoma pruinosa* (Fries) v. Höhn.) has been associated with bark lesions and dieback of young ash in America (Silverborg and Brandt 1957), but none of the species of *Cytospora* occurring on ash in Britain has been regarded as parasitic.

Diseases of leaves and shoots

There appear to be no important leaf diseases of ash in Britain. *Phyllosticta fraxinicola* Sacc. (Fungi Imperfecti, Sphaeropsidales) causes roundish or irregular brown spots with darker margins on ash leaves; minute black pycnidia develop on the brown spots. Wolf (1939) has recorded *P. viridis* Ell. & Kellerm., the perfect stage of which is *Mycosphaerella fraxinicola* (Schw.) House (Ascomycetes, Sphaeriales), on *F. americana* and *F. pennsylvanica* in the United States. Trees were sometimes defoliated four to six weeks before their normal time. *Septoria fraxini* Desm. (Fungi Imperfecti, Sphaeropsidales) also causes roundish brown spots in Britain. Neither *P. fraxinicola* nor *S. fraxini* causes economically important damage in Britain. The latter may in fact be the imperfect stage of a *Mycosphaerella* (Ascomycetes, Sphaeriales).

Mildew, *Phyllactinia corylea* (Ascomycetes, Erysiphales) (Chapter 18) occurs on ash as well as on other broadleaved trees, but is not damaging.

Venturia fraxini (Fr.) Aderh. (Ascomycetes, Sphaeriales) causes withering of the leaves and premature leaf-fall of ash on the Continent, but has not apparently been recorded in Britain. *Gloeosporium aridum* Ell. & Hollw. (Fungi Imperfecti, Melanconiales) attacks leaves and young shoots of ash in America (Coe and Wagener 1949), occasionally doing serious damage. Another species, *G. fraxineum* Peck., occurs rarely in Britain but has not been associated with injury.

In America ash is attacked by the rust *Puccinia sparganioides* Ell. & Barth (*P. peridermiospora* (Ell. & Tr.) Arth.). The damage is occasionally serious, since trees affected for several seasons tend to die back considerably (Creelman 1956). The symptoms are conspicuous, since the leaves are distorted and the twigs swollen. The yellow cup-like aecia appear on the leaves and on the swollen twigs. The alternate host carrying the uredial and telial stages is Cord grass, *Spartina*. Characteristically this is a salt marsh genus, but at least one American species occurs in fresh-water ditches, which presumably enables the fungus to exist at a considerable distance from the sea.

There is one native species of *Spartina* in Britain, *S. maritima*, and one introduced American species, *S. alterniflora*, occurs locally. The hybrid between these two, *S. townsendii*, has proved a rampant colonizer. All three are restricted to tidal mud flats, and have not invaded fresh water. This would lessen their danger as alternate hosts if the rust were introduced. In any case there is some doubt whether *S. alterniflora* is one of the hosts in America. It is reported by Creelman in America to be resistant, while Partridge and Rich (1957) consider it susceptible to the fungus. Nothing is known of the susceptibility of *Fraxinus excelsior* to this fungus. The American records have referred to their native species. Partial control can be obtained by spraying with various fungicides (Partridge and Rich 1957).

DISEASES OF BIRCH

It is now generally agreed that three species of birch, *Betula verrucosa*, *B. pubescens*, and *B. nana*, are native to Britain. In practice the first two are seldom separated, and pathological records generally refer to both species. Only a few differences in disease resistance between the two have been recorded. *B. nana* is a comparatively rare dwarf species, and little is known about its diseases in Britain. A number of other species of birch, mainly North American, have been tried on a small scale for forestry. Many others occur in arboreta.

Diseases due to non-living agencies

Birch in its youth is able to stand very adverse soil and climatic conditions. No other tree in Britain can grow on such a wide range of sites. It is extremely hardy to both winter cold and spring frost. In Britain, however, it becomes liable to dieback from about thirty years of age onwards, so that healthy old birches are comparatively rare. To what extent this dieback is fungal or physiological is not known. It occurs on a very wide range of sites. It is frequently associated with the fungus *Polyporus betulinus*, a common cause of decay in birch. It is therefore discussed later in this chapter under Stem Diseases.

In recent years dieback of *B. lutea* and *B. papyrifera* in the New England states of the U.S.A. and in the neighbouring parts of Canada has attracted a great deal of attention. It has been suggested that it is mainly due to site alterations (Chapter 5) and particularly to high soil temperatures, resulting from reduction in stand densities, and lethal to fine roots (Redmond 1959). Braathe (1957) is inclined to associate the trouble in some way with the very sudden premature thaw that occurred in the affected region in Canada in March 1936. He found that the general area covered by this thaw coincides very closely with that of birch dieback. The virus-like symptoms associated with the disease have stimulated the idea, as yet unproven, that a virus may be involved (Hansbrough and Stout 1947; Berbee 1957). Clark and Barter (1958) failed to find any association between the disease and climate, and concluded that birch dieback must be an infectious disease. This serious disease still awaits a proven explanation.

Nursery diseases

Rust

The most important leaf disease of birch in Britain is that caused by the rust fungus *Melampsoridium betulinum* (Desm.) Kleb. (Basidiomycetes, Uredinales). Although it occurs in the forest, it is important only in the nursery. The uredia and telia are borne on birch leaves, the aecia on larch needles. On larch the disease is unimportant, but on birch, particularly in the nursery, defoliation can be severe. The uredia appear as small orange spots mainly on the underside, but occasionally on the upperside, of birch leaves. They occur in quantity in the latter part of the summer, and are followed later by the dark-brown telia, the spores from which can infect only larch. The fungus frequently occurs on birch in the absence of larch, and the method of overwintering in this case has been the subject of some investigation. In Portugal (D'Oliveira and Pimentel 1953) it has been found to overwinter in the buds, while in the United States (Weir and Hubert 1918) it has been suggested that it remains alive as mycelium in fallen leaves, producing uredia in the spring.

In severe attacks, defoliation can be nearly complete, and is often followed in the autumn by dieback of the shoots or even death of the whole plant. This may be due to secondary infection by other fungi such as *Melanconium betulinum* (see below), or to frost injury of unripened shoots. It is a common observation that premature defoliation of deciduous broadleaved trees results in incomplete ripening and consequent autumn frost injury.

The rust fungus has been recorded on a considerable number of birch species. In Britain, *B. pubescens* appears to be worse affected than *B. verrucosa*. Isolated observations in Britain suggest that *B. lenta*, *B. papyrifera*, and *B. japonica szechuanica* are all more resistant than our native birches, while *B. ermani* and *B. utilis* are highly resistant. But in view of the known occurrence among the rusts of races of different varietal virulence, it would be rash to assume that this order of resistance holds generally, especially in other countries.

No work appears to have been done on control, although the damage in birch seed-beds would justify efforts, presumably by spraying, to reduce the severity of attack. If birch were a more important tree, this disease would certainly be regarded much more seriously.

Root diseases

Armillaria mellea (Chapter 16) occurs occasionally on birch, but is not an important factor in the dieback of older birch mentioned above. *Fomes annosus* grows strongly on birch, particularly on stumps and suppressed trees. Where birch scrub is succeeded by conifers it may serve as a source of infection for the latter.

Decay

Birch timber is very prone to attack by decay-causing fungi, so that the dead parts of still living trees are very rapidly invaded, especially by *Polyporus betulinus* (see below).

Stem diseases—bark and cambium

Dieback

Deterioration of older birch and the appearance of the fruit-bodies of *Polyporus betulinus* (Bull.) Fries. (Basidiomycetes, Aphyllophorales) on the dead stems are commonly observed phenomena. *P. betulinus* can invade and decay birch wood very rapidly after the death of the tree, but the evidence for its pathogenicity is not conclusive. Wound inoculation experiments have given results varying from failure to a few inches of spread. J. A. Macdonald (1937), whose own inoculation experiments failed to cause infection, nevertheless considered that the fungus was a parasite, basing this opinion on observations of its spread into living tissue from natural infections. Rozanova (1925), on the other hand, who also failed to obtain satisfactory artificial infection, was inclined to rate *P. betulinus* as a very weak parasite and to associate *Fomes igniarius* (L. ex Fr.) Kickx. (Basidiomycetes, Aphyllophorales) with birch dieback. In Holland dieback of birches and invasion by *P. betulinus* has followed lowering of the water table (Vliet, Verbrugge, and Boer 1955).

Without further work on the age and conditions under which dieback starts we cannot correctly evaluate the role of fungi, or suggest what other factors enter into the disease complex. For the time being, if birch is used as a forest crop, it is best to assume that the trees will have a relatively short healthy life.

Other diseases of bark and cambium

A few other fungi have been associated with minor injuries to the stems of birch. These include *Melanconium betulinum* Schm. & Kze. (Fungi Imperfecti, Melanconiales), which produces small erumpent black fruit-bodies, usually on dead bark, but which has been associated both in the United States and in Britain with dieback in young birch of several species, including *B. papyrifera*, *B. verrucosa*, and *B. populifolia* (Carter 1936*a*; Dimbleby 1958). However, there is no proof of its pathogenicity.

In Canada a *Phomopsis* (Fungi Imperfecti, Sphaeropsidales), the perfect stage of which is a form of *Diaporthe eres* Nits. (Ascomycetes, Sphaeriales), has been found to cause canker and dieback of *B. lutea* (Horner 1955). *D. eres* has been recorded on birch in Eire; but the *Phomopsis* disease has not so far been found in Europe.

Massee (1914) described *Plowrightia virgultorum* (Fr.) Sacc. (Ascomycetes, Dothidiales) as the cause of a disease which he called Black knot of birch. The fungus formed small black lesions, some of which spread and girdled the stems. Recovery shoots, many of which were attacked and killed back in turn, and which arose from below the point of girdling, gave an appearance suggestive of witches' brooms. The disease was said to occur commonly in Scotland and also in other parts of Britain. Although the fungus certainly still occurs on birch, it has not recently been associated with active disease.

Diseases of leaves and shoots

Witches' brooms

Taphrina (*Exoascus*) *turgida* Sadeb. (Ascomycetes, Taphrinales) causes proliferation of the buds on birch twigs, resulting in dense 'witches' brooms'. The mycelium of the fungus is perennial in the infected twigs, which in the spring bear pale-green leaves on which the asci are borne. Many twigs in the brooms die, but others remain alive and the broom gradually increases in size (Blomfield 1924). The fungus attacks *B. pubescens* more than *B. verrucosa*. It was once suggested that the fungi on these two hosts were different, that on *B. pubescens* being named *T. betulina* Rostr. Mix (1949), however, showed that these two supposed species were indistinguishable. The disease has been recorded on other species of *Betula*, but on none of them is it serious.

Other leaf and shoot diseases

Gloeosporium betulae Fckl. (Fungi Imperfecti, Melanconiales) and a closely related, or possibly identical, species *G. betulinum* Westend., occur on birch leaves in Britain and may cause very minor injury. Another species of the same genus, *G. betularum* Ell. & Mart. occurs on birch in North America, and is considered there a minor cause of damage.

Phyllosticta betulina Sacc. (Fungi Imperfecti, Sphaeropsidales) occurs fairly commonly in Britain as a cause of leaf spots on birch, but it is unimportant. *Marssonina betulae* (Lib.) Magn. (Fungi Imperfecti, Melanconiales) causes similar unimportant injury. The mildew *Phyllactinia corylea* (Pers.) Karst. (Chapter 18) is found on birch as well as on several other hardwoods, but it does not cause serious damage. Several other fungi have been recorded abroad as parasites on birch leaves, but none appears sufficiently important to warrant consideration here.

On the Continent *Sclerotinia betulae* Wor. (Ascomycetes, Helotiales) destroys the fruits of birch and forms black sclerotia on them (Malençon 1924). It has been recorded in Britain by Batko, but is probably rare and certainly does little damage here.

DISEASES OF HAZEL

The native British species of hazel is *Corylus avellana*. It was at one time planted widely as a coppice understory to oak. It is still very common and in places has a limited value for minor products such as pea sticks and hurdles. *C. maxima*, a native of south-east Europe and west Asia, is sometimes planted for its nuts.

Root diseases

Armillaria mellea (Chapter 16) occurs rather commonly on hazel. It is occasionally damaging in nut orchards, but it is much more serious when infected coppice is planted with other more valuable species, or even when hazel coppice is cleared and planted with fruit trees. *Armillaria* persists on the old stumps or on portions of roots in the soil, and spreads thence to the new crop.

The parasitic flowering plant, *Lathraea squamaria* (Chapter 13), occurs fairly commonly on the roots of hazel, but there is no evidence that it causes any appreciable harm.

Stem diseases—bark and cambium

Bacterial blight

Bacterial blight of *Corylus*, caused by *Xanthomonas corylina* Mill. *et al.*, is a very serious disease of commercially grown filberts (*C. avellana* and *C. maxima*) in the north-western United States (Miller, Bollen, and Simmons 1949). It also occurs on American native species of *Corylus*, but is much less serious. *C. colurna*, the Turkish hazel, which is also grown on a small scale for its nuts in America, is said to be resistant.

The bacterium does not kill the bushes completely, but numerous leaf and flower-buds, as well as twigs and branches, are killed. Cankers are formed on the larger branches, which are often girdled. Leaf infection results in reddish-brown spots, but this aspect of the disease is not serious. The bacterium over-winters in the cankers, and infection is spread by rain. In nut orchards it may also be spread on pruning tools, or by movement of infected bushes.

Cutting out all infected branches about mid-summer, and spraying with Bordeaux mixture just before the leaves fall, have been suggested as control measures (Šutić 1956). They would, of course, be economically possible only in orchards.

Fortunately this disease has not reached Britain. If it did appear here, it might very well be spread considerably on tools in the normal course of coppice cutting.

Other diseases of bark and cambium

Cytospora corylicola Sacc. (Fungi Imperfecti, Sphaeropsidales) has been associated with a serious dieback of *C. avellana* in Spain, Italy, and Sicily (Pupillo and Canova 1952). Reddish lesions are formed on the bark; sometimes these spread and girdle the shoots, in other cases they heal, but remain as points of weakness at which the stems often break. The orange-red masses of conidia are produced from the lesions in damp weather. The fungus is considered to be a wound parasite, and it has been suggested that it normally follows sunscorch (Servazzi 1950). Inoculations made by Graniti (1957) gave uniformly negative results, and he concluded it could only spread on bushes weakened by climatic or edaphic causes or by cultural mismanagement. *C. corylicola* has not been recorded in Britain. There is no suggestion that *C. fuckelii* Sacc., which does occur on hazel in Britain, is a parasite.

In the eastern United States *Cryptosporella anomala* (Peck.) Sacc. (Ascomycetes, Sphaeriales) is said to be very destructive on *C. avellana* (Slate 1930). It forms small oval warts, 6 to 8 mm. long, on the bark, and these are followed by dieback of affected branches, and eventually by the death of the whole bush. The fungus does not appear to have been recorded in Britain.

Diseases of leaves and shoots

A large number of fungi occur on the leaves of hazel, but normally none is seriously damaging. Mildew, *Phyllactinia corylea* (Chapter 18), is probably more serious on hazel than on its other broadleaved hosts, and occasionally causes quite severe defoliation.

Septoria avellanae B. & Br., *Phyllosticta coryli* Westend., *Labrella coryli* Sacc. (all Fungi Imperfecti, Sphaeropsidales), and *Gnomoniella coryli* (Batsch.) Sacc. (Ascomycetes, Sphaeriales) have all been found in association with leaf-spots on hazel in Britain, but none is important.

Ribaldi (1953) reported a leaf-spot of *C. avellana* in Italy caused by *Leptothyrium coryli* Fckl. (Fungi Imperfecti, Sphaeropsidales). This species of *Leptothyrium* has not apparently been recorded in Britain.

Diseases of fruit

Sclerotinia fructigena Aderh. & Ruhl. (Ascomycetes, Helotiales) attacks hazel nuts when they are partially developed, causing them to rot or to fall prematurely. The fungus is much better known as the cause of brown rot in many fruits such as apple and plum. It appears to be mainly a wound parasite on hazel, infecting punctures made by the nut weevil, *Balaninus nucum* (Wormald 1944; Moore, M. H. 1950). Sclerotia are formed by the fungus on the rotted nuts.

Moore also ascribed rotting of nuts to *Botrytis cinerea* Fr. (Fungi Imperfecti, Moniliales) and *Gibberella baccata* (Wallr.) Sacc. (Ascomycetes, Hypocreales), the imperfect stage of which is *Fusarium lateritium* Nees. (Fungi Imperfecti, Moniliales), and considered that they probably entered through wounds. They are, however, much less important than *Sclerotinia fructigena*.

It will be noted that there are no serious diseases of hazel in Britain, but that in America and southern Europe several very damaging species occur on *C. avellana*.

DISEASES OF HORNBEAM

Carpinus betulus occurs as a native constituent of mixed broadleaved woodland in the south-east of England, and has been planted occasionally elsewhere. In some places it is of limited forest importance, in others it is regarded as a weed species.

Stem diseases—bark and cambium

A few fungi have been recorded as apparent parasites on the living stems of *Carpinus*, but there is no evidence that any of them are important. *Dermatea carpinea* (Pers.) Rehm. (Ascomycetes, Helotiales) has been described as a wound parasite in Germany. It is probably the same fungus as *Pezicula fagi* (Phill.) Boud., which has been found in Britain, but not apparently as a cause of disease. *Stilbospora angustata* Pers. (Fungi Imperfecti, Melanconiales) has been associated with branch dieback of *Carpinus* in France. This fungus has also been recorded in Britain, but with no suggestion that it is a parasite here. Two species of *Melanconium* (Fungi Imperfecti, Melanconiales), *M. magnum* Berk. and *M. stromaticum* Corda, have been found in Britain on hornbeam twigs. Though the latter occurred on living twigs, it was not suggested that it was a parasite. Thus all the stem fungi associated with *Carpinus* are no more than very weak parasites.

Diseases of leaves and shoots

Hornbeam, like birch, is attacked by a fungus causing witches' brooms, in this case *Taphrina carpini* (Rostr.) Joh. (Ascomycetes, Taphrinales). The mycelium overwinters in infected buds. The leaves borne on the brooms are somewhat curled and bear the asci of the fungus on their under surfaces. The disease is not so common on hornbeam as is *T. turgida* on birch, and it is of no consequence in the cultivation of hornbeam.

Gloeosporium carpini (Lib.) Desm. (Fungi Imperfecti, Melanconiales) has been recorded in Britain causing brown spots on the leaves of hornbeam. It is possibly the conidial stage of *Gnomonia fimbriata* Auersw. (Ascomycetes, Sphaeriales), also called *Mamiana fimbriata* Ces. et de Not. or *Gnomoniella fimbriata* Sacc. *G. fimbriata* produces small black stromata on hornbeam leaves on the Continent.

The mildew, *Phyllactinia corylea* (Chapter 18), occurs on hornbeam as well as on other broadleaved hosts, but does little damage.

A rust fungus *Melampsoridium carpini* (Fckl.) Diet. (Basidiomycetes, Uredinales) occurs on hornbeam on the Continent and in America, but has not been recorded in Britain. The uredial and telial stages are on hornbeam, the aecidial host is unknown.

DISEASES OF LIME

Tilia cordata, the Small-leaved lime, is a native of Britain, and *T. platyphyllos*, the Large-leaved lime, though probably not native, is widely naturalized. The common lime, *T.* × *vulgaris*, though widely planted in parks and gardens, is rare as a woodland tree. A few other species of *Tilia* are quite commonly planted for ornamental purposes. Among these are *T. euchlora* and *T. petiolaris*.

Stem diseases—wilts

Lime is occasionally attacked by *Verticillium* wilt, caused by the fungus *V. albo-atrum* Reinke & Barth. (Fungi Imperfecti, Moniliales) (Chapter 30), but much less commonly than sycamore and maple.

Stem diseases—bark and cambium

Tilia is a common host for the bark fungus *Nectria* (Ascomycetes, Hypocreales) and a number of species have been recorded (Chapter 18). It is not known to what extent *Nectria* on lime is a primary parasite.

Exosporium tiliae Link. (Fungi Imperfecti, Moniliales) has been recorded occasionally as a suggested parasite on lime bark.The conidia are formed in black pustules arising below the outer bark. This peels off exposing the bast, which soon becomes characteristically shredded. It is almost certain that this fungus is normally a saprophyte, and if it does cause dieback, as has been suggested, it is probably on trees weakened by other causes.

Discella desmazierii B. & Br. (Fungi Imperfecti, Sphaeropsidales) has been occasionally recorded on live stems of *Tilia* in Britain and on the Continent, but it is probably only a very weak parasite. *Cryptodiaporthe hranicensis* (Petr.) Wehm. and *Diaporthe eres* Nits. (Ascomycetes, Sphaeriales) have both been recorded on live as well as dead stems, but they also seem to have little pathological significance. Petrak (1938*b*) considered that *Hercospora tiliae* (Pers.) Fr. (Ascomycetes, Sphaeriales) could be a parasite on lime branches on the Continent, but it has been recorded only as a saprophyte in Britain.

On the Continent *Pyrenochaeta pubescens* Rostr. (Fungi Imperfecti, Sphaeropsidales) causes deeply sunken lesions on the bark of young limes, especially on unsuitable sites. These lesions are later covered with small black pustular pycnidia. In severe attacks the twigs may be girdled, and the recovery shoots produce an appearance of excessively bushy growth (Příhoda 1950*a*). The disease has not been recorded in Britain.

Botryosphaeria ribis Gross. & Dugg. (Ascomycetes, Sphaeriales), which is also absent from Britain, is well known in America as the cause of a dieback of currants. It has also been recorded there as the probable cause of large conspicuous cankers on *T. neglecta* (Davidson, Wester, and Fowler 1949). Inoculations indicated that the European species of *Tilia* were less susceptible to the fungus.

Diseases of leaves and shoots

The commonest cause of leaf disease in lime is *Gnomonia tiliae* Kleb. (Ascomycetes, Sphaeriales), the conidial stage of which is *Gloeosporium tiliae* Oud. (Fungi Imperfecti, Melanconiales). In the early stages this causes brown spots with dark edges on the leaves and sometimes on the petioles and young shoots. Withering and defoliation may result. Defoliation is partially due to weakening of the petioles by the fungus, so that they tend to break in the wind while the leaves are still green (Lorrain Smith 1904; Salmon and Wormald 1915). This normally happens fairly early in the summer and lightly affected trees may recover completely before the end of the season; in more severe cases, however, the early defoliation may have serious effects on the general health of the tree. The disease occurs occasionally in Britain, but seldom causes serious defoliation. Nothing is known of the relative resistance of the various varieties of *Tilia*. Perithecia are formed in the spring on fallen leaves which have overwintered, so that sweeping them up from beneath infected trees should lessen the attack in the following season.

Cercospora microsora Sacc. (Fungi Imperfecti, Moniliales) causes small brown spots, 1 to 3 mm. across, on leaves of most species of lime, both in Europe and America. In severe attacks it can cause appreciable defoliation, but it is much less serious than *Gnomonia*. Its status in Britain is not known, though it certainly occurs.

Elsinoë tiliae Creelman (Ascomycetes, Myriangiales) has caused severe spotting and defoliation of the leaves of *T. europaea* and *T. platyphyllos* in Nova Scotia. The fungus overwinters in the buds and can be controlled by early spraying (Creelman 1958*b*).

Two other fungi, *Phyllosticta tiliae* Sacc. & Speg. and *Ascochyta tiliae* Kab. & Bub. (both Fungi Imperfecti, Sphaeropsidales), have been recorded causing leaf-spots on *Tilia* in Britain, but neither is important.

Tilia, particularly *T. vulgaris*, is particularly prone to attack by aphids. These secrete 'honey-dew', which serves as a substrate for sooty moulds, such as *Fumago* and *Capnodium* spp. (Chapter 21). These are not parasitic on the leaves, but they may add to the damage done by the aphids by reducing the amount of light reaching the foliage.

DISEASES OF *NOTHOFAGUS*

Nothofagus has possibilities as a forest tree, at any rate in the warmer parts of Britain. Two species from southern Chile, *N. procera* and *N. obliqua*, have so far been tried as timber trees. Other Chilean species, notably *N. dombeyi* and *N. antarctica*, have been planted in parks and gardens, but do not appear suitable for forest purposes. The New Zealand species of the genus are much less successful in Britain.

Diseases due to non-living agencies

As would be expected from a genus which extends into climates milder than that of Britain, frost injury is of considerable importance. The most susceptible to winter cold of the species mentioned above is probably *N. dombeyi*, but *N. procera* runs it very close. *N. obliqua* is much less damaged by winter cold, while *N. antarctica*, in most parts of the country at any rate, seems perfectly winter hardy. In the spring, *N. procera* is by far the most frost susceptible; on the very small amount of evidence available the other three species appear quite resistant. On *N. procera* winter and spring injury takes the form of extensive dieback, canker development, and occasionally death of the whole tree. On the other species the injuries are less pronounced, and death from frost does not occur.

Stem diseases—bark and cambium

Little is yet known of the fungal diseases of *Nothofagus* in Britain. Cankers, mostly thought to be initiated by frost, are rather common. *Nectria ditissima* Tul. (Ascomycetes, Hypocreales) has been recorded on cankers, very possibly following frost injury.

At Lael in west Scotland a twig dieback occurred on *N. procera*, disfiguring the trees but not causing serious injury. Frost is unlikely in such a maritime climate, and a species of *Phomopsis* (Fungi Imperfecti, Sphaeropsidales), which was consistently isolated from the twigs, seems the most probable cause of the injury. It is possible that this may be *P. diaporthes-macrostomae* (Nits.) Trav., the conidial stage of *Diaporthe medusaea* Nits. (Ascomycetes, Sphaeriales), which has been recorded on *Fagus*.

The most important disease of native *Nothofagus*, both in Chile and in New Zealand, is caused by species of the genus *Cyttaria* (Ascomycetes, Pezizales) (Palm 1932; Santisson 1945; Marchionatto 1949; Rawlings 1956). These fungi cause galls which restrict the growth of the branches on which they are formed; occasionally dieback occurs (Herbert 1930). The ascocarps, which develop during the summer, are very large—for instance in *C. gunnii* Berk. they are up to 40 mm. in diameter and are of a conspicuous orange in colour (White, N. H. 1954). These large brightly coloured ascocarps have probably attracted more attention to *Cyttaria* than its pathological importance warrants. Nevertheless, its absence from introduced *Nothofagus* in Britain is a cause for satisfaction.

35

DISEASES OF ORNAMENTAL TREES AND SHRUBS

THE information available on the diseases of ornamental trees and shrubs is very inadequate. Their comparatively low economic importance and the scattered nature of their planting have tended to restrict the attention paid to them. For Britain, a rather limited host range is included as part of a book on diseases of garden plants by Beaumont (1956), while rose diseases, which are not dealt with here, are covered in most of the more comprehensive books on the genus. On the Continent, diseases of ornamental plants have been specifically dealt with by Gram and Weber (1952) and Pape (1955). North American diseases of ornamentals are well covered by Dodge and Rickett (1948). They give full descriptions of the symptoms of many of the diseases only briefly mentioned below. They also quote many other American diseases without giving any indication of their importance; most of these have been omitted from this chapter. Pirone (1948) also deals with the diseases of some ornamental trees and shrubs. Much more is known about the diseases of those genera which include fruit trees or bushes (Wormald 1955), but in this chapter only the ornamental species of such genera are dealt with. There are, of course, many records of fungi found on ornamental species, but these are of little value without some indication of the pathological status of the fungi involved.

Certain fungi such as *Armillaria mellea*, which has an enormous host range on ornamental trees and shrubs, *Verticillium albo-atrum*, and canker-causing species of *Nectria*, are mentioned only if any particular host genus is unusually susceptible or resistant. Damping off, which affects almost all kinds of seedlings, is also not specifically referred to. All these diseases are dealt with in the general Chapters, 15, 16, and 18, with the exception of *Verticillium* wilt, which is covered in Chapter 30, since it is most important on maples.

A considerable number of virus diseases have been recorded on ornamental trees and shrubs, though few are damaging. They are listed in Chapter 12.

Much information is available on the frost susceptibility of various ornamental trees and shrubs, because the very obvious damage caused by particularly cold winters, or unusually severe late frosts, has led to the compilation of lists of relative susceptibility and resistance. Some of this information is summarized in Chapter 4, so that few references to frost are made below.

The omission of a host genus does not means that it is disease free, but merely that no specific pathological information is available about it. In most cases it is stated whether or not a fungus occurs in Britain. Where nothing is said, it can be assumed that the fungus is found in this country. Very little is known about varietal susceptibility of ornamental trees and shrubs. When this matter is not mentioned it should be assumed that the omission indicates lack of knowledge, rather than absence of varietal differences.

Diseases of palms and of bamboos are not included in this book. The importance of the former in Britain is not sufficient to justify the space that would have to be devoted to their diseases. There is a large literature on palm diseases in the tropics and sub-tropics. The larger bamboos are certainly woody plants, but pathologically they are better considered together with their near relatives the grasses.

Little specific work has been done on the control of these minor diseases, and therefore very few recommendations on that subject are made below. When control measures are required, the general suggestions made in Chapter 36 should be applied according to the nature of the disease.

AESCULUS

The most important disease of Horse chestnut, *A. hippocastaneum*, and of other species of *Aesculus* is a leaf spot which has been attributed to one or more species of *Phyllosticta* (Fungi Imperfecti, Sphaeropsidales) under a variety of names, the most frequently quoted being *P. paviae* Desm. These are almost certainly the conidial stage or stages of *Guignardia aesculi* (Peck) Stew. (Ascomycetes, Sphaeriales), which has also been associated with similar symptoms. Small, irregular, slightly discoloured water-soaked spots appear on the leaves in the spring. Later the spots become brown with yellow margins, and merge so that the whole leaf withers (Stewart, V. B. 1916). The small black pustules of the conidial stage appear in the centres of the spots while the leaves are still on the tree and cause further spread. The spores of the perfect stage are produced on leaves on the ground the following spring, so that sweeping up fallen leaves is a possible means of control for park or garden trees. Fungicidal spraying of the expanding foliage has been suggested in America, where the disease has attracted more attention than in Europe. This could be carried out in nurseries, where the disease sometimes occurs. It has been suggested that the disease may be seed-borne (Orton 1931). It occurs in Europe including Britain, as well as in America, occasionally doing enough damage to attract attention but never apparently reaching epidemic proportions.

Septoria hippocastani B. & Br. (Fungi Imperfecti, Sphaeropsidales), the perfect stage of which is *Mycosphaerella hippocastani* Kleb. (Ascomycetes, Sphaeriales), also causes leaf-spot of Horse chestnut in Britain, but appears to be much less damaging than the disease described above. The spots are small and do not have a yellow margin.

Glomerella cingulata (St.) S. & v. Schr. (Ascomycetes, Sphaeriales) has been associated in the United States (Luttrell 1949) with death of leaves, dieback of branches, and cankers on the branches and trunk of *Aesculus* species, the lesions on the leaves being smaller than those due to *Guignardia aesculi*. It has not been recorded on this host in Britain, though it has a wide host range, particularly on fruit trees.

Horse chestnut appears to be more liable to slime flux than any other tree commonly grown in Britain (Chapter 20).

AMELANCHIER

Three fungi, which occur on *A. canadensis* and other species abroad, have not yet been recorded on it in Britain, although they occur here on other hosts. These are *Fabraea maculata* (Lev.) Atk. (Ascomycetes, Helotiales), causing a leaf spot, and two rust fungi, *Gymnosporangium clavariaeforme* (Pers.) DC. and *Ochropsora ariae* (Fckl.) Ramsb. (Basidiomycetes, Uredinales). *G. clavariaeforme*, which occurs on *Crataegus* and *Pyrus* in Britain, with juniper as its alternate host, would produce swellings, studded with small orange fruit-bodies, on the leaves, petioles, fruits, and shoots of *Amelanchier*. *O. ariae* occurs commonly on *Anemone nemorosa*, but the alternate stage has not been recorded in Britain with certainty on any of its possible host plants; if it did occur on *Amelanchier* it would produce yellowish to reddish spots on the underside of the leaves. Other species of *Gymnosporangium* occur on *Amelanchier* in North America.

Fire blight, *Erwinia amylovora* (see *Crataegus*), has been recorded on *Amelanchier* in the United States.

ARBUTUS

Griphosphaeria corticola (Fckl.) Höhn. (Ascomycetes, Sphaeriales), which causes canker and dieback of roses in Britain, has been associated with leaf-spot and dieback of *A. unedo* on the Continent (Basile 1954).

In the United States *A. menziesii* is attacked by *Phytophthora cactorum* (Leb. & Cohn.) Schroet. (Phycomycetes, Peronosporales), which causes lesions of the main stem and upper roots and results in extensive crown-dieback (Stuntz and Seliskar 1943). This fungus has not been recorded on *Arbutus* in Britain. *Rhytisma arbuti* Phil. (Ascomycetes, Phacidiales), which causes black spots on the leaves of *A. menziesii* on the Pacific coast of North America, has also not been recorded in Britain. Several species of *Phyllosticta* (Fungi Imperfecti, Sphaeropsidales) have been associated with leaf spotting of *Arbutus*— on the Continent, but not in Britain.

ARCTOSTAPHYLOS

Macrosporium sarcinale Berk. (*Stemphylium botryosum* Wallr.) (Fungi Imperfecti, Moniliales) has been associated with a leaf-spot on *A. manzanita* in Britain by Briant and Martyn (1929). The perfect stage is *Pleospora herbarum* (Fr.) Rabenh. (Ascomycetes, Sphaeriales), which has a very wide host range.

ARONIA (Chokeberry)

Several species of *Gymnosporangium* (Basidiomycetes, Uredinales) have been recorded on Chokeberry, *A. melanocarpa* and other species, in the United States. The alternate host is *Juniperus*. None of the British species of *Gymnosporangium* has been recorded on *Aronia*. Fire blight (*Erwinia amylovora*) (see *Crataegus*) has been recorded on *Aronia* in the United States.

ARTEMESIA

Puccinia absinthii DC. (Basidiomycetes, Uredinales) causes brown spots on the leaves and shoots of several wild species of *Artemesia*, but is uncommon. It is quite likely to appear occasionally on cultivated varieties.

AUCUBA

Phomopsis aucubicola Grove (Fungi Imperfecti, Sphaeropsidales) has been associated with a twig dieback of *A. japonica* in Britain (Grove 1935), but it is certainly not a common disease.

Phyllosticta aucubae S. & S. (Fungi Imperfecti, Sphaeropsidales) has been associated with brown and black spotting of the margins of the leaves of *A. japonica* in North America. It sometimes follows infestation by scale insects. There is only a rather doubtful record of this fungus on *Aucuba* in Britain.

AZALEA (see *Rhododendron*)

BERBERIS AND *MAHONIA*

Rusts

The greatest pathological significance of *Berberis* lies in the fact that it is the alternate host for Black rust of wheat and other cereals, *Puccinia graminis* Pers. (Basidiomycetes, Uredinales). A comprehensive review of this disease has been given by Lehmann, Kummer, and Dannenmann (1937). In many parts of the world, particularly in North America, *P. graminis* is a devastating disease; but in Britain it is now only occasionally serious on wheat and oats where *B. vulgaris* occurs wild, particularly in parts of south Scotland (Bates, C. C. V. 1949) and in Wales. The winter in most parts of Britain is cold enough to kill the uredospores, which in warmer climates allow the fungus to over-winter without the necessity of infecting *Berberis*, so that in Britain the disease develops seriously on cereals only in the neighbourhood of the shrub. On *Berberis* the fungus produces orange aecia, 2 to 5 mm. in diameter, on the leaves and fruit, and causes comparatively little damage (Fig. 92). The uredo- and teleutospores occur on wheat, oats, and other cereals, and also on a number of grasses.

B. vulgaris is probably the only species that need be regarded seriously as a host for the fungus. Some of the innumerable species and hybrids of *Berberis* planted in gardens, including most of the evergreen varieties, appear to be resistant. Others are susceptible, but nevertheless they are seldom, if ever, naturally infected. The fungus can infect *M. aquifolium*, but does not develop freely on it. Lists of susceptible and resistant varieties have been given by Levine and Cotter (1932) and by Buchwald (1937).

In Britain the eradication of wild *B. vulgaris* from the neighbourhood of fields is desirable, and it should never be planted in gardens. There is no need to restrict the use of other species of *Berberis* or *Mahonia*. In America, where the disease is much more serious, *Berberis* eradication has been attempted on a large scale (Bulger 1954). *Berberis* plays a particularly important role there, not only because in the colder regions it is required for the successful over-wintering of the disease, but also because the fungus hybridizes on it to produce new strains, to which hitherto resistant wheats may prove susceptible (Craigie 1942).

Another species of *Puccinia*, *P. arrhenatheri* (Kleb.) Erikss., occurs very rarely on *B. vulgaris*, which is the aecial host. The aecia, which occur on leaves

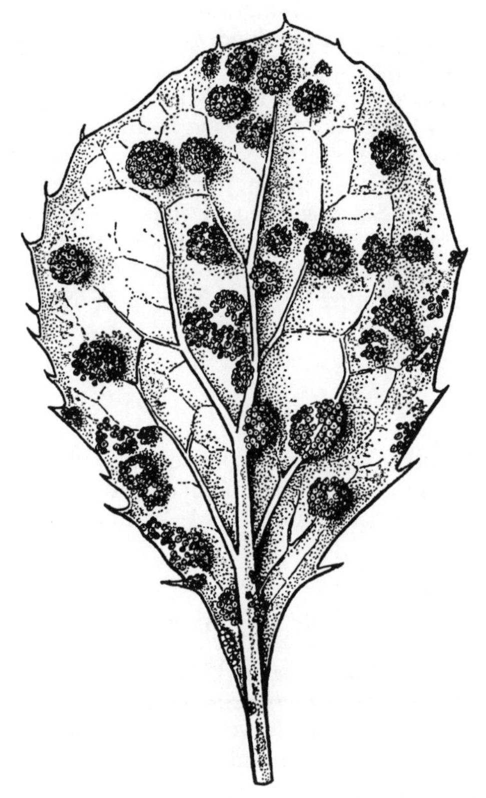

FIG. 92. Aecia of *Puccinia graminis* on *Berberis vulgaris* (×2). (J. C. from photograph by Department of Agriculture for Scotland, Science Service.)

arising from infected buds, are much smaller than those of *P. graminis*. The uredial and telial stages are found rather more frequently on Oat grass, *Arrhenatherum elatius*. The fungus has never been regarded as damaging in Britain, but in Germany it has been associated with dieback and even death of *Berberis* bushes (Ulbrich 1931).

Puccinia (*Cumminsiella*) *mirabilissima* Peck, which produces all its stages on *Mahonia*, was introduced from America, and has now spread widely both in Britain (Wilson, M. 1930*a*) and on the Continent (Poeverlein 1932). The aecidia, which are found on the younger leaves, are rare, but the uredial and telial stages, which occur as conspicuous reddish spots on the older leaves (Hammarlund 1932), are fairly common in some parts of Britain.

Other diseases

A number of species of *Berberis*, including *B. vulgaris* and *B. thunbergii*, are attacked by a bacterium, *Pseudomonas berberidis* Thornb. & And. This disease occurs in the United States (Thornberry and Anderson 1931), Denmark, and Britain. Affected shoots bear dead buds here and there and succulent shoots may be killed, but much more conspicuous are dark spots, 2 to 4 mm. across, on the leaves. The disease is not common in Britain.

Phoma berberidicola Vestergr. (Fungi Imperfecti, Sphaeropsidales) has been recorded in both England and Scotland on living twigs of *B. vulgaris*. It is doubtful if it is really a pathogen.

Mildew, *Microsphaera berberidis* (DC.) Lév. (Ascomycetes, Erysiphales), has been recorded on the leaves of *B. vulgaris, B. hookeri*, and *M. aquifolium* in Britain. Its host range on species and hybrids of *Berberis* and *Mahonia* is certainly wider than this. It is seldom, if ever, damaging.

Phyllosticta mahoniae Sacc. & Speg. (Fungi Imperfecti, Sphaeropsidales) causes spots on the leaves of *M. aquifolium* and *M. japonica. Septoria berberidis* Niessl. (Fungi Imperfecti, Sphaeropsidales) causes somewhat similar damage to *B. vulgaris*. Neither of these fungi is of any pathological significance.

BUDDLEIA

B. variabilis and other species can be seriously damaged by Cucumber mosaic virus, which causes malformation and mottling of the leaves and may reduce the amount of flower (Smith, K. M. 1952) (Chapter 12).

BUXUS (Box)

Puccinia buxi DC. (Basidiomycetes, Uredinales) is found occasionally on the leaves of *B. sempervirens*. Only the telial stage occurs, brown telia being produced on both sides of the leaves. Severe attacks have occasionally occurred in Britain. The removal and burning of infected shoots has been suggested as a means of control.

Volutella buxi (Cda.) Berk. (Fungi Imperfecti, Moniliales), the perfect stage of which is probably *Pseudonectria* (*Nectriella*) *rousseliana* (Mont.) Wollenw. (Ascomycetes, Hypocreales), causes withering of the foliage, dieback of twigs and branches, and occasionally cankers, both in North America (Dodge and Swift 1930; Andrus 1933; Weiss and Baumhofer 1940) and in Europe (Juel, O. 1925). The pink spore masses of the *Volutella* are produced on leaves still on the tree and later the larger perithecia, which vary in colour from yellow through red to green, form on fallen leaves. *Verticillium buxi* Link., Aersw. & Fleisch. (Fungi Imperfecti, Moniliales) and *Hyponectria buxi* (Desm.) Sacc. (Ascomycetes, Hypocreales) have also been associated with somewhat similar injuries to box (Dodge, B. O. 1944*a*, 1944*b*). The interrelationships of these fungi and their relative importance as causes of disease on box still require further investigation.

The fungi concerned all occur in Britain, but their status here is quite unknown. They may account for some of the fairly frequently observed deaths of twigs and branches of clipped box. This is quite probable, since they are usually considered to be wound parasites.

In the United States leaf-spotting has also been attributed to *Macrophoma candollei* (Berk. & Br.) Berl. & Vogl. (Dodge and Swift 1930; Andrus 1933; Weiss and Baumhofer 1940). This is possibly the same fungus as *M. mirbelii* B. & V., which together with *M. candollei* has been recorded in Europe (Juel, O. 1926), including Britain.

Two species of *Phyllosticta* (Fungi Imperfecti, Sphaeropsidales), *P. buxina* Sacc. and *P. limbalis* Pers., also cause leaf-spots of box in Britain; neither is important.

CALLUNA (Heather)

The diseases of heather are considered in Chapter 13, in connexion with the unfortunately remote possibility that they might exercise a controlling influence on one of Britain's most widespread forest-weed species.

CAMELLIA

The commonest disease of *C. japonica* and other species of *Camellia* in Britain is Leaf blotch caused by *Pestalozzia guepini* Desm. (Fungi Imperfecti, Melanconiales). The brown or yellow leaf blotches, which occur particularly on greenhouse-grown plants, are sprinkled with minute black fruit-bodies. Less commonly, pale spots with a narrow red border are caused by *Phyllosticta camelliae* Westend. (Fungi Imperfecti, Sphaeropsidales).

Exobasidium camelliae Shir. (Basidiomycetes, Exobasidiales) causes swellings covered with a whitish bloom on the leaves and flower buds. Shirai (1896) considered that there were two separate varieties of the fungus, one attacking the leaves and the other the flower-buds. The deformation caused by this fungus has been fully described by Wolf and Wolf (1952). The disease has once been recorded in Britain (Dennis and Wakefield 1946).

Glomerella cingulata (Stonem.) Spauld. & Schrenk. (Ascomycetes, Sphaeriales) causes dieback and canker of camellias in the United States (Baxter and Plakidas 1954) and Japan (Itô *et al.* 1956). It is a wound parasite, entering through leaf scars, grafting wounds, etc. It can be controlled by spraying (Anzalone and Baxter 1956; Anzalone and Plakidas 1956). In Britain this fungus has a wide host range on fruit trees, causing among other diseases Bitter rot of apples, but it has not been recorded on *Camellia.* It is known to possess different strains, possibly that on *Camellia* has not yet reached Britain.

Sclerotinia camelliae Hara. (Ascomycetes, Helotiales) occurs in Japan, where camellias are native, and in the United States, where they are grown on a large scale; but it is not present in Britain. It destroys the flowers, but can be controlled by removal of diseased blooms and collection of fallen ones before the apothecia are formed on them (Thomas and Hanson 1946). Control by spraying is also possible (Anzalone and Plakidas 1958*b*).

CARYA (Hickory)

Carya is comparatively little planted in Britain, and there are no disease records for it here. In North America, however, it is an important native forest genus, and in the southern United States *C. pecan* Engl. & Graebn. provides a valuable nut crop. A considerable number of diseases have been recorded on cultivated pecan, but there are comparatively few important ones on the forest species (Osborn *et al.* 1954; Campbell and Verrall 1956). *Poria spiculosa* Camp. & Dav. (Basidiomycetes, Aphyllophorales), mainly important as a cause of decay, is also associated with widespread and destructive canker formation (Campbell and Davidson 1942).

Other damaging diseases are leaf blotch caused by *Mycosphaerella dendroides* (Cke.) Dem. & Cole (Ascomycetes, Sphaeriales) (Demaree and Cole 1930), anthracnose caused by *Gnomonia caryae* Wolf (Ascomycetes, Sphaeriales), mildew caused by *Microstroma juglandis* (Bereng.) Sacc. (Fungi Imperfecti, Melanconiales), and scab caused by *Cladosporium effusum* (Wint.) Dem. (Fungi Imperfecti, Moniliales). The last is the most serious disease on cultivated pecans in the southern States. None of these fungi has been recorded in Britain.

CATALPA

Paclt (1951) has produced a comprehensive list of the many fungi occurring on *C. bignonoides* and other species of *Catalpa*. Few of these are significant, however, and none have been recorded in Britain. In the United States two fungi, *Phyllosticta catalpae* Ell. & Mart. (Fungi Imperfecti, Sphaeropsidales) and *Macrosporium catalpae* Ell. & Mart. (Fungi Imperfecti, Moniliales), are associated with quite serious leaf spotting. Pirone (1948) suggested, however, that the initial cause is bacterial and that the fungi are secondary. Dodge and Rickett (1948) considered that *M. catalpae* follows attack by a midge, *Itonida catalpae* Comst.

CEANOTHUS

This genus is particularly susceptible to frost injury, but appears to be free from important fungal diseases, both in Britain and abroad.

CELASTRUS

Phyllactinia corylea (Pers.) Karst. (Ascomycetes, Erysiphales), Powdery mildew, a fungus with a wide host range (Chapter 18), has been recorded in the United States, but not in Britain, causing leaf-fall of *C. scandens* (Dodge and Rickett 1943).

CELTIS

Although the genus *Celtis* has a wide distribution in northern temperate regions, it is not a native of, and is seldom planted in, Britain. Little is known of its diseases, but *C. australis*, which is scarcely hardy in Britain, is commonly attacked in the Mediterranean regions, where it is native, by *Gyroceras celtidis* (Biv.) Mont. & Ces. (Fungi Imperfecti, Moniliales), which sometimes causes severe leaf-spotting (Killian 1925; Mezzetti 1950). In the United States *C. occidentalis*, which is the best species for planting in Britain, is often disfigured by a witches' broom disease, which seems to be caused by mites of the genus *Aceria*, together with the mildew fungus *Sphaerotheca phytoptophila* Kell. & Swingle (Ascomycetes, Erysiphales) (Snetsinger and Himelick 1957).

There is no record of *Ceratocystis ulmi*, the cause of Elm disease, on the genus *Celtis*, although it is closely related to *Ulmus* (Chapter 31).

CERCIS (Judas Tree)

The Judas tree, *C. siliquastrum*, often displays considerable twig and branch dieback in Britain. The dead bark is rapidly colonized by *Nectria cinnabarina* (Tode) Fr. (Ascomycetes, Hypocreales). This fungus may play a part in the dieback, but the causes are probably complex and may well originate in the climatic unsuitability of the tree to our cool moist summers.

In the United States *Botryosphaeria ribis* Gross. & Dugg. (Ascomycetes, Sphaeriales) causes sunken oval cankers, which often result in girdling of twigs (Davidson, Wester, and Fowler 1949). *Mycosphaerella cercidicola* (Ell. & Kell.) Wolf (Ascomycetes, Sphaeriales), the conidial stage of which is *Cercospora cercidicola* Ell. (Fungi Imperfecti, Moniliales), causes severe leaf-spotting (Wolf 1940*a*). Neither fungus occurs in Britain.

CHIMONANTHUS (Winter Sweet)

A species of *Alternaria* Nees. (Fungi Imperfecti, Moniliales) and a species of *Ascochyta* Lib. (Fungi Imperfecti, Sphaeropsidales) have been recorded causing leaf-spots of *C. fragrans* in Scotland (Foister 1961). Nothing further is known of their distribution or importance.

CHOISYA

A species of *Botryodiplodia* (Fungi Imperfecti, Sphaeropsidales) has been found by Woodward (1926) to be the cause of a twig dieback of *C. ternata* Humb. in Britain. This Mexican shrub is frequently damaged during cold winters, and most occurrences of fungi on it are likely to be secondary to winter injury.

CISTUS

Many species of *Cistus* are damaged in Britain whenever there is a cold winter. Young bushes usually recover rapidly, but in older bushes winter injury may lead to progressive dieback. It is not known what fungi, if any, are associated with this.

CLEMATIS

Dieback often affects the large-flowered, garden varieties of *Clematis*. This is usually known as Clematis wilt, and has been attributed to many different causes, such as bad drainage, overfeeding, and graft failure. In the United States, Gloyer (1915) associated *Ascochyta clematidina* Thüm. (Fungi Imperfecti, Sphaeropsidales) with a disease of this nature. In this case the fungus usually attacked the leaves and passed down the petioles into the stems, but it also caused independent lesions on the stems. The dieback in this disease moved progressively down the plant, which is substantially different from the disease in Britain, where the main stem dies at, or near, ground-level, while the top is still healthy. A species of *Ascochyta* agreeing closely with Gloyer's description has been found on *Clematis* in Britain (Moore, W. C. 1947), but it is certainly not the usual cause of the dieback disease, which for the moment remains unexplained.

Mildew, caused by *Erysiphe polygoni* DC. (Ascomycetes, Erysiphales), is not uncommon, but seldom does much damage. Like other mildews it is probably controllable by sulphur spraying. *Septoria clematidis* Rob. & Desm. (Fungi Imperfecti, Sphaeropsidales) causes an unimportant leaf spot, mainly on the wild *C. vitalba*.

COLUTEA

Mildew, caused by a species of *Oidium* (Fungi Imperfecti, Moniliales), has been recorded in Britain on *C. arborescens* (Moore, W. C. 1959). The rust, *Uromyces laburni* (DC.) Fckl., which is found here on other leguminous shrubs, has not been recorded on *Colutea* in Britain, although it occurs on it on the Continent.

CORNUS

Mildew, caused by *Erysiphe tortilis* (Wallr.) Fr. (Ascomycetes, Erysiphales), is common on Wild dogwood, *C. sanguinea*, covering the leaves with white mycelium. It probably occurs also on planted species. Leaf-spots, 6 to 10 mm. across, caused by *Phyllosticta cornicola* Rab. (Fungi Imperfecti, Sphaeropsidales) and *Septoria cornicola* Desm. (Fungi Imperfecti, Sphaeropsidales) have been recorded quite commonly on *C. sanguinea* and the frequently planted *C. alba*. *P. cornicola* causes reddish spots, eventually with a pale centre; *S. cornicola* causes slightly larger, greyish spots, with a dark-purple border. The two fungi often occur together.

Leaf-spots on native Flowering dogwood, *C. florida*, in the United States are caused by *S. floridae* Tehon & Daniels and *Ascochyta cornicola* Sacc. (both Fungi Imperfecti, Sphaeropsidales). The latter sometimes causes withering of the foliage over the whole tree (Jenkins, Miller, and Hepting 1953). By far the most serious disease in the United States, however, is spot anthracnose caused by *Elsinoë corni* Jenkins & Bitanc. (Ascomycetes, Myrangiales). This fungus, which has not been recorded in Britain, produces lesions on the leaves, flowering bracts, berries, and twigs, and has recently attracted considerable attention in the south-eastern United States, where Flowering dogwood is esteemed as an ornamental tree both wild and cultivated. On cultivated trees control by spraying with copper and other substances is possible (Cox and Heuberger 1953).

The so-called Crown-canker, caused by *Phytophthora cactorum* (Leb. & Cohn) Schroet. (Phycomycetes, Peronosporales), is also regarded as a serious disease of Flowering dogwood in the United States. Though the first detectable symptoms are a reddening and withering of the leaves in part or whole of the crown, followed by twig and branch dieback, the actual seat of the disease is in the slowly developing lesions on the lower stem or upper roots (Creager 1937; Stuntz and Seliskar 1943). There are no records of *P. cactorum* on *Cornus* in Britain.

COTONEASTER

This genus is comparatively free from diseases in Britain, but *Phyllosticta sanguinea* (Desm.) Sacc. (Fungi Imperfecti, Sphaeropsidales) has been reported causing round, brown spots with a reddish margin on the leaves of *C. frigida* and the Silver leaf fungus, *Stereum purpureum* (Fr.) Fr., has been found causing dieback of *C. horizontalis.*

Cotoneaster is one of the genera susceptible to Fire blight, caused by the bacterium *Erwinia amylovora*, and it has been recorded on *C. salicifolia* in Britain (see *Crataegus* below). *C. horizontalis*, another species commonly planted in Britain, is known to be susceptible to Fire blight, but *C. simonsii* is probably immune (Thomas and Thomas 1931).

Fabraea maculata (Lév.) Atk. (Ascomycetes, Helotiales), causing a leaf-spot, and the rust *Gymnosporangium clavariaeforme* (Pers.) DC., both known to occur on *Cotoneaster* abroad, have not yet been recorded on it in Britain, though both fungi are present in the country on other hosts (see *Amelanchier* and *Crataegus*).

CRATAEGUS (Hawthorn)

Two species of *Crataegus*, *C. monogyna* and *C. oxycanthoides*, occur wild in Britain; others are planted.

Rusts

Gymnosporangium clavariaeforme (Jacq.) DC. and *G. confusum* Plowr. (both Basidiomycetes, Uredinales) occur in Britain with their aecial stages on hawthorn and other rosaceous plants, and with their telial stages on *Juniperus communis* and on *J. sabina* respectively (Chapter 26). There are no uredial stages. The aecia are borne on yellow or orange swellings on the young shoots, leaves or fruits, and are rather conspicuous. *G. clavariaeforme* is not uncommon in Britain, but does little damage. Montemartini (1925) found that *G. clavariaeforme* was able to overwinter as mycelium in hawthorn without the intervention of juniper in Italy. This may be possible in Britain, in which case the disease could persist on hawthorn, after the juniper from which it was originally infected had been removed.

Other species of *Gymnosporangium* with their aecia on *Crataegus* and other rosaceous hosts and their telia on *Juniperus* species occur abroad. As far as *Crataegus* is concerned, the most damaging of these is probably the North American *G. globosum*, which has been reported by Young and McNew (1947) severely damaging small plants of *C. mollis*.

Other diseases of leaves and shoots

Phleospora oxycanthae (Kunze & Schm.) Wallr. (Fungi Imperfecti, Sphaeropsidales) forms crowded yellowish spots on the leaves, which spread until the foliage assumes a scorched appearance. Small brown pycnidia are borne mainly on the underside of the leaves. It is quite common in Britain and occasionally damaging.

Mildew, caused by *Podosphaera oxycanthae* (DC.) de Bary (Ascomycetes, Erysiphales), which causes a typical white bloom on the foliage and exercises an appreciable check on growth, is very common. Control with sulphur sprays is possible, and in the case of hedges, summer trimming to remove most of the infected shoots should lessen future attacks.

Leaf blotch caused by *Sclerotinia crataegi* Magn. (Ascomycetes, Helotiales) is not particularly common in Britain. Dark-brown patches appear on the leaves in the spring. The affected leaves, which are normally scattered among healthy ones, later hang withered and black. Under damp conditions the fungus appears as a powdery grey, sweetly scented mould on the leaf surfaces (Dowson and Dillon-Weston 1937; Wormald 1937).

In the United States *Entomosporium thuemenii* (Cke.) Sacc. (Fungi Imperfecti, Melanconiales) attacks the leaves and young twigs of hawthorn (Stowell 1956). It can cause almost complete defoliation and appreciable dieback, but can be controlled by spraying (Dochinger and Bachelder 1954). *Crataegus monogyna* is said to be particularly susceptible. Some of the American species are comparatively resistant. A very similar disease, attributed to a species of *Entomosporium*, has been described in Germany (Gante 1927). There has been some confusion between *E. thuemenii* and *E. maculatum* Lév., which does occur in Britain, and which is responsible for a severe leaf blight of quince and to a lesser extent of pear. The perfect stage of the latter is *Fabraea maculata* Atk. (Ascomycetes, Helotiales), but only the *Entomosporium* stage has been recorded in Britain. There is no doubt, however, that they are different species.

In the United States and New Zealand, Fire blight, caused by the bacterium *Erwinia amylovora* (Burr.) Winsl. *et al.*, attacks various species of *Crataegus*. The disease has been described by Parker, Fisher, and Mills (1956). The bacterium enters through the blossoms, which are infected by pollinating insects contaminated with bacterial slime. Affected flowers become blackened and shrivelled, and this damage rapidly spreads to the leaves on flowering twigs. Injury is progressive, cankers being formed where infected twigs join larger branches, and substantial dieback taking place. Bacterial slime appears on the affected parts. The withered leaves are often retained after healthy leaves have fallen. The disease is serious chiefly on pear and apple, and the particular significance of hawthorn lies in its possible use for hedges in the neighbourhood of orchards, where it could act as a source of infection to fruit trees (Cockayne 1921). The disease has recently been recorded on pear, and on *Crataegus*, in Britain and has already done considerable damage to the former (Lelliott 1959; Glasscock unpubl.). Under 'The Fire Blight Disease Order, 1958' owners can be compelled to fell and destroy infected trees. The severity of the disease can be lessened by spraying; streptomycin has proved especially successful (Ark 1958).

CYDONIA

Most of the information available on the diseases of this genus refers to the Common quince, *C. oblonga* (Wormald 1955). Much less is known about the diseases of the varieties grown primarily for flower, which are now often separated in the genus *Chaenomeles*.

Entomosporium maculatum Lév. (Fungi Imperfecti, Melanconiales), which causes leaf-blight and fruit-spot of quince and is probably its most serious disease in Britain, has occasionally been recorded on other species of *Cydonia*. Its perfect stage, *Fabraea maculata* Atk. (Ascomycetes, Helotiales), has not been recorded in Britain. The much rarer *Sclerotinia cydoniae* Schellenb. (Ascomycetes, Helotiales), which is responsible for leaf-blotch of quince, has not been recorded on other species of *Cydonia*. Only the imperfect stage, *Monilia necans* Briosi & Cav., has so far been found in Britain. *S. fructigena* Aderh. & Ruhl. causes brown rot of the fruit of the flowering varieties, as well as of quince and many other fruits (see *Prunus* and *Pyrus*). Only the imperfect stage, *M. fructigena* Pers. (Fungi Imperfecti, Moniliales), has so far been found in Britain. *S. laxa* Aderh. & Ruhl., which attacks the blossoms and shoots of many kinds of fruit trees (see *Prunus* and *Pyrus*), has been recorded on quince in Britain, and quite frequently causes blossom wilt of *C. lageneria*, the commonest ornamental species.

Mildew caused by *Podosphaera leucotricha* (Ell. & Everh.) Salm. (Ascomycetes, Erysiphales), which is common on apple, also occurs on quince, some varieties being very susceptible. It has not been recorded on the decorative species of *Cydonia*. The rust fungus *Gymnosporangium confusum* Plowr. (Basidiomycetes, Uredinales) (see *Crataegus*) has also been recorded rarely on quince in Britain, but not on the ornamental species. In the Mediterranean region this fungus is quite damaging to quince (Fischer, E. 1931).

Members of this genus are known from American experience to be susceptible to *Erwinia amylovora*, the cause of Fire blight (see *Crataegus*).

CYTISUS AND *GENISTA* (Broom)

Ceratophorum (*Pleiochaeta*) *setosum* Kirchn. (Fungi Imperfecti, Moniliales), which causes a leaf-spot of lupins and laburnum, is also responsible for die-back of brooms. The disease first shows as dark spots on the leaves and stems, and can cause very rapid destruction of seedlings or cuttings (Raabe, A. 1938; Green and Hewlett 1949). It is probably best to burn infected plants.

The rust *Uromyces laburni* (DC.) Fckl. (Basidiomycetes, Uredinales) produces its uredial and telial stages on *Cytisus scoparius*, *Genista tinctoria*, *G. anglica* and *G. sagittalis*, as well as on laburnum and gorse. No aecial host has been recorded in Britain. It may well have a wider host range on *Cytisus* and *Genista* than has yet been recorded. It is of no practical importance.

Cylindrocladium scoparium Morg. (Fungi Imperfecti, Moniliales), which is particularly associated with shoot dieback of *Eucalyptus* (see below), has been found in Japan causing dieback of broom (Terashita and Itô 1956). The fungus occurs in Britain on various hosts, but has not been associated here with any disease of *Cytisus* or *Genista*.

DAPHNE

The only fungal disease of *Daphne* known in Britain is Leaf-spot caused by *Marssonina daphnes* Magn. (Fungi Imperfecti, Melanconiales). *Gloeosporium mezerei* Cke. is certainly identical with *M. daphnes*. It produces small spots about 1 mm. across on the leaves and petioles, and can cause appreciable defoliation (Green, D. E. 1935). It has been recorded only on *D. mezereum*. Virus mosaic (Chapter 12 is the most serious disease of *Daphne* in Britain.

Several species of *Daphne* tend to be rather short-lived in Britain and often die suddenly without any apparent cause.

DIERVILLA (Weigelia)

Gloeosporium diervillae Grove (Fungi Imperfecti, Melanconiales) has been found causing pale spots with a reddish border, 3 to 4 mm. across, on the leaves of *D. florida*. It is certainly not common in Britain.

ELAEAGNUS

Leaf-spots, caused by *Phyllosticta argyrea* Speg. (Fungi Imperfecti, Sphaeropsidales), have been recorded on *E. pungens* and other species in Britain. *P. elaeagni* (Sacc.) Allesch. may also be involved.

ERICA (Heath)

Attacks by *Phytophthora cinnamomi* Rands. (Phycomycetes, Peronosporales) and also by *P. cactorum* (Leb. & Cohn.) Schroet. have been reported on a number of occasions on the greenhouse species of heaths cultivated in pots for market (Oyler and Bewley 1937), as well as on *Calluna*. It was found that the compost used for the heaths was a very favourable medium for the growth of the fungi, and that rain-water tanks used for watering the plants were often heavily contaminated. It was suggested, in view of the known susceptibility of beech to *Phytophthora*, that beech-leaf mould should not be used in the compost. The disease can probably be avoided by soil sterilization, though this might raise problems with a genus so highly mycorrhizal as *Erica*. In Oregon the fungus has been recorded in nurseries, on *Calluna* and also on *Erica carnea* (Torgeson 1954), but in Britain it has not been found on hardy heaths.

The rhizomorph-forming fungus *Marasmius androsaceus* Fr. (Basidiomycetes, Agaricales), which is associated with dieback of heather (Chapter 13), has also been found attacking wild *E. tetralix*.

Mildew caused by *Oidium ericinum* Erikss. (Fungi Imperfecti, Moniliales), which is sometimes very harmful abroad, has been recorded only once in Britain. *Phyllosticta ericae* All. (Fungi Imperfecti, Sphaeropsidales), which was suspected of causing disease on *E. carnea* in Germany, has once been associated with leaf browning of *Erica* in Britain. *Stemphylium ericoctonum* Br. & de B. (Fungi Imperfecti, Moniliales), which has been associated on the Continent with leaf-fall and dieback of greenhouse heaths, has been recorded in Britain on *Calluna* as a Sooty-mould but not as a cause of disease. *Cladosporium herbarum* (Pers.) Link. (Fungi Imperfecti, Moniliales) has occurred more than once in Cornwall on cultivated *E. lusitanica* as a disfiguring sooty-mould on the foliage (Moore, W. C. 1959).

EUCALYPTUS

For the growth of eucalypts in Britain, the question of hardiness outweighs all other pathological considerations. D. Martin (1948) gave *E. vernicosa*, *E. parvifolia*, *E. niphophila*, and the hardier variants of *E. gunnii* as the species most likely to withstand the British climate. Wood (Macdonald *et al.* 1957), who collected together all the available data, suggested the addition of *E. pauciflora* to this list. Even in parts of the Mediterranean region, frost injury to eucalypts occurs in hard winters. Considerable damage was done in February 1956, and a great deal of valuable information on relative susceptibility was collected by Magnani (1957). His list does not include by any means all the varieties grown in Britain, but he also regards *E. gunnii* and *E. niphophila* as hardy.

Little is known of diseases of *Eucalyptus* in Britain, and even abroad most of the pathological literature is concerned with wood-destroying fungi. A mildew, caused by a species of *Oidium* (Fungi Imperfecti, Moniliales), has been recorded by Glasscock and Rosser (1958) on nursery plants of *E. gunnii* and *E. perrineana* in Britain. A similar *Oidium* had earlier been recorded on *E. camaldulensis* in Italy (Grasso 1948). *Cylindrocladium scoparium* Morg. (Fungi Imperfecti, Moniliales) has been associated in Britain chiefly with shoot wilt of *Prunus*. In the Argentine, as well as attacking cherries, it causes a collar rot of *Eucalyptus* seedlings (Jauch 1943). In Brazil it has also been reported causing dieback of mature trees by Batista (1951), who considered the fungus on *Eucalyptus* to be a separate variety. A species of *Alternaria* (Fungi Imperfecti, Moniliales) has been reported by Magnani (1958*b*) causing brown spots with a red edge on the leaves of eucalypts in several nurseries in Italy. *Physalospora eucalyptorum* Turc. (Ascomycetes, Sphaeriales) has been reported in Italy causing shoot dieback and eventually death of young trees (Turconi 1924). Possibly the most widespread disease is shoot dieback caused by *Botrytis cinerea* (Chapter 15).

EUONYMUS

Mildew, caused by *Oidium euonymi-japonicae* (Arcang.) Sacc. (Fungi Imperfecti, Moniliales), is now generally distributed in the south and west of

England and in Wales. The perfect stage has not been found, even in Japan. The fungus, which was probably introduced from Japan, attacks only the evergreen *E. japonicus*, overwintering as mycelium in sheltered parts of the bushes. It has been described by Salmon (1905). It is easily controlled by sulphur spraying. Much less important is the mildew on *E. europaeus*, caused by *Microsphaera euonymi* (DC.) Sacc. (Ascomycetes, Erysiphales).

Several fungi cause leaf spots on *Euonymus* in Britain. *Phyllosticta euonymi* (*bolleana*) Sacc. has been recorded on *E. japonicus* and *E. europaeus*, and *P. subnervisequa* All. on *E. japonicus* and *E. latifolius*. Both produce similar irregular spots and can be separated only under the microscope. Damaging attacks have been recorded. *Septoria euonymi* Rabenh. has also been recorded as a rare cause of leaf spots on *E. japonicus*.

The rust *Melampsora euonymi-caprearum* Kleb., with its uredial and telial stages on *Salix* spp. (Chapter 33), produces orange aecia on the leaves of *E. europaeus* in August and September, but is not common.

FORSYTHIA

Bacterial blight, caused by a bacterium very closely related to, if not identical with, *Pseudomonas syringae* van Hall, shows as blackening and dieback of the shoots (Metcalfe, G. 1939). *P. syringae* is more familiar as the cause of lilac blight. The disease is more prevalent on *F. intermedia spectabilis* than on *F. suspensa*. It seldom, if ever, kills, but causes repeated dieback and is very difficult to eradicate, even if all the apparently infected wood is cut out.

Sclerotinia sclerotiorum (Lib.) de Bary. (Ascomycetes, Helotiales), a fungus with a wide host range mostly on herbaceous plants, has been associated with twig dieback of *Forsythia* in the United States (Wolf 1947) and occasionally in Britain. Small lesions appear on the twigs and may girdle them.

Botrysphaeria ribis Gross. & Dugg. (Ascomycetes, Sphaeriales), well known in the United States as a cause of dieback of currants, but occurring on a large number of other hosts, many of them woody, can also cause twig dieback of *Forsythia* (Marshall, B. E. 1952). It does not occur in Britain.

Galls or nodules on the stems of *Forsythia* have several times been reported in Britain (Moore, W. C. 1948). They were regarded by J. M. Thompson (1946) as root-producing growths. They are not, however, of universal occurrence and there is probably some special physiological or pathological reason for their appearance. In America, similar galls have been attributed to a species of *Phomopsis* (Dodge and Rickett 1948).

Leaf-spot, caused by *Phyllosticta forsythiae* Sacc., has been rarely recorded in Britain on *F. suspensa*.

GARRYA

Phyllosticta garryae Cke. & Hark. (Fungi Imperfecti, Sphaeropsidales) has been recorded in Britain causing large grey spots with a purple border on *G. elliptica*. It is a native disease of the shrub in California and may well have been imported on nursery stock.

GENISTA (see *Cytisus*)

GLEDITSCHIA (Honey Locust)

This genus is not very commonly planted in Britain and there are no British records of disease on it. In the United States *Thyronectria austro-americana* (Speg.) Seeler (Ascomycetes, Hypocreales) causes sunken bark cankers, chiefly on Honey locust, *G. triacanthos*. These result in girdling and branch dieback. On Honey locust the progress of the disease is slow, but trees of *G. japonica* can be killed in a single season (Seeler 1940; Crandall 1942).

HEDERA

Ivy, *H. helix*, is subject to a number of diseases, but the only serious one is that caused by *Bacterium hederae* Arnaud (White and McCulloch 1934). It produces both leaf-spots and stem cankers. The leaf-spots are at first light green and water-soaked; later they turn brown or black with reddish margins. The petioles also become black and shrivelled. It has been recorded in France, Germany, and the United States, but not in Britain.

Calonectria hederae Arnaud (Ascomycetes, Hypocreales), the imperfect stage of which is *Cylindrocladium macrosporum* var. *hederae* (Fungi Imperfecti, Moniliales), has been recorded in Surrey, apparently killing patches of ivy several metres across, growing under fairly heavy shade. The stems were attacked as well as the leaves. Perithecia and conidia were borne abundantly on the infected parts (Booth and Murray 1960). The fungus has also been recorded on ivy on the Continent. There is no definite proof of its pathogenicity.

In Britain leaf-spot caused by *Mycosphaerella hedericola* (Desm.) Lind. (Ascomycetes, Sphaeriales), the imperfect stage of which is *Septoria hederae* Desm. (Fungi Imperfecti, Sphaeropsidales), is common, particularly in the damper districts of the south and west. It causes brown spots, 3 to 12 mm. in diameter, which eventually become whitish in the centre, but retain a purplish-brown margin. Three rather doubtfully separate species of *Phyllosticta* (Fungi Imperfecti, Sphaeropsidales) have also been recorded causing spots on ivy leaves in Britain. The commonest and best established species is *P. hedericola* Dur. & Mont., which causes spots difficult to distinguish from those due to *Septoria*. Both fungi produce minute black pycnidia on the lesions.

Much larger lesions, up to 2 cm. in diameter, are caused by *Colletotrichum trichellum* (Fr.) Duke (Fungi Imperfecti, Melanconiales) (Bongini 1933). This fungus has been reported causing considerable damage in Italy and in the United States, but though it occurs in Britain, the damage here is comparatively slight. None of these leaf-spots in Britain has any appreciable effect on the vigour of ivy or on its ability to spread.

HYDRANGEA

Mildew, caused by *Microsphaera polonica* Siem. (Ascomycetes, Erysiphales), is widely distributed in Britain, but seldom attracts attention. It is mainly a disease of *H. hortensis* grown under glass. The perfect stage of the fungus has not been found in Britain.

Septoria hydrangeae Bizz. (Fungi Imperfecti, Sphaeropsidales), causing brown spots with a marked red border on the leaves, very occasionally damages *H. hortensis*.

HYPERICUM (St. John's wort)

A rust, *Melampsora hypericorum* Wint. (Basidiomycetes, Uredinales), has been recorded on *H. androsaemum* and *H. elatum*, among the shrubby species. There is evidence that *H. calycinum*, *H. patulum*, and *H. moserianum* are resistant (Moore, W. C. 1948). The orange uredial and the brown telial stages occur on *Hypericum*; the aecial stage is unknown.

ILEX (Holly)

Two leaf-spotting fungi affect holly in Britain. *Phyllosticta aquifolina* Grove (Fungi imperfecti, Sphaeropsidales) causes grey spots with a dark brown border, and *Coniothyrium ilicis* Sm. & Ramsb. (Fungi Imperfecti, Sphaeropsidales) whitish spots, on the leaves of *I. aquifolium*. Neither is common, but the second is known to have reached epidemic level on one occasion.

Phacidium curtisii (B. & Rav.) Luttr. (Ascomycetes, Phacidiales) causes large brown spots on the foliage of American holly, *I. opaca*, in the United States. Later flat, black stromata develop. The disease is known as Tar-spot of holly (Luttrell 1940*a*).

Recently a serious disease of English holly (*I. aquifolium*) caused by a new species of *Phytophthora* (Phycomycetes, Peronosporales), *P. ilicis* Budd. & Young, has been described in the north-west United States (Buddenhagen and Young 1957). The fungus causes black spotting of the leaves, defoliation, twig dieback, and stem cankers.

A disfiguring twig and branch dieback of American holly in New Jersey has been attributed to *Fusarium solani* var. *martii* (App. & Wr.) Wr. (Fungi Imperfecti, Moniliales) (Bender 1941).

JUGLANS (Walnut)

The two species of importance in Britain are *J. regia*, the Common walnut, grown chiefly for its nuts, and the American species *J. nigra*, the Black walnut, which forms a very handsome tree. The wood of both species is valuable, but they are seldom planted specifically for timber. They are not easy to establish, and when young can be badly damaged by spring frosts.

The toxic effect of the roots of walnuts on other plants growing near them is dealt with in Chapter 8.

Root diseases

Both species are particularly susceptible to attack by Honey fungus, *Armillaria mellea*. The establishment of walnut on old coppice sites is therefore fraught with considerable risk, though such sites may often provide the fertile soil which the trees demand.

Several species of *Phytophthora* (Phycomycetes, Peronosporales) have been associated with the so-called 'Crown rot' of walnut in North America (Smith and Barrett 1931; Gravatt 1954), in the Argentine (Pontis Videla 1943), in Australia (Cookson 1929), and in Europe, particularly in Italy (Curzi 1933; Cristinzio 1942). Despite its name, it is essentially a disease of the upper roots and collar, exactly like the so-called 'Ink disease' caused by fungi of the same genus on *Castanea* and *Fagus* (Chapter 29). The same association with badly drained soils is suggested. The disease on walnut has not been recorded in Britain; possibly it has escaped attention, or been masked by Honey fungus.

Stem diseases—bark and cambium

Twig dieback of walnut is common. It is often primarily due to frost, and the fungi which colonize the dead twigs are secondary. Such fungi recorded in Britain include *Nectria cinnabarina* (Tode) Fr. (Ascomycetes, Hypocreales), *Cytospora juglandina* Sacc. (Fungi Imperfecti, Sphaeropsidales), and a *Phomopsis* stage of *Diaporthe perniciosa* March (Ascomycetes, Sphaeriales) (Hamond 1931).

Nectria galligena Bres. is considered in America a serious wound parasite on the branches of walnut (Ashcroft 1934). Eventually very large target cankers may be formed. This fungus causes apple and pear canker in Britain, but has not been suggested here as a cause of disease in walnut.

Hendersonula toruloidea Natt. (Fungi Imperfecti, Sphaeropsidales) causes severe dieback of *J. regia* in California, following damage to the stems by heat. Otherwise it is only a weak parasite (Sommer 1955). It does not occur in Britain.

A dieback caused by *Melanconis juglandis* (E. & E.) Graves (Ascomycetes, Sphaeriales) occurs in the United States, chiefly on the Butternut (*J. cinerea*) (Graves 1923). The imperfect stage of this fungus is *Melanconium oblongum* Berk. (Fungi Imperfecti, Melanconiales). A closely related fungus, *M. juglandinum* Kze., the perfect stage of which is *Melanconis carthusiana* Tull., has been recorded commonly on dying branches of *J. regia* in Britain and on the Continent, where it is said also to cause dieback in walnut grafts. There is strong evidence that *M. juglandis* in America is only a weak parasite causing slow deterioration of attacked trees; the same is probably the case with *M. juglandinum* in Europe. *Melanconium juglandinum* forms on the twigs black pustular fruit bodies, from which very large numbers of spores are extruded in black tendrils.

Recently a disease characterized by extensive, shallow, irregularly shaped cankers on the bark of the trunk and main branches has been described in California (Wilson, Starr, and Berger 1957). The cause is a bacterium, *Erwinia nigrifluens* Wilson, Starr, and Berger, which is very closely related to *E. amylovora*, the cause of Fire blight in fruit trees (see *Crataegus*). *E. amylovora* was known to be capable of infecting walnuts into which it had been inoculated (Smith, C. O. 1931).

Graft disease

When special varieties of walnuts are propagated by grafting under greenhouse conditions, the graft union is often attacked by *Chalaropsis thielavioides* Peyr. (Fungi Imperfecti, Moniliales). A mass of brownish-black spores is produced all over the graft junction. The disease can be avoided by sterilization of the propagating houses before use, treatment of the cut end of the stock with weak formalin, and periodic sterilization of the grafting knife (Hamond 1935). The fungus is known to occur on other hosts, attacking rose graft-unions in the United States, and being found on carrots and peach seedlings in Britain. It is not known whether it can be transferred from one host genus to another.

Diseases of leaves and shoots

Bacterial blight

The most serious disease of walnut is the bacterial blight caused by *Xanthomonas juglandis* (Pierce) Dowson. This has been known variously as *Bacterium*, *Phytomonas*, and *Pseudomonas juglandis*. It now occurs almost everywhere walnut is grown, having spread considerably in the last thirty years. Irregular black spots appear on the leaves and petioles; young fruits may be killed or spotted, and large black spots are formed on those that are nearly ripe. Black patches appear on the shoots, which are sometimes girdled and killed. Rain is the chief means of spread, but the disease can also be carried by contaminated pollen (Ark 1944). Infection is therefore intense only over short distances, and the disease is most serious where walnuts are grown in quantity. In Britain, where walnut is seldom grown on a commercial scale, this disease is not very serious, being seldom found except on young plants in nurseries. The disease might well become troublesome if walnut were being grown in plantations for timber.

The bacteria overwinter in the stem lesions, and pruning out infected wood is one method of control, but this may well lead to excessive mutilation and in any case walnut does not respond well to heavy pruning. When pruning is carried out, tools must be wiped clean after every cut, otherwise it may spread the disease. It can also be controlled by two or more sprayings in the spring with Bordeaux mixture, which is occasionally phytotoxic to *Juglans*, or with other copper compounds (Miller, P. A. 1940; Ark and Scott 1951). Streptomycin is also effective (Miller, P. A. 1958).

Phytomonas syringae has recently been recorded, under the name of Walnut blast, attacking the leaves, fruits, and buds of *J. nigra* and *J. mandshurica* in California (Ark and Bell 1959). The symptoms closely resembled those of Walnut blight, except that the withered nutlets and leaf spots were somewhat blacker and drier in appearance than those caused by *Xanthomonas juglandis*.

Leaf-blotch

Leaf-blotch, caused by *Gnomonia leptostyla* (Fr.) Ces. & De Not. (Ascomycetes, Sphaeriales), the conidial stage of which is *Marssonina juglandis* (Lib.) Magn. (Fungi Imperfecti, Melanconiales), causes brown blotches on the leaves and fruit. In severe attacks it can cause serious defoliation. It occurs commonly in England and Wales, and has been recorded in Scotland. The acervuli occur as small black specks on the infected leaves. The conidia can cause extensive secondary infections, especially in wet summers. The perithecia are produced on the fallen leaves, and the ascospores from them are responsible for fresh infections in the spring. Under orchard or garden conditions, therefore, the disease can be controlled by sweeping up and burning the fallen leaves (Castellani 1948; Wormald 1955). Spraying during early and midsummer is also a possible means of control on small trees (Carter and Hoffman 1953).

Leaf-spots

A number of fungi cause leaf-spots on walnut. One is *Microstroma juglandis* (Ber.) Sacc. (Fungi Imperfecti, Melanconiales), which occurs in south and central Europe and in North America, but not in Britain. Another is *Ascochyta juglandis* Bolts. (Fungi Imperfecti, Sphaeropsidales), which produces numerous greyish-brown spots with a darker border, 1 to 10 mm. in diameter, and which does occur, but not as a serious disease in this country.

KALMIA

K. latifolia and other species are subject to a number of leaf-spotting fungi which, with one exception *Pestalozzia macrotricha* Kleb. (Fungi Imperfecti, Melanconiales), are still confined to North America, where the genus *Kalmia* is native. *P. macrotricha* usually occurs on *Rhododendron*, causing brown spots which turn black as the fungus fructifies. In Britain it has been recorded only on *Rhododendron*, though in Italy it occurs commonly, though not seriously, on *Kalmia* (Servazzi 1936).

In North America *Phyllosticta kalmicola* (Sch.) E. & E. (Fungi Imperfecti, Sphaeropsidales) causes small silvery spots with a red or purple border; while larger brown spots are produced by *Phomopsis kalmiae* Enlows (Fungi Imperfecti, Sphaeropsidales). Both these diseases are common and occasionally destructive (Dodge and Rickett 1948; Pirone 1948).

KERRIA

K. japonica, especially the double form, is subject to considerable twig dieback when grown in Britain. In the main this occurs during the winter and indicates a certain lack of hardiness. There is no evidence that it is fungal in origin, or even that fungi play any significant part in its development.

LABURNUM

L. anagyroides and the other cultivated species are very prone to attack by wood-destroying fungi, which enter through pruning wounds, dead branches, etc. Among these is the parasite *Stereum purpureum* (Fr.) Fr. (Basidiomycetes, Aphyllophorales), the cause of Silver-leaf in plums. On laburnum it does not produce silvering, but it can cause severe dieback. *Cucurbitaria laburni* (Pers.) Ces. & De Not. (Ascomycetes, Sphaeriales) is also a wound parasite, but only on trees seriously weakened by other causes (Green, F. M. 1932). The conspicuous black crust-like fruit-bodies emerge through on the dead bark. Generally an old laburnum which has started to die back for any reason is best replaced. The easy entry of so many fungi greatly lessens the chances of recovery.

A species of *Fusarium* has been found associated with an otherwise unexplained dieback of laburnum in a Gloucestershire nursery. The disease was persistent and rather serious in this one locality, but has not been met with elsewhere. Wollenweber (1931) recorded three species of *Fusarium* on laburnum.

Leaf-spots on Laburnum have been attributed to a considerable number of fungi. *Ascochyta cytisi* Lib. (Fungi Imperfecti, Sphaeropsidales) causes roundish spots, up to 2 cm. in diameter, visible on both sides of the leaf and margined indistinctly by a thin brown line. This fungus is probably identical with *Phyllosticta cytisi* Desm., *Gloeosporium cytisi* B. & Br., *Septoria cytisi* Desm., and *Ascochyta kabatiana* Trott. (Grove 1935; Arx 1957*c*). Several of these fungi have been separately described as causes of leaf spots on laburnum.

Ceratophorum (*Pleiochaeta*) *setosum* Kirchn. (Fungi Imperfecti, Moniliales) occasionally attacks laburnum leaves, particularly in nurseries, causing small brown spots, which enlarge into irregular brown blotches. In severe attacks, dieback may take place, but in Britain this fungus is probably more important on lupin and *Cytisus*.

Two mildews occur on laburnum. *Peronospora cytisi* Magn. (Phycomycetes, Peronosporales) produces brown spots on the leaves. The underside of the spots becomes covered with grey mycelium. The disease appears to be quite rare in Britain, and may have been only recently introduced. A species of *Oidium* (Fungi imperfecti, Moniliales) has also been recorded occasionally on laburnum in Britain and elsewhere (Fischer, R. 1956). It causes typical whitening and distortion of the leaves and young shoots and may interfere with flower development. The perfect stage is unknown.

The rust *Uromyces laburni* (DC.) Fckl. (Basidiomycetes, Uredinales) very rarely produces its uredial and telial stages on laburnum; it is commoner on *Cytisus*, *Ulex*, etc.

LAURUS

The true laurel, *L. nobilis*, is not very commonly planted in Britain, and the only disease recorded on it, a leaf-spot caused by *Phyllosticta lauri* Westend. (Fungi Imperfecti, Sphaeropsidales), does not appear to be of any importance. In native stands of *L. nobilis* in southern Europe and of *L. canariensis* in the Canary Islands and the Azores, *Exobasidium lauri* (Bory.) Geyl. (Basidiomycetes, Exobasidiales) stimulates the production of antler-like outgrowths, up to 20 cm. in length, on the stems.

LAVANDULA

The only serious disease of lavender, *L. spica*, is 'Shab' caused by *Phoma lavandulae* Gab. (Fungi Imperfecti, Sphaeropsidales). Wilting of some of the shoots in May is the first symptom of the disease, being followed by death of the shoots and sometimes of the whole bush. Pycnidia are formed abundantly on the dead stems, portions of which may be blown about, spreading the disease over wider areas than is possible by the normal splashing about of the spores by rain. Pycnidia can also develop on dead stems of the weed *Chenopodium album*. Invasion takes place at the point of attachment of the leaves, or through wounds caused by picking the flowers. The disease is perpetuated mainly by the use of slightly infected cutting material for propagation. The use of very small cuttings, consisting entirely of young wood and not more than 7 cm. long, should result in healthy stock. The Dwarf French variety is immune, but it is of no commercial value (Metcalfe, C. R. 1931).

Lavender is often damaged by winter cold and sometimes by spring frost, especially if it has recently been pruned. Injury due to these causes can be confused with that due to 'Shab'.

Septoria lavandulae Desm. (Fungi Imperfecti, Sphaeropsidales) causes small circular spots on the leaves of lavender, often occurring on dying leaves on plants damaged by 'Shab' disease or frost. Wilt of the developing blossoms is sometimes caused by *Botrytis cinerea* Pers. ex Fr. (Fungi Imperfecti, Moniliales) (Ware 1931).

LAVATERA

The rust fungus, *Puccinia malvacearum* Mont. (Basidiomycetes, Uredinales), which is often serious on hollyhocks, has been recorded on Tree mallow, *L. arborea*.

LEDUM

L. palustre is of some importance on the Continent as the alternate host of the spruce rust *Chrysomyxa ledi* (A. & S.) De Bary (Basidiomycetes, Uredinales) (Chapter 23). Although *L. palustre* and other species occur occasionally in Britain, the fungus has not so far been recorded.

Another fungus on *Ledum*, *Sclerotinia heteroica* Woron. & Nawas. (Ascomycetes, Helotiales), is of interest because it is heteroecious, a phenomenon primarily associated with the Rust fungi. Conidia formed on the leaves of *Vaccinium uliginosum* infect the flowers of *L. palustre*, which become mummified and form apothecia the following spring. The ascospores then reinfect the leaves of *Vaccinium* (Woronin and Nawaschin 1896). This fungus also is absent from Britain.

LIGUSTRUM (Privet)

Root diseases

A considerable number of root diseases have been recorded on privet (*L. vulgare* and *L. ovalifolium*). Its apparent susceptibility to attack of this nature may rest to some extent on its frequent use for hedges. Death of part of a hedge, or even of one individual bush in it, must inevitably attract attention. *Armillaria mellea* (Fr.) Quél. (Basidiomycetes, Agaricales) has frequently been recorded damaging privet hedges. *Rosellinia necatrix* Prill. (Ascomycetes,

Sphaeriales), White root rot (Chapter 16), which occurs mainly on fruit trees and herbaceous plants in the south-west of England, has also been recorded on this genus. A *Phytophthora* species (Phycomycetes, Peronosporales) has been associated with the death of privet hedges in Somerset. *Verticillium dahliae* Kleb. (Fungi Imperfecti, Moniliales) has been found on the roots of privet in Yorkshire. *Polystictus velutinus* Fr. (Basidiomycetes, Aphyllophorales), usually regarded as a saprophyte, has been associated with the death of privet roots in Scotland.

Other diseases

Glomerella cingulata (Stonem.) Spauld. & Schrenk. (Ascomycetes, Sphaeriales), primarily important as the cause of Bitter rot of apples, has been recorded causing dieback and canker of privet twigs in the United States (Mix 1930) and more recently in Britain (Brooks 1953; Roberts 1957), as well as in Germany, Denmark, and Russia. It is certainly not a common disease in Britain. *L. ovalifolium* is said to be resistant (Dodge and Rickett 1943). Mix found that the fungus from apple could infect privet, so it is probably a mistake to plant privet hedges in the neighbourhood of orchards.

N. A. Brown (1936) reported nodular galls, 1 to 3 cm. in diameter, on privet in the United States, caused by a species of *Phomopsis* (Fungi Imperfecti, Sphaeropsidales). These caused inconsiderable dieback of the bushes. In Britain *Phoma ligustrina* Sacc. (Fungi Imperfecti, Sphaeropsidales) and *Phomopsis brachyceras* Grove have both been recorded on *Ligustrum*, but were not regarded as pathogens and did not produce galls.

Mycosphaerella ligustri Lind. (Ascomycetes, Sphaeriales), the pycnidial stage of which is *Phyllosticta ligustri* Sacc. (Ascomycetes, Sphaeropsidales), causes a leaf-spot on privet. The brown spots of variable shape and size later become paler and develop a brown or reddish border. *Septoria ligustri* (Desm.) Fckl. (Fungi Imperfecti, Sphaeropsidales) causes somewhat similar spots, but is less common in Britain.

LIQUIDAMBAR

No diseases have been recorded in Britain on *L. styraciflua*, though it is fairly commonly planted for autumn colour. In North America, where it is native, the tree is very free from fungal attack, but recently a serious disease, so far not completely explained, has caused considerable losses locally, though throughout the whole range of the tree the number of deaths is still negligible (Hepting 1955*b*). The disease, known as Sweetgum blight, is a progressive dieback of the whole crown, twigs which do not die immediately, producing small, sparse foliage. Trees frequently die within two years of the first appearance of disease symptoms (Miller and Gravatt 1952). In some regions, at any rate, drought appears to be a major factor in this disease (Toole 1959; Toole and Broadfoot 1959). Leader dieback, which may be a different disease or merely one aspect of Sweetgum blight, has been described by Garren (1954), who isolated *Diplodia theobromae* (Pat.) Now. (Fungi Imperfecti, Sphaeropsidales). Despite successful inoculation experiments, Garren (1956) regards this fungus only as a weak parasite.

LIRIODENDRON (Tulip Tree)

In North America, where it is native, *L. tulipifera* is subject to a number of fungal leaf-spots, none of any significance; but in Britain the only fungus recorded on living leaves has been *Rhytisma liriodendri* Wallr. (Ascomycetes, Phacidiales), causing typical black tar-spots, similar to those produced by *R. acerinum* on sycamore.

LONICERA (Honeysuckle)

Honeysuckle, *L. periclymenum*, appears to have few significant diseases. A form of *Ascochyta vulgaris* Kab. & Bub. (Fungi Imperfecti, Sphaeropsidales) causes roundish spots, often up to 1 cm. across, on the leaves. The spots are initially brown, but later turn pale and dry, often becoming torn. It is common, but seldom damaging. *Phyllosticta lonicerae* Westend (Fungi Imperfecti, Sphaeropsidales) may be the same fungus. *Leptothyrium periclymeni* Sacc. (Fungi imperfecti, Sphaeropsidales) causes much smaller spots, which are greenish in the early stages of development. It also is widely distributed.

A leaf-spot, caused by *Lasiobotrys lonicerae* Kunze (Ascomycetes, Erysiphales), occurs occasionally, but does no appreciable injury. Small black fruit-bodies are visible on the spots. The mildew *Microsphaera lonicerae* (DC.) Wint. (Ascomycetes, Erysiphales), which occurs on *Lonicera* on the Continent, has been recorded in Britain, but is certainly of no practical importance here. One of the grass rusts, *Puccinia festucae* Plowr. (Basidiomycetes, Uredinales), bears its aecia on yellow or brownish spots on the leaves of *L. periclymenum*, and its uredial and telial stages on species of *Festuca*. It is not common in Britain on either host.

MAGNOLIA

In Britain *Phyllosticta magnoliae* Sacc. (Fungi Imperfecti, Sphaeropsidales) causes irregular pale spots on the leaves of the evergreen *M. grandiflora*. *Microdiplodia solitaria* Bub. (Fungi Imperfecti, Sphaeropsidales) has been reported as a cause of leaf-spotting on the same host in Italy (Spaulding 1956), but in Britain *M. magnoliae* Grove has been found only on dead leaves and twigs. *Glomerella cingulata* (Stonem.) Spauld. & Schrenk. (Ascomycetes, Sphaeriales), commonly associated with rotting of apples and other fruit in Britain, has been recorded causing leaf spot of *M. grandiflora* in the United States (Fowler 1947). It has not been found on *Magnolia* in Britain. In Florida velvety, reddish-brown patches on the leaves and twigs are caused by the parasitic alga *Cephaleuros virescens* Kunze (Wolf 1929–30; Fowler 1947) (Chapter 21).

Canker and dieback of magnolias in the United States is comparatively common, but the cause has not been definitely established. However, a species of *Phomopsis* (Fungi Imperfecti, Sphaeropsidales) is constantly associated with the disease (Pirone 1948). There is no suggestion that *P. magnoliicola* Died., which is found on dead twigs and branches of several species of *Magnolia* in Britain, is a parasite.

MAHONIA (see *Berberis*)

MALUS (see *Pyrus*)

MORUS (Mulberry)

Though essentially a fruit tree, *M. nigra* is often grown for amenity in Britain in parks and gardens, while *M. alba* is planted abroad for purposes such as fodder, so that their diseases are dealt with here.

Root diseases

Rosellinia necatrix Prill. (Ascomycetes, Sphaeriales), White root rot, which has a wide host range but is particularly serious on apple and vine (Chapter 16), is regarded seriously on young mulberry plants in Italy (Voglino 1929) and Japan (Matuo and Sakurai 1954). In Britain the fungus has been found on a number of hosts, chiefly in south-west England, but not so far including mulberry. *Rosellinia aquila* (Fr.) De Not., which has been reported in Britain on spruce seedlings (Chapter 23), has been found on mulberry roots in France. Violet root rot, caused by *Helicobasidium mompa* Tan., which occurs on a wide range of host plants in Japan (Chapter 16), is said to be particularly serious on mulberry.

Diseases of bark and cambium

Hendersonula toruloidea Nattrass (Fungi Imperfecti, Sphaeropsidales), which does not occur in Britain, causes severe dieback of mulberry in Pakistan (Khan 1955). The same fungus has been associated with dieback of walnut in California, but only following initial sunscorch.

Diseases of leaves and shoots

Pseudomonas mori (Boyer & Lamb.) Stevens, more commonly known as *Bacterium mori* (Boyer & Lamb.) Smith, causes general dieback and stunting of young mulberry trees. It also causes angular, black spots, surrounded by a yellow zone on the leaves. In severe attacks, elongated lesions are formed on older twigs. In wet weather bacteria ooze from infected shoots and twigs. There are varietal differences in susceptibility (Zaprometov and Mikhailov 1937), which have not been fully worked out. The disease occurs in Britain, but is seldom serious (Wormald 1924). There is general agreement that pruning out infected twigs, accompanied by spraying with Bordeaux mixture or other fungicides, and possibly by sterilization of the pruning wounds, is the best method of control (Passinetti 1928; Zaprometov and Mikhailov 1937).

Gibberella moricola (Ces. & De Not.) Sacc. (Ascomycetes, Hypocreales) attacks the young shoots during the summer, girdling and killing them. The minute reddish-brown pustules of the conidial stage, *Fusarium lateritium* Nees. (Fungi Imperfecti, Moniliales), are borne near the base of the diseased shoots. The nearly black perithecia, which are rare in Britain, occur later (Wormald 1955). The disease, which has a very wide distribution abroad on *Morus nigra*, *M. alba*, and other species, is uncommon in Britain.

The genus *Sclerotinia* (Ascomycetes, Helotiales) has occasionally been associated with disease in mulberry. *S. sclerotiorum* (Lib.) de Bary caused dieback of young shoots on an epidemic scale in Bulgaria, possibly predisposed by a wet spring and summer (Christoff 1932), while *S.* (*Ciboria*) *carunculoides* Sieg. and Jenk. infected the flowers and destroyed the fruit of *M. alba* in North America (Siegler and Jenkins 1923). Since these diseases have not attracted further attention, they may be dependent on particular climatic circumstances for epidemic development.

Septogloeum mori (Lév.) Briosi & Cav. (Fungi Imperfecti, Melanconiales) (syn. *Phleospora mori* Sacc.) causes a leaf-spot of mulberry. The perfect stage, which has not been recorded in Britain, is *Mycosphaerella mori* Wolf. (Ascomycetes, Sphaeriales). The disease has been fully described by Cass Smith and Stewart (1947). Dark spots, 2 mm. in diameter, are formed on the leaves. Later the spots become larger and paler and the pycnidial pustules appear as small brown dots on the surface. In epidemic attacks, such as have occurred in Australia, the fungus can cause quite serious defoliation. The disease seems to occur almost everywhere mulberry is grown, having been reported among other places from Italy, Russia, and Australia, as well as Britain. There are differences in susceptibility between different species and between different varieties of the same species (Masera 1933; Zaprometov and Mikhailov 1937). It can be controlled by spraying with Bordeaux mixture or lime sulphur (Cass Smith and Stewart 1947).

Wolf (1936) has described another species of *Mycosphaerella*, *M. arachnoidea* Wolf., causing a false mildew of *M. rubra* in the United States. The leaves are covered with small patches of white mycelium, which later turns yellow. The disease ultimately leads to necrotic lesions and the death of the leaves. It has been suggested that there are other species of *Mycosphaerella* on mulberry, but there has obviously been confusion, and the whole situation requires further investigation.

A true mildew on the underside of the leaves, caused by a form of *Phyllactinia corylea* (Pers.) Karst. (Ascomycetes, Erysiphales), is one of the most widespread diseases of mulberry (Zaprometov 1945). In Britain the fungus is not uncommon on hazel, ash, and other trees (Chapter 18), but has not been recorded on *Morus*. Two other mildews, *Uncinula mori* Miyake (Ascomycetes, Erysiphales) in Japan, and *U. geniculata* Gerv. in North America, occur mainly on the upper side of the leaves; neither is found in Europe.

MYRICA (Bog Myrtle)

Two American rusts, *Cronartium comptoniae* Arth. and *Gymnosporangium ellisii* (Berk.) Farl. (Basidiomycetes, Uredinales), have *Myrica* spp. as alternate hosts. The native British *M. gale* can serve in this capacity, but neither fungus has been recorded in Britain. The first of these diseases on pine is discussed briefly in Chapter 22. The second has its telial stage (the uredial stage is missing) on *Chamaecyparis thyoides*. In the immediate neighbourhood of *Chamaecyparis*, the damage to *Myrica* may be serious.

MYRTUS (Myrtle)

M. communis and other species of *Myrtus*, which are barely hardy in most parts of Britain, flourish near the south and west coasts, and are there subject to attack by a number of leaf-spotting fungi. The most serious is *Pestalozzia decolorata* Speg. (Fungi Imperfecti, Melanconiales). This causes minute dark-red or reddish-purple spots, which increase in size and become pale in the centre. The acervuli appear as minute black spots, visible with a lens, in the centre of the lesions. Affected leaves wither, and defoliation can reach epidemic proportions (Hewlett 1952). Another leaf-spot, caused by *Cercospora myrticola* Speg. (Fungi Imperfecti, Moniliales) (syn. *C. myrti* Eriks.), is easily distinguished because the spots are dark in the centre and pale at the edges. The spores are borne on conidiophores protruding from the underside of the lesions, which are covered with grey downy superficial mycelium. This disease sometimes occurs separately, sometimes on bushes also affected by *Pestalozzia*. It is the less serious disease of the two, since it does not result in such rapid defoliation (Hewlett 1952). Both fungi have been recorded abroad, *P. decolorata* in Germany, and *C. myrticola* in Cyprus and the United States, but they have attracted little attention.

Spots similar to those caused by *Pestalozzia* have been attributed to *Phyllosticta nuptialis* Thüm. (Fungi Imperfecti, Sphaeropsidales), which is not apparently damaging.

NYSSA

N. sylvatica is occasionally planted as an ornamental in Britain, but so far no diseases have been recorded on it here. In its native North America it is attacked by *Phyllosticta nyssae* Cke. (Fungi Imperfecti, Sphaeropsidales), the perfect stage of which is *Mycosphaerella nyssaecola* (Cke.) Wolf (Ascomycetes, Sphaeriales). It causes irregular, purplish blotches, 2 cm. or more across, on the leaves, and severe attacks can result in premature defoliation. The perithecia of the *Mycosphaerella* stage are borne on fallen leaves the following spring (Wolf 1940*b*).

PAULOWNIA

The more serious diseases of *P. imperialis* do not appear to have spread beyond Japan, where it is widely cultivated, and possibly China where it is native. Brooming, a serious virus disease, has been reported only from Japan. This is also the case with its most serious fungal disease, Anthracnose caused by *Gloeosporium kawakamii* Miyake (Fungi Imperfecti, Melanconiales). This fungus attacks the leaves, particularly along the veins and on the petioles, and also the young shoots. It overwinters in the petioles of fallen leaves. In the nursery the disease can be fatal; on older trees it is much less damaging (Yoshii 1933; Itô and Chiba 1954–5). Serious dieback in northern Japan has also been associated with *Valsa paulowniae* Miyabe and Hemmi (Ascomycetes, Sphaeriales) (Togashi and Uchimura 1933), and *Physalospora paulowniae* Itô and Kob. (Ascomycetes, Sphaeriales) (Itô and Kobayashi 1951).

Only leaf-spotting fungi are present in Europe. Paclt (1948) reports *Ascochyta paulowniae* Sacc. & Brun., *Phyllosticta paulowniae* Sacc., and *Septoria paulowniae* Thüm. (all Fungi Imperfecti, Sphaeropsidales) in Czechoslovakia. Only the first two have been recorded in Britain, and neither is common. *Ascochyta* causes irregular yellowish-brown lesions, which later turn grey, and *Phyllosticta* causes spots with a dark, sinuous margin.

PHILADELPHUS

Circular leaf-spots, 1 cm. or more across, and yellowish with a darker border, are caused by *Ascochyta philadelphi* Sacc. & Sp. (Fungi Imperfecti, Sphaeropsidales) on *P. coronarius* and other species. The disease has been recorded in a number of places in Britain, but has not proved very damaging.

PHOTINIA

No diseases have been reported on *Photinia* in Britain, although *P. villosa* and other species are quite commonly planted. In the United States serious leaf-spotting of *P. serrulata* and other species is caused by several species of *Cercospora* (Fungi Imperfecti, Moniliales) (Anzalone and Plakidas 1957). It can be controlled by copper sprays (Anzalone and Plakidas 1958*a*).

Fabraea maculata Atk. (Ascomycetes, Helotiales), the pycnidial stage of which is *Entomosporium maculatum* Lév. (Fungi Imperfecti, Melanconiales), has been recorded in America causing purplish-brown leaf-spots on the foliage of *P. glabra*, and doing appreciable damage. Infection appeared to be from nearby pears (Plakidas 1957). Later it was suggested that the fungus on *Photinia* is a separate race from that on pear and loquat (*Eriobotrya*) (Stathis and Plakidas 1959). In Britain this fungus has been found on the nearly related, half-hardy loquat, *Eriobotrya japonica*, and on other Rosaceous shrubs, but not on *Photinia*.

In the United States *Photinia* is a host for *Erwinia amylovora*, the cause of Fire blight (see *Crataegus*), but it has not been recorded on this host in Britain.

PLATANUS (Plane)

Anthracnose

By far the most widespread and serious disease of plane is anthracnose caused by *Gnomonia veneta* (Sacc. & Speg.) Kleb. (Ascomycetes, Sphaeriales). This fungus has two conidial stages, the best known of which is *Gloeosporium nervisequum* (Fckl.) Sacc. (Fungi Imperfecti, Melanconiales). The disease appears in the spring when the developing leaves and shoots turn brown and die (Fig. 93). At this stage the damage is easily confused with late-frost injury. But later, the symptoms on expanded leaves are characteristic, well-defined patches of brown tissue bounded by the veins appearing in contrast to the green of the rest of the leaf. Later the entire leaf is killed. Small, pin-head, cream-coloured fruit-bodies, which are the *Gloeosporium* stage, appear on the underside of infected leaves. Small twigs are also killed, while cankers may be formed on twigs up to 2·5 cm. in diameter. The following spring, black pimples appear on the bark; these are the pycnidia of the second conidial stage, *Discula platani* Sacc. The perfect stage develops on dead fallen leaves. The fungus can spread considerably in the twigs, lesions on small branches usually being at points where twigs join them (Sempio 1933). Though repeated attacks may weaken trees, the disease is seldom fatal. It detracts from the amenity value of planes planted for ornamental purposes, although a fresh crop of shoots is usually produced in mid-summer.

Schenk (1926) suggested that the disease developed after frost injury. There is much more evidence associating attacks with damp weather in the spring

FIG. 93. *Gnomonia veneta* on *Platanus acerifolia* ($\times \frac{1}{2}$). (J. C.)

and early summer (Weisse 1925). The severe attacks which are common in California may be influenced by the frequent coastal mists. In Britain repetitive attacks are uncommon, though the disease may become epidemic locally for a single year (Day, W. R. 1947).

It has been suggested that *P. orientalis* is more resistant than the North American *P. occidentalis* or the widely planted London plane, *P. acerifolia* (Sempio 1938). Walther (1935) went so far as to suggest that certain individuals of *P. acerifolia* which showed higher resistance might be second-generation hybrids in which *P. orientalis* genes were predominant. However, Sprau (1951) regarded *P. orientalis* as equally susceptible.

Various methods of control are possible on ornamental trees of small or moderate size; pruning off infected shoots, spraying, and the collection and destruction of fallen leaves have all been suggested (Sempio 1938). Schuldt (1951) considered Bordeaux mixture one of the best substances for spraying; but Schneider and Campana (1955) have suggested that more modern sprays may be superior and Snyder (1957) has achieved satisfactory control with a single application of phenyl-mercury-acetate at the end of April. Early spraying is essential for successful control, the first application being given before the buds burst.

Canker stain

'Canker stain', caused by a form of *Endoconidiophora fimbriata* (Ell. & Halst.) Davidson (Ascomycetes, Sphaeriales), is locally serious on street trees, mainly *P. acerifolia*, in the United States. The form on plane is morphologically indistinguishable from the type which is known chiefly as the cause of Black rot of Sweet potatoes. The two can be separated only by host reactions (Walter, Rex, and Schreiber 1952).

Canker stain has been fully described by J. M. Walter (1946). It is similar to 'Sooty bark' of sycamore (Chapter 30), since the fungus spreads up and down the tree in the wood, causing a reddish-brown or bluish-black stain, and produces lesions where it reaches the cambium. Nevertheless, the majority of the lesions arise from individual infections. Dieback results from the fungus girdling the limbs. Leaves do not usually show symptoms until the branch is almost completely girdled. Infection is nearly always through wounds, especially those caused by contaminated tools. Pruning tools and even contaminated wound dressings are a ready means of spread in street trees (Walter and Mook 1940). Indeed the disease never reaches epidemic proportions in natural stands because of the relative absence of wounds and the complete absence of tool transfer. On the other hand, losses of over 80 per cent. can take place under town conditions (Walter, Rex, and Schreiber 1952). Winter pruning, sterilization of tools, and avoidance of unnecessary mutilation are all valuable means of control. Infected sawdust and fragments of bark, which may fall into normal asphalt wound dressings, are another means of transmission, so that the admixture of a fungicide with the dressing is a necessary precaution (Walter, J. M. 1946).

Neither *E. fimbriata* nor its form on plane has been recorded in Britain.

Other diseases

In North America Wolf (1938) has described a complex leaf disease of *P. occidentalis*. In mid-June minute, brown, necrotic spots appear on the leaves. These are due to *Cercospora platanifolia* Ell. & Ev. (Fungi Imperfecti, Moniliales), the perfect stage of which is *Mycosphaerella platanifolia* Cke. (Ascomycetes, Sphaeriales). The minute fruit-bodies of this fungus appear on both surfaces of the leaves. This is followed towards the end of July by the development of much larger pale-green areas due to the fungus *Stigmina platani* (Fckl.) Sacc. (Fungi Imperfecti, Moniliales), the perfect stage of which is *M. stigmina-platani* Wolf. A thin black web of mycelium, among which the fruit-bodies are produced, eventually covers the underside of the leaves. *Stigmina* is the more vigorous fungus and tends to mask the effects of *Cercospora*. *C. platanifolia* has also been recorded in Japan (Itô and Hosaka 1958). Neither of these fungi occurs in Britain.

Pirone (1952) recorded a species of *Dothiorella* (Fungi Imperfecti, Sphaeropsidales) causing bark cankers on *P. acerifolia* in New York City. This fungus may be the conidial stage of *Botryosphaeria ribis* Grossenb. & Dugg. (Ascomycetes, Sphaeriales), which has a wide host range on woody plants.

POLYGONUM

The vigorous climber *P. baldschuanicum*, which appears to be more or less disease-free in Britain, is a rare example of a woody plant attacked by a Smut fungus. *Ustilago raciborskiana* Siem. (Basidiomycetes, Ustilaginales) has been recorded in Poland and in Italy. The fungus causes a form of brooming, the internodes being shortened, and the growth more rigid than usual, while the flowers are crowded into spikes (Verona and Bozzini 1956).

PRUNUS (Cherry and Plum)

The bulk of the information on diseases of *Prunus* is derived from work on the cultivated fruits, in particular cherry and plum. The information available on the occurrence of these diseases on ornamental members of the genus is very incomplete. Comparatively little space, therefore, will be devoted here to the description or to the control of the diseases, since information is already available in other publications (Brooks 1953; Wormald 1955).

Virus diseases, which are of considerable importance on this genus, are dealt with in Chapter 12.

Diseases of bark and cambium

Silver leaf

Silver leaf, caused by *Stereum purpureum* (Fr.) Fr. (Basidiomycetes, Aphyllophorales), is one of the most serious diseases of plums in Britain (Brooks 1953). The fungus enters the wood through pruning wounds and broken branches, causing girdling and dieback. The silvering of the foliage is a secondary symptom produced by the toxic action of the fungus, which causes the mesophyll cells in the leaf to separate from each other and from the epidermis, thus altering the light-reflecting qualities of the leaf. The silvering symptom is typical on plums, apples, and some other members of the wide host range, but on some hosts it is absent. The small bracket-like fruit-bodies of the fungus, brown above and purple beneath, are borne on the diseased wood, usually at a rather late stage in the development of the disease.

Stereum purpureum has been recorded on a number of ornamental and wild varieties of *Prunus*, including *P. avium* (Gean), *P. laurocerasus* (Cherry laurel), *P. mahaleb*, *P. spinosa* (Blackthorn), and *P. triloba*. It has recently been reported in Britain doing serious damage to Japanese cherries and almonds planted along roads (Salter 1958). It can probably attack most varieties of *Prunus*, but there will certainly be large differences in susceptibility, as there are among plums and cherries grown for fruit. Control lies in the protective treatment of pruning wounds and the sterilization of tools used for pruning, in the removal of diseased wood from trees in the early stages of the disease, and in the destruction of badly infected trees.

Black knot

Dibotryon morbosum (Schw.) T. & S. (Ascomycetes, Sphaeriales) causes Black knot of plums and cherries, wild and cultivated, in North America (Koch, L. W. 1933, 1935). It causes elongated black swollen cankers on the branches, and results in considerable branch and twig dieback. It is not present in Europe.

Bacterial canker

Bacterial canker, leaf-spot, and shoot wilt of cherry and plum are caused mainly by *Pseudomonas mors-prunorum* Wormald. Less commonly, very similar damage is produced by *P. prunicola* Wormald, now regarded as a form of *P. syringae*, the cause of Lilac blight (Crosse 1954). Lesions appear on the twigs and branches during the winter and early spring. If these girdle the stem, as they often do, dieback takes place in the early summer, when in any case there is usually copious exudation of gum from the lesions. In the spring bacteria from the bark lesions infect the leaves, causing spots. These spots are of little importance to the health of the tree, but the bacterium passes the summer in them, dying out in the cankers in the spring. In the autumn fresh infection of the stems takes place through wounds and leaf scars. In the past, much of the damage caused by the bacterium was attributed to the fungus *Valsa leucostoma* (Pers. ex Fr.) Fr. (Ascomycetes, Sphaeriales), which in its conidial stage, *Cytospora leucostoma* Sacc. (Fungi Imperfecti, Sphaeropsidales), produces abundant reddish spore-tendrils on the dead bark. It is now considered that this fungus is at the most a weak secondary parasite. The disease has been found on *P. cerasifera pissardi*, almond (*P. amygdalus*) (Wormald 1938), and wild *P. avium* in several places in southern England (Peace unpubl.).

The disease can be controlled by spraying with Bordeaux mixture at the end of August, and again at the end of September. The antibiotic streptomycin has also given quite promising results (Crosse 1956; Dye 1956). Sprayings should kill a large proportion of the bacteria on the leaves before invasion of the shoots and stems has taken place. The method could be used to control local outbreaks on small wild *P. avium*; but destruction of all plants showing bark lesions during the winter, before leaf infection has taken place, is probably more generally practicable.

Diseases of leaves and shoots

Several species of *Taphrina* (Ascomycetes, Taphrinales) cause distortion of the leaves and shoots of *Prunus*. *T. cerasi* (Fckl.) Sadeb. produces witches' brooms on cherry, including quite frequently wild *P. avium*. It has also been recorded on Japanese cherries (Rathbun-Gravatt 1927). The brooms seldom flower, but bear leaves which are reddish, particularly when unfolding. The brooms are thus best removed from trees that are being grown for flower or fruit. *T. deformans* (Berk.) Tul. causes leaf-curl of peach and almond (Fig. 94). Affected leaves are extensively deformed and eventually reddish in colour. *T. minor* Sadeb. causes thickening, curling, and reddening of the leaves of cherries. It can attack ornamental varieties. *T. insititiae* (Sadeb.) Johans., which causes witches' brooms on plums, has also been recorded on *P. subhirtella autumnalis*, *T. pruni* (Fckl.) Tul., which distorts the fruit of plums, has been recorded on Bird cherry (*P. padus*), under the name *T. padi* (Jacz.) Mix.

FIG. 94. *Taphrina deformans* on almond, Farnham, Surrey, June (×½). (J. C.)

Two rust fungi (Basidiomycetes, Uredinales) affect the genus *Prunus* in Britain. *Tranzschelia pruni-spinosae* (Pers.) Diet. (*Puccinia pruni-spinosae* Pers.) has its uredial and telial stages on plums, and the aecial stage on *Anemone*. British wild species of *Prunus* are attacked, but the damage is never serious; many American species are also susceptible (Smith and Cochran 1939). The fungus causes orange spots on the leaves and sometimes on the fruits. *Thekopsora areolata* (Fr.) Magn. has the aecial stage on the cones of spruce (Chapter 23), while the uredial and telial stages cause small brown spots on the leaves of *P. padus* and other Bird cherries. It is not common.

Sclerotinia fructigena Aderh. & Ruhl. (Ascomycetes, Helotiales) causes brown rot of many different kinds of fruit. Only the imperfect stage, *Monilia fructigena* Pers. (Fungi Imperfecti, Moniliales), has been found in Britain. It can attack the fruits of ornamental varieties of *Prunus* (Wormald 1940). Much more important is *Sclerotinia laxa* Aderh. & Ruhl., which causes wilt of the developing blossoms and dieback of twigs on a wide range of *Prunus* species (Wormald 1940). Infection takes place through the flowers, and the fungus grows down into the twigs. Cushion-like tufts of conidia appear on affected parts. In *S. fructigena* these are yellowish and about 8 mm. across, while in *S. laxa* they are grey and smaller. On ornamental trees, infected twigs should be pruned out well below the apparently diseased portion. *S. laxa* has also been associated with the wilting of plum rootstocks earthed up for layering. The basal portion covered by the soil was infected. *Cylindrocladium scoparium* Morg. (Fungi Imperfecti, Moniliales) causes similar damage. *Botrytis cinerea* Fr. (Fungi Imperfecti, Moniliales) (Chapter 15) has been reported causing shoot wilt of *P. triloba* (Moore, W. C. 1939). Similar damage on *P. besseyi* and *P. pumila* has also been ascribed to it.

Mildew caused by *Podosphaera oxycanthae* (DC.) de Bary (Ascomycetes, Erysiphales), which is uncommon on plum and cherry including ornamental varieties, occurs rather more often on *P. laurocerasus* (Salmon 1906). The form on this species is now separated as *Podosphaera tridactyla* (Wallr.) De Bary from the type which occurs on hawthorn. It attacks particularly the young shoots, causing distortion and sometimes dieback (Montemartini 1930), so that frequent pruning of Cherry laurel hedges, with consequent repeated production of young shoots, is likely to encourage attacks. The disease can be controlled by dusting with sulphur.

Leaf-spots

A leaf-spot of Cherry laurel, *P. laurocerasus*, is caused by *Trochila laurocerasi* (Desm.) Fr. (Ascomycetes, Helotiales) (Gregor 1936). The spots, which are yellow and indefinite at first, later become well defined and brown, and finally fall out, leaving holes in the leaves. Frequent summer pruning of laurel hedges appears to favour the fungus by providing young wounded leaves, which it is best able to infect. A very similar disease is caused by *Coryneum laurocerasi* Prill. & Delacr. (Fungi Imperfecti, Melanconiales).

Leaf-spots of other species of *Prunus* are caused by *Clasterosporium carpophilum* (Lév.) Aderh. (Fungi Imperfecti, Moniliales) (syn. *Coryneum beijerinckii* Oudem.). The spots usually fall out, leaving shot-holes. The fungus can also cause lesions on the young twigs, but this aspect of the disease appears to be commoner in North America than in Europe (Ogawa, Nichols, and English 1955). It is most serious on fruiting varieties, including almond, but has been recorded on a wide range of ornamental species and varieties (Wilson, E. E. 1937; Smith and Smith 1942).

Polystigma rubrum (Pers. ex Fr.) (DC.) (Ascomycetes, Hypocreales) causes a leaf blotch of plums, attacking particularly Blackthorn, *P. spinosa*, and Bullace, *P. domestica insititia*. It produces on leaves large red stromata, which are covered with the small black punctures of the spermogonia and which become black after leaf fall.

Gnomonia padicola (Lib.) Kleb. (Ascomycetes, Sphaeriales) causes brownish-purple leaf-spots on Bird cherry, *P. padus*. It is known in Britain only as the imperfect stage, *Actinonema padi* Fr. (Fungi Imperfecti, Melanconiales).

Coccomyces hiemalis Higg. (Ascomycetes, Phacidiales), which causes leaf-spot of cherry, and can attack ornamental varieties, occurs in North America, western Europe, and other parts of the world, but not in Britain. In wet seasons it can cause extensive defoliation (Keitt *et al.* 1937).

PYRACANTHA

Only one serious disease affects this genus, Scab caused by *Fusicladium pyracanthae* (Otth.) Rostr. (Fungi Imperfecti, Moniliales). This fungus has sometimes been referred to as the variety *pyracanthae* of the Pear scab organism, *F. pirinum* (Lib.) Fckl.; but they differ in spore size, and cross inoculations do not succeed (McKay 1944). The perfect stage of the Pear scab organism is *Venturia pirina* Aderh. (Ascomycetes, Sphaeriales), but no *Venturia* stage has yet been found on *Pyracantha*.

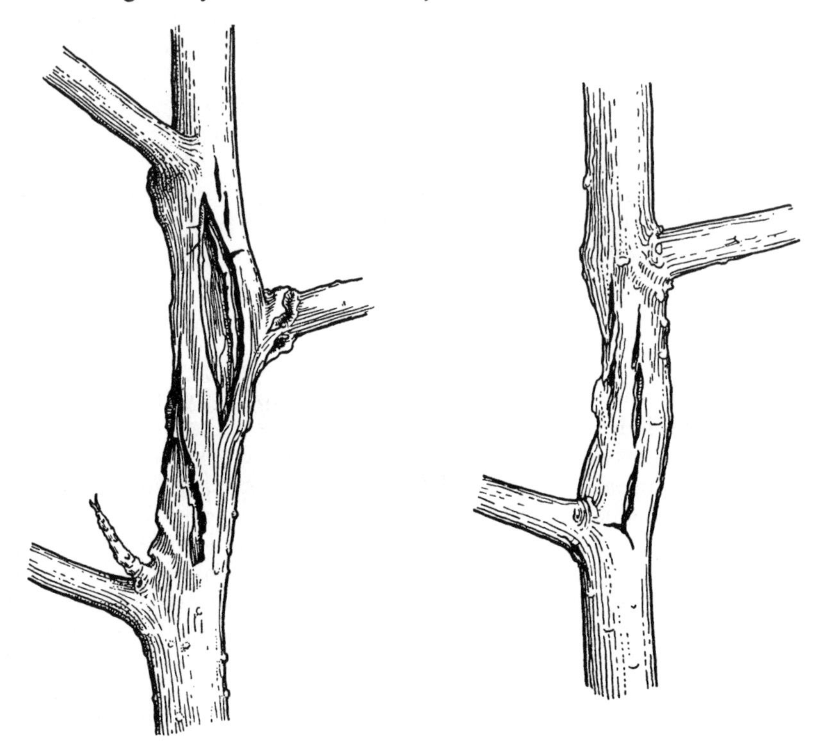

FIG. 95. *Fusicladium pyracanthae* on *Pyracantha*, Westonbirt, Gloucestershire, May (× 1⅓). (J. C.)

The fungus attacks the leaves, shoots, flowers, and fruit. Its greatest significance lies in its disfigurement of the fruit, which is the main beauty of the genus. The berries may be completely covered by the olive-brown coating of the fungus. Attack on the leaves occurs particularly at the base, or may even be confined to the petioles, and results in premature leaf fall. In addition scabby lesions appear on the twigs (Fig. 95). The fungus overwinters mainly on still living leaves infected the previous summer and also in the buds (McKay 1944; Cox 1951; Arx 1957*a*). The attack is worse in wet summers. It can be controlled by spraying with lime sulphur (McKay 1944) and by other sprays (Cox 1951, 1953*a*). Fire blight (*Erwinia amylovora*) (see *Crataegus*) has been recorded on *Pyracantha* in Britain.

PYRUS (Crab)

Most of the information available on diseases of the genus *Pyrus* refers, of course, to cultivated apples and pears, and is dealt with in other publications (Brooks 1953; Wormald 1955). The Mountain ashes, which are sometimes included in this genus, are here dealt with separately under *Sorbus*. On the other hand, the apples, often now separated as the genus *Malus*, are here brought back under *Pyrus*, largely as a matter of pathological convenience. Comparatively little information is available on the occurrence of even the principal diseases of apple and pear on wild or ornamental species of *Pyrus*.

The well-known Apple scab, *Venturia inaequalis* (Cke.) Wint. (Ascomycetes, Sphaeriales), the imperfect stage of which is *Fusicladium dendriticum* (Wallr.) Fckl. (Fungi Imperfecti, Moniliales), has been recorded on a number of ornamental crabs, including the Siberian crab, *P. baccata*, the Japanese *P. floribunda* and *P. prunifolia*, and the Chinese *P. spectabilis*. *Venturia pirina* Aderh., Pear scab, the imperfect stage of which is *F. pirinum* (Lib.) Fckl., has been found on *P. salicifolia* and other species. The spray techniques advised for use on apple, which are constantly being revised and improved, should be adaptable in the event of either of these fungi becoming seriously damaging on an ornamental species. *Sclerotinia laxa* Aderh. & Ruhl. (Ascomycetes, Helotiales), cause of blossom wilt and spur blight of apple, has been found on *P. purpurea*.

Two rust fungi, *Gymnosporangium clavariaeforme* (Pers.) DC. and *G. fuscum* DC. (Basidiomycetes, Uredinales), have been recorded in Britain on Wild pear (*P. communis*). The aecia are produced on conspicuous, orange swellings on the young shoots, leaves, and fruits of pear. The telia are borne on species of *Juniperus* (Chapter 26). Neither fungus has a uredial stage. *G. juniperi-virginianae* Schw., a North American species with similar hosts, has been recorded on *P. baccata*, *P. coronaria*, *P. ioensis*, and other American species. Crowell (1935*b*) lists the species of *Pyrus* which are resistant to this fungus, to *G. clavariaeforme*, and to other American species of *Gymnosporangium*.

Fire blight, caused by the bacterium *Erwinia amylovora* (Burr.) Winsl. *et al.*, is known to attack the American wild crab *P. coronaria* and ornamental flowering crabs (Dodge, B. O. 1936). Fire blight is particularly serious on pears. Hitherto confined to the United States and New Zealand, it has recently appeared on cultivated pears in Britain. It can attack many Rosaceous genera, including particularly *Crataegus*, under which heading it is more fully described above.

RHAMNUS (Buckthorn)

This genus is pathologically important chiefly because it is the alternate host of the Crown rust of oats, *Puccinia coronata* Corda (Basidiomycetes, Uredinales). Various other fungi have been listed by Godwin (1943) on the two native species, *R. frangula* and *R. cathartica*, but he considered that most of them were saprophytic. One of them, *Phyllosticta rhamni* Westend (Fungi Imperfecti, Sphaeropsidales), causes brown spots with a dark border on living leaves of *Rhamnus*, but does not appear to be damaging. It occurs on the Continent, and has been recorded in Britain on *R. frangula*.

The aecia of *Puccinia coronata*, which are orange with white peridia, are borne on the leaves and petioles, and less frequently on the young shoots, of Buckthorn in May and June. The aeciospores infect oats and grasses, on which the uredo- and teleutospores are later produced. In Britain, especially in the west and north, oats sometimes suffer very severely from this disease. In the warmer parts of the country the uredospores can overwinter, so that *Rhamnus* is not always essential for the perpetuation of the disease. *P. coronata* comprises a number of varieties with specific host ranges on grasses and oats, and there is evidence that these varieties are also specific as regards their occurrence on *Rhamnus* species (Dietz 1926; Brown, M. R. 1938). Other varieties of *Rhamnus*, besides our two native species, can be attacked (Dietz 1926). The fungus does no appreciable damage to Buckthorn in Britain.

RHODODENDRON (including Azalea)

This genus is subject to a large number of diseases, few of which, however, are of any great importance. The popularity of rhododendrons and azaleas as ornamental shrubs has no doubt focused more attention on their diseases than has been given to most other ornamental shrub genera.

In many forest areas, *R. ponticum* has become a weed species and drastic measures may be required to clear it for afforestation. It is unfortunate, therefore, that none of the diseases discussed below are likely to exercise any appreciable control on its growth or spread.

General reviews of the diseases of *Rhododendron* have been published in the United States by R. P. White (1933) and R. D. Raabe (1954), in Germany by Kaven (1934), and in Britain by Beaumont (1954, 1956).

Nursery diseases

Wilt of rhododendrons in the United States is caused by *Phytophthora cinnamomi* Rands. (White, R. P. 1937). It is purely a nursery disease, and is particularly common on *R. ponticum*. The wilt results, as is usual with *Phytophthora* diseases, from death of the roots. A similar disease, associated with a species of *Phytophthora*, was found by Barthelet (1934) on hybrid rhododendrons growing under damp shady conditions in a nursery in France. *P. cactorum* (Leb. & Cohn.) Schroet., on the other hand, causes large blotches on the leaves and sometimes very damaging branch cankers on hybrid rhododendrons in the United States. There are no records of *Phytophthora* on *Rhododendron* in Britain.

Rhizoctonia solani Köhn., the mycelial stage of *Corticium solani* (Prill. & Delacr.) Bourd. & Galz. (Basidiomycetes, Aphyllophorales), has been found causing Damping off and death of foliage on young greenhouse azaleas in Britain (Storey 1955) and in America (Raabe, R. D. 1954).

Root diseases

Armillaria mellea (Valh.) Quél. (Basidiomycetes, Agaricales) (Chapter 16) is frequently recorded on rhododendrons. There is no reason to suppose that the genus is specially susceptible. Probably it is due to the frequency with which rhododendrons are planted in heavily thinned or partially cleared woodland, where there are many stumps and root remains to act as centres of infection.

Stem diseases—bark and cambium

Gram (Gram and Weber 1952) has described wilting of greenhouse azaleas in Denmark, which he attributed to the common soil-fungus *Cylindrocarpon radicicola* Wollenw. (Fungi Imperfecti, Moniliales). The foliage wilted and the main stem near soil-level turned brown. Similar symptoms have been observed in Belgium and in Britain, but the causal agent was not isolated.

Diseases of leaves, shoots, buds, and flowers

Bud blast

This disease is caused by *Pycnostysanus azaleae* (Peck.) Mason (Fungi Imperfecti, Moniliales) (syn. *Sporocybe azaleae* (Peck.) Sacc.). In Britain the

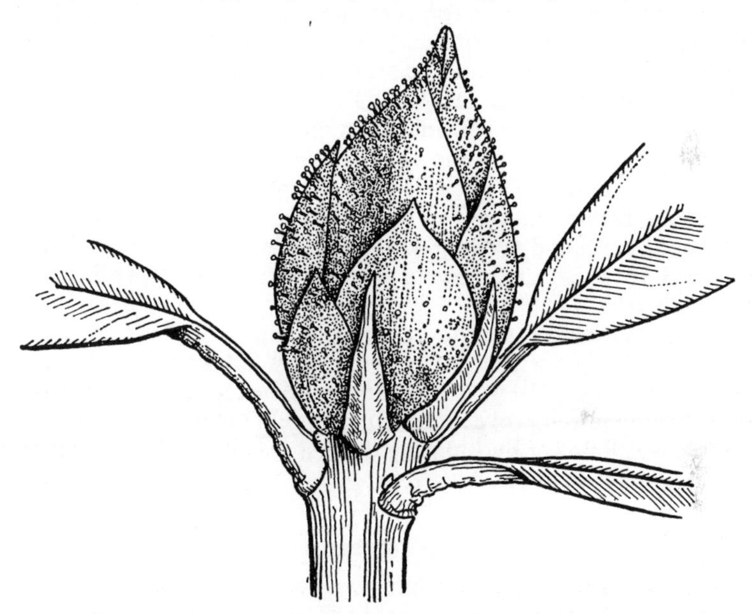

FIG. 96. Bud blast, *Pycnostysanus azaleae*, on a hybrid rhododendron, Tilford, Surrey (×2). (J. C.)

disease has attracted attention only in south-east England on evergreen rhododendrons, particularly on some of the hybrids. The fungus infects the developing buds in the summer, grows in them during the winter, and prevents the development of flowers in the spring. At this time the buds turn first greyish or brownish and then black as the small pin-like coremia of the fungus are formed (Fig. 96). Infection on some varieties can be so heavy as to practically prevent flowering, whereas other varieties in the immediate vicinity are unaffected (Baillie and Jepson 1951). The fungus is probably carried by a leaf-hopper, *Graphocephala coccinea* Farst. In America the fungus causes dieback of twigs as well as death of buds on azaleas (Davis, W. H. 1939), but seems, as in Britain, to attack buds only on evergreen rhododendrons. Various suggestions for control have been put forward. Street (1950) recommended lessening of overhead cover and heavy pruning of affected bushes. In addition he suggested insecticidal spraying, a course also proposed by Baillie and Jepson (1951), who found, however, that fungicidal sprays of Bordeaux mixture applied fortnightly from mid-June to mid-October gave very good control. Removal of dead buds in the early spring, before the fungus has fruited, has also been suggested, but they are not always easy to differentiate, and the task is hardly possible on large bushes. Further work on combined sprays directed against the insect and the fungus appears desirable, together with more exact information than is yet available, on resistant varieties.

Petal blight

This disease, caused by *Ovulinia* (*Sclerotinia*) *azaleae* Weiss (Ascomycetes, Helotiales), has been known for some years, mainly in the south-eastern United States and in California, attacking the petals of cultivated azaleas and rhododendrons (Raabe and Sciarone 1955). The disease has also been described in Switzerland (Terrier 1950). The fungus has recently been recorded in south-west Scotland and at Dunkeld on rhododendrons and azaleas (Paton 1954). Small round spots on the petals rapidly develop into large irregular brown blotches, until finally the flowers collapse. Infected buds may fail to open. Conidia formed in the diseased flowers can be transmitted to healthy ones by bees (Smith and Weiss 1942). Small, black, disc-shaped sclerotia are later formed in the infected flowers and fall with these to the ground where they overwinter.

Control by spraying is said to be possible, but difficult (Riley and Daigle 1948). Removal of infected flowers before they fall would be valuable, and heavy mulching during the winter to bury the sclerotia and prevent sporulation in the spring might well prove successful. It would obviously be a wise precaution to remove all flower-buds during the first season on plants received from an area known to be infected.

Other diseases of leaves and shoots

Hypertrophy of the leaves and sometimes of the buds and flowers of greenhouse azaleas and some species of hardy rhododendrons is caused by one or more species of *Exobasidium* (Basidiomycetes, Exobasidiales). In Britain the disease is attributed to *E. vaccinii* (Fckl.) Woron., which occurs commonly on wild species of *Vaccinium*. The galls, which are formed on the rhododendron leaves, buds, and flowers, and which are sometimes very irregular in shape, vary in size from that of a pea to a small plum (Fig. 97). When young they may be reddish in colour, but later, when the spores are formed, they become covered with a white bloom. They cause general interference with growth and reduce flowering (Chittenden 1908). It is usually possible to control the disease by the removal of the affected organs. Others have separated the fungi on *Rhododendron* from that on *Vaccinium*, and have described the one on greenhouse azaleas under the name *E. japonicum* Shir., and the one on hardy rhododendrons, which is known to attack *R. ferrugineum*, *R. hirsutum*, and a few other species, as *E. rhododendri* Cram. Savile (1959) considers that more than two species are involved.

The most serious disease of forced greenhouse azaleas is a leafscorch caused by *Septoria azaleae* Vogl. (Fungi imperfecti, Sphaeropsidales). Dark-brown patches, on which the small black pycnidia are produced, are formed on the leaves. Severe defoliation often occurs. There is a considerable importation into Britain from the Continent of plants for forcing, and it seems likely that

FIG. 97. *Exobasidium vaccinii* on a greenhouse azalea, Farnham, Surrey (×2). (J. N.)

these are sometimes infected before dispatch (Salmon and Ware 1927). Some varieties are much more susceptible than others (Pape 1955). So far the disease has been reported only occasionally in Britain, but it is becoming more common (Moore, W. C. 1959). Regular application of a copper spray from mid-summer onwards has been suggested as the best means of control. This is considered more effective than attempts to control the fungus after active damage has started the following spring. Other fungicides may be more satisfactory, however, since azaleas are occasionally damaged by copper sprays. Various proprietary substances have been used with success (Raabe and Lang 1958).

Several species of *Phyllosticta* (Fungi Imperfecti, Sphaeropsidales) cause leaf-spots on rhododendrons. *P. rhododendri* Westend. has attracted most attention in Britain, causing rusty brown spots mostly on the margins of the leaves. It is of no great pathological importance. *Septoria rhododendri* Cke. has been recorded causing leaf-spot and subsequent leaf-fall on *R. ponticum* in Cornwall, but it appears to be quite unimportant.

Commoner in Britain are large, irregular leaf-spots caused by *Gloeosporium rhododendri* Bri. & Cav. (Fungi Imperfecti, Melanconiales). These occur on other species besides *R. ponticum*. Arx (1957*c*) considers this fungus to be the same as *G. fructigenum* (syn. *Colletotrichum gloeosporioides*), the conidial stage of *Glomerella cingulata* (Ascomycetes, Sphaeriales), the cause of Bitter rot of apples and of many other diseases of woody plants. Stathis and Plakidas (1958) have recently described an anthracnose of 'Indian' and Kurume azaleas in Louisiana, associated with a species of *Colletotrichum* 'typical of the conidial stage of *Glomerella cingulata*'. In this disease the leaf-spots, which are so numerous as to cause severe defoliation, are very small (0·5–3·0 mm. in diameter). Both copper and organic fungicides were found effective in controlling the disease.

Pestalozzia guepini Desm. (Fungi Imperfecti, Melanconiales), better known as a cause of leaf blotch of *Camellia*, has also been found on *Rhododendron* leaves, while *P. macrotricha* Kleb. and *P. rhododendri* Guba. have been recorded, not only as causes of leaf-spot, but also of stem lesions on a number of species. The diseases caused by those species of *Pestalozzia* are characterized by so profuse a production of fruit-bodies and spores that the surface of the lesions is blackened. *P. rhododendri* has not been recorded in Britain. R. P. White (1930), however, considered that these fungi were only comparatively weak wound parasites, a view supported by Howarth and Chippindale (1931), who found *P. macrotricha* and *Diplodia eurhododendri* Voss. (Fungi Imperfecti, Sphaeropsidales) associated with graft failures in young Pink Pearl rhododendrons. They considered, however, that the primary cause was faulty grafting technique and the use of unsuitable stocks. *D. eurhododendri* has been found as a cause of leaf-spotting in Britain.

Several other fungi have been recorded both in Britain and abroad on dead spots on rhododendron leaves. Probably they are secondary to insect and climatic injury, having time for development because the leaves of evergreen rhododendrons persist on the bush even when quite severely damaged.

The rust *Chrysomyxa rhododendri* de Bary (Basidiomycetes, Uredinales) is more important in Britain as a cause of disease on its alternate host, spruce, than on *Rhododendron* (Chapter 23). The chief rhododendron hosts in Europe are *R. hirsutum* and *R. ferrugineum*, but the rust occurs on a number of other species, and has been recorded on *R. ponticum* (Wilson and Bisby 1954). Infection of rhododendrons takes place in late summer from the aeciospores liberated from spruces. Brown patches appear on the leaves, and yellow pustular uredia appear on their underside. The teliospores in their turn reinfect spruce. At the moment this disease seems unlikely to be dangerous on *Rhododendron* in Britain.

RHUS (Sumach)

In Britain the Stag's horn sumach, *R. typhina*, is often damaged by spring frost, but the genus as a whole is fairly free from diseases. In the United States, however, *R. typhina* is subject to serious wilts caused by a form of *Fusarium oxysporum* Schlecht (Fungi Imperfecti, Moniliales) (Toole 1949), and by *Botryosphaeria ribis* Gross. & Dugg. (Ascomycetes, Sphaeriales) (Bragonier 1949). *F. oxysporum* occurs in Britain on potatoes and other hosts, but presumably the form on *Rhus* is absent. *B. ribis* is confined to America.

Endothia parasitica, the cause of Chestnut blight (Chapter 29), has been recorded as a weak parasite on *R. typhina* in America.

RIBES (Currant)

The genus *Ribes* in Britain is important as a fruit crop in the form of currants and gooseberries. Its diseases are therefore covered by books on fruit-tree diseases (Wormald 1955). Comparatively little information is available on these diseases on flowering currants. Some of them do occur on ornamental and wild species, but normally without serious effects.

American gooseberry mildew, *Sphaerotheca mors-uvae* (Schw.) Berk. (Ascomycetes, Erysiphales), is often serious on gooseberries and is occasionally damaging to Red and Black currants. A typical, white, mycelial bloom is found on the leaves. Sulphur sprays give satisfactory control. The mildew can attack a number of wild and planted species of *Ribes* found in Britain, though most of such records are American or Continental. Susceptible species include *R. alpinum*, *R. aureum*, *R. cereum*, *R. fasciculatum chinense*, and *R. lacustre*.

Leaf-spot of currants caused by *Mycosphaerella ribis* (Fckl.) Kleb. (Ascomycetes, Sphaeriales), the conidial stage of which is *Septoria ribis* Desm. (Fungi Imperfecti, Sphaeropsidales), is occasionally damaging on Black currants in Britain. It has been reported in America on a number of wild species (Stone, R. E. 1916). The fungus causes angular spots with pale centres on which the small black pycnidia of the *Septoria* stage are borne.

A more serious leaf disease of Black currants is caused by *Pseudopeziza ribis* Kleb. (Ascomycetes, Helotiales), the conidial stage of which is *Gloeosporium ribis* (Lib.) Mont. & Desm. (Fungi Imperfecti, Melanconiales). It has been recorded on a number of wild species, including *P. aureum*, in the United States. The fungus causes small brown spots, initially about 2 mm. across, on the leaves. Severe defoliation sometimes occurs. Bordeaux mixture applied in May and June gives effective control.

It is known that some of the flowering currants, including the commonly planted *R. sanguineum*, carry the uredial and telial stages of *Cronartium ribicola* Fisch. (Basidiomycetes, Uredinales), the serious Blister rust of Five-needled pines (Chapter 22). Details of the susceptibility of different *Ribes* species and varieties have been given by Schellenberg (1923) and Kimmey (1938). In Britain the flowering currants are not nearly so frequently or heavily infected as the Black currant, *R. nigrum*. Nevertheless, they should not be planted in the vicinity of Five-needled pines.

Two British rust fungi on willows have aecial stages on *Ribes* species, but in the case of *Melampsora ribesii-purpureae* Kleb. (Basidiomycetes, Uredinales) the British records on *Ribes* are doubtful (Wilson and Bisby 1954), though it has been found abroad on *R. aureum*, *R. alpinum*, and other species. The aecial stage of *M. ribesii-viminalis* Kleb. has not been found in Britain, but it also has been noted abroad on *R. aureum*, *R. alpinum*, and other species. The aecial stage of another rust, *Puccinia pringsheimiana* Kleb., the uredial and telial hosts of which are *Carex* spp. (sedges), has been recorded on *R. sanguineum* in Britain and on *R. aureum* in Denmark. It is of no importance on ornamental species, but can be damaging to gooseberries, forming red or orange blotches on the leaves and fruits, on which the small cup-like fruit-bodies are produced in clusters.

ROBINIA (False acacia)

R. pseudacacia was planted on a moderate scale as a forest tree in the early nineteenth century, but was generally a failure. Its growth in Britain, compared with that on the Continent and in its native North America, suggests that it is not really suited to our oceanic climate. It is nevertheless still commonly planted in parks and gardens or as street trees. Several other species of *Robinia* are met with occasionally as garden shrubs or small trees.

Robinia is relatively free from diseases in Britain. A mildew, *Erysiphe polygoni* DC. (Ascomycetes, Erysiphales), has been recorded occasionally, but is unimportant. Brown leaf-spots attributed to *Phleospora robiniae* (Lib.) Hoehn. (Fungi Imperfecti, Sphaeropsidales) have been recorded in Britain as well as on the Continent and in America. *Septoria curvata* (Rab. & Braun) Sacc., and *Ascochyta robiniae* Sacc. & Speg. are probably identical.

Twig and branch dieback of *Robinia* on the Continent has been associated with *Diaporthe oncostoma* (Duby) Fckl. (Ascomycetes, Sphaeriales) (Arnaud and Barthelet 1933). The fungus is frequently found as the pycnidial stage, usually referred to as *Phomopsis oncostoma* (Thüm.) v. Höhn. (Fungi Imperfecti, Sphaeropsidales). Ribaldi (1954), however, regarded it only as a weak parasite favoured by adverse conditions. Wehmeyer (1933) reported that the fungus is common in Britain wherever the host tree is found; but there is no suggestion that it is the cause of death of the twigs and branches on which it occurs. Dieback and leaf-spot have also been associated on the Continent with a number of species of *Fusarium* (Spaulding 1956).

Wilts and root dieback caused by species of *Phytophthora* (Phycomycetes, Peronosporales) have been reported from the Continent as well as from North America (Lambert and Crandall 1936).

Brooming, a quite serious virus disease, occurs in North America and on the Continent (Chapter 12).

ROSA

Detailed accounts of diseases are included in many books on rose cultivation (Harvey, N. P. 1953; Mansfield 1953; Park 1956; Wright *et al.* 1957). There is therefore no need to deal with them here.

RUBUS (Raspberry and Blackberry)

This genus has naturally attracted most attention as a fruit crop, and its diseases are therefore dealt with in books covering fruit diseases (Brooks 1953; Wormald 1955). The wild blackberry is in places a serious forest weed, but all the pathological records refer to the cultivated forms derived from it. The purely ornamental varieties of *Rubus* are not sufficiently important for any specific information to be available on their diseases. It can be assumed that most of the diseases of cultivated blackberries and raspberries are present on their wild counterparts, since the wild and cultivated species are essentially the same. The extent of attacks on the wild forms is unknown; but certainly no disease exercises any controlling influence on the growth or spread of the wild forms. Both blackberry and raspberry, particularly the latter, are very subject to virus diseases (Chapter 12).

SAMBUCUS (Elder)

By far the most familiar fungus on this host is *Auricularia auricula-judae* (L. ex Fr.) Schröt. (Basidiomycetes, Auriculariales). The gelatinous, brown, ear-like fruit-bodies occur frequently on dead stems of the Common elder, *Sambucus nigra*, sometimes in enormous quantities, particularly on bushes shaded-out by the canopy closure of trees above them. Generally the fungus is a saprophyte, but its behaviour is occasionally suggestive of weak parasitism, and successful inoculation experiments have been carried out (Le Goc 1914; Banerjee 1957).

Leaf-spots due to *Cercospora depazeoides* (Desm.) Sacc. (Fungi Imperfecti, Moniliales) have been recorded in several places in south-east England. The spots, which are 2 to 5 mm. across, are dull grey in colour, turning whitish with age. Attacks can result in almost complete defoliation (Moore, W. C. 1946). Several other fungi, including *Phyllosticta sambucicola* Kalchbr. (Fungi Imperfecti, Sphaeropsidales) and *Ramularia sambucina* Sacc. (Fungi Imperfecti, Moniliales), have been associated with leaf-spots on elder. Both these fungi occur in Britain, but nothing is known of their importance.

SORBUS (Mountain Ash)

The Mountain ash and Whitebeam have few important diseases. Probably the commonest disease is caused by the rust fungus, *Gymnosporangium juniperi* Link. (Basidiomycetes, Uredinales), which has its aecial stage on Mountain ash, *S. aucuparia*, and its telial stage on *Juniperus communis*. It occurs fairly widely in Scotland, but is rare in England. The orange aecia, which are horn-shaped and about 2 mm. long, are borne in clusters on the leaves in late summer. These have given the disease its name of 'Cluster-cups'. Abroad, particularly in North America, there are a number of other species of *Gymnosporangium* with aecial stages on various species of *Sorbus*, and telial stages on junipers. Another rust, *Ochropsora ariae* (Fckl.) Ramsb. with the aecial stage on *Anemone* and the uredial and telia stages on *S. aucuparia* and other species of *Sorbus*, is fairly common in Britain on the aecial host, but has not yet been found on Mountain ash.

Eutypella sorbi Sacc. (Ascomycetes, Sphaeriales), the conidial stage of which is *Cytospora rubescens* Tul. *f. sorbi* Sacc. (Fungi Imperfecti, Sphaeropsidales), has been associated with dieback of *S. aucuparia* in Scotland. It has also been recorded in England, but possibly only as a saprophyte. It occurs frequently on dead bark, but its pathogenicity is rather doubtful. The spores of the pycnidial stage emerge as dark-red tendrils. On the Continent *Valsa leucostoma* (Pers. ex Fr.) Fr. (Ascomycetes, Sphaeriales), the pycnidial stage of which is *Cytospora leucostoma* Sacc., has been recorded on *S. aria* and *S. aucuparia*. This fungus occurs commonly on *Prunus* spp. in Britain, though it is doubtful if it is a parasite, and has also been recorded here on *S. aucuparia*. *C. chrysosperma* (Pers.) Fr., which is more important on poplar, and *C. massariana* Sacc. sometimes occur as apparent parasites on species of *Sorbus*, especially on trees weakened by other causes, both in the United States and on the Continent.

The mildew *Podosphaera oxycanthae* (DC.) de Bary (see *Crataegus* above) can attack *Sorbus* species, but does not seem damaging on them in Britain.

Venturia aucupariae (Lasch) Rostr. (Ascomycetes, Sphaeriales), which is closely related to *V. inaequalis* (Cke.) Wint. the cause of Apple scab, can attack various species of *Sorbus*, including *S. aucuparia*, *S. aria* (Whitebeam), and *S. torminalis*, our three principal native species. The attack is mainly on the leaves, though also on young twigs, and may result in appreciable defoliation. The disease has not, however, attracted much attention in Britain, though the fungus has been recorded here on *S. aucuparia*.

Leaf-spots are caused in Britain by a number of fungi, including *Septoria sorbi* Lasch (Fungi Imperfecti, Sphaeropsidales), but they generally occur on fading leaves and are of no pathological importance.

In America *Sorbus* spp. are sometimes severely attacked by Fire blight, caused by the bacterium *Erwinia amylovora* (Burr.) Winsl. *et al.* (see *Crataegus* above). This disease has recently been found on *Sorbus* in Britain (Glasscock unpubl.), where it is locally serious on pears.

STAPHYLEA (Bladdernut)

Some members of this genus are planted in gardens in Britain, but no diseases have been reported. In North America, where it is native, two dieback diseases have been reported on *S. trifolia*. One was caused by a variety of *Coryneum microstictum* (Berk. & Br.) (Fungi Imperfecti, Melanconiales). It was recorded by W. H. Davis (1931), who carried out successful inoculation experiments. The fungus attacked mainly the current year's growth, but was capable of killing small shrubs outright. Some forms of the ubiquitous *C. microstictum* are present in Britain, including one which causes quite serious canker and dieback of cultivated roses, but the fungus has not been recorded in Britain on *Staphylea*. Very similar damage was done by *Hypomyces ipomoeae* (Halst.) Wr. (syn. *Nectria ipomoeae* Halst.) (Ascomycetes, Hypocreales) (Davis, W. H. 1934). This is generally a fungus of the tropics and sub-tropics and does not occur in Britain.

SYMPHORICARPUS (Snowberry)

Ascochytula symphoricarpi (Pass.) Poteb. (Fungi Imperfecti, Sphaeropsidales) has been associated with stem dieback of *S. albus* in Britain. A closely related, or possibly identical, fungus, *Ascochyta vulgaris* Kab. & Bub. var. *symphoricarpi* Grove, causes large roundish brown spots on the leaves.

SYRINGA (Lilac)

Root diseases

Lilac is particularly prone to attack by Honey fungus, *Armillaria mellea* (Fr.) Quél. (Basidiomycetes, Agaricales) (Chapter 16), though bushes are often only weakened, not killed, by its attacks.

Stem diseases—wilts

Wilt caused by *Verticillium albo-atrum* Reinke and Barth (Fungi Imperfecti, Moniliales), which is particularly important on *Acer* (Chapter 30), is also common on lilac.

When grafted on privet, no longer a common practice, lilac is liable to suffer eventually from graft-incompatibility and as a result to die back (Chapter 10).

Diseases of leaves and shoots

The most serious disease of lilac is Bacterial blight caused by *Pseudomonas syringae* v. Hall. The first signs of the disease are usually wilting and blackening of the young shoots and inflorescences. Older twigs may also be attacked and girdled. Small angular brown spots appear on the leaves (Fig. 98), which may also be crinkled at the edges or in the region of the mid-rib; but the foliar symptoms are, however, less obvious than the shoot and flower dieback. Later, elongated cankers develop on attacked branches which have not been girdled. The disease is widely distributed and occasionally serious, having been described in the United States by Bryan (1928), on the Continent by Laubert (1927) and Pape (1928), and in Britain by Wormald (1932). Some control can be exercised by cutting out affected branches, but the disease is very hard to eradicate. Spraying with Bordeaux mixture, as soon as the disease appears, has been recommended (Dodge and Rickett 1943).

FIG. 98. *Pseudomonas syringae* attacking a leaf of lilac, Guernsey, August (× ⅔). (J. C.)

Rather similar damage is caused by *Phytophthora syringae* (Kleb.) Kleb. (Phycomycetes, Peronosporales), more familiar as a cause of root disease, particularly on *Castanea*. The symptoms tend to appear earlier than those of Bacterial blight, buds often being killed before they have flushed. The disease, which is much commoner on forced lilacs than in the garden, has been fully described by De Bruyn (1924). Stem infections often take place from diseased leaves in the autumn, so that artificial premature defoliation was suggested as a possible control. This cannot be done with safety, however, since it interferes with growth and flowering the following spring (De Bruyn 1928). Infection also takes place direct from the soil, and especially if the plants are heeled-in with their branches near the ground or in contact with it (Chester 1932). Where the disease is troublesome, use of a sterile mulch under the plants while they are in the open ground might be considered as a protection against infection from the soil. There are said to be large differences in susceptibility between horticultural varieties (Gram and Weber 1952).

Ascochyta syringae Bres. (Fungi Imperfecti, Sphaeropsidales) has been associated on the Continent with a less severe shoot dieback. In light attacks only the leaves are infected (Curzi 1927). A species of *Ascochyta* has been recorded on lilac in Britain, but it has not been definitely identified as *A. syringae*.

Botrytis cinerea Fr. (Fungi Imperfecti, Moniliales) (Chapter 15) is also associated with dieback of shoots and blossoms in some seasons. It is probably not always a primary parasite, and may follow previous frost injury.

Heterosporium syringae Oudem. (Fungi Imperfecti, Moniliales) causes large, irregular, grey-brown blotches on lilac leaves. Small, blackish, tufted fruit-bodies are borne on the diseased areas (Massee 1911). It is fairly common, but not serious. Another leaf-spot, caused by *Phyllosticta syringae* Westend. (Fungi Imperfecti, Sphaeropsidales), can be distinguished from that due to *Heterosporium* because the large pale spots have a brown border. Small black pycnidia occur on the upper surface of infected leaves.

Mildew caused by a species of *Oidium* (Fungi Imperfecti, Moniliales) occurs fairly commonly on lilac in Britain, but is seldom damaging. It may be the imperfect stage of *Microsphaera alni* (Wallr.) Wint. (Ascomycetes, Erysiphales), which causes mildew of lilac in North America, where Crowell (1937) has described in detail the susceptibility of lilac species and varieties to this disease. If necessary, application of sulphur should control this disease.

Virus diseases of lilac, which are of some importance, are dealt with in Chapter 12.

TAMARIX

Pathologically the most significant feature of *Tamarix* is its extreme resistance to salt winds (Chapter 8). In Britain, where it is not native, it has not yet acquired any significant diseases. In the arid regions of the southern United States a bacterium, as yet unnamed, attacks and kills the foliage of *T. pentandra*, which was introduced from Europe and has become an obnoxious weed. The bacterial attack results in the death of the bushes, and has therefore been welcomed (Brown and Scandone 1953). In the Argentine *Botryosphaeria tamaricis* (Cke.) Theiss. & Syd. (Ascomycetes, Sphaeriales), a fungus occurring in Europe but not in Britain, attacked and destroyed a young planting of *T. gallica* (Frezzi 1942). This fungus has not been recorded as a cause of serious damage in Europe.

ULEX (Gorse)

The Common gorse, *U. europaeus*, is often severely damaged in cold winters. This in no way affects its ability to survive; but the resultant dead tops make it one of our most dangerous fire hazards.

The rust, *Uromyces laburni* (DC.) Fckl. (Basidiomycetes, Uredinales), which also occurs on *Cytisus* and other leguminous trees and shrubs, produces its uredia and telia on *U. europaeus*. No aecial stage has been found in Britain. It is of no practical importance. The parasitic flowering plant, *Cuscuta epithymum*, Common Dodder (Chapter 13), is not uncommon on gorse, but does little damage (Fig. 34).

UMBELLULARIA (Californian Laurel)

U. californica, the only species of the genus, is occasionally planted in the milder parts of Britain for ornament. So far it has remained disease-free in Britain, but a number of pathogens have been recorded in its native California, where it is widely planted as a street tree. Death of leaves, dieback of shoots, and the production of small branch cankers have been attributed to a species of *Macrophoma* (Fungi Imperfecti, Sphaeropsidales) (Barrett, J. T. 1948). A form of *Kabatiella phorodendri* (Darling) Harvey (Fungi Imperfecti, Moniliales) was found to be responsible for brown necrotic patches on the leaves, leading to defoliation (Harvey, J. M. 1951). Smaller angular black spots on the leaves are caused by a bacterium, *Pseudomonas lauracearum* Harvey (Harvey 1952). All these diseases are locally damaging, but there is no suggestion that any of them is epidemic.

VACCINIUM

Species of *Vaccinium*, in particular *V. myrtillus* the Bilberry, are often a dominant constituent of acid vegetation on forest soils. A few are grown in gardens in Britain, mainly for their autumn colour. Some have edible fruits, and in the United States horticultural varieties of several species are cultivated. As a result of this commercial interest, several papers on diseases of *Vaccinium* have been published in the United States (Shear, Stevens, and Bain 1931; Demaree and Wilcox 1947). The diseases include a considerable number of leaf-spotting fungi. Some species are now separated into the genus *Oxycoccus*, but they are here referred to under the old name *Vaccinium*.

Distortion of the leaves and inflorescences of wild species is caused by *Exobasidium vaccinii* (Fckl.) Woron. (Basidiomycetes, Auriculariales). This fungus also attacks greenhouse azaleas and is described under diseases of *Rhododendron* (above). It has been found on *V. vitis-idaea*, *V. myrtillus*, *V. uliginosum*, and *V. oxycoccus*, among the British wild species. Spraying has proved ineffective as a means of control: the only satisfactory method being the destruction of diseased bushes (Lockhart 1958). Other species of *Exobasidium* occur on *Vaccinium* on the Continent and in North America (Zeller 1934; Savile 1959).

Swellings on the stems, which lack the white bloom associated with those caused by *Exobasidium*, are produced by the rust fungus *Calyptospora goeppertiana* Kühn. (Basidiomycetes, Uredinales). The aecia are borne on species of *Abies*, and the telia on *V. vitis-idaea*. The infected branches of *Vaccinium* stand very erect, and are longer and thicker than normal. The disease is common on the Continent, but in Britain there is doubt about the few records that do exist. The rust can attack other species of *Vaccinium*, and occurs on a wide range of *Abies* and *Vaccinium* species in North America (Faull 1939).

Another rust, *Thekopsora vacciniorum* Karst. (syn. *Pucciniastrum myrtilli* Arth.) produces its uredial stage as orange spots on the leaves of *V. vitis-idaea*, *V. myrtillus*, *V. uliginosum*, and *V. oxycoccus*. In Britain the uredial stage is widespread but not common. There are doubtful records of the telial stage. The fungus occurs on the Continent and in North America, where it is found on a wide range of *Vaccinium* species and can cause serious defoliation on cultivated crops. The aecial host there is *Tsuga canadensis*, but, as would be expected from the behaviour of the disease in Britain, *Tsuga* is not necessary for the continued existence of the fungus on *Vaccinium*.

Several species of *Sclerotinia* (Ascomycetes, Helotiales) are found on *Vaccinium*. None, however, has been recorded in Britain. The most interesting is the heteroecious species *S. heteroica* Woron. & Nawas., which alternates between the leaves of *Vaccinium* and the flowers of *Ledum* (see *Ledum* above).

Phomopsis vaccinii Shear *et al.* (Fungi Imperfecti, Sphaeropsidales) causes shoot dieback of *V. corymbosum* in the United States (Wilcox 1939). A species of *Phomopsis* has been associated with similar damage on American hybrid blueberries in Scotland (Foister 1961). Shoot dieback and death of young plants of cultivated blueberries (*V. corymbosum*) is serious in British Columbia and Nova Scotia. The causal fungus is *Godronia cassandrae* Peck (Ascomycetes, Helotiales) (McKeen, W. E. 1958). Creelman (1958*a*), dealing with the disease in Nova Scotia, gave the imperfect stage of the fungus as *Fusicoccum putrifaciens* Shear, already known as a cause of berry-rot in cranberries. Nothing is known of the occurrence of this fungus on wild species; it has not been recorded in Britain.

Botrytis cinerea Pers. ex Fr. (Fungi Imperfecti, Moniliales) causes death of the blossoms and twigs of cultivated blueberries in the United States (Pelletier and Hilborn 1954). A species of *Botrytis* has been found causing twig dieback on American blueberries in Scotland (Foister 1961).

Leaf-spotting of cultivated *Vaccinium* in North Carolina has been associated with *Gloeosporium minus* Shear (Fungi Imperfecti, Melanconiales) and *Dothichiza caroliniana* Dem. & Wilcox (Fungi Imperfecti, Sphaeropsidales). *G. minus* causes distortion of the leaves and small lesions on the stems as well as leaf-spots. *D. caroliniana* attacks only the leaves, but can cause severe defoliation (Taylor and Clayton 1959). Neither has been recorded in Britain.

VERONICA (including Hebe)

Leaf-spot caused by *Septoria exotica* Speg. (Fungi Imperfecti, Sphaeropsidales) is not uncommon in south-west England and Wales on shrubby species of *Veronica*. The spots are conspicuous, being whitish with a regular brown margin. Mildew caused by *Peronospora grisea* Unger (Phycomycetes,

Peronosporales) occurs in the south-west of England. It has been definitely recorded on *V. hulkeana*, but most probably occurs on other shrubby species. Greyish mycelium develops on the underside of pale blotches on the leaves.

VIBURNUM

Phyllosticta tinea Sacc. (Fungi Imperfecti, Sphaeropsidales) occasionally causes pale spots on the leaves of *V. tinus* in Britain. *Phyllosticta opuli* Sacc. causes similar damage to the leaves of native *V. opulus*. On the Continent *Ascochyta tini* Sacc. (Fungi Imperfecti, Sphaeropsidales) causes spots, which eventually fall out giving a shot-hole appearance, on the leaves of *V. tinus* and *V. davidii* (Nicolas and Aggéry 1931). None of these diseases is at all important.

VITIS (Vine, Virginia Creeper)

For convenience, the garden climbers once placed in the genus *Ampelopsis*, and more recently in *Parthenocissus*, are here treated as species of *Vitis*. The genus is, of course, primarily of importance for fruit, for which purpose a large number of species and varieties of *V. vinifera* are grown. The vine pathogens are therefore dealt with in books on fruit diseases or on the cultivation of vines (Hyams 1952; Wormald 1955). It can be assumed that most of the principal diseases attacking commercial vines grown in the open are also found on the true garden vines, such as *V. coignetiae* or *V. vinifera purpurea*. More specific information is available about some of the diseases of the popular climbers *V. inconstans* and *V. vitacea*.

Possibly the most serious disease of vines is Downy mildew caused by *Plasmopara viticola* (Berk. & Curt.) Berl. & de Toni (Phycomycetes, Peronosporales). Yellowish or brownish spots appear on the lower leaves, and from the underside of these the conidia develop as a delicate white down. Severe defoliation often results. The fungus was introduced into Europe from the United States about 1878, and is now periodically serious in most of the main vine-growing areas. In the past it was rare in Britain, but has been recorded more frequently of recent years, being found on *V. coignetiae* and *V. vinifera purpurea* as well as on Grape vines (Moore, W. C. 1959). Details of the susceptibility of these and various other species are given by Viennot-Bourgin (1949). *V. inconstans* is only slightly attacked. The disease can be controlled by systematic spraying with a copper fungicide; indeed it was to combat this fungus in France that Bordeaux mixture was first used.

Powdery mildew, *Uncinula necator* (Schw.) Burr. (Ascomycetes, Erysiphales), is commoner than Downy mildew in Britain. Typical white powdery patches appear on the leaves, young shoots, flowers, and developing fruit. The perithecial stage is rare in Europe and the fungus survives the winter in the form of mycelium in the buds or young shoots. It occurs on other species of *Vitis* besides *V. vinifera*, but does not appear to have been recorded on any of the important ornamental varieties. It can be controlled by sulphur sprays.

Guignardia bidwellii (Ellis) Viala & Ravaz (Ascomycetes, Sphaeriales) and *Elsinoë ampelina* Shear (Ascomycetes, Myrangiales) attack the fruit, leaves, and young stems of vines. The conidial stages of these two fungi have often been confused. *Gloeosporium ampelophagum* Sacc. (*Sphaceloma ampelinum* de Bary) (Fungi Imperfecti, Melanconiales), the conidial stage of the latter, has in the past been erroneously associated with the former, the conidial stage of which is really *Phoma uvicola* Berk. & Curt. (Fungi imperfecti, Sphaeropsidales). This has led to considerable confusion between the two diseases. It is probable that only *Elsinoë* is present in Britain and then only in its conidial stage. This fungus is not at all serious in Britain, where it causes leaf-spots, which are greyish but surrounded by a dark border; eventually they fall out leaving shot-holes. Both diseases are quite damaging on vines on the Continent and in America. An *Elsinoë*, described as *E. parthenocissi* Jenk. & Bitan., has been recorded causing severe withering of the leaves, shoots, and fruit of the true Virginia creeper, *V. quinquefolia*, in the United States (Jenkins and Bitancourt 1942). A form of *Guignardia bidwellii* also attacks *V. quinquefolia* and *V. inconstans* in the United States, causing considerable damage to the leaves (Luttrell 1948).

Red scorch of vine leaves, caused by *Pseudopeziza tracheiphila* Müller-Thurgau (Ascomycetes, Helotiales), occurs on the Continent, causing lesions which are mainly on the edges of the leaves. The colour of the lesions varies with the variety of the host. *P. tracheiphila* also attacks *V. vitacea* and *V. inconstans*, but has not been recorded in Britain.

WISTARIA

Profuse spotting of wistaria leaves has been observed in Britain, the damage being sufficient to disfigure the foliage in late summer, but little defoliation resulted (Batko unpubl.). The causal fungus appears to be *Phyllosticta wistariae* Sacc. (Fungi Imperfecti, Sphaeropsidales), which has also been recorded on the Continent.

YUCCA

A leaf-spot is commonly caused by *Coniothyrium concentricum* (Desm.) Sacc. (Fungi Imperfecti, Sphaeropsidales). The large spots are light brown in the centre and have purplish margins, the small black pycnidia being produced on them in more or less concentric circles. The disease is said to be controllable with copper sprays (Dodge and Rickett 1943). It occurs in Britain on most of the varieties grown, but no information is available on varietal resistance or on the amount of damage done.

ZELKOVA

Z. serrata was found by Wollenweber (1931*b*) to be highly susceptible to inoculation with *Ceratocystis ulmi*, the cause of Elm disease (Chapter 31). *Zelkova* is closely related to *Ulmus*. Although several species, notably *Z. carpinifolia*, have been planted quite widely in parks and gardens in Britain, there are no records of the disease on them. A tree of *Z. serrata* was observed in close proximity to a large number of diseased elms, but remained unharmed.

36

THE CONTROL OF FUNGAL AND BACTERIAL DISEASES OF TREES

THE control of diseases caused by non-living agencies has already been covered in Chapter 9. There are both similarities and differences between the methods used for the control of living and non-living agencies, but both are alike in the fact that their cost must be considered in relation to the comparatively low annual increment in value of a tree crop and in relation to the financial saving they achieve. These matters have been considered in more detail in Chapter 9 and the arguments used there will not be repeated.

The whole subject of forest disease control has been reviewed by J. S. Boyce (1954*a*), who laid great stress on indirect as against direct methods. This attitude is inevitable, since indirect control by choice of site or species, or by adaptations of silvicultural treatment, is almost always cheaper, and therefore more applicable to a low-value crop, than direct-control methods. Nevertheless, direct methods are often used, especially in nursery practice, and they are therefore considered in some detail below.

DIRECT METHODS OF CONTROL

Eradication

Some organisms, for instance *Botrytis cinerea*, are common saprophytes on dead and dying vegetable matter, and there is thus no possibility of eradicating them, but it should theoretically be possible to exterminate organisms of more limited distribution. In the case of obligate parasites in particular, eradication of all diseased trees should achieve this end, since the pathogen is limited to the host tree. In practice the difficulties of observation, especially of early stages or of slight cases of development, make detection, which is a necessary precursor of eradication, very difficult. Survey costs alone may render an eradication programme uneconomic.

Few serious efforts have been made to eradicate tree diseases. The policy of removing trees attacked by Chestnut blight, *Endothia parasitica*, in the United States was soon abandoned, and the more sustained efforts to stamp out *Ceratocystis ulmi*, the cause of Elm disease, were eventually given up. There is no doubt that a tree disease can be eradicated if it is detected at a very early stage, when its spread is still extremely localized. This was the case with spot outbreaks of Chestnut blight in the western United States. These were successfully stamped out by early detection and prompt action. Efforts to reduce a disease by removing all the more obviously diseased trees have been rather more successful, and are generally much less costly, because the accompanying survey can be done less thoroughly, or even dispensed with. Bacterial canker of poplar has been minimized in Holland and is being reduced in France by this means, accompanied by the substitution of resistant for susceptible varieties in new plantings (Anon. 1951). In Britain, Watermark disease of Cricket-bat willow has been lessened, though not eradicated, in the main willow-growing districts by inspection and subsequent felling of diseased willows. In this case there is a strong economic justification, since diseased willows, however slightly affected, cannot be used for cricket-bat making and therefore fetch much lower prices than disease-free ones. In some other diseases, such as Elm disease caused by *C. ulmi*, the position is complicated by possibilities of recovery, so that wholesale felling of diseased elms could lead to the destruction of many trees that would regain their health if left standing. Eradication should obviously be attempted where a newly introduced pathogen has been detected before it has really started to spread. It should never be adopted for tree diseases which have become established, for it will certainly then prove both costly and unsuccessful. Removal of diseased trees without any attempt at complete eradication is more often justified, especially if it can be done without much additional cost as a part of normal silvicultural practice.

In the nursery, removal of infected plants may sometimes be used to eradicate a newly introduced disease, or to stem the attacks of one like Oak mildew, which is known to spread from an initially small number of infected plants. With most nursery diseases, however, removal of infected plants will prove to be a continuous process, leading eventually to the loss of nearly as many trees as would normally have developed the disease, and possibly of more than would have been seriously harmed by it.

Many of the rust fungi have two distinct host plants, spending part of their life cycle on one and part on the other. This introduces possibilities of eradicating the alternate host. This has been done on a large scale in North America with *Cronartium ribicola*, White pine blister rust, where the eradication of *Ribes*, the alternate host, protects susceptible Five-needled pines. Two-needled pines can be guarded from the Twisting rust, *Melampsora pinitorqua*, by the removal of the alternate host, aspen poplar, or more probably by the early cutting each year of the aspen coppice which is the chief source of infection. In some cases, however, trees or shrubs may have to be removed for the sake of more valuable crops, for instance some species of *Berberis* must be destroyed to protect wheat from the rust *Puccinia graminis* (Anon. 1933*b*; Bulger 1954), and in parts of Europe junipers must be removed because they act as alternate hosts to *Gymnosporangium* rusts on apple (Anon. 1934, 1936*a*). Eradication campaigns of this nature have been reviewed by Fulling (1943).

Destruction of diseased trees or of undesirable alternate hosts does not necessarily involve felling; poisoning is a commonly used alternative. The various methods used for the eradication of *Ribes* in the case of White pine blister rust have been summarized by Offord *et al.* (1952).

Fungicidal treatments

There are obvious difficulties in spraying large trees, but with modern equipment it is possible, without undue wastage, to treat isolated or near-isolated trees up to 30 metres in height. Successful treatment of plantations with such apparatus is much more difficult, and it can be taken that successful spraying or dusting of close growing trees is only possible from the air (Anon. 1950*c*), or by cutting frequent, wide racks from which ground machinery can be operated. So far practical spraying has been limited to street and park trees, mostly those of smaller size and to tree nurseries, where the relatively high value of nursery crops better justifies the cost, and where application to small trees is comparatively easy. The expense of spraying larger trees, considering that it may cost as much as £1 per year to apply two sprays to a tree 10 metres high, usually places an impossibly high charge against the timber or even the amenity value of the trees concerned.

Progress in the development of fungicidal chemicals, and in their methods of application, is so rapid that it is no longer possible to deal with them adequately in a general textbook. In any case such incomplete information as could be given here would soon be out of date. Recent information on materials and methods of application has been given by a number of authors (Brown, A. W. A. 1951; De Ong 1954; Dimond and Horsfall 1956; Horsfall 1957; Martin, H. 1957, 1958).

Many tree diseases have never been the subject of experiments on control by spraying or dusting, and with others long periods have elapsed since such experiments were made. Thus no specific information is available on them, or the existing information is out of date. Only where specific recommendations based on research are available have actual control measures been quoted in the text, despite the fact that they often appear to be blatantly out of date. In the absence of more modern work, they do offer opportunities of control and they may serve as a guide to other fungicides which could be tried. The information given in the earlier chapters on control by spraying is admittedly very incomplete, but this is a true reflection of the state of our knowledge. In most cases, even where specific materials have been mentioned, details of the concentrations are omitted. Often there is little evidence that the concentrations used in trials were really the best, and in any case the dilutions required often vary for different formulations of the same basic material. With most standard fungicides it may be best to follow or adapt the current directions given by the makers. Where no recommendations are available, trials should be based on the use of fungicides known to be successful for similar diseases on other trees or even on other crops.

In some cases spraying as soon as the disease is first noticed may give satisfactory control, as it does with Oak mildew, *Microsphaera alphitoides*. With many diseases, however, infection has taken place before the symptoms start to appear, so that post-appearance spraying is far too late to prevent damage. *Lophodermium pinastri* on pine is a good example of a disease where spraying, to be successful, must be done long before the symptoms appear. In such cases, of course, the timing of applications requires some knowledge of the life history and development of the fungus, with particular regard to the time at which infection takes place. Control attempts made without this knowledge may fail merely because they were carried out at the wrong time of year.

Indirect control by spraying can sometimes be achieved by insecticidal treatment against vector insects. Spraying against *Scolytus* beetles on elms has been used to prevent infection by *Ceratocystis ulmi*, the cause of Elm disease. No doubt similar methods could be used with other tree diseases, known or suspected to be carried by insects, once the nature of the vector and the time of infection were known.

Spraying against insect vectors shares one serious defect with direct pre-appearance spraying against fungi. The sprays have to be applied on the assumption that, if they were omitted, the disease would in fact develop. Such treatments should be adopted only if there is clear evidence that the frequency and the severity of the disease justify the repeated annual costs. With Elm disease only trees of the highest amenity value are worth this high annual expenditure. In nurseries, where spraying small plants is obviously much cheaper, it can often be justified, especially against diseases such as *Keithia thujina*, where attack is severe in most years. But even here some diseases present a problem. On highly susceptible species such as the *Sequoias*, annual protective spraying against *Botrytis cinerea* is certainly justified, while other species such as Scots pine are so seldom and so lightly attacked that there is certainly no need to spray. The problem of whether to spray species of intermediate susceptibility to *Botrytis*, such as Sitka spruce, could only be solved if we had better knowledge of the climatic factors favouring infection (Chapter 15). If such knowledge were available, a decision on whether to spray could be based on the local collection of meteorological data, as is now done with Potato blight.

Fungicides can, of course, be used against tree diseases in many other ways than direct application to growing plants as sprays or dusts. Fungicidal treatment of seed to prevent storage losses or to lessen pre-emergence attack by soil fungi has been suggested in Chapters 14 and 15. Superficial treatment of cuttings is of particular value to lessen the risk of transporting diseases from one country to another. A successful method of surface sterilization for poplar cuttings, using hydroxymercurichlorophenol, has been worked out in the United States (Waterman and Aldrich 1952, 1954; Ford and Waterman 1954). No doubt this method could be adapted for use with a wider range of propagation material.

Partial sterilization of the soil with fungicides has been used both against Damping off in the nursery (Chapter 15) and against root fungi such as *Armillaria mellea* (Chapter 18). Generally such treatments can be applied only when the soil is fallow, for the chemicals used are toxic to tree roots as well as to fungi. Their value lies in the fact that they render the soil safer for continued nursery use or for replanting. It is possible, however, to apply fungicides of sufficient strength to kill soil fungi, but which do not damage tree roots. These are obviously valuable for the treatment of Damping off in nurseries.

A good deal of work has been done on chemotherapy, the injection or absorption into the tree, of chemicals which will kill pathogenic fungi or bacteria without damage to the tree itself. Such methods are obviously attractive in 'wilt' diseases, and much of the work on trees has been done in connexion with a typical 'wilt', Elm disease caused by *Ceratocystis ulmi*. Reviews of recent progress in this field have been made by Stoddard and Dimond (1949), Horsfall and Dimond (1951), Crowdy (1952), and Dimond (1953). Cost of materials and difficulties of application have both operated against its success. With Elm disease, chemotherapy worked only protectively and did not eradicate the fungus, so that annual applications were involved. Nevertheless, recent advances in systemic fungicides and in methods of applying them will probably make chemotherapy more practicable (Brian 1954; Young *et al.* 1955; Darpoux, Halmos, and Leblanc 1958). Antibiotics seem particularly promising as systemic fungicides and may eventually prove of special value for the control of 'wilt' diseases (Crowdy 1954; Skolko 1954; Ark and Alcorn 1956). They have already been used successfully against Damping off in nurseries (Gregory *et al.* 1952; Vaartaja 1957).

Comparatively little work has been done on the fungicidal treatment of actual lesions. Specific recommendations have been made for a few diseases, notably for Ink disease, *Phytophthora* spp., on *Castanea* (Chapter 29), and for *Armillaria mellea* (Chapter 16). For these diseases exposure of the upper roots, followed by fungicidal treatment, has been suggested; indeed exposure of the infected roots without fungicidal treatment has been advocated, especially in hot climates. Recent successful work by V. D. Moss (1957) on the fungicidal treatment of Blister rust cankers with cycloheximide suggests that the treatment of above-ground lesions may have been unduly neglected.

The fungicidal treatment of stumps to prevent their colonization by pathogenic fungi and their subsequent use as centres of spread has received particular attention in connexion with *Fomes annosus* (Chapter 16). Stumps from early thinnings were found to be the chief means whereby this pathogen became established in afforested areas, and its chief means of spread thereafter. Similar methods should be applicable to other root diseases which act in the same manner.

The treatment of pruning and other wounds to prevent the entry of fungi is dealt with in Chapter 37. Wounds form one of the commonest means whereby decay-causing fungi enter the stem, and their protection, if it can be achieved cheaply, is obviously of considerable value.

Surgical methods

The removal of infected portions is often advocated as a means of control and, though laborious, is sometimes very successful. Where a disease is very localized in its attack, as for instance *Pycnostysanus azaleae*, Rhododendron bud blast, which normally attacks only the buds, there is comparatively little difficulty in ensuring that all infected material is removed. Localized lesions can sometimes be successfully removed, especially if they are on branches. Work of this kind has been carried out on pine infected with *Cronartium ribicola* (Martin and Gravatt 1954; Stewart 1957). With many other diseases, however, it is very difficult to discover the exact extent of invasion and cautious surgery may easily leave enough infected material for the disease quickly to redevelop. This is particularly the case with wilt diseases, and although pruning-out infected branches has been suggested as a remedial measure for Elm disease, such treatment usually involves too great a mutilation to be justified.

Removal of roots is a more serious matter, because it is liable to affect the stability of the tree. Occasionally the excision of badly diseased roots may be justified on ornamental trees of high value, for such roots are likely to die quickly and thereafter play little part in the tree's anchorage, but generally such treatments have no place in normal forest practice. Indeed all forms of surgical treatment must normally be restricted to individual trees of special value, unless they can be adapted as part of a normal operation, such as pruning. These matters are discussed more fully in Chapter 37.

When work of this kind is undertaken the possibility of the tools acting as a means of transmission should be borne in mind. Tools proved to be the chief means by which *Endoconidiophora fimbriata*, the cause of Canker stain of planes, was spread from tree to tree in the United States (Walter, Rex, and Schreiber 1952). Periodic sterilization of tools is always desirable, and is essential if the work is concerned with disease control.

Other direct methods

Trenching has sometimes been advocated to stop the spread of root fungi. Successful control of *Paxillus giganteus*, a Fairy-ring fungus spreading through shallow, sandy soil over chalk and killing young pines, has been achieved in this way (Peace 1936*a*). *Rhizina inflata* appears unable to cross comparatively shallow drains in wet peat soils, where the fungus is unlikely to penetrate deeply into the excessively wet peat. If a trench is deep enough to stop the fungus passing underneath, or reaches to a layer of rock, chalk, or clay which the fungus cannot penetrate, and if it is kept deep and open and really includes all the infected area, further spread should be successfully controlled. In practice, however, satisfactory trenches are seldom achieved and fungi cross them, often by debris in the bottom. In addition the construction of trenches often involves cutting a large number of roots. Not only may the severed ends of live roots act as means of entry for disease fungi, but the detached portions on the opposite side of the trench may be colonized by root fungi and used as centres of subsequent invasion of live roots. The cutting of roots also increases the risk of windblow. In arboreta or gardens trenching is sometimes justified against *Armillaria mellea*, but the trench must be maintained until the danger has ceased.

Diseases such as *Fomes annosus* and *A. mellea*, which spread chiefly from stumps, can obviously be controlled by stump removal. Indeed the removal of stumps from the site of any particularly valuable planting of trees or shrubs is always a wise precaution, particularly if the trees to be planted include genera known to be susceptible to root fungi. In the past, stump removal has proved expensive and rather slow, but the use of heavy machinery may make possible its use on a forest scale. It might well prove the most efficient means of reducing the incidence of a fungus like *F. annosus* on an infected site. There are other possible methods of achieving the same end. Girdling before felling gave good control of *Armillaria* root rot in tea bushes planted in cleared jungle (Leach, R. 1937, 1939). The food reserves in the roots of the girdled trees were used up before felling, so that the stumps were no longer suitable for centres of spread. Poisoning of standing trees or of stumps are other possible, though not altogether easy, methods of achieving the same object. The chief difficulty here lies in getting a substance that will penetrate sufficiently far down into the roots.

Recovery from root disease can be encouraged by banking soil around the base to encourage the production of new roots (Rhoads 1942).

INDIRECT METHODS OF CONTROL

Choice of site

Given a proper knowledge of the behaviour of a disease it may well prove possible to effect partial control by a proper choice of site, avoiding those places where the disease is likely to develop. This is particularly true for those diseases which only attack trees growing under adverse conditions. Ink disease, caused by *Phytophthora* spp., on *Castanea* occurs seriously only on sites with a periodically high water table; Larch canker, associated with *Dasyscypha willkommii*, is much worse on frosty sites; *Brunchorstia destruens* on Corsican pine is particularly associated with high-elevation plantations; while drought predisposes pine to attack by *Diplodia pinea* and poplars to *Dothichiza populea*. All these diseases can be mitigated by restricting the host trees to sites where these adverse influences do not operate.

Avoidance of the pathogen by choice of site is more difficult. Most diseases are very generally distributed in the countries where they occur. Efforts to raise *Thuja* in nurseries free from *Keithia thujina* met with only very limited success (Peace 1958*b*), and *K. thujina* is a fungus which still has a somewhat incomplete distribution in Britain. Ground which has not previously carried a forest crop will almost always be found to be free of root-inhabiting fungi, such as *Fomes annosus* or *Armillaria mellea*, and in plantations on such sites the fungus can be more or less excluded by fungicidal stump treatment. Thus a species like *Tsuga heterophylla*, which is highly susceptible to *Fomes*, and cannot be planted with any degree of confidence on old woodland sites, can be safely used in afforestation projects on virgin ground.

Detailed work on the ecology of a disease may well disclose site preferences which are not merely a matter of the direct effect of the site on the resistance of the host tree. *F. annosus* is known to be particularly virulent on old plough-land and on alkaline soils, so that sites of this nature should not be used for highly susceptible species. Although the occurrence of Ink disease on soils with a periodically high water table has usually been attributed to the damaging effect of the variable soil water content on the roots of the host tree, it may well be that in part its occurrence is due to the presence of the wet soil required for spore germination.

It is obviously desirable to avoid planting susceptible trees in the immediate neighbourhood of a source of infection, or alternatively creating a source of infection near susceptible trees. It would be fatal to plant *Tsuga heterophylla* on an area containing stumps infected with *Fomes annosus*. Dying elm wood should not be left or stacked in the neighbourhood of standing elms, lest the emerging *Scolytus* beetles carry *Ceratocystis ulmi* and start a local outbreak. *Thuja* should never be raised in a nursery with a *Thuja* hedge, lest *Keithia thujina* spread from it to the young plants. Indeed the ideal nursery should never be surrounded by trees of any age or size of those genera that are going to be raised in it. Extremely severe local attacks of *Lophodermium pinastri* on

Scots pine have occurred on young trees in the neighbourhood of infected older ones (Murray and Young 1956).

Where a fungus has two alternate hosts, the chances of control by avoidance are greatly improved. In some cases the distance which the fungus can spread is very limited. Scots pine are safe a few hundred metres from the nearest aspen poplars infected by *Melampsora pinitorqua*, as are spruce from the nearest rhododendrons infected by *Chrysomyxa rhododendri*. Even when safe distances are much greater, as is the case with *Cronartium ribicola* on *Ribes* and on Five-needled pines, the intensity of infection falls off with distance, and it may be possible to plant pines on the outer part of the infection zone, on the assumption that only a small percentage will become infected.

Choice of species

Choice of species immediately raises the vexed question of the use of exotics. This has been argued more fully in Chapter 3, 'Trees in Relation to their Environment'. But whether one agrees with J. S. Boyce (1954*a*) that exotics are generally undesirable, or with the author (Peace 1957*b*) that there are no insuperable obstacles to their use, for British forestry the argument is somewhat academic, since as far as conifers are concerned we are compelled by paucity of native species to use selected exotics. It has already been pointed out that we must avoid using certain species on sites where they will prove particularly prone to disease. In the case of *Pinus strobus*, one of the pines susceptible to *Cronartium ribicola*, this has meant the virtual abandonment of this species as a forest tree in Britain. But generally, given a sensible choice of sites, no other species suited to the British climate need be eschewed on grounds of disease susceptibility.

At the moment about twelve coniferous species play a major part in British forestry. From the point of view of rational utilization this is too many, since there is often difficulty in getting a sufficient and sustained supply of timber of uniform sawmilling and processing qualities. Unless one assumes that in the future utilization can be adapted to deal economically with a mixed product, this suggests that widespread planting ought to be limited to a smaller range of species. This is not, however, an extension of the suggestion, made in Chapter 3 and again below, that pure plantations are pathologically admissible. If Britain were virtually limited to, say, four species, or four groups of species with similar timber properties, much larger areas would be involved in any major disease catastrophe, many opportunities of local disease control by the substitution or initial use of resistant species would be lost, and the choice of species for sites, where the use of one or more of the selected few was barred because of disease, would be drastically narrowed. It is possible that a reduction in a number of major species could be justified economically, even if it did lead to some loss of total increment due to site unsuitability and disease. Certainly its pathological implications would have to be seriously considered, before such action was taken.

On the Continent and elsewhere considerable stress has been placed on the desirability of mixed plantations, particularly admixtures of broadleaved species among conifers (Anon. 1939*b*; Rennerfelt 1946; Leibundgut 1947;

Boyce, J. S. 1954*a*; Braun 1958). They have suggested that mixed plantations are more 'natural' and therefore generally more beneficial to the tree, tending to lessen pathogenic infections and to limit disease development, if such infections do take place. It is obvious that tree-to-tree infection is likely to be more serious in pure plantations than in mixed ones, though even here unexpected effects may intervene. In the case of *Mycosphaerella laricina* on larch, admixtures with spruce increase the severity of the disease, because the infected larch needles lodge on the spruce branches through the winter and are in a good position to reinfect the larch in the spring. Whereas when larch are mixed with beech, the larch needles fall to the ground where they are covered by beech leaves and become innocuous. The question of mixtures has also been discussed more fully in Chapter 3. The idea has been put forward particularly as a means of avoiding attacks of the root-rot fungus *Fomes annosus* (Priehäusser 1935). This particular aspect of the general suggestion is dealt with in Chapter 16 on Root diseases. It is certainly true that mixed plantings can provide a partial control of some diseases. It is equally certain that mixed planting is not a panacea for all ills. In any case it may often be less desirable economically than pure planting. It should not be undertaken purely for pathological reasons unless there is definite evidence that it is desirable for a particular species on a particular site. In our present state of knowledge it is seldom possible to produce such definite evidence.

Adaptations of forest practice

It has already been pointed out in Chapter 2, 'The Symptoms and Diagnosis of Tree Diseases', that the final effect of a fungal or bacterial attack is often dependent as much on the powers of recovery of the tree as on the actual damage done by the pathogen. Vigorous recovery can sometimes overcome initially quite serious injury. For this reason anything that can be done to increase vigour is nearly always beneficial, not so much because it prevents attack, but because it lessens the final effect. Vigour in itself, however, does not necessarily prevent or even lessen attack, for instance *Ceratocystis ulmi* is often more severe on heavily foliaged, water-demanding elms, and *Fomes annosus* develops severely on conifers growing vigorously on old arable soils (Chapter 18). Nevertheless, any silvicultural measures, such as manuring or thinning, which increase vigour of growth are likely to promote recovery from disease.

Increases in plantation spacing sometimes induce this desirable increase in vigour, but the use of variations in spacing as a method of disease control is obviously more complex than this (Boyce, J. S. 1954*a*). Wide spacing may depress early growth rates, because it delays the closing of the canopy and the suppression of competing vegetation. On the other hand, while close spacing may increase the chances with some diseases of tree-to-tree infection, it may be desirable in order to allow for the removal of damaged individuals, while still leaving enough undiseased stems to form a final crop. This is particularly the case with canker diseases, where it is obviously desirable to remove all the cankered stems as thinnings.

Brashing can seldom be regarded as a control measure, though beneficial effects have followed the brashing of young *Thuja* plantations affected by *Keithia thujina*, when the operation resulted in the removal of the heavily infected, lower, live branches. The wounds which result from brashing certainly lead on occasion to fungal invasion, and evidence has been put forward in one case, that of *Phomopsis pseudotsugae* on Japanese larch (Chapter 24), to suggest that the season of brashing has a big effect on subsequent fungal injury. It may well be that a relationship of this kind holds for similar diseases on other trees.

Thinning provides an obvious method of removing diseased individuals from a stand, and gives the forester one of his best methods of control. In some instances the removal of diseased trees may involve a more irregular thinning than would normally take place, but this is worthwhile if it results in the early removal of trees which are a source of danger to their neighbours or unlikely because of disease to develop into good final crop trees themselves. The pathological condition of a stand should always be investigated before marking for thinning is started, and if, as a result, the thinning is to be directed partially against any disease, a decision should be reached as to the degree of injury that merits preferential removal. Thinning, of course, leaves stumps, which may provide *Fomes annosus* and other fungi with centres of spread. This danger can be met by appropriate fungicidal treatment of the stump surfaces, but delayed thinning, and even initial wide spacing so that delayed thinning may be more easily adopted, have both been suggested as feasible protection measures against *F. annosus*.

There are other more specialized ways in which forest techniques can effect disease control. In Group dying, caused by *Rhizina inflata*, it is desirable to restrict the lighting of fires in the forest, for it is from the sites of such fires that the fungus spreads (Murray and Young 1960). *F. annosus* is known to be able to spread from infected fence posts to neighbouring young trees, so that the use of partially decayed material for plantation fencing carries an appreciable risk (Jørgensen 1955).

Finally there is the general question of forest sanitation. Quite rightly the debris is not usually removed from the forest floor. Brashed branches and the lop and top from thinnings are left to rot down and add to the forest humus. Very occasionally disease outbreaks may originate from such sources. *Lophodermium pinastri* has been known to spread catastrophically to young pines from heavily infected, initially live, lop and top left on the ground under an immediately adjacent older plantation (Murray and Young 1956). In this case the infection did not extend far into the young crop, indicating that the spores of this fungus do not spread in quantity over very long distances. If the lop and top had been dragged back 20 metres into the plantation, away from the young pines, little damage would have occurred. It would obviously be wrong to leave branches of chestnut, infected with *Endothia parasitica*, lying in a chestnut stand; some of the fungi causing twig blights and canker on basket willows can be spread by leaving rejected rods lying around the stools from which they were cut. But such occurrences, individually important though they may be, do not make a general case for the removal and destruction of woody debris. Such material can normally be left lying with safety unless it is obviously infected with a dangerous disease.

Precautions in nurseries

If a nursery is remote from the forest, or surrounded by a wide belt of trees of a single species, it should be possible to keep out, at any rate for a considerable period, a number of the more specific diseases, not only those of incomplete distribution, such as *Keithia thujina*, but also others like *Lophodermium pinastri* or *Coleosporium tussilaginis*. This can be done by not allowing any plants or unsterilized manures into the nursery, and by limiting imports to seed and sterile forms of humus such as spent hops. It is a pity that this relatively simple form of nursery disease control is not more generally practised.

With some of the diseases restricted to a single genus, it seems possible that control could be achieved by using a number of nurseries in rotation, raising in each only one crop of plants from seed to planting size in three to four years. Using six nurseries this would leave a fallow period of two years between each crop. With many diseases this would allow a sufficient period for the fungus to die out. There would then be no carry over of infection from one set of plants to the next, and only the three or four years for any chance infection from a distance to build up on any batch of plants.

With some root diseases there are limited possibilities of control by grafting susceptible scions on to resistant stocks (Bond 1936). However, apart from the limitations imposed by the cost and difficulty of grafting, some root diseases may infect the stock without seriously damaging it and then pass upwards to destroy the scion. This is apt to happen with *Verticillium* wilt of grafted maples. Equally, disease may pass from a resistant, but not immune, scion to a susceptible stock. This sometimes happens with elms resistant to *Ceratocystis ulmi* grafted on susceptible stocks.

Biological control

Deliberate biological methods, involving the use of one organism to attack another or to prevent attacks by another, have so far played no real part in the control of tree diseases. Orłoś (1957) has suggested control of *Armillaria mellea* by infecting stumps with harmless, competitive fungi (Chapter 38). It is doubtful if such measures would be practicable on a large scale. There are several instances of fungi attacking pathogenic fungi or damaging mistletoes, but in no instance to they exert an appreciable control. None of the fungus diseases affecting weed plants such as bracken or heather, or climbers such as honeysuckle, is sufficiently serious to offer any hopes at all of control.

Selection and hybridization

Numerous examples have already been given of specific variations in host resistance to fungal and bacterial diseases. Variability in the resistance of Five-needled pines to *Cronartium ribicola*, or *Castanea* species to *Endothia parasitica*, or of *Ulmus* species to *Ceratocystis ulmi* are obvious examples. Within species there are often provenance differences, for instance in the resistance of Douglas fir to *Rhabdocline pseudotsugae*, or of Scots pine to *Lophodermium pinastri*. There are also variations between individual trees. For instance, there are clonal, as well as provenance, differences in the resistance of Scots pine to *L. pinastri*. Other examples of clonal variation in disease resistance are given by Bacterial canker of poplar, and by *Cronartium ribicola* on *Pinus strobus*. This whole subject has been reviewed on a number of occasions (Hartley 1927; Clapper and Miller 1949; Lindgren 1951; Clapper 1952; Larsen 1953; Riker 1954).

J. S. Boyce (1957) and Schreiner (1957) have called attention to some of the limitations of this method of disease control. Trees resistant to one pathogen may be abnormally susceptible to another. *Ulmus* '*Christine Buisman*', one of the Dutch elm selections resistant to *Ceratocystis ulmi* (Chapter 31), proved unusually susceptible to *Nectria cinnabarina*. Selected trees must be suitable in other ways besides merely disease resistance. The Asian species of *Castanea*, resistant to *Endothia parasitica* (Chapter 29), are not satisfactory, either as timber trees or for nuts, in most of the area affected by Chestnut blight. There is also the possibility of genetic variation in the pathogen, so that a tree resistant to the pathogen in its original form may prove susceptible to a mutation from it.

It must be admitted that in this field of work the possibilities are so far much greater than the actual accomplishments. Resistant selections have been made and propagated for a considerable number of diseases, but field use has so far been virtually limited to those trees where vegetative propagation, either by grafting or by cuttings, is a feasible and reasonably economic method of increase. Thus elms resistant to *Ceratocystis ulmi*, poplars resistant to Bacterial canker and to *Dothichiza populea*, and chestnuts resistant to *Endothia parasitica* have all reached the stage of field trials. The production of tree seed strains resistant to specific diseases is still a matter for the future. There is no doubt that this is the more desirable, though admittedly more laborious, method of introducing resistance into tree plantings. Clonal stocks selected for resistance to one disease may eventually display susceptibility to another. The large-scale use of single clones, however desirable they may appear to be, is quite unjustified on pathological grounds (Hartley 1939). It is essential, if clonal stocks are being used for planting, to ensure either that a number of clones are planted in mixture, or that the area occupied by any one clone is limited and that four or more clones at least are used in each forest or planting.

37

PRUNING AND TREE SURGERY

PRUNING

Natural pruning

UNDER forest or plantation conditions the lower branches of trees eventually die, primarily from lack of light. When dead they gradually decay and fall off, and the wounds which they leave finally heal over. From the time of death until the completion of healing they form an obvious means of entry for fungi, and are therefore of pathological importance.

Natural pruning has been fully discussed by Mayer-Wegelin (1936). It will only be briefly dealt with here. Some trees can shed small twigs at definite abscission layers, and therefore without danger of wound infection, but this ability is limited to the crown and has no effect on the shedding of larger branches. The rate at which dead lower branches are shed depends on a large number of factors. There are big differences between species not only in their shade-bearing capacity, and therefore in the length of time their lower branches remain alive, but also in the period for which dead branches are retained. Beech, for instance, loses its branches sooner than oak, and oak sooner than most conifers. Warm moist climates, because they encourage the growth of fungi, hasten natural pruning. There is a very marked retention of branches in cold countries. The diameter of the branch naturally affects the rate of decay, and larger branches generally take longer to fall than small ones. They do not usually fall under their own weight, but as a result of wind, snow, branches falling on them from above, and other external stimuli; the greater weight of larger branches does not have much influence on shedding. Since so many variable influences are involved, the time for which a dead branch is retained may vary from two or three up to over 100 years (Romell 1937).

A very wide range of fungi, not all capable of causing decay, colonize dead branches. Many are restricted to certain tree genera, while others are found much more commonly on some genera than on others, for instance, *Auricularia auricula-judae* on elder. The relationships between the different fungi causing decay on branches are very complex, and Köster (1934) has pointed out that some of them are much more efficient in this role than others. Further study is required on the fungi involved in the natural removal of branches.

As soon as a broadleaved branch dies a protective layer of gum-filled cells normally starts to form at the base. This phenomenon has been studied in detail in beech by Gelinsky (1933); it also occurs in most, but not all, other broadleaved species (Schönigh 1935). Such protective zones can be formed only where living tissue is present. In small branches this will be over the whole cross-section of the branch, but in larger ones, where heartwood is present, the protective zone is limited to the sapwood and the heartwood constitutes an unprotected entrance to the main stem for decay-causing fungi. Conifers form protective zones of resin at the base of dead branches (Köster 1934). This resin accumulation tends to spread, so that in conifers dead wood may be protected to some extent as well as live. The resin, however, diffuses up the dead branch, delaying its decay and consequently lengthening its period of retention. Included branches, appearing in converted timber as resin-soaked knots, are a common feature of coniferous timber.

The forester is concerned in the first place with the rate of decay of branches, because this affects the length of time they remain on the tree, and therefore the length of branch included in the wood of the growing stem. This aspect of the problem has been dealt with briefly above. Secondly, he is concerned with the importance of dead branches as means of entry of decay fungi; this has been covered in Chapter 19.

Artificial pruning

Pruning of trees which are being grown for timber is directed almost entirely to a single end, the improvement of timber quality by the elimination of included branches. This may involve the removal of live or dead branches. In the forest, and particularly in plantations, pruning may also be used to promote access by removing the lower, usually dead, branches, an operation usually referred to as 'brashing'. The subject of artificial pruning has been fully reviewed (Anon. 1940; Mayer-Wegelin 1936, 1952).

Trees in parks and gardens, and along streets and roads, may be pruned for other purposes, for instance removal of diseased or dead limbs, increase of safety, reduction of shade, or clearance for overhead wires. Pruning of this kind, sometimes involving the removal of large limbs or quite drastic crown reduction, requires rather different techniques and is dealt with separately below.

Pruning timber trees

Much has been written on the economics of pruning; this will not be considered here. Much has also been published on the effect of live-pruning on growth rate. According to species and conditions, varying amounts of the live crown can be removed without any serious effect on growth; if this amount is greatly exceeded death may sometimes take place (Helmers 1946; Slabaugh 1957). However, the amount of crown removal required to cause death lies far beyond anything likely to be done in practice.

There is also a very large literature on the design and use of pruning tools. From the pathological aspect, any tool is satisfactory which will remove the branch with a reasonably clean cut, and without undue crushing of the tissues; in this connexion damage to the tissues round the branch may be more important than the cutting of the branch itself (Vloten 1952). There is no evidence that the slight roughness of a saw-cut, compared with the cleaner surface produced by a straight-edged pruning tool, is any disadvantage (Donald 1936; Pudden 1957); but the saw is certainly the safer with unskilled labour.

It is generally considered that close pruning is desirable to get rapid healing and the shortest length of included branch. The cut should always be parallel to the main stem, even though this greatly increases the exposed area, particularly with acute-angled branching.

It is doubtful whether the raised collar of tissue which often surrounds the base of a branch should be removed or not. R. T. Anderson (1937), Adams and Schneller (1939), E. R. Roth (1948), and many others, advocate its removal. It has also been suggested that its partial elimination gives the best healing (Anon. 1948*c*). On the other hand, its removal greatly increases the size of the wound, and in some species where the branches in a whorl are very close together, elimination of the collars would practically girdle the stem (Hawley and Clapp 1935; Jacobs 1938). There is fairly general agreement that stubs should never be left; apart from other objections no knot-free wood will be formed till they have been occluded. Even a 6-mm. stub may delay healing for several years (Skilling 1958). Mayer-Wegelin (1936, 1952), however, suggested, on somewhat theoretical grounds, that in some cases stub-pruning would enable the tree to lay down protective barriers before the stubs were removed two or three years later, a method condemned for beech by Zimmerle (1943). In any case this dual operation is too expensive and need not be further considered. J. D. Curtis (1937) stressed that even a small saw snag will seriously delay healing, but Donald (1936) considered small snags of little importance.

Pruning trees in streets, parks, etc.

The general principles of close pruning, outlined above, apply also to tree pruning under roadside or park conditions. But here the necessity to remove larger limbs and to extend pruning into the crown, coupled with the degree to which the ultimate shape and beauty of the trees have to be considered, require the adoption of certain additional techniques.

When heavy limbs are cut they are apt to fall before the cut is completed, tearing away a strip of bark and wood from the base of the wound. A small upward cut at the base, made before the main downward cut is started, will usually prevent this. Ideally, however, the limb should be removed in two stages, the first cuts, upper and lower, being made some 30 cm. from the main stem, and the second cut close, when there is no weight to make a tear (Fig. 99).

Crown reduction should be made, as far as possible, by shortening stems back to points where smaller side branches emerge (Fig. 100). The activity of these side branches will help to promote healing of the wounds, and their growth will give the trees a more natural appearance. This type of crown reduction has been practised very successfully in the larger London parks. After a few years it is only detectable because a branch pruned in this way has more abrupt and irregular taper and more sudden changes of direction than a normal branch.

Lopping or pollarding of trees is pathologically very undesirable. However well the tree recovers, large wounds are left, and the junctions of the new shoots with the main stems from which they arise are never satisfactory. Such shoots cannot safely be left to reach large size, so that once lopping has been started it becomes a recurrent procedure. When it has to be done, every effort should be made to do the work as carefully as possible. Particular care should be taken to see that no cuts are horizontal, lest water lie on the surface and encourage decay. Some species recover from lopping much better than others (see below).

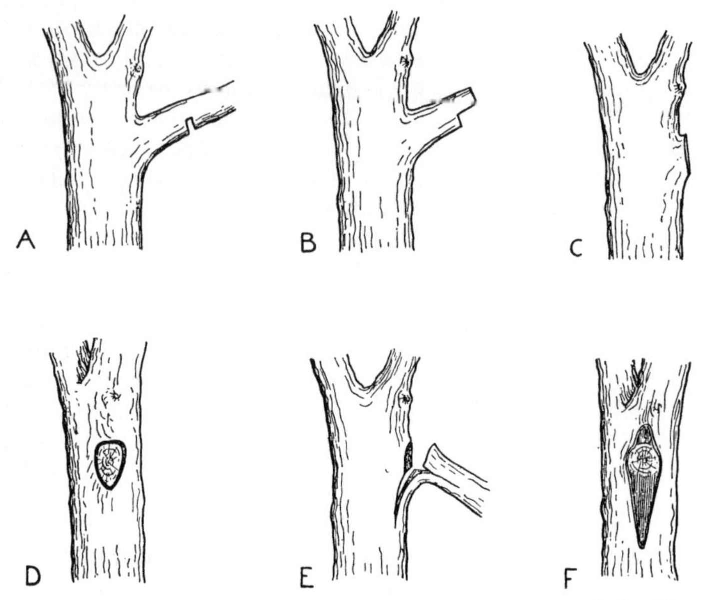

FIG. 99. Correct and incorrect methods of pruning (see text). (J. C.)

FIG. 100. Crown reduction in an over-large tree. Left, before pruning; right, after pruning. (A. C.)

Detailed information on the pruning of street and park trees is given by Pirone (1948), Le Sueur (1949), and Baxter (1952). The pruning of ornamental shrubs, as well as trees, is dealt with by Osborn and Bagenal (1952).

Rate of healing of pruning wounds

It is obviously desirable that pruning wounds, whether of live or dead branches, should heal as quickly as possible. This is not only because fungi can enter through unhealed wounds, but also because in timber trees the deposition of high quality, knot-free wood cannot start till healing is completed. The time of healing depends considerably on the size of the wound. In pruning timber trees, it is usually considered advisable not to exceed a certain limit. With most species, pruning of branches more than 5 cm. in diameter should be avoided. With ornamental trees it may sometimes be necessary to remove much larger limbs and to treat the wounds with some form of protective coating (see Wound Dressings below). It is generally agreed, however, that the vigour of the tree has more influence on the time of healing than the size of the wound (Anderson, E. A. 1951; Nylinder 1955).

The time of year at which pruning is carried out has an effect on wound healing, though possibly not so great a one as is often supposed. Many writers agree that a vigorous tree can safely be pruned at any time of year; but others consider that the winter is the logical time (Mayer-Wegelin 1936; Jacobs 1938; Hawley 1946; Strehlke 1952). This view is sometimes based on the theory that less damage will be done if the tissues are dormant at the time of pruning, but Rohmeder (1953) actually found that deliberate wounds on spruce were appreciably less infected by decay-causing fungi if they were made during the winter. Many consider late winter and early spring best, mainly because pruning at that time appears to stimulate immediate growth of healing tissue (Beach 1939; Roth 1948; Winterfeld 1956). Swarbrick (1926), working with sycamore, plum, and apple, found that in wounds made between May and August the tracheids were rapidly blocked by wound gum, that those made in September and October were only partially blocked, and that those made between November and April remained unblocked until the spring. R. T. Anderson (1937) claimed that small wounds made in the spring heal during the following summer, whereas those made earlier, during the previous winter, may not do so. Pudden (1957), working on cypresses in Kenya, found that wounds made when the soil was moist healed more rapidly than those made during drought periods. There is a strong belief also that species which 'bleed' freely in the spring, such as birches and maples, should be pruned later, in early summer (Pirone 1948), though it has not been established that the bleeding is always damaging. Winter brashing of Japanese larch is open to objection on pathological grounds (see below).

When pruning is designed mainly for the removal of diseased limbs, summer has obvious advantages, for disease can best be detected then. However, large limbs of deciduous trees are best removed in winter, when the foliage does not add its weight to that of the branch, and when the clear view through the branches makes it easier to carry out the operation and to appraise its final effect. However, it is not always easy to arrange pruning at a particular time of year, and it is probably best to regard it as something that can be done at any time of year, except in special cases where there are definite reasons for avoiding one season or selecting another.

Pruning and disease

Pruning wounds are often said to provide means of entry for fungi, particularly those causing wood decay; but it is sometimes overlooked that fungi may also enter through dead branches, and through the scars which side-branches leave when they fall. Nevertheless, it is probably true that pruning slightly increases the risk of fungal infection (Jacobs 1938; Paterson 1938; Romell 1940).

Certain diseases are specifically connected with pruning wounds. Examples are the infection of Japanese larch by *Phomopsis pseudotsugae* following winter brashing (Chapter 24), Silver leaf caused by *Stereum purpureum* on plum (Brooks 1953), and Canker stain caused by *Endoconidiophora fimbriata* on plane (Chapter 35).

Secondly, there is the entry of decay, though here the relative importance of artificial compared with natural pruning is harder to evaluate. Cases of decay arising from infection of artificial pruning wounds tend to attract notice, while decay associated with branches or branch scars is often disregarded as a natural phenomenon. Paterson (1938), working with Norway spruce, and Donald (1936), with various conifers, failed to find any evidence that pruning increased the incidence of decay. Indeed, some authors have suggested that pruning small branches of conifers and hardwoods lessens the risk of entry of decay (Staeger 1928; Platzer 1937; Andrews 1955; Skilling 1958). On the other hand, pruning, especially of branches over 5 cm. in diameter, does sometimes lead to increased infection. This was found with *Stereum sanguinolentum* on pruned *Pinus strobus* (Spaulding, McAloney, and Cline 1935). Rohmeder (1953) found 57 per cent. decay infection of deliberate wounds, 4–8 by 80–100 cm. in size, on spruce, but these, of course, were larger than normal pruning wounds, and were not around branch origins. Often fungal attack, following infection of pruning wounds, is confined to the region of the wound (Hawley and Clapp 1935; Winterfeld 1956). Such localized pockets of rot, however, lower the value of the timber. Hawley and Clapp considered the pruning of small live branches a lesser risk than dead-branch pruning, but Childs and Wright (1956), working with Douglas fir, took the opposite view, associating entry of decay with live pruning. In Britain and abroad, instances have been observed where careless brashing has led to infection by fungi such as *S. sanguinolentum* and even by *Fomes annosus* (Risby and Silverborg 1958; Peace unpubl.). Pruning of broadleaved trees, with the exception of poplar and willow, is generally considered to involve a high decay risk, though Nylinder (1955) considered that this was slight in the case of oak, and Winterfeld (1956) found that beech, from which small branches had been pruned, were no worse decayed than those left to prune naturally.

Thirdly, pruning may increase liability to injury by non-living agencies, such as frost. W. R. Day (1950*c*) reported frost as the probable cause of extensive lesions around pruning wounds in spruce. Wounding may stimulate premature cambial activity in the spring, rendering the wounds abnormally sensitive to frost.

On the evidence available it appears that pruning of live or dead branches, properly carried out, is sufficiently safe pathologically to be done without hesitation, if other considerations render it desirable.

The pruning of different genera

It is impossible to consider in detail here the different methods and times of pruning applicable to various trees and shrubs. For this, reference should be made to books devoted to the subject. Advice on the pruning of various forest genera has been given by Mayer-Wegelin (1936, 1952). Many non-pathological considerations naturally affect the treatment of different genera.

From the decay point of view, conifers are generally safer to prune than hardwoods, and trees of naturally rapid growth, such as poplar, than slow ones. Chadwick and Nank (1949) compared the healing rates of a number of different trees, and associated rapid healing with normally rapid growth. Trees such as beech, the timber of which is prone to decay, obviously are particularly dangerous to prune. Mayer-Wegelin (1936) has paid particular attention to beech.

If drastic pruning in the form of lopping or pollarding is unavoidable, careful consideration should be given to species. Most species tolerate heavy pruning when young, but older trees of birch, sycamore, ash, and beech, particularly the latter, make a very weak recovery. Elm, lime, and plane are notable examples of trees which can stand heavy pruning even when quite old. Drastic topping of conifers is a risky undertaking with most genera. The cypresses, *Thuja*, yew, and a few other conifers can stand lopping, but most conifers are likely to die if deprived of the bulk of their live crown. If conifers have to be topped at least two whorls of active growing branches should be left. With both broadleaved trees and conifers, species which produce strong coppice shoots from the base when felled, for example Spanish chestnut, poplar, and *Sequoia sempervirens*, are more likely to recover from over-heavy pruning.

Pruning by bud removal

Pruning by the removal of the lateral buds of conifers, instead of the later removal of branches, has been suggested as a method of getting lengths of knot-free timber. It is, in fact, generally used in the cultivation of Cricket bat willow, where only a comparatively short timber length is required. There is no evidence that it leads directly to disease, but it inevitably renders the loss of the leader by wind, fungal attack, or any other factor a much more serious matter than on a normal tree, where replacement by a side-branch or dormant bud is comparatively easy (Curtis, J. D. 1946; Paul, B. H. 1946). It also causes serious weakening, both structurally and physiologically, of the pruned trees.

Wound protectants

Large pruning wounds should obviously be protected against fungal or bacterial infection. An ideal protective coating should combine persistent fungicidal action and cambial stimulation with its protective properties. Wound dressings are essential where large limbs have been removed from trees of high individual value. They have been little used in forest practice, though even there they might be of value, especially on decay-prone species such as beech. Any wound more than 4 cm. in diameter on a broadleaved species, other than willow or poplar, ought to be protected.

There are several reviews of the literature on wound protectants (Lesser 1946; Pirone 1948; Le Sueur 1949); in addition various workers have carried out comparative tests on a substantial scale (Atkinson 1937; Tilford 1940; York 1941). Among the established substances, white-lead paint gives a satisfactory, but not very lasting, cover. It has no effect on cambial growth, and no strong fungicidal properties. On amenity trees an inconspicuous colour, such as dark brown, can be used. Varnish gives a rather more durable covering than paint, and can be used over it. Orange shellac stimulates early cambial growth and has very definite fungicidal properties, but it does not provide a permanent protective coating. It is best used as an undercoat to some other substances, such as a bitumen paint (Cooley 1942; McQuilkin 1950). Bordeaux paint, made by mixing Bordeaux paste with linseed oil, gives only a very short-lived cover. Its sole advantage is its fungicidal action.

Coal-tar has long been used as a wound protectant, often with good results, but it is not easy to apply cold and tends to crack with age. It has little or no fungicidal action. The volatile creosotes are too phytotoxic for use near cambium, and give no permanent covering although they may be highly fungicidal. A considerable number of bituminous substances are available, the less volatile of which are sometimes referred to as asphalts. These vary greatly in ease of application, durability, and phytotoxicity. Those which have the best fungicidal properties tend to be toxic to cambium. Some are completely innocuous to fungi. There is no great difficulty in finding forms which are easy to apply and give a good cover. Indeed a number of proprietary brands, produced especially for tree-wound treatment, are now on the market; some of these are very satisfactory (Atkinson 1937). It has been pointed out that the degree of protection given by substances of this nature depends to some extent on the thickness of the coverings, and it is suggested that bitumen paints should be applied thickly in the centre of the wound, but more thinly round the edges (Wenzl and Müller-Fembeck 1951).

Lanolin promotes a rapid start to healing, probably by keeping the wound moist. Its physical properties can be improved by incorporating resin and pine gum (McQuilkin 1950; Crowdy 1953). No information appears to be available on its protective qualities. Resin mixed with sardine oil and copper soap is said to be a satisfactory dressing (Cooley 1942). Another complex but successful dressing was beeswax, lubricating oil, and chalk, with 1 per cent. 2,4–D and 2 per cent. zinc oxide (Samish, Tamer, and Spiegel 1957).

Experiments on the incorporation of growth-promoting substances in wound dressings, in the hope of increasing cambial growth, have proved rather disappointing (Davis 1949; McQuilkin 1950); but Crowdy (1953) was able to increase the stimulation to early growth exerted by lanolin by the addition of growth substances to it.

York (1941) has suggested an ideal treatment, variations of which might be used to get maximum protection on trees of very high value. He proposes application of shellac to the edges of the wound to protect the cambium from the copper-sulphate solution used to sterilize the exposed wood surface.

This is followed by a complete coat of shellac, a coat of paint, and finally a protective layer of varnish.

Wound dressings which are not toxic to fungal spores can sometimes be dangerous. Walter and Mook (1940) reported that *Endoconidiophora fimbriata* on plane was transmitted in infected asphalt wound dressings, as well as by contaminated pruning tools. Marsden (1952) found that spores of *Ceratocystis ulmi*, the cause of Elm disease, could remain viable for at least a day in white-lead paint, and up to ten days in some forms of bituminous paints; both Bordeaux paint and shellac gave an immediate kill. J. M. Walter (1944) found that phenyl-mercury-nitrate could be mixed with certain varnish protectants to promote fungicidal action, but it was too phytotoxic when used with bituminous paints. Copper sulphate has sometimes been used as an additive, but may lead to chemical instability. There is room for more research on fungicidal additives to wound protectants.

There is some evidence that shading of wounds by a loose covering, such as sacking, hastens healing. Polythene has also been used to protect large wounds (Leiser 1958). Such treatments are obviously impracticable in the forest, but might sometimes be justified on ornamental trees.

TREE SURGERY

Tree surgery may be defined as treatment designed to improve the health, safey, beauty, or longevity of ornamental trees. Some of the techniques are discussed elsewhere in this book. A full but brief description of a tree surgeon's work has been given by C. Cook (1942). Few can afford to do much thorough work, even on their most valuable trees, and unnecessary operations are sometimes done in the name of tree surgery. Here we are only concerned with three aspects: cavity treatment, bracing and cabling, and the surgical treatment of roots.

Old trees often become seriously decayed, or even hollowed, by wood-destroying fungi. Sometimes the damage is so extensive that no treatment is possible; then it becomes necessary to decide whether the shell of live wood is strong enough safely to support the crown, possibly after it has been lightened by pruning. It is often suggested that smaller cavities should be cleaned out and filled. It is important to remember that it is almost impossible to remove all the decayed wood, let alone all the wood containing fungal hyphae, so that even after the best treatment the fungus may remain active behind the filling, especially if it is not fully waterproof. Unless the cavity is properly drained and is sealed watertight, the filling may do more harm than good. The filling, of course, adds nothing to the strength of the tree. It is certainly desirable to remove rotten wood from cavities, wherever it can be done without undue mutilation; it is highly desirable that cavities should be drained, even if this involves boring slanting auger holes up into them through living wood; the actual filling is much less important, and may well be omitted in favour of free air circulation, unless the open hole is considered unsightly, or filling is the only way of stopping water collecting in the cavity. Callus seldom grows right over a filling, though it may if the hole is very small or very narrow and the tissue round it in good health. Cavity treatment should be undertaken only on valuable trees in reasonably good health and with a reasonably long life expectation. General recommendations for cavity treatment are given by Pirone (1948), Le Sueur (1949), and Baxter (1952).

Trees which have forked low may split as the limbs become heavy. This is particularly the case with trees which have several large branches at the same level. They can often be rendered much safer by connecting the branches in pairs, or in groups, and thus using one branch as a counterpoise to another. In extreme cases all the branches may be connected to a single central point. Any attempt to do this by putting the connecting wire ropes round the branches, or by fastening them to bands encircling the branches, is bound to cause girdling. The only safe method is to bore through each branch and attach the wire rope to an eye-bolt, which goes right through the branch and through a small metal plate on the far side. For small branches, hooks screwed into the wood can be used. Careful observation of the movements of the branches in the wind should be made before the position and tension of the cables are decided on. The cables must be high enough to take the strain, but not so high that they restrict normal bending of individual branches. If a tree has already started to split at a crotch, it can sometimes be repaired by passing very long bolts or threaded lengths of rod with a plate on each end through both halves, and tightening. Rods can also be used instead of cable for two branches which are less than 0·5 metre apart. Cabling and bracing, however, should be carried out before the damage occurs. Their use should be considered on any old multi-branched tree which is in a potentially dangerous position or is of especial value. The techniques employed are fully described by Pirone (1948), Le Sueur (1949), and Baxter (1952).

Root pruning

In forestry root pruning may be used to induce flowering in seed-orchard trees. In such cases only comparatively small roots are likely to be severed. There is no evidence that this is likely to lead to disease. Root pruning as an alternative to transplanting in nurseries is even safer. More drastic cutting of roots may be inevitable in the course of road construction. Deep cuttings in the soil should be at least 2 metres from a tree. The distance can be less for a very young tree, but should be more for a really large one. If a trench is involved, it may be possible to dig it nearer, leaving some of the major roots unbroken across the trench. If this is done the roots should be wrapped in bits of sacking to protect them until the trench can be refilled. Severance of roots near the tree, however, is bound to make it unstable, and if such work has to be done the tree is better removed.

Where the cutting can be made at 2 or more metres from the tree the roots should be severed carefully, and the root-ends covered with at least 20 cm. of soil. This should be sloped, so that it does not wash away, and preferably covered with rock or supported by dry-stone walling. These precautions will enable the roots to regenerate, and by degrees restabilize the tree. Erosion of the soil round the severed roots will stop root-regeneration and eventually lead to windblow (Fig. 101).

It may occasionally be desirable to remove dead roots from ornamental trees attacked by *Armillaria mellea* or other root fungi. The removal of such roots from a tree which is not badly attacked may also lessen the infection reservoir from which the fungus could spread farther. It may also occasionally be necessary to remove a live root which is actually girdling the tree from which it arises (Watson, H. 1937). Root removal is naturally much more difficult than aerial pruning. Attempts to deal with the deeper roots probably involve more disturbance than is worth while. Large wounds, especially if they involve live wood, should be treated with a preservative coating (see above).

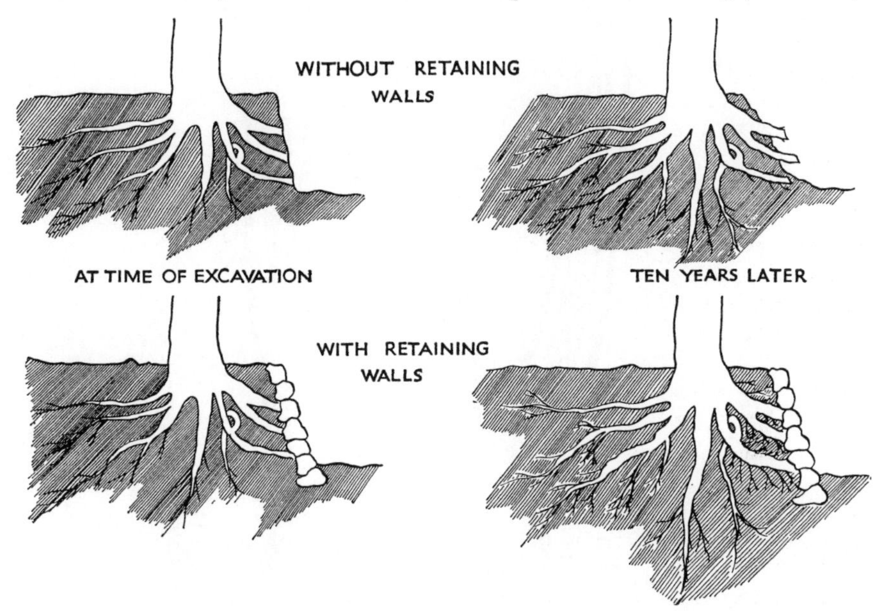

FIG. 101. The treatment of excavated tree roots (see text). (A. C.)

The dumping of soil over roots and the precautions that can be taken to lessen the damage caused thereby are discussed in Chapter 6.

Sterilization of tools

There is one clearly authenticated instance of the spread of a tree disease by pruning tools: *Endoconidiophora fimbriata* on plane in the United States (Walter, J. M. 1946). Probably other vascular diseases could be spread in this way. Tools should be surface sterilized after use on any tree known or suspected to be diseased, regardless of the nature of the disease. Sterilization under other circumstances is probably unnecessary. The best substance for use in the field is methylated spirits.

COPPICE CONVERSION AND DECAY

The practice of cutting hardwoods on short rotations and allowing them to sprout up again from the base to produce small poles has long been a part of English silviculture. With a few exceptions it is now limited to Spanish chestnut, which produces particularly durable poles; but the problem remains of converting existing coppice of other species to high forest, if possible without replanting. This operation of coppice conversion has a strong pathological aspect, which has been investigated for oak in the United States and published in a series of papers (Roth and Sleeth 1939; Roth and Hepting 1943*a*, 1943*b*; Roth, E. R. 1956).

Certain points emerge clearly from the American work. Stems arising from large stumps are more liable to decay than those from small ones. Stems

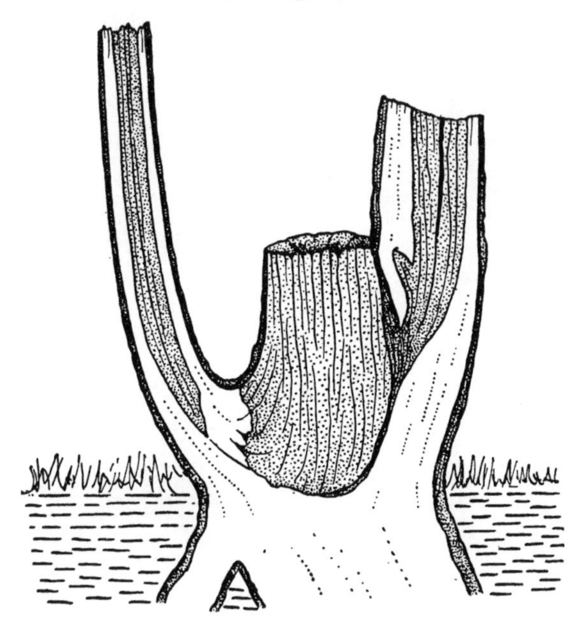

FIG. 102. The heartwood connexions to low and high sprouts on a coppice stump. The high sprout has such a connexion, the low sprout has not. (J. C. after Roth and Sleeth.)

arising low down, and especially from below ground-level, on the parent stump or stool, are less liable to decay than those emerging higher up (Fig. 102). In oak this was found to be merely a question of position affecting the continuity of heartwood between stump and sprout; there was no question of the low stem making its own root system. With chestnut in Britain, however, low sprouts certainly do form semi-independent root systems of their own. In the American oak stands the greatest freedom from decay was found where there had been fire. This had killed all the sprouts and buds above ground and therefore encouraged low sprout production.

They also found that where two stems arose close together, and especially if they were joined at the base, the removal of one often, but not always, led eventually to decay in the other (Fig. 103). This was particularly so if the stem removed was more than 7·5 cm. in diameter, if a stub was left or if it was axe-cut. Close cutting with a saw produced appreciably less decay in the remaining stem. Girdling unwanted stems did not significantly reduce decay in the remainder. Where a stem forked above ground, removal of one half almost invariably resulted in the entry of decay into the other.

This work can be used as a guide in the treatment of coppice conversion or to the raising of high forest from stump sprouts. Those of low origin should be encouraged at the expense of high ones, and companion sprouts, i.e. those nearly in contact at the base, singled if one of them appears at all likely to figure in the final crop. The earlier this is done the better; it might well be started when the stems are only 2 to 3 metres high. This effort to

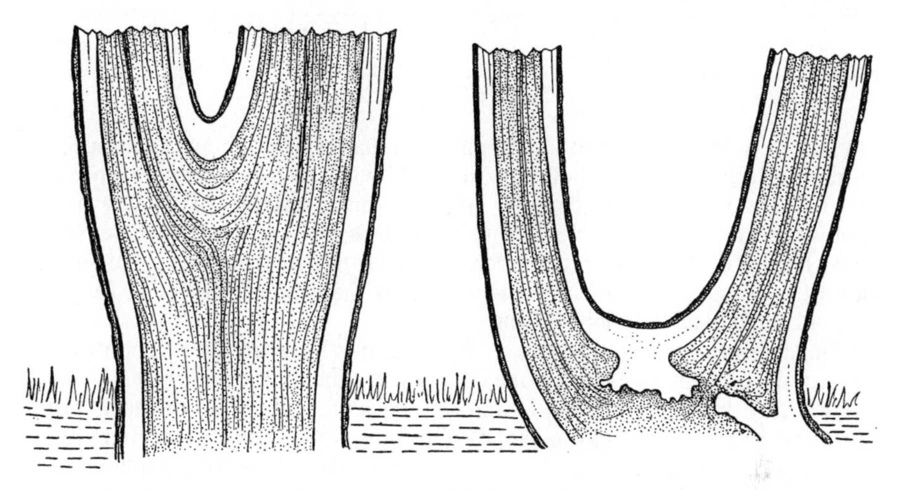

FIG. 103. The heartwood connexions in a tree forking above ground and at ground-level. The connexion is less definite in the latter. (J. C. after Roth and Sleeth.)

encourage low stems should continue throughout all the early thinnings, but as soon as the stems have reached 7·5 cm. average diameter, companion and forked stems should either be removed together or left alone, never singled. The use of a saw as opposed to an axe is a more debatable point. Where there are a lot of shoots from a stump or stool close cutting with a saw may be almost impossible, and an axe may well prove the better tool in practice.

38

QUARANTINE AND OTHER PHYTOSANITARY MEASURES IN RELATION TO TREE DISEASE

IT will be obvious from the foregoing pages that relatively few of the parasitic fungi and bacteria described have a world-wide distribution or even cover all those areas where availability of host and suitability of climate would allow their development. Numerous instances have been quoted where pathogens are known to have spread from one country, or from one continent, to another. In some cases it is known, and in many it is suspected, that this spread took place by man's agency.

On continuous land masses, and where the host tree has a continuous distribution, spread by natural means is almost inevitable and extremely hard to stop. Certainly no normal quarantine measures would be of much value. But where there are breaks in host distribution, for instance a mountain range or a desert, and particularly where there is a sea barrier, natural dissemination becomes much less important, or even non-existent, and man takes the primary position as an agent of spread (Orton and Beattie 1923). Where natural barriers do exist, the introduction of quarantine measures can be expected to strengthen them. Since Britain is entirely surrounded by water, it is obvious that such measures should be particularly useful for preventing, or at any rate postponing, entry of the very large number of pathogens which have not yet reached the country.

A long time ago, Hahn (1927) drew attention to the danger of introduced tree diseases, but unfortunately long before then many of the diseases which now plague exotic conifers in Britain had entered the country. The possibility that they may bring new diseases with them is indeed one of the arguments put forward against the use of exotics (Boyce, J. S. 1954*b*).

Quarantine can be applied in a variety of ways. General reviews of these have been made by Güssow (1936), McCubbin (1946, 1955), and Soraci (1957). The means used fall into three categories, (*a*) measures taken in the exporting country, (*b*) measures taken in the importing country, and (*c*) embargoes whereby passage of specified plants from one country to another is completely forbidden. Either of the first two methods may involve disinfection of plant material, a subject which is dealt with in more detail in Chapter 14 on Seed and in Chapter 36 on the Control of Diseases.

As far as trees are concerned there are three main types of material on which fungi and bacteria are likely to enter, namely seed, plants, and timber. Seed is certainly the smallest risk. Although a good case can be made for the routine treatment of many agricultural seeds (McCubbin 1936; Porter 1941), there is so little evidence of disease transmission on forest tree seeds that no general action can be justified (Chapter 14). Only in the case of *Endothia parasitica* on *Castanea* is there a strong case for seed treatment.

Timber is certainly much more dangerous, especially when it is unbarked. *Ceratocystis ulmi*, the cause of Elm disease, was taken from Europe to America on unbarked logs, and it is possible that *Endothia parasitica*, the cause of Chestnut blight, may have reached Europe from America in the same way. There has been a very general tendency, except in the case of fruit and vegetables, to neglect the risks involved in the movement of final products, partly no doubt because restrictions on them would almost invariably involve interference with trade. Yet in the case of some diseases, for instance *Endothia parasitica* on *Castanea* and on *Quercus*, or *Hypoxylon pruinatum* on poplar, transmission on timber is much more likely than on small plants. Morgan and Byrne (1957) have recently stressed the danger inherent in uncontrolled movements of timber from country to country or even from continent to continent.

There are, of course, examples of the transmission of tree diseases on plants. The most notable of these is the importation of *Cronartium ribicola*, the cause of White pine blister rust, into North America from Europe on nursery stock. Most of the quarantine regulations designed to prevent the entry of tree diseases have referred to living plants, rather than to seed or timber.

Although embargoes obviously give the best chances of success, there is rather general objection to their use (Güssow 1936; Moore, W. C. 1952). Where trade exists, they are bound to be restrictive, and there is always the possibility of their use for economic rather than phytosanitary purposes. A complete embargo on seed would obviously impose a severe restriction on any country, such as Britain, which was largely dependent on exotic conifers. A complete embargo on young trees would cause difficulties with trees such as poplars that are normally raised as vegetatively propagated clonal stocks. A complete embargo on timber would only be possible for a country that possessed a sufficiency of all kinds of timber for its own use. Conditions vary for each country and so, therefore, does the extent to which they can use this, the most potent weapon in the phytosanitary armoury. The extent of its adoption in Britain and the underlying reasons will be discussed later. Even embargoes do not provide complete safety. Spores may enter on aircraft, and passengers may carry prohibited material knowingly or unknowingly in their baggage (Sherman 1957).

It is generally admitted that inspection of imported material on arrival is a very unreliable way of preventing the entry of diseases (Gravatt and Parker 1949; Gram 1955). In the case of large consignments, only a sample can be inspected, and unless the incidence of disease is very high, it may well go undetected. In some cases the disease may be in a stage which is not detectable even by careful visual inspection. For instance, *Chrysomyxa rhododendri* was imported into the United States as mycelium in azalea leaves. By the time fructifications appeared and the disease became detectable, the infected plants had been widely distributed (Gould, Eglitis, and Doughty 1955). In fact inspection on entry is probably of little value, except in so far as it may occasionally lead to the rejection of a very badly diseased consignment or to the detection of material that is the subject of embargo.

It is obvious that there is a better chance of achieving satisfactory inspection in the country of origin. If inspection is started in the producing nursery, it can be extended over a period of time and carried out with a knowledge of some of the possibilities of infection to which the crop is subject. Even then it is impossible to state honestly that any plant is entirely free from all diseases. There is no doubt, however, that efficient inspection, especially in the nursery, can greatly lessen the risk, so that the danger of importing from any particular country depends not only on the pathogens established there, but on the efficiency of its phytopathological inspection service.

The principles underlying quarantine regulation have been discussed by W. C. Moore (1952, 1955) and Soraci (1957). There is a strong tendency to base such legislation on knowledge of specific diseases, in particular on their potential danger and on their distribution, and to condemn 'blanket' embargoes, on the ground that they are based on unreasonable fears of the unknown. However, the almost complete lack of knowledge of tree pathogens in some of the phytopathologically less advanced countries, and the obvious possibility that a known pathogen may behave quite differently when moved to a new environment, both support the use of embargoes, except where there are strong economic reasons against them.

It would be beyond the scope of this book to discuss all the quarantine restrictions on trees which have been imposed by various countries. They have been fully summarized by Ling (1952 and after). The United States, with a very long and detailed list of tree and shrub genera and species, and with their restrictions applying only to specified regions, provides a good example of a painstaking effort to relate quarantine measures to the existence and distribution of known diseases. The British regulations (Anon. 1952*f*) provide, as far as conifers are concerned, an example of the widespread embargo, designed to protect a group of genera against any fungi known or unknown from any region. Apart from occasional special licences used only for scientific purposes, the principal coniferous genera can only be imported into Britain in the form of seed. There should be no difficulty in raising in Britain all the coniferous nursery stock required therein, so that the embargo causes little interference to British forestry. A similar embargo on broadleaved trees, where even the most commonly planted genera only count as second-rank forest species, would have included many species of purely ornamental importance, restrictions on which would be very difficult for horticulturists. Restrictions on the importation of broadleaved trees into Britain have therefore been limited to oak and chestnut on account of *Endothia parasitica*, the cause of Chestnut blight, to elm on account of *Phloem necrosis*, and to poplar on account of *Septoria musiva* and the desirability of restricting importation of clones susceptible to Bacterial canker.

Diseases dangerous to Britain

Although in most countries quarantine regulations have been based on specific diseases, action has usually been taken piecemeal, often when a particular disease in some other country had become sufficiently violent to attract attention. On the other hand, if legislation was introduced against all non-native pathogenic fungi and bacteria found in other countries as soon as they were recorded, quarantine measures would soon cover so wide a field that they would amount to an embargo. In the United States an effort has been made to collect all the relevant information on the behaviour of pathogens on American tree species planted abroad (Spaulding 1956). In the course of writing the present book, similar information has emerged for many of the more important tree and shrub genera grown in Britain. Indeed the justification for describing so many diseases which did not occur in Britain lay mainly in the possibility of their one day reaching this country. This information forms the basis of the list of potentially dangerous pathogens, not yet present in Britain, given below.

The distribution given in the fourth column is usually expressed in very broad terms and in some cases is certainly incomplete. That fact alone indicates a deficiency in our knowledge of plant pathogens and their distribution. There is no doubt that the very high number listed for North America is considerably a reflection of the very active phytopathological work going on there. Other areas are much less well represented, partly because their population of fungal and bacterial pathogens has been less fully explored. Such regions undoubtedly contain pathogens at present unknown, a fact which provides the main justification for the comprehensive embargo which has been used to protect British conifers.

Dangerous pathogens not recorded in Britain

HOST	PATHOGEN	NATURE OF PATHOGEN	KNOWN DISTRIBUTION
Chapter 12			
Juglans	Brooming	Virus	N. America
Paulownia	Brooming	Virus	Japan
Robinia	Brooming	Virus	N. America
Syringa	Ring spot	Virus	Europe, N. America
Ulmus	Phloem necrosis	Virus	N. America
Chapter 13			
Conifers	*Viscum album* (vars. attacking conifers)	Mistletoe	Europe
Conifers	*Arceuthobium* spp.	Mistletoes	N. America
Broadleaved trees	*Phoradendron* spp.	Mistletoes	N. America
Chapter 16			
Conifers (especially Douglas fir)	*Poria weirii*	Fungus	N. America
Broadleaved trees and possibly conifers	*Phymatotrichum omnivorum*	Fungus	N. America
Chapter 18			
Broadleaved trees	*Botryosphaeria ribis*	Fungus	N. America
Chapter 22			
Pinus, including *P. contorta*	*Coleosporium solidaginis*	Fungus	N. America
Pinus, including *P. sylvestris* and *P. contorta*	*Cronartium harknessii*	Fungus	N. America

HOST	PATHOGEN	NATURE OF PATHOGEN	KNOWN DISTRIBUTION
Pinus, including *P. contorta*	*Cronartium stalactiforme*	Fungus	N. America
Pinus, including *P. contorta*	*Cronartium comptoniae*	Fungus	N. America
Pinus	*Atropellis* spp.	Fungus	N. America
Pinus	*Elytroderma deformans*	Fungus	N. America
Chapter 24			
Larix	*Hypodermella laricis*	Fungus	Europe, N. America
Larix	*Mycosphaerella laricina*	Fungus	Europe
Chapter 26			
Abies	*Bifusella faullii*	Fungus	N. America
Chamaecyparis, especially *C. lawsoniana*	*Phytophthora lateralis*	Fungus	N. America
Cupressus, especially *C. macrocarpa*	*Coryneum cardinale*	Fungus	Europe, N. America, S. America
Chamaecyparis and *Cupressus*, especially *C. macrocarpa*	*Monochaetia unicornis*	Fungus	Africa, New Zealand
Juniperus	*Phomopsis juniperovora*	Fungus	N. America
Juniperus (see also *Pyrus*)	*Gymnosporangium juniperi-virginianae* and other spp.	Fungi	N. America
Tsuga	*Melampsora farlowii*	Fungus	N. America
Chapter 27			
Quercus	*Ceratocystis fagacearum*	Fungus	N. America
Quercus (see also *Castanea*)	*Endothia parasitica*	Fungus	Asia, Europe, N. America
Quercus	*Strumella coryneoidea*	Fungus	N. America
Quercus (seed) (see also *Castanea*)	*Sclerotinia pseudo-tuberosa*	Fungus	Europe
Chapter 29			
Castanea	*Endothia parasitica*	Fungus	Asia, Europe, N. America
Castanea (seed) (see also *Quercus*)	*Sclerotinia pseudo-tuberosa*	Fungus	Europe
Chapter 30			
Acer	*Endoconidiophora virescens*	Fungus	N. America
Acer	*Eutypella parasitica*	Fungus	N. America
Acer	*Hypoxylon blakei*	Fungus	N. America
Chapter 31			
Ulmus	*Gnomonia ulmea*	Fungus	N. America
Ulmus	*Gnomonia oharana*	Fungus	Japan
Chapter 32			
Populus	*Septoria musiva*	Fungus	N. and S. America
Populus (especially aspen)	*Hypoxylon pruinatum*	Fungus	N. America
Populus	*Melampsora abietis-canadensis*	Fungus	N. America
Populus	*Melampsora albertensis*	Fungus	N. America, Argentine

HOST	PATHOGEN	NATURE OF PATHOGEN	KNOWN DISTRIBUTION
Populus	*Melampsora magnusiana*	Fungus	Europe, Japan
Populus	*Melampsora medusae*	Fungus	N. America, France?
Populus	*Melampsora occidentalis*	Fungus	N. America
Populus (White)	*Melampsora aecidioides*	Fungus	N. America, S. America, Near East, Europe
Populus	*Venturia populina*	Fungus	S. Europe
Populus	*Septotinia populiperda*	Fungus	N. America, Europe
Populus (particularly aspens)	*Sclerotinia bifrons*	Fungus	N. America
Populus	*Pestalozzia populinigrae*	Fungus	Japan
Chapter 34			
Alnus	*Didymosphaeria oregonensis*	Fungus	N. America
Fraxinus	*Puccinia sparganioides*	Fungus	N. America
Corylus	*Xanthomonas corylina*	Bacterium	N. America, Europe
Corylus	*Cytospora corylicola*	Fungus	Europe
Corylus	*Cryptosporella anomala*	Fungus	N. America
Carpinus	*Melampsoridium carpini*	Fungus	Europe, N. America
Tilia	*Pyrenochaeta pubescens*	Fungus	Europe
Nothofagus	*Cyttaria* spp.	Fungi	S. America, New Zealand
Chapter 35			
Camellia	*Sclerotinia camelliae*	Fungus	N. America, Japan
Cornus	*Elsinoë corni*	Fungus	N. America
Hedera	*Bacterium hederae*	Bacterium	Europe, N. America
Ilex	*Phytophthora ilicis*	Fungus	N. America
Myrica (see also *Pinus*)	*Cronartium comptoniae*	Fungus	N. America
Platanus	*Endoconidiophora fimbriata*	Fungus	N. America
Prunus	*Diobotryon morbosum*	Fungus	N. America
Prunus	*Coccomyces hiemalis*	Fungus	Europe, N. America, etc.
Pyrus (see also *Juniperus*)	*Gymnosporangium juniperi-virginianae* and other spp.	Fungi	N. America

International co-operation

The success of any phytosanitary measures, other than embargoes, obviously depends to a considerable extent on co-operation between importing and exporting countries. If it is agreed that inspection in the country of origin is more satisfactory than inspection on arrival, the importing country must obviously put considerable faith in the inspection service of the exporting one. It should be no reflection on this service if the importing country refuses to take any form of inspection as a complete guarantee of freedom from disease, since many diseases cannot be detected by visual inspection at some stage in their life. The most efficient co-operation in inspection can lessen, but cannot remove, the risk of disease transmission.

There are, however, other ways in which international co-operation can operate. Testing the trees of one country for their resistance to the diseases of another, whether by chance exposure, deliberate exposure, or inoculation, has considerable informative value. There is so much information already to be gleaned in many countries from existing plantations of exotics that it is doubtful if random plantings designed merely for disease-recording purposes would be justified, unless they could be placed in regions where the pathogen population was unexplored. Special plantations designed to test natural reactions to specific diseases, and therefore placed in regions of high infectivity, would be more rewarding, and there is certainly room for their extension. There is also room for expansion of inoculation tests by one country on tree species important to another (Gravatt and Parker 1949; Riker 1957).

Of course none of these methods will eliminate the difference in climate and in other environmental factors between one country and another. For this reason, the disease reactions of a species in one place cannot be taken as a certain indication of its behaviour towards the same pathogen elsewhere, but they do give the best information which can be gleaned without actually moving the pathogen, a method which must naturally be completely eschewed. There is a danger even in moving cultures of pathogens from one country to another (Wheeler 1957). Transference of cultures can be justified only if they serve some useful purpose and provided they are carefully guarded. But the use of an exotic pathogen, or even of an exotic strain of a native pathogen, for inoculation purposes is a pathological crime of grave significance.

Increasing speed and freedom in world travel may eventually lead to a much wider dissemination of plant diseases. Indeed present quarantine restrictions, which are fairly easily applied in a world divided by national barriers of frontier and trade, would be much harder to operate under those conditions of freedom of movement which may one day emerge. The remaining period of national division, when quarantine measures are still possible, should certainly be used to gain all the knowledge we can on the potential behaviour of the diseases that may eventually be spread far more widely than they are now. That information can only be gained by full and unrestricted exchange of knowledge gathered with a view to the requirements of other countries as well as one's own.

LIST OF HOST SPECIES

Abies alba Mill.	Common Silver fir	*A. pectinata* DC.
A. amabilis Forbes		
A. balsamea Mill.	Balsam fir	
A. cephalonica Loud.		
A. concolor Engelm.		
A. grandis Lindl.		
A. lasiocarpa Nutt.	Alpine fir	
A. magnifica Murr.		
A. nobilis Lindl.		*A. procera* Rehd.
A. nordmanniana Spach.		
A. numidica De Lan.	Algerian fir	
A. pindrow Spach.		
A. pinsapo Bois.		
Acer campestre L.	Field maple	
A. ginnala Maxim.		
A. macrophyllum Pursh.	Oregon maple	
A. negundo L.	Box elder	
A. palmatum Thunb.	Japanese maple	
A. platanoides L.	Norway maple	
A. pseudoplatanus L.	Sycamore	
A. rubrum L.	Red maple	
A. saccharinum L.	Silver maple	
A. saccharum Marsh.	Sugar maple	
A. trifidum Hook. & Arn.		*A. buergerianum* Miq.
Aesculus hippocastanum L.	Horse chestnut	
A. pavia L.	Red buckeye	
Ailanthus glandulosa Desf.	Tree of Heaven	*A. altissima* Swingle
Alnus cordata Desf.		
A. glutinosa (L.) Gaertn.	Common alder	
A. incana (L.) Moench.	Grey alder	
A. rubra Bong.	Oregon alder	*A. oregona* Nutt.
Amelanchier canadensis Med.	Snowy mespilus	
Anemone nemorosa L.	Wood anemone	
Araucaria cunninghamia Ait.		
A. excelsa R. Brown	Norfolk Island pine	
A. imbricata Pav.	Monkey puzzle	*A. araucana* (Mol.) K. Koch.
Arbutus menziesii Pursh.	Madrona	
A. unedo L.	Strawberry tree	
Arctostaphylos manzanita Parry		
Aronia melanocarpa Ell.	Black chokeberry	
Arrhenatherun elatius (L.) J. & C. Presl	Oat grass	
Asplenium ruta-muraria L.	Wall-rue	
Aucuba japonica Thunb.		
Berberis hookeri Lam.		
B. thunbergii DC.		
B. vulgaris L.	Common barberry	
Betula ermani Cham.		
B. japonica Sieb. var. szechuanica Schneid.		
B. lenta L.	Black birch	

Betula lutea Michx.	Yellow birch	
B. nana L.		
B. papyrifera Marsh.	Paper birch	
B. populifolia Ait.	Grey birch	
B. pubescens Ehrh.		
B. utilis D. Don	Himalayan birch	
B. verrucosa Ehrh.		*B. pendula* Roth.
Blechnum spicant (L.) Roth.	Hard fern	
Buddleia variabilis Hemsl.		
Buxus sempervirens L.	Box	
Calendula officinalis L.		
Callitris robusta R. Br.		
Calluna vulgaris (L.) Hull.	Heather, Ling.	
Camellia japonica L.		
Caragana arborescens Lam.		
Carpinus betulus L.	Hornbeam	
Carya pecan Engl. & Graebn.	Pecan	
Castanea crenata Zieb. & Zucc.	Japanese chestnut	*C. japonica* Bl.
C. dentata Borkh.	American chestnut	
C. henryi Rehd. & Wils.		
C. mollisima Bl.		
C. sativa Mill.	Spanish or Sweet chestnut	
C. seguinii Dode		
Catalpa bignonioides Walt.	Indian bean	
Cedrus atlantica Man.		
C. deodara Loud.	Deodar	
C. libani Loud.		*C. libanensis* Mirb.
Celastrus scandens L.		
Celtis australis L.		
C. occidentalis L.		
Cercis siliquastrum L.	Judas tree	
Chamaecyparis lawsoniana (Murr.) Parl.	Lawson cypress	
C. lawsoniana cv. 'lycopodioides' Beissn.		
C. nootkatensis Sudw.		
C. obtusa Endl.		
C. pisifera Endl.		
C. thyoides Brit.		
Chamaedaphne calyculata Moench.		
Chenopodium album L.		
Chimonanthus fragrans Lindl.	Winter sweet	
Choisya ternata Humb.		
Clematis vitalba L.	Traveller's Joy	
Colutea arborescens L.	Bladder senna	
Cornus alba L.		
C. florida L.	Flowering dogwood	
C. mas L.	Cornelian cherry	
C. sanguinea L.	Wild dogwood	
Corylus avellana L.	Hazel	
C. avellana cv. 'contorta' Bean		
C. colurna L.	Turkish Hazel	
C. maxima Mill.		
Cotoneaster frigida Wallr.		
C. horizontalis Dec.		
C. salicifolia Franch.		
C. simonsii Baker		

Crataegus mollis Sch.		
C. monogyna Jacq.	Hawthorn	
C. oxycanthoides Thuill.	Hawthorn	
Cryptomeria japonica (L.) D. Don		
Cupressocyparis leylandii Jacks. & Dall.		
Cupressus arizonica Gr.		
C. lusitanica Mill.		
C. lusitanica var. *benthami* Carr.		
C. macrocarpa Gord.	Monterey cypress	
C. sempervirens L.	Italian cypress	
C. sempervirens var. *stricta* Ait.		
C. torulosa Don		
Cydonia lagenaria Loisel.	Japanese quince	*Chaenomeles lagenaria* Koidz.
C. oblonga Mill.	Quince	
Cytisus scoparius L.	Common broom	*Sarothamnus scoparius* (L.) Wimm.
Daphne mezereum L.		
D. odora Thunb.		
Diervilla florida Zieb. & Zucc.		
Dryopteris felix-mas (L.) Schott.	Male fern	
Elaeagnus angustifolia L.		
E. pungens Thunb.		
Erica carnea L.		
E. lusitanica Rud.		
E. tetralix L.	Cross-leaved heath	
Eriobotrya japonica Max.	Loquat	
Eucalyptus gunnii Hook.		
E. niphophila Maid. & Blakeley		
E. parvifolia Camb.		
E. pauciflora Sieb.		
E. perriniana Rodw.		
E. camaldulensis Dehn.		*E. rostrata* Schlecht.
E. vernicosa Hook.		
Euonymus europaeus L.	Spindle-berry	
E. japonicus Thunb.		
E. latifolius Mill.		
Fagus grandifolia Ehrh.	American beech	
F. sylvatica L.	Common beech	
F. sylvatica var. *purpurea* Ait.	Copper beech	
Forsythia intermedia Zabel. cv. 'spectabilis' Kochne.		
F. suspensa Vahl.		
Fraxinus americana L.	White ash	
F. berlandieriana DC.	Arizona ash	
F. excelsior L.	Common ash	
F. excelsior cv. 'monophylla' Desf.		
F. ornus L.	Manna ash	
F. pennsylvanica Marsh.		
Garrya elliptica Dougl.		
Genista anglica L.		
G. sagittalis L.		
G. tinctoria L.	Dyer's greenweed	

Gentiana asclepiadea L.	Willow gentian	
Ginkgo biloba L.	Maidenhair tree	
Gleditschia triacanthos L.	Honey locust	
Hedera helix L.	Ivy	
Hevea braziliensis Muell. Arg.	Rubber	
Hydrangea arborescens L.		
H. hortensis Smith		
Hypericum androsaemum L.		
H. calycinum L.	Rose of Sharon	
H. elatum Ait.		
H. moserianum And.		
H. patulum Thunb.		
Ilex aquifolium L.	English holly	
I. opaca Ait.	American holly	
Juglans cinerea L.	Butternut	
J. mandshurica Maxim.		
J. nigra L.	Black walnut	
J. regia L.	Common walnut	
Juniperus chinensis L.		
J. communis L.	Common Juniper	
J. excelsa Bieb.		
J. oxycedrus L.		
J. procera Hoch.		
J. sabina L.		
J. virginiana L.		
Kalmia latifolia L.	Calico bush	
Kerria japonica DC.		
Laburnum alpinum Bercht. & Presl		
L. anagyroides Med.		*L. vulgare* Presl.
L. vossii Voss		
Larix decidua Mill.	European larch	*L. europaea* DC.
L. × eurolepis Henry	Hybrid larch	
L. gmelini (Rupr.) Gord.		
L. laricina K. Koch.	Tamarack	
L. leptolepis (Sieb. & Zucc.) Gord.	Japanese larch	
L. occidentalis Nutt.	Western larch	
L. sibirica Ladeb.		
Laurus canariensis Webb. & Berth.		
L. nobilis L.		
Lavatera arborea L.	Tree mallow	
Lavendula spica L.	Lavender	
Ledum palustre L.		
Libocedrus decurrens Torr.	Incense cedar	
Ligustrum lucidum Ait.		
L. ovalifolium Hassk.	Oval-leaved privet	
L. vulgare L.	Common privet	
Liquidambar styraciflua L.	Sweet gum	
Liriodendron tulipifera L.	Tulip tree, Yellow poplar	
Lonicera periclymenum L.	Honeysuckle	
Lycium halimifolium Mill.		
Magnolia grandiflora L.		
Mahonia aquifolium (Pursh.) Nutt.		

Mahonia japonica DC.		
Mespilus germanica L.	Medlar	
Metasequoia glyptostroboides Hu & Cheng.	Dawn redwood	
Morus alba L.	White mulberry	
M. nigra L.	Black mulberry	
M. rubra L.	Red mulberry	
Moneses uniflora (L.) A. Gray		
Myrica gale L.	Bog myrtle	
Myrtus communis L.	Myrtle	
Nothofagus antarctica Oerst.		
N. dombeyi Bl.		
N. obliqua Bl.		
N. procera Oerst.		
Nyssa sylvatica Marsh.	Tupelo	
Paulownia imperialis Sieb.		
Philadelphus communis L.	Mock orange, Syringa	
Photinia serrulata Lindl.		
P. villosa DC.		
Picea alba Link.	White spruce	*P. glauca* Voss
P. asperata Mass.		
P. engelmannii Englm.		
P. excelsa Link.	Norway spruce	*P. abies* (L.) Karst
P. jezoensis Maxim.		
P. omorika Purk.	Serbian spruce	
P. orientalis Carr.	Oriental spruce	
P. pungens Englm.	Blue spruce	
P. rubra Link.	Red spruce	
P. sitchensis Carr.	Sitka spruce	
Pinus armandi Franch.		
P. ayacahuite Ehr.	Mexican white pine	
P. banksiana Lamb.	Jack pine	
P. caribaea More.	Slash pine	*P. elliottii* Englm.
P. caribaea var. *hondurensis* Loock.		
P. cembra L.	Arolla pine	
P. contorta (Loud.) Doug.	Lodgepole pine	
P. densiflora Sieb. & Zucc.	Japanese red pine	
P. echinata Mill.	Shortleaf pine	
P. edulis Englm.		
P. excelsa Wall.	Himalayan pine	*P. griffithii* McCl.
P. flexilis James	Limber pine	
P. halipensis Mill.	Aleppo pine	
P. holfordiana Jacks.		
P. jeffreyi Balf.		
P. koraiensis Sieb. & Zucc.		
P. lambertiana Doug.	Sugar pine	
P. leiophylla Schl. & Cham.		
P. longifolia Roxb.		*P. roxburghii* Sarg.
P. merkusii Jungh. & d. Vr.		
P. montana Mill.	Mountain pine	*P. mugo* Turv.
P. monticola Lamb.	Western white pine	
P. muricata Don	Bishop pine	
P. nigra Arn.	Black pine	
P. nigra var. *austriaca* (Hoess.) Asch. & Graebn.	Austrian pine	
P. nigra var. *calabrica* Schn.	Corsican pine	*P. laricio* Poir.

Pinus palustris Mill.	Longleaf pine	
P. patula Schl. & Cham.		
P. peuce Griseb.	Macedonian pine	
P. pinaster Ait.	Maritime pine	*P. maritima* Poir.
P. pinea L.	Stone pine	
P. ponderosa Laws.		
P. pungens Lamb.		
P. radiata D. Don	Monterey pine	*P. insignis* Doug.
P. resinosa Ait.	Red pine	
P. rigida Mill.	Pitch pine	
P. strobus L.	Weymouth pine, Eastern white pine	
P. sylvestris L.	Scots pine	
P. taeda L.	Loblolly pine	
P. thunbergii Parl.		
P. virginiana Mill.		
Platanus acerifolia Willd.	London plane	
P. occidentalis L.	Buttonwood	
P. orientalis L.	Oriental plane	
Polygonum baldschuanicum Regel.		
Polypodium vulgare L.		
Polystichum setiferum (Forsk.) Waynar	Soft shield fern	
Populus adenopoda Maxim.		
P. Aigeiros Duby. (section)	Black poplars	
P. alba L.	White poplar	
P. alba cv. 'bolleana' Lauche		
P. ×'berolinensis' Dipp.	Russian poplar	
P. candicans Ait.		*P. gileadensis* Roul.
P. canescens Sm.	Grey poplar	
P. deltoides Marsh.	American black poplar	
P. deltoides var. *monolifera* Henry		
P. ×*euramericana* (Dode) Guinier (section)	Hybrid black poplars	*P. canadensis* Moench.
P. ×*euramericana* cv. 'brabantica' Houtz.		
P. ×*euramericana* cv. 'eugenei' Simon Louis		
P. ×*euramericana* cv. 'gelrica' Houtz.		
P. ×*euramericana* cv. 'marilandica' Bosc.		
P. ×*euramericana* cv. 'pseudoeugenei'		
P. ×*euramericana* cv. 'regenerata' Henry		
P. ×*euramericana* cv. 'robusta' Schneid.		
P. ×*euramericana* cv. 'serotina' Hart.	Black Italian poplar	
P. ×*euramericana* cv. 'serotina erecta' Houtz.		
P. ×'generosa' Henry		
P. grandidentata Michx.		
P. koreana Rehd.		
P. laurifolia Ledeb.		
P. Leuce Duby. (section)	White and Aspen poplars	
P. leucoides Spach. (section)		
P. maximowiczii Henry		
P. nigra L.	Black poplar	
P. nigra cv. 'italica' Duroi	Lombardy poplar	
P. simonii Carr.		
P. tacamahaca Mill.		
P. tremula L.	Aspen	

Populus tremuloides Michx.		
P. trichocarpa Torr. & Gray		
Prunus amygdalus Stokes	Almond	
P. avium L.	Gean	
P. besseyi Bail.		
P. cerasifera Ehrh. cv. 'pissardi' Carr.		
P. domestica L. var *insititia* Poir.	Bullace	
P. laurocerasus L.	Cherry laurel	
P. mahaleb L.		
P. padus L.	Bird Cherry	
P. pumila L.		
P. serotina Ehrh.		
P. spinosa L.	Blackthorn	
P. subhirtella Miq. cv. 'autumnalis' Mak.		
P. triloba Lind.		
P. virginiana L.		
Pseudolarix amabilis (Nels.) Rehd.		
Pseudotsuga taxifolia (Poir.) Brit.	Douglas fir	*P. douglasii* Carr.
Pteridium aquilinum L.	Bracken	
Pyrus baccata L.	Siberian crab	*Malus baccata* Borkh.
P. communis L.	Wild pear	
P. coronaria Mill.		*Malus coronaria* Mill.
P. floribunda Kirchn.		*Malus floribunda* Sieb.
P. ioensis Brit.		*Malus ioensis* Brit.
P. prunifolia Willd.		*Malus prunifolia* Borkh.
P. purpurea Hill.		
P. salicifolia Pall.		
P. spectabilis Ait.		*Malus spectabilis* Borkh.
Quercus alba L.	White oak	
Q. cerris L.	Turkey oak	
Q. coccinea Moench.	Scarlet oak	
Q. ilex L.	Evergreen oak	
Q. lusitanica Lam.		
Q. mirbeckii Dur.		
Q. nigra L.	Water oak	
Q. palustris L.	Pin oak	
Q. phellos L.	Willow oak	
Q. pubescens Willd.		*Q. lanuginosa* Thuill.
Q. robur L.	Pedunculate oak	
Q. robur cv. 'fastigiata' Kuntz.		
Q. robur var. *pectinata* K. Koch.		
Q. rubra Dur.	Red oak	*Q. borealis* Michx.
Q. sessiliflora Salisb.	Sessile oak	*Q. petraea* (Matt.) Liebl.
Q. stellata Wangh.	Post oak	
Q. toza Bosc.		
Q. velutina Lam.	Black oak	
Q. virginiana Mill.	Live oak	
Rhamnus cathartica L.	Purging buckthorn	
R. frangula L.	Breaking buckthorn	*Frangula alnus* Mill.
Rhododendron ferrugineum L.		
R. hirsutum L.		
R. ponticum L.		
Rhus typhina L.	Sumach	
Ribes alpinum L.		
R. aureum Pursh.		

Ribes cereum Doug.		
R. fasiculatum Sieb. var. *chinense* Max.		
R. lacustre Poir.		
R. nigrum L.	Black currant	
R. sanguineum L.		
R. uva-crispa L.	Gooseberry	
Robina pseudacacia L.	False acacia	
Salix alba L.	White willow	
S. alba cv. 'caerulea' (Sm.) Sm.	Cricket bat willow	
S. alba cv. 'vitellina' Stokes	Golden willow	
S. ×'americana' hort.		
S. atrocinerea Brot.	Common sallow	
S. aurita L.		
S. babylonica L.	Weeping willow	
S. caprea L.	Goat willow	
S. cinerea L.		
S. fragilis L.	Crack willow	
S. hookeriana Benth.		
S. matsudana Koid. cv. 'tortuosa' Rehd.		
S. pentandra L.	Bay willow	
S. purpurea L.		
S. repens L.		
S. scouleriana Barr.		
S. triandra L.	Almond-leaved willow	*S. amygdalina* L.
S. viminalis L.	Common osier	
Sambucus canadensis L.		
S. nigra L.	Elder	
S. nigra cv. 'heterophylla' Endl.		
Santalum album L.	Sandal	
Scolopendrium vulgare Sm.	Hart's-tongue fern.	
Senecio jacobaea L.	Ragwort	
S. smithii DC.		
S. sylvaticus L.	Wood groundsel	
S. vulgaris L.	Groundsel	
Sequoia gigantea Dec.		*Wellingtonia gigantea* Lindl.; *Sequoiadendron giganteum* Buch.
S. sempervirens (Lamb.) Endl.	Redwood	
Sorbus aria (L.) Crantz.	Whitebeam	
S. aucuparia L.	Mountain ash, Rowan	
S. torminalis (L.) Crantz.	Wild service	
Spartina alterniflora Lois.		
S. maritima (Curt.) Fern.		
S. townsendii H. & J. Groves		
Stachyurus praecox Sieb. & Zucc.		
Staphylea trifolia L.	Bladdernut	
Symphoricarpus albus Blake	Snowberry	
Syringa japonica Decne.		
S. vulgaris L.	Common lilac	
Tamarix gallica L.		
T. pentandra Pall.		
Taxodium distichum Rich.	Swamp cypress	
Taxus baccata L.	Yew	
T. brevifolia Nutt.		

Thelypteris phegopteris (L.) Sloss.	Beech fern	
Thuja orientalis L.		
T. occidentalis L.		
T. plicata Don	Western red cedar	
T. standishii (Gord.) Carr.		
Tilia cordata Mill.	Small-leaved lime	*T. parvifolia* Ehrh.
T. euchlora C. Koch.		
T. neglecta Spach.		
T. petiolaris Hook.		
T. platyphyllos Scop.	Large-leaved lime	
T. × *vulgaris* Hayne	Common lime	
Tsuga canadensis (L.) Carr.	Eastern hemlock	
T. caroliniana Englm.		
T. heterophylla (Raf.) Sarg.	Western hemlock	
Tussilago farfara L.	Coltsfoot	
Ulex europaeus L.	Common gorse	
Ulmus alata Michx.		
U. americana L.		
U. glabra Huds.	Wych elm	*U. scabra* Mill. *U. montana* With.
U. hollandica Mill. var. *belgica* Rehd.	Dutch elm	
U. hollandica var. *vegeta* (Loud.) Rehd.	Huntingdon elm	
U. parvifolia Jacq.	Chinese elm	
U. pumila L.		
U. pumila var. *pinnato-ramosa* Henry		
U. stricta Lindl.	Cornish elm	
Umbellularia californica Nutt.	Californian laurel	
Vaccinium corymbosum L.		
V. macrocarpon Ait.		
V. myrtillus L.	Bilberry	
V. oxycoccus L.	Cranberry	*Oxycoccus palustris* Pers.
V. uliginosum L.	Bog whortleberry	
V. vitis-idaea L.	Cowberry	
Veronica hulkeana v. Muel.		
Viburnum davidii Franch.		
V. opulus L.	Guelder rose	
V. tinus L.	Laurustinus	
Vincetoxicum officinale Moench.		
Vitis coignetiae Oull.		
V. inconstans Miq.		*Ampelopsis veitchii* hort. *Parthenocissus tricuspidata* Planch.
V. quinquefolia Lam.		*Parthenocissus quinquefolia* Planch.
V. vinifera L.	Grape vine	
V. vitacea Bean		*Parthenocissus vitacea* Hitch.
Zelkova carpinifolia K. Koch.		*Z. crenata* Spach.
Z. serrata Mak.		*Z. keaki* Mayr.

GLOSSARY

Many of the fungal terms are dealt with more fully in Chapter 11, 'The Organisms associated with the Diseases of Trees'.

Abscission layer. A layer of specialized cells formed at the base of the leaf stalk or sometimes at the base of a twig, before that leaf or twig is shed.

Acervulus (*-i*). An erumpent, cushion-like mass of hyphae, bearing conidia. Characteristic of the Melanconiales.

Adventitious bud. A bud produced from any part of the stem or roots, which is not connected with the strands of bud-bearing tissue arising from the axils of the leaves.

Adventitious root. A root produced on the stem.

Aecial. Pertaining to aecia (see below).

Aecium (*-ia*). A cup-like fruit body in the Uredinales.

Aeciospore. A spore formed in an aecium (see above).

Aerial. Above ground, or outside the substrate in which an organism is growing.

Annulus. A band or ring round the stem of a toadstool resulting from the breaking of the partial veil from the margin of the cap.

Anthracnose. A disease causing limited lesions.

Antibiotic. A chemical compound produced by one organism, which is harmful to another.

Apothecium (*-ia*). A cup or saucer-like ascocarp (see below) of the Discomycetes.

Ascigerous. Bearing asci (see below).

Ascocarp. An ascus-producing structure of the Ascomycetes (see below).

Ascospore. A spore produced in an ascus (see below).

Ascus (*-i*). A sac-like body of the perfect stage of the Ascomycetes, in which the ascospores (usually 8 in number) are produced.

Asexual. Having no sex organs or sexually produced spores; pertaining to spores, etc., produced without a previous sexual stage.

Axillary. Growing in the axil, e.g. buds in the axils of the leaves.

Basidium (*-ia*). The structure on which the sexually produced spores (basidiospores) of the Basidiomycetes develop.

Basidiospore. A spore from a basidium (see above).

Biocenosis. A balanced community of organisms, such as the soil flora and fauna.

Blaze. A shallow excision, removing a portion of the bark with or without wood, from a tree or timber, to leave a readily visible mark.

Blight. A loose term for a disease causing rapid dieback.

Bound water. Water held in close association by the cell contents and therefore not readily extractable.

Brash. The non-utilizable residue left after felling, pruning, or brashing (see below).

Brashing. Removing dead, or sometimes live, branches from trees within a man's reach in a young crop.

Britain. Used throughout to denote England, Scotland, and Wales.

Callus. Tissue that forms over a cut or damaged plant surface.

Cambial. Pertaining to the cambium (see below).

Cambium. Tissue from which secondary growth arises in stems and roots.

Canopy. The cover of branches and foliage formed by tree crowns.

Canopy, closed. A canopy (see above) in which the individual tree crowns are generally in contact with each other.

Carbohydrates. Compounds of carbon, hydrogen, and oxygen formed initially by photosynthesis (see below) in the leaves.

Check. A state in which a tree or crop, while remaining reasonably healthy, grows extremely slowly.

Chemotherapy. The use of chemicals, injected into or absorbed by the plant, against fungi or bacteria in the plant tissues.

Chlamydospore. A thick walled, asexual resting-spore.

Chlorophyll. The green colouring matter found in plants and concerned with photosynthesis (see below).

Chlorosis. An unseasonable yellowing of the foliage, often associated with mineral deficiency.

Chlorotic. Showing symptoms of chlorosis (see above).

Cleaning. Cutting back weed species, climbers, etc., in a sapling crop, either before or at the first thinning.

Cleistocarp. A fungal fruit-body having no special opening.

Clone. Plants derived from one individual by vegetative reproduction and therefore identical.

Clonal. Belonging to a clone (see above).

Collar. The transition zone between stem and root.

Conidium (*-ia*). An asexual spore.

Conidial. Pertaining to conidia (see above).

Conidiophore. A simple or branched hypha (see below) on which the conidia (see above) are produced.

Conk. The large, often bracket-like fruit-body of a Basidiomycete.

Continent, The. Used throughout to denote Europe, omitting Britain.

Conversion. The process of converting wood into any kind of product, particularly logs into timber.

Coppice. Growth arising from trees cut near the ground so that shoots develop from adventitious buds (see above).

Coremium (*-ia*). A sheaf-like group of hyphae (see below), sometimes joined together, generally upright and producing spores at the tips; characteristic of some groups of the Fungi Imperfecti.

Cortex. The outer or more superficial part of an organ, the bark.

Cotyledon. A seed leaf.

Cull. To reject inferior plants in the nursery, or defective logs or parts of logs in the forest.

Culture. A growth of an organism under artificial conditions.

Culturing. Growing an organism under artificial conditions.

Cutting. An unrooted piece of stem, inserted with more than half its length in the ground, with the intention of rooting it (see Set).

Damping off. Rotting of seedlings before or soon after emergence, often at soil-level.

Defoliate. To remove the leaves.

Dioecious. Having male and female flowers on different individuals.

Edaphic. Pertaining to the soil.

Epidemic. A marked and generally rapid increase in a disease.

Epicormic shoot. A shoot arising from a dormant or adventitious bud on a trunk or branch.

Epidermis. The thin outermost protective layer on a plant.

Epiphytic. Growing on, but not parasitic on, another plant.

Erumpent. Breaking through suddenly and irregularly.

Establish. To get a crop to the stage where it no longer needs special protection or tending operations, other than cleaning, pruning, and thinning.

Facultative parasites. Plants (including bacteria, fungi, etc.) having the power to live saprophytically, as well as parasitically, and therefore able to be cultured (see above).

Fasciation. Coalescent development of the branches of a shoot system, often producing a flattened shoot.

Fascicle. A small bundle or tuft, e.g. needle fascicles in pines.

Felling face. An exposed face of a crop left standing when part of it has been clear-felled.

Fluting. Vertical grooves, clefts, or hollows on the lower stem of a tree.

Foliation. Development of leaves.

Frond. A leaf, especially of the compound type, such as those of ferns, etc.

Fructification. A fruit-body of a fungus.

Girdle. To cut through or kill the bark and outer living layers of the wood round the stem or branch of a tree.

Globose. Spherical or globe-shaped.

Grain. The general direction of the fibres or wood elements in timber.

Haustorium (*-ia*). A special hyphal branch for the absorption of food, especially one within a living cell of the host.

Heteroecious. Passing different stages of its life history in different hosts.

Histology. The study of the detailed structure of animal or plant tissues.

Hymenium. The fertile surface of a fungal fruit-body, on which spores are formed.

Hypertrophy. Excessive, and often abnormal growth, due mainly to increase in size of cells.

Hypha (*-ae*). One of the threads of a mycelium (see below).

Hypocotyl. The portion of the stem below the seed-leaves in a seedling.

Immune. Unable to be attacked by an organism.

Imperfect stage. A stage in the life cycle of a fungus in which imperfect (asexual) spores are produced.

Increment. Increase in girth, height, volume, or even value of a tree or crop.

Increment borer. An auger-like instrument with a hollow bit, used to extract cores from timber.

Indigenous. Belonging to the locality; native.

Inoculation. Transference of an organism into or on to a potential host, in order to test pathogenicity.

Inoculum. Material used to make an inoculation.

Insolation. Exposure to the sun's rays.

Intercellular. Between the cells.

Internodal. Between the nodes or main-branch origins on a stem.

Interveinal. Between the veins on a leaf.

Intracellular. Inside a cell, or cells.

Inversion (climatic). A reversal of the normal temperature gradient in the air, so that the temperature increases, instead of decreases, with the height above the ground.

Inversion layer. The layer of static cold air near the ground during an inversion (see above).

Katabatic wind. The movement of cold air downhill under its own weight.

Lamella (-ae), middle. The layer between adjacent cell walls.

Lammas shoot. A continuation of an annual shoot, formed after a pause in growth in the summer.

Leaf primordium (-ia). The earliest stage in the formation of a leaf, while still in the bud.

Lenticel. A ventilating pore in the stem or roots.

Lesion. A circumscribed diseased area on a plant organ.

Lining-out. Transplanting seedlings from seed-beds to rows in the nursery.

Meristem. Tissue formed of cells all capable of multiplication, as found at growing points.

Mesophyll. The internal cells of a leaf.

Mildew. A fungal disease in which the bulk of the mycelium is produced on the surface of the host, looking whitish.

Monospore. Pertaining to a single spore, e.g. monospore culture, a culture originating from a single spore.

Mosaic. Patchy chlorosis of a leaf, usually caused by a virus.

Mulch. A covering over the ground around the base of a plant, to conserve moisture and check weed growth.

Mutant. An individual with transmissible characteristics different from those of the parent form.

Mutation. The production of a mutant (see above).

Mycelial. Appertaining to mycelium (see below).

Mycelial strands. Closely interwoven hyphae forming definite threads, but without any specialized covering.

Mycelium (-ia). A mass of hyphae, the vegetative body of a fungus.

Necrosis. Death of plant cells, usually resulting in darkening of the tissue.

Necrotic. See above.

Nematodes. Microscopic, free-moving, worm-like organisms, sometimes attacking plants, often living in soil; Eel worms.

Nucleus. A complex spheroidal mass, included in and essential to, the life of most cells.

Occlusion. Covering over; usually used in connexion with the healing of wounds.

Occluded. See above.

Obligate parasite. An organism able to exist only as a parasite.

Oospore. A resting spore of the lower fungi (Phycomycetes).

Ostiole. An opening by which spores are released from an ascigerous or pycnidial fruit-body.

Paraphyses. Sterile hyphal elements, mixed with spore-bearing organs on the hymenium (spore-bearing layer) of a fungus.

Parasite. An organism living on, and nourished by, another living organism.

Pathogen. An organism able to cause disease.

Pathogenic. Able to cause disease.

Pathogenicity. Ability to cause disease.

Perfect stage. The stage in the life cycle of a fungus in which spores are formed after sexual fusion.

Peridium (-ia). The wall or limiting membrane of a fungal fruit-body.

Perithecium (-ia). A closed, sub-globose or flask-like fruit-body produced by certain Ascomycetes.

Petiole. A leaf-stalk.

pH. A measure of acidity or alkalinity based on the hydrogen-ion concentration.

Phloem. Tissue of the inner bark, serving for the transport of foodstuffs.

Photosynthesis. The synthesis of carbohydrates from carbon dioxide and water in plant tissue in the presence of light.

Physiological. Concerned with the functions and activities of organisms.

Physiological drought. Drought symptoms arising from disturbance of the water balance within the plant, and not associated with dry soil.

Physiologic specialization. Division of an organism into strains which differ only in their parasitic behaviour.

Phytotoxic. Toxic to plants.

Phytosanitary. For the protection of the health of plants.

Pinnate. Completely divided into separate leaflets.

Pocket-rot. A rot causing small hollow or mycelium-filled cavities in the wood.

Pole-stage. A stage in the development of a tree crop; from the time the lower branches start to die, up to timber size.

Pole-size. See above.

Pollard. A tree that has been cut high, usually above animal-browsing height, to obtain a head of shoots.

Protoplasm. The living contents of a cell.

Provenance. The geographical source or place of origin from which a given lot of seed or plants was collected.

Pycnidium (*-ia*). The fruit-body of the Sphaeropsidales, frequently globose or flask-like.

Pycniospore. A spore from a pycnium (see below).

Pycnium (*-ia*). A pycnidium-like fruit-body produced as one stage in the life history of the Uredinales.

Raceme. An elongated axis, bearing stalked flowers.

Radicle. The first small root produced when a seed germinates.

Recovery shoots. Shoots produced as a reaction to damage.

Resistance. Ability to withstand attacks by an organism, or damage by a non-living agency.

Resistant. Able to withstand, without serious injury, attack by an organism or damage by a non-living agency, but not immune (see above) from such attacks or damage.

Rhizomorph. A specialized thread- or cord-like structure made up of hyphae with a protective covering.

Ride. A forest track, wide enough for vehicular extraction of produce, but not usually metalled

Saprophyte. An organism using dead organic material as food.

Scion. An unrooted portion of a plant; used for grafting or budding on to a rooted stock.

Sclerotium (*-ia*). A firm, frequently rounded, often black mass of hyphae; often acting as a resting body.

Screefing. Scraping away patches of vegetation in order to plant young trees.

Senescence. Ageing.

Septum (*-a*). A wall or division.

Set. An unrooted piece of stem, inserted with less than half its length in the ground, with the intention of rooting it (see Cutting).

Sexual fusion. The fusion of the contents of two unlike cells, resulting, in the fungi, in the production of the fruit-bodies and spores of the perfect stage.

Shake. An internal crack in the heartwood of standing trees, sometimes radial, sometimes following the line of an annual ring.

Spermatium (*-ia*). A separate sexual cell.

Spermogonium (*-ia*). A fruit-body in which spermatia (see above) are produced.

Spore. The final reproductive structure in the Cryptogams. Usually the main means of spread.

Sporidium (*-ia*). A basidiospore (see above) in the Uredinales.

Stag-headed. With dead branches projecting from the upper crown.

Stand. A plantation of trees.

Stock. A rooted plant on to which an unrooted portion of another plant is budded or grafted.

Stomata. The pores on leaves, and sometimes on other parts of plants, through which gaseous exchange takes place.

Stroma (*-ata*). A dense mass of vegetative hyphae in or on which fruiting bodies are produced.

Stromatic. See above.

Substrate. The material on or in which a saprophyte is living.

Succulent shoot. A shoot which is kept rigid mainly by the turgidity of the cell contents.

Susceptibility. Liability to injury by organisms or to damage by non-living agencies.

Susceptible. Unable to withstand attack by an organism or damage by a non-living agency without serious injury.

Symbiotic. Able to live in close and continuous association with an unlike organism.

Systemic. Generally present throughout an

organism, or able to spread internally through an organism.

Target canker. A canker which, because of annual spread and annual healing, develops concentric ridges of tissue, reminiscent of a target.

Taxonomy. The study of the classification of organisms.

Taxonomist. One who studies the classification of organisms.

Telium (*-ia*). A fruit-body of the Uredinales, which produces teliospores; the so-called 'winter stage'.

Telial. See above.

Teliospore. The spore (commonly a winter or resting spore) of the Uredinales, on which the basidia are produced.

Temperature inversion. See 'Inversion (climatic)'.

Thicket stage. The stage in the early life of a plantation in which the branches near the ground are in close contact.

Thinning. A felling made in an immature plantation for the purpose of improving the growth and form of the remaining trees, but without permanently breaking the canopy.

Toadstool. A fruit-body with a rounded cap and a central stalk, produced particularly by the Agaricales.

Transpiration. The process by which water vapour escapes from the leaves and other parts of the living plant.

Tubercle. A small wart-like growth.

Tushing. The extraction of poles and logs by dragging them along the ground.

Tyloses. Irregular cells in cell cavities; cellular intrusions into vessels through the pits in the walls.

Uredium (*-ia*). A fruit-body of the Uredinales, producing uredospores (see below); the so-called 'Summer stage'.

Uredospore. A summer spore of the Uredinales.

Variegation. Irregular distribution of the colouring matter in leaves, leading to white, cream, or yellow patches.

Varietal control. Restrictions on, or control of, the use of plant varieties in an effort to lessen or control disease.

Vascular. Pertaining to the water-conducting tissue of the plant.

Water table. The level at which free water occurs in the soil.

Weeding (*plantation*). The removal of competing growth from plantations, usually by cutting.

Wilt. The collapse of those parts of the plant which are sustained by cell turgidity; a disease capable of causing such effect in a plant.

Winged cork. Corky processes forming irregular ridges on twigs and small branches.

Witches' broom. A dense conglomeration of twiggy growth, arising from abnormally profuse bud development.

Xerophytic. Adapted to dry conditions.

Zonate canker. A canker showing distinct zones of dieback and/or healing tissue.

Zone line. A narrow dark zone in decayed wood, seen as a line in cross-section; often sclerotial (see above).

Zoosporangium (*-ia*). Sexual fruit-body of the lower fungi (Phycomycetes).

Zoospore. Spore produced in a zoosporangium (see above).

BIBLIOGRAPHY

Additional abstract references to *Review of Applied Mycology* (*R.A.M.*) or to *Forestry Abstracts* (*For. Abstr.*) are given for most foreign and the less easily obtained English language journals.

AAGE, B. 1941, 'Vanris paa Eg', *Dansk Skovforen. Tidsskr.* **26,** 389–99. *Biol. Abstr.* 1949, No. 8759.

ABRAHÃO, J. 1948. '*Botrytis cinerea* Pers. parasitando mudas de *Eucalyptus* spp.', *Biológico,* **14,** 172. *R.A.M.,* 1949, p. 258.

ACKERMAN, R. F. 1957. 'The effect of various seedbed treatments on the germination and survival of White spruce and Lodgepole pine seedlings', *Can. Dept. N.A. & N.R. For. Br. Res. Div. Tech. Note,* **63,** 23 pp.

ADAMS, D. F., MAYHEW, D. J., GNAGY, R. M., RICHEY, E. P., KOPPE, R. W., and ALLEN, I. W. 1952. 'Atmospheric pollution in the *Ponderosa* pine blight area, Spokane County, Washington', *Industr. Engng. Chem.* **44,** 1356–65. *R.A.M.,* 1953, p. 47.

——, SHAW, C. G., and YERKES, W. D. 1956. 'Relationship of injury indexes and fumigation fluoride levels', *Phytopathology,* **46,** 587–91.

ADAMS, J. F. 1918. '*Keithia* on *Chamaecyparis thyoides*', *Torreya,* **18,** 157–60.

ADAMS, W. R., and SCHNELLER, M. R. 1939. 'Some physiological responses to close pruning of Northern white pine', *Bull. Vt. Agric. Exp. Sta.* **444,** 26 pp. *For. Abstr.* **2,** 139.

ADDICOTT, F. T., and LYNCH, R. S. 1955. 'Physiology of abcission', *Ann. Rev. Plant Physiol.* 211–38.

ADKIN, B. W. 1945. 'Trees adjacent to highways', *Quart. J. For.* **39,** 125–6.

—— 1946. 'Trees adjacent to highways', *Quart. J. For.* **40,** 108–9.

—— 1947. 'Trees adjacent to highways', *Quart. J. For.* **41,** 23–25.

AHLGREN, C. E., and HANSEN, H. L. 1957. 'Some effects of temporary flooding on coniferous trees', *J. For.* **55,** 647–50.

AICHELE, H. 1952. 'Untersuchungen über die Frostschutzwirkung eines Kalkanstrichs an Obstbäumen', *Ber. dtsch. Wetterdienstes U.S. Zone,* **32,** 70–73. *Hort. Abstr.,* 1953, No. 383.

AICHINGER, E. VON. 1932. 'Höhenstufenumkehr der Vegetation durch Frostlöcher der montanen Stufe in den Karawanken', *Forstarchiv.* **8,** 20–26.

AINSWORTH, G. C. 1937. *The Plant Diseases of Great Britain—a Bibliography.* Chapman & Hall, London, 273 pp.

—— and BISBY, G. R. 1954. *A dictionary of the fungi.* Imp. Mycol. Inst., 475 pp.

AKAI, S., and TAKEUCHI, T. 1954. 'On damping-off disease of some coniferous seedlings caused by soil-inhabiting fungi', *Ann. Phytopath. Soc. Japan,* **18,** 122–4. *For. Abstr.* **17,** No. 1733.

ALBEN, A. O. 1955. 'Preliminary results of treating rosetted Pecan trees with chelated zinc', *Proc. Amer. Soc. Hort. Sci.* **66,** 28–30. *R.A.M.,* 1956, p. 730.

—— and BOGGS, H. M. 1926. 'Zinc content of soils in relation to Pecan rosette', *Soil Sci.* **41,** 329–32.

—— COLE, J. R., and LEWIS, R. D. 1932. 'New developments in treating Pecan rosette with chemicals'. *Phytopathology,* **22,** 979–81.

ALBERTSON, F. W., and WEAVER, J. E. 1945. 'Injury and death or recovery of trees in prairie climate', *Ecol. Monogr.* 393–433. *For. Abstr.* **8,** No. 654.

ALCOCK, N. L. 1924. 'A dieback and bark disease of willows, attacking the young twigs', *Trans. Roy. Scot. Arbor. Soc.* **38,** 128–30.

ALCOCK, N. L. 1926. 'Successional disease in plants as shown by willow rods', *Trans. Brit. Mycol. Soc.* **11,** 161–7.

—— 1928. '*Keithia thujina,* Durand: a disease of nursery seedlings of *Thuja plicata*', *Scot. For. J.* **42,** 77–79.

—— and MAXWELL, I. 1925. 'Successional diseases on willow. (*a*) *Cryptomyces maximus* (Rehm.) Fries. (*b*) *Scleroderris fuliginosa* Karst. (*c*) *Myxosporium scutellatum* Otth. (Petrak)', *Trans. Roy. Scot. Arbor. Soc.* **39,** 34–37.

—— and WILSON, M. 1927. '*Armillaria mellea* on heather', *Scot. For. J.* **41,** 224–5.

ALDHOUS, J. R. 1959. 'Polythene bags for movements of forest nursery stock', *Emp. For. Rev.* **38,** 65–76.

ALEXOPOULOS, C. J., 1952. *Introductory mycology.* John Wiley, New York, 482 pp.

ALLEN, M. C., and HAENSELER, C. M. 1935. 'Antagonistic action of *Trichoderma* on *Rhizoctonia* and other soil fungi', *Phytopathology,* **25,** 244–52.

ALLEN, R. M. 1955. 'Foliage treatments improve survival of longleaf pine plantings', *J. For.* **53,** 724–7.

AMES, R. W. 1952. '*Gloeosporium ulmicolum* reported on fruit of Rock elm and variegated English elm', *Plant Dis. Reptr.* **36,** 301.

ANDERSEN, K. F. 1954. 'Gales and gale damage to forests, with special reference to the effects of the storm of 31st January 1953, in the north-east of Scotland', *Forestry,* **27,** 97–121.

ANDERSON, A. B., ZAVARIN, E., and SCHEFFER, T. C. 1958. 'Nature of some decay-retardant extractive components in Incense Cedar heartwood (*Libocedrus decurrens* Torrey)', *Nature, Lond.* **181,** 1275–6.

ANDERSON, E. A. 1951. 'Healing time for pruned Douglas fir', *Timberman,* **52,** 74–80. *For. Abstr.* **13,** No. 2138.

ANDERSON, M. L. 1930. 'A case of "damping-off" induced by the use of wood-ashes as a manure on seed-beds', *Scot. For. J.* **44,** 7–16.

—— 1956. 'Some biological aspects of forestry', *Quart. J. For.* **50,** 278–89.

ANDERSON, R. L. 1952. 'Factors influencing the incidence of Hypoxylon canker of aspen', *Phytopathology,* **42,** 463.

—— 1956. 'Hypoxylon canker of aspen', *For. Pest. Leafl.* (U.S. Dep. of Agric.), **6,** 1–3. *R.A.M.,* 1957, p. 286.

ANDERSON, R. T. 1937. 'Pruning of green branches of conifers', *Quart. J. For.* **31,** 29–30.

ANDERSSON, E., and STRAND, L. 1951. 'Nagra data från två jämförande försöksodlingar med asp.', *Svensk PappTidn.* **54,** 81–92. *Plant Breed. Abstr.,* 1951, No. 3035.

ANDERSSON, H. 1953–4. 'Kvarstående effekt av hormonderivat på träd och buskar', *Växtstkyddsnotiser Statens Växtskyddsanst., Stockholm,* 59–64. *For Abstr.* **16,** No. 3116.

ANDRÉN, F. 1946. 'Betningförsök med Lupinfrö 1946', *Växtskyddsnotiser, Växskyddsanst., Stockholm,* 91–93. *R.A.M.,* 1947, p. 303.

ANDREWS, S. R. 1955. 'Red rot of Ponderosa pine', *Agric. Monogr.* **23,** 34 pp. *For. Abstr.* **16,** No. 4224.

—— and GILL, L. S. 1939. 'Determining the time branches on living trees have been dead', *J. For.* **37,** 930–5.

—— —— 1941. 'Western red rot control for the Black Hills', *J. For.* **39,** 818–23.

ANDRUS, C. F. 1933. 'Fungous flora accompanying decline of Boxwood, *Plant Dis. Reptr.* **17,** 169–70.

ANGUS, A. 1958. 'Notes on disease of bracken (*Pteridium aquilinum*) in Scotland', *Trans. Bot. Soc. Edinb.* **37,** 209–13.

ANON. 1921. 'Damage by fumes from manufacturing plants in Connecticut', *Amer. J. For.* **19,** 367–73.

ANON. 1930. 'Chronique forestière. La maladie du rond', *Bull. Soc. Centr. for. Belg.* **37,** 522–6. *R.A.M.*, 1931, p. 355.

ANON. 1931. 'Insect pests and fungus diseases of basket willows', *Min. Agric. Bull.* **29,** 14 pp.

ANON. 1932. *Reports on the work of Agricultural Research Institutes and on certain other agricultural investigations in the United Kingdom.* H.M.S.O., London, 377 pp.

ANON. 1933*a*. 'Plantesygdommer i Danmark 1932. Oversigt, samlet ved Statens plantepatologiske Forsøg', *Tidsskr. Planteavl*, **39,** 453–506. *R.A.M.*, 1934, p. 151.

ANON. 1933*b*. 'Gesetze und verordnungen', *Nachr.Bl. dtsch. PflSchDienst.* **13,** 47.

ANON. 1933*c*. 'Amtliche Pflanzenschutzbestimmungen', *Nachr.Bl. dtsch. PflSch-Dienst.* **4,** 159, 165–70. *R.A.M.*, 1933, p. 832.

ANON. 1934. 'Amtliche Pflanzenschutzbestimmungen', *Nachr.Bl. dtsch. PflSchDienst.* **5,** 222–3, 236–43; **9,** 248–50, 261–3, 275. *R.A.M.*, 1934, p. 543.

ANON. 1936*a*. 'Legislative and administrative measures', *Int. Bull. Pl. Prot.* **10,** 263–4, 269–71. *R.A.M.*, 1937, p. 288.

ANON. 1936*b*. 'Gesetze und Verordnungen', *Nachr.Bl. dtsch. PflSchDienst.* **16,** 55. *R.A.M.*, 1936, p. 832.

ANON. 1937. 'Plant pathology', *Rep. Ariz. Agric. Exp. Sta.*, 77–91. *R.A.M.*, 1938, pp. 503–4.

ANON. 1939*a*. *Effect of sulphur dioxide on vegetation*, Nat. Res. Coun. Canada, Ottawa, 447 pp.

ANON. 1939*b*. 'The relation of stand composition to crop security', Report of the Committee on Silviculture, New England Section, Soc. Amer. For., *J. For.* **37,** 49–54.

ANON. 1939*c*. 'The Swiss Leaf-cast disease of Douglas fir', *For. Abstr.* **1,** 69–71.

ANON. 1940. 'Recent work on the pruning of forest trees with particular reference to *Pinus* spp.', *For. Abstr.* **2,** 97–107.

ANON. 1941. 'Disease of Silver Fir in underplanting', *Gdnrs.' Chron.*, Ser. 3, **109,** 236.

ANON. 1945*a*. 'Damage to vegetation due to coal gas leakage through soil', *Gas Jour.* **246,** 532.

ANON. 1945*b*. 'Spring frost damage in orchards and its possible prevention', *Imp. Bur. Hort. & Plant. Crops. Tech. Bull.* **15,** 22 pp.

ANON. 1947*a*. '*Ceratostomella ulmi* Buisman', *C.M.I. Dis. Maps of Plant Disease,* **36** (2nd ed.).

ANON. 1947*b*. 'Protection against forest insects and diseases in the United States', *U.S. For. Serv. Reappraisal Rept.* **5,** 39 pp.

ANON. 1948*a*. 'Larch Canker (*Dasyscypha calycina* Fuckel.)', *For. Comm. Leaflet,* **16,** 8 pp.

ANON. 1948*b*. 'Hail damage to exotic pines in S. Africa', *Rep. Dep. For. S. Afr.*, 1946–7, 4 pp. *For. Abstr.* **10,** No. 2917.

ANON. 1948*c*. 'Phomopsis disease of conifers', *For. Comm. Leaflet,* **14,** 4 pp.

ANON. 1948*d*. *Court of Session Cases*, p. 526.

ANON. 1949*a*. 'Parasitic plant mainly on willows and poplars', *Gdnrs.' Chron.* **126,** 138.

ANON. 1949*b*. 'Watermark disease of the Cricket bat willow', *For. Comm. Leaflet,* **20,** 5 pp.

ANON. 1949*c*. 'The healing of pruning wounds in Douglas fir', *Rep. B.C. For. Serv.*, 1948, 24 pp. *For. Abstr.* **11,** No. 1821.

ANON. 1950*a*. 'Type of thinnings vs. glaze damage in plantations', *Pa. For. Wat.* 86–87. *For. Abstr.* **12,** No. 2299.

ANON. 1950*b*. 'Fungi inhabiting conifers in the Tôhoku district. I. Fungi on Sugi (*Cryptomeria japonica* D. Don.), *Bull. For. Exp. Sta. Meguro, Tokyo,* 27–53. *For. Abstr.* **13,** No. 413.

ANON. 1950*c*. 'Aerial dusting from fixed wing aircraft', *World Crops,* **2,** 383. *R.A.M.,* 1950, p. 573.

ANON. 1951. 'Actes Officiels. Lutte contre le chancre suitant du peuplier', *Rev. for. Franc.* **11,** 722.

ANON. 1952*a*. 'Sooty bark disease of sycamore, *Cryptostroma corticale*', *For. Comm. Leaflet,* **30,** 7 pp.

ANON. 1952*b*. 'Study of fallen snow on the crowns of forest trees', *Bull. For. Exp. Sta. Meguro, Tokyo,* **54,** 115–64. *For. Abstr.* **15,** No. 37.

ANON. 1952*c*. 'Injury to seedlings from soil radiation', *Rep. Dep. Agric. Seychelles,* 1951, 21–22. *For. Abstr.* **14,** No. 2341.

ANON. 1952*d*. *The forest tree breeding station at Ruotsinkylä.* Helsinki, 4 pp. *Plant Breed. Abstr.,* 1954, No. 653.

ANON. 1952*e*. *Air pollution. Proceedings of the United States Technical Conference on air pollution.* McGraw-Hill, New York, 847 pp.

ANON. 1952*f*. *Destructive insect and pest, Great Btitain. The importation of forest trees (prohibition) order,* 1952. H.M.S.O., London, No. 1929.

ANON. 1953. 'International rules for seed testing', *Proc. Int. Nat. Seed testing Assoc., Copenhagen,* 69 pp.

ANON. 1954. 'Chestnut blight and resistant chestnuts', *U.S. Dept. Agric. Farmers' Bull.* **2068,** 21 pp. *R.A.M.,* 1956, p. 558.

ANON. 1955*a*. 'Leaf cast of larch', *For. Comm. Leaflet,* **21,** 4 pp.

ANON. 1955*b*. 'Drought pinpoints the drainage weak spots', *Comm. Grower,* 26 Aug., 403.

ANON. 1955*c*. 'Resistance of Poplar hybrids to rust', *Årsberätt. Fören. Växtförädl. Skogsträd,* 1954 (1955), 11. *For. Abstr.* **18,** No. 4257.

ANON. 1956*a*. 'Oak mildew', *For. Comm. Leaflet,* **38,** 6 pp.

ANON. 1956*b*. Note in *World Crops,* May. *For. Abstr.* **17,** 470.

ANON. 1956*c*. 'Leaf scorch of shade trees', *Control Plant Diseases, Agric. Ext. Serv. Purdue University, Lafayette, Indiana,* No. Mimeo BP 2–4, 2 pp. *For. Abstr.* **17,** 4, No. 4053.

ANON. 1956*d*. 'Two Leaf-cast diseases of Douglas fir', *For. Comm. Leaflet,* **18,** 5 pp.

ANON. 1957*a*. *Weed control handbook.* Brit. Weed. Cont. Coun., 163 pp.

ANON. 1957*b*. '*Fomes annosus*: a fungus causing butt-rot and death of conifers', *For. Comm. Leaflet,* **5,** 10 pp.

ANON. 1957*c*. 'Cypress canker in Tasmania', *Tasm. J. Agric.* **28,** 168–9. *R.A.M.,* 1957, p. 675.

ANON. 1958*a*. '*Keithia* disease of *Thuja plicata*', *For. Comm. Leaflet,* **43,** 5 pp.

ANON. 1958*b*. 'Chestnut blight caused by the fungus *Endothia parasitica*', *For. Comm. Booklet,* **3,** 2 pp.

ANON. 1958*c*. 'Honey fungus', *For. Comm. Leaflet,* **6,** 5 pp.

ANON. 1958*d*. 'Poplars in forestry and land use', *FAO For. & For. Prod. Stud.* **12,** 511 pp.

ANON. 1958*e*. 'Elm disease *Ceratostomella ulmi*', *For. Comm. Leaflet,* **19,** 7 pp.

ANTIPOV, V. G. 1957*a*. 'The effect of smoke and gases discharged by industrial plants on the seasonal development of trees and shrubs', *Bot. Z.* **42,** 92–95. *Hort. Abstr.,* 1957, No. 2752.

—— 1957*b*. 'Dejstvie gazov, vybrasyvaemyh promyslennymi predprijatijami, na semena derev′ev i kustarnikov', *Bot. Ž.* **42,** 1230–2. *For. Abstr.* **19,** No. 4347.

ANZALONE, L., and BAXTER, L. W. 1956. 'Protecting *Camellia* grafts with fungicides', *Phytopathology*, **46**, 6–7.

—— and PLAKIDAS, A. G. 1956. 'Fungicides for control of dieback of Camellias', *Phytopathology*, **46**, 7.

—— —— 1957. '*Cercospora* leaf spot of *Photinia serrulata*', *Phytopathology*, **47**, 515–16.

ANZALONE, L. Jnr., and PLAKIDAS, A. G. 1958*a*. 'Control of *Cercospora* leaf-spot of *Photinia serrulata* with fungicides', *Phytopathology*, **48**, 341.

—— —— 1958*b*. 'Control of flower blight of camellias in Louisiana with fungicides', *Plant Dis. Reptr.* **42**, 804–6.

AOSHIMA, K. 1951. 'Studies on Birch-wood-rotting fungus, *Fuscoporia obliqua* (Pers.) Aoshima, comb. nov.', *Bull. Tokyo Univ. For.* **39**, 185–207. *R.A.M.*, 1952, p. 359.

APPEL, O. 1904. 'Über bestandweises Absterben von Roterlen', *Naturw. Z. Land- u. Forstw.* **2**, 311–20

AREND, J. L. 1955. 'Tolerance of conifers to foliage sprays of 2, 4-D and 2, 4, 5-T in Lower Michigan', *Tech. Note Lake State For. Exp. Stat.* **437**, 2 pp. *Weed Abstr.*, 1956, **5**, 173.

ARK, P. A. 1939. 'Bacterial leaf spot of maple', *Phytopathology*, **29**, 968–70.

—— 1944. 'Pollen as a source of walnut bacterial blight infection', *Phytopathology*, **34**, 330–4.

—— 1958. 'Control of fire-blight of pear with Agri-mycin formulations', *Plant Dis. Reptr.* **42**, 1397–8.

—— and ALCORN, S. M. 1956. 'Antibiotics as bactericides and fungicides against diseases of plants', *Plant Dis. Reptr.* **40**, 85–92.

—— and BELL, M. R. 1959. 'Walnut blast, a new bacterial disease of Walnut', *Plant Dis. Reptr.* **43**, 272–5.

—— and SCOTT, C. E. 1951. 'Walnut blight; three compounds found effective in prebloom-postbloom spray program', *Calif. Agric.* **5**, 7, 14. *R.A.M.*, 1952, pp. 152–3.

ARNAUD, G., and BARTHELET, J. 1933. 'Les Chancres du *Cédrela* et du *Robinia*', *Rev. Path. vég.* **20**, 323–32. *R.A.M.*, 1934, p. 479.

—— —— 1936. 'Le nérume ou pourriture noire des Châtaignes (*Sclerotinia pseudotuberosa* et *Rhacodiella castaneae*)', *Ann. Épiphyt.* (N.S.) **1**, 121–46. *R.A.M.*, 1936, p. 616.

ARNOLD-FORSTER, W. 1952. 'Wind-shelter and wind-tolerance', *J. Roy. Hort. Soc.* **76**, 341–7.

ARSDEL, E. P. VAN., RIKER, A. J., and PATTON, R. F. 1956. 'The effects of temperature and moisture on the spread of White pine blister rust', *Phytopathology*, **46**, 307–18.

ARTAZA, J. E. Y. 1949. *El castaño en España*. Instituto Forestal de Investigaciones y Experiencias, 303 pp.

ARVIDSON, B. 1954. 'En studie av granrötrotans (*Polyporus annosus* Fr.) ekonomiska konsekvenser', *Svenska SkogsvFören. Tidskr.* **52**, 381–412. *For. Abstr.* **17**, No. 1738.

ARX, J. A. VON. 1957*a*. 'Schurft op *Pyrancantha*', *Tijdschr. PlZiekt.* **63**, 198–9. *Biol. Abstr.*, 1958, No. 9717.

—— 1957*b*. 'Über *Fusicladium saliciperdum* (All. et Tub.) Lind', *Tijdschr. PlZiekt.* **63**, 232–6. *R.A.M.*, 1958, p. 252.

—— 1957*c*. 'Revision der zu *Gloeosporium* gestellten Pilze', *Verh. Akad. Wet. Amst.* **51**, 153 pp. *R.A.M.*, 1958, p. 26.

—— 1957*d*. 'Kultur- und Infektionsversuche mit einigen *Colletotrichum*-arten', *Tijdschr. PlZiekt.* **63**, 171–88.

ASHCROFT, J. M. 1934. 'European canker of Black walnut and other trees', *Agric. Exp. Sta. West Virginia Bull.* **261,** 52 pp. *R.A.M.*, 1935, pp. 407–8.

ATANASOFF, D. 1935. 'Old and new virus diseases of trees and shrubs', *Phytopath. Z.* **8,** 197–223. *R.A.M.*, 1935, p. 462.

ATKINS, E. S. 1956. 'The use of chemicals to release White pine reproduction', *Tech. Note For. Br. Can.* **37,** 8 pp. *Weed Abstr.*, 1957, No. 325.

ATKINSON, J. D. 1937. 'Wound dressings for fruit trees', *N.Z. J. Sci. Tech.* **19,** 313–16. *R.A.M.*, 1938, p. 187.

AYTOUN, R. S. C. 1953. 'The genus *Trichoderma*; its relationship with *Armillaria mellea* (Vahl. ex Fries.) Quel. *Polyporus schweinitzii* Fr., together with preliminary observations on its ecology in woodland soils', *Trans. Bot. Soc. Edinb.* **36,** 99–114.

BACHE-WIIG, S. 1940. 'Contributions to the life history of a systematic fungus parasite, *Cryptomycina pteridis*', *Mycologia*, **32,** 214–50. *R.A.M.*, 1940, p. 485.

BADOUX, H. 1922*a*. 'Le *Cenangium abietis*, Duby. un nouveau champignon parasitaire de l'épicéa', *J. For. Suisse*, **73,** 101–4.

—— 1922*b*. 'Ennemis du pin Weymouth', *J. For. Suisse*, **63,** 101–4. *R.A.M.*, 1923, p. 246.

BAGCHEE, K. 1929. 'Investigations on the infestations of *Peridermium complanatum*, Barclay, on the needles, and of *Peridermium himalayense* n. sp., on the stem of *Pinus longifolia*, Roxb. Part I. Distribution, pathological study of the infections, and morphology of the parasites', *Indian For. Rec.* (*Bot. Ser.*), **14,** 24 pp. *R.A.M.*, 1930, p. 146.

—— 1933. 'Investigations on the infestation of *Peridermium himalayense*, Bagchee, on *Pinus longifolia*. Part II. *Cronartium himalayense*, n. sp., on *Swertia* spp. Distribution, morphology of the parasite, pathological study of the infection, biological relationship with the Pine rust, and control', *Indian For. Rec.* (*Bot. Ser.*), **18,** 66 pp. *R.A.M.*, 1934, p. 340.

BAILLIE, A. F. H., and JEPSON, W. F. 1951. 'Bud blast disease of the *Rhododendron* in its relation to the leaf-hopper *Graphocephala coccinea* Forst.', *J. Roy. Hort. Soc.* **76,** 355–65.

BAKER, C. 1953. 'Baled straw for orchard heating', *Gdnrs.' Chron.* **134,** 6.

BAKER, F. S. 1929. 'Effect of excessively high temperatures on coniferous reproduction', *J. For.* **27,** 949–75.

BAKER, W. L. 1948. 'Transmission by leaf hoppers of the virus causing Phloem necrosis of American elm', *Science*, **108,** 307–8. *R.A.M.*, 1949, p. 256.

—— 1949. 'Studies on the transmission of a virus causing Phloem necrosis of American elm, with notes on the biology of its insect vector', *J. Econ. Ent.* **42,** 729–32. *For. Abstr.* **11,** No. 3306.

—— 1950. 'Recent developments on transmission of Elm phloem necrosis disease', Abstr. in *Proc. Ent. Soc. Washington*, **52,** 52. *For. Abstr.* **12,** No. 2431.

BAKSHI, B. K. 1950. 'Fungi associated with Ambrosia beetles in Great Britain', *Trans. Brit. Mycol. Soc.* **33,** 111–20.

BALDACCI, E., and ORSENIGO, M. 1952. 'Chestnut blight in Italy', *Phytopathology*, **42,** 38–39.

BALDASSINI, C. 1948. 'Di un'alterazione di polloni di *Castagno dovuta* ad una specie di *Endothia*', *Ann. Sper. Agri.*, (N.S.) **2,** 677–86. *R.A.M.*, 1949, p. 427.

BALDWIN, H. I. 1928. 'Relative humidity over snow crust and transpiration of conifers', *Amer. Meteorol. Soc. Bull.*, p. 81.

—— 1942. *Forest tree seed of the North temperate regions*, Chron. Bot., 240 pp.

BALFOUR, F. R. S. 1941. 'Low temperatures in January 1941 in S.E. Scotland and their effect on shrubs and trees', *J. Roy. Hort. Soc.* **66,** 230–6.

Bancroft, K. 1911. 'A pine disease, *Diplodia pinea* Kickx', *Kew Bull.* **1**, 60–62.

Banerjee, S. 1956. 'An oak (*Quercus robur* L.) canker caused by *Stereum rugosum* (Pers.) Fr.', *Trans. Brit. Mycol. Soc.* **39**, 267–77.

—— 1957. 'On the biology of *Auricularia auricula-judae* (Linn.) Schroet. causing rot in elder (*Sambucus niger* L.)', *Proceedings of the National Institute of Sciences of India, New Delhi*, **6**, 316–34. *For. Abstr.* **20**, No. 724.

Banks, C. H. 1953. 'Spiral grain and its effect on the strength of South African grown pines', *J. S. Afr. For. Assoc.* **23**, 45–50.

Barber, J. C., Dorman, K. W., and Bauer, E. 1957. 'Slash pine progeny tests indicate genetic variation in resistance to rust', *Res. Note Stheast. For. Exp. Sta.* **104**, 2. *R.A.M.*, 1958, pp. 424–5.

Barnett, H. L., and True, R. P. 1955. 'The oak wilt fungus *Endoconidiophora fagacearum*', *Trans. N.Y. Acad. Sci.*, ser. II, **17**, 552–9. *R.A.M.*, 1955, p. 686.

Barrett, D. K. 1958. 'Cracking in the main stem of Noble fir at Lethen, Nairnshire', *Scot. For.* **12**, 187–90.

Barrett, J. T. 1948. 'A leaf and twig blight of California laurel, *Umbellularia californica* Nutt', *Phytopathology*, **38**, 912–13.

Barriéty, L., Jacquiot, C., Moreau, C., and Moreau, M. 1951. 'La Maladie de l'encre du Chêne rouge (*Quercus borealis* Michx.)', *Rev. Path. vég.* **30**, 253–62. *R.A.M.*, 1953, p. 595.

Bart, G. J. 1957. 'Susceptibility of non-oak species to *Endoconidiophora fagacearum*', *Phytopathology*, **47**, 3.

Barthelet, J. 1934. 'Sur une maladie des Rhododendrons', *Rev. Path. vég.* **21**, 31–35. *R.A.M.*, 1935, p. 314.

—— and Vinot, M. 1944. 'Notes sur les maladies des cultures méridionales', *Ann. Épiphyt.*, (n.s.) **10**, Fasc. Unique., 11–23. *R.A.M.*, 1947, pp. 328–9.

Bartholomew, C. R. 1955. 'Effect of an agricultural spray on forest trees', *Quart. J. For.* **49**, 139–40.

Basile, R. 1954. '*Coryneum microstictum* Berkeley et Br. su foglie di *Arbutus unedo*', *Boll. Staz. Pat. veg., Roma*, ser. 3, **11**, 261–7. *R.A.M.*, 1955, p. 371.

Batchelor, L. D. 1923. 'Methods of harvesting, and irrigation in relation to moldy walnuts', *Calif. Agric. Exp. Sta. Bull.* **367**, 677–96. *R.A.M.*, 1924, p. 437.

Bates, C. G. 1930. 'The frost hardiness of geographic strains of Norway pine', *J. For.* **28**, 327–33.

—— and Roeser, J. 1924. 'Relative resistance of tree seedlings to excessive heat', *U.S. Dept. Agric. Dep. Bull.* **1263**, 1–16.

Batista, A. C. 1951. '"*Cylindrocladium scoparium*" Morgan "var. *brasiliensis*" Batista e Ciferri, um novo fungo do Eucalipto', *Bol. Sec. Agric. Pernambuco*, **18**, 188–91. *R.A.M.*, 1952, p. 524.

Batko, S. 1956. '*Meria laricis* on Japanese and Hybrid larch in Britain', *Trans. Brit. Mycol. Soc.* **39**, 13–16.

—— Murray, J. S., and Peace, T. R. 1958. '*Sclerophoma pithyophila* associated with needle-cast of pines and its connexion with *Pullularia pullulans*', *Trans. Brit. Mycol. Soc.* **41**, 126–8.

Batts, C. C. V. 1949. 'Black rust in south-east Scotland', *Nature, Lond.* **158**, 107.

Bavendamm, W. 1928. 'Neue Untersuchungen über die Lebensbedingungen holzzerstörender Pilze. Ein Beitrag zur Frage der Krankheitsempfänglichkeit unserer Holzpflanzen. I. Mitteilung: Gasversuche', *Centralbl. für Bakt.* **75**, 426, 503. *R.A.M.*, 1929, p. 210.

—— 1935. '*Dothidea noxia* an Amerikanischen Eichen', *Tharandt. forstl. Jb.* **86**, 273–5. *Biol. Abstr.* **10**, No. 12806.

—— 1936. 'Der Rindenbrand der Pappeln', *Tharandt. forstl. Jb.* **87**, 177–9. *R.A.M.*, 1936, p. 471.

BAWDEN, F. C. 1943. *Plant viruses and virus diseases.* Chronica Botanica, Waltham, Mass., 294 pp.

BAXTER, D. V. 1931. 'A preliminary study of *Coleosporium solidaginis* (Schw.) Thüm. in forest plantations in the region of the Lake States', *Pap. Mich. Acad. Sci.* **14,** 245–57. *R.A.M.*, 1931, p. 497.

—— 1933. 'Observations on forest pathology as a part of forestry in Europe', *Univ. Mich. Sch. For. & Cons. Bull.* **2,** 39 pp.

—— 1952. *Pathology in forest practice.* Wiley & Sons, New York, 601 pp.

—— and WADSWORTH, F. H. 1939. 'Forests and fungus succession in the Lower Yukon Valley', *Bull. Sch. For. Mich.* **9,** 52 pp. *R.A.M.*, 1939, p. 642.

BAXTER, L. W., and PLAKIDAS, A. G. 1954. 'Dieback and canker of camellias caused by *Glomerella cingulata*', *Phytopathology*, **44,** 129–33.

BAZZIGHER, G. 1953. 'Beitrag zur Kenntnis der *Endothia parasitica* (Murr.) And., dem Erreger des Kastaniensterbens', *Phytopath. Z.* **21,** 107–32. *R.A.M.*, 1954, p. 569.

—— 1956. 'Pilzschäden an Kastanien nördlich der Alpen', *Schweiz. Z. Forstw.* **107,** 694–5. *R.A.M.*, 1957, p. 673.

BEACH, J. E. G. 1939. 'The pruning of Scots pine and Sitka spruce', *Scot. For. J.* **53,** 85–89.

BEADLE, N. C. W. 1940. 'Soil temperatures during forest fires and their effect on the survival of vegetation', *J. Ecol.* **28,** 180–92.

BEALE, J. H., and BEALE, H. P. 1952. 'Transmission of a ringspot-virus disease of *Syringa vulgaris* by grafting,' *Phytopathology*, **42,** 463.

BEAN, W. J. 1950–1. *Trees and shrubs hardy in the British Isles.* John Murray, London, 3 vols.

BEAR, F. E., *et al.* 1957. 'Chelates in plant nutrition', *Soil Sci.* **84,** 1–97.

BEATTIE, R. K. 1933. 'How the Dutch elm disease reached America', *Proc. 9th Shade Tree Conf.*, 101–5.

—— and DILLER, J. D. 1954. 'Fifty years of Chestnut blight in America', *J. For.* **52,** 323–9.

BEAUMONT, A. 1954. 'Diseases of Rhododendron', *Gdnrs.' Chron.* **136,** 15.

—— 1956. *Diseases of garden plants.* W. H. & L. Collingridge, London, 152 pp.

BECKER-DILLINGEN, J. VON. 1939. 'Die Ernährung des Waldes', *Handbuch des Forstdüngung*, 589 pp.

—— 1940. 'Die Magnesiafrage im Waldbau', *Forstarchiv.* **16,** 88–92.

BEILMANN, A. P. 1938. 'The behaviour of a basswood during an artificial windstorm', *Missouri. Bot. Gard. Bull.*, pp. 50–56.

BENDER, T. R. 1941. '*Fusarium* die-back of American holly', *Plant. Dis. Reptr.* **225,** 403–6.

BENNETT, W. H. 1936. 'An abnormal spruce', *Quart. J. For.* **30,** 133–5.

BENSEND, D. W. 1943. 'Effect of nitrogen on growth and drought resistance of Jack pine seedlings', *Tech. Bull. Minn. Agric. Exp. Sta.* **163,** 63 pp. *For. Abstr.* **7,** 172.

BENZIAN, B. 1955. 'Nutrition problems in forest nurseries, summary report for 1953', *For. Comm. Rep. For. Res.*, 1954, pp. 38–50.

—— 1957. 'Copper deficiency in Poplar', *For. Comm. Rep. For. Res.*, 1957, p. 98.

—— and WARREN, R. G. 1956. 'Copper deficiency in Sitka spruce seedlings', *Nature, Lond.* **178,** 864–5.

BERBEE, J. G. 1957. 'Virus symptoms associated with Birch die-back', *Bi-M. Progr. Rep. Div. For. Biol. Dep. Agric. Can.* **13,** 1. *R.A.M.*, 1957, pp. 795–6.

—— BERBEE, F., and BRENER, W. H. 1953. 'The prevention of damping off of coniferous seedlings by pelleting seed', *Phytopathology*, **43,** 466.

BERG, A. 1940. 'A rust-resistant Red Cedar', *Phytopathology*, **30,** 876–88.

BERK, S. 1948. 'Inoculation experiments with *Polyporus schweinitzii*', *Phytopathology*, **38**, 370–7.

BERNAUX, P. 1956. 'Contribution à l'étude de la biologie des *Gymnosporangium*', *Ann. Inst. Rech. agron. Sér. C* (*Ann. Épiphyt.*), **7**, 1–210. *R.A.M.*, 1957, pp. 278–80.

BERRY, F. H., and GRAVATT, G. F. 1955. 'Walnut bunch disease', *Bull. Calif. Dep. Agric.* **44**, 63–67. *R.A.M.*, 1956, pp. 245–6.

BERTOTTI, F. 1930. 'Ricerche sulle muffe delle Castagne', *Ann. R. Acad. Agric. Torino*, **72**, 47–58. *R.A.M.*, 1931, pp. 696–7.

BESSEY, E. A. 1950. *Morphology and taxonomy of fungi.* Blakiston, Toronto, 791 pp.

BETHLAHMY, N. 1952. 'Why do plants wilt in cold weather?', *Ecology*, **33**, 301–3.

BIER, J. E. 1939*a*. 'Septoria canker of introduced and native hybrid Poplars', *Canad. J. Res.*, pp. 195–204.

—— 1939. 'Hypoxylon canker of maple', *For. Chron.* **15**, 122–3.

—— 1940. 'Studies in Forest Pathology. III. Hypoxylon canker of Poplar', *Tech. Bull. Dep. Agric. Can.* **27**, 40 pp. *R.A.M.*, 1940, p. 505.

—— 1959. 'The relation of bark moisture to the development of canker diseases caused by native, facultative parasites', *Canad. J. Bot.* **37**, 229–38.

—— and FOSTER, F. E. 1946. 'The relation of research in forest pathology to the preparation of forest inventories. 1. Suggested aids for cruising overmature stands of Sitka Spruce on the Queen Charlotte Islands', *B.C. Lumberm.* **30**, 38–40, 64. *For. Abstr.* **8**, No. 1415.

—— and FOSTER, F. E., SALISBURY, P. J. 1946. 'Studies in Forest Pathology. IV. Decay of Sitka spruce on Queen Charlotte Islands', *Tech. Bull. Dep. Agric. Can.* **56**, 1–35. *For. Abstr.* **9**, No. 1274.

—— SALISBURY, P. J., and WALDIE, R. A. 1948. 'Studies in Forest Pathology. V. Decay in Fir, *Abies lasiocarpa* and *A. amabilis*, in the Upper Fraser Region of British Columbia', *Tech. Bull. Dep. Agric. Can.* **66**, 28 pp. *R.A.M.*, 1949, p. 430.

BIGGS, P. 1957. 'Studies on *Meria laricis*, needle-cast disease of Larch', *For. Comm. Rep. For. Res.*, 1956/7, pp. 102–4.

BILHAM, E. G. 1938. *The climate of the British Isles.* Macmillan, London, 347 pp.

BINGHAM, R. T., and EHRLICH, J. 1943. 'A *Dasyscypha* following *Cronartium ribicola* on *Pinus monticola* I, II', *Mycologia*, **35**, 294–311. *R.A.M.*, 1943, p. 412.

—— SQUILLACE, A. E., and PATTON, R. F. 1956. 'Vigor, disease resistance, and field performance in juvenile progenies of the hybrid *Pinus monticola* Dougl. × *Pinus strobus* L.', *Z. Forstgenet.* **5**, 104–12. *R.A.M.*, 1957, p. 147.

BIRAGHI, A. 1938. 'Alcune gravi lesioni prodotte da gas tossici su rami di Nocciolo', *Boll. Staz. Pat. veg. Roma*, **18**, 497–508. *R.A.M.*, 1939, p. 561.

—— 1948. 'Prove di inoculazione di Castagni con *Endothia* sp. isolata in Biscaglia', *Ann. Sper. agr.*, (N.S.) **2**, 687–91. *R.A.M.*, 1949, p. 427.

—— 1949*a*. 'Il disseccamento fogliare del castagno causato da "*Sphaerella maculiformis*" ', *Ital. For. Mont.* **4**, 21–24. *For. Abstr.* **10**, No. 674.

—— 1949*b*. 'Il disseccamento degli Abeti di Vallombrosa', *Ital. For. Mont.* **4**, 1–11. *R.A.M.*, 1950, pp. 184–5.

—— 1951. '*Endothia parasitica* e gen. *Quercus*', *Ital. For. Mont.* **6**, 15–16. *R.A.M.*, 1951, p. 550.

—— 1953*a*. 'Sul proposto raggruppamento di *Endothia fluens* (Sow.) S. et S. e di *Endothia parasitica* (Murr.) P. J. et H. W. And. in un'unica specie', *Boll. Staz. Pat. veg. Roma*, ser. 3, **9** (1951), 133–67. *R.A.M.*, 1953, p. 523.

BIRAGHI, A. 1953*b*. 'Ulteriori notzie sulla resistenza di *Castinea sativa* Mill. nei confronti di *Endothia parasitica* (Murr.) And.', *Boll. Staz. Pat. veg., Roma*, ser. 3, **11**, 149–57. *Hort. Abstr.*, 1955, No. 336.

—— 1954. 'Some important diseases of conifers in Italy', *FAO Plant Prot. Bull.* **2**, 166–7.

—— 1955. 'Il "cancro della corteccia" ed i suoi riflessi sulla crisi del castagno', *Ital. for. mont.* **10**, 3–11. *R.A.M.*, 1957, p. 218.

BIRCH, T. T. C. 1935. 'A *Phomopsis* disease of conifers in New Zealand', *Bull. N.Z. For. Serv.* **7**, 30 pp. *R.A.M.*, 1935, pp. 541–2.

—— 1936. '*Diplodia pinea* in New Zealand', *Bull. N.Z. For. Serv.* **8**, 32 pp. *R.A.M.*, 1937, p. 148.

—— 1937. '*Armillaria mellea* (Vahl.) Quél. in relation to New Zealand forests', *Rep. Aust. N.Z. Ass. Sci.* **23**, 276–9. *R.A.M.*, 1938, p. 714.

BISBY, G. R. 1944. 'The British Hysteriales II', *Trans. Brit. Mycol. Soc.* **27**, 20–28.

—— and MASON, E. W. 1940. 'List of Pyrenomycetes recorded for Britain', *Trans. Brit. Mycol. Soc.* **24**, 127–243.

BITTMANN. 1933. 'Wucherungen an der Schwarzpappel', *Wiener Allg. Forst- u. Jagdztg.* **52**, 50. *R.A.M.*, 1935, p. 134.

BJALLOVIČ, Ju. P. 1957. 'Škala ustojčivosti drevesnyh i kustarnikovyh porod k zatopleniju', *Bot. Z.* **42**, 734–41. *For. Abstr.* **19**, No. 4343.

BJÖRKMAN, E. 1940. 'Om mykorrhizans utbildning hos Tall- och Granplantor, odlade i näringsrika jordar vid olika kvavetillförsel och ljustillgang', *Medd. Skogsförsöksanst. Stockh.* **32**, 23–74. *R.A.M.*, 1940, pp. 422–3.

—— 1942. 'Über die Bedingungen der Mykorrhizabildung bei Kiefer und Fichte', *Symb. Bot. uppsaliens.* **6**, 190 pp. *R.A.M.*, 1945, p. 465.

—— 1944. 'Om skogsplanteringens markbiologiska förutsättningar', *Svenska SkogsvFören. Tidskr.* **42**, 333–55. *For. Abstr.* **8**, No. 61.

—— 1949. 'Studier över snöskyttesvampens (*Phacidium infestans* Karst.) biologi samt metoder for snöskyttets bekämpande', *Medd. SkogsforskInst. Stockh.* **37**, 43–136. *R.A.M.*, 1949, p. 495.

—— 1953. 'Om "Granens gulspetssjuka" i plantskolor', *Svenska SkogsvFören. Tidskr.* **51**, 212–27. *R.A.M.*, 1954, p. 326.

—— 1957. 'Norrländsk tallkräfta (*Dasyscypha fuscosanguinea*) i sådder, och planteringar i höjdlägen', *Skogen*, **44**, 522. *For. Abstr.* **19**, No. 617.

BLAIS, J. R. 1958. 'Effects of defoliation by spruce budworm (*Choristoneura fumiferana* Clem.) on radial growth at breast height of Balsam fir (*Abies balsamea* (L.) Mill.) and White spruce (*Picea glauca* (Moench) Voss.)', *For. Chron.* **34**, 39–47.

BLATTNÝ, C. 1933. 'Vertikálni rozšiřeni virových chorob', *Ochr. Rost.* **13**, 145. *R.A.M.*, 1934, p. 390.

—— 1938*a*. 'Poznámka o méně známých virových chorobách', *Ochr. Rost.* **14**, 86–87. *R.A.M.*, 1938, p. 543.

—— 1938*b*. 'Virová choroba "maly list" a pohárovitost listů Lipy', *Ochr. Rost.* **14**, 80–81. *R.A.M.*, 1938, p. 568.

—— 1956. 'Poznámky o virose smrku. Předběžné sdělení', *Sborn. čsl. Akad. zeměd* (*Lesn.*), **29**, 771–4. *For. Abstr.* **18**, No. 3045.

BLEASDALE, J. K. A. 1952. 'Atmospheric pollution and plant growth', *Nature, Lond.* **169**, 376–7.

—— 1953. 'Air pollution and plant growth', *Smokeless Air*, repr. in *J. Pk. Adm.* **18**, 300–1. *Hort. Abstr.* 1954, No. 1084.

—— 1957. 'Smoke pollution and the growth of plants', *Herb. Abstr.* **27**, 161–5.

BLISS, D. E. 1944. 'Controlling *Armillaria* root rot in *Citrus*', *Calif. Agric. Exp. Sta.* 7 pp. *R.A.M.*, 1945, pp. 225–6.

BLISS, D. E. 1951. 'The destruction of *Armillaria mellea* in *Citrus* soils', *Phytopathology*, **41**, 665–83.

BLOMFIELD, J. E. 1924. 'Witches' brooms', *J. R. Micr. Soc., Lond.*, pp. 190–4.

BLOOMBERG, W. J. 1959. 'Some anatomical evidence for resistance of *Populus* spp. to *Cytospora* canker disease', Abstr. *C.P.S. News*, **11**, 1. *R.A.M.*, 1959, p. 490.

BLUMER, S. 1933. 'Die Erysiphaceen Mitteleuropas mit besonderer Breücksichtigung der Schweiz', *Beitr. Kryptogamenfl. Schweiz*, **7**, 483 pp. *R.A.M.*, 1934, p. 127.

BLYTH, A. W. 1953. 'Reduction of growth in conifers caused by "red belt" in the subalpine region of Alberta', *Silv. Leaf*, **79** (*Dep. Res. & Dev. For. Br.*) *Canada*, 3 pp.

BOLLAND, G. 1957*a*. 'Resistenzuntersuchungen vor allem über Kienzopf und Schütte an der Kiefer', *Züchter*, **27**, 38–47. *R.A.M.*, 1957, p. 799.

—— 1957*b*. 'Untersuchungen über den Drehwuchs an Kiefernmutterbäumen und ihren Nachkommenschaften', *Silvae Genet.* **6**, 148–56. *Plant Breed. Abstr.*, 1958, No. 2245.

BONAR, L. 1942. 'Studies on some California fungi. II', *Mycologia*, **34**, 180–92. *For. Abstr.* **4**, 189.

BOND, G. 1955. 'An isotopic study of the fixation of nitrogen associated with nodulated plants of *Alnus*, *Myrica*, and *Hippophaë*', *J. Exp. Bot.* **6**, 303–11. *For. Abstr.*, 1956, No. 1333.

—— 1956. 'Evidence for fixation of nitrogen by root nodules of alder (*Alnus*) under field conditions', *New Phytol.* **55**, 147–53.

—— FLETCHER, W. W., and FERGUSON, T. P. 1954. 'The development and function of the root nodules of *Alnus*, *Myrica*, and *Hippophaë*', *Plant & Soil*, **5**, 309–23.

BOND, T. E. T. 1936. 'Disease relationships in grafted plants and chimaeras', *Biol. Rev.* **11**, 269–85.

—— 1956. 'Notes on *Taphrina*', *Trans. Brit. Mycol. Soc.* **39**, 60–66.

BONGINI, V. 1933. 'Macchie di seccherreccio delle foglie di Edera', *Difesa Piante*, **10**, 123–30. *R.A.M.*, 1934, p. 309.

BOODLE, L. A. 1924. 'Mistletoe on oaks', *Kew Bull.*, pp. 331–3.

—— and DALLIMORE, W. 1911. 'Report on investigations made regarding Beech coccus', *Kew. Bull.*, pp. 333–43.

BOOTH, C. 1959. *Studies of* Pyrenomycetes IV. *Nectria* (*Part I*). Commonwealth Mycological Institute, 115 pp.

BOOTH, C., and MURRAY, J. S. 1960. '*Calonectria hederae* Arnaud and its *Cylindrocladium* conidial state', *Trans. Brit. Mycol. Soc.* **43**, 69–72.

BORNEBUSCH, C. H. 1941. 'Meddelelser fra Frøudvalget. Agern og Bog fra Holland og Belgien. Afsvampningsforsøg med Agern', *Dansk Skovforen. Tidsskr.* **26**, 371–3. *For. Abstr.* **8**, No. 223.

BORTHWICK, A. W. 1909. 'A new disease of *Picea*', *Notes Roy. Bot. Gdn. Edin.* **20**, 259–61.

BORZINI, G. 1940. 'Sulle cause di un deperimento di piantine di Cipresso'. *Boll. Staz. Pat. veg., Roma* (N.S. 20), **4**, 330–5. *R.A.M.*, 1946, p. 531.

BOSSHARD, H. H. 1955. 'Zur Physiologie des Eschenbraunkerns', *Schweiz. Z. Forstw.* **106**, 592–612. *For. Abstr.* **17**, No. 2188.

BOTTOMLEY, A. M. 1919. 'A preliminary investigation into a disease attacking *Cupressus* plants', *S. Afr. J. Sci.* **15**, 613–17.

BOUDRU, M. 1945. 'La rouille des aiguilles du Thuya géant (*Thuja plicata* Don)', *Bull. Soc. Centr. for. Belg.* **52**, 69–75. *R.A.M.*, 1947, pp. 221–2.

—— 1947. 'La crise du Pin laricio de Corse en Belgique', *Bull. Soc. Centr. for. Belg.* **54**, 49–94. *R.A.M.*, 1947, p. 366.

BOUGHEY, A. S. 1938. 'Honey fungus as a disease of *Rhododendron*', *Gdnrs.' Chron.* **104**, 84. *R.A.M.*, 1938, p. 832.

BOULD, C., NICHOLAS, D. J. D., TOLHURST, J. A. H., and POTTER, J. M. S. 1953*a*. 'Zinc deficiency of fruit trees in Great Britain', *J. Hort. Sci.* **28,** 260–7.

—— NICHOLAS, D. J. D., TOLHURST, J. A. H., and POTTER, J. M. S. 1953*b*. 'Copper deficiency of fruit trees in Great Britain', *J. Hort. Sci.* **28,** 268–77.

BOURCHIER, R. J. 1953. 'Forest disease survey; yellow witches' broom of spruce', *Rep. For. Insect Dis. Surv. Can. 1952,* 123 pp. *For. Abstr.* **14,** No. 3494.

BOUYOUCOS, G. J. 1923. 'Movement of soil moisture from small capillaries to large capillaries of the soil upon freezing', *J. Agric. Res.* **24,** 427–31.

BOWEN, P. R. 1930. 'A maple leaf disease caused by *Cristulariella depraedans*', *Conn. Agric. Exp. Sta. Bull.* **316,** 625–47. *R.A.M.*, 1930, p. 690.

BOWER, D. C. 1954. 'Soil and salt', *Agric.* **61,** 11–15.

BOYCE, J. S. 1918. 'Perennial mycelium of *Gymnosporangium blasdaleanum*', *Phytopathology*, **8,** 161–2.

—— 1920. 'The dry rot of incense cedar', *U.S. Dep. Agric. Bull.* **871,** 58 pp.

—— 1926. 'Observations on the White pine blister rust in Great Britain and Denmark', *J. For.* **24,** 893–6.

—— 1927. 'Decay and seed trees in the Douglas fir region', *J. For.* **25,** 835–9.

—— 1930. 'Decay in Pacific Northwest Conifers', *Yale Univ. Osborn. Bot. Lab. Bull.* **1,** 51 pp. *R.A.M.*, 1931, p. 570.

—— 1933. 'A canker of Douglas fir associated with *Phomopsis lokoyae*', *J. For.* **31,** 664–72.

—— 1940. 'A needle-cast of Douglas fir associated with *Adelopus gäumannii*', *Phytopathology*, **30,** 649–59.

—— 1941. 'Exotic trees and disease', *J. For.* **39,** 907–13.

—— 1943. 'Host relationships and distribution of Conifer rusts in the United States and Canada', *Trans. Conn. Acad. Arts Sci.* **35,** 329–482. *R.A.M.*, 1944, p. 156.

—— 1948. *Forest Pathology*. McGraw-Hill Book Co., 550 pp.

—— 1952. '*Cucurbitaria pithyophila*, an entomogenous fungus', *Plant Dis. Reptr.* **36,** 62–63.

—— 1954*a*. 'Forest plantation protection against diseases and insect pests', *FAO For. Developm. Pap.* **3,** 41 pp. *R.A.M.*, 1956, p. 53.

—— 1954*b*. 'Introduction of exotic trees. Dangers from diseases and insect pests', *Unasylva*, **8,** 8–14. *R.A.M.*, 1954, p. 570.

—— 1957*a*. 'The fungus causing Western gall rust and Woodgate rust of pines', *For. Sci.* **3,** 225–34.

—— 1957*b*. 'Limitations of forest disease control by breeding and selection', *Proc. Soc. Amer. For. Syracuse, N.Y.*, pp. 54–55.

BOYCE, J. S., and WAGG, J. W. B. 1953. 'Conk rot of old-growth Douglas fir in Western Oregon', Joint Publ. *Ore. For. Prod. Lab. & Res. Div. Ore. For. Dep. Bull.* **4,** 96 pp.

BOYCE, J. S., Jr. 1951. '*Lophodermium pinastri* and needle browning of Southern pines', *J. For.* **49,** 20–24.

—— 1952. 'A needle blight of Loblolly pine caused by the brown-spot fungus', *J. For.* **50,** 686–7.

—— 1959. 'Brown spot needle blight on Eastern White Pine', *Plant Dis. Reptr.*, **43,** 420.

BOYD, J. D. 1950. 'Tree growth stresses. II. The development of shakes and other visual failures in timber', *Aust. J. Appl. Sci.* **1,** 296–312. *Biol. Abstr.*, 1953, No. 17716.

BOYER, M. G. 1958. 'Phytotoxic action of *Endoconidiophora fagacearum* Bretz.', *Dissert. Abstr.* **18,** 1954. *R.A.M.*, 1959, p. 230.

BRAATHE, P. 1950. 'Granas veksthemning på lyngmark', *Tidsskr. Skogbr.* **58,** 42–45. *For. Abstr.* **13,** No. 1948.

—— 1957. 'Is there a connection between the birch die-back and the March thaw of 1936?', *For. Chron.* **33,** 358–63.

BRAGONIER, W. H. 1949. 'Umbrella disease of *Rhus glabra* and *R. typhina* caused by *Botryosphaeria ribis*', *Phytopathology*, **39,** 3.

BRAID, K. W., and TERVET, I. W. 1937. 'Certain botanical aspects of the dying-out of heather (*Calluna vulgaris*, Hull)', *Scot. J. Agric.* **20,** 365–72.

BRAUCHER, O. L., and SOUTHWICK, R. W. 1941. 'Correction of manganese-deficiency symptoms of walnut trees', *Proc. Amer. Soc. Hort. Sci.* **39,** 133–6. *R.A.M.*, 1943, p. 185.

BRAUN, H., and HUBBES, M. 1957. 'Sporeninfektion und Antagonismus bei *Dothichiza populea*', *Naturwissenschaften*, **44,** 333. *For. Abstr.* **19,** 1, No. 625.

BRAUN, H. J. 1958. 'Untersuchungen über den Wurzelschwamm *Fomes annosus* (Fr.) Cooke', *Forstwiss. Cbl.* **77,** 65–88. *For. Abstr.* **20,** No. 710.

BRETT, I. 1954. 'Observations on the damage caused by sea water to plantations in Norfolk and Suffolk', *Quart. J. For.* **48,** 217–18.

BRETZ, T. W. 1950. 'Seed transmission of the Elm mosaic virus', *Phytopathology*, **40,** 3–4.

—— 1952*a*. 'The ascigerous stage of the Oak wilt fungus', *Phytopathology*, **42,** 435–7.

—— 1952*b*. 'New hosts for the Oak wilt fungus, *Chalara quercina* Henry', *Phytopathology*, **42,** 3.

—— 1955. 'Some additional native and exotic species of *Fagaceae* susceptible to Oak wilt', *Plant Dis. Reptr.* **39,** 495–7.

—— and BUCHANAN, W. D. 1957. 'Oak wilt fungus not found in acorns from diseased tree', *Plant Dis. Reptr.* **41,** 546.

—— and JONES, T. W. 1958. 'Oak wilt distribution through 1957', *Plant Dis. Reptr.* **42,** 710.

—— and LONG, W. G. 1950. 'Oak wilt fungus isolated from Chinese chestnut', *Plant Dis. Reptr.* **34,** 291.

BRIAN, P. W. 1954. 'Recent work on systemic fungicides and bactericides', *Rep. 5th Commonwealth Myc. Conf.*, pp. 86–96.

—— HEMMING, H. G., and MCGOWAN, J. C. 1945. 'Origin of a toxicity to mycorrhiza in Wareham heath soil', *Nature, Lond.* **151,** 637–8.

BRIANT, A. K., and MARTYN, E. B. 1929. 'A leaf-spot of *Arctostaphylos manzanita*', *Trans. Brit. Mycol. Soc.* **14,** 221–5.

BRICHET, J. 1950. 'La protection des arbres contre les violences solaires', *Fruits & Primeurs*, **20,** 204–6. *R.A.M.*, 1951, p. 4.

BRIERLEY, P. 1944. 'Viruses described primarily on ornamental or miscellaneous hosts', *Plant Dis. Reptr.*, Suppl. **150,** 410–82.

—— 1957. 'Virus-free hydrangeas from tip cuttings of heat-treated ringspot-affected stock plants', *Plant Dis. Reptr.* **41,** 1005.

—— and LORENTZ, P. 1957*a*. 'Hydrangea ringspot virus, the probable cause of "running-out" of the florists' hydrangea', *Phytopathology*, **47,** 49–53.

—— and LORENTZ, P. 1957*b*. 'Wisteria mosaic and Peony leaf curl, two diseases of ornamental plants caused by viruses transmissible by grafting but not by sap inoculation', *Plant Dis. Reptr.* **41,** 691–3.

BRINK, G., and ENDE, G. v. d. 1951. 'Verslag van het onderzoek naar de populierenkanker in 1948 en 1949', *Meded. Ned. Heidemaatschappij*, **13,** 15 pp. *For. Abstr.* **14,** No. 3493.

BROEKHUIJSEN, M. J. 1934. 'Wilgenkanker, veroorzaakt door *Discella carbonacea* (Fries) Berk et Br.' *Tijdschr. PlZiekt.* **40,** 62–63. *R.A.M.*, 1934, p. 550.

BROEKHUIZEN, S. 1929. '*Wondreaksies van hout. Het ontstaan van thyllen en wondgom in het biezonder in verband met de Iepenziekte*', thesis. Univ. Utrecht, 78 pp. *R.A.M.*, 1930, p. 80.

BRONCHI, P. 1956. 'Origine dell'abetina pura artificiale nella Foresta Domaniale di Badia Prataglia in relazione ai recenti danni da "*Fomes annosus*" su "*Abies alba*",' *Monti e Boschi*, **7**, 368–73. *R.A.M.*, 1957, p. 363.

BROOKS, C. E. P., and DOUGLAS, C. K. M. 1956. 'Glazed frost of January 1940', *Air Ministry Met. Office Geo. Mem.* **78**, 40 pp. *Nature, Lond.* **178**, 579.

BROOKS, F. T. 1910. '*Rhizina undulata* (attacking newly planted conifers)', *Quart. J. For.* **4**, 308–9.

—— 1953. *Plant Diseases*. Second ed. Oxford University Press, 457 pp.

—— and WALKER, M. M. 1935. 'Observations on *Fusicladium saliciperdum*', *New Phytol.* **34**, 64–67.

BROUWER, F. I. 1945. 'Over het voorkomen van de zwan *Pithya cupressi* (Batsch.) Fuckel in Nederland', *Tijdschr. PlZiekt.* **51**, 54–56.

BROWER, A. E. 1949. 'The beech scale and beech bark disease in Acadia National Park', *J. Econ. Ent.* **42**, 226–8. *For. Abstr.* **10**, No. 1501.

BROWN, A. B. 1930. 'Observations on leaf fall in the Douglas fir when infected with *Rhabdocline pseudotsugae* Sydow', *Ann. Appl. Biol.* **17**, 745–54.

BROWN, A. W. A. 1951. *Insect control by chemicals.* John Wiley, New York, 817 pp.

BROWN, J. G., and SCANDONE, T. G. 1953. 'A bacterial parasite of tamarisk', *Plant Dis. Reptr.* **37**, 524–5.

BROWN, M. R. 1938. 'A study of crown rust, *Puccinia coronata* Corda, in Great Britain. II. The aecidial hosts of *P. coronata*', *Ann. Appl. Biol.* **25**, 506–27.

BROWN, N. A. 1932. 'Canker of ash trees produced by a variety of the Olive-tubercle organism, *Bacterium savastanoi*', *J. Agric. Res.* **44**, 701–22.

—— 1936. 'Privet and jasmine galls produced by a species of *Phomopsis*', *Phytopathology*, **26**, 795–9.

BROWN, W. 1915. 'Studies in the physiology of parasitism. I. The action of *Botrytis cinerea*', *Ann. Bot., Lond.* **29**, 313–48.

—— 1916. 'Studies in the physiology of parasitism. III. On the relation between the "infection drop" and the underlying tissue', *Ann. Bot., Lond.* **30**, 399–406.

—— 1917. 'Studies in the physiology of parasitism. IV. On the distribution of cytase in cultures of *Botrytis cinerea*', *Ann. Bot., Lond.* **31**, 489–98.

BROWN, W. R. 1957. 'Ivy and trees', *Quart. J. For.* **51**, 161–2.

BRUCE, C. E. R., and GOLDE, R. H. 1949. 'Lightning', *Research*, **2**, 12–19.

BRUCE, D. 1956. 'Effect of defoliation on growth of Longleaf pine seedlings', *For. Sci.* **2**, 31–35.

BRUYN, H. L. G. DE. 1922. 'The saprophytic life of *Phytophthora* in the soil', *Meded. LandbHogesch. Wageningen*, **24**, 1–38. *R.A.M.*, 1922, p. 399.

—— 1924. 'The *Phytophthora* disease of lilac', *Phytopathology*, **14**, 503–17.

—— 1928. 'Is ontbladering als bestrijdingswijze tigen Phytoppthoraziekte van de Seringen gewenschr?', *Tijdschr. PlZiekt.* **34**, 233–8.

BRYAN, M. K. 1928. 'Lilac blight in the United States', *J. Agric. Res.* **36**, 225–35.

BRYŃSKI, K. 1930. 'Przyczynek do znajomości wpływu wiatrów na wzrost drzew', *Kosmos*, **55**, 734–6. *Biol. Abstr.*, 1932, No. 12647.

BUCHANAN, T. S. 1938. 'Blister rust damage to merchantable Western white pine', *J. For.* **36**, 321–8.

—— HARVEY, G. M., and WELCH, D. S. 1951. 'Pole blight of Western white pine: a numerical evaluation of the symptoms', *Phytopathology*, **41**, 199–208.

BUCHWALD, N. F. 1937. '*Berberis*-Arternes Modtageligheid for Sortrust (*Puccinia gramminis*)', *Gartnertidende*, 5 pp. *R.A.M.*, 1938, p. 603.

BUCKLAND, D. C. 1948. '*Fusarium* top blight of Douglas fir seedlings at Duncan, V.I.', *For. Chron.* **24**, 326.
—— 1953. 'Observations on *Armillaria mellea* in immature Douglas fir', *For. Chron.* **29**, 344–6.
—— and KULJT, J. 1957. 'Unexplained brooming of Douglas fir and other conifers in British Columbia and Alberta', *For. Sci.* **3**, 236–42.
—— MOLNAR, A. C., and WALLIS, G. W. 1954. 'Yellow laminated root rot of Douglas fir', *Canad. J. Bot.* **32**, 69–81.
—— REDMOND, D. R., and POMERLEAU, R. 1957. 'Definitions of terms in forest and shade tree diseases', *Canad. J. Bot.* **35**, 675–9.
BUDDENHAGEN, I. W. 1955. '*Phytophthora cinnamomi* (Rands) bij *Chamaecyparis lawsoniana* (Parl.) en *Taxus baccata* (L.) in Nederland', *Tijdschr. PlZiekt.* **61**, 17–18. *For. Abstr.* **16**, No. 3138.
—— and YOUNG, R. A. 1957. 'A leaf and twig disease of English holly caused by *Phytophthora ilicis* n. sp.,' *Phytopathology*, **47**, 95–101.
BUDDIN, W., and WAKEFIELD, E. M. 1929. 'Further notes on the connection between *Rhizoctonia crocorum* and *Helicobasidium purpureum*', *Trans. Brit. Mycol. Soc.* **14**, 97–99.
BUISMAN, C. 1931. 'Three species of *Botryodiplodia* (Sacc.) on Elm trees in the United States', *J. Arnold Arbor.* **12**, 289–96. *R.A.M.*, 1932, p. 212.
—— 1935. 'Sensibilité de diverses espèces et variétés d'Orme à *Ceratostomella ulmi*', *Rev. Path. vég.* **22**, 200–8. *R.A.M.*, 1936, p. 125.
BULGER, R. O. 1954. 'Barberry eradication', *Agric. Chem.* **9**, 60–61. *R.A.M.*, 1955, p. 711.
BULL, R. A. 1951. 'A new gall disease of *Sequoia sempervirens*', *Gdnrs.' Chron.*, ser. 3, **130**, 110–11. *R.A.M.*, 1951, p. 638.
BULLER, A. H. R. 1909. *Researches on the Fungi.* Longmans, Green & Co., London, 1, 274 pp.
BURGER, H. 1940. 'Physikalische Eigenschaften von Wald und Freilandböden. VI. Der Wald als Erholungsstätte und der Waldboden', *Mitt. schweiz. Zent. Anst. forstl. Versuchsw.* **21**, 223–49. *For. Abstr.* **3**, 102–3.
—— 1941. 'Der Drehwuchs bei den Holzarten. I. Drehwuchs bei Fichte und Tanne', *Mitt. schweiz. Zent. Anst. forstl. Versuchsw.* **22**, 142–63. *For. Abstr.* **4**, 110.
—— 1944. 'Johannistriebe der Lärche', *Zbl. ges. Forstw.* **76**, 10–12. *For. Abstr.* **6**, 142.
BURTT, DAVY J., and DAY, W. R. 1958. 'Cultivation of the Cricket bat willow', *For. Comm. Bull.* **17**, 34 pp.
BUSH, R. 1946. *Frost and the fruitgrower.* Cassell, London, 119 pp.
BUSNEL, R. G., DARFOUX, H., and RIDÉ, M. 1951. 'Utilisation de la chaleur transmise par le rayonnement infrarouge comme méthode de désinfection des Châtaignes contre les spores d'*Endothia parasitica* (Murrill) Anderson', *C.R. Acad. Agric. Fr.* **37**, 513–15. *R.A.M.*, 1952, p. 261.
BUTIN, H. 1955. 'Über den Einfluss des Wassergehaltes der Pappel auf ihre Resistenz gegenüber *Cytospora chrysosperma* (Pers.) Fr., *Phytopath. Z.* **24**, 245–64. *For. Abstr.* **17**, No. 4111.
—— 1956*a*. 'Beobachtungen über das vorjährige Aufreten der *Dothichiza*-Krankheit der Pappel', *NachrBl. dtsch. PflSchDienst.* **8**, 55–58. *R.A.M.*, 1956, p. 797.
—— 1956*b*. 'Untersuchungen über Resistenz und Krankheitsanfälligkeit der Pappel gegenüber *Dothichiza populea* (Sacc. et Br.)', *Phytopath. Z.* **28**, 353–74. *For. Abstr.* **19**, No. 622.
—— 1957*a*. 'Die blatt- und rindenbewohnenden Pilze der Pappel unter besonderer

Berücksichtigung der Krankheitserreger', *Mitt. biol. Bundesanst. Land- u. Forstw., Berl.* **91,** 64 pp. *For. Abstr.* **19,** No. 1934.

BUTIN, H. 1957*b*. 'Über die Hauptfruchtform von *Dothichiza populea* Sacc. et Briard', *NachrBl. dtsch. PflSchDienst.* **9,** 69–71. *R.A.M.*, 1958, p. 188.

—— 1957*c*. 'Die jahreszeitlichen Wassergehaltsschwankungen und die Verteilung des Wassers in Stecklingen und im Stamm 2-jähriger Pappeln', *Ber. dtsch. bot. Ges.* **70,** 157–66. *R.A.M.*, 1958, p. 684.

—— 1958. 'Untersuchungen über ein Toxin in Kulturfiltraten von *Dothichiza populea* Sacc. et Br.', *Phytopath. Z.* **33,** 135–46. *R.A.M.*, 1959, p. 103.

—— and SCHUEPP, H. 1959. 'Schäden an Kastanien nördlich der Alpen', *Mitt. schweiz. Zent. Anst. forstl. Versuchsw.* **35,** 267–72.

BUTLER, E. J. 1930. 'Some aspects of the morbid anatomy of plants', *Ann. Appl. Biol.* **17,** 175–212.

—— and JONES, S. G. 1949. *Plant pathology*. Macmillan, London, 979 pp.

BUXTON, A. 1942. 'Norfolk sea floods. General effects of the February 1938 flood seen in 1942', *Trans. Norfolk and Norwich Naturalists' Soc.* **15,** 332–41. *For. Abstr.* **10,** No. 2535.

BYLTERUD, A. 1956. 'Kjemiske midler mot teinung og ugras', *Norsk. Skogbr.* **2,** 333–6. *Weed Abstr.*, 1957, No. 337.

BYRAM, G. M., and NELSON, R. M. 1952. 'Lethal temperatures and fire injury', *Res. Notes Stheast. For. Exp. Sta.* **1,** 1 p. *For. Abstr.* **13,** No. 3957.

CABORN, J. M. 1957. 'Shelterbelts and microclimate', *For. Comm. Bull.* **29,** 135 pp.

CADMAN, C. H. 1940. 'Graft-blight of lilacs', *Gdnrs.' Chron.* **107,** 25.

CADMAN, W. A. 1953. 'Shelterbelts for Welsh hill farms', *For. Comm. For. Rec.* **22,** 32 pp.

CAIN, S. A. 1944. *Foundations of plant geography*. Harper Bros., New York, 556 pp.

CALAVAN, E. C., WAMPLER, E. L., SUFFICOOL, J. R., and ORMSBY, H. W. 1952. 'Grey mould infections', *Citrus Leaves*, **32,** 10–11. *R.A.M.*, 1953, p. 555.

CALLAHAN, K. L. 1957. 'Pollen transmission of Elm mosaic virus', *Phytopathology*, **47,** 5.

—— and MOORE, J. D. 1957. 'Prunus host range of Elm mosaic virus', *Phytopathology*, **47,** 5.

CALLAN, E. McC. 1939. '*Cryptorrhynchus lapathi* L. in relation to the Watermark disease of the Cricket-bat willow', *Ann. Appl. Biol.* **26,** 135–7.

CALLEN, E. O. 1938. 'Some fungi on the yew', *Trans. Brit. Mycol. Soc.* **22,** 94–106.

CAMBONIE, L. 1932. 'Nos Châtaigniers sont malades; la jaunisse ou maladie des taches de feuilles', *La Vie Agric. et Rurale*, **21,** 336. *R.A.M.*, 1933, p. 252.

CAMICI, L. 1948. 'La "mummificazione delle Castagne" da *Phomopsis viterbensis* sp.n.', *Ann. Sper. agr.*, (N.S.) **2,** 557–66. *R.A.M.*, 1949, p. 314.

CAMPBELL, A. H. 1934. 'Zone lines in plant tissues. II. The black lines formed by *Armillaria mellea* (Vahl) Quél.', *Ann. Appl. Biol.* **21,** 1–22.

—— and VINES, A. E. 1938. 'The effect of *Lophodermellina macrospora* (Hartig) Tehog on leaf-abscission in *Picea excelsa* Link.', *New Phytol.* **37,** 358–68.

CAMPBELL, W. A. 1948. '*Phytophthora cinnamomi* associated with the roots of Littleleaf-diseased Shortleaf pine', *Plant Dis. Reptr.* **32,** 472.

—— 1951. 'Fungi associated with the roots of Littleleaf-diseased and healthy Shortleaf pine', *Phytopathology*, **41,** 439–45.

—— and COPELAND, O. L. 1954. 'Littleleaf disease of Shortleaf and Loblolly pines', *Circ. U.S. Dep. Agric.* **940,** 41 pp. *R.A.M.*, 1947, p. 220.

CAMPELL, W. A., and DAVIDSON, R. W. 1940. 'Top rot in glaze-damaged Black cherry and Sugar maple on the Allegheny Plateau', *J. For.* **38,** 963–5.

——— 1942. 'A species of *Poria* causing rot and cankers of hickory and oak', *Mycologia,* **34,** 17–26.

—— and MILLER, J. H. 1953. 'Windthrow of root-rotted Oak shade trees', *Arborist's News, Wooster,* **18,** 18. *For. Abstr.* **16,** No. 3136.

—— and VERRALL, A. F. 1956. 'Fungus enemies of hickory', *Hickory Task Force Rep., U.S. Dep. Agric.* **3,** 8 pp. *R.A.M.,* 1957, p. 286.

CANNON, J. F. M., and CANNON, M. 1957. 'The stability of the epiphytic flora of pollarded willows', *Proc. Bot. Soc. Brit. Isles,* **2,** 226–33. *Biol. Abstr.,* 1958, No. 3933.

CANNON, W. A. 1932. 'Absorption of oxygen by roots, when the shoot is in darkness or light', *Plant Physiol.* **7,** 673–84.

CAPRETTI, C. 1956. '*Diplodia pinea* (Desm.) Kickx. agente del disseccamento di varie specie del gen. *Pinus* e di altre conifere', *Ann. Accad. ital. Sci. for.* **5,** 171–202. *R.A.M.,* 1957, p. 436.

CARDOSO, L. 1951. '*Diplodia pinea* (Desm.) Kickx. associada a uma "seca das pontas" de *Pinus radiata* D. Don.', *Tech. Pap. St. Flora Serv. São Paulo,* **17,** 18 pp.

CAROSELLI, N., and HOWARD, F. L. 1939. 'Bleeding canker of Maples', *Bull. Bartlett Trees Res. Lab.* **3,** 44–48. *For. Abstr.* **2,** 192.

CAROSELLI, N. E. 1954. '*Verticillium* wilt of maple', *Diss. Abstr.* **14,** 2186–7. *R.A.M.,* 1955, p. 496.

—— 1955*a*. 'Investigations of toxins in vitro by the Maple wilt fungus *Verticillium* sp.', *Phytopathology,* **45,** 183–4.

—— 1955*b*. 'The relation of soil-water content and that of sapwood-water to the incidence of maple wilt caused by *Verticilium* sp.', *Phytopathology,* **45,** 184.

—— 1956. 'The effect of various soil amendments on maple wilt incited by *Verticillium* sp.', *Phytopathology,* **46,** 240.

—— 1957. 'Juniper blight and progress on its control', *Plant Dis. Reptr.* **41,** 216–18.

—— and TUCKER, C. M. 1949. 'Pit canker of elm', *Phytopathology,* **39,** 481–8.

CARRANZA, J. M. 1949. 'Una "*Alternaria*" parásita de las coniferas en la Argentina', *Lilloa,* **21,** 279–89. *R.A.M.,* 1951, p. 352.

CARTER, J. C. 1935. '*Cytosporina* canker on American elm in Illinois nurseries', *Plant Dis. Reptr.* **19,** 14–16. *R.A.M.,* 1935, p. 537.

—— 1936*a*. '*Melanconium betulinum* on *Betula* in Illinois', *Phytopathology,* **26,** 88.

—— 1936*b*. '*Cytosporina ludibunda* on American elm', *Phytopathology,* **26,** 805–6.

—— 1939. 'Progress in the control of Elm diseases in nurseries', *Biol. Notes Ill. Nat. Hist. Surv.* **10,** 1–19. *R.A.M.,* 1940, p. 442.

—— 1944. 'Wetwood of elm', *Proc. 20th Nat. Shade Tree Conf.,* pp. 175–80. *For. Abstr.* **7,** 222.

—— 1947. '*Tubercularia* canker and dieback of Siberian elm (*Ulmus pumila* L.)', *Phytopathology,* **37,** 243–6.

—— and HOFFMAN, P. F. 1953. 'Results of fungicide tests for control of leaf diseases of Black walnut, *Catalpa,* and American elm in 1951 and 1952', *Plant Dis. Reptr.* **37,** 114–15.

CARTWRIGHT, K. ST. G. 1937. 'A reinvestigation into the cause of "brown Oak", *Fistulina hepatica* (Huds.) Fr.', *Trans. Brit. Mycol. Soc.* **21,** 68–83.

—— 1938. 'A further note on fungus association in the Siricidae', *Ann. Appl. Biol.* **25,** 430–2.

—— 1942. 'The variability in resistance to decay of the heartwood of home-grown European Larch, *Larix decidua,* Mill. (*L. europaea*) and its relation to position in the log', *Forestry,* **16,** 49–51.

CARTWRIGHT, K. ST. G., and FINDLAY, W. P. K. 1958. *Decay of timber and its prevention*. H.M.S.O., London, 327 pp.

CARVELL, K. L., TRYON, E. H., and TRUE, R. P. 1957. 'Effects of glaze on the development of Appalachian hardwoods', *J. For.* **55**, 130–2.

CASH, E. K., and WATERMAN, A. M. 1957. 'A new species of *Plagiostoma* associated with a leaf disease of hybrid Aspens', *Mycologia*, **49**, 756–60. *R.A.M.*, 1958, p. 317.

CASS-SMITH, W. P., and STEWART, R. 1947. 'Leaf spot disease of Black mulberry', *J. Dep. Agric. W. Aust.* Ser. 2, **24**, 69–74. *R.A.M.*, 1947, p. 551.

CASTELLANI, E. 1948. 'Ricerche morfo-bio-ecologiche sulla "*Gnomonia juglandis*" (DC.) Trav.', Rep. *Ann. Accad. Agric. Torino*, **90**, 19 pp. *R.A.M.*, 1949, p. 315.

CECCARELLI, V. 1950. 'Revisione dei criteri distintivi dei danni da gas tossici e insussistenza dei considetti "danni invisibili"', *Notiz. Malatt. Piante.* **10,** 1–19. *Biol. Abstr.*, 1953, No. 30793.

CHADWICK, L. C., and NANK, E. E. 1949. 'The effects of certain pruning practices and wound dressings on callusing of tree wounds', *Proc. Amer. Soc. Hort. Sci.* **53,** 226–32. *For. Abstr.* **12,** No. 2294.

CHALK, L., and BIGG, J. M. 1956. 'The distribution of moisture in the living stem in Sitka spruce and Douglas fir', *Forestry*, **29,** 5–21.

CHAMPION, H. G. 1925. 'Contributions towards a knowledge of twisted fibre in trees. I. The phenomenon of twisted fibre with special reference to *Pinus longifolia* Roxb.', *Indian For. Rec.* **11,** 70 pp.

—— 1930. 'Second interim report on the progress of the investigations into the origin of twisted fibre in *Pinus longifolia* Roxb.', *Indian For.* **56,** 511–20.

—— 1933. 'European silvicultural research. IV. Mixtures', *Indian For.* **59,** 22–28

CHANDLER, W. H. 1954. 'Cold resistance in horticultural plants: A review', *Proc. Amer. Soc. Hort. Sci.* **64,** 552–72.

CHESTER, K. S. 1930. 'Graft-blight of lilac', *J. Arnold Arbor.* **11,** 232–3. *R.A.M.*, 1931, p. 190.

—— 1931. 'Graft blight: a disease of lilac related to the employment of certain understocks in propagation', *J. Arnold Arbor.* **12,** 79–146. *R.A.M.*, 1931, p. 599.

—— 1932. 'A comparative study of three *Phytophthora* diseases of lilac and their pathogens', *J. Arnold Arbor.* **13,** 232–68. *R.A.M.*, 1932, p. 579.

—— 1950. 'Plant disease losses: their appraisal and interpretation', *Plant Dis. Reptr. Suppl.* **193,** 190–362.

CHEYNEY, E. G. 1927. 'The effect of position of roots upon the growth of planted trees', *J. For.* **25,** 1013–15.

CHIBA, S. 1955*a*. 'Selection of cold-resistant individuals of *Cryptomeria japonica* under the field condition', *J. Jap. For. Soc.* **37,** 409–12. *Biol. Abs.*, 1957, No. 2408.

—— 1955*b*. 'Selection of resistant stocks to the needle blight of *Cryptomeria japonica*', *J. Jap. For. Soc.* **37,** 510–12. *Biol. Abs.*, 1956, No. 32809.

CHILDS, T. W. 1955. 'Needle blight of Ponderosa pine', *Res. Note Pacif. Nthwest For. Exp. Sta.* **114,** 6 pp. *R.A.M.*, 1957, p. 220.

—— and BEDWELL, J. L. 1948. 'Susceptibility of some White pine species to *Cronartium ribicola* in the Pacific Northwest', *J. For.* **46,** 595–9.

—— and WRIGHT, E. 1956. 'Pruning and occurrence of heart rot in young Douglas fir', *Res. Note Pacif. Nthwest For. Exp. Sta.* **132,** 5 pp. *R.A.M.*, 1957, p. 437.

CHITTENDEN, F. J. 1908–9. 'Azalea gall (*Exobasidium japonicum* Shirai)', *J. R. Hort. Soc.* **34,** 45–46.

CHRISTENSEN, C. M. 1932. 'Cultural races of *Pestalozzia funerea* and the production of variants resembling *Monochaetia*', *Phytopathology*, **22,** 6.

CHRISTENSEN, C. M. 1937. '*Cephalosporium* canker of Balsam fir', *Phytopathology*, **27,** 788–91.

—— 1940. 'Observations on *Polyporus circinatus*', *Phytopathology*, **30,** 957–63.

CHRISTOFF, A. 1932. 'Sklerotinijata po černicata', *Renseignements Agricoles, Sofia*, **13,** 127–39. *R.A.M.*, 1932, p. 485.

CHUPP, C. 1954. *A monograph of the fungus genus Cercospora.* Publ. by author, Ithaca, N.Y., 667 pp. *R.A.M.*, 1954, p. 635.

CIESLAR, A. VON. 1923. 'Untersuchungen über die wirtschaftliche Bedeutung der Herkunft des Saatgutes der Stieleiche', *Cbl. ges. forstw.* **49,** 97–149.

CIFERRI, R. 1950. 'La "mummificazione" delle Castagne da *Phomopsis endogena* (Speg.) nobis, n. comb.', *Notiz. Malatt. Piante*, **8,** 36–37. *R.A.M.*, 1950, p. 483.

—— 1951. 'Malattie Crittogamiche del Pioppo', *Att. Cong. Naz. Piop. Pavia*, **25,** 31–56.

—— and BALDACCI, E. 1948. 'Le malattie crittogamiche e disfunzionali del Pioppo', *Ente Naz. Cell. & Cart. Rome, 2nd Sess. Int. Pop. Comm.*, pp. 91–176.

CLAPHAM, A. R., TUTIN, T. G., and WARBURG, E. F. 1952. *Flora of the British Isles.* Cambridge University Press, 1,591 pp.

CLAPPER, R. B. 1952. 'Breeding and establishing new trees resistant to disease', *Econ. Bot.* **6,** 271–93. *R.A.M.*, 1953, p. 156.

—— GRAVATT, G. F., and STOUT, D. C. 1946. '*Endothia* canker on Post oak', *Plant Dis. Reptr.* **30,** 381.

—— and MILLER, J. M. 1949. 'Breeding and selecting pest-resistant trees', *Yearb. Agric. U.S. Dep. Agric.*, pp. 465–71.

CLARK, A. F. 1933. 'The Horntail borer and its fungal association', *N.Z. J. Sci. Technol.* **15,** 188–90. *R.A.M.*, 1934, p. 340.

CLARK, J., and BARTER, G. W. 1958. 'Growth and climate in relation to dieback of Yellow Birch'. *For. Sci.* **4,** 343–68.

—— and GIBBS, R. D. 1957. 'Studies in tree physiology. IV. Further investigations of seasonal changes in moisture content of certain Canadian forest trees', *Canad. J. Bot.* **35,** 219–53.

CLARKE, R. W. 1956. 'Wind damage in planted stock and natural regeneration of *Pinus radiata* at Mount Stromlo, A.C.T.', *Aust. For.* **20,** 37–39.

CLAUSEN, V. H., and KAUFERT, F. H. 1952. 'Occurrence and probable cause of heartwood degradation in commercial species of *Populus*', *J. For. Prod. Res. Soc.* **2,** 62–67.

CLINTON, G. P., and MCCORMICK, F. A. 1929. 'The Willow scab fungus *Fusicladium saliciperdum*', *Conn. Agric. Exp. Sta. Bull.* **302,** 443–69. *R.A.M.*, 1929, p. 614.

—— —— 1936. 'Dutch elm disease—*Graphium ulmi*', *Conn. Agric. Exp. Sta. Bull.* **389,** 701–52. *R.A.M.*, 1937, p. 216.

COCHRAN, L. C., and PINE, T. S. 1958. 'Present status of information on host range and host reactions to peach mosaic virus', *Plant Dis. Reptr.* **42,** 1225–8.

COCKAYNE, A. H. 1921. 'Fireblight and its control—the hawthorn question', *N.Z. J. Agric.* **23,** 30–36. *R.A.M.*, 1922, p. 22.

COCKERILL, J. 1957. 'Experiments in the control of damping-off at the Nursery, Orono, Ontario', *For. Chron.* **33,** 201–4.

COE, D. M., and WAGENER, W. W 1949. 'Ash anthracnose appears in California', *Plant Dis. Reptr.* **33,** 232.

COLEMAN, L. C. 1923. 'The transmission of Sandal spike', *Indian For.* **49,** 6–9.

COLLAER, P. 1940. 'Le rôle de la lumière dans l'établissement de la limite supérieure des forêts', *Ber. schweiz. bot. Ges.* **50,** 500–16. *For. Abstr.* **3,** 297.

COLLINS, J. F. 1915. 'The Chestnut bark disease on freshly fallen nuts', *Phytopathology*, **5,** 233–5.

CONDER, E. M. 1957. 'Problems of forestry in industrial areas', *Quart. J. For.* **51,** 38–45.

COOK, C. 1942. 'Some aspects of tree surgery in Canada', *Ann. Appl. Biol.* **29,** 205–8.

COOK, D. B., and WELCH, D. S. 1957. 'Backflash damage to residual stands incident to chemipeeling', *J. For.* **55,** 265–7.

COOK, M. T. 1923. 'The origin and structure of plant galls', *Science,* (N.S.) **57,** 6–14. *R.A.M.*, 1923, p. 377.

—— 1931. 'New virus diseases in Porto Rico', *Phytopathology,* **21,** 124.

COOKSON, I. C. 1929. 'An account of a crown rot of English walnut trees in Victoria', *Proc. Roy. Soc. Victoria,* (N.S.) **42,** 5–25. *R.A.M.*, 1930, p. 567.

COOLEY, J. S. 1942. 'Wound dressings on apple trees', *Circ. U.S. Dep. Agric.* **656,** 18 pp. *R.A.M.*, 1943, p. 29.

—— 1948. 'Collar injury of apple trees associated with waterlogged soil', *Phytopathology,* **38,** 736–9.

COPELAND, O. L., Jr. 1949. 'Some relations between soils and the Littleleaf disease of pine', *J. For.* **47,** 566–8.

—— 1955. 'The effects of an artificially induced drought on Shortleaf pine', *J. For.* **53,** 262–4.

—— and MCALPINE, R. G. 1955. 'The interrelations of littleleaf, site index, soil, and ground cover in Piedmont Shortleaf pine stands', *Ecology,* **36,** 635–40.

CORNFORD, C. E. 1938. 'Katabatic winds and the prevention of frost damage', *Quart. J. R. Met. Soc.* **64,** 553–87.

COTTON, A. D. 1919. 'The occurrence of Oak mildew on beech in Britain', *Trans. Brit. Mycol. Soc.* **6,** 198–200.

COUDERC, M. 1936. 'Les porte-greffes du Châtaignier et la maladie de l'encre', *Progr. agric. vitic.* **106,** 305–8, 352–6, 423–7, 449–52. *R.A.M.*, 1937, p. 354.

COVE, D. J. 1956. 'The Distribution of Mistletoe', *The Starfish* (*J. Ass. Sch. Nat. Hist. Soc.*), **9,** 20–22. *Nature, Lond.* **178,** 779.

COX, R. S. 1951. 'Control and overwintering studies on *Pyracantha* scab disease, *Fusicladium pirinum* var. *pyracanthae*, in Delaware', *Phytopathology,* **41,** 560.

—— 1953*a*. 'Control of pyracantha scab', *Plant Dis. Reptr.* **37,** 7–10.

—— 1953*b*. 'Etiology and control of a serious complex of diseases of conifer seedlings', *Phytopathology,* **43,** 469.

—— and HEUBERGER, J. W. 1953. 'Control of spot anthracnose and *Septoria* leaf spot of Flowering dogwood', *Nat. Hort. Mag.*, p. 70. *R.A.M.*, 1955, p. 371.

CRAIGHEAD, F. C. 1927. 'Abnormalities in annual rings resulting from fires', *J. For.* **25,** 840–2.

—— 1940. 'Some effects of artificial defoliation on pine and larch', *J. For.* **38,** 885–8.

CRAIGIE, J. H. 1942. 'Heterothallism in the Rust fungi and its significance', *Trans. Roy. Soc. Canad.*, 3rd ser. **36,** 19–40.

CRAM, W. H., and VAARTAJA, O. 1956. 'Toxicity of eight pesticides to spruce and caragana seed', *For. Chron.* **31,** 247–9.

—— —— 1957. 'Rate and timing of fungicidal soil treatments', *Phytopathology,* **47,** 169–73.

CRANDALL, B. S. 1942. 'Thyronectria disease of Honey locust in the South', *Plant Dis. Reptr.* **26,** 376.

—— GRAVATT, G. F., and RYAN, M. M. 1945. 'Root disease of *Castanea* species and some coniferous and broadleaf nursery stocks, caused by *Phytophthora cinnamomi*', *Phytopathology,* **35,** 162–80.

—— and HARTLEY, C. 1938. '*Phytophthora cactorum* associated with seedling diseases in forest nurseries', *Phytopathology,* **28,** 358–60.

CREAGER, D. B. 1937*a*. 'The Cephalosporium disease of elms', *Contr. Arnold. Arbor.* **10,** 91 pp. *R.A.M.*, 1937, p. 783.

——1937*b*. 'Phytophthora crown rot of dogwood', *J. Arnold Arbor.* **18,** 344–8.

CREELMAN, D. W. 1956. 'The occurrence of Ash rust in Western Nova Scotia', *Plant Dis. Reptr.* **40,** 580.

—— 1958*a*. 'Fusicoccum canker of the highbush Blueberry especially with reference to its occurrence in Novia Scotia', *Plant Dis. Reptr.* **42,** 843–5.

—— 1958*b*. 'The control of spot anthracnose of Linden', *Plant Dis. Reptr.* **42,** 895–6.

CRESCINI, F. 1940 and 1941. 'Sulla respirazione dell'apparato radicale delle piante cultivate', *Nuovo G. bot. ital.* **47,** 47–118; **48,** 449–94.

CRISTINZIO, M. 1942. 'Le malattie crittogamiche del Noce (*Juglans regia* L.)', *Ric. Ossvz. Divulg. fitopat. Campania ed Mezzogiorno* (*Portici*), **9,** 17–64. *R.A.M.*, 1946, p. 427.

CROCKER, W. 1931. 'The effect of illuminating gas on trees', *Proc. 7th Nat. Shade Tree Conf.*, pp. 24–34.

—— and BARTON, L. V. 1953. *Physiology of seeds. An introduction to the experimental study of seed and germination problems.* Chronica Botanica Co., Waltham, Mass. 267 pp.

CROKER, T. C., Jr. 1958. 'Soil depth affects wind-firmness of Longleafpine', *J. For.* **56,** 432.

CROSSE, J. E. 1954. 'Bacterial canker, leaf spot, and shoot wilt of cherry and plum', *Ann. Rep. East Malling Res. Sta.*, 1953, pp. 202–7.

—— 1956. 'Trials with the antibiotic streptomycin for the control of bacterial canker of cherry', *Ann. Rep. East Malling Res. Sta.*, 1955, pp. 170–2.

—— 1958. 'Fire-blight—a serious threat to the fruit grower', *Commercial Grower*, p. 745.

—— 1959. 'Fire-blight of pear in Britain', *Commonw. Phytopath. News*, **5,** 4–5.

CROWDY, S. H. 1952. 'The chemotherapy of plant diseases', *Emp. J. Exp. Agric.* **20,** 187–94. *R.A.M.*, 1953, p. 27.

—— 1953. 'Observations on the effect of growth-stimulating compounds on the healing of wounds on apple trees', *Ann. Appl. Biol.* **40,** 197–207.

—— 1954. 'Antibiotics in the systemic control of plant disease', *Research*, **7,** 483–6.

CROWELL, I. H. 1934. 'The hosts, life history, and control of the Cedar-Apple rust fungus *Gymnosporangium juniperi-virginianae* Schw.', *J. Arnold Arbor.* **15,** 163–232.

—— 1935*a*. 'The hosts, life history and control of *Gymnosporangium clivipes* C. and P.', *J. Arnold Arbor.* **16,** 368–410.

—— 1935*b*. 'Ornamental apples and cedar rust', *Horticulture*, **13,** 515. *Plant Breed. Abstr.*, 1935, p. 306.

—— 1937. 'Relative susceptibility of lilac species and varieties to *Microsphaera alni*', *Plant Dis. Reptr.* **21,** 134–8.

—— 1940. 'The geographical distribution of the genus *Gymnosporangium*', *Canad. J. Res.*, sect. C, **18,** 469–88. *R.A.M.*, 1941, p. 23.

CROXTON, W. C. 1939. 'A study of the tolerance of trees to breakage by ice accumulation', *Ecology*, **20,** 71–73.

CUNNINGHAM, H. S. 1928. 'A study of the histologic changes induced in leaves by certain leaf-spotting fungi', *Phytopathology*, **18,** 717–51.

CURRY, J. R., and CHURCH, T. W., Jr. 1952. 'Observations on winter drying of conifers in the Adirondacks', *J. For.* **50,** 114–16.

CURTIS, J. D. 1936. 'Snow damage in plantations', *J. For.* **34,** 613–19.

—— 1937. 'Artificial forest pruning with special reference to the healing of branch cuts', *Scot. For. J.* **51,** 138–40.

—— 1943. 'Some observations on wind damage', *J. For.* **41,** 877–82.

CURTIS, J. D. 1946. 'Pruning forest trees by the finger-budding. method', *J. For.* **44,** 502–4.

CURTIS, K. M. 1926. 'A die-back of *Pinus radiata* and *P. muricata* caused by the fungus *Botryodiplodia pinea* (Desm). Petr.', *Trans. N.Z. Inst.* **56,** 52–57. *R.A.M.*, 1926, p. 708.

CURZI, M. 1927. 'Di uno speciale parassitismo dell'*Ascochyta syringae*', *Riv. Patol. Veg.* **17,** 22–23. *R.A.M.*, 1927, p. 555.

—— 1933. 'La "*Phytophthora* (*Blepharospora*) *cambivora*" Petri sul Noce', *Rendic. R. Acad. Lincei*, ser. 6a, **18,** 587–92. *R.A.M.*, 1934, p. 336.

CUSTER, A. VON. 1934. 'Über den Frostkern der Karpathenbuche', *Schweiz. Z. Forstw.* **85,** 231–6.

CZAJA, A. T. 1951. 'Pflanzenschaden durch staubförmiges Wasserglas', *Z. PflKrankh.* **58,** 54–61. *Hort. Abstr.*, 1952, No. 1330.

DALLIMORE, W. 1917. 'Natural grafting of branches and roots', *Kew Bull.*, pp. 303–6.

—— 1932. 'Abnormal growth on the branches of a lime tree', *Quart. J. For.* **26,** 242–6.

DAMON, S. C., and SNELL, W. H. 1948. 'Two unusual hosts', *Plant Dis. Reptr.* **32,** 447.

DANA, B. F., and WOLFE, S. E. 1931. 'The occurrence of Violet root rot in central Texas', *Phytopathology*, **21,** 557–8.

DANCE, B. W. 1957. 'A fungus associated with blight and dieback of hybrid Aspen', *Bi-m. Progr. Rep. Div. For. Biol. Dep. Agric. Can.* **13,** 1–2.

DARK, S. T. E. 1935. 'Trees struck by lightning', *British Thunderstorms, Huddersfield Thunderstorm Census Organ., 5th Ann. Rep.*, pp. 86–88. *Nature, Lond.* **135,** 144.

DARKER, G. D. 1932. 'The Hypodermataceae of conifers', *Contr. Arnold Arbor*, **1,** 131 pp.

DARLEY, E. F., and WILBUR, W. D. 1954. 'Some relationships of carbon disulfide and *Trichoderma viride* in the control of *Armillaria mellea*', *Phytopathology*, **44,** 485.

DARPOUX, H., HALMOS, E., and LEBLANC, R. 1958. 'Étude de l'action systémique de diverses substances, la plupart antibiotiques', *Ann. épiphyt.* **9,** 387–414. *R.A.M.*, 1959, p. 309.

—— and RIDÉ, M. 1952. 'Recherche de procédés de désinfection des Châtaignes contre l'*Endothia parasitica* (Murril) Anderson', *Phytiatrie-Phytopharm.* **1,** 17–20. *R.A.M.*, 1954, p. 646.

—— —— and BONDOUX, P. 1957. 'Le chancre du châtaigner causé par l'*Endothia parasitica*', *Bull. Tech. Inf. Ing. Serv. Agric.* **123,** 24 pp.

DAVIDSON, R. W. 1935. 'Forest Pathology Notes', *Plant Dis. Reptr.* **29,** 94–97.

—— and LORENZ, R. C. 1938. 'Species of *Eutypella* and *Schizoxylon* associated with cankers of maple', *Phytopathology*, **28,** 733–45.

—— WESTER, H. V., and FOWLER, M. E. 1949. '*Botryosphaeria* and *Diplodia* associated with cankers on Linden and Redbud', *Phytopathology*, **39,** 502.

DAVIS, D. E. 1949. 'Some effects of calcium deficiency on the anatomy of *Pinus taeda*', *Amer. J. Bot.* **36,** 276–82.

DAVIS, E. A. 1949. 'Effect of several plant growth-regulators on wound healing of Sugar maple', *Bot. Gaz.* **111,** 69–77.

DAVIS, W. C., and LATHAM, D. H. 1939. 'Cedar blight on wilding and forest tree nursery stock', *Phytopathology*, **29,** 991–2.

DAVIS, W. H. 1931. 'Corynose twig blight of the American bladder nut, *Staphylea trifolia*', *Phytopathology*, **21,** 1163–71.

Davis, W. H. 1934. 'Twig blight (*Hypomyces ipomoeae*) of the American bladder nut', *Phytopathology*, **24**, 6.
—— 1939. 'A bud and twig blight of Azaleas caused by *Sporocybe azaleae*', *Phytopathology*, **29**, 517–28.
Day, M. W. 1940. 'Snow damage to conifer plantations', *Quart. Bull. Mich. Agric. Exp. Sta.* **23**, 97–98. *For. Abstr.* **2**, 284.
Day, W. R. 1924. 'The Watermark disease of the Cricket bat willow (*Salix caerulea*)', *Oxford Forestry Mem.*, **3**, 30 pp.
—— 1927*a*. 'The parasitism of *Armillaria mellea* in relation to conifers', *Quart. J. For.* **21**, 9–21.
—— 1927*b*. 'The oak mildew *Microsphaera quercina* (Schw.) Burrill and *Armillaria mellea* (Vahl.) Quél. in relation to the dying back of oak', *Forestry*, **1**, 108–12.
—— 1927*c*. 'A leaf cast of the Douglas fir due to *Rhabdocline pseudotsugae* Syd.', *Quart. J. For.* **21**, 193–9.
—— 1929*a*. 'The heart rot of timber in relation to forest management', *Quart. J. For.* **23**, 242–51.
—— 1929*b*. 'Environment and disease. A discussion on the parasitism of *Armillaria mellea* Vahl. Fr.', *Forestry*, **3**, 94–103.
—— 1930. 'The "Javart" disease of Sweet Chestnut. *Cytodiplospora castaneae* Oudemans (= *Diplodina castaneae* Prillieux et Delacroix)', *Quart. J. For.* **24**, 114–17.
—— 1931. 'The relationship between frost damage and Larch canker', *Forestry*, **5**, 41–56.
—— 1934*a*. 'Wind damage on Corsican and Scots pine', *Quart. J. For.* **28**, 226–9.
—— 1934*b*. 'Development of disease in the living tree', *Brit. Wood Pres. Ass. J.* **4**, 25–44.
—— 1937. 'The dying of larch: A note on Professor E. Münch's monograph "Das Lärchensterben" ', *Forestry*, **11**, 109–16.
—— 1938. 'Root-rot of sweet chestnut and beech caused by species of *Phytophthora*. I. Cause and symptoms of disease: its relation to soil conditions', *Forestry*, **12**, 101–16.
—— 1939. 'Root-rot of sweet chestnut and beech caused by species of *Phytophthora*. II. Inoculation experiments and methods of control', *Forestry*, **13**, 46–58.
—— 1945. 'A discussion of causes of dying-back of Corsican pine, with special reference to frost injury', *Forestry*, **19**, 4–26.
—— 1946. 'The pathology of beech on chalk soils', *Quart. J. For.* **40**, 72–82.
—— 1947. 'The leaf scorch of planes, caused by *Gnomonia veneta* (Sacc. and Speg.) Kleb.', *Quart. J. For.* **41**, 22.
—— 1948*a*. 'A note on canker development in poplars and willows', *Ned. Boschb. Tijdschr.* **20**, 323–30. *R.A.M.*, 1950, p. 129.
—— 1948*b*. 'The penetration of conifer roots by *Fomes annosus*', *Quart. J. For.* **32**, 99–101.
—— 1949. 'Forest pathology in relation to land utilization', *Emp. For. Rev.* **28**, 110–16.
—— 1950*a*. 'Frost as a cause of die-back and canker of Japanese larch', *Quart. J. For.* **44**, 78–82.
—— 1950*b*. 'The soil conditions which determine wind-throw in forests', *Forestry*, **23**, 90–95.
—— 1950*c*. 'Cambial injuries in a pruned stand of Norway spruce', *For. Comm. For. Rec.* **4**, 11 pp.
—— 1951. 'The susceptibility to injury by experimental freezing of strains of European larch (*Larix decidua* Mill.) of varying geographical origin', *Forestry*, **24**, 39–56.

DAY, W. R. 1953. 'The growth of Sitka spruce on shallow soils in relation to root-disease and wind-throw', *Forestry*, **26**, 81–95.
—— 1954. 'Drought crack of conifers', *For. Comm. For. Rec.* **26**, 40 pp.
—— 1955. 'Forest hygiene in Great Britain', *For. Bull.* (*Univ. Toronto*), **4**, 26 pp.
—— 1958*a*. 'The distribution of mycelia in European Larch bark, in relation to the development of canker', *Forestry*, **31**, 63–86.
—— 1958*b*. 'Variations in susceptibility of European Larch of differing seed origin in Scotland to injury by experimental freezing', *Scot. For.* **12**, 143–6.
—— and PEACE, T. R. 1934. 'The experimental production and the diagnosis of frost injury on forest trees', *Oxford Forestry Mem.* **16**, 60 pp.
—— —— 1937*a*. 'The influence of certain accessory factors on frost injury to forest trees. II. Temperature conditions before freezing', *Forestry*, **11**, 13–29.
—— —— 1937*b*. 'The influence of certain accessory factors on frost injury to forest trees. IV. Air and soil factors', *Forestry*, **11**, 92–103.
—— —— 1946. 'Spring Frosts', *For. Comm. Bull.* **18**, 111 pp.
DEARNESS, J., and HANSBROUGH, J. R. 1934. '*Cytospora* infection following fire injury in western British Columbia', *Canad. J. Res.* **10**, 125–8.
DEBOER, R. H. 1957. 'Effect of night lights on plant growth being studied', *N.J. Agric.* **39**, 3–5. *Hort. Abstr.*, 1958, No. 659.
DECKENBACH, K. N. 1927. 'Grusevaja rzavcina *Gymnosporangium sabinae* (Dicks) Winter i sposoby bor′by s nej v uslovijah Kryma', *Materialy po Mikologii i Fitopatologii*, **6**, 68–91. *R.A.M.*, 1928, p. 251.
DÉFAGO, G. 1937. '*Cryptodiaporthe castanea* (Tul.) Wehmeyer, parasite du Châtaignier', *Phytopath. Z.* **10**, 168–77. *R.A.M.*, 1937, p. 845.
DE FLUITER, H. J., and THUNG, T. H. 1951. 'Waarnemigen omtrent de dwergziekte bij Framboos en wilde Braam', *Tijdschr. PlZiekt.* **57**, 108–14. *R.A.M.*, 1952, p. 288.
DEITSCHMANN, G. H., and HERRICK, D. E. 1957. 'Logging injury in Central States upland hardwoods', *J. For.* **55**, 273–7.
DE KONING, M. 1923. 'Een nieuw bestrijdingsmiddel tegen de wortelzwam', *Tijdschr. PlZiekt.* **29**, 1–4. *R.A.M.*, 1923, p. 430.
DELEVOY, G. 1926. 'La fonte des semis ou "Damping off"', *Bull. Soc. Centr. for. Belg.* **29**, 305–15, 364–77. *R.A.M.*, 1927, p. 450.
—— 1927. 'La fonte des semis ou "Damping off"', *Bull. Soc. Centr. for. Belg.* **30**, 497–505. *R.A.M.*, 1928, p. 413.
—— 1928. 'Chronique forestière. Bruyères et *Armillaria mellea*', *Bull. Soc. Centr. for. Belg.* **31**, 359–60. *R.A.M.*, 1929, p. 147.
DELFORGE, P. 1930. 'Le *Chrysomyxa abietis* (rouille des aiguilles de l'Epicéa)', *Bull. Soc. Centr. for. Belg.* **37**, 419–23. *R.A.M.*, 1931, p. 141.
DEMAREE, J. B., and COLE, J. R. 1930. 'Pecan leaf blotch', *J. Agric. Res.* **40**, 777–89.
—— and WILCOX, M. S. 1947. 'Fungi pathogenic to blueberries in the Eastern United States', *Phytopathology*, **37**, 487–506.
DEMORLAINE, J. 1927. 'Le dépérissement du Chêne dans nos forêts françaises', *La Vie Agric. et Rurale*, **31**, 145–6. *R.A.M.*, 1928, p. 206.
DENGLER, A. 1930. *Waldbau auf ökologischer Grundlage.* Springer, Berlin, 560 pp.
—— 1955. 'Schütterversuch mit finnischen und märkischen Kiefern', *Arch. Forstw.* **4**, 4–8. *R.A.M.*, 1956, p. 731.
DENNIS, R. W. G. 1931. 'The Black canker of willows', *Trans. Brit. Mycol. Soc.* **16**, 76–84.
—— and WAKEFIELD, E. M. 1946. 'New or interesting British Fungi', *Trans. Brit. Mycol. Soc.* **29**, 141–66.
DENYER, W. B. G. 1953. 'Cephalosporium canker of Western hemlock', *Canad. J. Bot.* **31**, 361–6.

DE ONG, E. R. 1954. *Insect, fungus, and weed control.* Thames & Hudson, London, 400 pp.

DEPOERK, R. 1946. 'Sur un nouveau procédé de lutte contre les pourridiés en hévéaculture', *Bull. Inst. colon. belge*, **17**, 980–6. *R.A.M.*, 1947, p. 466.

DETERS, M. E. 1939. 'Frost heaving of forest planting stock at the Kellog reforestation tract, near Battle Creek, Michigan', *Pap. Mich. Acad. Sci.* **25**, 171–7. *For. Abstr.* **2**, 242.

DETMERS, F. 1923. '*Dothichiza* canker on Norway poplar', *Phytopathology*, **13**, 245–7.

DEUBER, C. G. 1936. 'Effect on trees of an illuminating gas in the soil', *Plant Physiol.* **11**, 401–12.

DICK, C. R. 1950. 'Miscellaneous note "Damping-off" in France', *Scot. For.* **4**, 7.

DIERCKS, R. 1957. 'Über den "Rutenbrenner" (*Physalospora miyabeana*) an Korbweiden unter Berücksichtigung zweijähriger Bekämpfungsversuche', *PflSchBer.* **9**, 37–44. *R.A.M.*, 1957, p. 561.

DIETZ, S.M. 1926. 'Alternate hosts of *Puccinia coronata.* II', *Phytopathology*, **16**, 84.

DILLER, J. D. 1935. '*Atropellis* canker of Eastern pines', *Plant Dis. Reptr.* **19**, 17.

—— 1943. 'A canker of Eastern pines associated with *Atropellis tingens*', *J. For.* **41**, 41–52.

—— 1957. 'Testing American chestnuts for blight resistance', *For. Res. Notes (N.E. For. Exp. Sta.)*, **74**, 3 pp.

DIMBLEBY, G. W. 1958. 'Experiments with hardwoods on heathland', *Imp. For. Inst. Oxford Inst. Paper*, **33**, 42 pp.

DIMOND, A. E. 1953. 'Progress in plant chemotherapy', *Agric. Chemic.* **8**, 34–35, 123, 125. *R.A.M.*, 1955, p. 661.

—— 1955. 'Pathogenesis of the wilt diseases', *Annual Review of Plant Physiology*, **6**, 329–50.

—— and HORSFALL, J. G. 1955. 'Fifty years of fungicides', *Ann. Appl. Biol.* **42**, 282–7.

—— PLUMB, G. H., STODDARD, E. M., and HORSFALL, J. G. 1949. 'An evaluation of chemotherapy and vector control of insecticides for combating Dutch Elm Disease', *Bull. Conn. Agric. Exp. Sta.* **531**, pp. 69. *For. Abstr.* **11**, 560.

—— and WAGGONER, P. E. 1953. 'On the nature and role of vivotoxins in plant disease', *Phytopathology*, **43**, 229–35.

DOBBS, C. G. 1951. 'A study of growth rings in trees. I. Review and discussion of recent work', *Forestry*, **24**, 22–35.

—— 1952. 'A study of growth rings in trees. II. A ring pattern in European larch', *Forestry*, **25**, 104–25.

DOCHINGER, L. S. 1956. 'New concepts of the *Verticillium* wilt disease of maple', *Phytopathology*, **46**, 467.

—— and BACHELDER, S. 1954. 'Sprays for leaf spot on English hawthorn', *Phytopathology*, **44**, 110.

DODGE, A. W. 1936. 'Lightning damage to trees', *Proc. 12th Nat. Shade Tree Conf.*, pp. 43–55.

—— 1937. 'Installation of lightning protection on shade trees', *Proc. 13th Nat. Shade Tree Conf.*, pp. 95–109.

DODGE, B. O. 1934. 'Witches' brooms on Southern white cedars', *J. N.Y. Bot. Gdn.*, **35**, 41–45. *R.A.M.*, 1934, p. 555.

—— 1936. 'Notes on some bacterial and fungous diseases in our gardens', *J. N.Y. Bot. Gdn.* **37**, 29–33. *R.A.M.*, 1936, p. 443.

—— 1944*a*. 'Boxwood blights and *Hyponectria buxi*', *Mycologia*, **36**, 215–22. *R.A.M.*, 1944, p. 320.

DODGE, B. O. 1944*b*. '*Volutella buxi* and *Verticillium buxi*', *Mycologia*, **37**, 416–25. *R.A.M.*, 1944, p. 503.

—— 1947. 'The Brooming disease of walnut', *J. N.Y. Bot. Gdn.* **48,** 112–14. *R.A.M.*, 1947, p. 362.

—— and RICKETT, H. W. 1948. *Diseases and pests of ornamental plants.* The Jaques Cattell Press, Lancaster, Penn., 638 pp. (A third edition, by Pirons, Dodge, and Rickett has appeared in 1960, too late to be used here.)

—— and SWIFT, M. E. 1930. 'Notes on boxwood troubles', *J. N.Y. Bot. Gdn.* **31,** 191–8. *R.A.M.*, 1931, p. 34.

D'OLIVEIRA, A. L. BRANQUINHO, and PIMENTEL, A. A. LOPES, 1953. 'Infecções latentes de *Melampsoridium betulinum* Pers. Kleb. em gomos de *Betula celtiberica* Rothm. et Vasc.', *Estud. Inform. Serv. flor. aqüic., Portugal,* **9,** C, p. 3. *For. Abstr.* **15,** No. 2646.

D'OLIVEIRA, M. DE L. 1939. 'Inoculações experimentais com o *Bacterium savastanoi* E. F. Smith e o *Bacterium svastanoi* var. *fraxini* N. A. Brown', *Agron. lusit.* **1,** 88–102. *R.A.M.*, 1939, p. 825.

DOMAŃSKI, S. 1954. 'Badania nad biologia *Fomes igniarius* (Linn.) Fr. na Białodrzewie (*Populus alba* L.)', *Acta Soc. Bot. Polon.* **23,** 589–616. *R.A.M.*, 1957, p. 143.

DOMINIK, T. 1946. 'Znaczenie mikroflory glebowej dla rozwoju sadzonek sosnowych', *Przegląd lésniczy, Poznań*, pp. 7–12. *For. Abstr.* **9,** No. 1433.

DONALD, G. H. 1936. 'Pruning studies at Princes Risborough', *Quart. J. For.* **30,** 111–20.

DONANDT, S. 1932. 'Untersuchungen über die Pathogenität des Wirtelpilzes *Verticillium alboatrum* R. u. B.', *Z. Parasitenk.* **4,** 653–711. *R.A.M.*, 1933, p. 116.

DONAUBAUER, E. 1957. 'Über eine Blatt- und Zweigkrankheit der Kanadapappel', *Allg. Forstztg.* **68,** 341. *R.A.M.*, 1958, p. 560.

DOWNS, A. A. 1938. 'Glaze damage in the birch-beech-maple-hemlock type of Pennsylvania and New York', *J. For.* **36,** 63–70.

—— 1943. 'Minimizing glaze damage in pine', *Tech. Note Appal. For. Exp. Sta.* **55,** 5 pp. *For. Abstr.* **7,** 93–94.

DOWSON, W. J. 1937. '*Bacterium salicis* Day, the cause of the Watermark disease on the Cricket-bat willow', *Ann. Appl. Biol.* **24,** 528–44.

—— 1957. *Plant diseases due to bacteria.* Cambridge University Press, 232 pp.

—— and CALLAN, E. McC. 1937. 'The Watermark disease in the White willow', *Forestry,* **11,** 104–8.

—— and DILLON-WESTON, W. A. R. 1937. 'Brown rot of the hawthorn', *Gdnrs.' Chron.* **101,** 426.

DOYER, C. M. 1925. 'Untersuchungen über die sogenannten Pestalozzia-Krankheiten und die Gattung *Pestalozzia* de Not.', *Meded. Phytopath. Lab. 'Willie Commelin Scholten', Baarn, Holland,* **9,** 72 pp. *R.A.M.*, 1926, p. 391.

DUCHAUFOUR, P. 1957. 'Un cas de carence azotée de l'épicéa décelée par le diagnostic foliaire', *Rev. for. France,* **9,** 128–35. *For. Abstr.* **18,** No. 2962.

DUFRÉNOY, J. 1922*a*. 'Les maladies du châtaignier', *C.R. Congr. Régional à Brive,* pp. 45–63. *R.A.M.*, 1924, p. 7.

—— 1922*b*. 'Tumeurs de *Sequoia sempervirens*', *Bull. Soc. Path. vég. France,* **9,** 148–50. *R.A.M.*, 1923, p. 95.

—— 1922*c*. 'Biologie de l'*Armillaria mellea*', *Bull. Soc. Path. vég. France,* **9,** 277–81. *R.A.M.*, 1923, p. 431.

—— 1922*d*. 'Sur la tuméfaction et la tubérisation', *C.R. Acad. Sci.* **174,** 1725–7. *R.A.M.*, 1923, p. 135.

—— 1925*a*. 'La maladie des Châtaigniers en Corse', *Rev. Eaux et Forêts,* **63,** 149–56. *R.A.M.*, 1925, p. 577.

DUFRÉNOY, J. 1925*b*. 'Les tumeurs des résineux', *Ann. Inst. Nat. Agron.* **19**, 33–201. *R.A.M.*, 1925, p. 513.

—— 1926. 'Les rapports des Châtaigniers exotiques avec le milieu biologique', Office Agric. Rég. du Massif Central, Clermont-Ferrand, 31 pp. *R.A.M.*, 1927, p. 5.

—— 1930. 'La lutte contre la maladie des Châtaigniers', *Ann. des Épiphyties*, **16**, 25–49. *R.A.M.*, 1931, p. 6.

—— 1932. 'L'inégale susceptibilité des épiceas vis-à-vis du *Chrysomyxa rhododendri*', *C.R. Soc. Biol.* **109**, 352–3. *R.A.M.*, 1932, p. 553.

—— 1933. 'Reconstitution par les Châtaigniers japonais des Châtaigneraies détruites par la maladie de l'encre', *Congr. Morbihan. de la Forêt et du Châtaignier 1932*, pp. 56–63. *R.A.M.*, 1934, p. 63.

—— and GAUDINEAU, M. 1924. 'Sur une maladie causée par un *Coryneum* nouveau', *Rev. Path. vég.* **11**, 164–7. *R.A.M.*, 1925, p. 134.

DUGELAY, A. 1957. 'Observations générales dur la gelée de Février 1956 dans les départements des Alpes-Maritimes et du Var', *Rev. for. France*, **9**, 1–27.

DUNCAN, D. P. 1954. 'A study of some of the factors affecting the natural regeneration of Tamarack (*Larix laricina*) in Minnesota', *Ecology*, **35**, 498–521.

DURRIEU, G. 1957. 'Influence du climat sur la biologie de *Phaeocryptopus gäumannii* (Rohde) Petrak parasite du *Pseudotsuga*', *C.R. Acad. Sci., Paris*, **244**, 2183–5. *For. Abstr.* **19**, No. 633.

DYE, M. H. 1956. 'Recent work with horticultural formulations of streptomycin', *Orchard, N.Z.* **29**, 2–3. *R.A.M.*, 1957, p. 331.

EATON, F. M., MCCALLUM, R. D., and MAYHUGH, M. S. 1941. 'Quality of irrigation waters of the Hollister area of California with special reference to boron content and its effect on apricots and prunes', *Tech. Bull. U.S. Dep. Agric.* **746**, 59 pp. *R.A.M.*, 1942, p. 26.

EDGERTON, L. J. 1957. 'Effect of nitrogen fertilization on cold hardiness of apple trees', *Proc. Amer. Soc. Hort. Sci.* **70**, 40–45. *Biol. Abstr.*, 1958, No. 20955.

EDLIN, H. L. 1943. 'A salt storm on the South coast', *Quart. J. For.* **37**, 24–26.

—— 1957. 'Saltburn following a summer gale in south-east England', *Quart. J. For.* **51**, 46–50.

—— and NIMMO, M. 1956. *Tree injuries.* Thames & Hudson, 167 pp.

EDWARDS, M. V. 1952. 'The effects of partial soil sterilization with formalin on the raising of Sitka spruce and other conifer seedlings', *For. Comm. For. Rec.* **16**, 20 pp.

—— 1953. 'Frost damage to Sitka spruce plants in the nursery and its relation to seed origin', *Scot. For.* **7**, 51.

—— 1957. 'Third report on an investigation of 1931–32 into various races of European larch', *For. Comm. Rep. For. Res., 1956*, pp. 142–54.

EHLERS, J. H. 1915. 'The temperature of leaves of *Pinus* in winter', *Amer. J. Bot.* **2**, 32–70.

EHRLICH, J. 1934. 'The beech bark disease a *Nectria* disease of *Fagus*, following *Cryptococcus fagi* (Baer.)', *Canad. J. Res.* **10**, 593–692.

EIMERN, J. VAN, and LOEWEL, E. L. 1953. 'Haben die Wassergräben in der Marsch des Alten Landes eine Bedeutung für den Frostschutz?', *Mitt. ObstbVersuchsrings Jork*, **8**, 225–31. *Hort. Abstr.*, 1954, No. 264.

EKBOM, O. 1928. 'Bidrag till kännendomen om bleckningsskador pa gran', *Svenska SkogsvFören. Tidskr.* **26**, 659–84.

ELIASON, E. J. 1928. 'Comparative virulence of certain strains of *Pythium* in direct inoculation of conifers', *Phytopathology*, **18**, 361–7.

ELLIS, D. E. 1946. 'Anthracnose of dwarf Mistletoe caused by a new species of *Septogloeum*', *J. Elisha Mitchell Sci. Soc.* **62,** 25–50. *R.A.M.*, 1947, p. 86.

—— and GILL, L. S. 1945. 'A new *Rhabdogloeum* associated with *Rhabdocline pseudotsugae* in the Southwest', *Mycologia*, **37,** 326–32.

ENDE, G. VAN DEN. 1952. 'Een bladvlekkenziekte voorkomend op de populieren veroorzaakt door *Septotinia populiperda* Waterman and Cash', *Tijdschr. PlZierkt.* **58,** 54–59. *For. Abstr.* **13,** No. 4001.

—— 1953. 'Verslag van het onderzoek naar de populierenkanker in 1950 en 1951', *Meded. Ned. Heidemaatsch.* **16,** 19. *Plant Breed. Abstr.*, 1953, No. 659.

—— 1954. 'Het parasitaire karakter van *Septotinia populiperda*', *Tijdschr. PlZiekt.* **60,** 253–5. *For. Abstr.* **16,** No. 3151.

—— 1955. 'Verslag van het onderzoek naar de populierenkanker in 1952 en 1953, veroorzaakt door *Pseudomonas syringae* v. Hall. f. sp. *populea* Sabet', *Meded. Ned. Heidemaatsch.* **21,** 19 pp. *For. Abstr.* **17,** No. 4112.

—— 1957*a*. 'Het onderzoek over de populierenkanker veroorzaakt door *Pseudomonas syringae* v. Hall. f. sp. *populea* Sabet', *Meded. LandbHogesch. Gent.* **22,** 527–33.

—— 1957*b*. 'Verslag van het onderzoek naar de Populierenkanker in de jaaren 1954, 1955 en 1956 veroorzaakt door *Pseudomonas syringae* v. Hall f. sp. *populea* Sabet', *Meded. Ned. Heidemaatsch.* **24,** 12 pp.

—— 1958. 'Untersuchungen über den Pflanzenparasiten *Verticillium albo-atrum* Reinke et Berth', *Acta bot. neerl.* **7,** 665–740.

ENDO, Y., and KURASAWA, T. 1937. 'On a strange virosis of the mulberry tree', *Bull. Seric. Silk Indust., Uyeda,* **9,** 115–32. *R.A.M.*, 1938, p. 538.

ENGEL, A. 1939. 'Les effets de la grêle en forêt', *Bull. Soc. for. Franche-Comté,* **23,** 12–20. *For. Abstr.* **1,** 75.

ENGELBRECHT, M. 1928. 'Soll man gegen die Kiefernschütte spritzen?', *Illustr. landw. Ztg.* **48,** 341. *R.A.M.*, 1929, p. 2.

ENGELHARD, A. W. 1957. 'Host index of *Verticillium albo-atrum* Reinke & Berth (including *Verticillium dahliae* Kleb.)', *Plant Dis. Reptr.* **244,** 49 pp.

ENGLERTH, G. H. 1942. 'Decay of Western Hemlock in western Oregon and Washington', *Bull. Yale Univ. Sch. For.* **50,** 53 pp. *For. Abstr.* **4,** 247.

—— BOYCE, J. S., Jr., and ROTH, E. R. 1956. 'Longevity of the Oak wilt fungus in Red Oak lumber', *For. Sci.* **2,** 2–6.

ENTRICAN, A. R. 1954. 'Annual report of the Forest Research Institute (New Zealand Forest Service) for the year ended 31st March 1954', *For. Res. Notes, N.Z.* **1,** 30 pp. *R.A.M.*, 1955, p. 267.

ERNSTSON, M., and HADDERS, G. 1948. 'Skadergörelse å granplantor genom uttorkning under vårvintern 1947', *Svenska SkogsvFören. Tidskr.* **46,** 310–22. *For. Abstr.* **11,** No. 3243.

ERŠOV, M. F. 1957. 'O fotosinteze čistyh i zapylennyh list′ev lipy melkolistnoj i vjaza melkolistnogo', *Dokl. Akad. Nauk S.S.S.R.* **112,** 1136–8. *For. Abstr.* **19,** No. 537.

ESLYN, W. E. 1959. 'Radiographical determination of decay in living trees by means of the Thulium X-ray unit', *For. Sci.* **5,** 37–47.

ETHERIDGE, D. E. 1955. 'Comparative studies of North American and European cultures of the root rot fungus, *Fomes annosus* (Fr.) Cooke', *Canad. J. Bot.* **33,** 416–28. *For. Abstr.* **17,** No. 1261.

—— 1957. 'Moisture and temperature relations of heartwood fungi in subalpine Spruce', *Canad. J. Bot.* **35,** 935–44.

—— 1958*a*. 'The effect on variations in decay of moisture content and rate of growth in subalpine Spruce', *Canad. J. Bot.* **36,** 187–206.

ETHERIDGE, D. E. 1958*b*. 'Decay losses in subalpine spuce on the Rocky Mountain Forest Reserve in Alberta', *For. Chron.* **34,** 116–31.

ETTLINGER, L. 1945. *Über die Gattung Crumenula sensu Rehm mit besonderer Berücksichtigung des Crumenula-Triebsterbens der Pinus-Arten*, thesis, Fed. Tech. Coll., Zürich, 73 pp. *R.A.M.*, 1946, p. 587.

EVISON, J. R. B. 1954. 'A garden by the sea', *J. R. Hort. Soc.* **79,** 294–302.

—— 1957. 'A summer gale', *J. R. Hort. Soc.* **82,** 88–91.

EWERT, R. 1924. 'Rauchkranke Böden', *Angew. Bot.* **6,** 97–104. *R.A.M.*, 1924, p. 681.

EZEKIEL, W. N. 1945. 'Effect of low temperatures on survival of *Phymatotrichum omnivorum*', *Phytopathology*, **35,** 296–301.

—— and TAUBENHAUS, J. J. 1934. 'Comparing soil fungicides with special reference to Phymatotrichum root rot', *Science*, (N.S.) **79,** 595–6. *R.A.M.*, 1934, p. 698.

FABRICIUS, L. 1929. 'Neue Versuche zur Feststellung des Einflusses von Wurzelwittbewerb und Lichtentzug des Schirmstandes auf de Jungwuchs', *Forstwiss. Cbl.* **51,** 477–506.

—— 1930. 'Die Schäden des Winterwetters 1928–1929 an den fremdländischen Holzarten des forstlichen Versuchsgartens in Grafrath bei München', *Forstwiss. Cbl.* **52,** 33–47.

FALCK, R. 1924. 'Über das Eichensterben im Regierungsbezirk Stralsund nebst Beiträgen zur Biologie des Hallimaschs und Eichenmehltaus', *Allg. Forst- u. Jagdztg.* **100,** 298–317.

—— 1930. 'Neue Mitteilungen über die Rotfäule', *Mitt. Forstw. Forstwiss.*, pp. 525–67. *R.A.M.*, 1931, p. 354.

FANG, C.-T., YUEN, S.-Y., LEE, C.-T., and WANG, K.-M. 1956. 'Experiments on the control of the stem rot disease of *Ginkgo* caused by *Macrophomina phaseoli*', *Acta phytopath. sinica*, **2,** 43–54. *R.A.M.*, 1958, p. 60.

FARRALL, A. W., SHELDON, W. H., and HANSEN, C. 1946. 'Protection of crops from frost damage through the use of radiant energy', *Quart. Bull. Mich. Agric. Exp. Sta.* **29,** 53–63. *Hort. Abstr.*, 1947, No. 108.

FAULKNER, R., and ALDHOUS, J. R. 1956. 'Nursery investigations', *For. Comm. Rep. For. Res. for year ended March 1955*, pp. 16–32.

FAULL, J. H. 1930. 'The spread and control of *Phacidium* blight in spruce plantations', *J. Arnold Arbor.* **11,** 136–47. *R.A.M.*, 1931, p. 1.

—— 1934. 'The biology of Milesian rusts', *J. Arnold Arbor.* **15,** 50–85. *R.A.M.*, 1934, p. 412.

—— 1938. '*Pucciniastrum* on *Epilobium* and *Abies*', *J. Arnold Arbor.* **19,** 163–73. *R.A.M.*, 1938, p. 571.

—— 1939. 'A review and extension of our knowledge of *Calyptospora goeppertiana* Kuehn', *J. Arnold Arbor.* **20,** 104–13. *R.A.M.*, 1939, p. 491.

FAWCETT, G. L. 1941. 'Departamento de Botánica y Fitopatologia. Ex Memoria anual del año 1941', *Rev. industr. agríc. Tucumán*, **32,** 41–45. *R.A.M.*, 1942, p. 481.

FELDMAN, A. W., CAROSELLI, N. E., and HOWARD, F. L. 1950. 'Physiology of toxin production by *Ceratostomella ulmi*', *Phytopathology*, **40,** 341–54.

FELT, E. P. 1943. 'Delayed winter injury', *Proc. 19th Nat. Shade Tree Conf.*, pp. 19–22. *For. Abstr.* **7,** 93.

FENTON, E. W. 1943. 'Some observations on heart rot in conifers from an ecological point of view', *Forestry*, **17,** 55–60.

FERDA, J. 1954. 'Vliv kouřových plynů na množství a poškození jehlic Smrkových porostů', *Práce výzk. ust. lesn. ČSR.* **5,** 283–95. *R.A.M.*, 1955, p. 687.

FERDINANDSEN, C., and JØRGENSEN, C. A. 1938–9. *Skovtraernes Sygdomme.* Gyldendalske Boghandel, Nordisk Forlag, København, 570 pp.

FERGUS, C. L. 1956*a*. 'An unusual cone fasciculation of Table mountain pine', *Plant Dis. Reptr.* **40,** 752–3.

—— 1956*b*. 'Some observations about *Polyporus dryadeus* on oak', *Plant Dis. Reptr.* **40,** 827–9.

—— 1956*c*. 'Frost cracks on oak', *Phytopathology,* **46,** 297.

FERGUSON, E. R. 1957. 'Causes of first-year mortality of planted Loblolly Pines in East Texas', *Proc. Soc. Amer. For.,* 1956, pp. 89–92.

FERKL, F. 1951. 'Výsledky měření teplotý v ovocných stromech', *Sborn. Čsl. Akad. Zeměd.* **24,** 175–81. *Hort. Abstr.,* 1952, No. 2292.

FERNANDES, C. T. 1949. 'A campanha contro a "doença da tinta" dos Cástanheiros no ano de 1948', *Bol. Jun. Frut., Lisb.* **9,** 41–66. *R.A.M.,* 1950, p. 390.

—— 1951. 'Le Mal des Châtaigniers au Portugal', *Publ. Dir. Serv. flor. aquic.* **18,** 111–20. *R.A.M.,* 1953, p. 523.

FERRÉ, Y. DE. 1955. 'Les *Pseudotsuga* de l'Arboretum de Joueou et leur résistance au *Phaeocryptopus*', *Trav. Lab. for. Toulouse,* **6** (4), Art. 1, 6 pp. *For. Abstr.* **17,** No. 2982.

FERRELL, W. K. 1955. 'The relationship of Pole blight of Western white pine to edaphic and other site factors', *Res. Note For. Exp. Sta. Idaho,* **13,** 7 pp. *For. Abstr.* **17,** No. 3086.

FIELD, C. P. 1942. 'Low temperature injury to fruit blossom. II. A comparison of the relative susceptibility and effect of environmental factors on three commercial apple varieties', *East Malling Res. Sta. Ann. Rep. 1941,* pp. 29–35.

FIELDING, J. M. 1940. 'Leans in Monterey Pine (*Pinus radiata*) plantations', *Aust. For.* **5,** 21–25.

—— 1952. 'The moisture content of the trunks of Monterey Pine trees', *Aust. For.* **16,** 3–21.

FILLER, E. C. 1933. 'Blister rust damage to Northern white pine at Waterford Vt.', *J. Agric. Res.* **47,** 297–313.

FINCH, A. H., and KINNISON, A. F. 1933. 'Pecan rosette: soil, chemical, and physiological studies', *Arizona Agric. Exp. Sta. Tech. Bull.* **47,** 407–42. *R.A.M.,* 1933, p. 602.

FINDLAY, W. P. K. 1951. 'Sap stain of timber', *For. Abstr.* **20,** 1–7.

—— 1953. *Dry Rot and other timber troubles.* Hutchinson, 267 pp.

FISCHER, E. 1930. 'Über einige Kleinarten von *Gymnosporangium* und ihre Einwirkung auf den Wirt', *Z. Bot.* **23,** 163–82. *R.A.M.,* 1930, p. 754.

—— 1931. 'Die Beziehungen zwischen *Gymnosporangium confusum* Plowr. auf *Juniperus phoenicea* und *J. sabina*', *Ber. schweiz. bot. Ges.* **40,** 8 pp. *R.A.M.,* 1932, p. 145.

—— 1933. 'Die Rostepidemie der Rottanne in den Alpen im Herbst 1932', *Mitt. Naturforsch. Ges. Bern, 1932,* pp. 20–21. *R.A.M.,* 1934, p. 201.

FISCHER, H. 1939. 'Die Douglasienschütte', *Blumen- u. PflBau ver. Gartenwelt,* **42,** 331–2. *R.A.M.,* 1939, p. 3.

FISCHER, R. 1956. 'Ein neuartiges Mehltauauftreten an Goldregen', *PflSchBer.,* **16,** 173–88. *R.A.M.,* 1956, p. 771.

FISCHER, W. 1957. 'Zur Föhrenschütte *Lophodermium pinastri*', *Schweiz. Z. Forstw.* **108,** 260–70. *For. Abstr.* **18,** No. 4255.

FISHER, P. L. 1941. 'Germination reduction and radicle decay of conifers caused by certain fungi', *J. Agric. Res.* **62,** 87–95.

FISHER, R. C. 1931. 'Notes on the biology of the Large elm bark-beetle, *Scolytus destructor*', *Forestry,* **5,** 120–31.

FISHER, R. C. 1937 'The genus *Scolytus* in Great Britain, with notes on the structure of *S. destructor* Ol.', *Ann. Appl. Biol.* **24,** 110–30.

—— THOMPSON, G. H., and WEBB, W. E. 1953. 'Ambrosia beetles in forest and sawmill. Their biology, economic importance and control. I. Biology and economic importance', *For. Abstr.* **14,** 381–9.

FLANDER, A. 1913. 'Hitzerisse an Fichten', *Forstwiss. Cbl.* **35,** 124–7.

FLURY, PH. 1926. 'Über Zuwachs und Ertrag reiner und gemischte Bestände', *Schweiz. Z. Forstw.* **77,** 337–42.

FOISTER, C. E. *The Economic Plant Diseases of Scotland,* Dept. Agric. Fish. Scotland, Tech. Bull. 1, 209 pp.

FORD, D. H., and RAWLINS, T. E. 1956. 'Improved cytochemical methods for differentiating *Cronartium ribicola* from *Cronartium occidentale* on *Ribes*', *Phytopathology,* **46,** 667–8.

FORD, H. F., and WATERMAN, A. M. 1954. 'Effect of surface sterilization on survival and growth of field planted hybrid poplar cuttings', *Plant Dis. Reptr.* **38,** 101–5.

FOSTER, A. A. 1956. 'Diseases of the forest nurseries of Georgia', *Plant Dis. Reptr.* **40,** 69–70.

—— CAIRNS, E. F., and HOPPER, B. 1956. 'Modifications in soils of Southern pine nurseries produced by fungicidal and nematocidal chemicals', *Phytopathology,* **46,** 12.

FOSTER, R. E., and A. T. 1951. 'Studies in Forest Pathology. VIII. Decay of Western hemlock on the Queen Charlotte Islands, British Columbia', *Canad. J. Bot.* **29,** 479–521.

FOSTER, R. E., CRAIG, H. M., and WALLIS, G. W. 1954. 'Studies in Forest Pathology. XII. Decay of Western hemlock in the Upper Columbia Region, British Columbia', *Canad. J. Bot.* **32,** 145–71.

FOWLER, D. P., and HEIMBURGER, C. 1958. 'The hybrid *Pinus peuce* Griseb. × *Pinus strobus* L.', *Silvae Genet.* **7,** 81–86.

FOWLER, M. E. 1936*a*. '*Sphaeropsis malorum* on *Abies concolor*', *Plant Dis. Reptr.* **20,** 30–31.

—— 1936*b*. 'Fasciation of *Betula pendula dalecarlica*', *Phytopathology,* **26,** 390, 392.

—— 1938. 'Twig cankers of Asiatic chestnuts in the eastern United States', *Phytopathology,* **28,** 693–704.

—— 1947. '*Glomerella* leaf spot of magnolia', *Plant Dis. Reptr.* **31,** 298.

—— 1953. 'Surveys for larch canker', *Phytopathology,* **43,** 405–6.

—— 1958. 'Oak wilt', *U.S. Dep. Agric. For. Serv. Pest Leafl.* **29,** 7 pp.

—— and ALDRICH, K. F. 1953. 'Resurvey for European larch canker in the United States', *Plant Dis. Reptr.* **37,** 160–1.

—— and BERRY, F. H. 1958. 'Blossom-end rot of Chinese Chestnuts', *Plant Dis. Reptr.* **42,** 91–96.

—— GRAVATT, G. F., and THOMPSON, A. R. 1945. 'Reducing damage to trees from construction work', *U.S. Dep. Agric. Farmer's Bull.,* No. 1967, 26 pp.

FRANCKE-GROSMANN, H. 1939. 'Über das Zusammenleben von Holzwespen (*Siricinae*) mit Pilzen', *Z. angew. Ent.* **25,** 647–80.

—— 1948. 'Rotfäule und Riesenbastkäfer, eine Gefahr für die Sitkafichte auf Öd- und Ackerlandaufforstungen Schleswig-Holsteins', *Forst-u. Holzw.* **3,** 232–5. *R.A.M.,* 1950, p. 67.

FRANKEN, E. 1956. 'Witterungsverlauf im Frühjahr 1955 und Pilzkrankheiten der Pappeln', *Holzzucht,* **9,** 30. *R.A.M.,* 1957, p. 288.

FRANKLIN, T. B. 1919–20. 'The cooling of the soil at night with especial reference to spring frosts', *Proc. Roy. Soc. Edin.* **40,** 10–22.

FRANSEN, J. J. 1939. *Iepenziekte, Iepenspintkevers en beider bestrijding,* thesis, Wageningen Agric. Coll., 118 pp. *R.A.M.,* 1939, p. 557.

FRANZ, J. 1955. 'Tannenstammläuse (*Adelges piceae* Ratz.) unter einer Pilzdecke von *Cucurbitaria pithyophila* (Kze. et Schm.) de Not., nebst Beobachtungen an *Aphidoletes thompsoni* Mohn (Dipt., Itonididae) und *Rabocerus mutilatus* Beck (Col., Pythidae) als Tannenläusfeinde', *Z. PflKrankh.* **62,** 49–61. *R.A.M.*, 1956, p. 251.

FRASER, J. W., and FARRAR, J. L. 1957. 'Frost hardiness of White spruce and Red pine seedlings in relation to soil moisture', *Tech. Note For. Br. Can.* **59,** 5 pp. *For. Abstr.*, 1958, No. 172.

FRED, E. B., BALDWIN, I. L., and MCCOY, E. 1932. 'Root nodule bacteria and leguminous plants', *Univ. Wisconsin Studies*, **52,** *Science* 5, 343 pp. Rev. *Soil Sci.*, 1933, pp. 167–8.

FREDERICK, L., and HOWARD, F. L. 1951. 'Comparative physiology and pathogenicity of eight isolates of *Ceratostomella ulmi*', *Phytopathology*, **41,** 12–13.

FRERICH. 1957. 'Sonnenseitige Kambium-Frostschäden an jungen Pappeln', *Allg. ForstZ.* **12,** 296–9.

FRESA, R. 1936. 'Argentine Republic: *Melampsora larici-populina* in the Delta of Parana', *Int. Bull. Pl. Prot.* **10,** 145–6. *R.A.M.*, 1937, p. 5.

FREZZI, M. J. 1942. 'Muerte del Tamarisco, ocasionado por *Botryosphaeria tamaricis*, en Corrientes, Argentina', *Rev. argent. agron.* **9,** 110–13. *R.A.M.*, 1944, p. 301.

—— 1950. 'Las especies de *Phytopathora* en la Argentina', *Rev. invest. agríc. B. Aires*, **4,** 47–133. *R.A.M.*, 1951, p. 433.

—— 1952. 'Presencia de *Botryosphaeria ribis* en la República Argentina y su importancia económica', *Rev. invest. agríc. B. Aires*, **6,** 247–62. *For. Abstr.* **18,** No. 3007.

FRIES, N. 1943. 'Zur Kenntnis des winterlichen Wasserhaushalts der Laubbäume', *Svensk bot. Tidskr.* **37,** 241–65. *For. Abstr.* **6,** 8.

FRITH, H. J. 1951. 'Frost protection in orchards using air from the temperature inversion layer', *Aust. J. Agric. Sci.* **2,** 24–42. *Hort. Abstr.*, 1952, No. 279.

FRITZSCHE, K. 1933. 'Sturmgefahr und Anpassung', *Tharandt. forstl. Jb.* **84,** 1–104.

FROHBERGER, P. E. 1956. 'Untersuchungen über die Wirkung von Chinonoximbenzoyl-hydrazon gegen Keimlingskranhkeiten verschiedener Kulturpflanzen', *Phytopath. Z.* **27,** 427–55. *R.A.M.*, 1957, p. 200.

FROHLICH, J. 1949. 'Zur Tannenfrage in Oesterreich', *Allg. Forstzt.* **60,** 186–8.

FRYER, L. D. 1947. 'Damage to *Pinus radiata* by climatic agents', *Aust. For.* **11,** 57–64.

FULLER, C. E. K., and NEWHOOK, F. J. 1954. 'A report on cypress canker in New Zealand', *N.Z. J. Agric.* **88,** 211, 213, 215, 217, 219–20. *R.A.M.*, 1955, p. 4.

FULLER, G. D., and LEADBEATER, M. R. 1935. 'Some effects of fuel oil on plants', *Plant Physiol.* **10,** 817–20. *Biol. Abstr.*, 1936, No. 12936.

FULLING, E. H. 1943. 'Plant life and the law of man. IV. Barberry, currant and gooseberry, and cedar control', *Bot. Rev.* **9,** 483–592. *For. Abstr.* **5,** 276.

FURNEAUX, B. S., and KENT, W. G. 1937. '"The Death": a trouble of fruit trees due to root suffocation', *Sci. Hort.* **5,** 67–77. *R.A.M.*, 1937, p. 540.

GAIL, F. W., and LONG, E. M. 1935. 'A study of root development and transpiration in relation to the distribution of *Pinus contorta*', *Ecology*, **16,** 88–100.

GAISBERG, E. VON. 1937. 'Über die Adelopus-Nadelschütte in Württembergischen Douglasienbeständen mit Hinweis auf die bisher hier bekanntgewordene Verbreitung von Rhabdocline', *Silva*, **25,** 37–42, 45–48. *R.A.M.*, 1938, pp. 84–86.

GALLOY, A. 1925. 'De la pourriture rouge consécutive aux dégâts de cerf dans certaines pessières de l'Hertogenwald', *Bull. Soc. Centr. for. Belg.* **28,** 400–5.

Galoux, A. 1951. 'Les dégâts de neige en forêt', *Bull. Soc. Centr. for. Belg.* **58,** 281–90. *For. Abstr.* **13,** No. 1285.

Gambogi, P., and Verona, O. 1958. 'Presenza in Italia di *Pestalotia populi-nigrae* Sawada et K. Itô, causa di una malattia ("shoot blight") del Pioppo', *Ann. Sper. agr.*, (n.s.) **12,** 1–3. *R.A.M.*, 1959, p. 282.

Gante, T. 1927. 'Eine Blattfallkrankheit des Rotdorns', *Gartenwelt,* **31,** 505. *R.A.M.*, 1928, pp. 99–100.

Gants, G. V. 1940. 'The silvicultural importance of the Grey Alder as an accumulator of nitrogen', *Mitt. Kirov. forsttech. Akad.* **58,** 178–89. *For. Abstr.* **3,** 285.

Garbowski, L. 1936. 'Przyczynek do znajomości mikroflory grzybnej nasion drzew leśnych', *Prace Wydz. Chor. Rośl. państw. Inst. Nauk. Gosp. wiejsk. Bydgoszczy,* **15,** 5–30. *R.A.M.*, 1937, p. 147.

Garcia, F., and Rigney, J. W. 1914. 'Hardiness of fruit buds and flowers to frost', *N. Mex. Agric. Exp. Sta.* **89,** 52 pp.

García Salmerón, J. M., and Breis, F. del B. 1958. 'Dos hongos aislados en los lechos de germinación de semillas de *Pinus halepensis* Mill.', *Montes, Madrid,* **14,** 183–7. *For. Abstr.* **20,** No. 411.

Gard, M. 1925. 'Le pourridié du Noyer. Principe du traitement curatif', *Rev. bot. appl.* **5,** 217–22. *R.A.M.*, 1925, p. 577.

Gardner, R. C. B. 1940. 'The ice storm of January 27th–29th 1940', *Quart. J. For.* **34,** 63–72.

Gardner, V. R. 1944. 'Winter hardiness in juvenile and adult forms of certain conifers', *Bot. Gaz.* **105,** 408–10. *For. Abstr.* **6,** 112.

Garren, K. H. 1941. 'Fire wounds on Loblolly pine and their relation to decay and other cull', *J. For.* **39,** 17–22.

—— 1954. 'Second report on leader dieback of sweetgum', *Plant Dis. Reptr.* **38,** 91–92.

—— 1956. 'Possible relation of *Diplodia theobromae* to leader dieback of sweetgum suggested by artificially induced infections', *Plant Dis. Reptr.* **40,** 1124–7.

Garrett, S. D. 1953. 'Rhizomorph behaviour in *Armillaria mellea* (Vahl) Quél. I. Factors controlling rhizomorph initiation by *A. mellea* in pure culture', *Ann. Bot., Lond.*, (n.s.) **17,** 63–79.

—— 1956*a*. *Biology of root-infecting fungi.* Cambridge University Press, 293 pp.

—— 1956*b*. 'Rhizomorph behaviour of *Armillaria mellea* (Vahl). Quél. II. Logistics of infection', *Ann. Bot., Lond.*, (n.s.) **20,** 193–200.

—— 1957. 'Effect of a soil microflora selected by carbon disulphide fumigation on survival of *Armillaria mellea* in woody host tissues', *Canad. J. Microbiol.* **3,** 135–49. *For. Abstr.* **19,** No. 593.

Gäumann, E. 1927. 'Über eine *Pestalozzia*-Krankheit der Nussbäume', *Mitt. schweiz. ZentAnst. forstl. Versuchsw.* **14,** 195–200. *R.A.M.*, 1928, p. 38.

—— 1948. 'Der Einfluss der Meereshöhe auf die Dauerhaftigkeit des Lärchenholzes', *Mitt. schweiz. ZentAnst. forstl. Versuchsw.* **25,** 327–93. *For. Abstr.* **13,** No. 652.

—— 1949. *Die Pilze; Grundzüge ihrer Entwicklungsgeschichte und Morphologie.* Birkhäuser, Basel, 382 pp.

—— 1950. *Principles of plant protection.* Eng. edn., trans. W. Brierley. Crosby Lockwood, London, 543 pp.

—— 1951. 'Über das Kastaniensterben in Tessin', *Schweiz. Z. Forstw.* **102,** 1–20. *R.A.M.*, 1951, p. 550.

—— 1952. *The fungi—A description of their morphological features and evolutionary development.* Hafner, New York, 420 pp.

GÄUMANN, E. 1954. 'Toxins and plant diseases', *Endeavour*, **13,** 198–204.

—— and JAAG, O. 1937. 'Über eine neue Erkrankung der Tanne (*Abies alba* Hill.) und der Fichte (*Picea excelsa* (Lam.) Ling.)', *Phytopath. Z.* **10,** 1–16. *R.A.M.*, 1937, p. 717.

—— and PÉTER-CONTESSE, J. 1951. 'Neuere erfahrungen über die Mistel', *Schweiz. Z. Forstw.* **102,** 108–19. *Biol. Abstr.*, 1953, No. 14025.

—— ROTH, C., and ANLIKER, J. 1934. 'Über die Biologie der *Herpotrichia nigra*', *Z. Pflanzenkrank. Pflanzenschutz*, **44,** 99–116. *Biol. Abstr.*, 1937, No. 10459.

GAUT, A. 1907. *Seaside planting of trees and shrubs.* Country Life, 101 pp.

GAVRIS, V. P. 1939. 'Selekcionnyj otbor immunyh form sosny obyknovennoj', *Lesn. Khoz.* **8,** 5–8. *R.A.M.*, 1940, p. 375.

GEIGER, R. 1950. *The climate near the ground.* A translation by M. N. Stewart and others of the 2nd German edition of *Das Klima der bodennahen Luftschicht*, with revisions and enlargements by the author. Harvard University Press, Cambridge, Mass., 482 pp. *For. Abstr.* **13,** No. 10.

GELINSKY, H. 1933. 'Die Astreinung der Rotbuche', *Z. Forst- u. Jagdw.* **6,** 289–329.

GENÈVES, L. 1957. 'Sur le rôle des écailles dans la résistance au froid des bourgeons de marronier: *Aesculus hippocastanum*', *C.R. Acad. Sci., Paris*, **244,** 2083–5. *For. Abstr.* **19,** No. 173.

GEORGE, E. J. 1936. 'Growth and survival of deciduous trees in shelter-belt experiments at Mandan, N. Dak., 1915–1934', *U.S. Dept. Agric. Tech. Bull.* **496,** 48 pp.

GEORGESCU, C., and GASMET, V. 1954. 'Un atac de *Rosellinia byssiseda* (Tode) Schroet., la puietii de Molid', *Rev. Pădurilor*, **68,** 31–34. *R.A.M.*, 1955, p. 195.

GEORGESCU, C. C., and BADEA, M. 1935. 'Căderea acelor de *Juniperus*, cauzată de o ciupercă nouă *Camarosporium juniperinum* Georgescu et Badea nov. sp. *Comunicare* prealabilă', *Rev. Pădurilor*, **47,** 155–62. *R.A.M.*, 1936, p. 473.

—— and Mocanu, V. V. 1956. 'Un atac de *Diplodia pinea* (Desm.) Kickx. pe lujeri si ace de pini vătămati de ingheturi tîrzii', *Rev. Pădurilor*, **71,** 383–6. *For. Abstr.* **18,** No. 4238.

—— and PETRESCU, M. 1954. 'Un parazit al fructelor de Ulm: *Gloeosporium ulmicola* Miles', *Rev. Pădurilor*, **69,** 106. *R.A.M.*, 1955, p. 191.

——, TEODORU, I., and BADEA, M. 1945. 'Uscarea în massă a stejarului', *Rev. Pădurilor*, **57,** 65–79. *For. Abstr.* **9,** No. 1921.

GEORGÉVITCH, P. 1926. '*Armillaria mellea* (Val.) Quél., cause du desséchement des forêts de Chêne en Yougoslavie', *C.R. Acad. Sci.* **182,** 289–491. *R.A.M.*, 1926, p. 456.

GERLACH, C. VON. 1928. 'Bestätigung von Eisenbahnrauchschäden im Walde', *Cbl. ges. Forstw.* **54,** 284–96.

GERLINGS, J. H. J. 1939. 'Herkomstonderzoek van den Douglasspar aan de afdeeling houtteelt van het Instituut voor Boschbouwkundig Onderzoek', *Ned. Boschb.Tijdschr.* **12,** 405–32. *R.A.M.*, 1940, p. 178.

GEYR, H. VON. 1932. 'Nochmals: Rhabdocline', *Forstarchiv.* **8,** 241–5. *R.A.M.*, 1933, p. 66.

GIBSON, I. A. S. 1955*a*. 'Trials of fungicides for the control of damping-off in pine seedlings', *E. Afr. Agric. J.* **21,** 96–102. *R.A.M.*, 1956, p. 798.

—— 1955*b*. 'Potassium permanganate as a seed bed treatment against damping-off in pines', *E. Afr. Agric. J.* **20,** 176–7. *R.A.M.*, 1955, p. 760.

—— 1956*a*. 'An anomalous effect of soil treatment with ethylmercury phosphate on the incidence of damping-off in pine seedlings', *Phytopathology*, **46,** 181–2.

—— 1956*b*. 'Trials of fungicides for the control of damping-off in pine seedlings. II. Field trials', *E. Afr. Agric. J.* **21,** 165–6. *For. Abstr.* **17,** No. 3873.

GIBSON, I. A. S. 1956*c*. 'Sowing density and damping-off in pine seedlings', *E. Afr. Agric. J.* **21,** 183–8. *For. Abstr.* **17,** No. 3872.

—— 1957. 'Saprophytic fungi as destroyers of germinating pine seeds', *E. Afr. Agric. J.* **22,** 203–6. *R.A.M.*, 1957, p. 739.

GILGUT, C. J. 1937. '*Cytospora* canker of Spruces', *Proc. Nat.* (*U.S.*) *Shade Tree Conf.*, pp. 113–19. *R.A.M.*, 1937, p. 648.

GILL, L. S. 1935. '*Arceuthobium* in the United States', *Trans. Conn. Acad. Arts & Sci.* **32,** 111–245.

GILMOUR, J. W. 1958. 'Chlorosis of Douglas fir', *N.Z. J. For.* **7,** 94–106.

GLASER, T., and SOSNA, Z. 1956. 'Badania porównawcze huby korzeniowej (*Fomes annosus* Fr.) pochodązcej z sosny świerka i brzozy na sztucznych pożywkach', *Acta Soc. Bot. Polon.* **25,** 285–303. *For. Abstr.* **18,** No. 574.

GLASSCOCK, H. H., and ROSSER, W. R. 1958. 'Powdery mildew on *Eucalyptus*', *Plant Path.* **7,** 152. *R.A.M.*, 1959, p. 229.

GLOCK, W. S. 1951. 'Cambial frost injuries and multiple growth layers at Lubbock, Texas', *Ecology*, **32,** 28–36.

GLOCKER, K., and KRÜSSMANN, G. 1957. 'Beiträge zur Kenntnis der Industriefestigkeit der immergrünen Gehölze', *Deutsche Baumschule, Aachen,* **9,** 85–92, 124–8, 150–3. *For. Abstr.* **19,** No. 3112.

GLOYER, W. O. 1915. 'The cause of stem-rot and leaf-spot of clematis', *J. Agric. Res.* **4,** 331–42.

GODWIN, H. 1943. 'Biological flora of the British Isles. Rhamnaceae: *Rhamnus cathartica* L., *Frangula alnus* Miller (*Rhamnus frangula* L.)', *J. Ecol.* **31,** 66–68.

GOETZ, J. 1951. 'Szablastość modrzewia polskiego na Górze Chelmowej', *Acta Soc. Bot. Polon.* **21,** 181–90. *For. Abstr.* **16,** No. 179.

GOIDÀNICH, G. 1934. 'La moria degli Aceri', *Italia Agric.* **71,** 1043–55. *R.A.M.*, 1935, p. 264.

—— 1935. 'Nuovi casi di tracheomicosi da "*Verticillium*" in Italia. Osservazioni su una specie nuova di "*Verticillium*" tracheicolo', *Boll. Staz. Pat. Veg. Roma.*, (N.S.) **15,** 548–54. *R.A.M.*, 1936, p. 474.

—— 1938. 'Nuove osservazioni sul "disseccamento dei germogli" dei Pioppi', *R.C. Accad. Lincei*, **27,** 3 pp. *R.A.M.*, 1938, p. 779.

—— 1940. 'La "necrosi corticale" del Pioppo causata da *Chondroplea populea* (Sacc. et Br.) Kleb.', *Riv. cellulosa*, **18,** 29 pp. *R.A.M.*, 1947, p. 431.

GOODALL, D. W., and GREGORY, F. G. 1947. 'Chemical composition of plants as an index of their nutritional status', *Imp. Bur. Hort. Plantat. Crops, Tech. Commun.* **17,** 167 pp.

GOODDING, L. N. 1930. '*Didymosphaeria oregonensis,* a new canker organism on alder', *Phytopathology*, **20,** 854.

GOODE, M. J. 1953. 'Control of oak leaf-blister in Mississippi', *Phytopathology,* **43,** 472.

GOOR, C. P. VAN. 1956. 'Kaligebrek als oorzaak van gelepuntziekte van Groveden (*Pinus sylvestris*) en Corsicaanse den (*Pinus nigra* var. *corsicana*)', *Ned. Boschb. Tijdschr.* **28,** 21–31. *R.A.M.*, 1957, p. 437.

—— *et al.* 1954. *I.U.F.R.O. Section 24, Forest Protection. Special conference on root and butt-rots of forest trees by Fomes annosus.* Bosbouwproefstation T.N.O. Wageningen, 22–26 July, 30 pp.

GOSS, R. W., and FRINK, P. R. 1934. '*Cephalosporium* wilt and die-back of the White elm', *Univ. Nebraska Agric. Exp. Sta. Res. Bull.* **70,** 24 pp. *R.A.M.*, 1934, p. 478.

GOSSELIN, R. 1944. 'Studies in *Polystictus circinatus* and its relation to butt rot of Spruce', *Bull. Serv. for. Québec*, (N.S.) **10,** 44 pp. *For. Abstr.* **7,** 532–3.

GOULD, C. J., EGLITIS, M., and DOUGHTY, C. C. 1955. 'European rhododendron rust (*Chrysomyxa ledi* var. *rhododendri*) in the United States', *Plant Dis. Reptr.* **39,** 781–2.

GOURLAY, W. B. 1951. 'Town trees and telephone wires', *J. R. Hort. Soc.* **76,** 64–65.

GOVI, G. 1952. 'Due specie di *Cylindrocarpon* isolate da fruttiferi', *Ann. Sper. agr.*, (N.S.) **6,** 793–804. *R.A.M.*, 1953, p. 261.

GRAEBNER, P. 1906. 'Beiträge zur Kenntnis nichtparasitärer Pflanzenkrankheiten an forstlichen Gewächsen. I. Absterbender Fichtenbestand des Schutzbezirks Woltöfen bei Lübberstedt', *Z. Forst- u. Jagdw.*, pp. 705–14. Abstr. in *J. For.*, 1907, p. 427.

GRAHAM, S. A. 1943. 'Causes of hemlock mortality in northern Michigan', *Bull. Mich. Sch. For.* **10,** 61 pp. *For. Abstr.* **6,** 19–20.

—— and HARRISON, R. P. 1954. 'Insect attacks and *Hypoxylon* infections in Aspen', *J. For.* **52,** 741–3.

GRAINGER, J., and ALLEN, A. L. 1936. 'The internal temperatures of fruit-tree buds', *Ann. Appl. Biol.* **23,** 1–10.

GRAM, E. 1955. 'Barriers and by-passes in plant trade', *Ann. Appl. Biol.* **42,** 76–81.

—— and ROSTRUP, S. 1924. 'Oversigt over Sygdomme hos Landbrugets og Havebrugets Kulturplanter i 1923', *Tijdschr. PlZiekt.* **30,** 361–414. *R.A.M.*, 1924, p. 506.

—— and WEBER, A. 1952. *Plant diseases in orchard nursery and garden crops.* Trans. Dennis, Macdonald, London, 618 pp.

GRANITI, A. 1957. 'Risultati di inoculazioni artificiali con ceppi di *Cytospora corylicola* Sacc., isolati da noccioli colpiti da mal dello stacco in Sicilia', *Ital. for. mont.* **12,** 93–98. *For. Abstr.* **19,** No. 750.

GRANT, T. J., and CHILDS, T. W. 1940. '*Nectria* canker of north-eastern hardwoods in relation to stand improvement', *J. For.* **38,** 797–802.

—— and SPAULDING, P. 1939. 'Avenues of entrance for canker-forming Nectrias of New England hardwoods', *Phytopathology*, **29,** 351–8.

—— STOUT, D. C., and READEY, J. C. 1942. 'Systemic brooming, a virus disease of Black Locust', *J. For.* **40,** 253–60.

GRASSO, V. 1948. 'L'oidio dell'Eucalipto', *Nuovo G. bot. ital.*, (N.S.) **55,** 581–4. *For. Abstr.* **11,** No. 2273.

—— 1951. 'Un nuovo agente patogeno del *Cupressus macrocarpa* Hartw. in Italia', *Ital. for. mont.* **6,** 62–65. *R.A.M.*, 1951, p. 352.

—— 1952. 'Conifere suscettibili ed immuni al *Coryneum cardinale* Wag.', *Ital. for. mont.* **7,** 148–9. *For. Abstr.* **14,** No. 439.

—— 1955. 'Un nuovo ospite del *Coryneum cardinale* Wag. in Italia', *Boll. Staz. Pat. veg. Roma*, ser. 3, **12,** 209–11. *R.A.M.*, 1956, p. 498.

—— 1957. 'Nuovi rinvenimenti di ospiti con *Coryneum cardinale*', *Boll. Staz. Pat. veg. Roma*, ser. 3, **14** (1956), 239–42. *R.A.M.*, 1958, p. 320.

—— and CAPRETTI, C. 1955. 'Un nuovo ospite di *Keithia tetraspora* (Phill.) Sacc. e prima segnalazione in Italia', *Ital. for. mont.* **10,** 273–5. *For. Abstr.* **17,** No. 2976.

GRATKOWSKI, H. J. 1956. 'Windthrow around staggered settings in old growth Douglas fir', *For. Sci.* **2,** 60–74.

GRAVATT, G. F. 1952. 'Blight on chestnut and oaks in Europe in 1951', *Plant Dis. Reptr.* **36,** 111–15.

—— 1954. 'Potential danger to the Persian walnuts, Douglas fir, and Port Orford cedar of the Pacific coast from the Cinnamon *Phytophthora*', *Plant Dis. Reptr.* **38,** 214–16.

GRAVATT, G. F., and PARKER, D. E. 1949. 'Introduced tree diseases and insects', *Yearb. Agric. U.S. Dept. Agric.*, pp. 446–51. *R.A.M.*, 1949, p. 652.

—— and POSEY, G. B. 1918. 'Gipsy moth larvae as agents in dissemination of the White pine blister rust', *J. Agric. Res.* **12,** 459–62.

GRAVES, A. H. 1923. 'The Melanconis disease of the butternut (*Juglans cinerea* L.)', *Phytopathology,* **13,** 411–34.

—— 1926. 'The cause of the persistent development of basal shoots from blighted chestnut trees', *Phytopathology,* **16,** 615–21.

—— 1950. 'Relative blight resistance in species and hybrids of *Castanea*', *Phytopathology,* **40,** 1125–31.

GRAY, E. 1940. 'The Willow wood wasp and watermark disease of Willows', *Vet. J.* **96,** 370–3.

GRAY, W. G. 1931. 'An instance of "damping-off" retarded by the use of basic slag', *Forestry,* **5,** 132–5.

GRAYBURN, A. W. 1957. 'Hail damage to exotic forests in Canterbury', *N.Z. J. For.* **7,** 50–57. *For. Abstr.* **19,** No. 4342.

GRAYSON, A. J. 1956. 'Effects of atmospheric pollution in forestry', *Nature, Lond.* **178,** 719–21.

GREEN, D. E. 1935. 'Leaf spot of *Daphne mezereum* caused by *Marssonina daphnes* (Desm. et Rob.) Magn.', *J. R. Hort. Soc.* **60,** 156–8.

—— and HEWLETT, M. A. 1949. 'Die-back of *Cytisus* cuttings', *J. R. Hort. Soc.* **74,** 310–12.

GREEN, F. M. 1932. 'Observations on *Cucurbitaria laburni* (Pers.) de Not.', *Trans. Brit. Mycol. Soc.* **16,** 289–303.

GREEN, R. J., Jr. 1954. 'A preliminary investigation of toxins produced in vitro by *Verticillium albo-atrum*', *Phytopathology,* **44,** 433–7.

GREEN, W. E. 1947. 'Effect of water impoundment on tree mortality and growth', *J. For.* **45,** 118–20.

GREENHAM, C. G., and BROWN, A. G. 1958. 'The control of mistletoe by trunk injection', *J. Aust. Inst. Agric. Sci.* **23,** 308–18. *Hort. Abstr.*, 1958, No. 1333.

GREGOR, M. J. F. 1932*a*. 'Observations on the structure and identity of *Tulasnella anceps* Bres. et Syd.', *Ann. Mycol.* **30,** 463–5. *R.A.M.*, 1933, p. 132.

—— 1932*b*. 'The possible utilisation of disease as a factor in bracken control', *Scot. For. J.* **46,** 52–59.

—— 1935. 'A disease of bracken and other ferns caused by *Corticium anceps* (Bres. et Syd.) Gregor', *Phytopath. Z.* **8,** 401–18. *R.A.M.*, 1935, p. 797.

—— 1936. 'A disease of Cherry laurel caused by *Trochila laurocerasi* (Desm.) Fr.', *Ann. Appl. Biol.* **23,** 700–4. *See also* Wilson, M. J. F.

GREGORY, K. R., ALLEN, O. N., RIKER, A. J., and PETERSON, W. H. 1952. 'Antibiotics as agents for the control of certain damping-off fungi', *Amer. J. Bot.* **39,** 405–15.

GREGORY, P. H., and WALLER, S. 1951. '*Cryptostroma corticale* and Sooty bark disease of Sycamore (*Acer pseudoplatanus*)', *Trans. Brit. Mycol. Soc.* **34,** 579–97.

GREMMEN, J. 1954. 'Op *Populus* en *Salix* voorkomende *Melampsora*-soorten in Nederland', *Tijdschr. PlZiekt.* **60,** 243–50. *For. Abstr.* **16,** No. 3148.

—— 1956. 'Een blad- en twijgziekte van populieren veroorzaakt door *Venturia tremulae* en *Venturia populina*', *Tijdschr. PlZiekt.* **62,** 236–42. *For. Abstr.* **18,** No. 1815.

—— 1958*a*. 'Een afsterven van *Pinus*-soorten in Nederland en de vermoedelijke oorzaak', *Ned. Bosb. Tijdschr.* **30,** 199–208. *For. Abstr.* **20,** No. 733.

—— 1958*b*. 'Bijdrage tot de biologie van *Cryptodiaporthe populea* (Sacc.) Butin (*Dothichiza populea* Sacc. et Bri.), *Ned. Bosb. Tijdschr.* **30,** 251–60. *For. Abstr.* **20,** No. 2019.

GRENTE, J. 1952. 'Le *Phytophthora cinnamomi* parasite du Châtaignier en France', *C.R. Acad. Sci., Paris,* **234,** 2226–8. *R.A.M.*, 1953, p. 106.

—— and SOLIGNAT, G. 1952. 'La maladie de l'encre du Châtaignier et son évolution', *C.R. Acad. Agric. Fr.* **38,** 126–9. *R.A.M.*, 1953, p. 106.

GRIES, G. A. 1943. 'Juglone (5-hydroxy-1, 4-naphthoquinone)—a promising fungicide', *Phytopathology,* **33,** 1112.

GRIFFIN, D. M. 1956. 'Fungal damage to roots of seedlings in forest nurseries', *For. Comm. Rep. For. Res. 1955*, pp. 75–76.

—— 1957. 'Fungal damage to roots of Sitka spruce seedlings in forest nurseries', *For. Comm. Rep. For. Res. 1956*, pp. 86–87.

—— 1958. 'Influence of pH on the incidence of Damping-off', *Trans. Brit. Mycol. Soc.* **41,** 483–90.

GRIMAL, R. 1956. 'Les dangers de maladies et d'invasions d'insectes par l'introduction d'arbres exotiques', *Chêne-liège,* **62,** 1525, pp. 25, 27, 29; 1526, pp. 31, 33, 35; 1527, pp. 17, 19. *R.A.M.*, 1956, p. 855.

GROSSH, P. VON. 1906. 'Grappenweise Beschädigung von Eichen in folge Blitzschlages', *Allg. Forst- u. Jagdztg.* **82,** 355–6.

GROVE, W. B. 1913. *The British rust fungi.* Cambridge University Press, 412 pp.

—— 1935–7. *British stem and leaf fungi.* Cambridge University Press, vol. I, 488 pp., vol. II, 405 pp.

GROVES, J. W. 1946. 'Variations in *Botrytis cinerea*', Abstr. in *Proc. Canad. phytopath. Soc.* **14,** 13. *R.A.M.*, 1947, p. 406.

—— and LEACH, A. M. 1949. 'The species of *Tympanis* occurring on *Pinus*', *Mycologia,* **41,** 59–76. *R.A.M.*, 1949, p. 489.

GRUENHAGEN, R. H. 1945. '*Hypoxylon pruinatum* and its pathogenesis on Poplar', *Phytopathology,* **35,** 72–89.

GUBA, E. F. 1934. 'Slime flux', *Proc. 10th Nat. Shade Tree Conf., U.S.A.*, pp. 56–60.

GÜDE, J. 1954. 'Die Feststellung der durch Steinkohlenrauch verursachten Zuwachsminderung in Fichtenbestanden', *Z. Weltfw.* **17,** 87–93.

GUILLEBAUD, W. H. 1930. 'Some experimental studies on the artificial regeneration of oak', *Forestry,* **4,** 113–21.

GUINIER, P. 1931. 'Note sur deux *Pucciniastrum* nuisibles aux conifères', ex *Travaux Cryptog.* dédiés à L. Mangin, Muséum National d'Hist. Nat., Paris, pp. 373–5. *R.A.M.*, 1932, p. 339.

GÜNZL, L. 1953. 'Forstkulturschäden durch Eis bei Hochwasser', *Allg. Forstztg.* **64,** 252–3. *Biol. Abstr.*, 1954, No. 24485.

GUPTA, S. L. 1950. 'Occurrence of Chestnut-blight in the Kumaon Hills', *Curr. Sci.* **19,** 13–14. *R.A.M.*, 1950, p. 441.

GÜSSOW, H. T. 1929. 'Needle blight of white pine', *Rept. Dom. Bot. for the year 1928*, Div. Bot. Canada Dept. Agric., pp. 31–33. *R.A.M.*, 1930, p. 500.

—— 1936. 'Plant quarantine legislation—a review and a reform', *Phytopathology,* **26,** 465–82.

GUSTAFSON, R. O. 1946. 'Forest fires, basal wounds and resulting damage to timber in an eastern Kentucky area', *Bull. Ky. Agric. Exp. Sta.* **493,** 15 pp. *R.A.M.*, 1947, p. 518.

GUTHRIE-SMITH, H. 1936. '*Elytranthe flavida*, the red mistletoe', *Gdnrs.' Chron.* **99,** 282–3.

GUYOT, A. L. 1934. 'Note sur une maladie chancreuse du Pin sylvestre dans le Nord de la France', *Rev. Path. vég.* **21,** 33–38. *R.A.M.*, 1934, p. 665.

GUYOT, M. 1921. 'Notes de Pathologie végétale', *Bull. Soc. Path. vég. France,* **8,** 132–6. *R.A.M.*, 1922, p. 334.

GUYOT, R. 1933. 'De la maladie du rond: de l'influence des foyers ou des foyers d'incendie dans sa propagation', *Rev. gén. des Sciences*, **44**, 239–47. *R.A.M.*, 1933, p. 798.

Gwynne-Vaughan, H. C. I., and Barnes, B. 1937. *The structure and development of the fungi*. Cambridge University Press, 449 pp.

Győrfi, J. 1957. 'Az erdei fák rácos niegbetegedései', *Erdész. Kutatás*. **1–2**, 83–94. *R.A.M.*, 1958, p. 559.

Haasis, F. W. 1923. 'Frost heaving of Western yellow pine seedlings', *Ecology*, **4**, 378–90.

Haddock, P. G., and Gessel, S. P. 1951. 'Soil compaction, soil aeration, and tree growth', *Proc. 27th Nat. Shade Tree Conf.*, pp. 266–72.

Haddow, W. R. 1938. 'The disease caused by *Trametes pini* (Thore) Fries in White Pine', *Roy. Canad. Inst. Trans.* **22**, 21–80. *For. Abstr.* **1**, p. 85.

—— 1941. 'Needle blight and late fall browning of Red pine (*Pinus resinosa* Ait.) caused by a gall midge (*Cecidomyiidae*) and the fungus *Pullularia* (De Bary) Berkhout', *Trans. Roy. Canad. Inst.* **23**, 161–89. *R.A.M.*, 1942, p. 356.

—— and Newman, F. S. 1942. 'A disease of the Scots Pine caused by *Diplodia pinea* associated with the Pine spittle-bug', *Trans. R. Canad. Inst.* **24**, 18 pp. *For. Abstr.* **4**, 126–7.

Hadfield, M. 1956. 'Does ivy damage trees?', *Country Life*, pp. 670–1.

Hagem, O. 1926. 'Schütteskader paa Furuen (*Pinus sylvestris*)', *Vestland. forstl. Forsøkssta. Medded.*, **7**, 133 pp. *R.A.M.*, 1927, p. 452.

Hahn, G. G. 1927. 'Fungi as an international problem', *Trans. Proc. Bot. Soc. Edin.* **29**, 342–8.

—— 1928. '*Phomopsis conorum* (Sacc.) Died.—an old fungus of the Douglas fir and other conifers', *Trans. Brit. Mycol. Soc.* **13**, 278–86.

—— 1930. 'Life-history of the species of *Phomopsis* occurring on conifers', *Trans. Brit. Mycol. Soc.* **15**, 32–93.

—— 1933. 'An undescribed *Phomopsis* from Douglas fir on the Pacific coast', *Mycologia*, **25**, 369–75. *R.A.M.*, 1934, p. 199.

—— 1939. 'Immunity of a staminate clone of *Ribes alpinum* from *Cronartium ribicola*', *Phytopathology*, **29**, 981–6.

—— 1940. '*Dasyscyphae* on conifers in North America. IV. Two new species on Douglas Fir from the Pacific Coast', *Mycologia*, **32**, 137–47.

—— 1943*a*. 'Blister rust on Red currant', *Phytopathology*, **33**, 341–53.

—— 1943*b*. 'Taxonomy, distribution, and pathology of *Phomopsis occulta* and *P. juniperovora*', *Mycologia*, **35**, 112–29. *R.A.M.*, 1943, pp. 281–2.

—— 1948. 'Immunity of Canadian Black currant selections from Blister rust', *Phytopathology*, **38**, 453–6.

—— 1957*a*. 'A new species of *Phacidiella* causing the so-called Phomopsis disease of conifers', *Mycologia*, **49**, 226–39. *R.A.M.*, 1957, p. 625.

—— 1957*b*. '*Phacidiopycnis* (Phomopsis) canker and dieback of conifers', *Plant Dis. Reptr.* **41**, 623–33.

—— and Ayers, T. T. 1934*a*. '*Dasyscyphae* on conifers in North America. I. The large-spored, white-excipled species', *Mycologia*, **26**, 73–101. *R.A.M.*, 1934, p. 482.

—— —— 1934*b*. '*Dasyscyphae* on conifers in North America. II. *D. ellisiana*', *Mycologia*, **26**, 167–80. *R.A.M.*, 1934, p. 553.

—— —— 1943. 'Role of *Dasyscypha willkommii* and related fungi in the production of canker and die-back of larches', *J. For.* **41**, 483–95.

—— Hartley, C., and Pierce, R. G. 1917. 'A nursery blight of cedars', *J. Agric. Res.* **10**, 533–40.

HALL, T. F., and SMITH, G. E. 1955. 'Effects of flooding on woody plants, West Sandy dewatering project, Kentucky reservoir', *J. For.* **53,** 281–5.

HAMBRIDGE, G., *et al.* 1941. *Hunger signs in crops—a symposium.* Amer. Soc. Agr. & Nat. Fertilizer Ass. Washington, 327 pp.

HAMILTON, J. R., and JACKSON, L. W. R. 1951. 'Treatment of Shortleaf pine and Loblolly pine seed with fungicidal dusts', *Plant Dis. Reptr.* **35,** 274–6.

HAMMARLUND, C. 1925. 'Zur Genetik, Biologie und Physiologie einiger Erysiphaceen', *Hereditas (Lund, Sweden),* **6,** 1–126. *R.A.M.*, 1925, p. 431.

—— 1932. 'Zur biologie des Mahonia-Rostes (*Puccinia mirabilissima* Peck). (Vorläufige Mitteilung)', *Bot. Notiser,* **6,** 401–16. *R.A.M.*, 1933, p. 293.

HAMOND, J. B. 1931. 'Some diseases of walnuts', *Ann. Rept. East Malling Res. Sta., 1928, 1929, 1930, II. Suppl.*, pp. 143–9.

—— 1935. 'A graft disease of walnuts caused by a species of *Chalaropsis*', *Trans. Brit. Mycol. Soc.* **19,** 158–9.

HANSBROUGH, J. R. 1934. 'Occurrence and parasitism of *Aleurodiscus amorphus* in North America', *J. For.* **32,** 452–8.

—— 1936. 'The Tympanis canker of Red pine', *Yale Univ. Sch. For. Bull.* **43,** 58 pp. *R.A.M.*, 1937, p. 289.

—— 1951. 'Strumella canker of oaks', *Tree Pest Leafl., S.A.F. New England Section,* **32,** 3 pp.

—— JENSEN, V. S., MACALONEY, H. J., and NASH, R. W. 1950. 'Excessive birch mortality in the northeast', *Tree Pest Leafl., S.A.F. New England Section,* **52,** 4 pp.

—— and STOUT, D. C. 1947. 'Viruslike symptoms accompanying birch dieback', *Plant Dis. Reptr.* **31,** 327.

HANSEN, H. N. 1956. 'The perfect stage of *Coryneum cardinale*', *Phytopathology,* **46,** 636–7.

—— and SMITH, R. E. 1937. 'A bacterial gall disease of Douglas fir, *Pseudotsugae taxifolia*', *Hilgardia,* **10,** 569–77. *R.A.M.*, 1937, p. 718.

HANSEN, T. S., *et al.* 1923. 'A study of the damping-off disease of coniferous seedlings', *Minn. Agri. Exp. Sta. Tech. Bull.* **15,** 35 pp.

HARDING, R. B., MILLER, M. P., and FIREMAN, M. 1956. 'Sodium chloride absorption by citrus leaves from sprinkler-applied water', *Citrus Leaves,* **36,** pp. 6–8 & 33. *Biol. Abstr.*, 1957, No. 32610.

HARLEY, J. L. 1956. 'The mycorrhiza of forest trees', *Endeavour,* **15,** 43–48. *R.A.M.*, 1956, p. 703.

—— 1957. 'The ecology of tree mycorrhiza', Abstr. in *Trans. Brit. Mycol. Soc.* **40,** 167–8.

—— 1959. *The biology of mycorrhiza.* Leonard Hill, London, 233 pp.

—— and BRIERLEY, J. K. 1955. 'The uptake of phosphate by excised mycorrhizal roots of the beech. VII. Active transport of ^{32}P from fungus to host during uptake of phosphate from solution', *New Phytol.* **54,** 296–301.

HARPER, A. G. 1913. 'Defoliation; its effects upon the growth and structure of the wood of *Larix*', *Ann. Bot.* **27,** 621–42.

HARRIS, H. A. 1932. 'Initial studies of American elm diseases in Illinois', *Bull. Illinois Dept. Registr. & Educ., Div. Nat. Hist. Survey,* **20,** Art. 1, 70 pp. *R.A.M.*, 1933, p. 124.

HARRIS, J. M. 1953. 'Heartwood formation in *Pinus radiata*', *Nature, Lond.* **172,** 552.

HARRIS, T. M. 1946. 'Zinc poisoning of wild plants from wire-netting', *New Phytol.* **45,** 50–55.

HARRISON, A. T. 1932. 'The mistletoe and its hosts', *Gdnrs.' Chron.* **91,** 189.

HARROW, R. L. 1948. 'The effect of frost of the winter of 1946–47 on vegetation', *J. R. Hort. Soc.* **73,** 389–415; 439–48.

HARTIG, R. 1894. *Text book of the diseases of trees.* Trans. W. Somerville, Macmillan, London, 331 pp.
—— 1900. *Lehrbuch der Pflanzenkrankheiten.* Springer, Berlin, 3rd ed., 324 pp.
HARTIGAN, D. 1949. 'Control of mistletoe', *Aust. J. Sci.* **11,** 174. *For. Abstr.* **11,** No. 1451.
HARTLEY, C. 1915. 'Injury by disinfectants to seeds and roots in sandy soils', *U.S. Dep. Agric. Bull.* **169,** 35 pp.
—— 1918. 'Stem lesions caused by excessive heat', *J. Agric. Res.* **14,** 595–604.
—— 1921. 'Damping-off in forest nurseries', *U.S. Dep. Agric. Bull.* **934,** 99 pp.
—— 1927. 'Forest genetics with particular reference to disease resistance', *J. For.* **25,** 667–86.
—— 1939. 'The clonal variety for tree planting; Asset or liability?', *Phytopathology,* **29,** 9.
—— and CRANDALL, B. S. 1935. 'Vascular disease in poplar and willow', *Phytopathology,* **25,** 18–19.
—— and DAVIDSON, R. W. 1950. 'Wetwood in living trees', *Phytopathology,* **40,** 871.
—— MERRILL, T. C., and RHOADS, A. S. 1918. 'Seedling diseases of conifers', *J. Agric. Res.* **15,** 521–38.
HARTLEY, I., and ELLIS, H. 1931. 'Can *Lathraea squamaria* have a saprophytic existence?', *Northwestern Naturalist,* **6,** 29–30. *Biol. Abstr.,* 1932, No. 3439.
HARVEY, J. M. 1951. 'An anthracnose disease of *Umbellularia californica*', *Madroño, Calif.* **11,** 162–71. *R.A.M.,* 1952, p. 38.
—— 1952. 'Bacterial leaf spot of *Umbellularia californica*', *Madroño, Calif.* **11,** 195–8. *R.A.M.,* 1952, p. 261.
HARVEY, N. P. 1953. *The rose in Britain.* Plant Protection Ltd., 181 pp.
HARVEY, R. B. 1923. 'Cambial temperatures of trees in winter and their relation to sun scald', *Ecology,* **4,** 261–5.
—— 1930. 'Length of exposure to low temperatures as a factor in the hardening process in tree seedlings', *J. For.* **28,** 50–53.
HASELHOFF, E. 1932. *Grundzüge der Rauchschädenkunde.* Borntraeger, Berlin, 162 pp.
—— BREDEMANN, G., and HASELHOFF, W. 1932. *Entstehung, Erkennung und Beurteilung von Rauchschäden.* Borntraeger, Berlin, 472 pp. *R.A.M.,* 1933, p. 524.
HASEN, C. M. 1951. 'The helicopter rotor as a means of controlling frost damage in fruit orchards', *Quart. Bull. Mich. Agric. Exp. Sta.* **34,** 182–8. *Hort. Abstr.,* 1952, No. 1325.
HASKINS, C. P., and MOORE, C. N. 1933. 'The physiological basis of the twisting habit in plant growth', *Science,* **72,** 283.
HATCH, A. B. 1936. 'The role of mycorrhizae in afforestation', *J. For.* **34,** 22–29.
HATFIELD, W. C. 1946. 'Shoot blight of Aspen and Poplar caused by species of *Fusicladium*', *Univ. Wyo. Publ. Sci.* **12,** 73–74. *R.A.M.,* 1947, p. 571.
HAVELIK, K. 1935. 'Odumíráni Jedle', *Ann. Acad. tchécosl. Agric.* **10,** 124–8. *R.A.M.,* 1935, p. 481.
HAWKE, E. L. 1944. 'Thermal characteristics of a Hertfordshire frost-hollow', *Quart. J. Roy. Met. Soc.* **70,** 22–48. *Hort. Abstr.,* 1947, No. 1287.
HAWKER, L. E., and FRAYMOUTH, J. 1951. 'A re-investigation of the root-nodules of species of *Elaeagnus, Hippophäe, Alnus,* and *Myrica,* with special reference to the morphology and life histories of the causative organisms', *J. Gen. Microbiol.* **5,** 369–86. *R.A.M.,* 1952, p. 36.
HAWKES, C. 1953. 'Planes release tree plantation', *J. For.* **51,** 345–8.

HAWKSWORTH, F. G. 1958. 'Rate of spread and intensification of dwarf mistletoe in young Lodgepole pine stands', *J. For.* **56,** 404–7.

—— and LUSHER, A. A. 1956. 'Dwarf mistletoe survey and control on the Mescalero-Apache Reservation, New Mexico', *J. For.* **54,** 384–90.

HAWLEY, R. C. 1946. *Practice of Silviculture.* John Wiley. New York, 5th ed., 354 pp.

—— and CLAPP, R. T. 1935. 'Artificial pruning in coniferous plantations', *Yale Univ. Sch. For. Bull.* **39,** 36 pp.

—— and STICKEL, P. W. 1948. *Forest protection.* John Wiley, New York, 355 pp.

HEALD, F. D., GARDNER, M. W., and STUDHALTER, R. 1915. 'Air and wind dissemination of ascospores of the Chestnut-blight fungus', *J. Agric. Res.* **3,** 493–526.

—— and STUDHALTER, R. A. 1914. 'Birds as carriers of the Chestnut-blight fungus', *J. Agric. Res.* **2,** 405–22.

HEDGCOCK, G. G. 1927. '*Dothichiza populae* and its mode of infection', *Phytopathology,* **17,** 545–7.

—— 1928. 'A key to the known aecial forms of *Coleosporium* occurring in the United States and a list of the host species', *Mycologia,* **20,** 97–100. *R.A.M.,* 1928, p. 603.

—— 1929. 'The large leaf spot of chestnut and oak associated with *Monochaetia desmazierii*', *Mycologia,* **21,** 324–5. *R.A.M.,* 1930, p. 353.

—— BETHEL, E., and HUNT, N. R. 1918. 'Piñon blister-rust', *J. Agric. Res.* **14,** 411–24.

—— HAHN, G. G., and HUNT, N. R. 1922. 'Two important pine cone rusts and their new Cronartial stages', *Phytopathology,* **12,** 199–222.

—— and LONG, W. H. 1915. 'A disease of pines caused by *Cronartium pyriforme*', *Bull. U.S. Dep. Agric.* **247,** 20 pp.

—— and SIGGERS, P. V. 1949. 'A comparison of the Pine-oak rusts', *Tech. Bull. U.S. Dep. Agric.* **978,** 30 pp.

HEIBERG, S. O., and WHITE, D. P. 1951. 'Potassium deficiency of reforested pine and spruce stands in northern New York', *Proc. Soil. Sci. Soc. Amer.* **15,** 369–76. *R.A.M.,* 1952, p. 307.

HEIM, R., and JACQUES-FÉLIX, M. 1953. 'Études expérimentales sur la spécificité des hyménomycètes. Nutrition et balancement des caractères morphologiques culturaux chez *Armillariella mellea* et *Clitocybe tabescens.* Les rapports de parenté entre *Armillariella mellea* et *Clitocybe tabescens*', *C.R. Acad. Sci., Paris,* **236,** 167–70. *R.A.M.,* 1953, p. 698.

—— and LAMI, R. 1950. 'La disparition des Peupliers de Paris sous les atteintes de l'*Ungulina inzengae* (de Not.) Pat.', *C.R. Acad. Agric. Fr.* **36,** 257–8. *R.A.M.,* 1950, p. 542.

HEINICKE, A. J. 1932. 'The effect of submerging the roots of apple trees at different seasons of the year', *Proc. Amer. Soc. Hort. Sci.* **29,** 205–7.

HEINRICHER, E. 1906. 'Ein bemerkenswerter Standort der *Lathraea squamaria*', *Naturw. Z. Forst- u. Landw.* **4,** 274–6.

—— 1914. 'Ein Hexenbesen auf *Juniperus communis* verursacht durch *Arceuthobium oxycedri,* (DC.) M. Bieb.', *Naturw. Z. Forst- u. Landw.* **12,** 36–39.

—— 1930. 'Über *Arceuthobium oxycedri* (DC.) M. Bieb. auf *Chamaecyparis sphaeroidea* Spach. *pendula* Hort. und einen Hexenbesen der durch den Einfluss des *Arceuthobium* auf dieser Cupressinee entstand', *Z. Wiss. Biol. Abt. E. Planta.* **10,** 374–80. *Biol. Abstr.,* 1934, No. 10503.

HELMERS, A. E. 1946. 'Effect of pruning on growth of Western white pine', *J. Fr.* **44,** 673–6.

HEMBERG, T. 1949. 'Growth-inhibiting substances in terminal buds of *Fraxinus*', *Physiol. Plant., Copenhagen*, **2,** 37–44. *For. Abstr.* **11,** No. 77.

HENDERSON, D. M. 1954. 'The genus *Taphrina* in Scotland', *Notes Roy. Bot. Gard. Edin.* **21,** 165–80.

—— 1957. 'The *Melampsora epitea* complex on mountain willows in Scotland', *Notes Roy. Bot. Gard., Edin.* **22,** 201–6. *Biol. Abstr.*, 1958, No. 5951.

HENDRIX, J. W. 1956. 'Variation in sensitivity to atmospheric fluorides among Ponderosa pine seedlings', *Phytopathology*, **46,** 637.

HENRIKSEN, H. A. 1951. 'Røntgenfotografering som diagnostisk hjaelpemiddel ved undersøgelse af traeer', *Dansk. Skovforen. Tiddskr.* **36,** 515–20. *For. Abstr.* **14,** 1, No. 430.

—— and JØRGENSEN, E. 1954. 'Rodfordaerverangreb i relation til udhugningsgrad', *Forstl. Forsøgsv. Danm.* **21,** 215–51. *R.A.M.*, 1955, p. 194.

HENRY, B. W. 1950. 'Control of a root rot of pine seedlings by soil fumigation', *Phytopathology*, **40,** 788.

HENSON, W. R. 1952. 'Chinook winds and red-belt injury to Lodgepole pine in the Rocky Mountain parks area of Canada', *For. Chron.* **28,** 62–64. *For. Abstr.* **13,** No. 3941.

HEPTING, G. H. 1935. 'Decay following fire in young Mississippi Delta hardwoods', *Tech. Bull. U.S. Dep. Agric.* **494,** 32 pp. *R.A.M.*, 1936, p. 470.

—— 1941. 'Prediction of cull following fire in Appalachian oaks', *J. Agric. Res.* **62,** 109–20.

—— 1944. 'Sapstreak, a new killing disease of Sugar maple', *Phytopathology*, **34,** 1069–76.

—— 1954. 'Gum flow and pitch-soak in Virginia pine following Fusarium inoculation', *S.E. For. Exp. Sta. Station Paper*, **40,** 9 pp.

—— 1955*a*. 'The current status of oak wilt in the United States', *For. Sci.* **1,** 95–103.

—— 1955*b*. 'A southwide survey for Sweetgum blight', *Plant Dis. Reptr.* **39,** 261–5.

—— and DAVIDSON, R. W. 1937. 'A leaf and twig disease of hemlock caused by a new species of *Rosellinia*', *Phytopathology*, **27,** 305–10.

—— and DOWNS, A. A. 1944. 'Root and butt rot in planted White pine at Biltmore, North Carolina', *J. For.* **42,** 119–23.

—— GARREN, K. H., and WARLICK, P. W. 1940. 'External features correlated with top rot in Appalachian Oaks', *J. For.* **38,** 873–6.

—— and ROTH, E. R. 1950. 'The fruiting of heart-rot fungi on felled trees', *J. For.* **48,** 332–3.

—— —— 1953. 'Host relations and spread of the pine Pitch canker disease', *Phytopathology*, **43,** 475.

—— —— and SLEETH, B. 1949. 'Discolorations and decay from increment borings', *J. For.* **47,** 366–70.

—— and TOOLE, E. R. 1939. 'The hemlock rust caused by *Melampsora farlowii*', *Phytopathology*, **29,** 463–73.

—— —— and BOYCE, J. S., Jr. 1952. 'Sexuality in the Oak wilt fungus', *Phytopathology*, **42,** 438–42.

HERBERT, D. A. 1931. '*Cyttaria septentrionalis*, a new fungus attacking *Nothofagus moorei* in Queensland and New South Wales', *Proc. Roy. Soc. Queensland*, **41,** 158–61. *R.A.M.*, 1931, p. 697.

HERRICK, E. H. 1932. 'Further notes on twisted trees', *Science*, **76,** 406–7.

HESS, E. 1933. 'Races des pins et cris de neige', *J. For. Suisse*, **84,** 269–77. *Biol. Abstr.*, 1936, No. 16683.

HESSELINK, E. 1927. 'Onder welke omstandigheden doet *Lophodermium pinastri* Chev. te Appelscha, Exloo en Odoorn schade in Dennenbeplantingen?', *Tijdschr. PlZiekt.* **33,** 105–24. *R.A.M.*, 1927, p. 705.

HESTERBERG, G. A. 1957. 'Deterioration of Sugar Maple', *Sta. Pap. Lake St. For. Exp. Sta.* **51,** 58. *For. Abstr.* **19,** No. 3162.

HEWITT, J. L. 1936. 'A survey concerning a native pathogen—*Armillaria mellea*', *Bull. Dept. Agric. Calif.* **25,** 226–34.

HEWLETT, M. A. 1952. 'Two leaf spots of Myrtle new to Great Britain, caused by the fungi *Pestalotia decolorata* and *Cercospora myrti*', *J. R. Hort. Soc.* **77,** 413–18.

HEY, T. 1956. '"Sweat of Heaven." The nature and causes of Honeydew', *Gdnrs.' Chron.* **140,** 522.

HEYBROEK, H. M. 1957. 'Iepenveredeling in Nederland', *Ned. Boschb. Tijdschr.* **29,** 96–100.

HICKMAN, C. J. 1958. '*Phytophthora*—Plant destroyer', *Trans. Brit. Mycol. Soc.* **41,** 1–13.

HIKSCH, F. 1934. 'Beitrag zur forstlichen Schädenfrage der arsenigen Säure im weissen Hüttenrauch der Arsenikhüttenwerke', *Tharandt. forstl. Jb.* **85,** 117–66. *R.A.M.*, 1935, p. 725.

HILDEBRAND, E. M. 1954. 'Camellia variegation in Texas', *Plant Dis. Reptr.* **38,** 566–7.

HILEY, W. E. 1919. *Fungal diseases of the common larch.* Clarendon Press, Oxford, 204 pp.

HIMELICK, E. B., and CURL, E. A. 1955. 'Experimental transmission of the Oak wilt fungus by caged squirrels', *Phytopathology*, **45,** 581–4.

—— —— 1958. 'Transmission of *Ceratocystis fagacearum* by insects and mites', *Plant Dis. Reptr.* **42,** 538–45.

—— SCHEIN, R. D., and CURL, E. A. 1953. 'Rodent feeding on mycelial pads of the Oak wilt fungus', *Plant Dis. Reptr.* **37,** 101–3.

HINTIKKA, T. J. 1933. 'Muutamia havaintoja Männyn tuudenpesistä'. *Acta Forest. Fennica*, **39,** 15 pp. *R.A.M.*, 1934, p. 283.

HIRATA, T., and MAEZEWA, K. 1956. 'Wind damage in the Tokyo University forest, Hokkaido', *Misc. Inform. Tokyo Univ. For.* **11,** 1–8. *For. Abstr.* **19,** No. 1854.

HIRATSUKA, N. 1936–7. '*Gymnosporangium* of Japan', *Bot. Mag. Tokyo*, **51,** 587, pp. 481–4; 598, pp. 549–55; 599, pp. 593–9; 600, pp. 661–8; 601, pp. 1–8. *R.A.M.*, 1937, p. 411.

HIRT, R. R. 1932. 'On the biology of *Trametes suaveolens* (L.) Fries', *New York State Coll. For. Tech. Publ.* **37,** 36 pp. *R.A.M.*, 1932, p. 811.

—— 1938. 'Relation of stomata to infection of *Pinus strobus* by *Cronartium ribicola*', *Phytopathology*, **28,** 180–90.

—— 1939. 'Canker development by *Cronartium ribicola* on young *Pinus strobus*', *Phytopathology*, **29,** 1607–76.

—— 1940. 'Relative susceptibility to *Cronartium ribicola* of 5-needled Pines planted in the east', *J. For.* **38,** 932–7.

—— 1944. 'Distribution of Blister-rust cankers on Eastern white pine according to age of needle-bearing wood at time of infection', *J. For.* **42,** 9–14.

—— 1948. 'Evidence of resistance to blister rust by Eastern white pine growing in the northeast', *J. For.* **46,** 911–13.

—— 1956.' Fifty years of White pine blister rust in the northeast', *J. For.* **54,** 435–8.

—— and ELIASON, E. J. 1938. 'The development of decay in living trees inoculated with *Fomes pinicola*', *J. For.* **36,** 705–9.

HOBBS, C. H. 1944. 'Studies on mineral deficiency in pine', *Plant Physiol.* **19,** 590–602. *R.A.M.*, 1945, p. 215.

HOCKING, G. H. 1949. 'Compression failure in *Pinus radiata* stems exposed to strong wind', *N.Z. J. For.* **6,** 65–66. *For. Abstr.* **12,** No. 1541.

HOEPFFNER, A. VON. 1910. 'Beobachtungen über elektrische erscheinungen im Walde', *Naturw. Z. Forst- u. Landw.* **8,** 411–16.

HÖFKER, H. 1919. 'Über den Einfluss der Winterwitterung auf die Gehölze mit besonderer Berücksichtigung des strengen Frostes im Winter 1916–1917', *Mitt. Deutsch. Dendrol. Ges.*, **28,** 196–207. *Bot. Abstr.*, 1925, No. 7376.

HOLDSWORTH, L. W. 1956. 'A note on new methods of treating trace element deficiencies in plants', *J. R. Hort. Soc.* **81,** 318–21.

HOLLY, K. 1954. 'Morphological effects on plants due to damage by growth regulator weed-killers', *Plant Path.* **3,** 1–5.

HOLMES, F. W. 1958. 'Recorded Dutch elm disease distribution in North America as of 1957', *Plant Dis. Reptr.* **42,** 1299–1300.

HOLMES, G. D. 1953. 'Recent developments in the control of weeds in forest nurseries', *Proc. Brit. Weed Control Conf.*, pp. 282–9.

—— and BUSZEWICZ, G. M. 1953–7. 'Forest tree seed investigations', *For. Comm. Rep. For. Res.*, 1952, pp. 12–14; 1953, pp. 14–17; 1954, pp. 1–4; 1955, pp. 13–16; 1956, pp. 15–19.

—— —— 1958. 'The storage of seed of temperate forest tree species', *For. Abstr.* **19,** 313–22, 455–76.

HOLSTENER-JØRGENSEN, H., and KLUBIEN, E. 1957. 'Diagnosticering af kvaelstoff-mangel i bøgekulturer', *Dansk Skovforen. Tidsskr.* **42,** 593–600. *For. Abstr.* **19,** No. 1869.

HOLTZMANN, O. V. 1955. 'Organisms causing damping-off of coniferous seedlings, and their control', *Diss. Abstr.* **15,** 2392. *R.A.M.*, 1956, p. 731.

HOPP, P. J. 1957. 'Zur Kenntnis des Lärchenkrebses', *Forstwiss. Cbl.* **76,** 334–54.

HOPFFGARTEN, E. H. von 1933. 'Beiträge zur Kenntnis der Stockfäule (*Trametes radiciperda*)', *Phytopath. Z.* **6,** 1–48. *R.A.M.*, 1933, p. 738.

HORD, H. H. V., GROENEWOUD, H. VAN, and RILEY, C. G. 1957. 'Low temperature injury to roots of White elm', *For. Chron.* **33,** 156–63.

HORNER, R. M. 1956. 'A Diaporthe canker of *Betula lutea*', Abstr. in *Proc. 22nd Sess. Canad. Phytopathological Soc., Edmonton 1955,* **23,** 16–17. *For. Abstr.* **18,** No. 1808.

HORNIBROOK, E. M. 1950. 'Estimating defect in mature and overmature stands of three Rocky Mountain conifers', *J. For.* **48,** 408–17.

HORSFALL, J. G. 1945. *Fungicides and their action.* Chronica Botanica, 239 pp.

—— 1956. *The principles of fungicidal action.* Chronica Botanica, 280 pp.

—— and DIMOND, A. E. 1951. 'Plant chemotherapy', *Trans. N.Y. Acad. Sci.*, ser. 2, **13,** 338–41. *R.A.M.*, 1952, p. 133.

HORTON, G. S., and HENDEE, C. 1934. 'A study of rot of Aspen in the Chippewa National Forest', *J. For.* **32,** 493–4.

HOSOKAWA, T., and ODANI, N. 1957. 'The daily compensation period and vertical ranges of epiphytes in a beech forest', *J. Ecol.* **45,** 901–15.

HOTSON, J. W., and STUNTZ, D. E. 1934. 'Canker on *Chamaecyparis lawsoniana*', *Phytopathology*, **29,** 1145–6.

HOUTZAGERS, G., *et al.* 1941. *Handboek voor de Populierenteelt.* H. Veenman & Zonen, Wageningen, 231 pp.

HOWARD, A. 1925. 'The effect of grass on trees', *Proc. Roy. Soc., Lond.* B. **97,** 284–321. *Bot. Abstr.* **15,** 519, No. 3386.

HOWARD, F. L. 1941. 'The bleeding canker disease of hardwoods and possibilities of control', *Proc. 8th W. Shade Tree Conf.*, 10 pp. *R.A.M.*, 1942, p. 394.

HOWARD, N. F. 1932. 'Twisted trees', *Science*, **75,** 132–3.

HOWARD, N. O. 1929. 'A new disease of Douglas fir', *Science*, **69,** 651–2. *R.A.M.*, 1929, p. 744.

HOWARTH, W. O., and CHIPPINDALE, H. G. 1931. 'On some fungi occurring on Rhododendrons', *Mem. & Proc. Manchester Lit. & Phil. Soc.*, 1930–1, **7**, 95–103. *R.A.M.*, 1931, p. 798.

HUBBES, M. 1959. 'Untersuchungen über *Dothichiza populea* Sacc. et Briard, den Erreger des Rindenbrandes der Pappel', *Phytopath. Z.* **35**, 58–96. *R.A.M.*, 1959, p. 490.

HUBERMAN, M. A. 1943. 'Sunscald of Eastern white pine, *Pinus strobus* L.', *Ecology*, **24**, 456–71.

HUBERT, E. E. 1918. 'Fungi as a contributory cause of windfall', *J. For.* **16**, 696–714.

—— 1935*a*. 'Observations on *Tuberculina maxima*, a parasite of *Cronartium ribicola*', *Phytopathology*, **25**, 253–61.

—— 1935*b*. 'A disease of conifers caused by *Stereum sanguinolentum*', *J. For.* **33**, 485–9.

HUBERT, F. P., McCUBBIN, W. A., and WHEELER, W. H. 1952. '2, 4-D injury to common horsechestnut causing virus-like symptoms', *Plant Dis. Reptr.* **36**, 65.

HUCKENPAHLER, B. J. 1936. 'Amount and distribution of moisture in a living shortleaf pine', *J. For.* **34**, 399–401.

HUET, M. 1936. 'La Maladie du rond (*Polyporus annosus*)', *Bull. Soc. for. Belg.* **48**, 349–71. *R.A.M.*, 1937, p. 145.

HUIKARI, O. 1955. 'Experiments on the effect of anaerobic media upon birch, pine, and spruce seedlings', *Commun. Inst. Forest Fennial*, **42**, 1–13. *Biol. Abstr.*, 1957, No. 2422.

HULBARY, R. L. 1941. 'A needle blight of Austrian pine', *Bull. Ill. nat. Hist. Surv.* **21**, 231–6. *R.A.M.*, 1942, p. 275.

HULSHOF, H. J., and ZEGERS, H. J. M. 1951. 'Wortelsterfte en slechte bladstand bij peren', *Fruitteelt*, **41**, 758–9. *Hort. Abstr.*, 1952, No. 2286.

HUMPHREY, C. J., and SIGGERS, P. V. 1933. 'Temperature relations of wood destroying fungi', *J. Agric. Res.* **47**, 997–1008. *R.A.M.*, 1934, p. 413.

HUNTER, L. M. 1927. 'Comparative study of spermagonia of rusts of *Abies*', *Bot. Gaz.* **83**, 1–23. *R.A.M.*, 1927, p. 588.

—— 1936. 'The genus *Milesia* in Great Britain and Ireland', *Trans. Brit. Mycol. Soc.* **20**, 116–19.

—— 1948. 'A study of the mycelium and haustoria of the rusts of *Abies*', *Canad. J. Res.*, sect. C, **26**, 219–38. *R.A.M.*, 1949, p. 201.

HUNTER-BLAIR, H. 1946. 'Frost damage to woodlands on Blairquhan estate in April 1945', *Scot. For. J.* **60**, 38–43.

HUSS, E. 1951. 'Skogsforskningsinstitutets metodik vid fröundersökningar', *Medd. Skogsforskn. Inst. Stockh.* **40**, 1–82. *R.A.M.*, 1953, p. 45.

HUTCHINS, L. M. 1947. 'Soil aeration—its importance to plant growth and disease control', *Proc. 23rd Nat. Shade Tree Conf.*, pp. 17–25.

HUTCHINSON, M. T., and VARNEY, E. H. 1954. 'Ringspot—a virus disease of cultivated Blueberry', *Plant Dis. Reptr.* **38**, 260–2.

HYAMS, E. 1952. *Grapes under cloches.* Faber & Faber, 133 pp.

HYLANDER, N. 1957*a*. 'On cut-leaved and small-leaved forms of *Alnus glutinosa* and *A. incana*', *Svensk bot. Tidskr.* **51**, 437–53. *For. Abstr.* **19**, No. 2733.

—— 1957*b*. 'On cut-leaved and small-leaved forms of Scandinavian Birches', *Svensk bot. Tidskr.* **51**, 417–36. *For. Abstr.* **19**, No. 2734.

—— JØRSTAD, I., and NANNFELDT, J. A. 1953. 'Enumeratio Uredinearum Scandinavicarum', *Opera bot.* (*Bot. Notiser Suppl.*), **1**, 102 pp. *R.A.M.*, 1956, p. 791.

IGMÁNDY, Z. 1956. 'Fagyrepedés okozta károk csereseinkben', *Erdőmérn. Főisk. Közl.* **2**, 81–101. *For. Abstr.* **19**, No. 1862.

IGMÁNDY, Z., MILINKÓ, I., and SZATALA, Ö. 1954. 'Investigations and control experiments to combat the damping-off of conifer seedlings', *Erdész. Tud. Int. Évk.* **2** (1952), 210–26. *R.A.M.*, 1956, p. 133.

IIZUKA, H., TAMATE, S., TAKAKUWA, T., and SATO, T. 1950. 'Experiments with model windbreaks (1st report). Effect on salt-laden wind', *Bull. For. Exp. Sta., Meguro, Tokyo,* **45,** 1–15. *For. Abstr.* **13,** No. 3852.

IIZUKA, N. 1956. 'Study on the judgment method of inner rot of the living tree (Todo-Fir) by X-ray', *Bull. Tokyo Univ. For.* **52,** 153–63. *For. Abstr.* **19,** No. 3152.

IJJÁSZ, E. 1933. 'Die meteorologischen Faktoren des Bodenauffrierens und die Rolle der Frostlinien bei Auffrierungschäden', *Erdész. Kisérletek,* **35,** 318–19. *Imp. For. Inst. Trans.*, No. 21.

INGESTAD, T. 1957. 'Studies on nutrition of forest tree seedlings. 1. Mineral nutrition of Birch', *Physol. Plant. (Lund.)*, **10,** 418–39.

INOUE, K., and MASUDA, H. 1956. 'Trees for fire mantle in Hokkaido', *For. Exp. Sta., Hokkaido,* **5,** 150–61. *For. Abstr.* **18,** No. 555.

ISAAC, L. A. 1930. 'Seedling survival on burned and unburned surfaces', *J. For.* **28,** 569–71.

ISHIDA, S. 1950. 'A study on frost crack in trees', *Low Temperature Science, Hokkaido,* **5,** 61–73. *For. Abstr.*, 1953, No. 1282.

ITÔ, K. 1949. 'Studies on "murasaki-mompa" disease caused by *Helicobasidium mompa* Tanaka', *Bull. For. Exp. Sta., Meguro, Tokyo,* **43,** 126 pp. *R.A.M.*, 1951, p. 337.

—— 1950. 'Contributions to the diseases of poplars in Japan. I. Shoot blight of poplars caused by a new species of *Pestolotia*', *Bull. For. Exp. Sta., Meguro, Tokyo,* **45,** 135–46. *R.A.M.*, 1951, p. 549.

—— and CHIBA, O. 1954–5. 'Studies on some anthracnoses of woody plants. I. Overwintering of *Gleosporium kawakamii.* II. *Glomerella* parasitic on *Paulownia* trees', *Bull. For. Exp. Sta., Meguro, Tokyo,* **70,** 93–101; **81,** 43–62. *R.A.M.*, 1956, p. 249.

—— —— ONO, K., and HOSAKA, Y. 1956. 'Studies on some anthracnoses of woody Plants. III. Anthracnose affecting the fruit of *Camellia japonica* L.', *Bull. For. Exp. Sta., Meguro, Tokyo,* **83,** 65–88.

—— and HOSAKA, Y. 1950. 'Notes on some leaf-spot diseases of broadleaved trees. I. Cercospora leaf spot of plane trees', *Bull. For. Exp. Sta., Meguro, Tokyo,* **46,** 17–32. *R.A.M.*, 1951, p. 549.

—— —— 1951. 'Grey mold and sclerotial disease of "Sugi" (*Cryptomeria japonica* D. Don) seedlings, the causes of the so-called "snow molding"', *Bull. For. Exp. Sta., Meguro, Tokyo,* **51,** 1–27. *R.A.M.*, 1952, p. 582.

—— —— 1953. 'Gray moulds of Japanese larch (*Larix kaempferi* Sarg.) seedlings', *Bull. For. Exp. Sta., Meguro, Tokyo,* **59,** 33–44. *R.A.M.*, 1955, p. 412.

—— and KOBAYASHI, T. 1951. '*Physalospora paulowniae* sp.nov. causing a die-back of the *Paulownia* tree and its conidial stage, *Macrophoma*', *Bull. For. Exp. Sta., Meguro, Tokyo,* **49,** 79–87. *R.A.M.*, 1952, p. 153.

—— and KOBAYASHI, T. 1953. 'Contributions to the diseases of poplars in Japan. II. The Cercospora leaf spot of poplars with special reference to the life history of the causal fungus', *Bull. For. Exp. Sta., Meguro, Tokyo,* **59,** 1–32. *R.A.M.*, 1955, p. 497.

—— and KONTANI, S. 1951. 'Notes on damping-off and root rot of woody seedlings. I. Relation between the mycelial growth of causal fungi and the concentration of fungicides. II. A comparative study of isolates of *Rhizoctonia solani* with special reference to their parasitism', *Rep. For. Exp. Sta., Meguro, Tokyo,* **60,** 65–91. *R.A.M.*, 1952, p. 153.

Itô, K., and Kontani, S. 1954. '*Pestalotia* parasitic on seedlings of *Chamaecyparis obtusa* Sieb. et Zucc.', *Bull. For. Exp. Sta., Meguro, Tokyo*, **76**, 63–72. *R.A.M.*, 1955, p. 416.

—— —— and Kondô, H. 1955. 'Web-blight fungus of Japanese larch seedlings', *Bull. For. Exp. Sta., Meguro, Tokyo*, **79**, 43–70. *R.A.M.*, 1955, p. 688.

—— Satô, K., and Ôta, N. 1957. 'Studies on the needle-cast of Japanese larch. I. Life history of the causal fungus, *Mycosphaerella larici-leptolepis* sp. nov.', *Bull. For. Exp. Sta., Meguro, Tokyo*, **96**, 69–88.

—— Shibukawa, K., and Kobayashi, T. 1952. 'Etiological and pathological studies on the needle blight of *Cryptomeria japonica*. I. Morphology and pathogenicity of the fungi inhabiting the blighted needles', *Bull. For. Exp. Sta., Meguro, Tokyo*, **52**, 79–152. *R.A.M.*, 1954, p. 327.

—— —— and Terashita, T. 1954. 'Etiological and pathological studies on the needle blight of *Cryptomeria japonica*. II. Physiological and ecological characters of *Cercospora cryptomeriae* Shirai, the most important pathogen of the disease', *Bull. For. Exp. Sta., Meguro, Tokyo*, **76**, 27–61. *R.A.M.*, 1955, p. 415.

Itô, T. 1958. '*Pellicularia koleroga* Cooke causing the thread blight of *Ginkgo biloba*', *Bull. For. Exp. Sta., Meguro, Tokyo*, **105**, 11–18. *R.A.M.*, 1958, p. 604.

Ivanoff, S. S. 1958. 'The water soak method of plant disease control in relation to microbial activities, oxygen supply, and food availability', *Phytopathology*, **48**, 502–8.

Iyengar, A. V. V. 1958. 'Spike disease of Sandal: A retrospect', *Indian For.* **84**, 603–12.

Jacks, H. 1956. 'Seed disinfection. XII. Glasshouse tests for control of damping-off of *Pinus radiata* seed', *N.Z. J. Sci.*, Technol., sect. A, **37**, 427–31. *R.A.M.*, 1956, p. 730.

Jackson, H. S. 1914. 'A new Pomaceous rust of economic importance, *Gymnosporangium blasdaleanum*', *Phytopathology*, **4**, 261–70.

Jackson, L. W. R. 1940*a*. 'Lightning injury to black locust seedlings', *Phytopathology*, **30**, 183–4.

—— 1940*b*. 'Effects of H-ion and Al-ion concentrations on damping-off of conifers and certain causative fungi', *Phytopathology*, **30**, 563–79.

—— 1948. '"Needle curl" of Shortleaf pine seedlings', *Phytopathology*, **38**, 1028–9.

Jacobs, H. L. 1950. '2, 4-D, friend or foe', *Proc. 26th Nat. Shade Tree Conf.*, pp. 23–30.

Jacobs, M. R. 1936. 'The effect of wind on trees', *Aust. For.* **1**, 25–32.

—— 1938. 'Notes on pruning *Pinus radiata*. I. Observations on factors which influence pruning', *Bull. For. Bur. Aust.* **23**, 47 pp.

—— 1954. 'The effect of wind sway on the form and development of *Pinus radiata* D. Don.', *Aust. J. Bot.* **2**, 35–51. *Biol. Abstr.*, 1954, No. 26941.

Jacobsen, —. 1956. 'Spätfrostbekämpfung', *Forst- u. Holzw.* **11**, 297. *For. Abstr.* **18**, No. 540.

Jacques, J. E. 1944. 'Un Chancre de l'Orme de Sibérie (*Ulmus pumila*) causé par *Nectria cinnabarina*', *Ann. Assoc. canad.-franç. Sci.* **10**, 89. *R.A.M.*, 1946, p. 85.

Jacquiot, C., and Viney, R. 1954. 'La Graisse du chêne', *Rev. Path. vég.* **33**, 66–79. *For. Abstr.* **16**, No. 4358.

Jahnel, H., and Junghans, B. 1957. 'Über eine wenig bekannte Kiefernkrankheit (*Sclerophoma pithyophila*)', *Forstwiss. Cbl.* **76**, 129–32.

Jamalainen, E. A. 1956. 'A test on the control of black snow mould (*Herpotrichia nigra* Hartig) in spruce seedlings by the use of pentachloronitrobenzene', *Valt. Maatalousk. Julk.* **148**, 68–72. *R.A.M.*, 1957, p. 740.

JAMES, N. D. G. 1939. 'Lightning damage in trees', *Quart. J. For.* **33,** 16–18.

JAŇČAŘÍK, V., and UROŠEVIĆ, B. 1958. 'Hromadný výskyt houby *Coniothyrium quercinum* (Bonord.) Sacc. na semenáčcích Dubu červeného (*Quercus rubra* L.)', *Preslia,* **30,** 370–1. *R.A.M.,* 1959, p. 166.

JANSON, A. 1925. 'Rauchempfindlichkeit der Ziergehölze', *Dtsch. ObstbZtg.* **70,** 578–9. *R.A.M.,* 1926, pp. 67–68.

JAUCH, C. 1943. 'La presencia de *Cylindrocladium scoparium* en la Argentina', *Rev. argent. Agron.* **10,** 355–60. *For. Abstr.,* 1945, p. 259.

—— 1952. 'La "antracnosis" de los Sauces cultivados en el Delta del Paraná', *Rev. Fac. Agron. B. Aires,* **13,** 285–308. *R.A.M.,* 1954, p. 455.

JAYME, G., HARDES-STEINHÄUSER, M., and MOHRBERG, W. 1951. 'Einfluss des Zugholzanteils auf die technologische und chemische Verwendbarkeit von Papperhölzern', *Das Papier, Darmstadt,* **5,** 12 pp.

JEMISON, G. M. 1943. 'Effect of litter removal on diameter growth of Shortleaf pine', *J. For.* **41,** 213–14.

—— 1944. 'The effect of basal wounding by forest fires on the diameter growth of some southern Appalachian hardwoods', *Duke Univ. Sch. For. Bull.* **9,** 63 pp. *For. Abstr.* **8,** No. 1398.

—— and SCHUMACHER, F. X. 1948. 'Epicormic branching in old growth Appalachian hardwoods', *J. For.* **46,** 252–5.

JENKINS, A. E. 1944. 'Two new records of *Sphaceloma* diseases in the United States', *Phytopathology,* **34,** 981–3.

—— and BITANCOURT, A. A. 1939. 'Illustracões das doenças causadas por *Elsinoe* e *Sphaceloma* conhecidas na America do Sul até Janiero de 1936', *Arq. Inst. Biol. São Paulo,* **10,** 31–60. *R.A.M.,* 1940, p. 366.

—— —— 1942. 'An *Elsinoe* causing an anthracnose of Virginia Creeper', *Phytopathology,* **32,** 424–7.

—— and GRODSINKSY, L. 1943. '*Sphaceloma* on willow in New Zealand', *Trans. Brit. Mycol. Soc.* **26,** 1–3.

—— MILLER, J. H., and HEPTING, G. H. 1953. 'Spot anthracnose and other leaf and petal spots of flowering dogwood', *Nat. Hort. Mag.* **32,** 57–69. *R.A.M.,* 1955, p. 371.

JENKINS, P. T. 1952. '*Armillaria* on fruit trees', *J. Dep. Agric. Vict.* **50,** 88–90. *Hort. Abstr.,* 1952, No. 2329.

JENNINGS, O. E. 1934. 'Smoke injury to shade trees', *Proc. 10th Nat. Shade Tree Conf.,* pp. 44–48.

JESSEN, W. 1938. 'Phosphorsäuremangelerscheinungen bei verschiedenen Holzarten', *Phosphorsäure,* **7,** 263–70. *R.A.M.,* 1938, p. 573.

—— 1939. 'Kalium- und Magnesiummangelerscheinungen und Wirkung einer Düngung mit Kaliumchlorid und Kalimagnesia auf das Wachstum verschiedener Holzarten', *Ernähr. Pfl.* **35,** 228–30. *R.A.M.,* 1940, p. 52.

JOACHIM, H. F. 1957*a*. 'Über den Einfluss des Sandrohrs (*Calamagrostis epigeios*) auf das Wachstum der Pappel', *Forst- u. Jagd. Spec. No. Die Pappel, II,* pp. 7–12. *For. Abstr.* **19,** No. 1494.

—— 1957*b*. 'Über Frostschäden an der Gattung *Populus*', *Arch. Forstw.* **6,** 601–78. *For. Abstr.* **19,** No. 3105.

—— 1958. 'Über den Braunfleckengrind', *Allg. ForstZ.* **13,** 548–51. *For. Abstr.* **20,** No. 2129.

JOHNSON, E. M. 1937. 'Distribution of *Cephalosporium* and *Verticillium* on Elm in Massachusetts', *Plant Dis. Reptr.* **21,** 58–59.

JOHNSON, L. P. V., and LINTON, G. M. 1942. 'Experiments on chemical control of damping-off in *Pinus resinosa* Ait.', *Canad. J. Res.,* sect. C, **20,** 559–71. *R.A.M.,* 1943, p. 158.

JOHNSSON, H. 1957. 'Aktuellt om hybridasp.', *Skogen*, **44**, 798, 801. *For. Abstr.* **19**, No. 2895.

JONES, A. T. 1931. 'Trees with twisted bark', *Science*, **73**, 341.

JONES, E. W. 1944. 'Biological flora of the British Isles, *Acer* L.', *J. Ecol.* **32**, 215–52.

—— 1952. 'Some observations on the lichen flora of tree boles, with special reference to the effect of smoke', *Rev. bryol. lichénol., Paris*, **21**, 96–115. *For. Abstr.*, 1953, No. 2347.

JONES, F. H., and PEACE, T. R. 1939. 'Experiments with frost heaving', *Quart. J. For.* **33**, 79–89.

JONES, S. G. 1925. 'Life-history and cytology of *Rhytisma acerinum* (Pers.) Fries', *Ann. Bot. Lond.* **39**, 41–75.

—— 1935. 'The structure of *Lophodermium pinastri* (Schrad.) Chev.', *Ann. Bot. Lond.* **49**, 699–728.

JONES, T. W. 1958. 'Mortality in wilt infected oaks', *Plant Dis. Reptr.* **42**, 552–3.

—— and BRETZ, T. W. 1955. 'Transmission of Oak wilt by tools', *Plant Dis. Reptr.* **39**, 498–9.

JØRGENSEN, C. A., LUND, A., and TRESCHOW, C. 1939. 'Undersøgelser over Rodfordaerveren, *Fomes annosus* (Fr.) Cke.', *K. VetHøjsk. Arsskr.*, pp. 71–128. *R.A.M.*, 1939, p. 772.

JØRGENSEN, E. 1955. 'Trametesangreb i laehegn', *Dansk. Skovforen. Tidsskr.* **40**, 279–85. *R.A.M.*, 1956, p. 731.

—— and PETERSEN, B. B. 1951. 'Angreb af *Fomes annosus* (Fr.) Cke. og *Hylesinus piniperda* L. på *Pinus silvestris* i Djurslands plantager', *Dansk. Skovforen. Tidsskr.* **36**, 453–79. *For. Abstr.* **13**, No. 3979.

JØRGENSEN, H. A. 1952. 'Studies on *Nectria cinnabarina.* Hosts and variations', *Årsskr. Vet.-Landbohøjsk.*, pp. 57–120. *For. Abstr.* **14**, No. 106.

JØRSTAD, I. 1925. 'Norske skogsykdommer. I. Nåletre-sykdommer bevirket av Rustsopper, Ascomyceter og Fungi Imperfecti', *Medd. Norske Skogfor søksvesen*, **6**, 186 pp. *R.A.M.*, 1926, p. 196.

JUEL, C. 1940. 'Grentørrens Smittefarlighed', *Dansk Skovforen. Tidsskr.* **25**, 154–9. *For. Abstr.* **8**, No. 684.

JUEL, H. O. 1925. 'Mycologische Beiträge IX', *Ark. Bot.* **19**, 20, pp. 1–10.

JUEL, O. 1926. 'Några parasiter på *Buxus*', *Svensk Bot. Tidsskr.* **20**, 493–4. *R.A.M.*, 1927, p. 618.

JUMP, J. A. 1937. 'Further notes on the disease of Himalayan pines', *Bull. Morris Arbor. Univ. Pa.* **1**, 97–98. *R.A.M.*, 1938, p. 281.

—— 1938*a*. 'A new disturbance of Red pine', *Science*, (N.S.) **87**, 138–9.

—— 1938*b*. 'A study of forking Red pine', *Phytopathology*, **28**, 798–811.

KÄÄRIK, A., and RENNERFELT, E. 1957. 'Investigations on the fungal flora of spruce and pine stumps', *Medd. SkogsForsknInst. Stock.* **47**, 88 pp. *R.A.M.*, 1958, p. 253.

KADAMBI, K. 1954. 'On *Loranthus* control', *Indian For.* **80**, 493–4.

—— and DABRAL, S. N. 1955. 'On Twist in Chir, *Pinus longifolia* Roxb.', *Indian For.* **81**, 58–64.

KALANDRA, A. 1938. 'Nová sypavka u nás působená houbou *Hypodermella sulcigena* (Rostr.) Tub. na Borovici obecné a Kleči v Tatrách a na Šumavě', *Ochr. Rost.* **14**, 38–46. *R.A.M.*, 1938, p. 570.

KAMEI, S. 1932. 'On new species of heteroecious Fern rusts', *Trans. Sapporo Nat. Hist. Soc.* **12**, 161–74. *R.A.M.*, 1932, p. 813.

—— 1940. 'Studies on the cultural experiments of the fern rusts of *Abies* in Japan', *J. Fac. Agric. Hokkaido Univ.* **47**, 91 pp. *R.A.M.*, 1949, p. 605.

KANGAS, E. 1952. Über die Brutstätternwahl von *Dendroctonus micans* Kug. (Col., Scolytidae) auf Fichten', *Suomen Hyonteistieteellinen Aikakauskirja*, **18,** 154–70.

KARNATZ, H. 1957. 'Das Kalken der Stämme — eine wichtige Schutzmassnahme im Obstbau', *Mitt. Obstb. Versuchsringes Jork.* **12,** 195–9. *Hort. Abstr.*, 1958, No. 1225.

KATSER, A. 1938. 'Ein Beitrag zur Anwendung des Antagonismus als biologische Bekämpfungsmethode unter besonderer Berücksichtigung der Gattungen *Trichoderma* und *Phytophthora*', *Boll. Staz. Pat. veg. Roma*, (N.S.) **18,** 195–217, 221–330. *R.A.M.*, 1939, p. 485.

KATWIJK, W. VAN. 1953. 'Mozaiek bij gouden regen', *Tijdschr. PlZiekt.* **59,** 237–9. *Hort. Abstr.*, 1954, No. 1807.

—— 1955. 'Ringvlekkenmozaiek bij Seringen in Nederland', *Meded. Dir. Tuinb.* **18,** 823–8. *R.A.M.*, 1956, p. 190.

KATZ, M. 1949. 'Sulfur dioxide in the atmosphere and its relation to plant life', *Industr. Engng. Chem.* **41,** 2450–65. *For. Abstr.* **11,** No. 2237.

KAUFERT, F. 1937. 'Factors influencing the formation of periderm in aspen', *Amer. J. Bot.* **24,** 24–30.

KAVEN, G. 1934. 'Krankheiten und Schädlinge an *Rhododendron*', *Die Kranke Pflanze*, **11,** 123–6. *R.A.M.*, 1935, p. 173.

KEITT, G. W., BLODGETT, E. C., WILSON, E. E., and MAGIE, R. O. 1937. 'The epidemiology and control of cherry leaf spot', *Res. Bull. Wis. Agric. Exp. Sta.* **132,** 117 pp. *R.A.M.*, 1938, p. 328.

KELLEY, A. P. 1948. *Further studies on the continued growth of American chestnut*, Landenberg Lab., Landenberg, Penn., 6 pp. *R.A.M.*, 1949, p. 200.

KENDRICK, J. B., DARLEY, E. F., MIDDLETON, J. T., and PAULUS, A. O. 1956. 'Plant response to polluted air', *Calif. Agric.* **10,** 9–10. *R.A.M.*, 1957, p. 48.

KENDRICK, J. B., Jr., MIDDLETON, J. T., and DARLEY, E. F. 1954. 'Chemical protection of plants from ozonated olefin (smog) injury', *Phytopathology*, **44,** 494–5.

KENNEDY, R. W., and ELLIOTT, G. K. 1957. 'Spiral grain in red alder', *For. Chron.* **33,** 238–51.

KERR, A. 1956. 'Some interactions between plant roots and pathogenic soil fungi', *Aust. J. Biol. Sci.* **2,** 45–52.

KESSELL, S. L., and STOATE, T. N. 1936. 'Plant nutrients and pine growth', *Aust. For.* **1,** 4–13.

KHAN, A. H. 1955. 'A new disease of Mulberry in the forest plantations of the Punjab', *Pakist. J. Sci. Res.* **7,** 92–99. *R.A.M.*, 1956, p. 647.

KIENHOLZ, R. 1932. 'Fasciation in red pine', *Phytopathology*, **22,** 15.

KIKUCHI, T. 1955. 'A study of the protective effects of irrigation against frost-injury in mulberry plants', *Proc. Crop Sci. Soc., Japan*, **24,** 52. Abstr. in *Soils & Fertil.*, 1956, No. 982.

KILLIAN, C. 1925. 'Le *Gyroceras celtidis* Mont. et Ces. parasite du *Celtis australis* L.', *Bull. Soc. Hist. Nat. Afrique du Nord*, **16,** 271–81. *R.A.M.*, 1926, p. 268.

KIMMEY, J. W. 1938. 'Susceptibility of *Ribes* to *Cronartium ribicola* in the west', *J. For.* **36,** 312–20.

—— 1946. 'Notes on visual differentiation of White pine blister rust from Pinyon rust in the telial stage', *Plant Dis. Reptr.* **30,** 59–61.

—— 1950. 'Cull factors for forest tree species in Northwestern California', *U.S. Dep. Agric.* (*Calif. For. Range Exp. Sta.*) *For. Survey Release*, **7,** June.

—— and HORNIBROOK, E. M. 1952. 'Cull and breakage factors and other tree measurement tables for Redwood', *U.S. Dep. Agric.* (*Calif. For. Range Exp. Sta.*) *For. Survey Release*, **13,** May.

KLÁŠTERSKÝ, I. 1951. 'A cowl-forming virosis in Roses, Lime-trees and Elm-trees', *Stud. bot. Čechosl.* **12,** 73–171. *R.A.M.*, 1952, p. 38.

KLEBAHN, H. 1918. *Haupt- und Nebenfruchtformen der Askomyzeten.* I. Borntraeger, Leipzig, 395 pp.

—— 1924. 'Kulturversuche mit Rostpilzen. XVII. Bericht (1916–1924)', *Z. Pfl-Krankh.* **34,** 289–303. *R.A.M.*, 1925, p. 375.

—— 1934. 'Eine Blattkrankheit der Edelkastanie und einige sie begleitende Pilze', *Z. PflKrankh. u. Pflanzenschutz*, **44,** 1–23. *R.A.M.*, 1934, p. 407.

—— 1939. 'Untersuchungen über *Cronartium gentianeum* v. Thümen', *Ber. dtsch. bot. Ges.* **57,** 92–98. *R.A.M.*, 1939, p. 644.

KLEIN, L. 1913. 'Forstbotanik', *Forstwissenschaft, Tübingen*, **1,** 299–584.

KLEPAC, D. 1955. 'Utjecaj imele na prirast jelovih šuma', *Šum. List*, **79,** 231–44. *For. Abstr.* **17,** No. 2947.

KLIKA, J. 1922. 'Einige Bemerkungen über die Biologie des Mehltaues', *Ann. mycol.* **20,** 74–80. *R.A.M.*, 1922, p. 335.

KLOTZ, K. 1956. 'Praktische Auswertung des Mykorrhiza-Problems in der Forstwirtschaft', *Allg. ForstZ.* **11,** 362–3. *R.A.M.*, 1957, p. 364.

KLOTZ, S. J., and SOKOLOFF, V. P. 1943. 'The possible relation of injury and death of small roots to decline and collapse of Citrus and Avocado', *Calif. Citrogr.* **28,** 86–87. *R.A.M.*, 1943, p. 319.

KNOWLSON, H. 1939. 'A long-term experiment on the radial growth of the oak—structural changes in the wood of earthed-up trunks', *The Naturalist*, April, pp. 93–99.

KOBAYASHI, T. 1957. 'Studies on the shoot blight disease of Japanese Cedar. *Cryptomeria japonica* D. Don caused by *Guignardia cryptomeriae* Sawada', *Bull. For. Exp. Sta. ,Meguro, Tokyo*, **96,** 17–36. *For. Abstr.* **19,** No. 635.

—— and ITO, K. 1956. 'Notes on the genus *Endothia* in Japan. I. Species of *Endothia* collected in Japan', *Bull. For. Exp. Sta., Meguro, Tokyo*, **92,** 81–98.

KOBENDZA, R. 1955. 'Dalsze studia nad zarastaniem ściętych pni drzew', *Roczn. dendrol. polsk. Tow. Bot., Warsz.* **10,** 1–37. *For. Abstr.* **17,** No. 2561.

KOCH, J. 1957. 'Mykorrhizernes betydning for skovtræernes ernæring belyst ved radioaktive isotoper', *Dansk Skovforen. Tidsskr.* **42,** 310–15. *R.A.M.*, 1958, p. 76.

KOCH, L. W. 1933. 'Investigations on black knot of plums and cherries. I', *Sci. Agric.* **13,** 576–90.

—— 1935. 'Investigations on black knot of plums and cherries. II, III, IV', *Sci. Agric.* **15,** 80–95, 411–23, 729–44.

KOCHMAN, J. 1929. 'Studja biologiczne nad pasorzytem Wierzby *Fusicladium saliciperdum* (All. et Tub.) Lind.', *Mém. Inst. Nat. Polonais d'Écon. Rur. à Puławy*, **10,** 555–73. *R.A.M.*, 1931, p. 139.

KÖCK, G. 1935. 'Eichenmehltau und Rauchgasschäden', *Z. PflKrankh.* **45,** 44–45. *R.A.M.*, 1935, p. 406.

KOEHLER, A. 1931. 'More about twisted grain in trees', *Science*, **73,** 477.

—— 1933. 'A new hypothesis as to the cause of shakes and rift cracks in green timber', *J. For.* **31,** 551–6.

KOHL, E. J. 1933. 'An explanation of the cause of spiral grain in trees', *Science*, **78,** 58–59.

KOKKONEN, P. 1926. 'Beobachtungen über die Struktur des Bodenfrostes', *Acta For. Fenn.* **30,** 56 pp.

—— 1933. 'Die Verteilung des Wassers im Boden infolge der Einwirkung des Bodenfrostes', *Trans. 6th Commission Int. Soc. Soil Sci.*, vol. B, pp. 13–17.

KONEV, G. I. 1951. '*Lophodermium pinastri* Chev. na khvoe kedra v Pribayakal′e', *Bot. Zh.* **35,** 664–6. *R.A.M.*, 1952, p. 40.

KONING, H. C. 1938. 'The bacterial canker of Poplars', *Meded. Phytopath. Lab. Willie Commelin Scholten*, **14,** 5–42. *R.A.M.*, 1939, p. 67.

—— 1939. 'Verslag van het onderzoek naar den populierenkanker over 1938', *Tijdschr. Ned. Heidemaatsch.*, 12 pp.

KORSTIAN, C. F. 1923. 'Control of snow moulding in coniferous nursery stock', *J. Agric. Res.* **24,** 741–7.

—— 1924. 'Density of cell sap in relation to environmental conditions in the Wasatch Mountains of Utah', *J. Agric. Res.* **28,** 845–907.

—— and BRUSH, W. D. 1931. 'Southern white cedar', *U.S. Dept. Agric. Tech. Bull.* **251,** 13–15.

—— and FETHEROLF, N. J. 1921. 'Control of the girdle of spruce transplants caused by excessive heat', *Phytopathology*, **11,** 485–90.

—— HARTLEY, C., WATTS, L. F., and HAHN, G. G. 1921. 'A chlorosis of conifers corrected by spraying with ferrous sulphate', *J. Agric. Res.* **21,** 153–71.

KOŠČEEV, A. L. 1952. 'Značenie drevesnoj ratitel′nosti v uvlažneni ii issušenii počv na vyrubkah lesa', *Dokl. Akad. Nauk S.S.S.R.* **87,** 297–9. *For. Abstr.* **14,** No. 3006.

KÖSTER, E. 1934. 'Die Astreinigung der Fichte', *Mitt. Forstwirt. Forstwiss.* **5,** 393–416.

KÖSTLER, J. N. 1956*a*. *Silviculture*. English trans. by M. L. Anderson, Oliver & Boyd, Edinburgh, 416 pp.

—— 1956*b*. 'Waldbauliche Beobachtungen an Wurzelstöcken sturmgeworfener Nadelbäume', *Forstwiss. Cbl.* **75,** 65–91. *For. Abstr.* **18,** No. 2953.

KRAHL-URBAN, J. 1955. 'Winterfrostschäden an Trauben-, Stiel-, und Roteichen', *Forst- u. Holzw.* **10,** 111–13. *For. Abstr.* **17,** No. 2911.

KRAMER, P. J. 1937. 'Photoperiodic stimulation of growth by artificial light as a cause of winter killing', *Plant Physiol.* **12,** 881–3.

—— 1939. 'The effect of drops of water on leaf temperatures', *Amer. J. Bot.* **26,** 12–14.

—— 1949. *Plant and Soil Water Relationships.* McGraw-Hill Book Co., New York, 347 pp.

—— 1950. 'Soil aeration and tree growth', *Proc. 26th Nat. Shade Tree Conf.*, pp. 51–58.

——, COSTING, H. J., and KORSTIAN, C. F. 1952. 'Survival of pine and hardwood seedlings in forest and open', *Ecology*, **33,** 427–30.

KRAMPE, O., and REHM, H. J. 1952. 'Untersuchungen über den Befall von *Pseudotsuga taxifolia viridis* mit *Adelopus gäumanni* Rohde', *NachrBl. dtsch. PflSch-Dienst., Berl.* **6,** 208–12. *R.A.M.*, 1953, p. 409.

KRASILNIKOV, N. A., and RAZNITSINA, E. A. 1946. 'A bacterial method of controlling damping-off of Scots pine seedlings caused by *Fusarium*', *Agrobiologiya*, **5–6,** 109–21. *R.A.M.*, 1949, p. 259.

KRAUSS, H. H. 1956. 'Die Bedeutung der Waldstreunutzung für den land- und Forstwirt', *Dtsch. Landwirtsch.* **7,** 458–62. *For. Abstr.* **18,** No. 1934.

KRSTIČ, M. 1952. 'Rak kestenove kore ustven je kod nas y na chrastu', *Zašt. Bilja. (Plant Prot.) Beograd*, **10,** 83–84. *R.A.M.*, 1952, p. 638.

KRÜGER, E. 1951. 'Die Verhütung von Rauchschäden als pflanzenzuchterisches Problem', *Freiberg, Forschungsh.* **2,** 3 pp. *R.A.M.*, 1953, p. 394.

KRUPENIKOV, I. A. 1947. 'O soleustoychivosti tamariksa v svyazi s ego sposobnost′yu k izbiratel′nomu "nakopleniyu" soley v assimiloruyushchikh organakh', *Dokl. Akad. Nauk S.S.S.R.* **56,** 765–8. *For. Abstr.* **12,** No. 96.

—— 1951. 'Tamariks i ego soleustoĭčivostj', *Priroda, Moskva*, **40,** 65–66. *For. Abstr.* **13,** No. 1940.

KÜHNE, H. 1954. 'Über die Festigkeits- und Verformungseigenschaften des braunen Kernholzes der Esche', *Schweiz. Z. Forstw.* **105,** 733–45. *For. Abstr.* **16,** No. 3550.

KUIJT, J. 1955. 'Dwarf Mistletoes', *Bot. Rev.* **21,** 569–627.

KUJALA, V. 1948. 'Murrayn mäntyä uhkaavista tuhosienistä', *Metsät. Aikak.* **3,** 42–44. *For. Abstr.* **10,** No. 2238.

KUNTZ, J. E., and RIKER, A. J. 1949. 'Winter injury versus disease in Wisconsin poplar plantings', *Phytopathology,* **39,** 12.

—— and RIKER, A. J. 1950*a*. 'Root grafts as a possible means for local transmission of Oak wilt', *Phytopathology,* **40,** 16–17.

—— —— 1950*b*. 'The translocation of poisons between trees through natural root grafts', *Northcent. Weed Control Conf. Res. Rept., 7th Ann. Meet.,* p. 242. *For. Abstr.* **13,** No. 1208.

—— —— 1956. 'The use of radio-active isotopes to ascertain the role of root grafting in the translocation of water, nutrients, and disease-inducing organisms among forest trees', *Proc. Int. Conf. Peaceful Uses of Atomic Energy, Geneva,* **12,** 144–8. *For. Abstr.* **18,** No. 1283.

KURAUCHI, I. 1956. 'Salt spray damage to the coastal forest', *Jap. J. Ecol.* **5,** 213–17. *Biol. Abstr.,* 1957, No. 5885.

KÜSTER, E. 1930. 'Anatomie der Gallen', *Handb. Pflanat.* **5,** 197 pp.

LACHAUSSÉE, E. 1941. 'La technique du boisement des trous à gelées', *Rev. Eaux For.* **79,** 506–9. *For. Abstr.* **8,** No. 1053.

—— 1953. 'Note sur la roulure et la gélivure du chêne', *Bull. Soc. for. Franche-Comté,* **26,** 655–9. *For. Abstr.* **15,** No. 2598.

LACHMUND, H. G. 1934*a*. 'Seasonal development of *Ribes* in relation to spread of *Cronartium ribicola* in the Pacific Northwest', *J. Agric. Res.* **49,** 93–114.

—— 1934*b*. 'Growth and injurious effects of *Cronartium ribicola* cankers on *Pinus monticola*', *J. Agric. Res.* **48,** 475–503.

LADEFOGED, K. 1943. 'Indre Brud i og Valkdannelser paa staaende Granstammer', *Dansk Skovforen. Tidskr.* **28,** 504–10. *For. Abstr.* **8,** No. 657.

LAGERBERG, T. 1919. 'Snöbrott och toppröta hos granen', *Medd. SkogsforsknInst.* **16,** 115–62.

—— 1933. '*Ascochyta parasitica* (Hartig), en skadesvanp på Granplantor', *Svenska SkogsvFören. Tidskr.* **30,** 1–10. *R.A.M.,* 1933, p. 667.

LAING, E. V. 1929. 'Notes from the Forestry Department, Aberdeen University', *Scot. For. J.* **43,** 48–52.

—— 1932. 'Studies on tree roots', *For. Comm. Bull.* **13,** 72 pp.

—— 1947. 'Preliminary note on a disease of Sitka spruce in Cairnhill Plantation, Durris, Kincardineshire (*Picea sitchensis* Carr.)', *Forestry,* **21,** 217–20.

LAMB, H., and SLEETH, B. 1941. 'Distribution and suggested control measures for the Southern pine fusiform rust', *Occ. Pap. Sth. For. Exp. Sta.* **91,** 5 pp. *R.A.M.* 1941, p. 186.

LAMBERT, E. B., and CRANDALL, B. S. 1936. 'A seedling wilt of Black locust caused by *Phytophthora parasitica*', *J. Agric. Res.* **53,** 467–76.

LAMPRECHT, H. 1950. 'Über den Einfluss von Umweltsfaktoren auf die Frostrissbildung bei Stiel- und Traubeneiche im nordostschweizerischen Mittelland', *Mitt. schweiz. ZentAnst. forstl. Versuchsw.* **26,** 360–418. *For. Abstr.* **12,** No. 3340.

LANE, R. D., and MCCOMB, A. L. 1948. 'Wilting and soil moisture depletion by tree seedlings and grass', *J. For.* **46,** 344–9.

LANGHOFFER, A. 1929. 'Le dépérissement du Chêne en Yougoslavie, spécialement dans la Slavonie', *Rev. Eaux For.* **67,** 763–5. *R.A.M.,* 1930, p. 278.

LANGLET, O. 1934. 'Om variationen hos tallen (*Pinus sylvestris* L.) och dess samband med klimatet', *Svenska SkogsvFören. Tidskr.* **1–2,** 87–110. Abstr. in *Forestry*, **8,** 80–82.

LANGNER, W. 1933. 'Über die Schüttekrankheit der Kiefernnadeln (*Pinus silvestris* und *Pinus strobus*)', *Phytopath. Z.* **5,** 625–40. *R.A.M.*, 1933, p. 604.

—— 1936. 'Untersuchungen über Lärchen-, Apfel- und Buchenkrebs', *Phytopath. Z.* **9,** 111–45. *R.A.M.*, 1936, pp. 693–4.

LANTELMÉ, W. 1951. 'Der Barfrost', *Forstwiss. Cbl.* **70,** 628–38. Abstr. in *Soils & Fertil.* **16,** No. 2187.

LANTERNIER, M. 1944. 'Le gui sur sapin pectiné dans le département de la Savoie', *Rev. Eaux For.* **82,** 460–9. *For. Abstr.* **8,** No. 1403.

LANZA, FELICE, 1950. 'Sulla conservazione delle Castagne destinate all'esportazione. Nota II. Ricerche sperimentali di lotta contro le infezioni crittogamiche (disinfezione chimica). L'ossido di etilene ed il bromuro di metile come fungicidi', *Ann. Sper. agr.*, (N.S.) **4,** 321–8. *R.A.M.*, 1951, p. 4.

LARSEN, C. S. 1953. 'Studies of diseases in clones of forest trees', *Hereditas, Lund,* **39,** 179–92. *Plant Breed. Abstr.*, 1953, No. 2215.

LARSEN, P. 1943. 'Die Bedeutung der Winterkälte für die Kernbildung der Buche', *Schweiz. Z. Forstw.* **94,** 265–72. *R.A.M.*, 1944, p. 201.

LASSCHUIT, J. A. 1952. 'Alterations in spirally curved trees of *Pinus merkusii*', *Tectona,* **41,** 179–88. *For. Abstr.* **13,** No. 3699.

LATHAM, D. H., DOAK, K. D., and WRIGHT, E. 1939. 'Mycorrhizae and Pseudomycorrhizae on Pines', *Phytopathology,* **29,** 14.

LATOUR, J. M. 1950. 'Intervention de la gelée dans la formation du chancre du Mélèze d'Europe (*Larix decidua* Miel.)', *Bull. Soc. for. Belg.* **57,** 239–41. *R.A.M.*, 1950, p. 591.

LAUBERT, R. 1926. 'Beobachtungen und Bemerkungen über das seuchenhafte diesjährige "Zweigspitzensterben" der Kiefern', *Illus. Landw. Zeit.* **46,** 543–4. *R.A.M.*, 1927, p. 264.

—— 1927. 'Die Fliederseuche', *Gartenwelt,* **31,** 374–5. *R.A.M.*, 1928, p. 246.

LAUDER, A. 1909. 'The Loganburn Smoke case', *Trans. Roy. Scot. Arb. Soc.* **22,** 15–20.

LAUGHTON, E. M. 1937. 'The incidence of fungal disease on timber trees in South Africa', *S. Afr. J. Sci.* **33,** 377–82. *R.A.M.*, 1937, p. 787.

LAUGHTON, F. S. 1937. 'The effects of soil and climate on the growth and vigour of *P. radiata* D. Don in South Africa', *S. Afr. J. Sci.* **33,** 589–604.

LEACH, J. G. 1940. *Insect transmission of plant diseases.* McGraw-Hill, 615 pp.

LEACH, R. 1937. 'Observations on the parasitism and control of *Armillaria mellea*' *Proc. Roy. Soc.* B, **121,** 561–73. *R.A.M.*, 1937, p. 564.

—— 1939. 'Biological control and ecology of *Armillaria mellea* (Vahl) Fr.', *Trans. Brit. Mycol. Soc.* **23,** 320–9.

LEAPHART, C. D., COPELAND, O. L., and GRAHAM, D. P. 1957. 'Pole blight of Western white pine', *For. Pest. Leafl. U.S.D.A.* **16,** 4 pp. *R.A.M.*, 1958, p. 321.

—— and GILL, L. S. 1955. 'Lesions associated with Pole blight of Western white pine', *For. Sci.* **1,** 232–9.

LEBEN, C., SCOTT, R. W., and ARNY, D. C. 1956. 'On the nature of the mechanism of the water-soak method for controlling diseases incited by certain seed-borne pathogens', *Phytopathology,* **46,** 273–7.

LE GOC, M. J. 1914. 'Further observations on *Hirneola auricula-judae* Berk. (Jew's Ear)', *New Phytol.* **13,** 122–33.

LEHMANN, E., KUMMER, H., and DANNENMANN, H. 1937. *Der Schwarzrost, seine Geschichte, seine Biologie und seine Bekämpfung in Verbindung mit der Berberitzenfrage,* J.F. Lehmanns Verlag, Munich–Berlin. 584 pp. *R.A.M.*, 1938, p. 21.

LEIBUNDGUT, H. 1947. 'Über die Planung von Bestandesumwandlungen', *Schweiz. Z. Forstw.* **98,** 372–89. *For. Abstr.* **10,** No. 171.

—— 1955. 'Untersuchungen über Augusttrieb- und Zwieselbildung bei der Fichte', *Schweiz. Z. Forstw.* **106,** 286–90. *Biol. Abstr.*, 1956, No. 32551.

—— and FRICK, L. 1943. 'Eine Buchenkrankheit im schweizerischen Mittelland', *Schweiz. Z. Forstw.* **94,** 297–306. *R.A.M.*, 1944, p. 200.

LEISER, A. T. 1958. 'Polyethylene treatment for tree bark wounds', *Amer. Nurserym.* **107,** 13, 49. *For. Abstr.* **19,** No. 4209.

LEK, H. A. A. VAN DER. 1929. 'Over aparasitaire nitwassen aan houtige planten', *Tijdschr. PlZiekt.* **35,** 25–59.

LELLIOTT, R. A. 1959. 'Fire blight of Pears in England', *Agric. Lond.* **65,** 564–8.

LEONTOVYČ, R. 1958. 'Napadnutie jednotlivých klonov topol′ov hrdzou *Melampsora allii-populina* Kleb. v. Selekčnej Vel′koškôlke Gabčikovo roku 1956', *Lesn. čas.* **4,** 30–45. *For. Abstr.* **19,** No. 3855.

LEROY, M. 1957. 'Enseignements à tirer d'une attaque de rouge cryptogamique dans les régénérations de Pin sylvestre de la Forêt de Haguenau', *Rev. for. franç.* **9,** 745–9. *R.A.M.*, 1958, p. 562.

LEROY, P. 1956. 'Dégats occasionnés par la glace dans les forêts de l'inspection de Haguenau', *Rev. for. franç.* **8,** 647–57. *For. Abstr.* **18,** No. 1739.

LESSER, M. A. 1946. 'Tree wound dressings', *Agric. Chem.* **1,** 26–28, 60–61. *Biol. Abstr.*, 1949, No. 12990.

LE SUEUR, A. D. C. 1931. 'An example of sewage poisoning', *Quart. J. For.* **25,** 336–7.

—— 1949. *The care and repair of ornamental trees.* Country Life, London, 224 pp.

LEVEN, G. 1932. 'A die-back disease on Pines (*Brunchorstia destruens* Erikss.)', *Quart. J. For.* **26,** 225–31.

LEVINE, M. N., and COTTER, R. U. 1932. 'Susceptibility and resistance of *Berberis* and related genera to *Puccinia graminis*', *U.S. Dept. Agric. Tech. Bull.* **300,** 26 pp. *R.A.M.*, 1932, p. 628.

LEVISOHN, I. 1952. 'Forking in pine roots', *Nature, Lond.* **169,** 715.

—— 1953. 'Growth response of tree seedlings to mycorrhizal mycelia in the absence of a mycorrhizal association', *Nature, Lond.* **172,** 316–17.

—— 1954. 'Test for the pseudomycorrhizal group of soil fungi', *Nature, Lond.* **174,** 408–9.

—— 1956. 'Growth stimulation of forest tree seedlings by the activity of free living mycorrhizal mycelia', *Forestry,* **29,** 53–59.

—— 1957*a*. 'Differential effects of root-infecting mycelia on young trees in different environments', *Emp. For. Rev.* **36,** 281–6.

—— 1957*b*. 'Antagonistic effects of *Alternaria tenuis* on certain root-fungi of forest trees', *Nature, Lond.* **179,** 1143–4.

—— 1958. 'Effects of mycorrhiza on tree growth', *Soils & Fertil.* **21,** 73–82.

LEVITT, E. C. 1947. '*Armillaria* root rot control', *Agric. Gaz. N.S.W.* **58,** 67–71. *R.A.M.*, 1947, p. 337.

LEVITT, J. 1956. *The hardiness of plants.* American Society of Agronomy, 278 pp. Rev. in *Nature, Lond.* **179,** 223–4.

LEVY, B. F. G. 1946. 'Tree injection. I. The estimation of dosage in relation to tree size. II. Methods for overcoming resistance to absorption of liquids. III. Reinvigoration of debilitated trees', *Rep. E. Malling Res. Sta., 1946*, pp. 99–103, 104–6, 107–12.

LEYTON, L. 1957. 'The diagnosis of mineral deficiencies in forest crops', Paper for *7th Brit. Commonw. For. Conf.* (*Aust. & N.Z.*), 6 pp.

—— and ROUSSEAU, L. Z. 1958. 'Root growth of tree seedlings in relation to aeration', *Physiology For. Trees,* Ronald Press, N.Y., pp. 467–75.

LIERNUR, A. G. M. 1927. 'Hexenbesen: ihre Morphologie, Anatomie und Entstehung', Thesis, Univ. Utrecht, 57 pp. *R.A.M.*, 1927, p. 706.

LIESE, J. 1926. 'Achtet auf den Buchenkeimlingspilz!', *Forstarchiv*, **2,** 217–18. *R.A.M.*, 1926, p. 705.

—— 1931. 'Zur Rhabdoclinekrankheit der Douglasie', *Forstarchiv*, **7,** 341–6. *R.A.M.*, 1932, p. 141.

—— 1932. 'Zur Biologie der Douglasiennadelschütte', *Z. Forst- u. Jagdw.* **64,** 680–93. *Biol. Abstr.*, 1934, No. 10471.

—— 1933. 'Vererbung der Hexenbesenbildung bei der Kiefer', *Z. Forst- u. Jagdw.* **65,** 10, pp. 541–4. *R.A.M.*, 1934, p. 202.

—— 1936*a*. 'Zur Frage der Vererbbarkeit der rindenbewohnenden Blasenrostkrankheiten bei Kiefer', *Z. Forst- u. Jagdw.* **68,** 602–9. *R.A.M.*, 1937, p. 288.

—— 1936*b*. 'Beiträge zum Kiefernbaumschwammproblem', *Forstarchiv.* **12,** 37–48. *R.A.M.*, 1936, p. 472.

—— 1936*c*. 'Die Douglasienrassen und ihre Anfälligkeit gegenüber der Douglasiennadelschütte (*Rhabdocline pseudotsugae*),' *Mitt. dtsch. dendrol. Ges.* **48,** 259–63. *R.A.M.*, 1937, p. 507.

—— 1939. 'The occurrence in the British Isles of the Adelopus disease of Douglas fir', *Quart. J. For.* **33,** 247–52.

—— 1953. 'Pflanzenphysiologische Betrachtungen zum Ulmensterben', *Arch. Forstw.* **1,** 59–70. *For. Abstr.* **15,** No. 2640.

LIGHTLE, P. C. 1954. 'The pathology of *Elytroderma deformans* on Ponderosa pine', *Phytopathology*, **44,** 557–69.

LIHNELL, D. 1942. '*Coenococcum graniforme* als Mykorrhizabildner von Waldbäumen', *Symb. bot. uppsaliens.* **5,** 19 pp. *R.A.M.*, 1943, p. 104.

LILPOP, J. 1923. 'Łuskiewnik (*Lathraea squamaria* L.) na świerku w Tatrach', *Acta Soc. Bot. Pol.* **1,** 60–61. *Bot. Abstr.*, 1924, No. 6698.

LIMING, O. N., REX, E. G., and LAYTON, K. 1951. 'Effects of a source of heavy infection on the development of Dutch elm disease in a community', *Phytopathology*, **41,** 146–51.

LINDE, R. J. VAN DER, and MEIDEN, H. A. VAN DER. 1954. 'Enkele voorlopige resultaten van een onderzoek naar de invloed van de inundatie 1953 op houtsoorten', *Tijdschr. Ned. Heidemaatsch.* **65,** 1–13. *Biol. Abstr.*, 1954, No. 22124.

LINDEIJER, E. J. 1932. 'De bacterie-ziekte van de Wilg veroorzaakt door *Pseudomonas saliciperda* n. sp.', Thesis, Univ. Amsterdam, 82 pp. *R.A.M.*, 1933, p. 60.

LINDGREN, R. M. 1932. 'Field observations of needle rusts of Spruce in Minnesota', *Plant Dis. Reptr.* **16,** 126–9.

—— 1933. 'Decay of wood, and growth of some hymenomycetes as affected by temperature', *Phytopathology*, **23,** 73–81.

—— 1948. 'Care needed in thinning pines with heavy Fusiform rust infection', *For. Fmr.* **7,** 3. *R.A.M.*, 1949, p. 604.

—— 1950. 'Damage from heavy Fusiform cankering in '49 can be reduced', *For. Fmr.* **9,** 9. *R.A.M.*, 1950, p. 591.

—— 1951. *The disease problem in relation to tree improvement.* Forest Genetics 'Tree Improvement' Conf., Atlanta, Georgia, Jan. 9, 1951, 5 pp. *Plant Breed. Abstr.*, 1952, No. 3009.

—— and HENRY, B. W. 1949. 'Promising treatments for controlling root disease and weeds in a southern pine nursery', *Plant Dis. Reptr.* **33,** 228–31.

LINES, R. 1953. 'The Scottish gale damage', *Irish For.* **10,** 3–15.

LING, L. 1952. *Digest of plant quarantine regulations*, F.A.O. Rome, 164 pp., and subsequent supplements.

LINNEMANN, G., and MEYER, H. 1958. 'Mykorrhiza und Phosphormangel', *Forst- u. Holzw.* **13,** 8–10. *For. Abstr.* **19,** No. 1484.

LINZON, S. N. 1958*a*. 'The effect of artificial defoliation of various ages of leaves upon White pine growth', *For. Chron.* **34,** 50–56.

—— 1958*b*. *The influence of smelter fumes on the growth of White pine in the Sudbury region.* Canad. Dep. Agric. For. Biol. Div. Science Service, Toronto, 45 pp.

LIOU, TCHEN-NGO. 1929. 'La végétation épiphytique des bois de conifères', *Bull. Soc. bot. France*, **66,** 21–30.

LITTLE, S. J., MOHR, J. J., and SPICER, L. L. 1958. 'Salt-water storm damage to Loblolly pine forests', *J. For.* **56,** 27–28.

LJUBIČ, F. P. 1955. 'Vzaimadojstvie kornevyh sistem raznyh vidov derevzev pri sovmestrom proizrastanii', *Agrobiologija,* **1,** 112–16. *For. Abstr.* **17,** No. 231.

LOBANOV, N. V. 1951. '"Mikotrofnost" glavneyshikh drevesnȳkh i kustarnikovȳkh porod v usloviyakh evropeyskoy chasti SSSR', *Agrobiologiya,* **4,** 49–62. *R.A.M.*, 1952, p. 622.

LOCKHART, C. L. 1958. 'Studies on red leaf disease of lowbush blueberries', *Plant Dis. Reptr.* **42,** 764–7.

LOEHWING, W. F. 1937. 'Root interactions of plants', *Bot. Rev.* **3,** 195–239.

LØFTING, E. C. L. 1937. 'Hedeskovenes Foryngelse meddelelser fra det af den forstlige forsøgskommission, skovriderne for statens hededistrikter og det danske hedeselskab nedsatte udvalg', *Forstl. Forsøgsv. Danm.* **14,** 133–60.

LOHMAN, M. L., and CASH, E. K. 1940. '*Atropellis* species from pine cankers in the United States', *J. Wash. Acad. Sci.* **30,** 255–62. *R.A.M.*, 1940, p. 629.

—— and WATSON, A. J. 1943. 'Identity and host relations of *Nectria* species associated with diseases of hardwoods in the United States', *Lloydia,* **6,** 77–108. *R.A.M.*, 1944, p. 47.

LOHWAG, H. 1931. 'Zur Rinnigkeit der Buchenstämme', *Z. PflKrankh.* **41,** 371–85. *R.A.M.*, 1932, p. 81.

LONG, W. H. 1924. 'The self pruning of Western yellow pine', *Phytopathology,* **14,** 336–7.

—— 1930. '*Polyporus dryadeus,* a root parasite of White fir', *Phytopathology,* **20,** 758–9.

LOONEY, W. S., and DUFFIELD, J. W. 'Proliferated cones of Douglas fir', *For. Sci.* **4,** 154–5.

LOPATIN, M. I. 1936. 'Poražaemost rastenij vozbuditelem kornevogo raka rastenij *Bacterium tumefaciens*', *Mikrobiol.* **5,** 716–24. *R.A.M.*, 1937, p. 235.

LORENTZ, P., and BRIERLEY, P. 1953. 'Graft transmission of lilac witches-broom virus', *Plant Dis. Reptr.* **37,** 555.

LORENZ, R. C. 1944. 'Discolorations and decay resulting from increment borings in hardwoods', *J. For.* **42,** 37–43.

LORRAIN SMITH, A. 1904. 'Diseases of plants due to fungi—*Gloeosporium tiliae* Oud.', *Trans. Brit. Mycol. Soc.* **2,** 55–56.

—— 1931. *Lichens.* Cambridge University Press, 464 pp.

—— 1933. 'Recent lichen literature', *Trans. Brit. Mycol. Soc.* **18,** 93–126.

LOW, J. D., and GLADMAN, R. J. 1960. '*Fomes annosus* in Great Britain', *For. Comm., For. Rec.* **41,** 22 pp.

LÜCKHOFF, H. A. 1955. 'Two hitherto unrecorded fungal diseases attacking pines and eucalypts in South Africa', *J. S. Afr. For. Ass.* **26,** 47–61. *R.A.M.*, 1957, p. 73.

LUCKWILL, L. C. 1950. 'Some virus diseases of fruit trees in England', *Fruit Yearb.* **4,** 84–88. *R.A.M.*, 1951, p. 167.

LUDBROOK, W. V. 1940. 'Boron deficiency symptoms on pine seedlings in water culture', *J. Coun. Sci. Industr. Res. Aust.* **13,** 186–90. *R.A.M.*, 1941, p. 40.

—— 1942. 'Fertilizer trials in southern New South Wales pine plantations', *J. Coun. Sci. Industr. Res. Aust.* **15,** 307–14. *R.A.M.*, 1943, p. 283.

LUTTRELL, E. S. 1940*a*. 'Tar spot of American holly', *Bull. Torrey Bot. Cl.* **67,** 692–704. *For. Abstr.* **3,** 77.

—— 1940*b*. '*Morenoella quercina*, cause of leaf spot of oaks', *Mycologia*, **32,** 652–66. *R.A.M.*, 1941, p. 95.

—— 1948. 'Physiologic specialization in *Guignardia bidwellii*, cause of black rot of *Vitis* and *Parthenocissus* species', *Phytopathology*, **38,** 716–23.

—— 1949*a*. 'Horse Chestnut anthracnose in Missouri', *Plant Dis. Reptr.* **33,** 324–7.

—— 1949*b*. '*Scirrhia acicola*, *Phaeocryptopus pinastri*, and *Lophodermium pinastri* associated with the decline of Ponderosa pine in Missouri', *Plant Dis. Reptr.* **33,** 397–401.

—— 1950. 'Botryosphaeria stem canker of elm', *Plant Dis. Reptr.* **34,** 138–9.

LUTZ, H. J. 1936. 'Scars resulting from glaze on woody stems', *J. For.* **34,** 1039–41.

—— 1952. 'Occurrence of clefts in the wood of living White spruce in Alaska', *J. For.* **50,** 99–102.

LYR, H. 1955. 'Untersuchungen zur Pathologie der Douglasie', *Arch. Forstw.* **4,** 533–44. *R.A.M.*, 1957, p. 74.

LYUBARSKY, L. V. 1952. 'Lesohozjajstvennoe značenie napl′ȳvov na vet′vjah sosnȳ obȳknovennoj', *Lesn. Hoz.* **5,** 78–79. *R.A.M.*, 1955, p. 827.

MCCALLUM, A. W. 1928. 'Studies in forest pathology. I. Decay in Balsam fir (*Abies balsamea* Mill.)', *Canad. Dep. Agric. Bull.* (N.S.) **104,** 25 pp.

—— 1929. 'Woodgate rust in Canada', *Phytopathology*, **19,** 414.

—— and STEWART, K. E. 1958. 'Dutch Elm Disease', *Canad. Agric. Sci. Serv. For. Biol. Div.* **1010,** 12 pp.

MCCOMB, A. L. 1943. 'Mycorrhizae and phosphorus nutrition of pine seedlings in a prairie soil nursery', *Res. Bull. Ia. Agric. Exp. Sta.* **314,** 582–612. *R.A.M.*, 1944, p. 238.

—— and GRIFFITH, J. E. 1946. 'Growth stimulation and phosphorus absorption of mycorrhizal and non-mycorrhizal northern White pine and Douglas fir seedlings in relation to fertilizer treatment', *Plant Physiol.* **21,** 11–17.

MCCOOL, M. M. and MEHLICH, A. 1938. 'Soil characteristics in relation to distance from industrial centres', *Contrib. Boyce Thompson Inst.* **9,** 353–70.

MCCORMACK, H. A. 1936. 'The morphology and development of *Caliciopsis pinea*', *Mycologia*, **28,** 188–96. *R.A.M.*, 1936, p. 620.

MCCORMICK, F. A. 1939. '*Phaeocryptopus gäumannii* on Douglas fir in Connecticut', *Plant Dis. Reptr.* **23,** 368–9.

MCCUBBIN, W. A. 1936. 'Analysis of typical plant diseases from the quarantine standpoint', *Phytopathology*, **26,** 991–1006.

—— 1946. 'Preventing plant disease introduction', *Bot. Rev.* **12,** 101–39. *R.A.M.*, 1946, p. 320.

—— 1954. *The plant quarantine problem: A general review of the biological, legal, administrative and public relations of plant quarantine with special reference to the United States situation.* Chronica Botanica Co., 255 pp.

MCCULLOCH, W. F. 1943. 'Ice breakage in partially cut and uncut second growth Douglas fir stands', *J. For.* **41,** 275–8.

MACDONALD, J. 1949. 'Notes on the die-back of European larch', *J. For. Comm.* **20,** 192–8.

—— 1951. 'Climatic limitations in British forestry', *Quart. J. For.* **45,** 161–8.

—— WOOD, R. F., EDWARDS, M. V., and ALDHOUS, J. R. 1957. 'Exotic forest trees in Great Britain', *For. Comm. Bull.* **30,** 167 pp.

MACDONALD, J. A. 1937. 'A study of *Polyporus betulinus* (Bull.) Fries', *Ann. Appl. Biol.* **24,** 289–310.

—— 1948. 'Heather rhizomorph fungus in Scotland', *Scot. Agric.* **28,** 99–101.

—— and RUSSELL, J. R. 1937. '*Phomopsis scobina* (Cke.) v. Höhn. and *Phomopsis controversa* (Sacc.) Trav. on Ash', *Trans. Bot. Soc. Edinb.* **32,** 341–52.

MACDONALD, J. M. 1952. 'Wind damage in middle-aged crops of Sitka spruce and its prevention', *Scot. For.* **6,** 82–85.

MACGINITIE, H. D. 1933. 'Redwoods and frost', *Science,* **78,** 190.

MACINTIRE, W. H. 1957. 'Fate of air-borne fluorides and attendant effects upon soil reaction and fertility', *J. Ass. Agric. Chem. Wash.* **40,** 958–73. *Soils & Fertil.*, 1957, No. 2241.

M'INTOSH, C. 1915. '*Cucurbitaria pithyophylla* Fries', *Trans. Roy. Scot. Arb. Soc.* **29,** 209–10.

MCKAY, R. 1944. 'Scab on Pyracanthas and its control', *J. R. Hort. Soc.* **69,** 204–7.

—— and CLEAR, T. 1955. 'A further note on group dying of Sitka spruce and *Rhizinia inflata*', *Irish For.* **12,** 58–63.

—— —— 1957. 'Violet root rot (*Helicobasidium purpureum*) on Douglas fir (*Pseudotsuga taxifolia*) and *Pinus contorta*', *Irish For.* **14,** 90–97.

MCKEEN, W. E. 1958. 'Blueberry canker in British Columbia', *Phytopathology,* **48,** 277–80.

MCKELL, C. M., and FINNIS, J. M. 1957. 'Control of soil moisture depletion through use of 2, 4-D on a mustard nurse crop during Douglas fir seedling establishment', *For. Sci.* **3,** 329–35.

MCKELLAR, A. D. 1942. 'Ice damage to Slash pine, Longleaf pine, and Loblolly pine plantations in the Piedmont section of Georgia', *J. For.* **40,** 794–7.

MCKENZIE, M. A., and JOHNSON, E. M. 1939. '*Cephalosporium* elm wilt in Massachusetts', *Bull. Mass. Agric. Exp. Sta.* **368,** 24 pp. *For. Abstr.* **2,** 4, p. 327.

MACLACHLAN, J. D. 1935*a*. 'The hosts of *Gymnosporangium globosum* Farl. and their relative susceptibility', *J. Arnold Arbor.* **16,** 98–142. *R.A.M.*, 1935, p. 368.

—— 1935*b*. 'The dispersal of viable basidiospores of the *Gymnosporangium* rusts', *J. Arnold Arbor.* **16,** 411–22. *R.A.M.*, 1936, p. 160.

—— 1936. 'Studies on the biology of *Gymnosporangium globosum* Farl.', *J. Arnold Arbor.* **17,** 1–24. *R.A.M.*, 1936, p. 512.

MACLEAN, H., and GARDNER, J. A. F. 1958. 'Distribution of fungicidal extractives in target pattern heartwood of Western Red Cedar', *For. Prod. J.* **8,** 107–8.

MACLEAN, N. A. 1950. 'Variation in monospore cultures of *Armillaria mellea*', *Phytopathology,* **40,** 968.

MCQUILKIN, W. E. 1946. 'Use of mulch, fertilizer, and large stock in planting clay sites', *J. For.* **44,** 28–29.

—— 1950. 'Effects of some growth regulators and dressings on the healing of tree wounds', *J. For.* **48,** 423–8.

MADGWICK, H. A. I., and OVINGTON, J. D. 1959. 'The chemical composition of precipitation in adjacent forest and open plots', *Forestry,* **32,** 14–22.

MAGERSTEIN, V. 1928. 'Der Wurzeltöter als Korbweidenschädling', *Der Deutsche Korbweidenzüchter*', pp. 80–81. *R.A.M.*, 1929, p. 143.

—— 1931. 'Rakovina Vrby', *Ochrana Rostlin,* **11,** 135–7. *R.A.M.*, 1932, p. 274.

MAGNANI, G. 1954. 'Alcuni casi di deperimento di piopelle in vivaio', *Cellulosa e Carta*, **5**, 14–15. *R.A.M.*, 1955, p. 684.

—— 1957. 'Danni da Freddo sugli Eucalitti nell'Inverno 1955–56. Memorie presentata al Congresso Mondiale dell'Eucalitto (Roma, Oct. 1956)', *Centro Sper. Agric. For. Roma*, pp. 159–84. *For. Abstr.* **19**, No. 1860.

—— 1958*a*. 'La diffusione e l'importanza della *Cytospora* sui pioppi', *Ital. for. mont.* **13**, 23–28. *For. Abstr.* **19**, No. 4432.

—— 1958*b*. 'Una alternariosi fogliare di piantine di eucalitto', *Monti e Boschi*, **9**, 181–7. *For. Abstr.* **19**, No. 4435.

MAGUIRE, W. P. 1955. 'Radiation, surface temperature and seedling survival', *For. Sci.* **1**, 277–85.

MAIRE, R. 1927. 'Champignons Nord-Africains nouveaux ou peu connus', *Bull. Soc. Hist. Nat. Afrique du Nord*, **18**, 117–20. *R.A.M.*, 1928, p. 59.

MALAISSE, F. 1957. 'Note sur la gélivure du chêne rouge d'Amérique en Campine', *Bull. Soc. Roy. for. Belg.* **64**, 439–61.

MALENÇON, G. 1924. 'Le *Sclerotinia betulae* Woronin', *Bull. Soc. mycol. France*, **40**, 177–80. *R.A.M.*, 1925, pp. 199–200.

MANGIN, M. 1912. 'Contribution à l'étude de la malades ronds du pin', *Compt. Rend.* **154**, 1525.

MAŃKA, K. 1956. *Fitopatologia leśna.* Poznań, 225 pp.

MANLEY, G. 1952 *Climate and the British scene.* Collins, London, 314 pp.

MANNERS, J. G. 1953. 'Studies on Larch canker. I. The taxonomy and biology of *Trichoscyphella willkommii* (Hart.) Nannf. and related species', *Trans. Brit. Mycol. Soc.* **36**, 362–74.

—— 1957. 'Studies on Larch canker. II. The incidence and anatomy of cankers produced experimentally either by inoculation or by freezing', *Trans. Brit. Mycol. Soc.* **40**, 500–8.

MANSFIELD, T. C. 1953. *Roses in colour and cultivation.* William Collins, 264 pp.

MANSHARD, E. 1927. 'Der Buchenkeimlingspilz *Phytophthora omnivora* de Bary und seine Bekämpfung', *Forstarchiv.* **3**, 84–86. *R.A.M.*, 1927, p. 326.

MARCELIN, P. 1951. 'Essai sur la Dépérissement de la Châtaigneraie. Semaine Internationale du Châtaignier, Sept. 1950', *Bull. Tech. Châtaignier*, **3**, 78–94.

MARCHIONATTO, J. B. 1949. 'Las citarias argentinas', *Alm. Minist. Agric. Argent.* **24**, 127–8. *For. Abstr.* **12**, No. 731.

—— 1952. '*Botryosphaeria ribis*, como hongo patógeno del sauce de adorno en la Argentina', *Rev. argent. Agron.* **19**, 238–9. *For. Abstr.* **14**, No. 2407.

MARINKOVIĆ, P., and MARINKOVIĆ, B. 1957. 'Uticaj drvne truleži na kvalitet jedne izdanačke sŭme hrasta Lŭznjaka u Sremu', *Šumarstvo*, **10**, 168–78. *R.A.M.*, 1958, p. 423.

MARSDEN, D. H. 1952. 'Are wound paints Trojan horses?', *Trees, Cleveland, Ohio*, **12**, 9 pp. *For. Abstr.* **13**, No. 3843.

—— 1953. 'Dutch elm disease; an evaluation of practical control efforts', *Plant Dis. Reptr.* **37**, 3–6.

MARSH, R. W. 1952. 'Field observations on the spread of *Armillaria mellea* in apple orchards and in a blackcurrant plantation', *Trans. Brit. Mycol. Soc.* **35**, 201–7.

MARSHALL, B. H., Jr. 1952. 'Cane blight of *Forsythia* in Maryland', *Plant Dis. Reptr.* **36**, 440–1.

MARSHALL, M., and MAKI, T. E. 1946. 'Transpiration of pine seedlings as influenced by foliar coatings', *Plant Physiol.* **21**, 95–101.

MARSHALL, R. P. 1939. '*Sphaeropsis* canker and die-back of shade trees', *Proc. Nat. Shade Tree Conf.* **15,** 65–69. *For. Abstr.* **2,** 245.

—— 1940. 'Black spot fungus of Elm (*Gnomonia ulmea* (Schw.) Thuem.)', *Tree Pest Leafl.* **41,** 4 pp. *R.A.M.*, 1941, p. 610.

—— 1941. 'Control of Cedar-Apple rust on Red cedar', *Trans. Conn. Acad. Sci.* **34,** 85–118. *R.A.M.*, 1944, p. 444.

MARTIN, D. 1948. '*Eucalyptus* in the British Isles, with some notes on records of frost resistance', *Aust. For.* **12,** 63–74. *For. Abstr.* **11,** No. 956.

MARTIN, E. I. 1957. 'Neoplastisches Wachstum bei *Sequoiadendron giganteum* Buchholz', *Phytopath. Z.* **30,** 342–3. *For. Abstr.* **19,** No. 1955.

MARTIN, H. 1957. *Guide to the chemicals used in crop protection.* Canada Dept. Agric. 3rd edn., 313 pp., mimeographed.

—— 1959. *The scientific principles of crop protection.* Edward Arnold & Co., London, 4th ed. 358 pp.

MARTIN, J. F. 1938. 'Some economic aspects of White pine blister rust control', *J. For.* **26,** 986–96.

—— and GRAVATT, G. F. 1954. 'Saving White pines by removing blister rust canker', *Circ. U.S. Dept. Agric.* **948,** 22 pp. *For. Abstr.* **16,** No. 3147.

MARTÍNEZ, J. B. 1933. 'Una grave micosis del Pino observada por primera vez en España', *Bol. Soc. Española Hist. Nat.* **33,** 25–30. *R.A.M.*, 1933, p. 480.

—— 1942. 'Las micosis del *Pinus insignis* en Guipúzcoa', *Inst. for. Invest. Exp., Madrid,* **13,** 72 pp. *R.A.M.*, 1943, p. 186.

MASERA, E. 1933. 'Osservazioni sulla "fersa" del Gelso', *Ann. Tecn. Agrar. Roma,* **6,** 178–84. *R.A.M.*, 1935, p. 265.

MASSEE, G. 1910. *Diseases of cultivated plants and trees.* Duckworth, London, 602 pp.

—— 1911. 'A disease of lilac', *Kew Bull.*, pp. 81–82.

—— 1914. 'Black knob of birch', *Kew Bull.*, pp. 322–3.

MASSEY, A. B. 1925. 'Antagonism of the walnuts *Juglans nigra* L. and *J. cinerea* L. in certain plant associations', *Phytopathology,* **15,** 773–83.

MATHUR, A. K. 1949. 'Angiospermic parasitesof our forests', *Indian For.* **75,** 449–56.

MATTSSON-MÅRN, L. 1944. 'Några skogliga synpunkter på snöskytteproblemet', *Norrlands SkogsvFörb. Tidskr.* **4,** 380–402. *For. Abstr.* **9,** No. 2552.

—— and NENZELL, G. 1941. 'Studier över snöskytteangrepp inom tall-föryngringar å Bergvik & Ala Nya Aktiebolags', *Norrlands SkogsvFörb. Tidskr.*, pp. 160–91. *For. Abstr.* **5,** 4, p. 278.

MATUO, T., and SAKURAI, Y. 1954. 'Hot-water-disinfection of the sapling of the Mulberry Tree infected by *Rosellinia necatrix* (Hart.) Berl. and the situation of the isolation ditch preventing the development of this disease in the Mulberry farm', *J. seric. Sci. Tokyo,* **23,** 271–7. *R.A.M.*, 1956, p. 647.

MATZKE, E. B. 1936. 'The effect of street lights in delaying leaf-fall in certain trees', *Amer. J. Bot.* **23,** 446–52. *Biol. Abstr.*, 1937, No. 8016.

MAXWELL, H. 1933. 'The sycamore fungus', *Nature, Lond.* **132,** 409, 752.

MAY, C., and BAKER, W. L. 1952. 'Insects and spread of forest-tree diseases', *U.S. Dept. Agric. Yearb.*, pp. 677–82.

—— and GRAVATT, G. F. 1951. 'Vascular diseases of hardwoods', *Tree Pest Leafl., Hillsboro', N.H.* **19,** 4 pp.

—— WALTER, J. M., and MOOK, P. V. 1941. 'Rosy canker of London plane associated with illuminating gas injury', *Phytopathology,* **31,** 349–51.

MAYER-WEGELIN, H. 1936. *Astung.* M. & H. Schaper, Hannover, 178 pp.

—— 1952. *Das Aufästen der Waldbäume.* M. & H. Schaper, Hannover, 92 pp.

—— 1956. 'Die biologische, technologische und forstliche Bedeutung des Drehwuchses der Waldbäume', *Forstarchiv,* **27,** 265–71. *For. Abstr.* **18,** No. 3476.

MEER, J. H. H. VAN DER. 1926. 'Verticillium-wilt of maple and elm-seedlings in Holland', *Phytopathology*, **16,** 611–14.

MEETHAM, A. R. 1952. *Atmospheric pollution, its origins and prevention.* Pergamon Press, London, 268 pp.

MEHLISCH, K. 1938. 'Ein pilzlicher Schädling an *Abies*', *Blumen- u. PflBau ver. Gartenwelt*, **42,** 92. *R.A.M.*, 1938, p. 493.

MEIDEN, H. A. VAN DER. 1957/8. 'Kernhout bij populieren en zijn praktische betekensis', *De Houtwereld*, **11,** 2, pp. 950–1; 6, pp. 101–3; 8, p. 169.

—— and VLOTEN, H. VAN. 1958. 'Roest en schorsbrand als bedreiging van de teelt van populier', *Ned. Boschb.-Tijdschr.* **30,** 261–73. *For. Abstr.* **20,** No. 2018.

MEIER, K. 1937. 'Über eine durch Kalimangel bedingte "Gelbsucht" an Thujapflanzen', *Landw. Jb. Schweiz*, **51,** 297–304. *R.A.M.*, 1937, p. 573.

MEIERHANS, J. 1951. 'Herpotrichia-Versuche 1950–51', *Schweiz. Z. Forstw.* **102,** 526–9. *R.A.M.*, 1952, p. 360.

MEINECKE, E. P. 1916. '*Peridermium harknessii* and *Cronartium quercuum*', *Phytopathology*, **6,** 225–40.

—— 1928. *A report upon the effect of excessive tourist travel on the California Red wood parks.* California State Printing Office, Sacramento, 20 pp. *R.A.M.*, 1929, p. 415.

MELEHOV, I. S. 1946. 'O povreždenijakh elovykh lesov severnoĭ tajgĭ ržavčinnym gribom *Chrysomyxa ledi*', *Sborn. naučno-issledov. Rabot Arkhangel. lesoteh. Inst.* **8,** 59–75. *R.A.M.*, 1948, p. 301.

MELIN, E. 1948. 'Recent advances in the study of tree mycorrhiza', *Trans. Brit. Mycol. Soc.* **30,** 92–99.

—— 1953. 'Physiology of Mycorrhizal relations in plants', *Ann. Rev. Plant Physiol.* **4,** 325–46.

—— and NILSSON, H. 1955. 'Ca^{45} used as indicator of transport of cations to pine seedlings by means of mycorrhizal mycelium', *Svensk. bot. Tidskr.* **49,** 119–22. *For. Abstr.* **16,** No. 3878.

MELLONI, M. 1936. 'Disseccamenti dei rami di Olmo provocati da due sferopsidali', *Boll. Staz. Pat. veg. Roma*, (N.S.) **16,** 208–13. *R.A.M.*, 1937, p. 354.

MELZER, H. 1931. 'Frostschäden des Winters 1928–29 in Österreich', *Cbl. ges. Forstw.* **57,** 49–75.

MERGEN, F. 1954. 'Mechanical aspects of wind-breakage and windfirmness', *J. For.* **52,** 119–25.

—— and WINER, H. I. 1952. 'Compression failures in the boles of living conifers', *J. For.* **50,** 677–9.

MERKLE, R. 1951. 'Über die Douglasien-Vorkommen und die Ausbreitung der Adelopus-Nadelschütte in Württemberg-Hohenzollern', *Forst- u. Jagdztg.* **122,** 161–91. *R.A.M.*, 1953, p. 410.

METCALFE, C. R. 1931. 'The "shab" disease of Lavender', *Trans. Brit. Mycol. Soc.* **16,** 149–76.

METCALFE, G. 1939. 'A bacterial disease of *Forsythia*', *Nature, Lond.* **144,** 1050.

—— 1940. 'The watermark disease of willows. I. Host-parasite relationships', *New Phytol.* **39,** 322–32.

—— 1941. 'The watermark disease of willows. II. Pathological changes in the wood', *New Phytol.* **40,** 97–107.

MEULI, L. J., and SHIRLEY, H. L. 1937. 'The effect of seed origin on drought resistance of Green Ash in the Prairie-Plains States', *J. For.* **35,** 1060–2.

MEYER, H. 1951*a*. 'Aufgaben und Wege der Douglasienzüchtung', *Allg. ForstZ.* **6,** 281–3. *For. Abstr.* **14,** No. 1946.

—— 1951*b*. 'Die Verbreitung der Douglasienschütten. Zum Stand der Krankheitserforschung', *Forstarchiv*, **22,** 5–11. *R.A.M.*, 1953, p. 410.

MEYER, H. 1957. 'Über einige Ergebnisse der Blasenrost-Resistenzzüchtung', *Silvae Genet.* **6,** 148–56. *Plant Breed. Abstr.*, 1958, No. 2251.

MEYFARTH, H. 1955. 'Schnee- und Sturmschäden im Thüringer Wald', *Forst- u. Jagdztg.* **5,** 53–56. *For. Abstr.* **18,** No. 2954.

MEZZETTI, A. 1950. 'Osservazioni sulla morfologia e sulla tassonomia del *Gyroceras celtidis* (Biv.) Mont. et Ces.', *Ann. Sper. agr.*, (N.S.) **4,** 857–67. *R.A.M.*, 1951, p. 203.

MICHIE, C. Y. 1885. *The Larch. A practical treatise on its culture and general management.* William Blackwood & Sons, Edinburgh. 280 pp.

MIDDLETON, J. T. 1943. 'The taxonomy, host range and geographic distribution of the genus *Pythium*', *Mem. Torrey Bot. Cl.* **20,** 1–171. *R.A.M.*, 1943, p. 373.

—— CRAFTS, A. S., BREWER, R. F., and TAYLOR, O. C. 1956. 'Plant damage by air pollution', *Calif. Agric.* **10,** 9–12.

MIELKE, J. L. 1935. 'Rodents as a factor in reducing aecial sporulation of *Cronartium ribicola*', *J. For.* **33,** 994–1003.

—— 1943. 'White pine blister rust in western North America', *Bull. Sch. For. Yale*, **52,** 155 pp. *R.A.M.*, 1944, p. 122.

—— 1952. 'The rust fungus *Cronartium filamentosum* in Rocky Mountain Ponderosa pine', *J. For.* **50,** 365–73.

—— 1956*a*. 'The rust fungus (*Cronartium stalactiforme*) in Lodgepole pine', *J. For.* **54,** 518–21.

—— 1956*b*. 'A needle-cast of Lodgepole pine caused by the fungus *Hypodermella concolor*', *Res. Note Intermt. For. Range Exp. Sta.* **27,** 3 pp. *R.A.M.*, 1958, p. 121.

——1957*a*. 'Aspen leaf blight in the Intermountain region', *Res. Note Intermt. For. Range Exp. Sta.* **42,** 5 pp. *R.A.M.*, 1958, p. 118.

—— 1957*b*. 'The *Comandra* blister rust in Lodgepole pine', *Res. Note Intermt. For. Range Exp. Sta.* **46,** 8 pp. *R.A.M.*, 1958, p. 189.

—— 1957*c*. 'The yellow witches' broom of subalpine fir in the intermountain region', *Res. Note Intermt. For. Range Exp. Sta.* **47,** 5 pp. *R.A.M.*, 1958, p. 188.

—— and KIMMEY, J. W. 1942. 'Heat injury to the leaves of California black oak and some other broadleaves', *Plant Dis. Reptr.* **26,** 116–19.

MIHIN, S. D. 1928. 'O zimneĭ zasuhe u drevesnykh porod. I. Isparenie vody počkami i morozoustojčivost', *Trudy Sib. inst. sel. Koz. Lesovod.* **10,** 123–45.

MIKOLA, P. 1952. 'Effect of forest humus on parasitic fungi causing damping-off disease of coniferous seedlings', *Phytopathology*, **42,** 202–3.

MILBRATH, J. A. 1940. 'Coryneum blight of Oriental arborvitae caused by *Coryneum berckmanii* n. sp.', *Phytopathology*, **30,** 592–602.

—— and YOUNG, R. A. 1949. 'Root rot of *Chamaecyparis*', *Newslett. Nurserym.* **60,** 4. *R.A.M.*, 1951, p. 132.

—— —— 1956. 'Cucumber mosaic virus and alfalfa mosaic virus isolated from *Daphne odora*', *Plant Dis. Reptr.* **40,** 279–83.

MILES, L. E. 1921. 'Leaf spots of the elm', *Bot. Gaz.* **71,** 161–96.

MILLER, E. J., NEILSON, J. A., and BANDEMER, S. L. 1937. 'Wax emulsions for spraying nursery stock and other plant materials', *Mich. Agric. Exp. Sta. Spec. Bull.* **282,** 39 pp.

MILLER, F. J. 1938. 'The influence of mycorrhizae on the growth of Shortleaf pine seedlings', *J. For.* **36,** 526–7.

MILLER, J. A., and ALDRICH, K. F. 1936. '*Pseudolarix amabilis*, a new host for *Dasyscypha willkommii*', *Science*, (N.S.) **83,** 499. *R.A.M.*, 1937, p. 220.

MILLER, J. K. 1935. 'A new species of *Keithia* on red cedar', *J. Elisha Mitchell Sci. Soc.* **51,** 167–71. *R.A.M.*, 1936, p. 68.

—— 1943. '*Fomes annosus* and red cedar', *J. For.* **41,** 37–40.

MILLER, J. M., PERRY, J. P., and BORGLAUG, N. E. 1957. 'Control of sunscald and subsequent Buprestid damage in Spanish cedar plantations in the Yucatan', *J. For.* **55,** 185–8.

MILLER, P. A. 1941. 'Phytophthora bleeding canker of *Quercus agrifolia*', *Phytopathology,* **31,** 863.

MILLER, P. R., and GRAVATT, G. F. 1952. 'The sweetgum blight', *Plant Dis. Reptr.* **36,** 247–52.

MILLER, P. W. 1940. 'Further studies on the comparative efficacy of Bordeaux mixture, copper oxalate, and some other "insoluble" copper sprays for the control of Walnut blight', *Ore. St. Hort. Soc.* 1939, pp. 127–34. *R.A.M.,* 1940, p. 373.

—— 1958. 'Recent studies on the effectiveness of agri-mycin 100 and agri-mycin 500 for the control of Walnut blight in Oregon', *Plant Dis. Reptr.* **42,** 388–9.

—— BOLLEN, W. B., and SIMMONS, J. E. 1949. 'Filbert bacteriosis and its control', *Tech. Bull. Ore. Agric. Exp. Sta.* **16,** 70 pp. *R.A.M.,* 1949, pp. 649–50.

—— and MCWHORTER, F. P. 1948. 'The use of vapor-heat as a practical means of disinfecting seeds', *Phytopathology,* **38,** 89–101.

MILNE, HOME J. 1952. 'Ivy as a forest weed', *Scot. For.* **6,** 86–87.

MINCKLER, L. S. 1951. 'Southern pines from different geographic sources show different responses to low temperatures', *J. For.* **49,** 915–16.

MINDERMAN, G. 1949. 'Een mogelijke oorzaak van het "ruien" van fijnsparren', *Ned. Boschb. Tijdschr.* **21,** 137–9. *For. Abstr.* **12,** No. 2296.

MIŠIČ, V. 1956. 'Masovna pojava sekundarnog lišća bukve u Jugoslaviji', *Zborn. Rad. Inst. Ekol. Beograd,* **6,** 14 pp. *For. Abstr.* **19,** No. 1859.

MITCHELL, H. L., FINN, R. F., and ROSENDAHL, R. O. 1937. 'The relation between mycorrhizae and the growth and nutrient absorption of coniferous seedlings in nursery beds', *Black Rock For. Pap. (N.Y.),* **1,** 58–73. *R.A.M.,* 1938, p. 53.

MITCHELL, R. L. 1954. 'Trace elements in Scottish peats', *Int. Peat Symposium, Dublin,* 9 pp. *Macaulay Inst. collected Papers,* 1951–4, **4.**

MIX, A. J. 1930. 'Further studies of privet anthracnose', *Phytopathology,* **20,** 257–61.

—— 1949. 'A monograph of the genus *Taphrina*', *Kans. Univ. Sci. Bull.* **33,** 3–167. *R.A.M.,* 1949, p. 548.

MIYAZAKI, S., OKINAGA, T., and HARATA, M. 1954. 'The effect of fertilizer on the growth of Black Locust (*Robinia pseudoacacia*) seedlings transplanted in Kosaka bare lands and injured by the strong sulphate smoke', *Bull. For. Exp. Sta. Meguro, Tokyo,* **74,** 177–90. *For. Abstr.* **16,** No. 3100.

MODESS, O. 1941. 'Zur Kenntnis der Mykorrhizabildner von Kiefer und Fichte', *Symb. bot. uppsaliens.* **5,** 3–147. *R.A.M.,* 1943, p. 104.

MODLIBOWSKA, I. 1953. 'Some experiments on "washing-off" the hoar frost', *Ann. Rep. East Malling Res. Sta. 1952,* **A 36,** 73–77. *Hort. Abstr.,* 1953, No. 2784.

MOLIN, N. 1955. 'Fallsjuka på groddplantor av barrträd', *Medd. SkogsforsknInst., Stockh.* **45,** 12 pp. *For. Abstr.* **16,** No. 4013.

—— 1957. 'Om *Fomes annosus* spridningsbiologi', *Medd. SkogsforsknInst., Stockh.* **47,** 36 pp. *For. Abstr.* **18,** No. 4228.

MOLNAR, A. C. 1954. 'A disease causing flagging on Ponderosa pine', *Bi-m. Progr. Rep. Div. For. Biol., Dep. Agric. Can.* **10,** (3) 4. *R.A.M.,* 1955, p. 117.

MOLOTKOV, P. I. 1956. 'Podvižnost′ pasoki v vodoprovodjašče sisteme dereva i soobščaemost′ meždu derevjami imejuščimi kornevuju svjaz′', *Bot. Ž.* **41,** 407–9. *For. Abstr.,* **18,** No. 212.

MONTEMARTINI, L. 1925. 'Svernamento del *Gymnosporangium clavariaeforme* (Jacq.) Rees sopra il *Crataegus ozyacantha* L.', *Riv. Pat. Veg.* **15,** 85–86. *R.A.M.*, 1925, p. 711.

—— 1930. 'Note di fitopatologia', *Riv. Pat. Veg.* **20,** 201–6. *R.A.M.*, 1931, p. 294.

MOOI, J. C. 1948. 'Kanker en takinsterving van der Wilg veroorzaakt door *Nectria galligena* en *Cryptodiaporthe salicina*', Thesis, Univ. Amsterdam, 119 pp. *R.A.M.*, 1949, p. 92.

MOORE, B. 1933. 'Unusual wind and soil effects', *J. For.* **31,** 97–98.

MOORE, M. H. 1945. 'A note on Medlar cluster-cup rust (*Gymnosporangium confusum* Plowr.) in Kent in 1943 and 1944', *Trans. Brit. Mycol. Soc.* **28,** 13–15.

—— 1950. 'Nut rotting in *Corylus avellana* L. in relation to the activities of the Nut weevil, *Balaninus nucum* L.', *J. Hort. Sci.* **25,** 213–24.

MOORE, W. C. 1939. 'New and interesting plant diseases. 3. A shoot wilt of *Prunus triloba* caused by *Botrytis cinerea* Pers.', *Trans. Brit. Mycol. Soc.* **23,** 313–15.

—— 1946. 'New and interesting plant diseases', *Trans. Brit. Mycol. Soc.* **29,** 250–8.

—— 1948. 'Report on fungus, bacterial and other diseases of crops in England and Wales for the years 1943–1946', *Minist. Agric., Lond., Bull.* **139,** 90 pp.

—— 1952. 'Principles underlying plant import and export regulations', *Plant Pathology,* **1,** 15–17.

—— 1954. 'Plant disease legislation against seed-borne diseases', *Rep. 5th Commonwealth Mycol. Conf.*, pp. 20–23.

—— 1955. 'The development of international co-operation in crop protection', *Ann. Appl. Biol.* **42,** 67–72.

—— 1959. *British Parasitic Fungi.* Cambridge University Press, 430 pp.

MOREAU, C. 1952. 'Coexistence des formes *Thielaviopsis* et *Graphium* chez une souche de *Ceratocystis major* (van Beyma) nov. comb. Remarques sur les variations des *Ceratocystis*', *Rev. Mycol.* **17,** Suppl. colon. **1,** 17–25. *R.A.M.*, 1953, p. 453.

—— and MOREAU, M. 1952. 'Étude mycologique de la maladie de l'encre du Chêne', *Rev. Path. vég.* **31,** 201–31. *R.A.M.*, 1953, p. 702.

—— —— 1953. 'Les Maladies du Châtaignier en Forêt de Marly', *Rev. for. franç.* **5,** 411–14. *R.A.M.*, 1956, p. 55.

—— —— 1954. 'Nouvelles observations sur le dépérissement des érables', *Bull. Soc. Linn. de Normandie,* ser. 9, **7,** 66–67.

MOREAU, F. 1952–3. *Les champignons, physiologie, morphologie, développement et systématique.* Lechevalier, Paris, vol. 1, 940 pp., vol. 2, 2,120 pp.

MOREILLON, H. 1918. 'The damage to oaks caused by the fungus *Diaporthe taleola* Tul.', *J. For. Suisse,* **69,** 1–2.

MOREILLON, M. 1929. 'Épicéas sur pied fendus pendant la période de sécheresse', *J. For. Suisse,* **80,** 128–9.

MORGAN, F. D., and BYRNE, J. 1957. *The desirability of international quarantine measures for exports and imports of timber and timber products.* Br. Comm. For. Conf., 7 pp.

MORIONDO, F. 1954. 'Osservazioni sul ciclo biologico della *Melampsora* sp. del pioppo in Italia', *Ital. for. mont.* **9,** 259–64. *For. Abstr.* **16,** No. 1931.

—— 1957. 'Osservazioni sulla biologia di *Melampsora pinitorqua* Rostr. sul litorale tirrenico', *Monti e Boschi,* **8,** 31–35. *R.A.M.*, 1958, p. 189.

MORITZ, O. 1930. 'Studien über Nectriakrebs. I. Infektionsversuch', *Z. Pfl-Krankh.* **40,** 251–61. *R.A.M.*, 1930, p. 669.

MORLING, R. J. 1954. *Trees in towns.* The Estates Gazette, London, 79 pp.

MORQUER, R. 1923. 'La maladie de "l'encre" du châtaignier', *Bull. Soc. Hist. Nat. Toulouse,* **1,** 255–99. *R.A.M.*, 1925, p. 249.

MORRIS, C. L. 1953. 'Chemical control of *Hypoderma lethale* on Pitch pine', *Plant Dis. Reptr.* **37**, 368–70.

MOSS, A. E. 1940. 'Effect on trees of wind-driven salt water', *J. For.* **38**, 421–5.

MOSS, V. D. 1957. 'Acti-dione treatment of Blister rust trunk cankers on Western white pine', *Plant Dis. Reptr.* **41**, 709–14.

—— 1958. 'Acti-dione stove oil treatment of blister-rust trunk cankers on reproduction and Pole western white pine', *Plant Dis. Reptr.* **42**, 703–6.

—— and WELLNER, C. A. 1953. 'Aiding Blister rust control by silvicultural measures in the Western white pine type', *Cir. U.S. Dep. Agric.* **919**, 32 pp. *For. Abstr.* **15**, No. 2645.

MOTT, D. G., NAIRN, L. D., and COOK, J. A. 1957. 'Radial growth in forest trees and effects of insect defoliation', *For. Sci.* **3**, 286–304.

MOULDS, F. R. 1957. 'Exotics can succeed in forestry as in agriculture', *J. For.* **55**, 563–6.

MÜHLDORF, A. 1928. 'Leaf-fall in frost', *Bull. Facult. Ştiinţe din Cernauti.* **2**, 267–304. Rev. in *Nature, London*, 1929, p. 505.

MÜHLSTEPH, W. 1942. 'Wege zum chemischen Nachweis von Abgas- (Rauch-) Schäden mit besonderer Rücksicht auf den Wald', *Tharandt. forstl. Jb.* **93**, 631–58. *For. Abstr.* **7**, 95.

MÜLDER, D. 1951. 'Neue Grundlagen für eine rationelle Abwehr des Kienzopfs (*Peridermium pini*)', *Forstw. Zbl.* **70**, 369–95. *R.A.M.*, 1952, p. 306.

—— 1955. 'Die Disposition der Kiefer für den Kienzopfbefall als Kernproblem waldbautechnischer Abwehr', *Schr. Reihe forstl. Fak. Univ. Göttingen*, **10**, 35 pp. *For. Abstr.* **17**, No. 1751.

MÜLLER, A. 1910. 'Der Kampf gegen den Kiefernbaumschwamm', *Z. Forst- u. Jagdw.* **42**, 129–46.

MÜLLER, E. 1950. 'Eine Knospenkrankheit der Stechfichte', *Schweiz. Beitr. Dendrol.* **2**, 69–72. *R.A.M.*, 1951, p. 639.

MÜLLER, K. 1912. 'Über das biologische Verhalten von *Rhytisma acerinum* auf verschiedene Ahornarten', *Ber. dtsch. bot. Ges.* **30**, 385–91.

MÜLLER, K. M. 1938. 'Untersuchungen über die Ursachen des Blitzeinschlages in der freien Natur insbesondere über die Blitzfrage im Walde', *Cbl. ges. forstw.* **64**, 287–300, 316–35.

MÜLLER, R. 1953. 'Zur Frage des Pappelrindentods', *Schweiz. Z. Forstw.* **104**, 408–28. *For. Abstr.* **15**, No. 404.

MÜLLER-KÖGLER, E. 1954. 'Bekämpfung des Eichenmehltaus', *Forstsch. Merkbl. niedersächs. forstl. VersAnst.*, Abt. B, **5**, 4 pp. *R.A.M.*, 1955, p. 495.

MÜLLER-STOLL, W. R., and HARTMANN, U. 1950. 'Über den *Cytospora*-Krebs. der Pappel (*Valsa sordida* Nitschke) und die Bedingungen für eine parasitäre Ausbreitung', *Phytopath. Z.* **16**, 443–78. *R.A.M.*, 1952, p. 523.

MÜNCH, E. 1928. 'Weitere Untersuchungen über Früh- und Spätfichten', *Z. Forst- u. Jagdw.*, pp. 129–77.

—— 1936*a*. 'Das Lärchensterben', *Forstwiss. Zbl.* **58**, 469–94, 537–62, 581–90, 643–71. *R.A.M.*, 1937, p. 76.

—— 1936*b*. 'Das Erlensterben', *Forstwiss. Zbl.* **58**, 173–94, 230–48. *R.A.M.*, 1936, p. 540.

MUNRO, J. W. 1926. 'British bark-beetles', *For. Comm. Bull.* **8**, 77 pp.

MURRAY, B. J. 1926. 'Three fungous diseases of *Salix* in New Zealand and some saprophytic fungi found on the same hosts', *Trans. N.Z. Inst.* **56**, 58–70. *R.A.M.*, 1926, p. 706.

MURRAY, J. S. 1953. 'A note on the outbreak of *Chrysomyza abietis* Unger (Spruce needle rust) in Scotland, 1951', *Scot. For.* **7**, 52–54.

—— 1955*a*. 'Rusts of British forest trees', *For. Comm. Booklet*, **4**, 15 pp.

MURRAY, J. S. 1955*b*. 'An exceptional frost in East Anglia', *Quart. J. For.* **49,** 120–5.

—— 1957. 'Top dying of Norway spruce in Great Britain', *Proc. 7th Brit. Comm. For. Conf.* 9 pp.

—— 1958. 'Lightning damage to trees', *Scot. For.* **12,** 70–71.

—— and YOUNG, C. W. T. 1956. 'The effect of brashing and thinning debris on the incidence of *Lophodermium pinastri*', *Quart. J. For.* **50,** 75–76.

—— —— 1961. 'Group dying of conifers', *For. Comm. For. Rec.* **46,** 19 pp.

MURSELL, P. 1949. 'Late frost and orchard heating (in Britain)', *Grower,* **32,** 25, 27, 29, 31. *Hort. Abstr.,* 1949, No. 2868.

MUTHANNA, M. A. 1956. 'Letters to the Editor. IV', *Indian For.* **82,** 268.

—— 1955. 'Spike disease of Sandal (*Santalum album*)', *Indian For.* **81,** 500–8.

NAGEL, C. M. 1949. 'Leaf rust resistance within certain species and hybrids of *Populus*', *Phytopathology,* **39,** 16.

NAPPER, R. P. N. 1932*a*. 'Observations on the root disease of rubber trees caused by *Fomes lignosus*', *J. Rubber Res. Inst. Malaya,* **4,** 5–33. *R.A.M.,* 1933, p. 52.

—— 1932*b*. 'A scheme of treatment for the control of *Fomes lignosus* in young rubber areas', *J. Rubber Res. Inst. Malaya,* **4,** 34–38. *R.A.M.,* 1933, p. 54.

NARASIMHAN, M. J. 1928. 'Note on the occurrence of intracellular bodies in Spike disease of Sandal (*Santalum album* Linn.)', *Phytopathology,* **18,** 815–17.

NATTRASS, R. M. 1928. 'The Physalospora disease of the Basket willow', *Trans. Brit. Mycol. Soc.* **13,** 286–304.

—— 1930. 'A note on two Marssonina diseases on willows', *Min. of Agric. Egypt* (*Plant Protect. Sect.*), *Bull.* **99,** 19 pp. *R.A.M.,* 1930, p. 813.

—— 1949. 'A Botrytis disease of *Eucalyptus* in Kenya', *Emp. For. Rev.* **28,** 60–61.

—— 1950. 'Annual Report of the Senior Plant Pathologist, 1948', *Dep. Agric. Kenya,* 1948, pp. 95–98. *R.A.M.,* 1951, p. 308.

—— and CICCARONE, A. 1947. 'Monochaetia canker of *Cupressus* in Kenya', *Emp. For. Rev.* **26,** 289–90.

NAZAROVA, E. S. 1936. 'Bolezn sosen, vyzyvaemaja *Sclerophoma pithyophila* v. H.', *Izv. Akad. Nauk S.S.S.R.* (Ser. biol.), **6,** 1191–1208. *R.A.M.,* 1937, p. 427.

NEČESANÝ, V. 1956. 'Tříděni bukových jader', *Dřevo,* **11,** 93–98. *For. Abstr.* **18,** No. 2251.

NEGER, F. W. 1918. 'Experimentelle Untersuchungen über Rußtaupilze', *Flora,* **10,** 67–139.

—— 1924. *Die Krankheiten unserer Waldbäume.* Ferdinand Enke, Stuttgart, 296 pp.

NEILSON-JONES, W. 1941. 'Fused needle disease of pines', *Emp. For. J.* **20,** 151–61.

—— 1945. 'Further field observations on Fused needle disease of pines', *Emp. For. J.* **24,** 235–9.

—— 1952. 'Fused needle disease in *Pinus palustris*', *Emp. For. Rev.* **31,** 325–6.

NĚMEC, A. 1936. 'Studie o karenčnich zjevech u Borovice v lesní školce v Řevnicích', *Ann. Acad. Tchécosl. Agric.* **11,** 531–4. *R.A.M.,* 1937, p. 357.

—— 1951. 'Příspěvek k otázce krění borovice na degradované hadcové půdě. (Intoxikace chromem, niklem a kobaltem.)', *Lesn. Práce,* **30,** 214–37. *For. Abstr.* **14,** No. 2344.

—— 1956. 'Vliv hrabání lesního steliva v monokulturách borovice na povahu půdy a na výživu borovice', *Práce výzkum. Úst. Lesn. ČSR,* **10,** 5–19. *For. Abstr.* **19,** No. 3252.

NENZELL, G. 1943. 'Redogörelse för undersökningar åren 1940–43 över snöskyttesvampen och dess bekämpande i tallföryngringar å Bergvik och Ala Nya A.-B: s marker', *Norrlands SkogsvFörb. Tidskr.* **4,** 293–314. *For. Abstr.* **9,** No. 2552.

NEVES, C. M. B. 1955. 'Nota sôbre a fasciação do Pinheiro bravo (*Pinus pinaster* Sol. ex Ait.) na Ilha da Madeira', *An. Inst. sup. Agron. Lisboa,* **19,** 103–6. *R.A.M.,* 1956, p. 132.

NEWBIGIN, M. I. 1909. 'The Glencorse Smoke Case', *Trans. Roy. Scot. Arb. Soc.* **22,** 221–7.

NEWHOOK, F. J. 1957*a*. 'Mortality of *Pinus radiata* in New Zealand', *Proc. Canad. phytopath. Soc.* **25,** 16.

——1957*b*, 'The relationship of saprophytic antagonism to control of *Botrytis cinerea* Pers. on Tomatoes', *N.Z.J. Sci. Technol.,* sect. A, **38,** 473–81. *R.A.M.,* 1958, p. 114.

NEWMAN, I. V. 1956. 'On fluting of the trunk in young trees of *Pinus taeda* L. (Loblolly pine)—with an appendix on the measurement of radial growth as ring-width', *Aust. J. Bot.* **4,** 1–12. *For. Abstr.* **17,** No. 3735.

NEWSAM, A. 1954. '*Fomes lignosus* in replanted areas', *Quart. Circ. Rubb. Res. Inst. Ceylon,* **29,** 78–84. *R.A.M.,* 1955, p. 61.

NICHOLSON, C. 1932. 'The mistletoe and its hosts', *Gdnrs.' Chron.* **91,** 102–4, 145–6.

NICOLAS, G., and AGGÉRY. 1931. 'Champignons observés sur deux *Viburnum* de la section *tinus* (*Viburnum davidi, Vib. tinus*)', *Bull. Soc. Hist. Nat. Toulouse,* **61,** 249–52. *R.A.M.,* 1932, p. 651.

NIENSTAEDT, H. 1958. 'Height growth is indicative of the relative frost resistance of Hemlock seed sources', *Tech. Note, Lake St. For. Exp. Sta.* **525,** 2 pp.

—— and GRAVES, A. H. 1955. 'Blight resistant chestnuts', *Circ. Conn. Agric. Exp. Sta.* **192,** 18 pp. *R.A.M.,* 1956, p. 853.

NIKKI, I., USHIYAMA, M., and TOMII, T. 1957. 'Studies on root-breaking in certain crops caused by icicle layers developing in the ground and protection measures against the damage. 4. The density of root development as a factor affecting the formation of icicle layers. 5. Planting methods in relation to root-breaking', *Proc. Crop Sci. Soc. Japan,* **26,** 75–77. *Soils & Fertil.,* 1958, No. 993.

NIKOLIĆ, V. 1951. 'Viroza na Jorgovanu', *Zašt. Bilja Beograd,* **3,** 71–72. *R.A.M.,* 1952, p. 328.

NISHIKADO, Y. 1944. 'Studies on the defoliation disease of the Japanese cypress caused by *Lophodermium chamaecyparisii*', *Rep. Chara Agric. Inst.* **36,** 351–60. *Biol. Abstr.,* 1949, No. 16901.

—— and MATSUMOTO, H. 1929. 'A new disease of elm, caused by *Gnomonia oharana* n. sp.', *Ber. Ohara Inst.* **4,** 279–87. *R.A.M.,* 1930, p. 351.

—— and WATANABE, K. 1953. 'The large leaf spot of Chestnut', *Ber. Ohara Inst.* **10,** 9–16. *R.A.M.,* 1954, p. 390.

NISSEN, T. V. 1956. 'Soil actinomycetes antagonistic to *Polyporus annosus* Fr.', *Friesia,* **5,** 332–9. *R.A.M.,* 1957, p. 221.

NOEL, P. 1941. 'Observations forestières en Lorraine pendant l'hiver 1940', *Rev. Eaux For.* **79,** 189–93. *For. Abstr.* **8,** No. 1378.

NOHARA, Y., and ZINNO, Y. 1955. 'Researches on the prevention of needle blight of "Sugi", *Cryptomeria japonica* D. Don. (III)', *Bull. For. Exp. Sta., Meguro, Tokyo,* **81,** 31–41. *R.A.M.,* 1956, p. 133.

NORDIN, V. J. 1953. 'A leaf-spot disease of Aspen', *Bi-m. Progr. Rep. Div. For. Biol. Dep. Agric. Can.* **9,** 4. *For. Abstr.* **14,** No. 3498.

—— 1956*a*. 'Heart rots in relation to the management of spruce in Alberta', *For. Chron.* **32,** 79–84.

NORDIN, V. J. 1956*b*. 'Rocky mountain region. An epidemic occurrence of needle-rust of white spruce in Alberta', *Can. Dep. of Agric. bi-m. Progr. Rep.* **12,** 3.

NORTHCOTT, P. L. 1957. 'Is spiral grain the normal growth pattern?', *For. Chron.* **33,** 335–52.

NYLINDER, P. 1951. 'Om patologiska hartskanaler', *Medd. SkogsforsknInst. Stockholm,* **40,** 12 pp. *For. Abstr.* **13,** No. 2412.

—— 1955. 'Kvistningsundersökningar, I. Grönkvistning av ek', *Medd. Skogs-forsknInst. Stockholm,* **45,** 44 pp. *For. Abstr.* **17,** No. 1578.

OBERLI, H. 1937. 'Einige Untersuchungen über den braunen Kern der Esche', *Schweiz. Z. Forstw.* **88,** 274–8.

OECHSLIN, M. 1949. 'Bespritzungs- und Bestäubungsversuche gegen die *Herpotrichia nigra.* (Vorläufige Mitteilung)', *Schweiz. Z. Forstw.* **100,** 229–31. *R.A.M.,* 1950, p. 183.

—— 1957. 'Schädigungen in Aufforstungen im Hochgebirge', *Schweiz. Z. Forstw.* **108,** 93–101. *For. Abstr.* **18,** No. 3039.

OFFORD, H. R., *et al.* 1952. 'Improvements in the control of *Ribes* by chemical and mechanical methods', *Circ. U.S. Dep. Agric.* **906,** 72 pp. *Hort. Abstr.* 1953, No. 2962.

——, QUICK, C. R., and MOSS, V. D. 1958. 'Blister rust control aided by the use of chemicals for killing *Ribes*', *J. For.* **56,** 12–18.

OGANESJAN, A. P. 1953. 'Salt tolerance of some fruit crops', *Bot. Ž.* **38,** 744–51. *Hort. Abstr.,* 1953, No. 2412.

OGAWA, J. M., NICHOLS, C. W., and ENGLISH, H. 1955. 'Almond scab', *Plant Dis. Reptr.* **39,** 504–8.

OGAWA, T. 1937. 'Shoot drooping disease of *Acer trifidum* Hook et Arn. caused by *Pseudomonas acernea* n.sp.', *Ann. phytopath. Soc. Japan,* **7,** 125–35. *R.A.M.,* 1938, p. 357.

OGILVIE, L. 1924. 'Observations on the "slime-fluxes" of trees', *Trans. Brit. Mycol. Soc.* **9,** 167–82.

—— 1932. 'Notes on the rusts of Basket willows and their control', *Ann. Rept. Agric. & Hort. Res. Sta., Long Ashton, 1931,* pp. 133–8. *R.A.M.,* 1932, p. 756.

—— and HICKMAN, C. J. 1938. 'Cuprous oxide and zinc oxide as seed protectants', *Gdnrs.' Chron.* **103,** 79–80.

—— and HUTCHINSON, H. P. 1933. '*Melampsora amygdalinae* the rust of Basket willows (*Salix triandra*). I. Observations and experiments in 1932. II. Spore germination experiments', *Ann. Rept. Agric. & Hort. Res. Sta., Long Ashton, 1932,* pp. 125–38.

OHLWEILER, W. W. 1912. 'Relation between the density of cell saps and the freezing point of leaves', *Ann. Rep. Missouri Bot. Gard.* **23,** 101–31.

OKANOUE, M. 1958. 'On an injury to Akamatsu (*Pinus densiflora* Sieb. et Zucc.) Forest in the vicinity of the smelting works at Hitachi', *Bull. For. Exp. Sta., Meguro, Tokyo,* **105,** 141–7. *R.A.M.,* 1958, p. 562.

OKSBJERG, E. 1956. 'Sommervejret 1955 på en midtjydsk lokalitet, og tørkens virkning på parceller med tidligt- og sentudspringende rødgraner', *Dansk. Skovforen. Tidskr.* **41,** 273–302. *For. Abstr.* **19,** No. 534.

OLBERG, A. 1955. 'Über die Kiefernschütte *Lophodermium pinastri* Schrad', *Forst- u. Holzw.* **10,** 307–8. *R.A.M.,* 1956, p. 562.

OLIVEIRA, A. L. F. 1944. 'Um fungo parasita do *Pinus halepensis* Miller', *Agros,* **27,** 158–64. *R.A.M.,* 1947, p. 272.

OLSON, D. S. 1952. 'Underground damage from logging in the Western white pine type', *J. For.* **50,** 460–2.

ONAKA, F. 1950. 'The effects of defoliation, disbudding, girdling, and other treatments on growth, especially radial growth, of conifers', *Bull. Kyoto Univ. For.* **18**, 55–95. *For. Abstr.* **14**, No. 175.

OPPLIGER, F. 1932. 'Stammbeschädigung durch reißerstriche', *Schweiz. Z. Forstw.* **83**, 59–61.

ORŁOŚ, H. 1957. 'Badania nad zwalczaniem opieńki miodowej (*Armillaria mellea* Vahl.) metoda biologiczna', *Roczn. Nauk leśn.* **15**, 195–235.

—— 1958. 'Studie o výsypu výtrusů z plodnic hub chorošovitých', *Česká mykologie*, **12**, 200–4.

—— and BRENNEJZEN, B. 1954. 'Badania nad zwalczaniem osutki sosnowej (*Lophodermium pinastri* Chev.) w szkółkach i uprawach sosny pospolitej', *Ministerstwo Leśnictwa Instytut Badawczy Leśnictwa*, **118**, 181–206.

—— —— 1957. 'Badania nad zwalczaniem hub drzewnych za pomoca zastrzyków środków grzybobójczych', *Roczn. Nauk leśn.* **19**, 3–42. *For. Abstr.* **20**, No. 713.

—— and DOMINIK, T. 1960. 'Z biologii huby korzeniowej — *Fomes annosus* (Fr.) Cooke', *Sylwan*, **104**, 1–12.

ORLOVA, A. A. 1954. 'Nekotorye dannye o mikaflore semjan drevesnykh i kustarnikovykh porod', *Trud. Inst. Les.* **16**, 281–96.

ORR, M. Y. 1925. 'The effect of frost on the wood of larch', *Trans. Roy. Scot. Arb. Soc.* **39**, 38–41.

ORSENIGO, M. 1949. 'Inchiesta sulla diffusione del "cancro Americano" del Castagno (*Endothia parasitica* (Murr.) And.) nell'Italia settentrionale', *Notiz. Malatt. Piante*, **2**, 41–42. *R.A.M.*, 1950, p. 236.

—— 1951. 'Contributi alla conoscenza del genere *Endothia*. II. Identificazione fra *E. parasitica*. (Murr.) P. J. et H. W. And. ed *E. fluens* (Sow.) S. et S. e loro fusione in *E. radicalis* (Schw.) DeNot. Considerazioni sulla patogenicità delle specie e sulla classificazione del genere', *Phytopath. Z.* **18**, 210–20. *Biol. Abstr.*, 1953, No. 7318.

—— and BONCOMPAGNI, T. 1950. 'Nota preliminare sui trattamenti antisporulanti nell'eradicazione del cancro del Castagno da *Endothia parasitica*', *Notiz. Malatt. Piante*, **10**, 52–57. *R.A.M.*, 1950, p. 545.

ORTON, C. R. 1931. 'Seed borne parasites. A bibliography', *Agric. Exp. Sta. West Virginia Univ. Bull.* **245**, 47 pp.

ORTON, W. A., and BEATTIE, R. K. 1923. 'The biological basis of foreign plant quarantines', *Phytopathology*, **13**, 295–306.

OSBORN, A., and BAGENAL, N. B. 1952. *Pruning—a practical handbook on the pruning of ornamental and flowering trees.* Ward Locke, London, 160 pp.

OSBURN, M. R., PHILLIPS, A. M., and PIERCE, W. C. 1954. 'Insects and diseases of the Pecan and their control', *Fmrs.' Bull. U.S. Dep. Agric.* **1829**, 56 pp. *R.A.M.*, 1955, p. 759.

OSMASTON, L. S. 1927. 'Mortality among oak', *Quart. J. For.* **21**, 28–30.

OW, L. VON. 1948. 'Über die Dürreempfindlichkeit der einzelnen Holzarten im Auwald', *Allg. ForstZ.* **3**, 219–21. *For. Abstr.* **11**, No. 3244.

OWEN, J. H. 1955. 'Fasciation of roots in Gallberry', *Plant Dis. Reptr.* **39**, 242.

OYLER, E., and BEWLEY, W. F. 1937. 'A disease of cultivated heaths caused by *Phytophthora cinnamomi* Rands', *Ann. Appl. Biol.* **24**, 1–16.

PACLT, J. 1948. 'O chorobné skvrnitosti listů Pavlonie císařské', *Ochr. Rost.* **21**, 34–37. *R.A.M.*, 1950, p. 588.

—— 1951. 'Fungus and the related diseases of the genus *Catalpa* (Bignoniaceae)', *Sydowia*, **5**, 160–8. *R.A.M.*, 1952, p. 153.

—— 1953. 'Kernbildung der Buche (*Fagus silvatica* L.)', *Phytopath. Z.* **20**, 255–9. *Biol. Abstr.*, 1955, No. 27156.

Palm, B. 1952. 'A die-back disease of *Metasequoia*', *Bot. Notiser*, **4,** 441. *R.A.M.*, 1953, p. 289.

Papajoannou, J. 1934. 'Die Temperaturverhältnisse unter Pflanzenschutzvorrichtungen im Forstgarten', *Forstw. Cbl.* **56,** 769–82.

Pape, H. 1928. 'Folgeerscheinungen der Fliederseuche', *Gartenwelt*, **32,** 303–4. *R.A.M.*, 1928, p. 725.

—— 1955. *Krankheiten und Schädlinge der Zierpflanzen und ihre Bekämpfung*. Paul Parey, Berlin, 559 pp.

Pardé, L. 1928. 'La régénération du chêne rouge d'Amérique dans le domaine des Barres', *Rev. Eaux Forêt*, **66,** 567–70. *R.A.M.*, 1929, p. 141.

Park, B. 1956. *Collins' Guide to Roses*. Collins, 288 pp.

Parker, A. K. 1953. 'Pole blight of Western white pine', *Bi-m. Progr. Rep. Div. For. Biol. Dep. Agric. Can.* **9,** 4. *For. Abstr.* **14,** No. 3581.

—— 1957. 'The nature of the association of *Europhium trinacriforme* with Pole blight lesions', *Canad. J. Bot.* **35,** 845–56.

Parker, J. 1955. 'Annual trends in cold hardiness of Ponderosa pine and Grand fir', *Ecology*, **36,** 377–80.

—— 1956*a*. 'Drought resistance in woody plants', *Bot. Rev.* **22,** 241–89.

—— 1956*b*. 'Variations in copper, boron, and manganese in leaves of *Pinus ponderosa*', *For. Sci.* **2,** 190–8.

Parker, K. G., Collins, D. L., Tyler, L. J., Connola, D. P., Ozard, W. E., and Dietrich H. 1947. 'The Dutch elm disease: association of *Ceratostomella ulmi* with *Scolytus multistriatus*, its advance into new areas, method of determining its distribution, and control of the disease', *Mem. Cornell Agric. Exp. Sta.* **275,** 44 pp. *R.A.M.*, 1949, p. 257.

——, Fisher, E. G., and Mills, W. D. 1956. 'Fire blight on Pome fruits and its control', *Cornell Ext. Bull.* **966,** 21 pp.

Parris, G. K. 1958. 'Soil fumigants and their use: A summary', *Plant Dis. Reptr.* **42,** 273–8.

Partridge, A. D., and Rich, A. E. 1957. 'A study of the ash leaf rust syndrome in New Hampshire, suscepts, incitant, epidemiology, and control', *Phytopathology*, **47,** 246.

Pasinetti, L. 1928. 'Suggerimenti terapeutici contro la batteriosi del Gelso', *Curiamo le Piante*, **6,** 2–6. *R.A.M.*, 1929, p. 142.

Paterson, A. 1938. 'The occlusion of pruning wounds in Norway spruce (*Picea excelsa*)', *Ann. Bot., London*, (n.s.) **2,** 681–98.

Paton, M. R. 1954. 'Petal blight of rhododendron in Scotland', *Plant Path.* **3,** 50.

Patton, R. F., and Riker, A. J. 1954. 'Needle droop and Needle blight of Red pine', *J. For.* **52,** 412–18.

—— —— 1958. 'Blister rust resistance in Eastern White Pine', *Proc. 5th Ntheast. For. Tree Impr. Conf., Orono, 1957*, pp. 46–51. *For. Abstr.* **20,** No. 1463.

Paul, B. H. 1946. 'Tree pruning by annual removal of lateral buds', *J. For.* **44,** 499–501.

—— 1957. 'Double branch whorls in White pine', *For. Sci.* **3,** 71–72.

Paul, W. R. C. 1929. 'A comparative morphological and physiological study of a number of strains of *Botrytis cinerea* Pers. with special reference to their virulence', *Trans. Brit. Mycol. Soc.* **14,** 118–35.

Pauley, S. S., and Perry, T. O. 1954. 'Ecotypic variation of the photoperiodic response in *Populus*', *J. Arnold Arbor.* **35,** 167–88.

Pawsey, R. G. 1957. 'The overwintering of *Keithia thujina*, the causal agent of Cedar leaf blight', *Trans. Brit. Mycol. Soc.* **40,** 166–7.

PEACE, T. R. 1932. 'The Dutch elm disease', *Forestry*, **6,** 125–42.

—— 1936*a*. 'Destructive fairy rings, associated with *Paxillus giganteus*, in young pine plantations', *Forestry*, **10,** 74–78.

—— 1936*b*. 'Spraying against *Meria laricis*, the leaf cast disease of larch', *Forestry*, **10,** 79–82.

—— 1938. 'Butt rot of conifers in Great Britain', *Quart. J. For.* **32,** 81–104.

—— 1940. 'An interesting case of lightning damage to a group of trees', *Quart. J. For.* **34,** 61–63.

—— 1944. 'The occurrence of *Malampsora pinitorqua* on Scots pine in south-eastern England', *Forestry*, **18,** 47–48.

—— 1948. 'The variation of Douglas fir in its native habitat (*Pseudotsuga taxifolia* Brit. syn. *P. douglasii* Carr.)', *Forestry*, **22,** 45–61.

—— 1952*a*. 'The effect of atmospheric pollution on forest trees in Great Britain', *Smokeless Air*, **83,** 12–16.

—— 1952*b*. 'Poplars', *For. Comm. Bull.* **19,** 50 pp.

—— 1953*a*. 'The defoliation of pines with particular reference to *Lophodermium pinastri*', *Scot. For.* **7,** 17–22.

—— 1953*b*. 'Asiatic chestnuts and American chestnut blight', *J. R. Hort. Soc.* **78,** 289–91.

—— 1954. 'Experiments on spraying with DDT to prevent the feeding of *Scolytus* beetles on elm and consequent infection with *Ceratostomella ulmi*', *Ann. Appl. Biol.* **41,** 155–64.

—— 1955. 'Sooty bark disease of Sycamore—a disease in eclipse', *Quart. J. For.* **49,** 197–204.

—— 1957*a*. 'Recent observations on the rusts of pine in Britain', *7th British Comm. For. Conf.*, 5 pp. *R.A.M.*, 1957, p. 799.

—— 1957*b*. 'Approach and perspective in forest pathology', *Forestry*, **30,** 48–56.

—— 1958*a*. 'A single case of fume damage', *Quart. J. For.* **52,** 41–45.

—— 1958*b*. 'The raising of *Thuja* in isolated nurseries to avoid infection by *Keithia thujina*', *Scot. For.* **12,** 7–10.

—— 1960. 'The status and development of elm disease in Britain', *For. Comm. Bull.* **33,** 44 pp.

—— and HOLMES, C. H. 1933. '*Meria laricis* the leaf cast disease of Larch', *Oxford For. Mem.* **15,** 29 pp.

PEARSON, G. A. 1931. 'Recovery of Western yellow pine seedlings from injury by grazing animals', *J. For.* **29,** 876–94.

PECHMANN, H. VON. 1949. 'Die Auswirkung eines Hagelschlages auf Zuwachsentwicklung und Holzwert', *Forstwiss. Cbl.* **68,** 445–56.

PECROT, A. 1956. 'Différentes relations entre l'apparition de la gélivure chez le peuplier euraméricain et les principaux caractères morphologiques du sol en Belgique', *Bull. Soc. Roy. for. Belg.* **63,** 1–19. *Soils & Fertil.*, 1956, No. 578.

PELLETIER, E. N., and HILBORN, M. T. 1954. 'Blossom and twig blight of low-bush Blueberries (*Botrytis cinerea*)', *Bull. Me. Agric. Exp. Sta.* **529,** 27 pp. *R.A.M.*, 1955, p. 42.

PELTIER, G. L. 1937. 'Distribution and prevalence of ozonium root rot in the shelterbelt zone of Texas', *Phytopathology*, **27,** 145–58.

PELZ, E. 1956. 'Gasförmige Luftverunreinigungen und Holzartenwahl in Gebieten mit Industrierauchschäden', *Forst- u. Jagdztg.* **6,** 347–9. *For. Abstr.* **19,** No. 539.

PENNINGSFELD, F. 1957. 'Kupfermangel bei Azaleen in Torf', *Süddtsch. ErwGärtn.* **11,** 2 pp. *Hort. Abstr.*, 1958, No. 734.

PERIŠIĆ, M. 1951. 'Nova virusna bolest na kanadskoj Topoli kod nas', *Šumarstvo*, **4**, 313. *R.A.M.*, 1953, p. 44.

PERSSON, A. 1955. 'Kronenmykose der Hybridaspe. I. Untersuchungen über Auftreten, selektive Wirkung und Pathogenität des Erregers', *Phytopath. Z.* **24**, 55–72. *R.A.M.*, 1956, p. 248.

PERSSON, E. 1957. 'Über den Stoffwechsel und eine antibiotisch wirksame Substanz von *Polyporus annosus* Fr.', *Phytopath. Z.* **30**, 45–86. *For. Abstr.* **19**, No. 1941.

PETCH, T. 1938. 'British Hypocreales', *Trans. Brit. Mycol. Soc.* **21**, 243–305.

PÉTER-CONTESSE, J. 1930. 'Du gui', *J. For. Suisse*, **81**, 217–23, 247–58.

—— 1937. 'Influence du gui sur la production du bois de service', *J. For. Suisse*, **7**, 145–51.

PETHYBRIDGE, G. H. 1911. 'The "Bladder rust" of Scots pine', *J. Dept. Agric. Ireland*, **11**, 500–2.

—— 1919. 'A destructive disease of seedling trees of *Thuja gigantea*', *Quart. J. For.* **13**, 93–97.

PETRAK, F. 1938*a*. 'Beiträge zur Systematik und Phylogenie der Gattung *Phaeocryptopus* Naoumov', *Ann. mycol. Berl.* **36**, 9–26. *R.A.M.*, 1938, p. 638.

—— 1938*b*. 'Beiträge zur Kenntnis der Gattung *Hercospora* mit besonderer Berücksichtigung ihrer Typusart *Hercospora tiliae* (Pers.) Fr.', *Ann. mycol. Berl.* **36**, 44–60. *R.A.M.*, 1938, p. 636.

PETRI, L. 1916. 'Studi sulla Malattia del Castagno detta "dell'inchiostro" ', *Ann. R. Ist. Sup. Forest. Naz. Firenze*, **1**, 360–93.

—— 1923. 'Sul modo di diffendersi del mal dell'inchiostro del Castagno e sui mezzi più efficaci per combatterlo', *Nuovi Ann. Min. Agric.* **3**, 3–19. *R.A.M.*, 1924, p. 245.

—— 1924*a*. 'Istruzioni pratiche per riconoscere e combattere la malattia del Castagno detta dell'inchiostro', *Nuovi Ann. Min. Agric.* **4**, 276–80. *R.A.M.*, 1925, p. 2.

—— 1924*b*. 'I tumori batterici del Pino d'Aleppo', *Ann. R. Ist. Sup. Forest. Naz. Firenze*, **9**, 43 pp. *R.A.M.*, 1925, p. 196.

—— 1928. 'Di un nuovo metodo di cura del "mal dell'inchiostro" del Castagno', *Boll. R. Staz. Pat. Veg.* (N.S.), **8**, 339–56. *R.A.M.*, 1929, p. 474.

PETRIE, S. M., and MACKAY, A. 1948. 'Seed-beds and the heat wave', *Scot. For.* **1**, 30–31.

PETRINI, S. 1946. 'Om granrötans inverkan på avverkningens rotvärde. Specialundersökningar i Lanforsbeståndet 1938 och 1941', *Medd. SkogsforsknInst.* Stockholm, **34**, 327–40. *R.A.M.*, 1947, p. 367.

PETROV. P. I. 1955. 'O temperaturnom režime drevesnyh stvolov', *Bot. Ž.*, **40**, 584–7. *For. Abstr.* **17**, No. 1221.

PETTINGA, J. J. 1950. 'De honingzwam (*Armillaria mellea* (Vahl.) Sacc.), *Fruitteelt*, **40**, 886. *R.A.M.*, 1951, p. 420.

PFEFFER, A., ŠKODA, B., and ZLATUŠKA, V. 1948. 'Vliv sucha v r. 1947 na lesní dřeviny', *Lesn. Práce*, **27**, 193–214. *For. Abstr.* **10**, No. 2215.

PHILLIPS, J. F. V. 1929. 'The influence of *Usnea* sp. (near *barbata* Fr.) upon the supporting tree', *Trans. Roy. Soc. S. Afr.* **17**, 101–7. *Biol. Abstr.*, 1930, No. 10951.

PHILLIPS, W. 1880. 'New British Discomycetes', *Gdnrs.' Chron.* **14**, 308.

PICCO, D. 1948. 'Ricerche sulla biologia dell'*Endothia parasitica* (Murr.) Anderson in Italia', *Ital. for. mont.* **3**, 254–8. *R.A.M.*, 1949, p. 428.

PICHEL, R. J. 1956. 'Les pourridiés de l'Hévéa dans la cuvette congolaise', *Publ. Inst. nat. agron. Congo. Belge, Sér. tech.* **49**, 480 pp. *R.A.M.*, 1956, p. 925.

PIERCE, A. S. 1934. 'Positive infection trials with Elm "wilt" fungi', *Science* (N.S.), **80**, 385. *R.A.M.*, 1935, p. 203.

PIERSON, R. K., and BUCHANAN, T. S. 1938*a*. 'Age of susceptibility of *Ribes petiolare* leaves to infection by aeciospores and urediospores of *Cronartium ribicola*', *Phytopathology*, **28**, 709–15.

—— —— 1938*b*. 'Susceptibility of needles of different ages on *Pinus monticola* seedlings to *Cronartium ribicola* infection', *Phytopathology*, **28**, 833–9.

PILLICHODY, A. 1936. 'Afforestation of frost hollows with Mountain pine', *Bull. Soc. For. de Franche-Comté*, **21**, 360–4.

PIRONE, P. P. 1938. 'The detrimental effect of Walnut to Rhododendrons and other ornamentals', *Plant Dis. Reptr.* **22**, 450–2.

—— 1948. *Maintenance of shade and ornamental trees.* Oxford University Press, New York, 2nd ed., 436 pp.

—— 1952. 'A bark canker disease of London plane, redbud, and sweet gum', *Phytopathology*, **42**, 16.

—— 1957. '*Ganoderma lucidum*, a parasite of shade trees', *Bull. Torrey Bot. Club*, **84**, 424–8. *Biol. Abstr.*, 1958, No. 28597.

PLAGNAT, F. 1950*a*. 'Le gui du sapin', *Ann. Éc. Eaux For. Nancy*, **12**, 155–231. *For. Abstr.* **12**, No. 2332.

—— 1950*b*. 'Sylviculture des sapinières à gui', *Rev. for. franç.* **7–8**, 365–78. *For. Abstr.* **12**, No. 2333.

PLAISANCE, G. 1956. 'Les couches aquifères perchées à éclipses', *VIe Congr. Int. Sci. Sol., Rapp. E,* pp. 553–60. *Soils & Fertil.*, 1956, No. 2280.

PLAKIDAS, A. G. 1942. '*Venturia acerina*, the perfect stage of *Cladosporium humile*', *Mycologia*, **34**, 27–37. *R.A.M.*, 1942, p. 274.

—— 1945. 'Blight of Oriental Arborvitae', *Phytopathology*, **34**, 181–90.

—— 1949. 'Witches' broom, a graft-transmissible disease of Arizona Ash (*Fraxinus berlandieriana*)', *Phytopathology*, **39**, 498–9.

—— 1954. 'Transmission of leaf and flower variegation in Camellias by grafting', *Phytopathology*, **44**, 14–18.

—— 1957. 'New or unusual plant diseases in Louisiana, II. Entomosporium leaf spot of *Photinia glabra*', *Plant Dis. Reptr.* **41**, 643–5.

PLATZER, B. 1937. 'Ergebnisse alter Kiefernastungen, III. Die Finowtaler Astungsversuche', *Forstarchiv*, **13**, 357–61.

POETEREN, N. VAN. 1937 and 1938. 'Verslag over de werkzaamheden van den Plantenziektenkundigen Dienst in het jaar 1936. Verslag over de werkzaamheden van den Plantenziektenkundigen Dienst in het jaar 1937', *Versl. PlZiekt. Dienst Wageningen*, 1937, **87**, 84 pp.; 1938, **89**, 82 pp. *R.A.M.*, 1939, p. 153.

POEVERLEIN, H. 1932. 'Die Gesamtverbreitung der *Uropyxis sanguinea* in Europa. (Nachtrag.)', *Ann. Mycol.* **30**, 402–4. *R.A.M.*, 1932, p. 580.

POLE EVANS, I. B. 1934. 'Aiming at better pastures and field crops. Annual Report of the Division of Plant Industry', *Fmg. S. Afr.* **9**, 539–48. *R.A.M.*, 1935, p. 426.

POMERLEAU, R. 1938. 'Recherches sur le *Gnomonia ulmea* (Schw.) Thüm.', *Contrib. Inst. Bot. Univ. Montreal*, **31**, 139 pp.

—— 1940. 'Studies on the ink-spot disease of Poplar', *Canad. J. Res. Sect. C.* **18**, 199–214.

—— 1942. 'Études sur la fonte des semis de conifères', *Rev. trim. canad.* **28**, 127–53. *R.A.M.*, 1943, p. 158.

—— and RAY, R. G. 1957. 'Occurrence and effects of summer frost in a conifer plantation', *Tech. Note For. Br. Can.* **51**, 15 pp. *For. Abstr.* **18**, No. 4171.

PONOMAREFF, N. V. 1938. 'The conidial stage of *Hypoxylon pruinatum*', *Phytopathology*, **28**, 515–18.

PONTIS, VIDELA R. E. 1943. 'El "mal de la tinta", del Nogal en la República Argentina', *Rev. B.A.P.* **27**, 31–33. *R.A.M.*, 1945, p. 39.

POPULER, CH. 1956. 'La pourriture rouge du cœur des résineux (*Fomes annosus* (Fr.) Cooke)', *Bull. Soc. for. Belg.* **63,** 297–329. *R.A.M.*, 1957, p. 221.

PORTER, R. H. 1941. 'Seed-borne organisms and plant quarantine', *J. Econ. Ent.* **34,** 543–8. *R.A.M.*, 1942, p. 112.

POSNETTE, A. F. 1954. 'Virus diseases of cherry trees in England. I. Survey of diseases present', *J. Hort. Sci.* **29,** 44–58.

—— 1957. 'Virus diseases of Pears in England', *J. Hort. Sci.* **32,** 53–61.

POTLAYCHUK, V. I. 1953. 'Vrednaja mikroflora želudei i ee razvitie v zavisimosti ot uslovii proizrastanija i khranenija'. *Bot. Ž.* **38,** 135–42. *R.A.M.*, 1954, p. 57.

POTTER, H. S. 1956. 'The use of chemicals to suppress symptoms of Dutch elm disease in two-year-old American elm trees inoculated with *Ceratostomella ulmi* (Schwarz) Buisman', *Diss. Abstr.* **16,** 2292. *R.A.M.*, 1957, p. 738.

POTZGER, J. E. 1939. 'Microclimate, evaporation stress, and epiphytic mosses', *Bryologist*, **42,** 53–61. *For. Abstr.* **1,** p. 228.

POVAH, A. H. 1921. 'An attack of poplar canker following fire injury', *Phytopathology*, **11,** 157–65.

—— 1935. 'The fungi of the Isle Royale, Lake Superior, including *Nectria galligena* on *Populus*', *Pap. Mich. Acad. Sci.* **20,** 113–56. *R.A.M.*, 1935, p. 794.

PRENTICE, I. W. 1950. 'Rubus stunt: a virus disease', *J. Hort. Sci.* **26,** 35–42. *R.A.M.*, 1951, p. 376.

PRESCOTT, G. W. 1956. 'A guide to the literature on ecology and life histories of algae', *Bot. Rev.* **22,** 167–240.

PRIEHÄUSSER, G. 1935. 'Beitrag zur Frage der Entstehung der Fichtenrotfäule' *Forstw. Zbl.* **57,** 649–55. *R.A.M.*, 1936, p. 184.

—— 1943. 'Über Fichtenwurzelfäule, Kronenform und Standort. Beitrag zur Kenntnis der Fichtenrotfäule', *Forstw. Zbl.* **6,** 259–73. *R.A.M.*, 1944, p. 417.

PRIESTLEY, J. H. 1932. 'The growing tree', *Forestry*, **6,** 105–12.

PŘÍHODA, A. 1950*a*. 'Keřovitý růst Lipových sazenic jako následik napadení houbon *Pyrenochaeta pubescens* Rostr.', *Ochr. Rost.* **23,** 366–8. *R.A.M.*, 1952, p. 38.

—— 1950*b*. '*Naemacyclus niveus* (Pers.) Sacc., houba na borových jehpicích', *Lesn. Práce*, **1–2,** 30–32.

—— 1955. 'Poškození Habrového semene při stratifikaci', *Ann. Acad. tchécosl. Agric.* **28,** 385–92. *R.A.M.*, 1957, p. 145.

—— 1957*a*. 'Hniloby stromů, poškozených přibližováním dřeva', *Lesn. Práce*, **36,** 271–3. *For. Abstr.* **19,** No. 3161.

—— 1957*b*. 'Nákaza živých smrků václavkou', *Les. Bratislava*, **13,** 173–6.

PRILLEUX. 1897. *Maladies des Plantes Agricoles et des Arbres Fruitiers et Forestiers.* Maison Didot, Paris, **2,** 592 pp.

PRIOR, E. M. 1913. 'Contribution to a knowledge of "The snap Beech" disease', *J. Econ. Biol.* **8,** 249–63.

PRODAN, I. 1935. '*Diplodia pinea* (Desm.) Kickx. in Rumänien', *Bul. Inf. Grăd. bot. Cluj*, 1934, **14,** 240–3. *R.A.M.*, 1935, p. 483.

PROTSENKO, E. P., and A. E. 1950. 'Kol'cevaja mozaika sireni — infekcionnoe zabolevanie', *Bull. Bot. Garden, Moscow*, **5,** 46–50. *R.A.M.*, 1953, p. 434.

PRYOR, L. D. 1937. 'Some observations on the roots of *Pinus radiata* in relation to wind resistance', *Aust. For.* **2,** 37–40.

—— 1940. 'The effect of fire on exotic conifers', *Aust. For.* **5,** 37–38.

—— 1956. 'Chlorosis and lack of vigour in seedlings of renantherous species of *Eucalyptus* caused by lack of mycorrhiza', *Proc. Linn. Soc. N.S.W.* **81,** 91–96. *R.A.M.*, 1957, p. 50.

PUDDEN, H. H. C. 1957. 'A problem of pruning in Kenya', *Brit. Commonw. For. Conf. Statement by Colony and Protectorate of Kenya*, 8 pp.

PUPILLO, M., and CANOVA, A. 1952. 'Contributo alla conoscenza del "mal dello stacco" dei Nocciòli in Sicilia', *Ann. Sper. agr.* (N.S.), **6,** 895–906. *R.A.M.*, 1953, p. 286.

PURNELL, H. M. 1957. 'Shoot blight of *Pinus radiata* Don caused by *Diplodia pinea* (Desm.) Kickx.', *7th Br. Comm. For. Conf. Melbourne, Bull.* **5,** 11 pp. *For. Abstr.* **19,** No. 1952.

PUTTERILL, V. A. 1923. 'Silver leaf disease of fruit trees, and its occurrence in South Africa', *Dep. Agric. S. Afr. Bull.* **27,** 19 pp. *R.A.M.*, 1923, p. 451.

QUANJER, H. M. 1924. '*Endothia parasitica*', *Phytopathology*, **14,** 535.

RAABE, A. 1938. '*Ceratophorum setosum* Kirchn. als Ursache eines Sämlingssterbens bei Ginster', *Z. PflKrankh.* **48,** 231–2. *R.A.M.*, 1938, p. 686.

RAABE, R. D. 1953. 'Cultural studies of *Armillaria mellea*', *Phytopathology*, **43,** 482.

—— 1954. 'Diseases of Rhododendrons and Azaleas', *Quart. Bull. Amer. Rhodo. Soc.* **8,** 82–86.

—— 1955. 'Variation in pathogenicity of isolates of *Armillaria mellea*', *Phytopathology*, **45,** 695.

—— 1958. 'Some previously unreported non-woody hosts of *Armillarea mellea* in California', *Plant Dis. Reptr.* **42,** 1025.

—— and LENZ, J. V. 1958. 'Septoria leaf scorch of Azalea', *Calif. Agric.* **12,** 11. *R.A.M.*, 1959, p. 148.

—— and SCIARONI, R. H. 1955. 'Petal blight disease of Azaleas', *Calif. Agric.* **9,** 7, 14. *R.A.M.*, 1956, p. 190.

RACK, K. 1953. 'Untersuchungen über die Bedeutung der Verwundung und über die Rolle von Wuchstoffen beim bakteriellen Pflanzenkrebs', *Phytopath. Z.* **21,** 1–44. *R.A.M.*, 1954, p. 470.

—— 1955. 'Über die Bedingungen und den Verlauf der Schütte-Infektion im Sommerhalbjahr 1954', *Forst- u. Holzw.* **10,** 224–5. *For. Abstr.* **17,** No. 2979.

—— 1957*a*. 'Untersuchungen über die Anfälligkeit verschiedener Eichenprovenienzen gegenüber dem Eichenmehltau', *Allg. Forst- u. Jagdztg.* **128,** 150–6. *For. Abstr.* **19,** No. 1979.

—— 1957*b*. 'Versuche zur Bekämpfung des Eichenmehltaus', *Forst- u. Holzw.* **12,** 5–6. *R.A.M.*, 1957, p. 672.

—— 1958. 'Erfahrungen mit Fungiziden zur Bekämpfung der Kiefernschütte, *NachrBl. dtsch. PflSchDienst, Stuttgart*, **10,** 54–58. *R.A.M.*, 1958, p. 744.

RADEMACHER, B. 1940. 'Kupfermangelerscheinungen bei Forstgewächsen auf Heideböden', *Mitt. Forstw. Forstwiss.*, pp. 335–44. *R.A.M.*, 1943, p. 82.

RAGGI, C. A. 1947. 'Nota sobre un interesante caso de parasitismo del *Botrytis cinerea* Pers. sobre *Eucalyptus* sps.', *Publ. misc. Minist. Agric., B. Aires*, ser. A, **3,** 11 pp. *R.A.M.*, 1947, p. 516.

RAMSAY, J. 1871–2. 'Notes on the injury done to vegetation by the severe frost of the 17th May 1871', *Proc. Nat. Hist. Soc. Glasgow*, **2,** 169–74.

RAMSBOTTOM, J., and BALFOUR-BROWNE, F. L. 1951. 'List of Discomycetes recorded from the British Isles', *Trans. Brit. Mycol. Soc.* **34,** 38–137.

RANGASWAMI, S., and GRIFFITH, A. L. 1939. 'Host plants and the Spike disease of Sandal', *Indian For.* **65,** 335–45. *R.A.M.*, 1939, p. 819.

—— and SREENIVASAYA, M. 1935. 'Insect transmission of spike disease of Sandal (*Santalum album* Linn.)', *Curr. Sci.* **4,** 17–19. *R.A.M.*, 1935, p. 802.

RANGONE-GALLUCCI, M. M., and PEROTTI, G. 1959. 'Osservazioni su una alterazione di pioppelle in vivaio', *Pioppicoltura*, **2,** 4–5.

RAO, H. S. 1954. 'The phenomenon of twisted trees', *Indian For.* **80,** 165–70.

RATHBUN, A. E. 1922. 'Root rot of pine seedlings', *Phytopathology*, **12,** 213–20.
—— 1923. 'Damping-off of taproots of conifers', *Phytopathology*, **13,** 385–90.
RATHBUN-GRAVATT, A. 1925. 'Direct inoculation of coniferous stems with damping-off fungi', *J. Agric. Res.* **30,** 327–39.
——1927. 'A witches' broom of introduced Japanese cherry tree', *Phytopathology*, **17,** 19–24.
—— 1931. 'Germination loss of coniferous seeds due to parasites', *J. Agric. Res.* **42,** 71–92.
RATTSJÖ, H., and RENNERFELT, E. 1955. 'Värdeförlusten på virkesutbytet till följd av rotröta', *Norrlands SkogsvFörb. Tidskr.* **3,** 279–98. *For. Abstr.* **17,** No. 1739.
RAULT, J. P., and MARSH, E. K. 1952. *The incidence and sylvicultural implications of spiral grain in* Pinus longifolia *Roxb. in South Africa and its effect on converted timber.* Forest Products Institute, Pretoria, 21 pp. *For. Abstr.* **13,** No. 4364.
RAUNECKER, H. 1956. 'Der Buchenrotkern nur eine Alterserscheinung?', *Allg. Forst- u. Jagdztg.* **127,** 16–31. *For. Abstr.* **18,** No. 2252.
RAUP, L. C. 1930. 'An investigation of the lichen flora of *Picea canadensis*', *The Bryologist*, **33,** 1–11. Rev. in *J. Ecol. Suppl.*, 1932, **9,** 149.
RAWLINGS, G. B. 1955. 'Epidemics in *Pinus radiata* forests in New Zealand', *N.Z. J. For.* **7,** 53–55. *R.A.M.*, 1956, p. 498.
RAWLINGS, G. H. 1956. 'Australasian Cyttariaceae', *Tech. Pap. N.Z. For. Serv.* **9,** 10 pp. *For. Abstr.* **18,** No. 3033.
RAY, W. W. 1936. 'Pathogenicity and cultural experiments with *Caliciopsis pinea*', *Mycologia*, **28,** 201–8. *R.A.M.*, 1936, p. 760.
—— 1939. 'Contribution to knowledge of the genus *Taphrina* in North America', *Mycologia*, **31,** 56–75. *R.A.M.*, 1939, p. 414.
RAYCHAUDHURI, S. P. 1952. 'Bayberry yellows', *Phytopathology*, **42,** 17.
RAYNER, M. C. 1929. 'The biology of fungus infection in the genus *Vaccinium*', *Ann. Bot. Lond.* **43,** 55–70.
—— 1930. 'Observations on *Armillaria mellea* in pure culture with certain conifers', *Forestry*, **4,** 65–77.
—— 1938. 'The use of soil or humus inocula in nurseries and plantations', *Emp. For. J.* **17,** 236–43.
—— 1947. 'Behaviour of Corsican pine stock following different nursery treatments (*Pinus nigra* var. *calabrica* Schneid.)', *Forestry*, **21,** 204–16.
—— and LEVISOHN, I. 1941. 'The mycorrhizal habit in relation to forestry. IV. Studies on mycorrhizal response in *Pinus* and other conifers', *Forestry*, **15,** 1–36.
REA, C. 1922. *British Basidiomycetae.* Cambridge University Press, 799 pp.
READ, R. A. 1957. 'Effect of livestock concentration on surface-soil porosity within shelterbelts', *J. For.* **55,** 529–30.
RECKENDORFER, P. 1952. 'Ein Beitrag zur Mikrochemie des Rauchschadens durch Fluor. Die Wanderung des Fluors im pflanzlichen Gewebe. I. Die unsichtbaren Schäden', *Pflanzenschutzber.* **9,** 33–55. *Biol. Abstr.*, 1953, No. 20689.
REDMOND, D. R. 1954. 'Condition of tree rootlets in Nova Scotia following a hurricane', *Bi-m. Progr. Rep. Div. For. Biol. Dep. Agric. Can.* **10,** 5, p. 1. *For. Abstr.* **16,** No. 1852.
—— 1958. 'Soil temperatures and birch decline', *Bi-m. Progr. Rep. Div. For. Biol. Dep. Agric. Can.* **14,** 5, p. 1.
REEVES, E. L., CHEYNEY, P. W., and MILBRATH, J. A. 1955. 'Normal-appearing Kwanzan and Shiro-fungen oriental flowering cherries found to carry a virus of little cherry type', *Plant Dis. Reptr.* **39,** 725–6.

Regler, W. 1957. 'Der Kieferndrehrost (*Melampsora pinitorqua*) eine wirtschaftlich wichtige Infektionskrankheit der Gattung *Pinus*', *Wiss. Abh. dtsch. Akad. LandwWiss., Berlin*, **27**, 205–34. *R.A.M.*, 1958, p. 561.

Rehder, A. 1949. *Manual of cultivated trees and shrubs hardy in North America.* Macmillan, 996 pp.

Reichert, I., and Palti, J. 1947. 'The control of lichens in citrus groves', *Palest. J. Bot.*, R. Ser., **6**, 222–4. *R.A.M.*, 1949, p. 63.

Reid, W. D. 1930. 'The diagnosis of fireblight in New Zealand', *N.Z. J. Sci. and Tech.* **12**, 166–72. *R.A.M.*, 1931, p. 318.

Rendle, B. J., Armstrong, F. H., and Nevard, E. H. 1941. 'The utilization of softwood timber damaged in the glazed frost, 1940', *Forestry*, **15**, 55–64.

Rennerfelt, E. 1943. 'Om vår nuvarande kunskap om törskatesvampen (*Peridermium*) och sättet för dess spridning och tillväxt', *Svenska SkogsvFören. Tidskr.* **41**, 305–24. *R.A.M.*, 1946, p. 373.

—— 1946. 'Om rotrötan (*Polyporus annosus* Fr.) I. Sverige. Dess Utbredning och sätt att uppträda', *Medd. SkogsforsknInst. Stockh.* **35**, 88 pp.

—— 1947. 'Några undersökningar över olika rötsvampars förmåga att angripa splint- och kärnved hos tall', *Medd. SkogsforsknInst. Stockh.* **36**, 24 pp. *For. Abstr.* **9**, No. 2277.

—— 1949. 'The effect of soil organisms on the development of *Polyporus annosus* Fr., the root rot fungus', *Oikos*, **1**, 65–78. *R.A.M.*, 1950, p. 595.

—— 1952. 'Om angrepp av rotröta på tall', *Medd. SkogsforsknInst. Stockh.* **41**, 1–39. *R.A.M.*, 1954, p. 190.

—— 1953. 'Biologische Untersuchungen uber den Kieferndreher *Melampsora pinitorqua* (Braun.) Rostr.', *Proc. Congr. Int. Union For. Res. Organ., Rome*, pp. 705–11. *R.A.M.*, 1955, p. 117.

—— 1956. 'The natural resistance to decay of certain conifers', *Friesia*, **5**, 361–5. *R.A.M.*, 1957, p. 290.

—— 1957. 'Untersuchungen über die Wurzelfäule auf Fichte und Kiefer in Schweden', *Phytopath. Z.* **28**, 259–74.

Rennie, P. J. 1956. 'The importance of shelter to early tree growth on upland moors', *Forestry*, **29**, 147–53.

—— 1957. 'The uptake of nutrients by timber forest and its importance to timber production in Britain', *Quart. J. For.* **51**, 101–15.

Reuss, H. 1928. 'Wesen, eigenschaften und wirtschaftliche bedeutung der früh- und spättreibenden fichtenform', *Clb. ges. Forstw.* **54**, 1–18.

Reynolds, E. S. 1939. 'Tree temperatures and thermostasy', *Mo. Bot. Gard. Ann.* **26**, 165–255.

Reynolds, R. R. 1940. 'Lightning as a cause of timber mortality', *South. Forest Exp. Sta. South. Forest Notes*, **31**, 1.

Rhine, J. B. 1924. 'Clogging of stomata of conifers by smoke', *Bot. Gaz.* **78**, 226–31. *Bot. Abstr.* **14**, No. 1431 (1925).

Rhoads, A. S. 1942. 'Growing new root systems by soil banking—a promising method of rejuvenating trees attacked by root diseases', *Phytopathology*, **32**, 529–36.

—— 1943. 'Lightning injury to oak and pine trees in Florida', *Plant Dis. Reptr.* **27**, 556–7.

—— 1945. 'A comparative study of two closely related root-rot fungi, *Clitocybe tabescens* and *Armillaria mellea*', *Mycologia*, **37**,741–66.*R.A.M.*, 1946, p. 186.

—— 1950. 'Clitocybe root rot of woody plants in the southeastern United States', *Circ. U.S. Dep. Agric.* **853**, 25 pp. *For. Abstr.* **13**, No. 1332.

—— and Wright, E. 1946. '*Fomes annosus* commonly a wound pathogen rather than a root parasite of Western hemlock in Western Oregon and Washington', *J. For.* **44**, 1091–2.

RIBALDI, M. 1948. 'Su di un seccume delle foglie dell'*Acer negundo* dovuto a *Macrophoma negundinis* nov. sp.', *Annali della Facoltà Agraria di Perugia*, **5,** 165–9. *For. Abstr.* **12,** No. 4415.

—— 1953. 'Maculatura fogliare del nocciòlo spontaneo causata da *Cylindrothyrium coryli* (*Leptothyrium coryli* Fuck.) n. comb., con osservazioni biologiche e sistematiche sul fungo', *Ann. Sper. agr.* (N.S.), **7,** 1941–56. *For. Abstr.* **15,** No. 2647.

—— 1954. 'Su un deperimento di *Robinia pseudo-acacia* L. var. *umbraculifera* DC. f. *bessoniana* Cowel, dovuto a *Phomopsis oncostoma* (Thüm.) v. Höhnel', *Ann. Sper. agr.* (N.S.), **8,** 1197–1212. *R.A.M.*, 1956, p. 250.

RICHARDS, P. W. M. 1928. 'Ecological notes on the Bryophytes of Middlesex', *J. Ecol.* **16,** 269–300.

RICHARDSON, A. D. 1923. 'Witches' broom on Silver fir', *Gard. Chron.* **73,** 11. *R.A.M.*, 1923, p. 430.

RICHARDSON, S. D. 1953. 'Root growth of *Acer pseudoplatanus* L. in relation to grass cover and nitrogen deficiency', *Meded. Landbouwk. Tijdschr. Wageningen*, **53,** 75–97.

—— 1955. 'Effects of sea-water flooding on tree growth in the Netherlands', *Quart. J. For.* **49,** 22–28.

RICHMOND, B. G. 1932. 'A Diaporthe canker of American elm', *Science* (N.S.), **75,** 110–11. *R.A.M.*, 1933, p. 125.

RICHTER, H. 1928. 'Die wichtigsten holzbewohnenden Nectrien aus der Gruppe der Krebserreger', *Z. Parasitenk.* **1,** 24–75. *R.A.M.*, 1928, p. 676.

—— 1933. 'Krebs und Rindenbrand der Pappel', *Mitt. dtsch. dendrol. Ges.* **45,** 262–7. *R.A.M.*, 1934, p. 479.

RIDÉ, M. 1958. 'Sur l'étiologie du chancre suintant du Peuplier', *C.R. Acad. Sci., Paris*, **246,** 2795–8. *R.A.M.*, 1958, p. 604.

RIEBEN, E. 1940. 'Un ennemi des forêts de montagne: *Usnea barbata*', *J. For. Suisse*, **91,** 62–64. *For. Abstr.* **2,** p. 20.

RIGGENBACH, A. 1956. 'Untersuchung über den Eschenkrebs', *Phytopath. Z.* **27,** 1–40. *R.A.M.*, 1957, p. 144.

RIKER, A. J. 1954. 'Opportunities in disease and insect control through genetics', *J. For.* **52,** 651–2.

—— 1957. 'The discovery of important diseases before they move from one country to another', *Phytopathology*, **47,** 388–9.

——, GRUENHAGEN, R. H., ROTH, L. F., and BRENER, W. H. 1947. 'Some chemical treatments and their influence on damping-off, weed control, and winter injury of Red pine seedlings', *J. Agric. Res.* **74,** 87–95.

—— KOUBA, T. F., BRENER, W. H., and BYAM, L. 1943. 'White pine selections tested for resistance to blister rust', *Phytopathology*, **33,** 11.

RILEY, C. G. 1948. 'Poplar disease in relation to thinning', *For. Chron.* **24,** 321. *For. Abstr.* **10,** No. 2947.

—— 1952. 'Studies in forest pathology. IX. *Fomes igniarius* decay of Poplar', *Canad. J. Bot.* **30,** 710–34.

—— 1953. 'Hail damage in forest stands', *For. Chron.* **29,** 139–42. *R.A.M.*, 1955, pp. 494–5.

—— and BIER, J. E. 1936. 'Extent of decay in poplar as indicated by the presence of sporophores of the fungus *Fomes igniarius* Linn.', *For. Chron.* **12,** 249–53.

RILEY, H. K., and DAIGLE, C. J. 1948. 'Spray tests for Azalea petal blight', *Proc. Amer. Soc. Hort. Sci.* **51,** 651–3. *R.A.M.*, 1949, p. 456.

RISHBETH, J. 1950. 'Observations on the biology of *Fomes annosus*, with particular reference to East Anglian pine plantations. I. The outbreaks of disease and ecological status of the fungus', *Ann. Bot., Lond.* (N.S.), **14,** 365–83.

RISHBETH, J. 1951*a*. 'Observations on the biology of *Fomes annosus*, with particular reference to East Anglian pine plantations. II. Spore production, stump infection, and saprophytic activity in stumps', *Ann. Bot., Lond.* (N.S.), **15**, 1–21.

—— 1951*b*. 'Observations on the biology of *Fomes annosus*, with particular reference to East Anglian pine plantations. III. Natural and experimental infection of pines, and some factors affecting severity of the disease', *Ann. Bot., Lond.* (N.S.), **15**, 221–46.

—— 1951*c*. 'Butt rot by *Fomes annosus* Fr. in East Anglian conifer plantations and its relation to tree killing', *Forestry*, **24**, 114–20.

—— 1952. 'Control of *Fomes annosus* Fr.', *Forestry*, **25**, 41–50.

—— 1955. 'Root diseases in plantations, with special reference to tropical crops', *Ann. Appl. Biol.* **42**, 220–7.

—— 1957. 'Some further observations on *Fomes annosus* Fr.', *Forestry*, **30**, 69–89.

—— 1958. 'Detection of viable air-borne spores in air', *Nature, London*, **181**, 1549.

—— and MEREDITH, D. S. 1957. 'Surface microflora of pine needles', *Nature, London*, **179**, 682–3.

RISLEY, J. A., and SILVERBORG, S. B. 1958. '*Stereum sanguinolentum* on living Norway spruce following pruning', *Phytopathology*, **48**, 337–8.

RITSCHL, A. 1937. 'Untersuchungen über *Gloeosporium fagicolum* Passerini, den Erreger der Blattfleckenkrankheit der Buche', *Z. PflKrankh.* **47**, 486–91. *R.A.M.*, 1928, p. 145.

ROACH, W. A., and ROBERTS, W. O. 1945. 'Further work on plant injection for diagnostic and curative purposes', *Imp. Bur. Hort. Plant. Crops, Tech. Comm.* **16**, 12 pp.

ROBAK, H. 1946. 'Tre skogsykdommer som hittil har vaert lite kjent eller påaktet i Norge', *Tidsskr. Skogbr.* **10–11**, 323–34. *R.A.M.*, 1947, p. 222.

—— 1951*a*. 'Noen iakttakelser til belysning av forholdet mellom klimatiske skader og soppangrep på nåletrær', *Medd. Vestl. forstl. Forsøksstа.* **27**, 43 pp. *R.A.M.*, 1952, p. 92.

—— 1951*b*. 'Om saprofyttiske og parasittiske raser af Lerkekreftsoppen, *Dasyscypha willkommii* (Hart.) Rehm.', *Medd. Vestl. forstl. Forsøksstа.* **29**, 117–211. *R.A.M.*, 1952, p. 406.

—— 1952*a*. '*Dothichiza pithyophila* (Cda.) Petr. the pycnidial stage of a mycelium of the type *Pullularia pullulans* (de B.) Berkhout', *Sydowia*, ser. 2, **6**, 361–2. *R.A.M.*, 1952, p. 598.

—— 1952*b*. '*Phomopsis pseudotsugae* Wilson—*Discula pinicola* (Naumov) Petr. as a saprophyte on coniferous woods', *Sydowia*, ser. 2, **6**, 378–82. *R.A.M.*, 1955, p. 419.

ROBERTS, E. G., and CLAPP, R. T. 1956. 'Effect of pruning on the recovery of ice-bent Slash pines', *J. For.* **54**, 596–7.

ROBERTS, E. T. 1957. '*Glomerella cingulata* causing a die-back of privet', *Plant Path.* **6**, 76.

ROBERTSON, N. F. 1955. 'Investigations of the biology of *Cryptostroma corticale* and the sooty bark disease of sycamore', *For. Comm. Rep. For. Res.*, 1953–4, pp. 57–58.

ROBINSON, R. L. 1927. 'Mortality among oak', *Quart. J. For.* **21**, 25–27.

ROESER, J. 1929. 'The occurrence of "pitch girdle" in sapling Douglas fir stands on the Pike National Forest', *J. For.* **27**, 813–20.

ROGERS, W. S., *et al.* 1954. 'Low temperature injury to fruit blossom. IV. Further experiments on water-sprinkling as an anti-frost measure', *J. Hort. Sci.* **29**, 126–41. *Hort. Abstr.*, 1954, No. 2423.

ROHDE, T. 1932. 'Das Vordringen der Rhabdocline-Schütte in Deutschland. Die Folgen des Rhabdocline-Befalls in deutschen Douglasienbeständen. Welche

Douglasien sind in Deutschland durch *Rhabdocline* gefährdet?', *Forstarchiv*, **14,** 247–9; **18,** 318–26; **22,** 389–92. *R.A.M.*, 1933, p. 65.
ROHDE, T. 1934. 'Zur Biologie der Douglasienschütte', *Z. Forst- u. Jagdw.* **66,** 151–6. *R.A.M.*, 1934, p. 606.
—— 1936. 'Schüttegefährdung verschiedener Douglasien-"Herkünfte"', *Z. Forst- u. Jagdw.* **68,** 610–16. *R.A.M.*, 1937, p. 289.
—— 1937. 'Erscheinungsformen und Erkennung der Schweizer Douglasienschütte', *Silva*, **25,** 69–77. *R.A.M.*, 1937, p. 507.
ROHMEDER, E. 1935. 'Zusammenhänge zwischen Baumklasseneinteilung und Wasserreiserbefall jungerer Eichenbestände', *Forstwiss. Cbl.* **79,** 205–10.
—— 1937. 'Die Stammfäule (Wurzelfäule und Wundfäule) der Fichtenbestockung', *Mitt. LandesForstverw. Bayerns*, **23,** 166 pp. *R.A.M.*, 1938, p. 86.
—— 1953. 'Wundschutz an verletzten Fichten. II. Teil', *Forstwiss. Cbl.* **72,** 321–5. *For. Abstr.* **15,** No. 2513.
—— 1956. 'Das Problem der Alterung langfristig vegetativ vermehrter Pappelklone', *Forstwiss. Cbl.* **75,** 380–407. *For. Abstr.* **18,** No. 3990.
RÖHRIG, H. 1934. 'Verbreitung und Bekämpfung des Kiefernbaumschwammes in den Staatsforsten des Regierungsbezirks Potsdam', *Forstarchiv*, **10,** 137–46. *R.A.M.*, 1934, p. 666.
ROL, R. 1951. 'Le chancre du Chêne Rouge d'Amérique', *Rev. for. franç.* **11,** 704–7. *R.A.M.*, 1952, p. 409.
ROLAND, G. 1945. 'Une nouvelle maladie du Chêne', *Bull. Soc. for. Belg.* **52,** 29–33. *R.A.M.*, 1945, p. 390.
ROLL-HANSEN, F. 1940. 'Undersøkelser over *Polyporus annosus* Fr., særlig med henblikk på dens forekomst i det sønnefjelske Norge', *Medd. norske Skogsforsøksv.* **24,** 100 pp. *R.A.M.*, 1946, p. 193.
—— 1947. 'Nytt om lokkrusten (*Pucciniastrum padi*)', *Medd. norske Skogselsk.* **34,** 503–10.
ROMBOUTS, J. 1936. 'Uma molestia de "*Eucalyptus*" e de "*Populus*", na Bahia, causada por "*Corticium salmonicolor*" B. et Bred.', *Rodriguésia*, **2,** 301–5. *R.A.M.*, 1937, p. 646.
ROMELE, L. 1923. 'Hänglavar och tillvaxt hos norrländskrgran', *Medd. Statens Skogsförsöksanst.* **19,** 405–38. Rev. in *Forstwiss. Cbl.* **67,** 354.
ROMELL, L.-G. 1937. 'Kvistrensning och övervallning hos okvistad och torrkvistad tall', *Svenska SkogsvFören. Tidskr.* **35,** 99–324. Rev. in *Forestry*, **12,** 50–51.
—— 1940. 'Kvistnings-studier å tall och gran', *Medd. Skogsförsöksanst. Stockh.* **32,** 143–94. *For. Abstr.* **3,** p. 47.
—— 1941. 'Localized injury to plant organs from hydrogenfluoride and other acid gases', *Svensk bot. Tidskr.* **35,** 271–86. *R.A.M.*, 1942, p. 151.
ROSENFELD, W. 1944. 'Erforschung der Bruchkatastrophen in den Ostschlesischen Beskiden in der Zeit von 1875–1942', *Forstwiss. Cbl. u. Tharandt. forstl. Jb.* **1,** 1–31. *For. Abstr.* **7,** p. 526.
ROTH, C. 1935. 'Untersuchungen über den Wurzelbrand der Fichte (*Picea excelsa* Link)', *Phytopath. Z.* **8,** 1–110. *R.A.M.*, 1935, p. 482.
ROTH, E. R. 1948. 'Healing and defects following oak pruning', *J. For.* **46,** 500–4.
—— 1950. 'Discolorations in living Yellow-poplar trees', *J. For.* **48,** 184–5.
—— 1954. 'Spread and intensification of the Little leaf disease of pine', *J. For* **52,** 592–6.
—— 1956. 'Decay following thinning of sprout oak clumps', *J. For.* **54,** 26–30.
—— and HEPTING, G. H. 1943*a*. 'Origin and development of oak stump sprouts as affecting their likelihood to decay', *J. For.* **41,** 27–36.
—— —— 1943*b*. 'Wounds and decay caused by removing large companion sprouts of oaks', *J. For.* **41,** 190–5.

ROTH, E. R., and SLEETH, B. 1939. 'Butt rot in unburned sprout oak stands', *Tech. Bull. U.S. Dep. Agric.* **684,** 42 pp. *R.A.M.*, 1940, p. 245.

ROTH, J. 1920. 'Maifrostschäden an Exoten', *Forstwiss. Cbl.* **46,** 151–61.

ROTH, L. F., and RIKER, A. J. 1943*a*. 'Life history and distribution of *Pythium* and *Rhizoctonia* in relation to damping-off of Red pine seedlings', *J. Agric. Res.* **67,** 129–48.

—— —— 1943*b*. 'Influence of temperature, moisture, and soil reaction on the damping-off of Red pine by *Pythium* and *Rhizoctonia*', *J. Agric. Res.* **67,** 273–93.

—— —— 1943*c*. 'Seasonal development in the nursery of damping-off of Red pine seedlings caused by *Pythium* and *Rhizoctonia*', *J. Agric. Res.* **67,** 417–31.

ROUPPERT, K. 1935. 'Blasenrost der Arve in der Hohen Tatra', *Bull. int. Acad. Cracovie*, sér. B, **1,** 241–52. *R.A.M.*, 1937, p. 75.

ROZANOVA, M. A. 1925. 'O rasprostranenii *Polyporus betulinus* Fr., *Fomes fomentarius* Fr., i *Fomes igniarius* Fr. v berezovyh roščah Zvenigorodskogo uezda Moskovskoj gubernii', *Zaščita Rastenij ot Vreditelej*, **2,** 24–25. *R.A.M.*, 1926, p. 335.

RUBIN, S. S., POPOVA, N. E., DANILEVSKIĬ, A. F., and KORZUNECKAJA, N. K. 1952. 'Vlijanie travjanistoĭ rastiteljnosti i ee kornevyh vydeleniĭ na rost drevesnyh rasteniĭ', *Lesn. Hoz.* **5,** 48–51. *For. Abstr.* **16,** No. 1517.

RUBNER, K. 1937. 'Schüttebefall an Kiefern verschiedener Herkunft', *Tharandt. forstl. Jb.* **88,** 289–93. *R.A.M.*, 1937, p. 847.

—— 1953. *Die pflanzengeographischen Grundlagen des Waldbaues.* Neumann, Berlin, 583 pp.

RUBNER, N. 1942. 'Die Gefährdung der Fichte durch Schnee- und Rauhreifbruch', *Mitt. H.-Göring-Akad. dtsch. Forstwiss.* **2,** 211–22. *For. Abstr.* **5,** 274.

RUDD JONES, D. 1953. 'Studies on a canker disease of Cypresses in East Africa, caused by *Monochaetia unicornis* (Cooke & Ellis) Sacc. I. Observations on the pathology, spread and possible origins of the disease', *Ann. Appl. Biol.* **40,** 323–43.

—— 1954*a*. 'Studies on a canker disease of Cypresses in East Africa, caused by *Monochaetia unicornis* (Cooke & Ellis) Sacc. II. Variation in the morphology and physiology of the pathogen', *Trans. Brit. Mycol. Soc.* **37,** 286–305.

—— 1954*b*. 'Studies on a canker disease of Cypresses in East Africa, caused by *Monochaetia unicornis* (Cooke & Ellis) Sacc. III. Resistance and susceptibility of species of *Cupressus* and allied genera', *Ann. Appl. Biol.* **41,** 325–35.

RUDMAN, P., and DA COSTA, E. W. B. 1958. 'The causes of natural durability in timber. II. The role of toxic extractives in the resistance of silvertop ash (*Eucalyptus sieberiana* F. Muell.) to decay', *Aust. Commonw. Sci. Ind. Res. Organ. Div. For. Prod. Tech. Pap.* **1,** 3–8.

RUDOLF, P. O. 1949. 'Recovery of winter injured pine', *Tech. Notes Lake St. For. Exp. Sta.* **323,** 1 p. *For. Abstr.* **11,** No. 2233.

RUHLAND, W. 1904. 'Ein neaer verderblisher Schädling der Eiche', *Cbl. Bakt.* **11,** 250–3.

RUMBOLD, C. 1915. 'Notes on chestnut fruits infected with the Chestnut blight fungus', *Phytopathology*, **5,** 64–65.

RUPERT, J. A., and LEACH, J. G. 1942. 'Willow blight in West Virginia', *Phytopathology*, **32,** 1095–6.

RUSHDI, M. K., and JEFFERS, W. F. 1952. 'Variation in *Rhizoctonia solani*', *Phytopathology*, **42,** 473–4.

—— —— 1956. 'Effect of some soil factors on efficiency of fungicides in controlling *Rhizoctonia solani*', *Phytopathology*, **46,** 88–90.

RUSHMORE, F. M. 1956. 'Beech root sprouts can be damaged by sodium arsenite treatment of parent tree', *For. Res. Note Ntheast. For. Exp. Sta.*, **57**, 4 pp. *For. Abstr.* **17**, No. 3957.

RŮŽIČKA, J. 1938*a*. 'Sealing-up of the topmost layer of forest soils is in most cases the true cause of trouble in forests', *Lesn. Práce*, **17**, 273–90. *For. Abstr.* **1**, p. 16.

—— 1938*b*. 'Doklad o škodlivosti nesprávného původu smrkového semene', *Lesn. Práce*, **17**, 533–9. *R.A.M.*, 1939, p. 219.

RYPÁČEK, V., TICHÝ, V., and HEJTMÁNEK, M. 1951. 'Teplotní poměry v trouchnivějícím dřevě na přirozeném stanovišti', *Acta Acad. Sci. nat. Morav.* **23**, 435–50. *R.A.M.*, 1957, p. 223.

SAARNIJOKI, S. 1955. 'Anatomisch-morphologische Untersuchungen über die Schlitzblättrigkeit bei einigen Bäumen und Sträuchern', *Commun. Inst. for. Fenn.* **44**, 118 pp. *For. Abstr.* **17**, No. 2460.

SABET, K. A. 1953*a*. 'Studies in the bacterial die-back and canker disease of Poplar. II. The relation between the bacterial slime and the causal organism', *Proc. Soc. Appl. Bact.* **16**, 45–55. *R.A.M.*, 1955, p. 1.

—— 1953*b*. 'Studies in the bacterial die-back and canker disease of Poplar. III. Freezing in relation to the disease', *Ann. Appl. Biol.* **40**, 645–50.

—— and DOWSON, W. J. 1952. 'Studies in the bacterial die-back and canker disease of Poplar. I. The disease and its cause', *Ann. Appl. Biol.* **39**, 609–16.

ŠAFAR, J. 1955. 'Srašćivanje korijenja. Biološko i ekonomsko značenje nekih odnosa drveća u šumskoj pedosferi', *Šum. List*, **79**, 563–78. *For. Abs.* **18**, No. 211.

SAHAROV, M. I. 1952. 'Zavisimosti temperatury stvolov sosny ot tipov lesa', *Dokl. Akad. Nauk. S.S.S.R.* **85**, 1373–6. *For. Abstr.* **15**, No. 2261.

SALISBURY, E. J. 1947. 'The span of life', *Proc. Roy. Soc. Med.* **40**, 638–41.

SALISBURY, P. J. 1953. 'Some aspects of conifer seed microflora', *Bi-m. Progr. Rep. Div. For. Biol. Dep. Agric. Can.* **9**, 3–4. *R.A.M.*, 1955, p. 194.

—— 1955. 'Parasitism of *Phytophthora* spp. isolated from root rots of Port Orford Cedar in British Columbia', *Bi-m. Progr. Rep. Div. For. Biol. Dep. Agric. Can.* **11**, 3–4. *R.A.M.*, 1956, p. 336.

—— 1957. 'Heavy damage to Chinese junipers, *Juniperus chinensis* L., associated with *Pestalozzia funerea*', *Bi-m. Progr. Rep. Div. For. Biol. Dep. Agric. Can.* **13**, 6, p. 4.

—— and LONG, J. R. 1955. 'A new needle blight of Douglas Fir seedlings caused by *Rosellinia herpotrichioides* Hepting and Davidson', *Proc. Canad. Phytopath. Soc.* **23**, 19. *R.A.M.*, 1956, p. 854.

SALMON, E. S. 1905. 'On a fungus disease of *Euonymus japonicus* Linn.', *J. R. Hort. Soc.* **29**, 434–42.

—— 1906. 'On a fungus disease of the cherry laurel (*Prunus laurocerasus* Linn.)', *J. R. Hort. Soc.* **31**, 142–6.

—— and WARE, W. M. 1927. 'Leaf scorch of Azalea', *Gdnrs.' Chron.* **81**, 286–8.

—— and WORMALD, H. 1915. 'Leaf-spot disease of lime', *Gdnrs.' Chron.* **58**, 193–4.

SALTER, R. G. 1958. 'Silver leaf disease', *Quart. J. For.* **52**, 335.

SAMISH, R. M., TAMIR, P., and SPIEGEL, P. 1957. 'The effect of various dressings on the healing of pruning wounds in apple trees', *Proc. Amer. Soc. Hort. Sci.* **70**, 5–9. *Biol. Abstr.*, 1958, No. 20967.

SAMPSON, H. C. 1901. 'Report on the effect of hailstorms on growing crops', *Trans. Scot. Arbor. Soc.* **16**, 467–9.

SANTESSON, R. 1945. '*Cytarria*, a genus of inoperculate Discomycetes', *Svensk bot. Tidskr.* **39,** 319–45.

SANZEN-BAKER, R. G., and NIMMO, M. 1941. 'Glazed frost 1940—damage to forest trees in England and Wales', *Forestry*, **15,** 37–54.

SARASOLA, A. A. 1944. 'Dos septoriosis de las Alamedas argentinas', *Rev. argent. Agron.* **11,** 20–43. *R.A.M.*, 1944, p. 365.

—— 1945. 'Nuevas observaciones sobre la cancrosis de los Alamos', *Rev. argent. Agron.* **12,** 115–19. *R.A.M.*, 1946, p. 16.

—— and MAGI, A. O. 1951. 'Algunos factores ambientales en correlación con la cancrosis de los Alamos (*Mycosphaerella populorum* Thomp.)', *Phyton*, **1,** 42–45. *R.A.M.*, 1951, p. 636.

SARAVÍ CISNEROS, R. 1950. 'El marchitamiento de los pinos provocado por *Diplodia pinea* Kickx en la provincia de Buenos Aires (Argentina)', *Rev. Fac. Agron., La Plata*, **27,** 163–79. *For. Abstr.* **13,** No. 1334.

—— 1953. 'Cancrosis de los Cipreses provocada por "*Coryneum cardinale*" Wagener en la provincia de Buenos Aires (Argentina)', *Rev. Fac. Agron. Eva Peron* (*formerly Rev. Fac. Agron. B. Aires*), **3,** 107–19. *R.A.M.*, 1955, p. 196.

SASAKI, T., and YOKOTA, S. 1955. 'Wood decay of *Abies sachalinensis* trees in Tokyo University forest in Hokkaido. I', *Misc. Inform. Tokyo Univ. For.* **10,** 15–21. *R.A.M.*, 1956, p. 251.

SATÔ, K. 1955. 'On the infection by fungi to "Sugi" seeds sown in soil, and the effects of the seed treatments with organic mercury compounds', *Bull. For. Exp. Sta. Meguro, Tokyo*, **81,** 63–74. *R.A.M.*, 1956, pp. 133–4.

——, ÔTA, N., and SHÔJI, T. 1955*a*. 'Influence of MH-30 treatment upon the control of overgrowth in "Sugi" seedlings. Especially on the effects of the frost damage and grey mold control', *J. Jap. For. Soc.* **37,** 533–7. *R.A.M.*, 1957, p. 438.

—— —— —— 1955*b*. 'Relation between weeds and damping-off of coniferous seedlings caused by *Rhizoctonia solani* in forestry nurseries', *Bull. For. Exp. Sta., Meguro, Tokyo*, **77,** 1–14. *R.A.M.*, 1955, p. 412.

SAVILE, D. B. O. 1959. 'Notes on *Exobasidium*', *Canad. J. Bot.* **37,** 641–56.

SAVORY, J. G. 1954. 'Breakdown of timber by ascomycetes and fungi imperfecti', *Ann. Appl. Biol.* **41,** 336–47.

SAVULESCU, T., and RAYSS, T. 1929. 'Un parasite des Pins peu connu en Europe, *Neopeckia coulteri* (Peck) Sacc.', *Ann. des Épiphyties*, **14,** 322–53. *R.A.M.*, 1930, p. 75.

SCARAMELLA, P. 1931. 'Sullo svernamento delle Melampsorae dei Salici in alta montagna', *Nuovo G. bot. ital.* (N.S.), **38,** 538–40. *R.A.M.*, 1932, p. 412.

SCARAMUZZI, G., and CIFERRI, R. 1957. 'Una nuova virosi: la "maculatura lineare" nel Nocciòlo', *Ann. Sper. agr.* (N.S.), **11,** 6, suppl. pp. lxi–lxxii. *R.A.M.*, 1958, p. 322.

—— and CORTE, A. 1957. 'La "maculatura lineare" del Ciliegio da Fiore', *Atti Ist. bot. Univ. Pavia*, ser. 5, **14,** 351–61. *R.A.M.*, 1958, p. 490.

SCHANTZ-HANSEN, T. 1945. 'The effect of planting methods on root development', *J. For.* **43,** 447–8.

SCHEFFER, T. C. 1957. 'Decay resistance of Western red cedar', *J. For.* **55,** 434–42.

—— and ENGLERTH, G. H. 1952. 'Decay resistance of second-growth Douglas fir', *J. For.* **50,** 439–42.

—— —— and DUNCAN, C. G. 1949. 'Decay resistance of seven native oaks', *J. Agric. Res.* **78,** 129–52.

—— and HEDGCOCK, G. G. 1955. 'Injury to northwestern forest trees by sulfur dioxide from smelters', *Tech. Bull. U.S. Dep. Agric.* **1117,** 49 pp. *For. Abstr.* **17,** No. 1700.

SCHEFFER, T. C., and HOPP, H. 1949. 'Decay resistance of black locust heartwood', *Tech. Bull. U.S. Dep. Agric.* **984,** 37 pp.

SCHEFFER-BOICHORST. 1934. 'Pappelkrebs', *Mitt. dtsch. dendrol. Ges.* **41,** 181. *R.A.M.*, 1935, p. 478.

SCHELLENBERG, H. C. 1923. 'Die Empfänglichkeit der Ribesarten für den Rost der Weymouthkiefer', *Schweiz. Z. Forstw.* **74,** 25–50. *R.A.M.*, 1923, p. 483.

SCHENK, P. J. 1926. 'Platanenziekte en koude', *Floralia*, **47,** 456–7. *R.A.M.*, 1927, p. 63.

SCHIMMLER, G. 1935. 'Rauchschäden an Laub- und Nadelgehölzen', *Gartenflora*, **84,** 271–2. *R.A.M.*, 1936, p. 267.

SCHINDLER, U. 1951. 'Das Buchensterben', *Forstarchiv*, **22,** 109–19. *For. Abstr.* **14,** No. 1389.

SCHIPPER, M. A. A., and HEYBROEK, H. M. 1957. 'Het toetsen van stammen van *Nectria cinnabarina* (Tode) Fr. op levende takken in vitro', *Tijdschr. PlZiekt.* **63,** 192–4. *R.A.M.*, 1958, p. 117.

SCHLUTER. 1956. 'Einige abnorme Triebbildungen der Kiefer und ihre waldbauliche Bedeutung', *Forst- u. Holzw.* **11,** 219–26. *For. Abstr.* **17,** No. 3733.

SCHMID, G. 1927. 'Zur Ökologie der Luftalgen', *Ber. dtsch. bot. Ges.* **45,** 518–33.

SCHMIDLE, A. 1953*a*. 'Die Cytospora-Krankheit der Pappel und die Bedingungen für ihr Auftreten', *Phytopath. Z.* **21,** 83–96. *R.A.M.*, 1954, p. 388.

—— 1953*b*. 'Zur Kenntnis der Biologie und der Pathogenität von *Dothichiza populea* Sacc. et Briard, dem Erreger eines Rindenbrandes der Pappel', *Phytopath. Z.* **21,** 189–209. *R.A.M.*, 1954, p. 568.

—— 1955. 'Über Infektionsversuche mit *Septotis populiperda* Waterman et Cash an *Populus deltoides*', *Angew. Bot.* **29,** 14–25. *R.A.M.*, 1955, p. 684.

SCHMIDT, R. 1938. 'Mischpflanzung als Schädlingsbekämpfungsmittel', *Blumen- u. PflBau ver. Gartenwelt*, **42,** 32 pp. *R.A.M.*, 1938, p. 420.

SCHMITZ, H. 1916. 'Some observations on witches' brooms of cherries', *Plant World*, **19,** 239–42.

—— and JACKSON, L. W. R. 1927. 'Heart rot of aspen with special reference to forest management in Minnesota', *Minn. Agric. Exp. Sta. Tech. Bull.* **50,** 43 pp.

SCHNEIDER, I. R., and CAMPANA, R. J. 1955. 'Fungicide tests in 1954 for the control of Sycamore anthracnose', *Plant Dis. Reptr.* **39,** 64–65.

SCHNEIDERHAN, F. J. 1927. 'The black walnut (*Juglans regia* L.) as a cause of death of apple trees', *Phytopathology*, **17,** 529–40.

SCHOBER, R. VON. 1954. 'Douglasien-Provenienzversuche', *Allg. Forst- u. Jagdztg.* **125,** 160–79.

—— 1958. 'Ergebnisse von Lärchen-Art- und Provenienzversuchen', *Silvae Genet.* **7,** 137–54. *Plant Breed. Abstr.* 1959, No. 1991.

—— and ZYCHA, H. 1948. 'Beobachtungen über Stockfäule in nordwestdeutschen Lärchenbeständen', *Forstwiss. Cbl.* **67,** 119–28.

SCHOENWALD, R. 1931. 'Wahrnehmungen über das Triebschwinden der Kiefer (*Cenangium abietis* (Pers.)) in den Jahren 1926–1928', *Dtsch. Forstztg.* **46,** 484–5. *R.A.M.*, 1931, p. 699.

SCHÖNHAR, S. 1951. 'Eichen- und Roteichenkrebs in Württemberg', *Allg. ForstZ.* **6,** 367–9. *For. Abstr.* **14,** No. 2404.

—— 1952. 'Untersuchungen über den Erreger des Pappelrindentodes', *Allg. ForstZ.* **7,** 509–12.

—— 1953. 'Untersuchungen über die Biologie von *Dothichiza populea* (Erreger des Pappelrindentodes)', *Forstw. Zbl.* **72,** 358–68. *R.A.M.*, 1954, p. 694.

—— 1956*a*. 'Braunfleckengrind und Rindentod der Pappel', *Allg. ForstZ.* **11,** 349–52. *For. Abstr.* **19,** No. 2101.

SCHÖNHAR, S. 1956*b*. 'Erfahrungsberichte aus der Forstlichen Versuchsanstalt, Stuttgart. Nr. 5. Spritzversuche gegen die Forchenschütte mit verschiedenen Fungiziden', *Allg. ForstZ.* **11,** 556–7. *R.A.M.*, 1958, p. 744.

—— 1957*a*. 'Ein Beitrag zur Frage der Anfälligkeit verschiedener Pappelarten und Pappelsorten gegen *Dothichiza populea*', *Mitt. Ver. forstl. Standortskunde ForstpflZücht.* **6,** 59–62. *For. Abstr.* **19,** No. 623.

—— 1957*b*. 'Zur Frage der Bekämpfung von *Dothichiza populea*', *Forst- u. Holzw.* **12,** 421–2. *For. Abstr.* **19,** No. 1962.

—— 1958. 'Bekämpfung der durch *Meria laricis* verursachten Lärchenschütte', *Allg. ForstZ.* **13,** 100. *For. Abstr.* **19,** No. 4398.

SCHÖNIGH, J. 1935. '*Ästung der Birke*', *Forstarchiv,* **11,** 261–7.

SCHREINER, E. J. 1931*a*. 'Two species of *Valsa* causing disease in *Populus*', *Amer. J. Bot.* **18,** 1–29.

—— 1931*b*. 'The rôle of disease in the growing of poplar', *J. For.* **29,** 79–82.

—— 1950. 'Can Black walnut poison pines?', *Shade Trees,* **23,** 2. *For. Abstr.* **11,** No. 3260.

—— 1957. 'The possibilities and limitations of selection and breeding for pest resistance in forest trees', *Proc. Soc. Amer. For.* 50–52.

SCHRÖCK, O. 1956. 'Das physiologische Alter und seine Bedeutung für die Wuchsleistung und Abgrenzung von Pappelklone', *Wiss. dtsch. Akad. Landwirtsch. Berlin,* **16,** 39–50. *For. Abstr.* **18,** No. 3991.

SCHULDT, P. H. 1951. 'Sprays for control of sycamore anthracnose', *Proc. Iowa Acad. Sci.* **58,** 201–7. *R.A.M.*, 1953, p. 109.

—— 1955. 'Comparison of anthracnose fungi on oak, sycamore, and other trees', *Contr. Boyce Thompson Inst.* **18,** 85–107. *R.A.M.*, 1955, p. 759.

SCHULZ, H. 1956. 'Untersuchungen an Frostrissen im Frühjahr 1956', *Forstwiss. Cbl.* **76,** 14–24. *For. Abstr.* **19,** No. 4339.

SCHÜTT, P. 1957. 'Untersuchungen über Individualunterschiede im Schüttebefall bei *Pinus silvestris* L.', *Silvae Genet.* **6,** 109–12. *For. Abstr.* **18,** No. 4254.

SCHWARTZ, H. 1932. 'Winter Sonnenbrand an der Buche', *Wien. allg. Forst- u. Jagdztg.* **50,** 39.

SCOTT, D. R. M., and PRESTON, S. B. 1955. 'Development of compression wood in Eastern white pine through the use of centrifugal force', *For. Sci.* **1,** 178–82.

SCURFIELD, G. 1955. 'Atmospheric pollution considered in relation to horticulture', *J. R. Hort. Soc.* **80,** 93–101.

SEEHOLZER, M. 1934. 'Zur wasserreiserfrage', *Forstwiss. Cbl.* **56,** 437–48.

SEELER, E. V. 1940. 'Two diseases of *Gleditsia* caused by a species of *Thyronectria*', *J. Arnold Arbor.* **21,** 405–27. *R.A.M.*, 1940, p. 734.

SEMPIO, C. 1933. 'Sulla progressiva distruzione delle alberato di Platani in alcune zone dell'Italia Centrale', *Riv. Pat. veg.* **23,** 129–70. *R.A.M.*, 1933, p. 734.

—— 1938. 'La cura di Platani fortemente colpiti della "*Discula platani*" (Peck.) Sacc.', *Riv. Pat. veg.* **18,** 365–75. *R.A.M.*, 1939, p. 213.

SERTZ, H. 1921. 'Über die Wirkung von Fluorwasserstoff und Fluorsilizium auf die lebende Pflanze', *Tharandt. Forstl. Jb.* **72,** 1–13.

SERVAZZI, O. 1934*a*. 'Su alcune *Pestalotia* parassite facoltative di piante ornamentali', *Difesa Piante,* **11** (1), 16–35. *R.A.M.*, 1934, p. 598.

—— 1934*b*. 'Contributi alla patologia dei Pioppi. I. La "fillostictosi" del Pioppo nero e del Pioppo del Canada', *Difesa Piante,* **11** (6), 185–207. *R.A.M.*, 1935, p. 478.

—— 1935*a*. 'Contributi alla patologia dei Pioppi. II. La "tafrinosi" o "bolla fogliare" dei Pioppi', *Difesa Piante,* **12** (2), 48–62. *R.A.M.*, 1935, p. 665.

—— 1935*b*. 'Contributi alla patologia dei Pioppi. III. La "defogliazione primaverile" dei Pioppi', *Difesa Piante,* **12** (5), 162–73. *R.A.M.*, 1936, p. 328.

SERVAZZI, O. 1935*c*. 'Intorno ad alcune *Pestalotia*', *Difesa Piante*, **12** (1), 22–32. *R.A.M.*, 1935, p. 608.

—— 1936. 'Sulla biologia di *Pestalotia macrotricha* Kleb.', *Boll. Lab. sper. e Reg. Oss. Fitopat., Torino* (formerly *Difesa Piante*), **13,** 72–92. *R.A.M.*, 1937, p. 467.

—— 1938*a*. 'Contributi alla patologia dei Pioppi. V. Segnalazione di tumori su Pioppo bianco', *Boll. Lab. sper. e Reg. Oss. Fitopat., Torino*, **15,** 30–33. *R.A.M.*, 1938, p. 780.

—— 1938*b*. 'Contributi alla patologia dei Pioppi. VI. Ricerche sulla cosidetta "defogliazione primaverile dei Pioppi" ', *Boll. Lab. sper. e Reg. Oss. Fitopat., Torino*, **15,** 49–152. *R.A.M.*, 1939, p. 639.

—— 1938*c*. 'Intorno ad un caso di disseccamento osservato su *Araucaria*', *Boll. Lab. sper. e Reg. Oss. Fitopat. Torino*, **15,** 34–37. *R.A.M.*, 1938, p. 748.

—— 1950. 'Brevi notizie sulla "moria" o "seccume" del Nocciòlo Gentile delle Langhe', *Nouvo G. bot. ital.* (N.S.), **57,** 679–82. *R.A.M.*, 1953, p. 287.

SETTERSTROM, C., and ZIMMERMANN, P. W. 1939. 'Factors influencing susceptibility of plants to sulphur dioxide injury', *Contr. Boyce Thompson Inst.* **10,** 155–81. *R.A.M.*, 1939, p. 467.

SHAW, C. G., FISCHER, G. W., ADAMS, D. F., and ADAMS, M. F. 1951. 'Fluorine injury to Ponderosa pine", *Phytopathology*, **41,** 943.

SHAW, L. 1933. 'The resistance of Rosaceous plants to fire-blight', *Phytopathology*, **23,** 32–33.

SHEAR, C. L. 1916. 'False blossom of cultivated cranberry', *U.S. Dept. Agric. Bull.* **444,** 7 pp.

——, STEVENS, N. E., and BAIN, A. F. 1931. 'Fungous diseases of the cultivated cranberry', *U.S. Dept. Agric. Tech. Bull.* **258,** 57 pp.

SHEARER, R. C., and MIELKE, J. L. 1958. 'An annotated list of the diseases of Western larch', *Intermt. For. Range Exp. Sta. Res. Note*, **53,** 6 pp.

SHERMAN, R. W. 1957. 'Co-operation of world tourists sought in plant quarantine enforcement', *FAO Plant Pro. Bull.* **5,** 89–90.

SHIRAI, M. 1896. 'Description of some new Japanese species of *Exobasidium*', *Bot. Mag., Tokyo*, **10,** 51–54.

SHIRLEY, H. L., and MEULI, L. J. 1939*a*. 'Influence of moisture supply on drought resistance of conifers', *J. Agric. Res.* **59,** 1–21.

—— —— 1939*b*. 'The influence of soil nutrients on drought resistance of two-year-old Red pine', *Amer. J. Bot.* **26,** 355–60.

SIEGLER, E. A., and JENKINS, A. E. 1923. '*Sclerotinia carunculoides*, the cause of a serious disease of the Mulberry (*Morus alba*)', *J. Agric. Res.* **23,** 833–6.

SIEVERS, F. J. 1924. 'Crop injury resulting from magnesium oxide dust', *Phytopathology*, **14,** 108–13.

SIGGERS, P. V. 1944. 'The brown spot needle blight of pine seedlings', *Tech. Bull. U.S. Dep. Agric.* **870,** 36 pp. *R.A.M.*, 1944, p. 505.

—— 1951. 'Spray control of the Fusiform rust in forest-tree nurseries', *J. For.* **49,** 350–2.

—— 1955. 'Control of the Fusiform rust of Southern pine', *J. For.* **53,** 442–6.

—— and LINDGREN, R. M. 1947. 'An old disease—a new problem', *Sth. Lumberm.* **175,** 172–5. *R.A.M.*, 1948, p. 398.

SILVÉN, F. 1944. 'Stämpelröta hos gran i Norrland', *Norrlands SkogsvFörb. Tidskr.* **2,** 135–58. *For. Abstr.* **9,** No. 2543.

SILVERBORG, S. B., and BRANDT, R. W. 1957. 'Association of *Cytophoma pruinosa* with dying ash', *For. Sci.* **3,** 75–78.

SIMMONDS, J. H. 1933. 'The work of the pathological branch', *Ann. Rep. Qd. Dept. of Agric. & Stock for the year 1932–3*, pp. 61–63. *R.A.M.*, 1934, p. 214.

SIMMONDS, J. H. 1940. 'Report of the Plant Pathological Section', *Dep. Agric. Qd., 1939–40*, pp. 10–11. *R.A.M.*, 1941, p. 151.

SIPKENS, J. 1952. 'De Droge Zomer 1947 en de Diktegroei van enige Houtsoorten in Drenthe', *Ned. Boschb. Tijdschr.* **24,** 85–89.

SIRÉN, G., and BERGMAN, F. 1951. 'Svamparna och våra skogsträd', *Skogbruket, Helsingfors* (Helsinki), **21,** 39–43. *R.A.M.*, 1953, p. 443.

SJÖSTRÖM, H. 1937. 'Iakttagelser och undersökningar över snöskyttets (*Phacidium infestans*) uppträdande pa Tallen i höjdlägen i Norrland och Dalarna', *Svenska SkogsvFören. Tidskr.* **35,** 205–49. *R.A.M.*, 1938, p. 86.

—— 1946. 'Om tallens föryngring och snöskyttet', *Norrlands SkogsvFörb. Tidskr.* **3,** 421–50. *For. Abstr.* **9,** No. 1886.

SKILLING, D. L. 1957. 'Is the epicormic branching associated with hardwood pruning wounds influenced by tree crown class?', *Tech. Note Lake States For. Exp. Sta.* **510,** 2 pp. *For. Abstr.* **19,** No. 1751.

—— 1958. 'Wound healing and defects following northern hardwood pruning', *J. For.* **56,** 19–22.

SKOLKO, A. J. 1954. 'Antibiotics in plant disease control', *Rep. 5th Commonwealth Mycol. Conf.*, pp. 78–81.

SKORIĆ, V. 1938. 'Jasenov rak i njegov uzročnik', *Ann. Exp. For., Zagreb,* **6,** 66–97. *R.A.M.*, 1939, p. 560.

SLABAUGH, P. E. 1957. 'Effects of live crown removal on the growth of Red pine', *J. For.* **55,** 904–6.

SLAGG, C. M., and WRIGHT, E. 1943. 'Diplodia blight in coniferous seedlings', *Phytopathology,* **33,** 390–3.

SLANKIS, V. 1949. 'Wirkung von β-Indolylessigsäure auf die dichotomische Verzweigung isolierter Wurzeln von *Pinus silvestris*', *Svensk bot. Tidskr.* **43,** 603–7. *R.A.M.*, 1950, p. 44.

—— 1955. 'Causality of morphogenesis of mycorrhizal pine roots', *Proc. Canad. Phytopath. Soc.* **23,** 20–21. *R.A.M.*, 1956, p. 914.

SLATE, G. L. 1930. 'Filberts', *New York* (*Geneva*) *Agric. Exp. Sta. Bull.* **588,** 32 pp. *R.A.M.*, 1931, p. 347.

SLATER, C. H. W., and RUXTON, J. P. 1955. 'The effect of soil compaction on incidence of frost', *E. Malling Res. Sta. Rep.*, 1954, pp. 88–91. *Soils & Fertil.*, 1955, No. 2004.

SLEETH, B., and LORENZ, R. C. 1945. 'Strumella canker of oak', *Phytopathology,* **35,** 671–4.

SLIPP, A. W. 1953. 'Survival probability and its application to damage survey in Western white pine infected with Blister rust', *Res. Notes For. Exp. Sta. Univ. Idaho,* **7,** 13 pp. *R.A.M.*, 1957, p. 74.

SLOCUM, G. K., and MAKI, T. E. 1956. 'Exposure of Loblolly pine planting stock', *J. For.* **54,** 313–15.

SMERLIS, E. 1957. 'Hylobius injuries as infection courts of root and butt rots in immature Balsam fir stands', *Bi-m. Progr. Rep. Div. For. Biol. Dep. Agric. Can.* **13,** 2, p. 1.

SMITH, C. O. 1931. 'Pathogenicity of *Bacillus amylovorus* on species of *Juglans*', *Phytopathology,* **21,** 219–23.

—— 1934. 'Inoculations showing the wide host range of *Botryosphaeria ribis*', *J. Agric. Res.* **49,** 467–76. *R.A.M.*, 1935, p. 196.

—— 1937. 'Crown gall on Incense cedar, *Libocedrus decurrens*', *Phytopathology,* **27,** 844–9.

—— 1939. 'Susceptibility of species of Cupressaceae to crown gall as determined by artificial inoculation', *J. Agric. Res.* **59,** 919–25.

SMITH, C. O., and BARRETT, J. T. 1931. 'Crown rot of *Juglans* in California', *J. Agric. Res.* **43,** 885–904.

—— and COCHRAN, L. C. 1939. 'Rust on the California native Pruni', *Phytopathology*, **29,** 645–6.

—— and SMITH, D. J. 1942. 'Host range and growth-temperature relations of *Coryneum beijerinckii*', *Phytopathology*, **32,** 221–5.

SMITH, F. F., and WEISS, F. 1942. 'Relationship of insects to the spread of Azalea flower spot', *Tech. Bull. U.S. Dep. Agric.* **798,** 43 pp. *R.A.M.*, 1943, p. 169.

SMITH, K. M. 1940. 'Graft-blight of Lilacs', *Gdnrs.' Chron.* **107,** 144.

—— 1951. *Recent advances in the study of plant viruses.* J. & A. Churchill, 300 pp.

—— 1952. 'Some garden plants susceptible to infection with the Cucumber mosaic virus', *J. R. Hort. Soc.* **77,** 19–21.

—— 1957. *Textbook of plant virus diseases.* J. & A. Churchill, 2nd ed., 652 pp.

SMITH, M. E. 1943. 'Micronutrients essential for the growth of *Pinus radiata*', *Aust. For.* **7,** 22–27. *For. Abstr.* **5,** p. 274.

SMITH, W. W., and TINGLEY, M. A. 1940. 'Frost rings in fall-fertilized McIntosh apple trees', *Proc. Amer. Soc. Hort. Sci.* **37,** 110–12.

SMITHSON, E. 1952. 'Development of winged cork in *Acer campestre* L.', *Proc. Leeds Phil. Lit. Soc.* (*Sci. Sect.*), **6,** 97–103. *For. Abstr.* **17,** No. 1225.

—— 1954. 'Development of winged cork in *Ulmus* × *hollandica* Mill.', *Proc. Leeds Phil. Lit. Soc.* (*Sci. Sect.*), **6,** 211–20. *For. Abstr.* **17,** No. 1226.

SMOLÁK, J. 1948. 'Virosa smrku', *Lesn. Práce*, **27,** 113–17. *R.A.M.*, 1948, p. 503.

—— and NOVÁK, J. B. 1950. 'Virové choroby Šeříku', *Ochr. Rost.* **23,** 285–304. *R.A.M.*, 1951, p. 520.

SMUCKER, S. J. 1935. 'Air currents as a possible carrier of *Ceratostomella ulmi*', *Phytopathology*, **25,** 442–3.

—— 1937. 'Relation of injuries to infection of American elm by *Ceratostomella ulmi*', *Phytopathology*, **27,** 140.

—— 1941. 'Susceptibility of *Planera* and several elm species to *Ceratostomella ulmi*', *Phytopathology*, **31,** 21.

SNELL, W. H. 1931. 'Forest damage and the White pine blister rust', *J. For.* **29,** 68–78.

—— and DICK, E. A. 1957. *A glossary of mycology.* Harvard University Press, 171 pp.

SNETZINGER, R., and HIMELICK, E. B. 1957. 'Observations on witches'-broom of Hackberry', *Plant Dis. Reptr.* **41,** 541–4.

SNYDER, H. D. 1957. 'Single spray control of *Gnomonia veneta* on *Platanus occidentalis* and *P. acerifolia*', *Phytopathology*, **47,** 246.

SNYDER, W. C., TOOLE, E. R., and HEPTING, G. H. 1949. 'Fusaria associated with Mimosa wilt, Sumac wilt, and Pine pitch canker', *J. Agric. Res.* **78,** 365–82.

SØEGAARD, B. 1956. 'Leaf blight resistance in *Thuja*. Experiments on resistance to attack by *Didymascella thujina* (Dur.) Maire (*Keithia thujina*) on *Thuja plicata* Lamb', *K. VetHøjsk. Aarsskr.*, pp. 30–48. *R.A.M.*, 1957, p. 438.

SOLBERG, R. A., and ADAMS, D. F. 1956. 'Histological responses of some plant leaves to hydrogen fluoride and sulfur dioxide', *Amer. J. Bot.* **43,** 755–60. *Hort. Abstr.*, 1957, No. 2014.

SOMERVILLE, W. 1909. '*Rhizoctonia violacea* causing a new disease of trees', *Quart. J. For.* **3,** 134–5.

—— 1914. 'The mistletoe in England', *Quart. J. For.* **8,** 20–25.

SOMMER, N. F. 1955. 'Sunburn predisposes walnut trees to branch wilt', *Phytopathology*, **45,** 607–13.

SORACI, F. A. 1957. 'Redefinition of the principles of plant quarantine and their relation to the current problems', *Phytopathology*, **47,** 381–2.

SOUTHAM, C. M., and EHRLICH, J. 1943. 'Decay resistance and physical characteristics of wood', *J. For.* **41,** 666–73.

SPAULDING, P. 1912. 'Notes upon tree diseases in the Eastern United States', *Mycologia*, **4,** 148–51.

—— 1929. 'White pine blister rust; a comparison of European with North American conditions', *U.S. Dep. Agric. Tech. Bull.* **87,** 59 pp. *R.A.M.*, 1929, p. 538.

—— 1948. 'The role of *Nectria* in the beech bark disease', *J. For.* **46,** 449–53.

—— 1952. 'A stem rust of *Pinus roxburghii* in India potentially dangerous to the United States', *Plant Dis. Reptr.* **36,** 159–61.

—— 1956. 'Diseases of North American forest trees planted abroad', *Agric. Handb. U.S. Dep. Agric.* **100,** 144 pp.

—— 1958. 'Diseases of foreign forest trees growing in the United States', *Agric. Handb. U.S. Dep. Agric.* **139,** 118 pp.

—— and BRATTON, A. W. 1946. 'Decay following glaze storm damage in woodlands of Central New York', *J. For.* **44,** 515–19.

—— GRANT, T. J., and AYERS, T. T. 1936. 'Investigations of Nectria diseases in hardwoods of New England', *J. For.* **34,** 168–79.

—— and HANSBROUGH, J. R. 1932. '*Cronartium comptoniae*, the Sweetfern blister rust of Pitch pines', *U.S. Dep. Agric. Circ.* **217,** 21 pp. *R.A.M.*, 1932, p. 615.

—— —— 1943. 'The needle blight of Eastern white pine', *U.S.D.A. Bureau of Plant Industry* (*Washington*), 2 pp. *For. Abstr.* **6,** p. 188.

—— —— 1944. 'Decay in Balsam fir in New England and New York', *U.S. Dep. Agric. Tech. Bull.* **872,** 30 pp. *R.A.M.*, 1945, p. 298.

——, MACALONEY, H. J., and CLINE, A. C. 1935. '*Stereum sanguinolentum* a dangerous fungus in pruning wounds on Northern white pine', *U.S. Dep. Agric. Tech. Note*, **19,** 2 pp.

—— and SIGGERS, P. V. 1927. 'The European larch canker in America', *Science* (N.S.), **66,** 480–1. *R.A.M.*, 1928, p. 285.

SPEGAZZINI, C. 1925. 'La "piptostelechia" del Álamo blanco', *Physis* (*Rev. Soc. argent. Cien. Nat.*), **8,** 1–11. *R.A.M.*, 1925, p. 575.

SPENCE, M. T. 1956. 'Damage to crops by lightning', *Agric.* **63,** 387–9. *Hort. Abstr.*, 1957, No. 1036.

SPIERENBURG, D. 1937. 'Bestrijding van het "vuur" in Eschdoorns', *Tijdschr. PlZiekt.* **43,** 150–1. *R.A.M.*, 1937, p. 783.

SPILKER, O. W., and YOUNG, H. C. 1955. 'Longevity of *Endoconidiophora fagacearum* in lumber', *Plant Dis. Reptr.* **39,** 429–32.

SPLETTSTÖSSER. 1957. 'Ästen von Eichen mit Wuchsstoffen', *Forst- u. Holzw.* **12,** 127–30. *For. Abstr.* **18,** No. 4075.

SPRAU, F. 1951. 'Starkes Auftreten der Platanen-Blattnervenkrankheit im Bodenseegebiet', *Pflanzenschutz*, **3,** 109–11. *R.A.M.*, 1952, p. 408.

SREENIVASAYA, M. 1930. 'Masking of spike-disease symptoms in *Santalum album* (Linn.)', *Nature, London*, **126,** 957.

—— 1948. 'The spike disease of Sandal', *Curr. Sci.* **17,** 141–5. *R.A.M.*, 1948, p. 585.

STAEGER, H. 1928. 'L'élagage des résineux, traitement complémentaire', *J. For. Suisse*, **79,** 185–92.

STAHL, E. 1912. *Die Blitzgefährdung der verschiedenen Baumarten. I. Häufigkeit starker Blitzbeschädigung der verschiedenen Baumarten.* Gustav Fischer, Jena, pp. 4–10.

STALLINGS, J. H. 1954. 'Soil-produced antibiotics—plant disease and insect control', *Bact. Rev.* **18,** 131–46. *R.A.M.*, 1955, p. 49.

STARK, R. W., and COOK, J. A. 1957. 'The effects of defoliation by the Lodgepole needle miner', *For. Sci.* **3,** 376–96.

STATHIS, P. D., and PLAKIDAS, A. G. 1958. 'Anthracnose of azaleas', *Phytopathology*, **48,** 256–60.

—— —— 1959. 'Entomosporium leaf spot of *Photinia glabra* and *Photinia serrulata*', *Phytopathology*, **49,** 361–5.

STECKI, K., and RADA, A. 1951. 'Szablastość strzał u modrzewi europejskich w nadleśnictwie Kwidzyń', *Acta Soc. Bot. Polon.* **21,** 165–79. *For. Abstr.* **14,** No. 178.

STEINBRENNER, E. C., and GESSEL, S. P. 1955. 'The effect of tractor logging on physical properties of some forest soils in southwestern Washington', *Proc. Soil Sci. Soc. Amer.* **19,** 372–6. *For. Abstr.* **17,** No. 676.

STEWART, D. M. 1957. 'Factors affecting local control of White pine blister rust in Minnesota', *J. For.* **55,** 832–7.

STEWART, V. B. 1916. 'The leaf blotch of horse chestnut', *Phytopathology*, **6,** 5–20.

STICKEL, P. W. 1933. 'Drought injury in Hemlock-hardwood stands in Connecticut', *J. For.* **31,** 573–7.

STIMPFLING, J. H., and SHUBERT, M. L. 1950. 'Iron fluctuations in *Acer saccharinum*', Abstr. in *J. Colo.-Wyo. Acad. Sci.* **4,** 55–56. *R.A.M.*, 1951, p. 592.

STOATE, T. N. 1951. 'Nutrition of the pine', *Bull. For. Timb. Bur. Aust.* **30,** 61 pp. *R.A.M.*, 1953, p. 45.

—— and BEDNALL, B. H. 1957. 'Disorders in conifer forests', *Brit. Commonw. For. Conf. Bull.* **11,** 9 pp.

STODDARD, E. M., and DIMOND, A. E. 1949. 'The chemotherapy of plant diseases', *Bot. Rev.* **15,** 345–76. *For. Abstr.* **11,** No. 1457.

STOECKELER, J. H. 1946. 'Alkali tolerance of drought-hardy trees and shrubs in the seed and seedling stage', *Proc. Minn. Acad. Sci. Minneapolis*, **14,** 79–83. *For. Abstr.* **14,** No. 3153.

—— 1948. 'Recovery of winter-injured conifers', *Amer. Nurserym.* **88,** 9, 54. *For. Abstr.* **10,** No. 2214.

—— 1951. 'Proper watering in the nursery produces drought-hardy Jack pine', *Tech. Note Lake St. For. Exp. Sta.* **348,** 1 p. *For. Abstr.* **13,** No. 255.

—— and RUDOLF, P. O. 1956. 'Winter coloration and growth of Jack pine in the nursery as affected by seed source', *Z. Forstgenet.* **5,** 161–5. *For. Abstr.* **18,** No. 1311.

STOKLASA, J. 1923. *Die Beschädigungen der Vegetation durch Rauchgase und Fabrikexhalationen*, Urban & Schwarzenberg, Berlin and Vienna, 487 pp.

STOLTENBERG, E. 1934. 'Snemugg (sneskytte)', *Tidsskr. for. Skogbruk.* **7–8,** 14 pp. *R.A.M.*, 1934, p. 814.

STONE, E. C. 1957. 'Dew as an ecological factor. I. A review of the literature. II. The effect of artificial dew on the survival of *Pinus ponderosa* and associated species', *Ecology*, **38,** 407–22.

STONE, E. L., Jr. 1953*a*. 'Magnesium deficiency in some northeastern pines', *Proc. Soil Sci. Soc. Amer.* **17,** 297–300. *For. Abstr.* **16,** No. 3096.

—— 1953*b*. 'The origin of epicormic branches in fir', *J. For.* **51,** 366.

STONE, E. L., and BAIRD, G. 1956. 'Boron level and boron toxicity in Red and White pine', *J. For.* **54,** 11–12.

——, MORROW, R. R., and WELCH, D. S. 1954. 'A malady of Red pine on poorly drained sites', *J. For.* **52,** 104–14.

STONE, G. E. 1914. 'Electrical injuries to trees', *Mass. Agric. Exp. Sta. Bull.* **156,** 19 pp.

STONE, R. E. 1916. 'Studies in the life histories of some species of *Septoria* occurring on *Ribes*', *Phytopathology*, **6,** 419–27.

STOREY, I. F. 1955. 'New or uncommon plant diseases and pests in England and Wales', *Plant Path.* **4,** 71–72. *R.A.M.*, 1956, p. 349.

STOWELL, E. A. 1956. 'A study of *Entomosporium* on *Crataegus*', *Diss. Abstr.* **16,** 222. *R.A.M.*, 1956, p. 796.

ŠTRAUCH-VALEVA, S. A. 1954. 'O biologii nekotorykh gribov, vyzyvajuščikh zabolevanija želudeĭ', *Trud. Inst. Les.* **16,** 269–80.

STREET, F. 1950. 'Some observations and notes on bud blast on Rhododendrons', *Rhododendron Yearb.* **5,** 72–77. *R.A.M.*, 1951, p. 162.

STREHLKE, E. G. 1952. 'Zur Ästung der Fichte: ältere Erfahrungen und neue Untersuchungen aus dem Forstamt Westerhof', *Forstarchiv*, **23,** 93–100. *For. Abstr.* **15,** No. 270.

STROBEL, J., and VERMILLION, M. T. 1956. 'A preliminary study of an apparent disease of Hemlock (*Tsuga canadensis* (L.) Carr.) in the nursery', *Lloydia*, **19,** 214–44. *R.A.M.*, 1957, p. 562.

STRONG, F. C. 1944. 'A study of calcium chloride injury to roadside trees', *Quart. Bull. Mich. Agric. Exp. Sta.* **27,** 209–24. *R.A.M.*, 1945, p. 294.

—— 1952. 'Damping-off in the forest tree nursery and its control', *Quart. Bull. Mich. Agric. Exp. Sta.* **34,** 285–96. *R.A.M.*, 1953, p. 111.

—— 1953. 'Spruce branch canker', *Proc. 29th Nat. Shade Tree Conf.*, pp. 30–35. *Biol. Abstr.*, 1955, No. 19893.

—— and KLOMPARENS, W. 1955. 'The control of Red cedar-apple and hawthorn rusts with acti-dione', *Plant Dis. Reptr.* **39,** 569.

STUDHALTER, R. A., and GLOCK, W. S. 1948. 'Artificial frost as a tool in dating growth layers and seasonal growth in the branches of trees', *Abstr. J. Colorado-Wyoming Acad. Sci.* **3,** 56. *For. Abstr.* **11,** No. 3242.

—— and RUGGLES, A. G. 1915. 'Insects as carriers of the Chestnut blight fungus', *Dept. For. Penn. Bull.* **12,** 33 pp.

STUNTZ, D. E., and SELISKAR, C. E. 1943. 'A stem canker of Dogwood and Madrona', *Mycologia*, **35,** 207–21. *R.A.M.*, 1943, p. 458.

SUDDS, R. H., and MARSH, R. S. 1943. 'Winter injury to trunks of young bearing apple trees in West Virginia following a fall application of nitrate of soda', *Proc. Amer. Hort. Sci.* **42,** 293–7.

SULIOTIS, M. 1936. 'Contributi alla patologia dei Pioppi. IV. Un disseccamento di piantine di Pioppo canadese e P. caroliniano intorno a *Physalospora populina* Maubl. ed una *Phoma* sp.', *Boll. Lab. sper. e Reg. Oss. Fitopat., Torino*, **8,** 62–72. *R.A.M.*, 1937, p. 422.

SŬTIĆ, D. 1956. 'Bakteriska pegavost lišća leske', *Zašt. Bilja*, **37,** 47–53. *Hort. Abstr.* 1958, No. 255.

SUTTON, O. G. 1947. 'The theoretical distribution of airborne pollution from factory chimneys', *Q. J. Roy. Met. Soc. (London)*, **73,** 426–36. *Biol. Abstr.*, 1949, No.1 84.

SWINGLE, R. U. 1942. 'Phloem necrosis: a virus disease of the American elm', *Circ. U.S. Dep. Agric.* **640,** 8 pp. *R.A.M.*, 1943, p. 45.

—— and BRETZ, T. W. 1950. 'Zonate canker, a virus disease of American elm', *Phytopathology*, **40,** 1018–22.

——, TILFORD, P. E., and IRISH, C. F. 1943. 'A graft transmissible mosaic of American elm', *Phytopathology*, **33,** 1196–1200.

——, WHITTEN, R. R., and YOUNG, H. C. 1949. 'The identification and control of elm phloem necrosis and Dutch elm disease', *Ohio Agric. Exp. Sta. Spec. Circ.* **80,** 11 pp.

SYLVÉN, N. 1944. 'Om ekens lövspricknings- och lövfällningsdata. Ett bidrag till kännedoman om ekens mångformighet', *Svensk Papp. Tidn.* **47,** 167–74. *For. Abstr.* **6,** 155.

Sylvén N. 1947. 'Årsberättelse över Föreningens för växtförädling av skogsträd verksamhet under år 1946', *Medd. Fören. Växtförädl. Skogsträd*, **44**, 38 pp. *For. Abstr.* **9**, No. 1455.

Szántó, I. 1948. 'A bükkfa rákja mint éghajlati betegség', *Erdész. kisérl.* **48**, 10–31. *R.A.M.*, 1950, p. 234.

Szpor, S. 1945. 'Elektrische Widerstände der Bäume und Blitzgefährdung', *Schweiz. Z. Forstw.* **96**, 209–19. *For. Abstr.* **7**, No. 2021.

Taber, S. 1920. 'Frost heaving', *J. Geol.* **37**, 428–61. Abstr. in *Nature*, 1929, **124**, 388.

—— 1930. 'The mechanics of frost heaving', *J. Geol.* **38**, 303–17. Rev. in *J. For.* 1931, **29**, 403–5.

Tabor, R. J., and Barratt, K. 1917. 'On a disease of beech caused by *Bulgaria polymorpha*', *Ann. Appl. Biol.* **4**, 20–27.

Tamm, C. O. 1951. 'Removal of plant nutrients from tree crowns by rain', *Physiol. Plant., Copenhagen*, **4**, 184–8. *For. Abstr.* **13**, No. 61.

—— 1953. 'Mera om granens gulspetssjuka', *Svenska Skogsvfören. Tidskr.* **51**, 390–6. *For. Abstr.* **17**, No. 1699.

Tanner, H. 1953. 'Hagelschäden an Waldbäumen und ihre Folgen', *Schweiz. Z. Forstw.* **104**, 232–7. *For. Abstr.* **15**, No. 358.

Taris, B. 1957. *Contribution à l'étude des maladies cryptogamiques des rameaux et des jeunes plantes de peuplier*, Imp. Alençonnaise Maison Poulet-Malassis, France, 232 pp.

Taylor, J., and Clayton, C. N. 1959. 'Comparative studies on Gloeosporium stem and leaf fleck and Dothichiza leaf spot of highbush blueberry', *Phytopathology*, **49**, 65–67.

Taylor, J. A. 1958. 'London's plane. Some notes and observations', *Gdnrs.' Chron.* **144**, 253.

Tengwall, T. A. 1924. 'Über einen bisher unbekannten Fall von Symbiose von Algen und Pilzen', *Meded. Phytopath. Lab. "Willie Commelin Scholten", Baarn (Holland)*, **6**, 52–57. *R.A.M.*, 1924, p. 731.

Terashita, T., and Itô, K. 1956. 'Some notes on *Cylindrocladium scoparium* in Japan', *Bull. For. Exp. Sta., Meguro, Tokyo*, **87**, 33–47. *R.A.M.*, 1957, p. 470.

Terrier, C. A. 1943. 'Über zwei in der Schweiz bisher wenig bekannte Schüttepilze der Kiefern: *Hypodermella sulcigena* (Rostr.) v. Tub. und *Hypodermella conjuncta* Darker', *Phytopath. Z.* **14**, 442–9. *For. Abstr.* **7**, No. 408.

—— 1947. 'Un nouveau champignon parasite des aiguilles du Sapin blanc: *Dimeriella terrieri* Petrak nov. spec. in litt.', *Ber. schweiz. bot. Ges.* **57**, 164–73. *R.A.M.*, 1949, p. 97.

—— 1950. 'La "Tacheture" des fleurs d'azalées provoquee par le champignon *Ovulinia azaleae* Weiss: La Presence de cette maladie en Suisse', *Stat. Fed. Vitic. Arbor. Chim. agric. Lausanne, Sect. Microbiol. Publ.* 390, 6 pp.

—— 1953. 'Note sur *Lophodermium macrosporum* (Hartig) Rehm.', *Phytopath. Z.* **20**, 397–404. *R.A.M.*, 1954, p. 456.

Thacker, D. G., and Good, H. M. 1952. 'The composition of air in trunks of sugar maple in relation to decay', *Canad. J. Bot.* **30**, 475–85.

Thoday, D. 1951. 'The haustorial system of *Viscum album*', *J. Exp. Bot.* **2**, 1–19.

—— 1958. 'Modes of union and interaction between parasite and host in the Loranthaceae. III. Further observations on *Viscum* and *Korthalsella*', *Proc. Roy. Soc.* B, **148**, 188–206. *Biol. Abstr.*, 1958, No. 38706.

Thomas, G. P. 1950. 'Two new outbreaks of *Phomopsis lokoyae* in British Columbia', *Canad. J. Res.* sect. C, **28**, 477–81.

THOMAS, G. P., and PODMORE, D. G. 1953. 'Studies in Forest Pathology. XI. Decay in Black cottonwood in the Middle Fraser Region, British Columbia', *Canad. J. Bot.* **31**, 675–92.

THOMAS, H. E. 1933. 'The Quince-rust disease caused by *Gymnosporangium germinale*', *Phytopathology*, **23**, 546–53.

—— 1934. 'Studies on *Armillaria mellea* (Vahl) Quél., infection, parasitism, and host resistance', *J. Agric. Res.* **48**, 187–218.

—— and HANSEN, H. N. 1946. 'Camellia flower blight', *Phytopathology*, **36**, 380–1.

—— and MILLS, W. D. 1929. 'Three rust diseases of the apple', *Cornell Agric. Exp. Sta. Mem.* **123**, 21 pp. *R.A.M.*, 1929, p. 582.

—— and RAWLINS, T. E. 1939. 'Some mosaic diseases of *Prunus* species', *Hilgardia*, **12**, 623–44. *R.A.M.*, 1940, p. 416.

—— and THOMAS, H. E. 1931. 'Plants affected by fire blight', *Phytopathology*, **21**, 425–35.

THOMAS, J. E., and LINDBERG, G. D. 1954. 'A needle disease of pines caused by *Dothistroma pini*', *Phytopathology*, **44**, 333.

THOMAS, M. D. 1951. 'Gas damage to plants', *Ann. Rev. Plant Physiol.* **2**, 293–322.

—— 1955. 'The invisible injury theory of plant damage' *J. Air. Pollut. Control Ass.* **5**, 4 pp.

THOMAS, R. W. 1953. 'Forest disease survey: Parasite on Jack pine mistletoe', *Rep. For. Insect Dis. Surv. Can.*, 1952, p. 94. *For. Abstr.* **14**, No. 3472.

THOMPSON, A. R. 1939. 'Grade change protection for valuable trees', *J. For.* **37**, 837–45.

—— 1943. 'Lightning struck tree survey', *Proc. 19th Nat. Shade Tree Conf.*, pp. 34–41. *For. Abstr.*, **7**, p. 95.

THOMPSON, G. E. 1939. 'A leaf blight of *Populus tacamahaca* Mill. caused by an undescribed species of *Linospora*', *Canad. J. Res.*, sect. C, **17**, 232–8.

—— 1941. 'Leaf-spot diseases of poplars caused by *Septoria musiva* and *S. populicola*', *Phytopathology*, **31**, 241–54.

—— 1954. 'The perfect stages of *Marssonina rhabdospora* and *Septogloeum rhopaloideum*', *Mycologia*, **46**, 652–9. *R.A.M.*, 1955, p. 261.

THOMPSON, J. MCLEAN, 1946. 'Some features of horticultural interest in the Forsythias', *J. R. Hort. Soc.* **71**, 166–72.

THOMSEN, M., BUCHWALD, N. F., and HAUBERG, P. A. 1949. 'Angreb af *Cryptococcus fagi*, *Nectria galligena* og andre parasiter paa Bøg i Danmark 1939–43', *Forstl. Forsøgsv. Danm.* **18**, 97–326. *R.A.M.*, 1950, p. 483.

THORN, W. A., and ZENTMYER, G. A. 1954. 'Hosts of *Phytophthora cinnamomi* Rands', *Plant Dis. Reptr.* **38**, 47–52.

THORNBERRY, H. H., and ANDERSON, H. W. 1931. 'A bacterial disease of Barberry caused by *Phytomonas berberidis*, n.sp.', *J. Agric. Res.* **43**, 29–36.

THRING, M. W. 1957. *Air Pollution.* Butterworth Scientific Publications, London, 248 pp.

TIFFANY, L. H., GILMAN, J. C., and MURPHY, D. R. 1955. 'Fungi from birds associated with wilted oaks in Iowa', *Iowa State Coll. J. Sci.* **30**, 21–32.

TILFORD, P. E. 1940. 'Tree wound dressings', *Proc. 16th Nat. Shade Tree Conf.*, pp. 41–51. *For. Abstr.* **2**, 325.

TIMONIN, M. I., and SELF, R. L. 1955. '*Cylindrocladium scoparium* Morgan on Azaleas and other ornamentals', *Plant Dis. Reptr.* **39**, 860–3.

TINT, H. 1945*a*. 'Studies in the Fusarium damping-off of conifers. I. The comparative virulence of certain Fusaria. II. Relation of age of host, pH, and some nutritional factors to the pathogenicity of Fusarium', *Phytopathology*, **35**, 421–57.

TINT, H. 1945*b*. 'Studies in the Fusarium damping-off of conifers. III. Relation of temperature and sunlight to the pathogenicity of Fusarium', *Phytopathology*, **35**, 498–510.

TOCCHETTO, A. 1954. 'Causas da podridão da Castanha japonesa e contrôle', *Rev. Agron. Pôrto Alegre*, **17**, 113–21. *R.A.M.*, 1955, p. 495.

TOGASHI, K., and UCHIMURA, K. 1933. 'A contribution to the knowledge of parasitism of *Valsa paulowniae*, in relation to temperature', *Jap. J. Bot.* **6**, 477–87. *R.A.M.*, 1933, p. 603.

TOKUSHIGE, Y. 1951. 'Witches' broom of *Paulownia tomentosa* L.', *J. Fac. Agric. Kyushu Univ.* **10**, 45–67. *R.A.M.*, 1952, p. 359.

—— 1955. 'Studies on the witches' broom of *Paulownia tomentosa*', *Sci. Bull. Fac. Agric. Kyushu Univ.* **15**, 287–331. *Biol. Abstr.*, 1957, No. 9149.

TOOLE, E. R. 1949. 'Fusarium wilt of Staghorn Sumac', *Phytopathology*, **39**, 754–9.

—— 1954. 'Rot and cankers on oak and honeylocust caused by *Poria spiculosa*', *J. For.* **52**, 941–2.

—— 1955. '*Polyporus hispidus* on southern bottomland oaks', *Phytopathology*, **45**, 177–80.

—— 1957. 'Twig canker of sweetgum', *Plant Dis. Reptr.* **41**, 808–9.

—— 1959. 'Sweetgum blight', *U.S. For. Serv. For. Pest Leafl.* **37**, 4 pp.

—— and BROADFOOT, W. M. 1959. 'Sweetgum blight as related to alluvial soils of the Mississippi River Floodplain', *For. Sci.* **5**, 2–9.

—— and FURNIVAL, G. M. 1957. 'Progress of heart rot following fire in bottomland Red oaks', *J. For.* **55**, 20–24.

TORGESON, D. C. 1951. 'The Phytophthora root rots of *Chamaecyparis lawsoniana*', *Phytopathology*, **41**, 944.

—— 1954. 'Root rot of Lawson Cypress and other ornamentals caused by *Phytophthora cinnamomi*', *Contr. Boyce Thompson Inst.* **17**, 359–73. *R.A.M.*, 1955, p. 4.

TOUMEY, J. W., and NEETHLING, E. J. 1924. 'Insolation a factor in the natural regeneration of certain conifers', *Yale Univ. Sch. For. Bull.* **11**, 63 pp.

TOWNROW, J. A. 1954. 'The biology of *Cryptostroma corticale* and the Sooty bark disease of Sycamore', *Rep. For. Res. For. Comm., Lond. 1952/3*, pp. 118–20.

TOWNSEND, B. B. 1954. 'Morphology and development of fungal rhizomorphs', *Trans. Brit. Mycol. Soc.* **37**, 222–33.

TRAUNMÜLLER, J. 1954. 'Hochwasserschäden in den oberösterreichischen Donau-Auen', *Allg. Forstztg.* **65**, 267–8. *For. Abstr.* **17**, No. 4055.

TRESCHOW, C. 1943. 'Undersøgelser over Brintjonkoncentrationens Indflydelse paa Vaeksten af Svampen *Polyporus annosus*', *Forstl. Forsøgsv. Danm.* **15**, 17–32. *R.A.M.*, 1946, p. 374.

—— 1958. 'Forsøg med rødgranracers resistens overfor angreb af *Fomes annosus* (Fr.) Cke.', *Forstl. Forsøgsv. Danm.* **25**, 1–23. *For. Abstr.* **20**, No. 711.

TRIONE, E. J. 1959. 'The pathology of *Phytophthora lateralis* on native *Chamaecyparis lawsoniana*', *Phytopathology*, **49**, 306–10.

—— and ROTH, L. F. 1956. 'Aerial infections of Port Orford cedar caused by *Phytophthora lateralis*', *Phytopathology*, **46**, 640.

TROUP, R. S. 1952. *Silvicultural systems.* Clarendon Press, Oxford, 2nd ed., 216 pp.

TRUE, R. P. 1938. 'Gall development on *Pinus sylvestris* attacked by the Woodgate Peridermium, and morphology of the parasite', *Phytopathology*, **28**, 24–49.

—— and SNOW, A. G. 1949. 'Gum flow from turpentine pines inoculated with the pitch-canker Fusarium', *J. For.* **47**, 894–9.

—— and TRYON, E. H. 1956. 'Oak stem cankers initiated in the drought year 1953', *Phytopathology*, **46**, 617–21.

TRUMAN, R. 1952. 'X-ray pole inspection', *Tech. Notes For. Comm.* (*Div. Wood Technol.*) *N.S.W.* **5**, 12–13.

TRUMBOWER, J. A. 1934. 'Control of elm leaf spot in nurseries', *Phytopathology*, **24**, 62–73.

TRUTER, S. J. 1947. 'Een voorlopig onderzoek naar de insterving van *Alnus glutinosa* (L.) Gaertner', Thesis, Univ. Utrecht, 110 pp. *R.A.M.*, 1947, p. 516.

TRYON, H. E. 1943. 'Stem girdling of coniferous nursery stock by frost-heaved soil', *J. For.* **41**, 768–9.

TSING, T., FANG, Y.-H., and WANG, W.-L. 1956. 'Salt tolerance of some popular trees in North Kiangsu', *Acta bot. Sinica*, **5**, 153–76. *For. Abstr.* **19**, No. 1476.

TUBEUF, C. VON. 1905. 'Absterben ganzer Baumgruppen durch den Blitz', *Naturw. Z. Forst- u. Landw.* **3**, 493–507.

—— 1906. 'Über sogennante Blitzlöcher im Walde', *Naturw. Z. Forst- u. Landw.* **4**, 344–51.

—— 1907. 'Die Varietäten oder Rassen der Mistel', *Naturw. Z. Forst- u. Landw.* **5**, 321–41.

—— 1908. 'Über die Verbreitung und Bedeutung der Mistelrassen in Bayern', *Naturw. Z. Forst- u. Landw.* **6**, 561–99.

—— 1910. 'Die Ausbreitung der Kiefernmistel in Tirol, und ihre Bedeutung als besondere Rasse?', *Naturw. Z. Forst- u. Landw.* **8**, 12–39.

—— 1913. 'Rassenbildung bei Ahorn-Rhytisma', *Naturw. Z. Forst- u. Landw.* **11**, 21–24.

—— 1915. 'Kann der Efeu den Bäumen schädlich werden?', *Naturw. Z. Forst- u. Landw.* **13**, 476–81.

—— 1927. 'Aufruf zum Anbau der rumelischen Strobe, *Pinus peuce* an Stelle der nordostamerikanischen Weymouthskiefer, *Pinus strobus* und der westamerikanischen Strobe, *Pinus monticola*', *Z. PflKrankh.* **37**, 6–8. *R.A.M.*, 1927, p. 522.

—— 1930. 'Biologische Bekämpfung des Blasenrostes der Weymouthskiefer', *Z. PflKrankh.* **40**, 177–81. *R.A.M.*, 1930, p. 691.

—— 1931. 'Ist *Pinus peuce* gegen den Blasenrostpilz immun oder für ihn nur wenig disponiert?', *Z. PflKrankh.* **41**, 369–70. *R.A.M.*, 1932, p. 81.

—— 1933. 'Studien über Symbiose und Disposition für Parasitenbefall sowie über Vererbung pathologischer Eigenschaften unserer Holzpflanzen. I. Das Problem der Hexenbesen', *Z. PflKrankh.* **43**, 193–242. *R.A.M.*, 1933, p. 663.

—— 1935. 'Ausführung der organisierten praktischen Bekämpfung des Blasenrostes fünfnadeliger Kiefern', *Z. PflKrankh.* **45**, 297–301. *R.A.M.*, 1935, p. 666.

—— 1936. 'Tuberkulose, Krebs und Rindergrind der Eschen-(*Fraxinus*) Arten und die sie veranlassenden Bakterien, Nektriapilze und Borkenkäfer', *Z. PflKrankh.* **46**, 449–83. *R.A.M.*, 1937, p. 216.

TUCKER, C. M. 1933. 'The distribution of the genus *Phytophthora*', *Missouri Agric. Exp. Sta. Res. Bull.* **184**, 80 pp. *R.A.M.*, 1933, p. 594.

—— and MILBRATH, J. A. 1942. 'Root rot of *Chamaecyparis* caused by a species of *Phytopathora*', *Mycologia*, **24**, 94–103. *R.A.M.*, 1942, p. 276.

TUINZING, W. D. J. 1946. 'Ziekten en plagen van den wilg in grienden', *Landbouwk. Tijdschr.* **58**, 639–44. *For. Abstr.* **9**, No. 2559.

TURCONI, M. 1924. 'Una moria di giovani piante di Eucalipti', *Atti Ist. Bot. R. Univ. di Pavia*, 3rd ser., **1**, 125–35. *R.A.M.*, 1925, p. 74.

TURRILL, W. B. 1920. '*Arceuthobium oxycedri* and its distribution', *Kew Bull.*, pp. 264–8.

TWYMAN, E. S. 1946*a*. 'The iron-manganese balance and its effect on the growth and development of plants', *New Phytol.* **45**, 18–24.

TWYMAN, E. S. 1946*b*. 'Notes on the die-back of Oak caused by *Colpoma quercinum* (Fr.) Wallr.', *Trans. Brit. Mycol. Soc.* **29,** 234–41.

TYLER, L. J., and PARKER, K. G. 1945. 'Pathogenicity of the Dutch Elm disease fungus', *Phytopathology*, **35,** 257–61.

ULBRICH, E. 1931. 'Über den Hexenbesenrost der Berberitze, *Puccinia arrenatheri* (Kleb.) Erikss. (*Aecidium graveolens* Shuttl.)', *Notizbl. bot. Gart. Mus. Berlin-Dahlem*, **11,** 124–8. *R.A.M.*, 1931, pp. 601–2.

ULMER, W. 1937. 'Über den Jahresgang der Frosthärte einiger immergrüner Arten der alpinen Stufe, sowie der Zerbe und Fichte', *Jb. Wiss. Bot.* **84,** 553–92.

UMANN, H. 1930. 'Birchenvorwald als Schutz gegen Spätfröste', *Forstwiss. Cbl.* **52,** 493–502, 581–92.

URI, J. 1948. 'Het parasitisme van *Nectria cinnabarina* (Tode) Fr.', *Tijdschr. PlZiekt.* **54,** 29–73. *R.A.M.*, 1949, p. 39.

UROŠEVIĆ, B. 1957. 'Mykoflora skladovaných Žaludů', *Praće výzkum. Ůst. lesn. ČSR.* **13,** 149–200. *R.A.M.*, 1958, p. 423.

URQUIJO LANDALUZE P. 1936. 'Hacia la solución del problema del Castaño', *La Coruña, Papelería e Imprenta Lombardero*, 38 pp. *R.A.M.*, 1936, p. 540.

—— 1947. 'Revisión taxonómica de los hongos productores de la enfermedad del Castaño llamada de la "tinta"', *Bol. Pat. Veg. Ent. agríc., Madrid* **16,** 253–70. *R.A.M.*, 1950, p. 234.

—— 1951. 'Recherches complémentaires concernant le traitement de la maladie de l'encre du châtaignier', *Semaine Internationale du Châtaignier, Sept. 1950, Bull. Techn. Châtaigner*, **3,** 117–26.

VAARTAJA, O. 1949. 'High surface soil temperatures: on methods of investigation and thermocouple observations on a wooded heath in the south of Finland', *Oikos*, **1,** 6–28. *For. Abstr.* **12,** No. 54.

—— 1952. 'Forest humus quality and light conditions as factors influencing damping-off', *Phytopathology*, **42,** 501–6.

—— 1954*a*. 'Puiden siemeniä ja sirkkataimia tuhoavista tekijöistä', *Acta For. Fenn.* **62,** 3–11. *Biol. Abstr.*, 1957, No. 25815.

—— 1954*b*. 'Photoperiodic ecotypes of trees', *Canad. J. Bot.* **32,** 392–9.

—— 1954*c*. 'Microflora on the surface of seedlings as affected by thiram', *Bi-m. Progr. Rep. Div. For. Biol. Dep. Agric. Can.* **10,** 4, p. 3. *R.A.M.*, 1955, p. 180.

—— 1956*a*. 'Notes on low temperature fungi', *Bi-m. Progr. Rep. Div. For. Biol. Dep. Agric. Can.* **12,** 4, p. 3. *R.A.M.*, 1957, p. 438.

—— 1956*b*. 'Screening fungicides for controlling damping-off of tree seedlings', *Phytopathology*, **46,** 387–90.

—— 1956*c*. 'Principles and present status of chemical control of seedling diseases', *For. Chron.* **32,** 45–48.

—— 1957*a*. 'Damping-off control by antibiotic plant substances', *Bi-m. Progr. Rep. Div. For. Biol. Dep. Agric. Can.* **13,** 2, p. 2. *R.A.M.*, 1957, p. 798.

—— 1957*b*. 'Effects of *Trichoderma* on tree seedlings and their pathogens', *Bi-m. Progr. Rep. Div. For. Biol. Dep. Agric. Can.* **13,** 5, p. 1. *R.A.M.*, 1958, p. 381.

—— 1957*c*. 'The susceptibility of seedlings of various tree species to *Phytophthora cactorum*', *Bi-m. Progr. Rep. Div. For. Biol. Dep. Agric. Can.* **13,** 2, p. 2. *R.A.M.*, 1957, p. 795.

—— and CRAM, W. H. 1956. 'Damping-off pathogens of conifers and of *Caragana* in Saskatchewan', *Phytopathology*, **46,** 391–7.

—— and WILNER, J. 1956. 'Field tests with fungicides to control damping-off of Scots pine', *Canad. J. Agric. Sci.* **36,** 14–18. *R.A.M.*, 1956, p. 730.

VALDER, P. G. 1958. 'The biology of *Helicobasidium purpureum* Pat.', *Trans. Brit. Mycol. Soc.* **41,** 283–308.

VARNEY, E. H. 1957. 'Mosaic and Shoestring, virus diseases of cultivated blueberry in New Jersey', *Phytopathology,* **47,** 307–9.

VEEN, B. 1954. 'De klimatologische eisen van de japanese lariks', *Ned. Boschb.-Tijdschr.* **26,** 311–19.

VERGNANO, O. 1953. 'Caratteristici effetti del boro su piante di Olmo e Pioppo nella zona dei soffioni boriferi di Travale (catena metallifera toscana)', *Nouvo G. bot. ital.* (N.S.), **60,** 225–9. *R.A.M.,* 1954, p. 189.

VERNEAU, R. 1953. 'Sterilizzazione parziale delle doghe di Castagno da esportazione', *Ann. Sper. agr.* (N.S.), **7,** 539–45. *R.A.M.,* 1955, pp. 410–11.

VERONA, O. 1950. 'Note sopra una malattia degli strobili di "*Pinus pinea*" prodotta da "*Sphaeropsis necatrix*"', *Ann. Fac. agr. Pisa* (N.S.), **11,** 193–236. *R.A.M.,* 1952, p. 308.

—— and BOZZINI, A. 1956. 'Sul "carbone" del *Polygonum baldschuanicum*', *Phytopath. Z.* **27,** 461–6. *R.A.M.,* 1957, p. 189.

VERPLANCKE, G. 1930. 'Une Maladie intéressante du Châtaignier', *Bull. Soc. Roy. Bot. de Belg.* **62,** 105–7. *R.A.M.,* 1930, p. 566.

VERRALL, A. F., and GRAHAM, T. W. 1935. 'The transmission of *Ceratostomella ulmi* through root grafts', *Phytopathology,* **25,** 11, pp. 1039–40.

VIENNOT-BOURGIN, G. 1949. *Les Champignons parasites des plantes cultivées,* Masson et Cie, Paris, vol. I, 755 pp., vol. II, 1,850 pp.

VIRTANEN, A. I. 1957. 'Investigations on nitrogen fixation by the Alder. II. Associated culture of Spruce and inoculated Alder without combined nitrogen', *Physiol. Plant., Copenhagen,* **10,** 164–9. *For. Abstr.* **18,** No. 3781.

VIVANI, W. 1957. 'Notes sur la "Batteriosi" du Peuplier', *Ist. Sperimentazione Pioppicoltura, Casale Monferrato, Italia,* 23 pp. *R.A.M.,* 1957, p. 674.

VLAD, I. 1944. 'Rezistenta la inundatii a speciilor forestiere din bazinul inferior al Ialometei', *Rev. Pădurilor,* **56,** 85–93. *For. Abstr.* **9,** No. 1861.

VLIET, J. I. VAN. 1931. 'Esschenkankers en hun bouw', Thesis, Univ. Utrecht, 73 pp. *R.A.M.,* 1932, p. 12.

VLIET, W. F. VAN, VERBRUGGE, E., and BOER, P. J. DEN. 1955. 'De aantasting van de berken door houtzwammen in het gebied van de duin-waterleiding van 's Gravenhage', *Levende Natuur,* **58,** 125–7. *Biol. Abstr.* 1957, No. 22496.

VLOTEN, H. VAN. 1929. '*Brunchorstia destruens* Erikss. on Scots pine in Scotland' *Scot. For. J.* **43,** 157–8.

—— 1932. '*Rhabdocline pseudotsugae* Sydow. oorzaak eener ziekte van Douglasspar', Thesis, Wageningen Agric. Coll., 168 pp. *R.A.M.,* 1933, p. 63.

—— 1936. 'Onderzoekingen over *Armillaria mellea* (Vahl) Quel.', *Fungus, Wageningen,* **8,** 20–23. *R.A.M.,* 1937, p. 215.

——1938. 'Het onderzoek naar de vatbaarheid van Populieren voer aantasting door *Dothichiza populea* Sacc. et Briard', *Tijdschr. ned. Heidemaatsch.* **3,** 18 pp. *R.A.M.,* 1938, p. 569.

—— 1941. 'Roest van Populieren', *Ned. Boschb.Tijdschr.* **7,** 347–50.

—— 1943. 'Verschillen in virulentie bij *Nectria cinnabarina*', *Tijdschr. PlZiekt.* **49,** 163–71. *R.A.M.,* 1946, p. 87.

—— 1946. 'Over de ziekteverschijnselen van den Corsicaanschen den, die van 1940 tot 1943 de aandacht trokken', *Ned. Boschb.Tijdschr.* **18,** 281–4. *For. Abstr.* **8,** p. 559.

—— 1949. 'Kruisingsproeven met rassen van *Melampsora larici-populina* Klebahn', *Tijdschr. PlZiekt.* **55,** 196–209. *R.A.M.,* 1950, p. 186.

—— 1952. 'Evidence of host-parasite relations by experiments with *Phomopsis pseudotsugae* Wilson', *Scot. For.* **6,** 38–46.

VLOTEN, H. VAN, and GREMMEN, J. 1953. 'Studies in the Discomycete genera *Crumenula* de Not. and *Cenangium* Fr.', *Acta Bot. neerl.* **2,** 226–41. *Biol. Abstr.* 1954, No. 26841.

VOGLINO, P. 1928. 'Il servizio di controllo fitopatologico sulle Castagne destinate agli Stati Uniti d'America nella campagna 1928 esercitato dal R. osservatorio di Fitopatologia di Torino', *Nuovi Annali dell'agricoltura,* **8,** 319–44.

—— 1929. 'Una cura pratica del marciume radicale del Gelso', *Difesa Piante,* **6,** 5, 1–2. *R.A.M.*, 1930, p. 213.

—— 1931. 'Il nerume delle Castagna', *Difesa Piante,* **8,** 2, 1–4. *R.A.M.*, 1931, p. 696.

—— 1932. 'La marcescenza del fusto nella *Araucaria imbricata*', *Difesa Piante,* **9,** 2, 17–20. *R.A.M.*, 1933, p. 63.

—— 1933. 'Sopra un deperimento dell'*Araucaria imbricata*', *Difesa Piante,* **10,** 3, 37–39. *R.A.M.*, 1934, p. 68.

VOIGT, G. K. 1951. 'Causes of injury to conifers during the winter of 1947–1948 in Wisconsin', *Trans. Wisconsin Acad. Sci. Arts and Letters,* **40,** 241–3. *Biol. Abstr.*, 1953, No. 4633.

—— 1954. 'Chemicals applied to soil may harm forest seedlings', *Wis. Agric. Exp. Sta.* 1952/53, **1,** 42–43.

VOLGER, C. VON. 1957. 'Untersuchungen zum Problem der chemischen Bekämpfung pilzparasitärer Keimlingskrankheiten unserer Nadelbäume', *Forstwiss. Cbl.* **76,** 294–305.

VÖRÖS, J. 1954. 'Antibiotikumok és antagonista mikroorganizmusok felhasználása fenyőcsemetedőlés ellen', *Növénytermelés,* **3,** 115–22. *R.A.M.*, 1956, p. 498.

VULTERIN, Z. 1952. 'Bioklimatické poměry Severočeské hnědouhelné pánve a přilehlé části hřebene. Rudohoří vzhledem k chřadnutí lesních porostů', *Lesn. Knihovna* (*velká Rada*), **5,** 228–41. *For. Abstr.* **14,** No. 2350.

WADE, I. W. 1953. 'Sounding damage on thinbarked trees', *For. Mimeo. Ohio Agric. Exp. Sta.* **12,** 1 p. *For. Abstr.* **16,** No. 732.

WADSWORTH, F. H. 1943. 'Lightning damage in Ponderosa pine stands of northern Arizona', *J. For.* **41,** 684–5.

WAGENER, W. W. 1939. 'The canker of *Cupressus* induced by *Coryneum cardinale* n. sp.', *J. Agric. Res.* **58,** 1–46.

—— 1940. 'The physical effects of drought on shade trees', *Proc. 7th West. Shade Tree Conf., Los Angeles, Calif.*, pp. 21–30.

—— 1948. 'Diseases of Cypresses', *El Aliso,* **1,** 255–321. *R.A.M.*, 1948, p. 545.

—— and CAVE, M. S. 1946. 'Pine killing by the root fungus, *Fomes annosus*, in California', *J. For.* **44,** 1–54.

—— CHILDS, T. W., and KIMMEY, J. W. 1949. 'Notes on some foliage diseases of forest trees on the Pacific slope', *Plant Dis. Reptr.* **33,** 195–7.

—— and DAVIDSON, R. W. 1954. 'Heart rots in living trees', *Bot. Rev.* **20,** 61–134.

WAHLENBURG, W. G. 1929. 'The relation of quantity of seed sown and density of seedlings to the development and survival of forest planting stock', *J. Agric. Res.* **38,** 219–27.

WAID, J. S., and WOODMAN, M. J. 1957. 'A non-destructive method of detecting diseases in wood', *Nature, London,* **180,** 47.

WAKEFIELD, E. M., and BISBY, G. R. 1941. 'List of Hyphomycetes recorded for Britain', *Trans. Brit. Mycol. Soc.* **25,** 49–126.

—— and DENNIS, R. W. G. 1950. *Common British Fungi.* P. R. Gawthorn, London, 290 pp.

WALDIE, J. S. L. 1926. 'A die-back disease of Pines', *Trans. R. Scot. Arbor. Soc.* **40,** 120–5.
—— 1930. 'An oak seedling disease caused by *Rosellinia quercina* Hartig', *Forestry,* **4,** 1–6.
WALES, B. H. 1931. 'A study of damage by tractor skidding', *J. For.* **27,** 495–9.
WALKER, L. C. 1956. 'Foliage symptoms as indicators of potassium-deficient soils', *For. Sci.* **2,** 113–20.
WALKER, R. B., GESSEL, S. P., and HADDOCK, P. G. 1955. 'Greenhouse studies in mineral requirements of conifers: Western red cedar', *For. Sci.* **1,** 51–60.
WALLACE, R. H., and MOSS, A. E. 1941. 'Salt spray damage from recent New England hurricane', *Proc. 15th Nat. Shade Tree Conf.*, pp. 112–19.
WALLACE, T. 1951. *The diagnosis of mineral deficiencies in plants by visual symptoms*', H.M.S.O., London, 107 pp.
WALLIS, G. W. 1954. 'Commercial thinning in Douglas fir in relation to control of *Poria weirii* root rot', *Bi-m. Progr. Rep. Div. For. Biol. Dep. Agric. Can.* **10,** 3, p. 4. *R.A.M.*, 1955, p. 420.
—— and BUCKLAND, D. C. 1956. 'The effect of trenching on the spread of yellow laminated root rot of Douglas fir', *For. Chron.* **31,** 356–9.
WALTER, B. 1932. 'Über den Unterschied der Blitzgefahr der Eiche und der Buche', *Physikal. Z.* **33,** 306–7.
WALTER, H. J. 1956. 'Variation in sensitivity to atmospheric fluorides among Ponderosa pine seedlings', *Phytopathology,* **46,** 637.
WALTER, J. M. 1937. 'Variation in mass isolates and monoconidium progenies of *Ceratostomella ulmi*', *J. Agric. Res.* **54,** 509–23.
—— 1939*a*. 'Effects of *Ceratostomella ulmi* on *Ulmus americana* and some types of European elm', *Phytopathology,* **29,** 23.
—— 1939*b*. 'Observations on fructification of *Ceratostomella ulmi* in England', *Phytopathology,* **29,** 551–3.
—— 1944. 'Wound dressing for London plane trees', *Plant Dis. Reptr.* **28,** 356.
—— 1946. 'Canker stain of Plane trees', *U.S. Dep. Agric. Circ.* **742,** 12 pp. *R.A.M.*, 1947, p. 85.
—— and MAY, C. 1937. 'Pathogenicity of a brown cultural variant of *Ceratostomella ulmi*', *Phytopathology,* **27,** 142–3.
—— ——, and COLLINS, C. W. 1943. 'Dutch elm disease and its control', *U.S. Dep. Agric. Circ.* **677,** 12 pp.
—— and MOOK, P. V. 1941. 'Transmission of the Planetree Ceratostomella in asphalt wound dressings', *Phytopathology,* **31,** 23–24.
——, REX, E. G., and SCHREIBER, R. 1952. 'The rate of progress and destructiveness of canker stain of plane trees', *Phytopathology,* **42,** 236–9.
WALTHER, E. 1935. 'Genetic constitution of host plant as a factor in pest control', *Mon. Bull. Calif. Dep. Agric.* **24,** 242–4. *Plant Breed. Abstr.* **6,** 88.
WARCUP, J. H. 1951. 'Effect of partial sterilization by steam or formalin on the fungus flora of an old forest nursery soil', *Trans. Brit. Mycol. Soc.* **34,** 519–32.
—— 1952. 'Effect of partial sterilization by steam or formalin on damping-off of Sitka spruce', *Trans. Brit. Mycol. Soc.* **35,** 248–62.
—— 1957. 'Chemical and biological aspects of soil sterilization', *Soils & Fertil.* **20,** 1–5.
WARE, W. M. 1931. 'A blossom wilt of Lavender caused by *Botrytis cinerea*', *J. S.E. Agric. Col. Wye, Kent,* **28,** 206–10.
WARENG, P. F. 1948. 'Photoperiodism in woody species', *Forestry,* **2 2,**211–21.
WARREN, G. L., and WHITNEY, R. D. 1951. 'Spruce root borer (*Hypomolyx* sp.), root wounds, and root diseases of White spruce', *Bi-m. Progr. Rep. Div. For. Biol. Dep. Agric. Can.* **7,** 4, pp. 2–3. *For. Abstr.* **13,** No. 1386.

WASSINK, E. C., and WIERSMA, J. H. 1955. 'Daylength responses of some forest trees', *Acta bot. Neerl.* **4,** 657–70. *Hort. Abstr.*, 1956, No. 2536.

WATERMAN, A. M. 1940. 'Sycamore and oak anthracnose. *Gnomonia veneta* (Sacc. & Speg.) Kleb.', *Tree Pest Leafl.* **48,** 4 pp. Rev. 1951, 3 pp. *R.A.M.*, 1941, p. 611.

—— 1943*a*. '*Diplodia pinea* and *Sphaeropsis malorum* on soft pines', *Phytopathology*, **33,** 828–31.

—— 1943*b*. '*Diplodia pinea*, the cause of a disease of hard pines', *Phytopathology*, **33,** 1018–31.

—— 1945. 'Tip blight of species of *Abies* caused by a new species of *Rehmiellopsis*', *J. Agric. Res.* **70,** 315–37.

—— 1947. '*Rhizosphaera kalkhoffi* associated with a needle cast of *Picea pungens*', *Phytopathology*, **37,** 507–11.

—— 1954. 'Septoria canker of poplars in the United States', *Circ. U.S. Dep. Agric.* **947,** 24 pp. *R.A.M.*, 1956, p. 646.

—— 1955. 'The relation of *Valsa kunzei* to cankers on conifers', *Phytopathology*, **45,** 686–92.

—— 1957. 'Canker and dieback of poplars caused by *Dothichiza populea*', *For. Sci.* **3,** 175–83.

—— and ALDRICH, K. F. 1952. 'Surface sterilization of Poplar cuttings', *Plant Dis. Reptr.* **36,** 203–7.

—— —— 1954. 'Additional information on the surface sterilization of Poplar cuttings', *Plant Dis. Reptr.* **38,** 96–100.

—— and CASH, E. K. 1950. 'Leaf blotch of poplar caused by a new species of *Septotinia*', *Mycologia*, **42,** 374–84. *R.A.M.*, 1951, p. 130.

—— and MARSHALL, R. P. 1947. 'A new species of *Cristulariella* associated with a leaf spot of Maple', *Mycologia*, **39,** 690–8. *R.A.M.*, 1948, p. 395.

WATSON, H. 1928. 'Notes on attack by *Rhizoctonia crocorum* on Sitka Spruce (*Picea sitchensis*)', *Scot. For. J.* **42,** 58–61.

—— 1933. 'Disease attacking common silver fir (*Abies pectinata*)', *Scot. For. J.* **47,** 71–72.

—— 1937. 'Tree felled by its own roots', *Scot. For. J.* **51,** 62.

WATSON, W. 1936. 'The bryophytes and lichens of British woods. I. Beechwoods. II. Other woodland types', *J. Ecol.* **24,** 139–61, 446–78.

WATT, A. S. 1927. '*Fomes annosus* on *Calluna vulgaris*', *Scot. For. J.* **41,** 225.

WEAN, R. E. 'The parasitism of *Polyporus schweinitzii* on seedling *Pinus strobus*', *Phytopathology*, **27,** 1124–42.

WEBER, G. F. 1941. 'Leaf blister of oaks', *Pr. Bull. Fla. Agric. Exp. Sta.* **558,** 2 pp. *R.A.M.*, 1942, p. 232.

WEBSTER, A. D. 1918. *Seaside planting*. T. Fisher Unwin, London, 156 pp.

WECK, J. 1952. *Ödlandaufforstung*. Fritz Haller, Berlin, 101 pp.

WEDDELL, D. J. 1942. 'Damage to Catalpa due to recreational use', *J. For.* **40,** 807.

WEHMEYER, L. E. 1933. 'The British species of the genus *Diaporthe* Nits. and its segregates', *Trans. Brit. Mycol. Soc.* **17,** 237–95.

WEINDLING, R. 1932. '*Trichoderma lignorum* as a parasite of other soil fungi', *Phytopathology*, **22,** 837–45.

—— and EMERSON, O. H. 1936. 'The isolation of a toxic substance from the culture filtrate of *Trichoderma*', *Phytopathology*, **26,** 1068–70.

—— and FAWCETT, H. S. 1934. 'Experiments in biological control of *Rhizoctonia* damping off', *Phytopathology*, **14,** 1142.

WEIR, J. R. 1915. 'Observations on *Rhizina inflata*', *J. Agric. Res.* **4,** 93–95.

—— 1916. '*Keithia thujina*, the cause of a serious leaf disease of Western red cedar', *Phytopathology*, **6,** 360–3.

WEIR, J. R. 1920. 'Note on the pathological effect of blazing trees', *Phytopathology*, **10,** 371–3.

—— 1921. '*Thelephora terrestris*, *T. fimbriata*, and *T. caryophyllea* on forest tree seedlings', *Phytopathology*, **11,** 141–4.

—— and HUBERT, E. E. 1918. 'Notes on the overwintering of forest tree rusts', *Phytopathology*, **8,** 55–59.

WEISS, F., and BAUMHOFER, L. G. 1940. 'Culture, diseases, and pests of the Box Tree', *Fmrs' Bull. U.S. Dep. Agric.*, **1855**, 18 pp. *R.A.M.*, 1941, p. 434.

WEISSE, A. 1925. 'Neue Beobachtungen über die Blattkrankheiten der Platanen', *Verh. Bot. Ver. Brandenburg*, **67,** 24–25. *R.A.M.*, 1926, p. 11.

WELCH, D. S. 1934. 'The range and importance of Nectria canker on hardwoods in the northeast', *J. For.* **32,** 997–1002.

—— and MATTHYSSE, J. G. 1955. 'Control of the Dutch elm disease in New York State', *Cornell Ext. Bull.* **932,** 14 pp. *R.A.M.*, 1956, p. 853.

WELLINGTON, W. G. 1954. 'Pole blight and climate', *Bi-m. Progr. Rep. Div. For. Biol. Dep. Agric. Can.* **10,** 6, pp. 2–4. *R.A.M.*, 1956, p. 132.

WELLS, B. W. 1939. 'A new forest climax: the salt spray climax of Smith Island, N.C.', *Bull. Torrey Bot. Cl.* **66,** 629–34.

WENDELKEN, W. J. 1955. 'Root development and wind-firmness on the shallow gravel soils of the Canterbury Plains', *N.Z. J. For.* **7,** 71–76.

WENGER, K. F. 1955. 'Light and mycorrhiza development', *Ecology*, **36,** 518–20.

WENNER, J. J. 1914. 'Morphology and life history of *P. funerea*', *Phytopathology*, **4,** 375–83.

WENT, J. C. 1947. 'Verslag van de onderzoekingen verricht in 1946 voor het Comité inzake bestudeering en bestrijding van de iepenziekte en andere boomziekten, uitgevoerd op het Phytopathologisch Laboratorium "Willie Commelin Scholten" te Baarn', *Mededeeling*, **42,** 12 pp. *For. Abstr.* **9,** No. 2560.

—— 1954. 'The Dutch elm disease. Summary of fifteen years' hybridization and selection work (1937–1952)', *Tijdschr. PlZiekt.* **60,** 109–27. *R.A.M.*, 1955, p. 1.

WENZL, H., and MÜLLER-FEMBECK, J. 1951. 'Der Einfluss der Schichtdicke von Wundverschlussmitteln auf die Wundüberwallung', *Bodenkultur*, **5,** 223–30. *Hort. Abstr.* 1952, No. 399.

WERFF, H. J. VAN DER. 1955. 'Bevloeiing van boomgaarden met zoet water ter bestrijding van verzilting', *Meded. Dir. Tuinb.* **18,** 288–96. *Hort. Abstr.* 1955, No. 3704.

WERNER, F., and ÄRMANN, J. 1955. 'Stormfällningens dynamik—en studie', *Svenska SkogsvFören. Tidskr.* **53,** 311–30. *For. Abstr.* **18,** No. 4167.

WESTERDIJK, J., LEDEBOER, M., and WENT, J. 1931. 'Mededeelingen omtrent gevoeligheidsproeven van Iepen voor *Graphium ulmi* Schwarz, gedurende 1929 en 1930', *Tijdschr. PlZiekt.* **37,** 105–10. *R.A.M.*, 1931, p. 695.

—— and LUIJK, A. VAN. 1924. 'Die Gloeosporien der Eiche und der Platane. II', *Meded. Phytopath. Lab. "Willie Commelin Scholten", Baarn, Holland*, **6,** 31–33. *R.A.M.*, 1925, p. 1.

WEST-NIELSEN, G., and OKSBJERG, E. 1954. 'Om rodfordaerverangreb. II. Fortsatte undersøgelser over Granens vaekst og sundhed', *Hedeselsk Tidsskr., Aarhus*, **75,** 336–43, 347–51. *R.A.M.*, 1955, p. 826.

WETTSTEIN-WESTERSHEIM, W. 1933. 'Zur Frage der Züchtung von Forstpflanzen', *Z. Züchtung*, A, **18,** 357–69. *Plant Breed. Abstr.* **3,** 216.

WHEELER, W. H. 1957. 'The movement of plant pathogens', *Phytopathology*, **47,** 386–8.

WHELDON, J. A., and TRAVIS, W. G. 1915. 'The Lichens of South Lancashire', *J. Linn. Soc. (Bot.)* **43,** 87–136.

WHITE, D. P., and LEAF, A. L. 1956. 'Forest Fertilization; a bibliography with abstracts on the use of fertilizers and soil amendments in forestry', *World For. Ser. N.Y. State Univ. Coll. For. Syracuse, Bull.* **2,** 303 pp.

WHITE, I. G. 1955. 'Toxin production by the oak wilt fungus, *Endoconidiophora fagacearum*', *Amer. J. Bot.* **42,** 759–64. *For. Abstr.* **17,** No. 1744.

WHITE, J. H. 1920. 'On the biology of *Fomes applanatus* (Pers.) Wallr.', *Trans. Roy. Canad. Inst.* **12,** 133–74.

WHITE, J. W. 1912. *Flora of Bristol.* John Wright & Son, Bristol, 722 pp.

WHITE, L. T. 1956. 'Biological control of seedling diseases', *For. Chron.* **32,** 49–52.

WHITE, N. H. 1954. 'The development of the ascocarp of *Cyttaria gunnii* Berk.', *Trans. Brit. Mycol. Soc.* **37,** 431–6.

WHITE, O. E. 1948. 'Fasciation', *Bot. Rev.* **14,** 319–58.

WHITE, R. P. 1929. 'Juniper blight', *50th Ann. Rep. New Jersey Agric. Exp. Sta.*, pp. 270–2. *R.A.M.*, 1931, p. 71.

—— 1930. 'Pathogenicity of *Pestalotia* spp. on Rhododendron', *Phytopathology*, **20,** 85–91.

—— 1933. 'The insects and diseases of the Rhododendron and Azalea', *J. Econ. Entom.* **26,** 631–40. *R.A.M.*, 1933, p. 696.

—— 1937. 'Rhododendron wilt and root rot', *Bull. N.J. Agric. Exp. Sta.* **615,** 32 pp. *R.A.M.*, 1939, p. 33.

—— and MCCULLOCH, L. 1934. 'A bacterial disease of *Hedera helix*', *J. Agric. Res.* **48,** 807–15.

WHITNEY, R. D. 1952. 'Relationship between entry of root-rotting fungi and root-wounding by *Hypomolyx* and other factors in White spruce', *Bi-m. Progr. Rep. Div. For. Biol. Dep. Agric. Can.* **8,** 1, pp. 2–3. *For. Abstr.* **13,** No. 3976.

WHITTEN, R. R., and SWINGLE, R. U. 1958. 'The Dutch Elm Disease and its control', *U.S.D.A. Agric. Inf. Bull.* **193,** 12 pp.

WIANT, J. S. 1929. 'The Rhizoctonia damping-off of conifers and the control by chemical treatment of the soil', *Cornell Agric. Exp. Sta. Mem.* **124,** 64 pp. *R.A.M.*, 1930, p. 7.

WICHMANN, H. E. 1953. 'Rindenbrüter und Hallimasch', *Forstwiss. Cbl.* **72,** 57–63. *For. Abstr.* **14,** No. 2489.

WIDDER, F. 1948. 'Untersuchungen über forstschädliche Cronartium-Arten', '*Carinthia II.*', *Mitteilungen des Naturwissenschaftlichen Vereines für Kärnten, Klagenfurt*, 137–8, pp. 82–93. *For. Abstr.* **11,** No. 665.

WIEDERMANN, E. 1935. 'Über den Schaden der Streunutzung im deutschen Osten', *Forstarchiv*, **11,** 386–90.

WIELER, A. 1922. 'Die Beteiligung des Bodens an den durch Rauchsäuren hervorgerufenen Vegetationsschäden', *Z. Forst- u. Jagdw.* **54,** 534–43. *R.A.M.*, 1923, p. 20.

WIJBRANS, J. R. 1957. 'Bliksemschade in jonge Hevea-aanplanten', *Bergcultures*, **26,** 291–7. *R.A.M.*, 1958, p. 180.

WILCOX, M. S. 1939. 'Phomopsis twig blight of Blueberry', *Phytopathology*, **29,** 136–42.

WILCOX, R. B. 1942. 'Blueberry stunt, a virus disease', *Plant Dis. Reptr.* **26,** 211–13.

WILDE, S. A. 'Mycorrhizal fungi: their distribution and effect on tree growth *Soil Sci.* **78,** 23–31. *R.A.M.*, 1956, pp. 534–5.

——, VOIGT, G. K., and PERSIDSKY, D. J. 1956. 'Transmitted effect of allyl alcohol on growth of Monterey pine seedlings', *For. Sci.* **2,** 58–59.

—— and WHITE, D. P. 1939. 'Damping-off as a factor in the natural distribution of pine species', *Phytopathology*, **29,** 367–9.

WILHELM, S., and FERGUSON, J. 1953. 'Soil fumigation against *Verticillium albo-atrum*', *Phytopathology*, **43,** 593–5.

WILKINS, W. H. 1936. 'Studies in the genus *Ustulina* with special reference to parasitism. II. A disease of the common lime (*Tilia vulgaris* Hayne) caused by *Ustulina*', *Trans. Brit. Mycol. Soc.* **20,** 133–56.

WILKINSON, R. E. 1952. 'Woody plant hosts of the tobacco ringspot virus', *Phytopathology*, **42,** 478.

—— 1953. '*Berberis Thunbergii*, a host of cucumber mosaic virus (*Marmor cucumeris*)', *Phytopathology*, **43,** 489.

WILL, G. N. 1955. 'Removal of mineral nutrients from tree crowns by rain', *Nature, London*, **176,** 1180.

WILLIS, J. C., and BURKHILL, I. H. 1884. 'Observations on the Flora of Pollard willows near Cambridge', *Proc. Camb. Phil. Soc.* **8,** 82–91.

WILNER, J., and VAARTAJA, O. 1958. 'Prevention of injury to tree seedlings during cellar storage', *For. Chron.* **34,** 132–8.

WILSON, C. C. 1953. 'The response of two species of Pine to various levels of nutrient zinc', *Science*, **117,** 231–3.

WILSON, E. E. 1937. 'The shot-hole disease of stone-fruit trees', *Bull. Calif. Agric. Exp. Sta.* **608,** 40 pp. *R.A.M.*, 1938, p. 256.

—— STARR, M. P., and BERGER, J. A. 1957. 'Bark canker, a bacterial disease of the Persian walnut tree', *Phytopathology*, **47,** 669–73.

WILSON, M. 1920. 'Two diseases new to Scotland caused by species of *Hypoderma*', *Trans. Roy. Scot. Arbor. Soc.* **34,** 222–3.

—— 1921*a*. 'A newly-recorded disease on Japanese larch', *Trans. Roy. Scot. Arbor. Soc.* **35,** 73–74.

—— 1921*b*. '*Armillaria mellea* as a potato disease', *Trans. Roy. Scot. Arbor. Soc.* **35,** 186–7.

—— 1922. 'Studies in the pathology of young trees and seedlings. I. The Rosellinia disease of the spruce', *Trans. Roy. Scot. Arbor. Soc.* **36,** 226–35.

—— 1924. 'Observations on some Scottish Uredineae and Ustilagineae. I', *Trans. Brit. Mycol. Soc.* **9,** 135–44.

—— 1925. 'The Phomopsis disease of conifers', *For. Comm. Bull.* **6,** 34 pp.

—— 1927. 'The host plants of *Fomes annosus*', *Trans. Brit. Mycol. Soc.* **12,** 147–9.

—— 1930*a*. 'The rust disease of *Berberis* (*Mahonia*) *aquifolium*', *Gdnrs.' Chron.* **87,** 132–3.

—— 1930*b*. 'The Phomopsis disease of cedars', *Gdnrs.' Chron.* **88,** 412–13.

—— 1937*a*. 'The occurrence of *Keithia tsugae* in Scotland', *Scot. For. J.* **51,** 46–47.

—— 1937*b*. 'The Phomopsis diseases of conifers', *Scot. For. J.* **51,** 39–44.

—— and BISBY, G. R. 1954. 'List of British Uredinales', *Trans. Brit. Mycol. Soc.* **37,** 61–86.

—— and MACDONALD, J. 1924. 'A new disease of the Silver firs in Scotland', *Trans. Roy. Scot. Arbor. Soc.* **38,** 114–18.

—— and WALDIE, J. S. L. 1926*a*. '*Rhizosphaera kalkhoffi* Bubák as a cause of defoliation of conifers', *Trans. Roy. Scot. Arbor. Soc.* **40,** 34–36.

—— —— 1926*b*. 'An epidemic disease of the oak', *Gdnrs.' Chron.* **80,** 106.

—— —— 1928. 'Notes on new or rare forest fungi', *Trans. Brit. Mycol. Soc.* **13,** 151–6.

—— and WILSON, M. J. F. 1926. '*Rhabdocline pseudotsugae* Syd.: a new disease of the Douglas fir in Scotland', *Trans. Roy. Scot. Arbor. Soc.* **40,** 37–40.

WILSON, M. J. F. 1928. 'A disease of the Douglas fir and other conifers', *Gdnrs.' Chron.* **83,** 105.

—— 1931. 'A comparative study of growth-forms within the species *Dermatea livida* (B. et Br.) Phillips', *Ann. Bot., Lond.* **45,** 73–90.

—— 1936. 'A disease of Cherry laurel caused by *Trochila laurocerasi* (Desm.) Fr.', *Ann. Appl. Biol.* **23,** 700–4. *See also* Gregor, M. J. F.

WINKLER, A. 1913. 'Über den Einfluss der Aussenbedingungen auf die Kälteresistenz ausdauernder Gewächse', *Jb. wiss. Bot.*, pp. 467–506.

WINTERFELD, K. 1956. 'Untersuchungen über die Auswirkung der Grünästung bei der Rotbuche. I. Beeinflussung des lebenden Holzgewebes durch den Grünästungsschnitt bei der Rotbuche. II. Holzzerstörungen durch Pilzbefall an Grünästungswunden der Rotbuche. III. Kann die Grünästung der Rotbuche die Rohholzqualität verbessern?', *Holz-Zbl.* **82,** 1053–5, 1089–90, 1115–17. *For. Abstr.* **18,** No. 438.

WOELFLE, M. 1936–7. 'Sturmschäden im Wald', *Forstwiss. Cbl.* **17,** 605–17; **18,** 77–92, 565–88.

WOESTE, U. 1956. 'Anatomische Untersuchungen über die Infektionswege einiger Wurzelpilze', *Phytopath. Z.* **26,** 225–72. *For. Abstr.* **18,** No. 1793.

WOFFENDEN, L. M., and PRIESTLEY, J. H. 1924. 'The toxic action of coal gas upon plants. II. The effect of coal gas upon cork and lenticel formation', *Ann. Appl. Biol.* **11,** 42–53.

WOLF, F. A. 1929–30. 'A parasitic alga *Cephaleuros virescens* Kunze on *Citrus* and certain other plants', *J. Elisha Mitchell Sci. Soc.* **45,** 187–205.

—— 1936. 'False mildew of Red mulberry', *Mycologia,* **28,** 268–77. *R.A.M.*, 1936, p. 762.

—— 1938. 'Life histories of two leaf-inhabiting fungi on Sycamore', *Mycologia,* **30,** 54–63.

—— 1939. 'Leafspot of ash and *Phyllosticta viridis*', *Mycologia,* **31,** 258–66. *R.A.M.*, 1939, p. 638.

—— 1940*a*. 'Cercospora leafspot of Red bud', *Mycologia,* **32,** 129–36. *For. Abstr.* **2,** 192–3.

—— 1940*b*. 'A leaf spot fungus on *Nyssa*', *Mycologia,* **32,** 331–5. *R.A.M.*, 1940, p. 626.

—— 1947. 'Twig blight of Golden Bell, *Forsythia viridissima* Lindl.', *Plant Dis. Reptr.* **31,** 325.

—— and BARBOUR, W. J. 1941. 'Brown-spot needle disease of pines', *Phytopathology,* **31,** 61–74.

WOLF, F. T., and WOLF, F. A. 1939. 'A study of *Botryosphaeria ribis* on Willow', *Mycologia,* **31,** 217–27. *R.A.M.*, 1939, p. 486.

—— —— 1952. 'Pathology of *Camellia* leaves infected by *Exobasidium camelliae* var. *gracilis* Shirai', *Phytopathology,* **42,** 147–9.

WOLLENWEBER, H. W. 1927. 'Das Ulmensterben und sein Erreger *Graphium ulmi* Schwarz', *NachrBl. deutsch. PflSchDienst.* **7,** 97–100. *R.A.M.*, 1928, p. 286.

—— 1929. 'Die Wirtelpilz-Welkekrankheit (Verticilliose) von Ulme, Ahorn und Linde, usw.', *Arb. Biol. Reichsanst. Land- u. Forstw.* **7,** 273–99. *R.A.M.*, 1930, p. 6.

—— 1931*a*. 'Fusarium-Monographie. Fungi parasitici et saprophytici', *Z. Parasitenk.* **3,** 269–516. *R.A.M.*, 1931, p. 626.

—— 1931*b*. 'La Maladie de l'orme', *Congr. Int. Bois et Sylvicult., Paris. Rapports Groupe II*, pp. 603–4.

WOOD, R. K. S. 1951. 'The control of diseases of lettuce by the use of antagonistic organisms. I. The control of *Botrytis cinerea* Pers.', *Ann. Appl. Biol.* **38,** 203–16.

WOODCOCK, A. H. 1955. 'Bursting bubbles and air pollution', *Sewage Indust. Wastes,* **27,** 1189–92. Abstr. in *Atmospheric Poll. Bull.* **24,** 49.

WOODFORD, E. K. 1957. 'The toxic action of Herbicides', *Outlook on Agriculture,* **1,** 145–54.

WOODS, R. V. 1955. 'Salt deaths in *Pinus radiata* at Mount Crawford Forest Reserve, S.A.', *Aust. For.* **19**, 13–19. *For. Abstr.* **17**, No. 1697.

WOODWARD, R. C. 1926. 'A note on *Botryodiplodia* sp. on *Choisya ternata* in England', *Trans. Brit. Mycol. Soc.* **11**, 281–3.

——, WALDIE, J. S. L., and STEVEN, H. M. 1929. Oak mildew and its control in forest nurseries', *Forestry*, **3**, 38–56.

WORLEY, C. L., LESSELBAUM, H. R., and MATTHEWS, T. M. 1941. 'Deficiency symptoms for the major elements in seedlings of three broadleaved trees', *J. Tenn. Acad. Sci.* **16**, 239–47.

WORMALD, H. 1924. 'The mulberry "blight" in Britain', *Ann. Appl. Biol.* **11**, 169–74.

—— 1930. 'Bacterial "blight" of Walnuts in Britain', *Ann. Appl. Biol.* **17**, 59–70.

—— 1932. 'A bacterial disease of Lilacs', *Gdnrs.' Chron.* **92**, 116–17.

—— 1937. 'Leaf blotch of Hawthorn', *Gdnrs'. Chron.* **102**, 47.

—— 1938. 'Two ornamental shrubs as hosts of the organism causing plum bacterial canker', *Rep. E. Malling Res. Sta., 1937*, pp. 198–200.

—— 1940. 'Host plants of the brown rot fungi of Britain', *Trans. Brit. Mycol. Soc.* **24**, 20–28.

—— 1944. 'Nut drop: a disease of cultivated hazel nuts', *Gdnrs.' Chron.* **115**, 60–61.

—— 1955. *Diseases of fruit and hops.* Crosby Lockwood and Son, London, 325 pp.

WORONIN, M., and NAWASCHIN, S. 1896. '*Sclerotinia heteroica*', *Z. PflKrankh.* **6**, 129–40.

WORSNOP, F. E. 1955. 'The growth of zinc-sensitive tree seedlings in tinplate and galvanized iron tubes', *Aust. For.* **19**, 74–86.

WÖSTMANN, E., and GOOSSEN, H. 1956. 'Bekämpfungsversuche gegen *Dothichiza populea* mit Fungiziden', *Forst- u. Holzw.* **11**, 371–2. *R.A.M.*, 1957, p. 143.

WRIGHT, E. 1941. 'Control of damping-off of broadleaf seedlings', *Phytopathology*, **31**, 857–8.

—— 1942. '*Cytospora abietis*, the cause of a canker of true Firs in California and Nevada', *J. Agric. Res.* **65**, 143–53.

—— 1944. 'Damping-off in broadleaf nurseries of the Great plains region', *J. Agric. Res.* **69**, 77–94.

—— 1945. 'Relation of macrofungi and micro-organisms of soils to damping-off of broadleaf seedlings', *J. Agric. Res.* **70**, 133–41.

—— 1957*a*. 'Influence of temperature and moisture on damping-off of American and Siberian elm, Black locust, and Desert willow', *Phytopathology*, **47**, 658–62.

—— 1957*b*. 'Cytospora canker of Rocky Mountain Douglas fir', *Plant Dis. Reptr.* **41**, 811–13.

—— 1957*c*. 'Cytospora canker of cottonwood', *Plant Dis. Reptr.* **41**, 892–3.

—— 1957*d*. 'Importance of Mycorrhizae to Ponderosa pine seedlings', *For. Sci.* **3**, 275–80.

—— and ISAAC, L. A. 1956. 'Decay following logging injury to Western hemlock, Sitka spruce and true firs (*Abies grandis* and *A. amabilis*)', *Tech. Bull. U.S. Dep. Agric.* **1148**, 34 pp. *R.A.M.*, 1957, p. 291.

—— and WELLS, H. R. 1948. 'Tests on the adaptability of trees and shrubs to shelterbelt planting on certain *Phymatotrichum omnivorum* root rot infested soils in Oklahoma and Texas', *J. For.* **46**, 256–62.

WRIGHT, R. C. M., *et al.* 1957. *Roses.* Ward, Lock, 160 pp.

WRÓBLEWSKI, A., KORCZYŃSKA, E., and WILUSZ, Z. 1952. 'Szkody mrozowe w Arboretum Kórnickim w czasie zimy 1939/40. II. Drzewa liściaste', *Pr. Zakład. Dend. i Pom. w Kórniku*, pp. 126–48.

WURGLER, W. 1955. 'Destruction des broussailles dans les pâturages de montaigne', *Landw. Jb. Schweiz*, **69,** 771–82. *Weed Abstr.*, 1956, No. 1226.

WYCOFF, H. B. 1952. 'Green manure crop causes seedling mortality', *Tree Plant Notes*, **12,** 9–10. *For. Abstr.* **14,** No. 2167.

—— 1955. 'Methyl bromide fumigation of an Illinois nursery soil', *J. For.* **53,** 811–15.

WYMAN, D. 1939. 'Saltwater injury of woody plants resulting from the hurricane of September 21, 1938', *Bull. Arnold Arbor.*, ser. 4, **7,** 45–52. *For. Abstr.* **2,** p. 118.

WYSONG, N. 1952. 'Salting icy pavement may affect trees', *Amer. Nurserym.* **96,** 30–31. *For. Abstr.* **14,** No. 1286.

YEAGER, L. E. 1949. 'Effect of permanent flooding in a river-bottom timber area', *Bull. Ill. Nat. Hist. Surv.* **25,** 65 pp. *For. Abstr.* **12,** No. 1543.

YEATMAN, C. W. 1955. 'Tree root development on upland heaths', *For. Comm. Bull.* **21,** 72 pp.

YLI-VAKKURI, P. 1954. 'Tutkimuksia puiden välisista elimellisistä juuriyhteyksistä männiköissä', *Acta Forest. Fennica,* **60,** 6, 1–117. *Biol. Abstr.* 1956, No. 2693.

YORK, H. H. 1929. 'The Woodgate rust', *J. Econ. Entom.* **22,** 482–4. *R.A.M.*, 1929, p. 682.

—— 1941. 'Tree-wound dressings', *Bull. Morris Arbor. Univ. Pa.* **3,** 73–77. *R.A.M.*, 1942, p. 276.

—— WEAN, R. E., and CHILDS, T. W. 1936. 'Some results of investigations on *Polyporus schweinitzii* Fr.', *Science* (N.S.), **84,** 160–1. *R.A.M.*, 1937, p. 76.

YOSHII, H. 1933. 'Studies on *Gloeosporium kawakamii* Miyabe. IV. On the anthracnose of *Paulownia tomentosa* caused by *Gloeosporium kawakamii* Miyabe', *Bull. Sci. Fakultato Terkultura Kyushu Imper. Univ.* **5,** 523–45. *R.A.M.*, 1934, p. 411.

YOUNG, G. Y. 1943. 'Root rots in storage of deciduous nursery stock and their control', *Phytopathology,* **33,** 656–65.

YOUNG, H. C., *et al.* 1955. 'Systemic chemicals', *Plant Dis. Reptr., Suppl.* **234,** 123–34.

YOUNG, H. E. 1936. 'The species of *Diplodia* affecting forest trees in Queensland', *Qd. Agric. J.* **46,** 310–27. *R.A.M.*, 1937, p. 219.

—— 1940. 'Fused needle disease and its relation to the nutrition of *Pinus*', *Bull. Qd. For. Serv.* **13,** 108 pp. *For. Abstr.* **2,** p. 322.

—— 1948. 'Rhizoctonia root rot of Hoop pine', *Qd. J. Agric. Sci.* **5,** 13–16. *For. Abstr.* **12,** No. 709.

—— 1952. 'Root disease in replanted areas', *Quart. Circ. Rubb. Res. Inst. Ceylon.* **27,** 19–23. *R.A.M.*, 1952, p. 629.

YOUNG, R. A., and MCNEW, G. L. 1947. 'Destruction of seedling *Crataegus mollis* by *Gymnosporangium globosum* in Iowa', *Plant Dis. Reptr.* **31,** 484–6.

ZAK, B. 1957. 'Littleleaf of pine', *U.S.D.A. For. Pest Leafl.* **20,** 4 pp.

—— and CAMPBELL, W. A. 1958. 'Susceptibility of southern pines and other species to the littleleaf pathogen in liquid culture', *For. Sci.* **4,** 156–61.

ZAMBETTAKIS, C. 1955. '*Rosellinia necatrix* (Hart.) Berl.', *Fiches de Phytopathologie Tropicale, Supplément Colonial de la Revue de Mycologie, Paris,* **13,** 8 pp. *For. Abstr.* **18,** No. 579.

ZAPROMETOV, N. G. 1932. 'Bakterioz i karlikovost' — novye bolezni šelkovicy v Sredneĭ Azii', *Šelkovodstvo*, pp. 36–38. *R.A.M.*, 1932, p. 756.

—— 1945. 'Bolezni šelkovicy', *State Publ. Dept. Tashkent,* 76 pp. *R.A.M.*, 1946, p. 16.

ZAPROMETOV, N. G., and MIKHAILOV, E. N. 1937. 'Bolezni šelkovicy', *Trud. sred.-aziat. nauč.-issled. Inst. šelkovodstva*, **14**, 50 pp. *R.A.M.*, 1937, p. 785.

ZEHETMAYR, J. W. L. 1954. 'Experiments in tree planting on peat', *For. Comm. Bull.* **22,** 110 pp.

ZELLER, S. M. 1934. 'Some new or noteworthy fungi on Ericaceous hosts in the Pacific Northwest', *Mycologia*, **26**, 291–304. *R.A.M.*, 1935, p. 65.

—— 1935. 'Some miscellaneous fungi of the Pacific Northwest', *Mycologia*, **27,** 449–66. *R.A.M.*, 1936, p. 117.

—— and GOODDING, L. N. 1930. 'Some species of *Atropellis* and *Scleroderris* on conifers in the Pacific Northwest', *Phytopathology*, **20,** 555–67.

—— and MILBRATH, J. A. 1952. 'Banded chlorosis, a transmissible disease of Cherry', *Phytopathology*, **32,** 634–5.

ZENTMYER, G. A., HORSFALL, J. G., and WALLACE, P. P. 1946. 'Dutch elm disease and its chemotherapy', *Bull. Conn. Agric. Exp. Sta.* **498,** 70 pp. *R.A.M.*,1947, p. 429.

ŽHURAVLEV, I. I. 1940. 'Materialy k metodike fitopatologičeskoĭ ekspertizy lesnykh semjan', *Sbornik Trudov, Central'nyĭ Naučno-Issled. Inst. Lesn. khoz.* **15,** 36–52.

—— 1952*a*. 'Novyĭ sposob dezinfekcii semjan', *Lesn. khoz.* **5,** 52–54. *For. Abstr.* **16,** No. 2885.

—— 1952*b*. 'Virulentnost' Fuzariumov vyzyvajuščikh poleganie vskhodov sosny', *Mikrobiologija*, **21,** 588–93. *R.A.M.*, 1955, p. 196.

—— and SKABICHEVSKAYA, T. P. 1953. 'Patogennost' griba *Alternaria* v otnošenii vskhodov khvojnykh porod v taežnoĭ zone', *Mikrobiologija*, **22,** 719–22. *R.A.M.*, 1956, p. 133.

ZIEGER, E. 1953–4. 'Rauschschäden im Walde', *Wiss. Z. Techn. Hochsch. Dresden*, **3,** 271–80. *R.A.M.*, 1955, p. 7.

—— 1955. 'Die heutige Bedeutung der Industrie-Rauchschäden für den Wald', *Arch. Forstw.* **4,** 66–79. *R.A.M.*, 1956, p. 856.

——, PELZ, E., and HORNIG, W. 1958. 'Ergebnisse einer Umfrage über Umfang und Art der Frostschäden des Winters 1955/56 in den Staatlichen Forstwirtschaftsbetrieben der DDR', *Arch. Forstw.* **7,** 316–37.

ZILLER, W. G. 1954. 'Studies of western tree rusts. I. A new cone rust on Sitka Spruce', *Canad. J. Bot.* **32,** 432–9. *R.A.M.*, 1955, p. 116.

—— 1955. 'Studies of western tree rusts. II. *Melampsora occidentalis* and *M. albertensis*, two needle rusts of Douglas fir', *Canad. J. Bot.* **33,** 177–88.

—— 1961. 'Pine twist rust (*Melampsora pinitorqua*) in North America', *Plant Dis. Reptr.* **45,** 5, 327–9.

ZIMMERLE, H. 1943. 'Über Ästungsversuche bei der Rotbuche', *Allg. Forst- u. Jagdztg.* **119,** 88–104. *For. Abstr.* **6,** 166.

—— 1952. 'Ertragszahlen für grüne Douglasie, japaner Lärche und Roteiche in Württemberg', *Mitt. Württemb. Forst. VersAnst.* **9,** 44 pp. *For. Abstr.* **14,** No. 2541.

ZIMMERMAN, G. A. 1936. 'A further report on induced immunity to chestnut blight', *Rep. Proc. 27th Ann. Meet. Nth. Nut. Gr. Assoc., Geneva N.Y.*, pp. 90–94. *Plant. Breed. Abstr.*, 1938, No. 255.

ZIMMERMAN, P. W. 1930. 'Oxygen requirements for root growth of cuttings in water', *Amer. J. Bot.* **17,** 842–61.

—— 1955. 'Chemicals involved in air pollution and their effects upon vegetation', *Prof. Pap. Boyce Thompson Inst.* **2,** 124–45. *R.A.M.*, 1956, p. 913.

—— and BERG, R. O. 1934. 'Effects of chlorinated water on land plants, aquatic plants, and goldfish', *Contrib. Boyce Thompson Inst.* **6,** 39–50.

ZOBEL, B. J., and GODDARD, R. E. 1955. 'Preliminary results on tests of drought hardy strains of Loblolly pine (*Pinus taeda* L.)', *Res. Note Tex. For. Serv.* **14,** 23 pp. *For. Abstr.* **17,** No. 1461.

ZOBRIST, L. 1950. 'Zehn Jahre Versuche zur Bekämpfung des schwarzen Schneeschimmels *Herpotrichia nigra* Hartig.', *Schweiz. Z. Forstw.* **101,** 632–42. *R.A.M.*, 1951, p. 498.

ZOGG, H. 1943. 'Untersuchungen über die Gattung *Hysterographium* Corda, insbesondere über *Hysterographium fraxini* (Pers.) de Not.', *Phytopath Z.* **14,** 310–84. *R.A.M.*, 1943, p. 504.

ZYCHA, H. 1937. 'Über das Wachstum zweier holzzerstörender Pilze und ihr Verhältnis zur Kohlensäure', *Zbl. Bakt. Abt.* **2,** 223–44. *R.A.M.*, 1938, p. 282.

—— 1951. 'Das Rindensterben der Buche', *Phytopath. Z.* **17,** 444–61. *R.A.M.*, 1952, p. 91.

—— 1955*a*. 'Eine Krebserkrankung der Sitka-Fichte (*Picea sitchensis* (Bong.) Carr.)', *Forstw. Zbl.* **74,** 293–304. *R.A.M.*, 1956, p. 562.

—— 1955*b*. 'Definition von Rindenbrand und Krebs bei Waldbäumen', *Meded. LandbHogesch. Wageningen,* **20,** 411–18. *R.A.M.*, 1956, p. 561.

INDEX

PLATE I

Norway spruce damaged by spring frost; the top, being above the cold air layer, is undamaged.

1. Multiple buds arising in larch after the original short-shoot bud had been killed by frost.

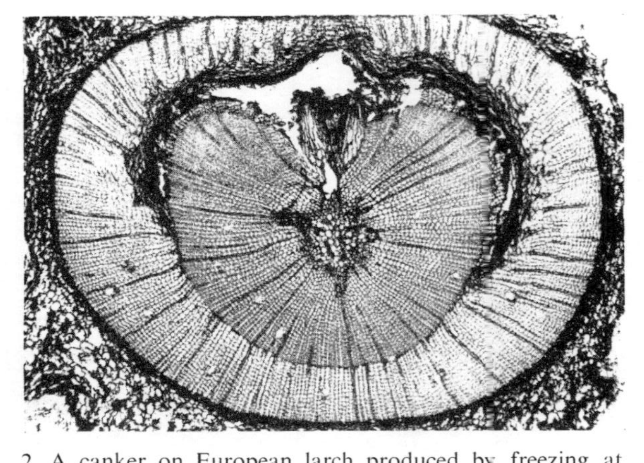

2. A canker on European larch produced by freezing at 22° F on April 22nd.

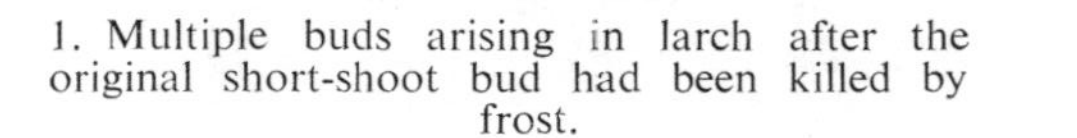

3. A section of Douglas fir stem showing two natural frost rings: (*a*) an autumn ring and (*b*) a spring ring.

4. An enormous witches' broom in the crown of a European larch, Honingham, Norfolk.

1. Ice deposition on trees.

2. Wind blast damage to young Norway spruce, Dartmoor. Later, as mutual shelter increased, these trees formed a normal crop.

PLATE IV

A stagheaded oak, the typical result of drought or waterlogged soil.

Top dying of Norway spruce, probably the result of excessive transpiration.

PLATE VI

Poplar beside a cart track at Bordil, Spain. The tree was damaged for a short period each year by carts removing poles from the plantations.

1. Partly successful graft union in a beech, showing the fan-like surface disclosed when the tree broke many years after grafting. Elmley Castle, Worcestershire.

2. Typical blanks left by damping off in a conifer seedbed ,Maelor Nursery, Flint, August.

PLATE VIII

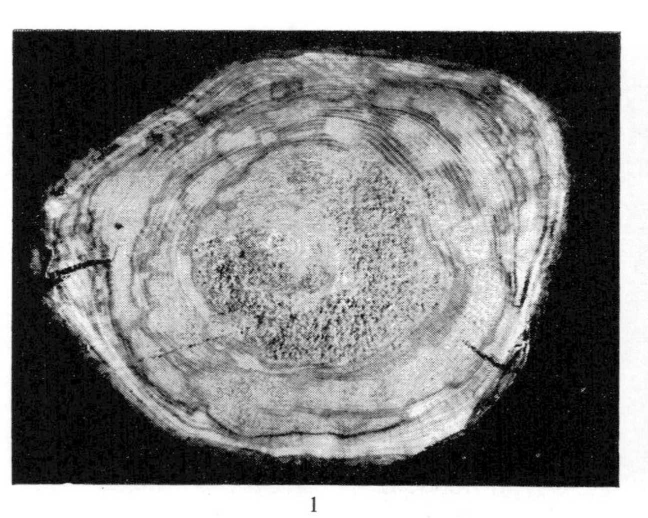

1

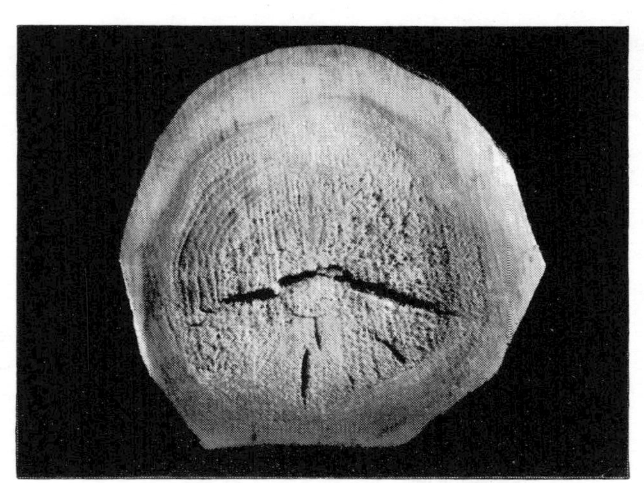

2

3

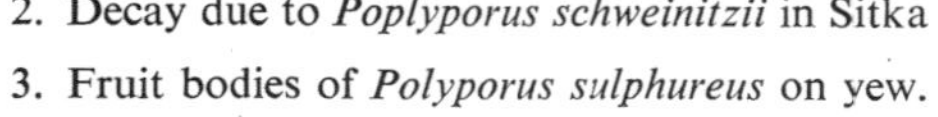

1. Decay due to *Fomes annosus* in European larch.
2. Decay due to *Poplyporus schweinitzii* in Sitka spruce.
3. Fruit bodies of *Polyporus sulphureus* on yew.

Hebeloma mesophaeum established as a mycorrhizal fungus on a potted oak.

PLATE X

Typical Forest of Dean oak, with large, dying lower branches, leading to infection by *Stereum gausapatum*.

1. Spiral grain in an old Scots pine, Harling, Norfolk.

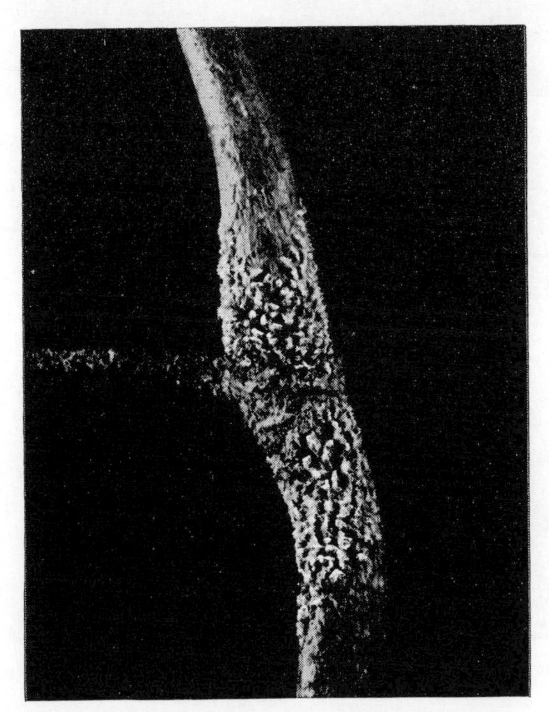

2. Aecia of *Peridermium pini* on Scots pine, Santon Downham, Norfolk, May.

Effect of *Cronartium ribicola* infection on a pole-stage *Pinus strobus*, Lynford, Norfolk, May.

Canker on Japanese larch possibly associated with *Phomopsis pseudotsugae*, Coed-y-Brenin Forest, Merioneth.

PLATE XIV

1. Witches' broom on *Abies numidica* caused by *Melampsorella caryophyllacearum*, Thirlmere, Cumberland.

2. Oak mildew, *Microsphaera alphitoides*.

Bark disease of beech, Steep, Hampshire.

PLATE XVI

Bacterial canker of *Populus* 'pseudoeugenei', beside an unaffected *P. nigra* 'italica', Aldershot, Hampshire.